Teacher's Edition

PRACTICAL MATHEMATICS

REVISED EDITION

SKILLS AND CONCEPTS

Holt, Rinehart and Winston, Inc.

Austin New York San Diego Chicago Toronto Montreal

Authors

Marguerite M. Fredrick
Computer Consultant, K-12
Pleasantville Union Free School District
Pleasantville, New York
Formerly Math Chairperson, 7-12
The Ursuline School
New Rochelle, New York

Steven J. Leinwand
Mathematics Consultant, K-12
Connecticut State Dept. of Education
Hartford, Connecticut

Robert D. Postman
Professor and Chairperson
Mathematics Education
Mercy College
Westchester, New York

Laurence R. Wantuck
Mathematics Supervisor, K-12
Broward County Schools
Fort Lauderdale, Florida

For Photo and Art Credits, see page 486.

We would like to thank the following teachers and administrators who reviewed this text or who field-tested material from this text in their classrooms.

Cindy Clark
Mathematics Teacher
Knox Doss Jr. High School
Hendersonville, Tennessee

Dan Hall
Mathematics Instructor
Yerba Buena High School
San Jose, California

Karen Hanna
Mathematics Consultant
Department of Education
Nashville, Tennessee

Ruth A. Hicks
Mathematics Teacher
Mira Mesa High School
San Diego, California

Debra Klein
Mathematics/Computer Teacher
Nova High School
Ft. Lauderdale, Florida

Mary Ann May
Mathematics Teacher
Williams High School
Plano, Texas

Wendell Meeks
Independent Mathematics Consultant
Springfield, Illinois

Judy Needham
Mathematics Teacher
La Porte High School
La Porte, Texas

F. W. Stanley
Mathematics Teacher
Shelby Junior High School
Shelby, North Carolina

Ruth Tiven
Mathematics Teacher
Hamden High School
Hamden, Connecticut

Nancy Vigliotti
Mathematics Teacher
Dobbs Ferry High School
Dobbs Ferry, NY

Printed in the United States of America

ISBN: 0-03-076789-X

1 2 3 4 5 6 7 8 9 036 96 95 94 93 92

CONTENTS

CHAPTER 1 NUMERATION

CHAPTER 2 WHOLE NUMBER COMPUTATION

CHAPTER 3 DECIMAL COMPUTATION

CHAPTER 4 STATISTICS

CHAPTER 5 FRACTIONS

CHAPTER 6 FRACTION COMPUTATION

CHAPTER 7 MEASUREMENT

CHAPTER 8 RATIO AND PROPORTION

CHAPTER 9 PERCENT

CHAPTER 10 PROBABILITY

CHAPTER 11 GEOMETRY

CHAPTER 12 PERIMETER, AREA, AND VOLUME

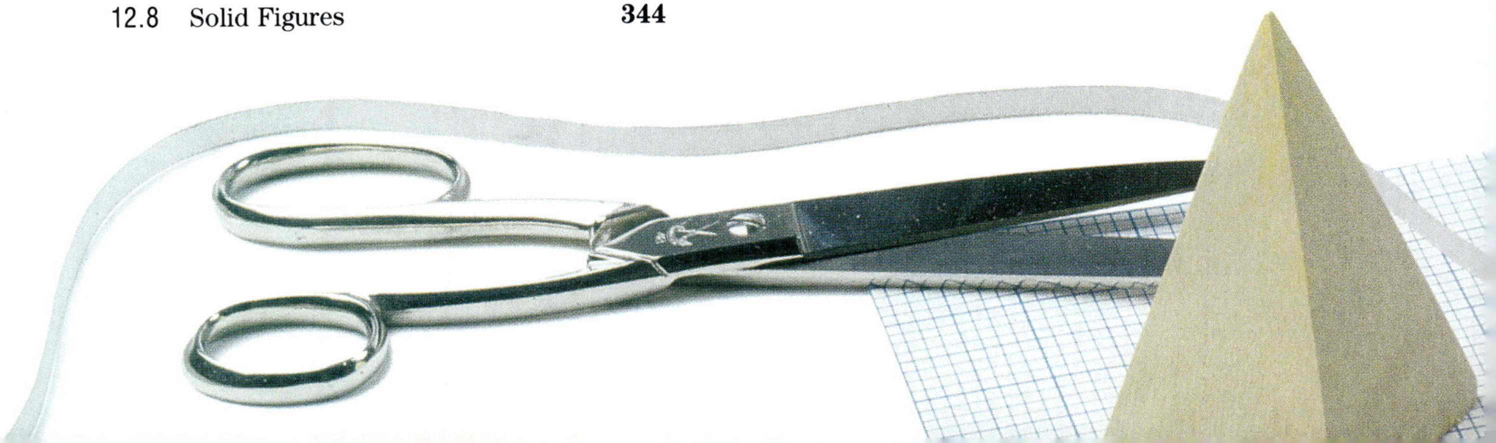

CHAPTER 13 RATIONAL NUMBERS

CHAPTER 14 EQUATIONS

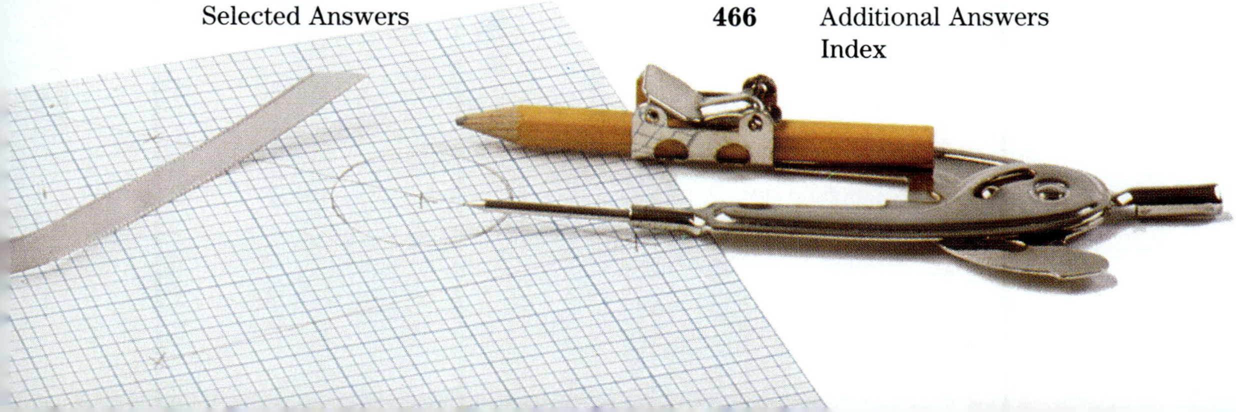

USING PRACTICAL MATHEMATICS
SKILLS AND CONCEPTS

Practical Mathematics **presents mathematical skills and concepts in meaningful, real-life situations. The varied examples, exercises, and applications help students become proficient, confident problem solvers.**

CLEAR INSTRUCTION AND EXAMPLES

Each lesson is introduced with a problem of interest to students.

Think offers helpful hints for solving problems and emphasizes the mental processes students should be using.

Examples illustrate the skill and relevant sub-skills in a step-by-step manner.

The **Checkpoint**—a lesson feature that contains exercises in multiple-choice format with common errors included among the answers—helps to evaluate student errors and teach correct techniques.

6.2 MULTIPLYING MIXED NUMBERS

Amir Johnson is building a cabin cruiser. The hull of the cruiser is 48 feet long. The owner wants $\frac{3}{4}$ of the hull painted white. How many feet of the hull should Amir paint white?

THINK: $\frac{3}{4}$ of 48 feet is the same as $\frac{3}{4} \times 48$.

EXAMPLE 1 Multiply: $\frac{3}{4} \times 48$

1. Rename the whole number as an improper fraction.

$48 = \frac{48}{1}$

2. Multiply.

$\frac{3}{4} \times \frac{48}{1} = \frac{36}{1} = 36$

Amir should paint 36 feet of the hull white.

To multiply mixed numbers and whole numbers, write each factor as an improper fraction.

EXAMPLE 2 Multiply:

a. $3 \times 2\frac{1}{6}$

$3 \times 2\frac{1}{6} = \frac{3}{1} \times \frac{13}{6} = \frac{13}{2} = 6\frac{1}{2}$

b. $4\frac{2}{3} \times 5\frac{1}{4}$

$4\frac{2}{3} \times 5\frac{1}{4} = \frac{14}{3} \times \frac{21}{4} = \frac{49}{2} = 24\frac{1}{2}$

CHECKPOINT Write the letter of the correct answer.

Multiply. What is the product in simplest form?

1. $\frac{3}{4} \times 12$ **a.** $\frac{36}{4}$ **b.** $12\frac{3}{4}$ **c.** 16 **d.** 9

2. $3 \times 4\frac{2}{3}$ **a.** $12\frac{2}{3}$ **b.** 14 **c.** $12\frac{6}{3}$ **d.** 6

3. $3\frac{2}{5} \times 1\frac{3}{4}$ **a.** $3\frac{6}{20}$ **b.** $3\frac{3}{10}$ **c.** $5\frac{19}{20}$ **d.** $4\frac{5}{9}$

148 CHAPTER 6

AMPLE AND VARIED PRACTICE

PRACTICE EXERCISES

Multiply. Write the product in simplest form.

1. $2 \times \frac{1}{2}$	**2.** $3 \times \frac{2}{3}$	**3.** $5 \times \frac{3}{5}$	**4.** $\frac{1}{2} \times 6$	**5.** $\frac{2}{3} \times 6$
6. $8 \times \frac{3}{4}$	**7.** $\frac{1}{3} \times 9$	**8.** $12 \times \frac{5}{6}$	**9.** $5 \times \frac{2}{3}$	**10.** $\frac{4}{5} \times 3$
11. $2 \times 1\frac{1}{2}$	**12.** $3 \times 2\frac{1}{3}$	**13.** $2 \times 3\frac{1}{2}$	**14.** $4 \times 3\frac{1}{4}$	**15.** $3 \times 2\frac{2}{3}$
16. $1\frac{3}{4} \times 8$	**17.** $2 \times 2\frac{5}{8}$	**18.** $3\frac{1}{6} \times 24$	**19.** $3\frac{7}{20} \times 5$	**20.** $4 \times 2\frac{1}{12}$
21. $\frac{7}{8} \times 2\frac{1}{2}$	**22.** $\frac{3}{5} \times 4\frac{1}{3}$	**23.** $7\frac{1}{4} \times \frac{8}{9}$	**24.** $6\frac{2}{3} \times \frac{4}{5}$	**25.** $\frac{5}{6} \times 5\frac{1}{3}$
26. $1\frac{1}{5} \times 2\frac{1}{2}$	**27.** $2\frac{2}{3} \times 1\frac{1}{3}$	**28.** $2\frac{1}{4} \times 2\frac{2}{3}$	**29.** $1\frac{3}{5} \times 2\frac{1}{2}$	**30.** $2\frac{2}{3} \times 3\frac{1}{2}$
31. $4\frac{3}{4} \times 2\frac{1}{3}$	**32.** $3\frac{3}{4} \times 1\frac{3}{4}$	**33.** $4\frac{2}{3} \times 1\frac{2}{7}$	**34.** $4\frac{1}{2} \times 2\frac{2}{3}$	**35.** $1\frac{1}{3} \times 3\frac{2}{7}$
36. $2\frac{1}{4} \times 3$	**37.** $3\frac{2}{3} \times 3\frac{3}{4}$	**38.** $\frac{2}{5} \times 3\frac{1}{2}$	**39.** $5\frac{1}{3} \times 2\frac{3}{8}$	**40.** $\frac{1}{2} \times 9\frac{2}{3}$
41. $5\frac{1}{5} \times 6\frac{1}{4}$	**42.** $7\frac{2}{3} \times 2$	**43.** $\frac{1}{3} \times 3\frac{2}{4}$	**44.** $6\frac{2}{5} \times 10$	**45.** $10\frac{1}{2} \times \frac{4}{7}$

46. $\frac{2}{5} \times \frac{3}{8} \times 10$	**47.** $2\frac{1}{2} \times \frac{1}{3} \times \frac{1}{4}$	**48.** $3\frac{2}{3} \times \frac{1}{4} \times 3$	**49.** $\frac{2}{3} \times \frac{1}{5} \times 2\frac{3}{4}$
50. $1\frac{1}{2} \times 1\frac{1}{4} \times 1\frac{1}{3}$	**51.** $4\frac{2}{5} \times 8\frac{1}{3} \times 2\frac{1}{4}$	**52.** $5\frac{2}{5} \times 7\frac{1}{3} \times 6\frac{1}{11}$	**53.** $9\frac{3}{5} \times 6\frac{1}{8} \times 2\frac{4}{7}$
54. $\left(2 + 1\frac{1}{2}\right) \times \frac{4}{7}$	**55.** $\left(1\frac{1}{4} + 3\frac{3}{8}\right) \times \frac{1}{2}$	**56.** $\left(\frac{3}{4} + \frac{7}{8}\right) \times \frac{2}{3}$	**57.** $\left(\frac{2}{3} + 1\frac{1}{2}\right) \times \frac{3}{4}$

Solve.

58. Maria's boat has a maximum speed of 36 miles per hour. Its cruising speed is $\frac{3}{4}$ of its maximum speed. What is the boat's cruising speed?

59. Latoya earns $18.00 an hour as a carpenter building boats. One week she works 42.5 hours. How much money does she earn that week?

60. Cesar's budget allows him to spend $78,000 on lumber. He has spent $45,963. How much more can he spend without going over budget?

61. A ship can travel at a rate of $18\frac{3}{4}$ miles per hour. How far can it travel in $3\frac{1}{5}$ hours?

TIME OUT A ship is sailing from New York City to Southhampton, England. The total distance is about 3,600 miles. The ship has traveled $\frac{1}{2}$ the distance to the halfway point. How far has the ship traveled?

FRACTION COMPUTATION **149**

The abundant exercises are patterned on the examples in the lesson.

Additional mixed-practice items keep students from simply following a pattern.

Some practice items go beyond the scope of the lesson and provide a challenge for more capable students. These are clearly identified in the *Teacher's Edition* Assignment chart that accompanies each lesson.

Each lesson concludes with an applications section that reinforces and builds on concepts and skills taught in this lesson and previous lessons.

Time Out offers a fun and challenging problem related to the lesson's theme.

OTHER IMPORTANT LESSON FEATURES

FOR DISCUSSION

Would a bar graph or a broken-line graph be better to show each of the following sets of data? Explain.

Favorite colors of different age groups
Hourly temperatures during a 24-h period

PRACTICE EXERCISES

Use each set of data to make a broken-line graph.

1. Jarene kept a record of the number of calls she received on her answering machine in 1 week.

DAILY CALLS: ANSWERING MACHINE

Day	Mon.	Tues.	Wed.	Thurs.	Fri.	Sat.	Sun.
Number of calls	3	2	6	8	5	10	2

2. The manager of the rock band "STAR" tracked the sales of the group's latest music video.

MONTHLY SALES— "FIREBALL" VIDEO MUSIC

Month	July	Aug.	Sept.	Oct.	Nov.	Dec.
Number sold	7,834	12,642	25,869	46,008	43,683	37,304

3. An airline company has gathered information about the number of mobile telephones installed in airplanes.

MOBILE TELEPHONE INSTALLATIONS

Year	1981	1982	1983	1984	1985	1986
Number installed	24	15	65	89	100	895

4. For the past 3 years, Green Ridge High School has made a videotape of the highlights of the school year in addition to the yearbook. Make a double broken-line graph to compare the sales of the video and the yearbook over the past 3 years. Use - - - - - to show the sales of the video. Use ——— to show the sales of the yearbook.

YEARBOOK/VIDEO SALES

Year	Yearbook	Video
1987	363	52
1988	287	101
1989	167	198

STATISTICS

FOR DISCUSSION

The **For Discussion** feature alternates with Checkpoint and stimulates student thinking and class discussion providing an additional opportunity to assess students' understanding.

CALCULATOR

Calculator activities are featured in every chapter. These lively boxed inserts give students numerous chances to use the calculator in a variety of everyday situations. This feature illustrates the calculator's advantages, in some circumstances, over pencil and paper.

PRACTICE EXERCISES

Divide. Write the quotient in simplest form.

1. $5 \div 1\frac{2}{3}$ **2.** $3 \div 2\frac{3}{4}$ **3.** $10 \div 3\frac{1}{3}$ **4.** $15 \div 2\frac{1}{7}$ **5.** $8 \div 2\frac{2}{5}$

6. $2\frac{1}{2} \div 5$ **7.** $1\frac{2}{3} \div 10$ **8.** $4\frac{3}{4} \div 3$ **9.** $5\frac{3}{5} \div 2$ **10.** $3\frac{3}{8} \div 3$

11. $\frac{2}{3} \div 1\frac{1}{2}$ **12.** $\frac{4}{9} \div 2\frac{2}{3}$ **13.** $\frac{2}{5} \div 4\frac{1}{2}$ **14.** $\frac{3}{4} \div 2\frac{1}{4}$ **15.** $\frac{7}{10} \div 4\frac{1}{5}$

16. $2\frac{2}{5} \div \frac{3}{5}$ **17.** $2\frac{1}{4} \div \frac{2}{3}$ **18.** $4\frac{3}{4} \div \frac{5}{6}$ **19.** $2\frac{2}{5} \div \frac{3}{10}$ **20.** $2\frac{5}{6} \div \frac{3}{4}$

21. $1\frac{3}{4} \div 2\frac{1}{3}$ **22.** $1\frac{2}{3} \div 2\frac{1}{2}$ **23.** $2\frac{2}{3} \div 1\frac{3}{5}$ **24.** $3\frac{3}{5} \div 4\frac{1}{4}$ **25.** $4\frac{1}{2} \div 2\frac{5}{8}$

26. $3\frac{3}{4} \div 4\frac{2}{3}$ **27.** $2\frac{3}{4} \div 2\frac{4}{5}$ **28.** $4\frac{2}{3} \div 5\frac{1}{6}$ **29.** $6\frac{1}{8} \div 1\frac{3}{4}$ **30.** $1\frac{2}{3} \div 2\frac{1}{12}$

31. $5\frac{1}{3} \div \frac{3}{4}$ **32.** $3 \div 4\frac{2}{3}$ **33.** $12\frac{1}{4} \div 1\frac{2}{5}$ **34.** $10\frac{5}{8} \div 15\frac{5}{6}$ **35.** $8\frac{2}{3} \div 4$

Solve.

36. The price of Fast Lane Records stock is $\$5\frac{3}{4}$ per share. How many shares can be purchased with $230?

37. 250 shares of Trans Air stock were purchased for $1,200. What was the price per share?

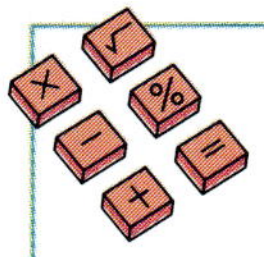

CALCULATOR

You can use what you know about changing a fraction to a decimal to help you multiply fractions and mixed numbers on a calculator.

Ground beef—$1.79 per lb Amount—$24\frac{3}{4}$ lb Cost—■

1. Change $24\frac{3}{4}$ to a decimal.

$24\frac{3}{4} = 24.75$

2. Use your calculator to multiply. Round to the nearest cent.

$24.75 \times \$1.79 = \$44.3025 \rightarrow \$44.31$

Copy and complete the table. Use your calculator to find the cost of each type of meat and the total cost.

	Chicken	Steak	Turkey	Bacon	
Number of pounds	$18\frac{1}{2}$	$21\frac{1}{4}$	$26\frac{1}{2}$	$15\frac{3}{4}$	
Cost per pound	$1.19	$4.99	$0.99	$1.09	Total cost
Cost	1. ■	2. ■	3. ■	4. ■	5. ■

FRACTION COMPUTATION 153

PROBLEM-SOLVING APPLICATIONS LESSONS

Two or more applications in each chapter—one career application and one or more consumer applications—emphasize the usefulness of math skills.

The realistic situation and practical problems relate math to people, careers, and the world around us.

Practice reinforces students' understanding and provides opportunities to use estimation and the calculator to solve problems.

PROBLEM Solving APPLICATION

9.6 CAREER: STORE OWNER

Rosa Aquino owns a neighborhood clothing store. She buys merchandise at a certain cost and then sells it to people in small quantities.

Rosa must pay certain expenses such as rent, telephone, and electricity. To meet these expenses and make a profit, she must sell her merchandise for a greater amount than its cost.

To do this, Rosa marks up the merchandise at a rate of 50%. That is, the **markup** is 50% of the cost. The **selling price** is found by adding the markup to the cost. What would Rosa sell a pair of slacks for if they cost her $18?

1. Find the amount of the markup.

THINK: 50% = 0.50

$0.50 \times \$18 = \$9$

2. Add to the find the selling price.

$\$18 + \$9 = \$27$

So, Rosa would sell the pair of slacks for $27.

Find the markup and the selling price. Remember to estimate whenever you use your calculator.

Item	Cost of merchandise	Percent of markup	Amount of markup	Selling price
1. A	$24	50%	■	■
2. B	$15	30%	■	■
3. C	$25	40%	■	■
4. D	$60	45%	■	■
5. E	$8.50	43%	■	■
6. F	$19.75	36%	■	■
7. G	$25.20	$33\frac{1}{3}\%$	■	■
8. H	$18.50	37.5%	■	■

Solve. Rosa buys one type of shirt at $10...

9. What is th...

PROBLEM Solving APPLICATION

9.11 CONSUMER: COMMISSION

You have a job as a real estate salesperson. You do not receive a weekly salary. Instead, you receive a **commission** on each house you sell. The **rate of commission** is a percent of the selling price of the house.

A. If your rate of commission is 3%, what is your commission on a selling price of $90,000?

You need to find the number that is 3% of 90,000.

1. Write a number sentence.

THINK: 3% = 0.03

What number is 3% of 90,000?

$n = 0.03 \times 90{,}000$

2. Find n

$n = 2{,}700$

So, you receive a commission of $2,700 on a selling price of $90,000.

B. Your friend is a salesperson for another real estate company. She sells a home for $150,000 and receives a $6,000 commission. What is her rate (percent) of commission?

You need to find what percent of 150,000 is 6,000.

1. Write a number sentence.

Let n represent the rate of commission.

What percent of 150,000 is 6,000?

$n \times 150{,}000 = 6{,}000$

2. Find n.

$n = 6{,}000 \div 150{,}000$

$n = 0.04$, or 4%

So, your friend's rate of commission is 4%.

Find the commission. Remember to estimate whenever you use your calculator.

	Selling price	Rate of commission	Commission
1.	$80,000	3%	
2.	$110,000	4%	
3.	$176,000	4%	
4.	$158,000	3%	
5.	$190,000	4%	
6.	$232,500	3.5%	

Find the *rate* of commission.

	Selling price	Rate of commission	Amount of commission
7.	$100,000		$3,000
8.	$120,000		$4,800
9.	$160,000		$5,600
10.	$400,000		$18,000

Solve.

11. Carl Marder sells boats. He receives a salary and a commission. His rate of commission is 2%. How much commission does he receive for a month in which he makes sales totaling $38,000?

12. Maria Gomez sells commercial real estate. Her rate of commission on sales is 5%. How much commission does she receive if she sells a building for $650,000?

13. Eddie Chang sells cars. On sales of $180,000, he receives commissions totaling $4,500. What is his rate of commission?

14. One month Kelly Mandera receives $2,500 in commissions. Her rate of commission is 4%. What is the total amount of her sales that month?

PROBLEM-SOLVING STRATEGY LESSONS

The Problem-Solving Strategy lessons provide an intensive, structured approach to teaching varied methods of solving problems.

The strategy is always identified.

Real-world situations show how math fills a need in everyday life.

The lesson shows how to apply the strategy in a step-by-step manner.

Practice Exercises reinforce students' ability to use the problem-solving strategy.

PROBLEM *Solving* STRATEGY

4.6 MAKING A TABLE

Situation:

Bill, Tom, Mary, and Sue each belong to one of the following after-school clubs: stamp, music, tennis, and hiking. Mary and Tom tried out for the tennis club, but did not make it. Sue does not like to participate in sports activities. Sue and Mary went to hear the music club perform. Which club does each person belong to?

Strategy:

Making a table can help you solve some problems.

Applying the Strategy:

Clue 1: Mary and Tom tried out for the tennis club but did not make it. Write *No* to show that Mary and Tom are not in the tennis club.

	Clubs			
	Stamp	Music	Tennis	Hiking
Bill				
Tom			No	
Mary			No	
Sue				

Clue 2: Sue does not like to participate in sports. Write *No* to show that Sue is not in the tennis or hiking clubs. Write *Yes* to show that Bill belongs to the tennis club.

	Clubs			
	Stamp	Music	Tennis	Hiking
Bill	No	No	Yes	No
Tom			No	
Mary			No	
Sue			No	No

Clue 3: Sue and Mary went to hear the music club perform. Write *No* to show that Sue and Mary are not in the music club. Why do you know that:

- Sue is in the stamp club?
- Mary is in the hiking club?
- Tom is in the music club?

	Clubs			
	Stamp	Music	Tennis	Hiking
Bill	No	No	Yes	No
Tom	No	Yes	No	No
Mary	No	No	No	Yes
Sue	Yes	No	No	No

102 CHAPTER 4

PRACTICE EXERCISES

Read the problem. Then decide which statement is true.

1. Jamie collects stamps either from foreign countries or from the United States. He does not collect stamps from foreign countries.
   a. Jamie collects U.S. stamps.
   b. Jamie collects stamps from foreign countries.

2. Some hikers like to camp. Amelia is a camper.
   a. Amelia is a hiker.
   b. Amelia may or may not be a hiker.

Solve by making a table.

3. Eli, Bert, and June exercise regularly. One jogs, one swims, and one plays golf. The jogger and Bert work together. Eli and the jogger went to see the golfer play in a tournament. Which kind of exercise does each person do?

4. Willy, Julia, Anne, and Sean all have carpentry as their hobby. They won the first four prizes in a carpentry contest. Sean won third prize. Julia did not win second prize. If Ann won first prize, which prize did Willy win?

5. Jason, Eve, Ben, and Kerry belong to a cooking club. They each baked different muffins: blueberry, banana, apple, and bran. Kerry needed to wash some berries for her muffins. Eve needed to remove the seeds from her fruit. Ben's muffins did not have any fruit. Which kind of muffins did each person bake?

6. Mr. Moore, Mrs. Donato, and Mr. Alen are the teachers for these clubs: photography, science, and acting. Mr. Moore and the photography club teacher are cousins. The science club teacher are cousins. The science club teacher and Mr. Alen have lunch together. Mr. Alen is not the photography club teacher. What does each person teach?

DECISION-MAKING LESSONS

The open-ended *Decision-Making* lessons reinforce logical reasoning and move beyond computation to a consideration of all factors involved in making sound decisions.

Problems are placed in a realistic situation.

After considering the **Decision-Making Factors**, students practice simple, efficient ways to organize the information necessary for **Decision-Making Comparisons.**

The open-ended problems promote decision making based on logical reasons and choices.

DECISION MAKING

3.13 WHERE TO SHOP

Most people have several stores to choose from when they go shopping. Choosing the right store can save time and money, but there are other benefits to be gained from shopping wisely.

PROBLEM

Carla and Tony have just moved from Ohio to Texas. They each need a new suit for the warmer climate that they can wear on job interviews. They also want to replace some of the items they sold before they moved. They made a list and checked department store ads for prices. They also made some notes about the stores and their locations.

	Downtown Department Store	Exclusive Department Store	Discount Department Store
Tony:			
Suit	\$ 159 – \$ 239	\$ 265 and up	\$ 75
Tie	\$ 12 – \$ 29	\$ 19 – \$ 79	2 for \$15
desklamp	\$ 39	\$ 39	not available
19" Television	\$ 297	\$ 499	\$ 249
Carla			
Suit	\$100 – \$ 250	\$265 – \$500	\$50
briefcase	\$150	\$275	not available
clock radio	\$ 79	\$ 79	\$ 13
Calculator	\$ 19	\$ 49	\$ 5
Location	2 mi away	30 mi away at Marvelous Mall	10 mi away at Mini-Mall
Hours	Mon.-Wed. 10am - 6pm Thurs. – Sat. 10 am - 9 pm	7 days 10 am – 9 pm	Mon.- Sat. 9am - 9 pm
Parking	none	3,000 Cars	100 Cars
Public Transportation	yes	none	none
Number of stores in area	30	150	10
Selection	average	extensive	limited
Quality	Well-Known moderate-priced brands, some better labels	Well-known brands, designer fashions, latest trends	Some brand names, mostly store labels
Other	Free delivery, free alternations on all clothing	Free delivery, free alterations on men's suits	Free delivery for furniture and major appliances, no alterations

DECISION-MAKING FACTORS

Cost Selection Convenience Quality

DECISION-MAKING COMPARISONS

Compare the three stores by completing the table.

	Downtown	Exclusive	Discount
Cost	1. ■	2. ■	Inexpensive
Selection	Average	3. ■	4. ■
Convenience			
Location	5. ■	Inconvenient	6. ■
Parking	7. ■	Extensive	8. ■
Shopping hours	9. ■	10. ■	Convenient
Quality	11. ■	Excellent	12. ■

MAKING THE DECISIONS

Where should Tony and Carla shop:

13. if cost were the only factor?

14. if selection were the only factor?

15. if convenience were the only factor?

16. if quality were the only factor?

17. How much could Carla save on a suit by shopping at Discount rather than Exclusive?

18. If Carla is looking for a suit that is very fashionable, where should she shop?

19. Tony wants a suit that will wear well and look good. If his total budget is \$500, where should he shop for the suit?

20. If Carla and Tony didn't have a car, where would they have to shop?

21. If the television set is the same brand and model at all three stores, where should Tony buy it? Why?

22. Would it make sense to buy the desk lamp and clock radio at Downtown or at Exclusive? Why?

23. The briefcases vary in price and quality. What other factors should Carla consider?

24. Why do people, even if they can't afford to shop there, visit large malls with expensive stores?

25. If you were Tony or Carla, where would you buy each of the items on his or her list?

REVIEW, TEST, AND MAINTENANCE FEATURES

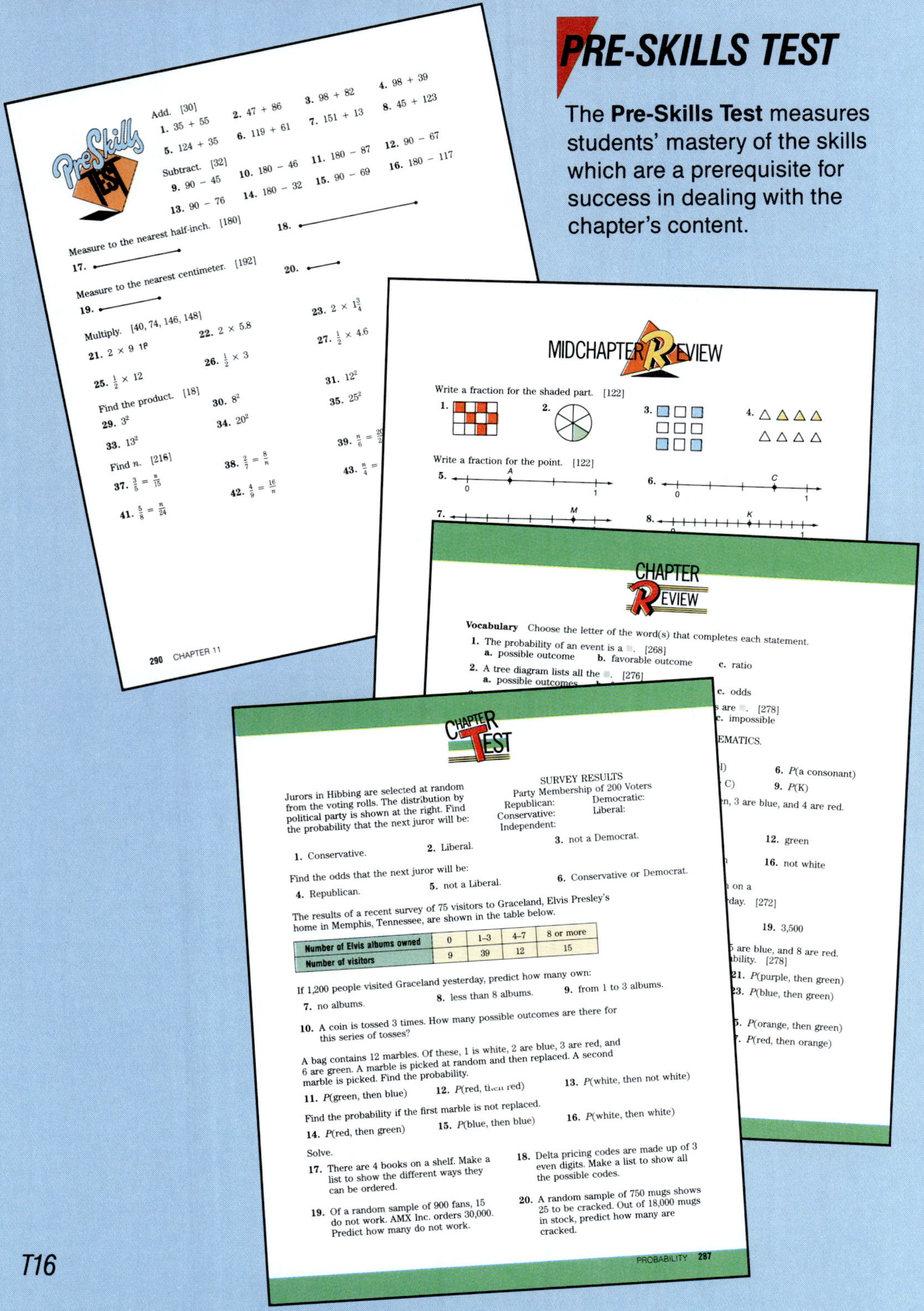
Add. [30]
1. 35 + 55 2. 47 + 86 3. 98 + 82 4. 98 + 39
5. 124 + 35 6. 119 + 61 7. 151 + 13 8. 45 + 123
Subtract. [32]
9. 90 − 45 10. 180 − 46 11. 180 − 87 12. 90 − 67
13. 90 − 76 14. 180 − 32 15. 90 − 69 16. 180 − 117
Measure to the nearest half-inch. [180]
17. 18.
Measure to the nearest centimeter. [192]
19. 20.
Multiply. [40, 74, 146, 148]
21. 2 × 9 22. 2 × 5.8 23. $2 \times 1\frac{3}{4}$
25. $\frac{1}{2} \times 12$ 26. $\frac{1}{2} \times 3$ 27. $\frac{1}{2} \times 4.6$
Find the product. [18]
29. 3^2 30. 8^2 31. 12^2
33. 13^2 34. 20^2 35. 25^2
Find n. [216]
37. $\frac{3}{5} = \frac{n}{15}$ 38. $\frac{2}{7} = \frac{8}{n}$ 39. $\frac{n}{6} = \frac{20}{2}$
41. $\frac{5}{8} = \frac{n}{24}$ 42. $\frac{4}{9} = \frac{16}{n}$ 43. $\frac{n}{4} =$
290 CHAPTER 11

MIDCHAPTER REVIEW
Write a fraction for the shaded part. [122]
1. 2. 3. 4.
Write a fraction for the point. [122]
5. 6.
7. 8.

CHAPTER REVIEW
Vocabulary Choose the letter of the word(s) that completes each statement.
1. The probability of an event is a ■. [268]
a. possible outcome b. favorable outcome c. ratio
2. A tree diagram lists all the ■. [276]
a. possible outcomes c. odds
...s are ■. [278]
c. impossible
...EMATICS.
6. P(a consonant)
9. P(K)
...n, 3 are blue, and 4 are red.
12. green
16. not white
...on a ...day. [272]
19. 3,500
...5 are blue, and 8 are red. ...bility. [278]
21. P(purple, then green)
23. P(blue, then green)
25. P(orange, then green)
27. P(red, then orange)

CHAPTER TEST
Jurors in Hibbing are selected at random from the voting rolls. The distribution by political party is shown at the right. Find the probability that the next juror will be:

SURVEY RESULTS
Party Membership of 200 Voters
Republican: Democratic:
Conservative: Liberal:
Independent:

1. Conservative. 2. Liberal. 3. not a Democrat.
Find the odds that the next juror will be:
4. Republican. 5. not a Liberal. 6. Conservative or Democrat.
The results of a recent survey of 75 visitors to Graceland, Elvis Presley's home in Memphis, Tennessee, are shown in the table below.

Number of Elvis albums owned	0	1–3	4–7	8 or more
Number of visitors	9	39	12	15

If 1,200 people visited Graceland yesterday, predict how many own:
7. no albums. 8. less than 8 albums. 9. from 1 to 3 albums.
10. A coin is tossed 3 times. How many possible outcomes are there for this series of tosses?
A bag contains 12 marbles. Of these, 1 is white, 2 are blue, 3 are red, and 6 are green. A marble is picked at random and then replaced. A second marble is picked. Find the probability.
11. P(green, then blue) 12. P(red, then red) 13. P(white, then not white)
Find the probability if the first marble is not replaced.
14. P(red, then green) 15. P(blue, then blue) 16. P(white, then white)
Solve.
17. There are 4 books on a shelf. Make a list to show the different ways they can be ordered.
18. Delta pricing codes are made up of 3 even digits. Make a list to show all the possible codes.
19. Of a random sample of 900 fans, 15 do not work. AMX Inc. orders 30,000. Predict how many do not work.
20. A random sample of 750 mugs shows 25 to be cracked. Out of 18,000 mugs in stock, predict how many are cracked.
PROBABILITY 287

PRE-SKILLS TEST

The **Pre-Skills Test** measures students' mastery of the skills which are a prerequisite for success in dealing with the chapter's content.

MIDCHAPTER REVIEW

The **Midchapter Review** assesses students' progress and their readiness to move on to more complex skills.

CHAPTER REVIEW

Each full-page **Chapter Review** reinforces key vocabulary, skills, and concepts.

CHAPTER TEST

The **Chapter Test** evaluates students' mastery of basic skills and concepts as well as applications and strategies. Each test item is keyed to a chapter objective in the *Teacher's Edition*.

MIXED REVIEW

The **Mixed Review** feature occurs several times in each chapter. The variety of exercises keeps students alert by reviewing a mixture of the skills taught previously.

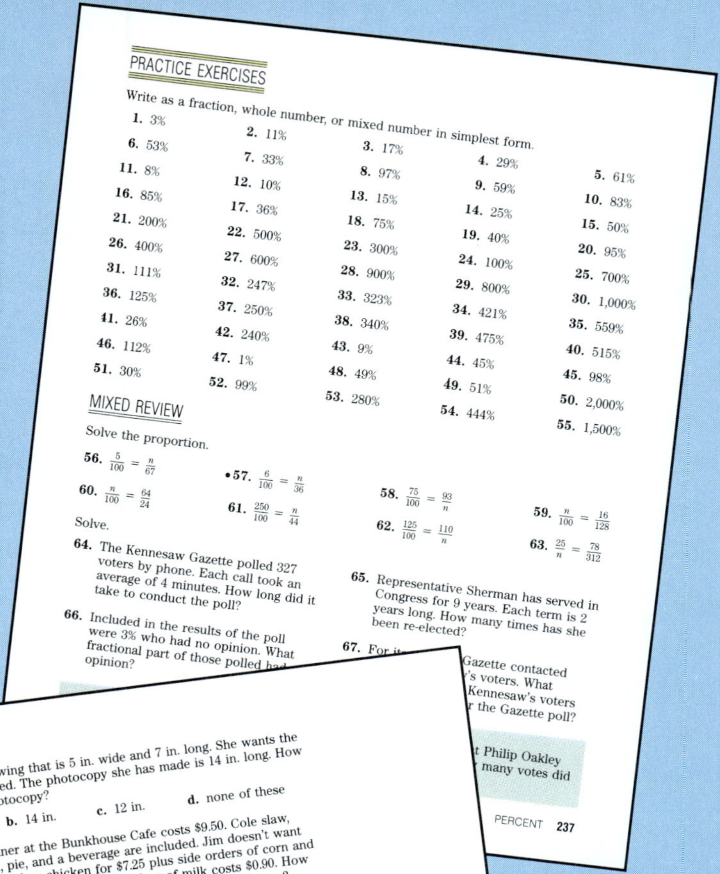

PRACTICE EXERCISES

Write as a fraction, whole number, or mixed number in simplest form.

1. 3%	2. 11%	3. 17%	4. 29%	5. 61%
6. 53%	7. 33%	8. 97%	9. 59%	10. 83%
11. 8%	12. 10%	13. 15%	14. 25%	15. 50%
16. 85%	17. 36%	18. 75%	19. 40%	20. 95%
21. 200%	22. 500%	23. 300%	24. 100%	25. 700%
26. 400%	27. 600%	28. 900%	29. 800%	30. 1,000%
31. 111%	32. 247%	33. 323%	34. 421%	35. 559%
36. 125%	37. 250%	38. 340%	39. 475%	40. 515%
41. 26%	42. 240%	43. 9%	44. 45%	45. 98%
46. 112%	47. 1%	48. 49%	49. 51%	50. 2,000%
51. 30%	52. 99%	53. 280%	54. 444%	55. 1,500%

MIXED REVIEW

Solve the proportion.

56. $\frac{5}{100} = \frac{n}{67}$ •57. $\frac{6}{100} = \frac{n}{36}$ 58. $\frac{75}{100} = \frac{93}{n}$ 59. $\frac{n}{100} = \frac{16}{128}$

60. $\frac{n}{100} = \frac{64}{24}$ 61. $\frac{250}{100} = \frac{n}{44}$ 62. $\frac{125}{100} = \frac{110}{n}$ 63. $\frac{25}{n} = \frac{78}{312}$

Solve.

64. The Kennesaw Gazette polled 327 voters by phone. Each call took an average of 4 minutes. How long did it take to conduct the poll?

65. Representative Sherman has served in Congress for 9 years. Each term is 2 years long. How many times has she been re-elected?

66. Included in the results of the poll were 3% who had no opinion. What fractional part of those polled had ... opinion?

67. For i... Gazette contacted ...'s voters. What ... Kennesaw's voters ...r the Gazette poll?

...t Philip Oakley ... many votes did

CUMULATIVE REVIEW

The **Cumulative Reviews** provide abundant practice for reviewing key concepts, skills, applications, and strategies. The reviews are carefully spaced throughout the book (after Chapter 3, 6, 9, and 12).

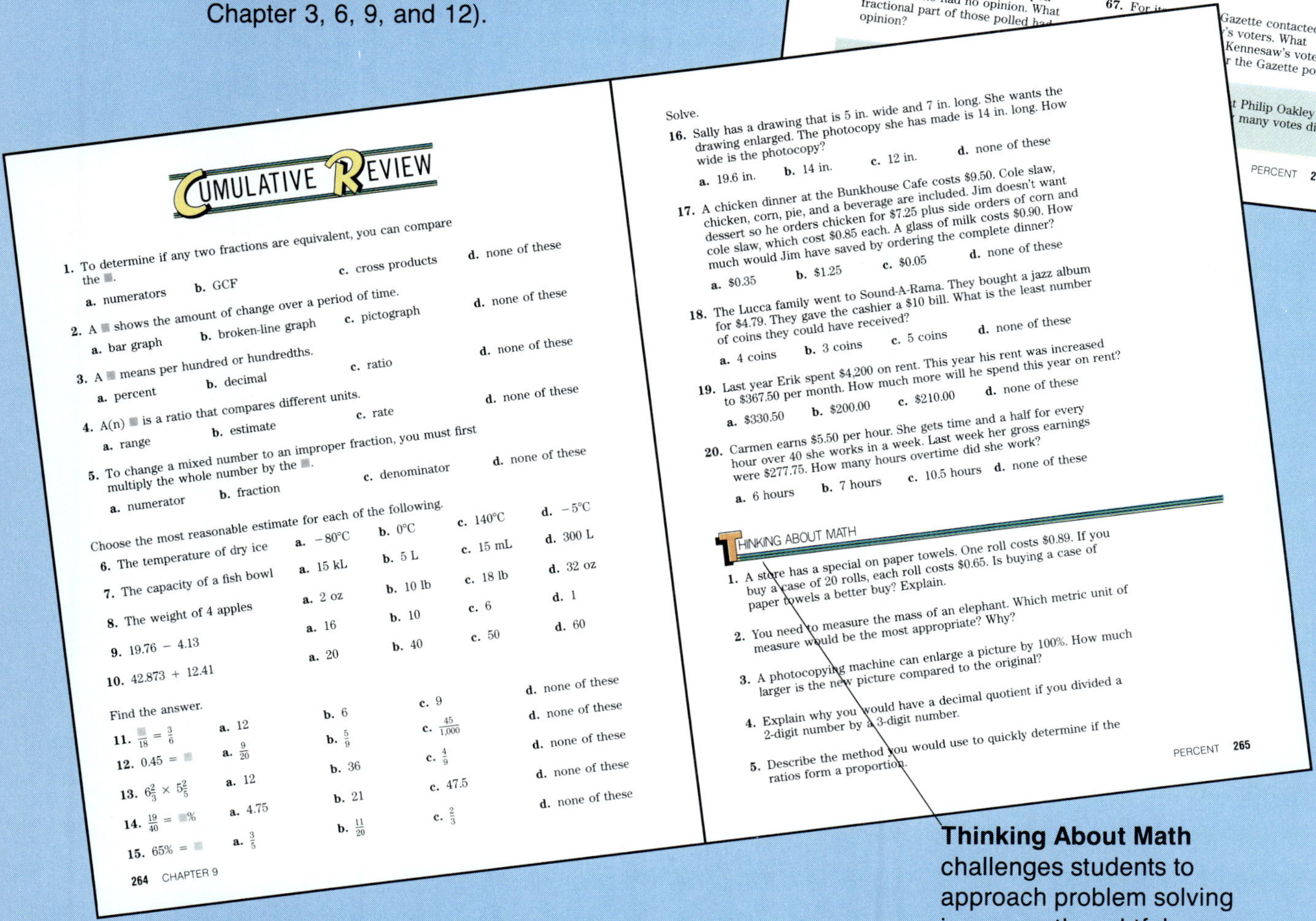

CUMULATIVE REVIEW

1. To determine if any two fractions are equivalent, you can compare the ■.
   a. numerators b. GCF c. cross products d. none of these
2. A ■ shows the amount of change over a period of time.
   a. bar graph b. broken-line graph c. pictograph d. none of these
3. A ■ means per hundred or hundredths.
   a. percent b. decimal c. ratio d. none of these
4. A(n) ■ is a ratio that compares different units.
   a. range b. estimate c. rate d. none of these
5. To change a mixed number to an improper fraction, you must first multiply the whole number by the ■.
   a. numerator b. fraction c. denominator d. none of these

Choose the most reasonable estimate for each of the following.

	a.	b.	c.	d.
6. The temperature of dry ice	−80°C	0°C	140°C	−5°C
7. The capacity of a fish bowl	15 kL	5 L	15 mL	300 L
8. The weight of 4 apples	2 oz	10 lb	18 lb	32 oz
9. 19.76 − 4.13	16	10	6	1
10. 42.873 + 12.41	20	40	50	60

Find the answer.

	a.	b.	c.	d.
11. $\frac{■}{18} = \frac{3}{6}$	12	6	9	none of these
12. 0.45 = ■	$\frac{9}{20}$	$\frac{5}{9}$	$\frac{45}{1,000}$	none of these
13. $6\frac{2}{3} \times 5\frac{2}{5}$	12	36	$\frac{4}{9}$	none of these
14. $\frac{19}{40}$ = ■%	4.75	21	47.5	none of these
15. 65% = ■	$\frac{3}{5}$	$\frac{11}{20}$	$\frac{2}{3}$	none of these

264 CHAPTER 9

Solve.

16. Sally has a drawing that is 5 in. wide and 7 in. long. She wants the drawing enlarged. The photocopy she has made is 14 in. long. How wide is the photocopy?
    a. 19.6 in. b. 14 in. c. 12 in. d. none of these
17. A chicken dinner at the Bunkhouse Cafe costs $9.50. Cole slaw, chicken, corn, pie, and a beverage are included. Jim doesn't want dessert so he orders chicken for $7.25 plus side orders of corn and cole slaw, which cost $0.85 each. A glass of milk costs $0.90. How much would Jim have saved by ordering the complete dinner?
    a. $0.35 b. $1.25 c. $0.05 d. none of these
18. The Lucca family went to Sound-A-Rama. They bought a jazz album for $4.79. They gave the cashier a $10 bill. What is the least number of coins they could have received?
    a. 4 coins b. 3 coins c. 5 coins d. none of these
19. Last year Erik spent $4,200 on rent. This year his rent was increased to $367.50 per month. How much more will he spend this year on rent?
    a. $330.50 b. $200.00 c. $210.00 d. none of these
20. Carmen earns $5.50 per hour. She gets time and a half for every hour over 40 she works in a week. Last week her gross earnings were $277.75. How many hours overtime did she work?
    a. 6 hours b. 7 hours c. 10.5 hours d. none of these

THINKING ABOUT MATH

1. A store has a special on paper towels. One roll costs $0.89. If you buy a case of 20 rolls, each roll costs $0.65. Is buying a case of paper towels a better buy? Explain.
2. You need to measure the mass of an elephant. Which metric unit of measure would be the most appropriate? Why?
3. A photocopying machine can enlarge a picture by 100%. How much larger is the new picture compared to the original?
4. Explain why you would have a decimal quotient if you divided a 2-digit number by a 3-digit number.
5. Describe the method you would use to quickly determine if the ratios form a proportion.

PERCENT 265

Thinking About Math challenges students to approach problem solving in a more thoughtful way.

OPTIONAL LEARNING AIDS AND EXTENSIONS

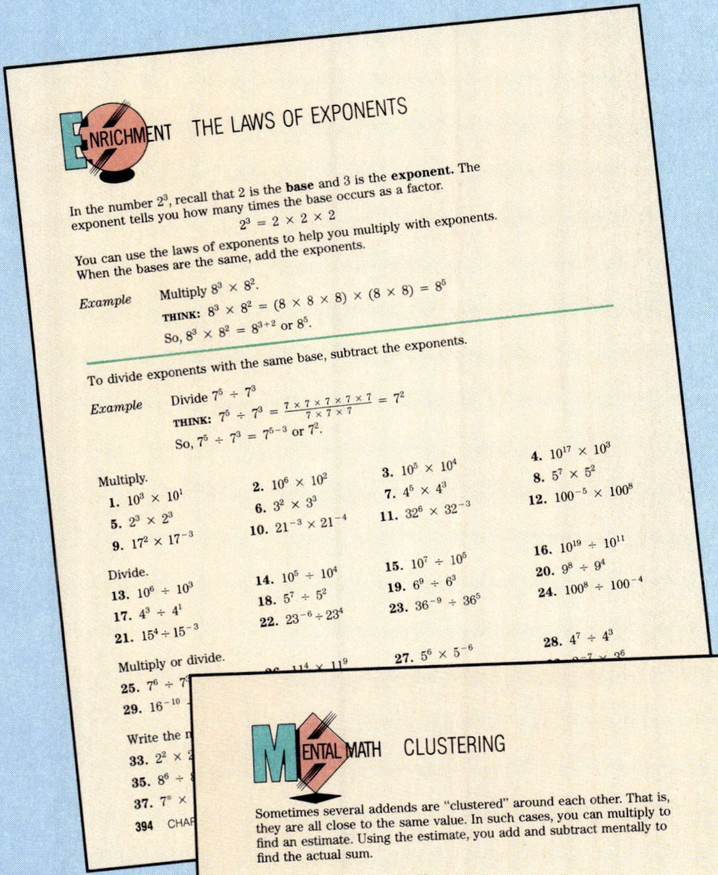

ENRICHMENT THE LAWS OF EXPONENTS

In the number 2^3, recall that 2 is the **base** and 3 is the **exponent.** The exponent tells you how many times the base occurs as a factor.

$$2^3 = 2 \times 2 \times 2$$

You can use the laws of exponents to help you multiply with exponents. When the bases are the same, add the exponents.

Example Multiply $8^3 \times 8^2$.
THINK: $8^3 \times 8^2 = (8 \times 8 \times 8) \times (8 \times 8) = 8^5$
So, $8^3 \times 8^2 = 8^{3+2}$ or 8^5.

To divide exponents with the same base, subtract the exponents.

Example Divide $7^5 \div 7^3$
THINK: $7^5 \div 7^3 = \frac{7 \times 7 \times 7 \times 7 \times 7}{7 \times 7 \times 7} = 7^2$
So, $7^5 \div 7^3 = 7^{5-3}$ or 7^2.

Multiply.

1. $10^3 \times 10^1$
2. $10^6 \times 10^2$
3. $10^5 \times 10^4$
4. $10^{17} \times 10^3$
5. $2^3 \times 2^3$
6. $3^2 \times 3^3$
7. $4^5 \times 4^3$
8. $5^7 \times 5^2$
9. $17^2 \times 17^{-3}$
10. $21^{-3} \times 21^{-4}$
11. $32^6 \times 32^{-3}$
12. $100^{-5} \times 100^8$

Divide.

13. $10^6 \div 10^3$
14. $10^5 \div 10^4$
15. $10^7 \div 10^5$
16. $10^{19} \div 10^{11}$
17. $4^3 \div 4^1$
18. $5^7 \div 5^2$
19. $6^9 \div 6^3$
20. $9^8 \div 9^4$
21. $15^4 \div 15^{-3}$
22. $23^{-6} \div 23^4$
23. $36^{-9} \div 36^5$
24. $100^8 \div 100^{-4}$

Multiply or divide.

25. $7^6 \div 7$
26. $11^4 \times 11^9$
27. $5^6 \times 5^{-6}$
28. $4^7 \div 4^3$
29. 16^{-10}

Write the

33. $2^2 \times 2$
35. $8^6 \div$
37. 7 ×

394 CHAP

ENRICHMENT

The **Enrichment** pages move beyond the core material covered in the chapter to give students an extra challenge and additional information.

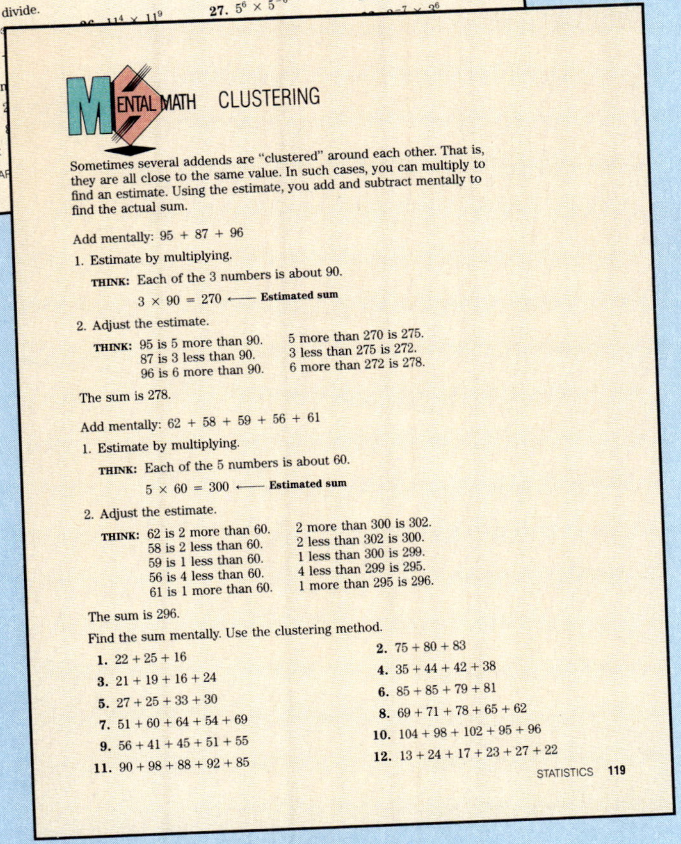

MENTAL MATH CLUSTERING

Sometimes several addends are "clustered" around each other. That is, they are all close to the same value. In such cases, you can multiply to find an estimate. Using the estimate, you add and subtract mentally to find the actual sum.

Add mentally: 95 + 87 + 96

1. Estimate by multiplying.
   THINK: Each of the 3 numbers is about 90.
   3 × 90 = 270 ⟵ **Estimated sum**
2. Adjust the estimate.
   THINK: 95 is 5 more than 90. 5 more than 270 is 275.
   87 is 3 less than 90. 3 less than 275 is 272.
   96 is 6 more than 90. 6 more than 272 is 278.

The sum is 278.

Add mentally: 62 + 58 + 59 + 56 + 61

1. Estimate by multiplying.
   THINK: Each of the 5 numbers is about 60.
   5 × 60 = 300 ⟵ **Estimated sum**
2. Adjust the estimate.
   THINK: 62 is 2 more than 60. 2 more than 300 is 302.
   58 is 2 less than 60. 2 less than 302 is 300.
   59 is 1 less than 60. 1 less than 300 is 299.
   56 is 4 less than 60. 4 less than 299 is 295.
   61 is 1 more than 60. 1 more than 295 is 296.

The sum is 296.

Find the sum mentally. Use the clustering method.

1. 22 + 25 + 16
2. 75 + 80 + 83
3. 21 + 19 + 16 + 24
4. 35 + 44 + 42 + 38
5. 27 + 25 + 33 + 30
6. 85 + 85 + 79 + 81
7. 51 + 60 + 64 + 54 + 69
8. 69 + 71 + 78 + 65 + 62
9. 56 + 41 + 45 + 51 + 55
10. 104 + 98 + 102 + 95 + 96
11. 90 + 98 + 88 + 92 + 85
12. 13 + 24 + 17 + 23 + 27 + 22

STATISTICS 119

MENTAL MATH

The **Mental Math** pages develop techniques to solve problems without using pencil and paper.

ESTIMATION

The **Estimation Skills** pages extend students' understanding of estimation techniques.

ESTIMATION SKILLS — ESTIMATING WITH CUSTOMARY MEASURES

When you estimate the sum or difference of measures expressed in two units, round each measure to the greater unit.

Estimate the difference: 15 ft 2 in. − 8 ft 11 in.

1. Round each measure to the nearest number of feet.

THINK: 2 in. < $\frac{1}{2}$ ft, so 15 ft 2 in. rounds to 15 ft.
11 in. > $\frac{1}{2}$ ft, so 8 ft 11 in. rounds to 9 ft.

2. Subtract.

15 − 9 = 6

The estimated difference is 6 ft.

Estimate the sum: 1 h 15 min + 5 h 37 min + 3 h 6 min

1. Round each measure to the nearest number of hours.

THINK: 15 min < $\frac{1}{2}$ h, so 1 h 15 min rounds to 1 h.
37 min > $\frac{1}{2}$ h, so 5 h 37 min rounds to 6 h.
6 min < $\frac{1}{2}$ h, so 3 h 6 min rounds to 3 h.

2. Add.

1 + 6 + 3 = 10

The estimated sum is 10 hours.

Estimate the sum or difference.

1. 18 ft 4 in. + 6 ft 9 in.
2. 9 ft 5 in. − 3 ft 3 in.
3. 4 h 8 min + 43 min
4. 16 h 40 min − 5 h 52
5. 9 lb 2 oz + 8 lb 3 oz
6. 24 lb 3 oz − 15 lb 6 oz
7. 2 gal 3 qt + 3 gal 1 qt
8. 7 gal − 2 gal 1 qt
9. 35 min 2 s + 20 min 8 s
10. 10 min 4 s − 5 min 53 s
11. 6 c 7 fl oz + 8 c 2 fl oz
12. 8 c 5 fl oz − 3 c 7 fl oz
13. 2 yd 32 ... 3 yd 27 in.
14. 10 yd 32 in. − 5 yd 5 in.
15. 1 ft 5 in ...
16. ... 4 min 42 s + 8 min 9 s
17. 2 lb 9 ...

CALCULATOR

The **Calculator** pages teach the keys and functions commonly available on calculators. Lessons emphasize the time-saving benefits of using these problem-solving tools correctly in a variety of situations.

CALCULATOR — DIFFERENCE BETWEEN CALCULATORS

Do you have a scientific calculator or a nonscientific calculator? Enter the following example to find out.

Enter: 3 + 4 × 5 =

A scientific calculator uses the rules for the order of operations. It gives the correct answer of 23.
A nonscientific calculator performs the operations in the order in which they are entered. It gives an incorrect answer of 35.

Compute: 38 + 91 ÷ 13

Scientific Calculator
Enter the numbers and the operation symbols in the order given.
Enter: 3 8 + 9 1 ÷ 1 3 =
Display: 45.

Nonscientific Calculator
Find the quotient 91 ÷ 13 first, then add 38.
Enter: 9 1 ÷ 1 3 = + 3 8 =
Display: 45.

Evaluate the expression 7.8 + 1.5n when n = 6.

Scientific Calculator
Mentally substitute 6 for n. Enter the numbers and operation symbols in the order given.
Enter: 7 . 8 + 1 . 5 × 6 =
Display: 16.8

Nonscientific Calculator
Mentally substitute 6 for n. Multiply 1.5 × 6 first. Then add 7.8.
Enter: 1 . 5 × 6 = + 7 . 8 =
Display: 16.8

Compute. Use a calculator.

1. 29 + 6 × 53
2. 37 + 54 ÷ 6
3. 360 + 19 × 82
4. 76 + 851 ÷ 37
5. 4.7 + 3.6 × 4.5
6. 8.4 + 31.9 ÷ 29
7. 0.28 + 0.32 × 140
8. 0.4 + 58.5 ÷ 6.5
9. 6.29 + 1.9 × 2.7

Evaluate each expression for the given value of n. Use a calculator.

10. $14 + 8n$; $n = 12$
11. $49 + n \div 2$; $n = 68$
12. $16 + 25n$; $n = 11$
13. $17 + n \div 14$; $n = 98$
14. $3.6 + 39n$; $n = 2.1$
15. $0.6 + n \div 8$; $n = 5$
16. $17 + 12 \div n$; $n = 2$
17. $22 + \frac{n}{19}$; $n = 152$
18. $7 + \frac{306}{n}$; $n = 36$
19. $12.8 + 7n$; $n = 4.2$
20. $3.6 + n \div 1.5$; $n = 6$
21. $20 + \frac{196}{n}$; $n = 7$

EQUATIONS 425

TEACHER'S EDITION

The *Teacher's Edition* is practical and flexible, with all the materials needed for a lesson presented in one place. The variety of suggestions minimizes preparation time and makes it easier to adapt each lesson to your teaching style and your students' needs.

Objectives—clearly stated goals that students should attain by the end of the lesson.

Warm-up—a brief activity for reviewing the lesson's prerequisite skills.

Introduction—material you can use to capture students' interest at the beginning of the lesson.

Instruction—useful teaching notes that help you derive maximum benefit from every page and every feature.

OBJECTIVE

- Multiply fractions, whole numbers, and mixed numbers.

TEACHING THE LESSON

WARM-UP Present the following exercises on the chalkboard or overhead transparency.

Write as an improper fraction.

1. $2\frac{2}{3}$ **2.** $4\frac{2}{5}$ **3.** $3\frac{3}{4}$

4. $8\frac{1}{3}$ **5.** $9\frac{3}{5}$ **6.** $1\frac{4}{7}$

(**1.** $\frac{8}{3}$ **2.** $\frac{22}{5}$ **3.** $\frac{15}{4}$ **4.** $\frac{25}{3}$ **5.** $\frac{48}{5}$ **6.** $\frac{11}{7}$)

INTRODUCTION Write the following on the chalkboard:

$\frac{1}{2} \times 6$

Explain that the word "of" can mean multiplication. Continue to illustrate the example:

$\frac{1}{2}$ of 6 (• • •)

$\frac{1}{2}$ of 6 = 3 • • •

INSTRUCTION Remind students that a whole number, such as 8, can be written as $\frac{8}{1}$. When discussing Example 1, point out that the word "of" is used to mean multiplication. Write the following on the chalkboard:

$\frac{3}{4}$ of 48 $\frac{3}{4} \times 48$

Elicit from students that when multiplying a fraction and a mixed number, the product will always be less than the mixed number.

As you discuss Example 2, remind students to write the mixed numbers as improper fractions before they begin to multiply. As students work through Example 2b, point out that the product of two mixed numbers is greater than either mixed number.

CHECKPOINT The incorrect answer choices include common errors.

148

Chapter 6 Fraction Computation

6.2 MULTIPLYING MIXED NUMBERS

Amir Johnson is building a cabin cruiser. The hull of the cruiser is 48 feet long. The owner wants $\frac{3}{4}$ of the hull painted white. How many feet of the hull should Amir paint white?

THINK: $\frac{3}{4}$ of 48 feet is the same as $\frac{3}{4} \times 48$.

EXAMPLE 1 Multiply: $\frac{3}{4} \times 48$

1. Rename the whole number as an improper fraction.

$48 = \frac{48}{1}$

2. Multiply.

$\frac{3}{4} \times \frac{48}{1} = \frac{36}{1} = 36$

Amir should paint 36 feet of the hull white.

To multiply mixed numbers and whole numbers, write each factor as an improper fraction.

EXAMPLE 2 Multiply:

a. $3 \times 2\frac{1}{6}$

$3 \times 2\frac{1}{6} = \frac{3}{1} \times \frac{13}{6} = \frac{13}{2} = 6\frac{1}{2}$

b. $4\frac{2}{3} \times 5\frac{1}{4}$

$4\frac{2}{3} \times 5\frac{1}{4} = \frac{14}{3} \times \frac{21}{4} = \frac{49}{2} = 24\frac{1}{2}$

CHECKPOINT Write the letter of the correct answer.

Multiply. What is the product in simplest form?

1. $\frac{3}{4} \times 12$ d **a.** $\frac{36}{4}$ **b.** $12\frac{3}{4}$ **c.** 16 **d.** 9

2. $3 \times 4\frac{2}{3}$ b **a.** $12\frac{2}{3}$ **b.** 14 **c.** $12\frac{6}{3}$ **d.** 6

3. $3\frac{2}{5} \times 1\frac{3}{4}$ c **a.** $3\frac{6}{20}$ **b.** $3\frac{3}{10}$ **c.** $5\frac{19}{20}$ **d.** $4\frac{5}{9}$

148 CHAPTER 6

Reduced Student Pages—clear, readable student pages. Dots next to some exercises tell the teacher which answers appear in the back of the student's book.

PRACTICE EXERCISES

Multiply. Write the product in simplest form.

• **1.** $2 \times \frac{1}{2}$ 1 **2.** $3 \times \frac{2}{3}$ 2 **3.** $5 \times \frac{3}{5}$ 3 **4.** $\frac{1}{2} \times 6$ 3 **5.** $\frac{2}{3} \times 6$ 4

6. $8 \times \frac{3}{4}$ 6 **7.** $\frac{1}{3} \times 9$ 3 **8.** $12 \times \frac{5}{6}$ 10 **9.** $5 \times \frac{2}{3}$ $3\frac{1}{3}$ **10.** $\frac{4}{5} \times 3$ $2\frac{2}{5}$

• **11.** $2 \times 1\frac{1}{2}$ 3 **12.** $3 \times 2\frac{1}{3}$ 7 **13.** $2 \times 3\frac{1}{2}$ 7 **14.** $4 \times 3\frac{1}{4}$ 13 **15.** $3 \times 2\frac{2}{3}$ 8

16. $1\frac{3}{4} \times 8$ 14 **17.** $2 \times 2\frac{5}{8}$ $5\frac{1}{4}$ **18.** $3\frac{1}{6} \times 24$ 76 **19.** $3\frac{7}{20} \times 5$ $16\frac{3}{4}$ **20.** $4 \times 2\frac{1}{12}$ $8\frac{1}{3}$

21. $\frac{7}{8} \times 2\frac{1}{2}$ $2\frac{3}{16}$ **22.** $\frac{3}{5} \times 4\frac{1}{3}$ $2\frac{3}{5}$ **23.** $7\frac{1}{4} \times \frac{8}{9}$ $6\frac{4}{9}$ **24.** $6\frac{2}{3} \times \frac{4}{5}$ $5\frac{1}{3}$ **25.** $\frac{5}{6} \times 5\frac{1}{3}$ $4\frac{4}{9}$

• **26.** $1\frac{1}{5} \times 2\frac{1}{2}$ 3 **27.** $2\frac{2}{3} \times 1\frac{1}{3}$ $3\frac{5}{9}$ **28.** $2\frac{1}{4} \times 2\frac{2}{3}$ 6 **29.** $1\frac{3}{5} \times 2\frac{1}{2}$ 4 **30.** $2\frac{2}{3} \times 3\frac{1}{2}$ $9\frac{1}{3}$

31. $4\frac{3}{4} \times 2\frac{1}{3}$ $11\frac{1}{2}$ **32.** $3\frac{3}{4} \times 1\frac{3}{4}$ $6\frac{9}{16}$ **33.** $4\frac{2}{3} \times 1\frac{2}{7}$ 6 **34.** $4\frac{1}{2} \times 2\frac{2}{3}$ 12 **35.** $1\frac{1}{3} \times 3\frac{2}{7}$ $4\frac{8}{21}$

36. $2\frac{1}{4} \times 3$ $6\frac{3}{4}$ **37.** $3\frac{2}{3} \times 3\frac{3}{4}$ $13\frac{3}{4}$ **38.** $\frac{2}{5} \times 3\frac{1}{2}$ $1\frac{2}{5}$ **39.** $5\frac{1}{3} \times 2\frac{3}{8}$ $12\frac{2}{3}$ **40.** $\frac{1}{2} \times 9\frac{2}{3}$ $4\frac{5}{6}$

41. $5\frac{1}{5} \times 6\frac{1}{4}$ $32\frac{1}{2}$ **42.** $7\frac{2}{3} \times 2$ $15\frac{1}{3}$ **43.** $\frac{1}{3} \times 3\frac{2}{4}$ $1\frac{1}{6}$ **44.** $6\frac{2}{5} \times 10$ 64 **45.** $10\frac{1}{2} \times \frac{4}{7}$ 6

46. $\frac{2}{5} \times \frac{3}{8} \times 10$ $1\frac{1}{2}$ **47.** $2\frac{1}{2} \times \frac{1}{3} \times \frac{1}{4}$ $\frac{5}{24}$ **48.** $3\frac{2}{3} \times \frac{1}{4} \times 3$ $2\frac{3}{4}$ **49.** $\frac{2}{3} \times \frac{1}{5} \times 2\frac{3}{4}$ $\frac{11}{30}$

50. $1\frac{1}{2} \times 1\frac{1}{4} \times 1\frac{1}{3}$ $2\frac{1}{2}$ **51.** $4\frac{2}{5} \times 8\frac{1}{3} \times 2\frac{1}{4}$ $82\frac{1}{2}$ **52.** $5\frac{2}{5} \times 7\frac{1}{3} \times 6\frac{1}{11}$ $241\frac{1}{5}$ **53.** $9\frac{3}{5} \times 6\frac{1}{8} \times 2\frac{4}{7}$ $151\frac{1}{5}$

54. $\left(2 + 1\frac{1}{2}\right) \times \frac{4}{7}$ 2 **55.** $\left(1\frac{1}{4} + 3\frac{3}{8}\right) \times \frac{1}{2}$ $2\frac{5}{16}$ **56.** $\left(\frac{3}{4} + \frac{7}{8}\right) \times \frac{2}{3}$ $1\frac{1}{12}$ **57.** $\left(\frac{2}{3} + 1\frac{1}{2}\right) \times \frac{3}{4}$ $1\frac{5}{8}$

Solve.

• **58.** Maria's boat has a maximum speed of 36 miles per hour. Its cruising speed is $\frac{3}{4}$ of its maximum speed. What is the boat's cruising speed? 27 mph

59. Latoya earns $18.00 an hour as a carpenter building boats. One week she works 42.5 hours. How much money does she earn that week? $765.00

60. Cesar's budget allows him to spend $78,000 on lumber. He has spent $45,963. How much more can he spend without going over budget? $32,037

61. A ship can travel at a rate of $18\frac{3}{4}$ miles per hour. How far can it travel in $3\frac{1}{5}$ hours? 60 mi

TIME OUT A ship is sailing from New York City to Southhampton, England. The total distance is about 3,600 miles. The ship has traveled $\frac{1}{2}$ the distance to the halfway point. How far has the ship traveled? 900 mi

FRACTION COMPUTATION 149

TIME OUT It may be helpful for students to draw a number line to illustrate that "$\frac{1}{2}$ the distance to the halfway point" is $\frac{1}{4}$.

COMMON ERROR

When multiplying mixed numbers, some students will multiply whole numbers and fractions instead of writing the mixed numbers as improper fractions. For example, in item 3 of Checkpoint, these students will select a. Have these students follow this procedure before they begin to multiply:

$$3\frac{2}{5} = \frac{(5 \times 3) + 2}{5} = \frac{17}{5}$$
$$1\frac{3}{4} = \frac{(1 \times 4) + 3}{4} = \frac{7}{4}$$

ASSIGNMENTS

Level 1	Odd 1–45, 58–59
Level 2	Even 2–48, 58–61, TO
Level 3	31–61

FOLLOW-UP ACTIVITY

APPLICATION Present the following recipe. Direct students to rewrite the recipe, to make $2\frac{1}{2}$ times the amount.

White Bread	
2 pkgs. dry yeast	2 tbsp salt
$\frac{1}{2}$ cup water	2 cups milk
$\frac{1}{3}$ cup sugar	$1\frac{1}{2}$ cups water
$\frac{1}{3}$ cup shortening	10 cups flour

SUPPLEMENTARY MATERIALS

TRP Practice, p. 60
TRP Reteaching, p. 37
TRP Transparency 30

149

Common Error—helpful hints that identify common errors and help you plan strategies to deal with them.

Assignments—guidelines for gearing lessons to students' varying abilities, making lessons challenging and interesting, yet attainable for each student.

Follow-up Activity—an additional activity for reinforcing or extending the skills taught in the lesson. The nature of the activity is identified in boldfaced type.

Supplementary Materials—correlations to the extra support materials in the *Teacher's Resource Package*.

TEACHER'S RESOURCE PACKAGE

One source provides
- **Maximum teaching flexibility.**
- **Materials for students of all ability levels.**
- **Reinforcement and extension of skills.**
- **All items in blackline-master format.**

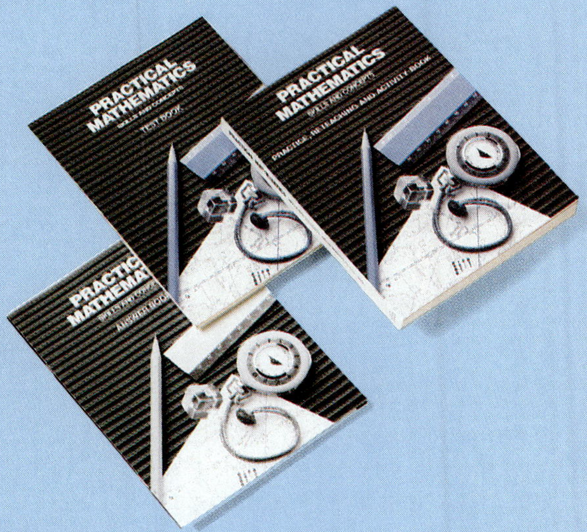

PRACTICE, RETEACHING AND ACTIVITY BOOK

- **Practice Sheets**—one page of practice exercises per lesson.
- **Reteaching Sheets**—one page of practice with instruction and worked examples for every skills lesson in the text.
- **Applications Sheets**—extensions of the real-life problem-solving applications in the text.
- **Group Projects**—cooperative-learning projects for each chapter.
- **Lesson Aids**—useful forms, such as grid paper, rulers, and protractors.

TRANSPARENCIES

- Includes a transparency for every lesson in the text, enabling you to work through problems as a class activity.
- Features two lessons on each transparency.

TEST BOOK

- **Skills Inventory**—a multiple-choice inventory test that evaluates students' initial skill proficiencies.
- **Posttests**—a two-page test for each chapter in the book—available in alternate A and B forms.
- **Cumulative Tests**—multiple-choice tests for every Cumulative Review in the book.
- **Final Test**—a multiple-choice end-of-year exam.
- **Management Forms**—tools for recording and tracking student progress.

THE ANSWER BOOK

- Contains the answers to every worksheet and test.

PACING CHART

This pacing chart is a general guide for using Holt Practical Mathematics during the school year. It offers a suggested number of days to spend on each chapter for 3 levels of student ability. Your own individual needs and those of your students may require alternate plans.

	Suggested Number of Days		
Chapter	Level 1 Below Average Students	Level 2 Average Students	Level 3 Above Average Students
1 Numeration	12	10	8
2 Whole Number Computation	16	14	10
3 Decimal Computation	15	13	10
4 Statistics	12	12	14
5 Fractions	12	9	9
6 Fraction Computation	16	14	12
7 Measurement	14	12	10
8 Ratio and Proportion	9	9	11
9 Percent	14	14	14
10 Probability	8	9	10
11 Geometry	12	15	16
12 Perimeter, Area, and Volume	12	14	16
13 Rational Numbers	10	13	15
14 Equations	8	12	15
Total	170	170	170

Chapter Objectives

Chapter 1 NUMERATION
A. Identify the uses of numbers.
B. Write word names for numbers and vice versa.
C. Identify the values of digits in whole numbers and decimals.
D. Compare and order whole numbers and decimals.
E. Round whole numbers and decimals.
F. Write numbers in exponent form and vice versa.
G. Identify needed information in solving problems.
H. Apply computational skills in real-life situations.

Chapter 2 WHOLE NUMBER COMPUTATION
A. Estimate by rounding.
B. Estimate by using front-end estimaton.
C. Estimate by using compatible numbers.
D. Add or subtract whole numbers.
E. Multiply or divide whole numbers.
F. Write subproblems to solve problems.
G. Apply computational skills in real-life situations.

Chapter 3 DECIMAL COMPUTATION
A. Estimate with decimals.
B. Add or subtract decimals.
C. Multiply or divide decimals by powers of 10.
D. Multiply or divide decimals.
E. Guess and check to solve problems.
F. Apply computational skills in real-life situations.

Chapter 4 STATISTICS
A. Make a grouped-frequency table.
B. Find statistical measures.
C. Read and interpret data from a graph.
D. Make a broken-line graph.
E. Make a table to solve problems.
F. Apply computational skills in real-life situations.

Chapter 5 FRACTIONS
A. Write fractions.
B. Find the GCF or LCM.
C. Write fractions in lowest terms.
D. Compare fractions.
E. Write improper fractions as whole numbers or mixed numbers and vice versa.
F. Write decimals as fractions or mixed numbers.
G. Write fractions or mixed numbers as decimals.
H. Interpret remainders to solve problems.
I. Apply computational skills in real-life situations.

Chapter 6 FRACTION COMPUTATION
A. Multiply fractions.
B. Divide fractions.
C. Add fractions.
D. Subtract fractions.
E. Work backward to solve problems.
F. Apply computational skills in real-life situations.

Chapter 7 MEASUREMENT
A. Change measurement units within the customary and metric systems.
B. Choose the most reasonable estimate of measure.
C. Add or subtract customary measures.
D. Compute with time.
E. Read temperatures.
F. Measure lengths to the nearest $\frac{1}{16}$ inch or millimeter.
G. Draw diagrams to solve problems.
H. Apply computational skills in real-life situations.

Chapter 8 RATIO AND PROPORTION
A. Write ratios.
B. Write ratios in simplest form.
C. Find unit rates.
D. Compare rates.
E. Determine if two ratios form a proportion.
F. Write and solve proportions.
G. Apply computational skills in real-life situations.
H. Solve problems using estimation.

Chapter 9 PERCENT
A. Write a percent as a decimal.
B. Write a percent as a fraction, whole number, or mixed number.
C. Write a decimal, fraction, or mixed number as a percent.
D. Compute, using the three cases of percent.
E. Estimate with percents.
F. Find the percent of increase or decrease.
G. Apply computational skills in real-life situations.
H. Solve problems by supplying missing information.

Chapter 10 PROBABILITY
A. Find the probabilities of simple events.
B. Find the odds in favor of or against events occurring.
C. Predict the number of times an event will occur.
D. Find the number of possible outcomes.
E. Find the probability of independent and dependent events.
F. Make an organized list to solve problems.
G. Apply computational skills in real-life situations.

Chapter 11 GEOMETRY
A. Draw geometric figures.
B. Classify and measure angles.
C. Find the complement or supplement of an angle.
D. Construct a congruent angle and an angle bisector.
E. Find the length of the diameter or radius of a circle.
F. Classify a triangle as equilateral, isosceles, or scalene.
G. Find the measure of the third angle of a triangle.
H. Find the missing side of a right triangle.
I. Find the missing sides of similar figures.
J. Find patterns to solve problems.
K. Apply computational skills in real-life situations.

Chapter 12 PERIMETER, AREA, AND VOLUME
A. Name solid figures.
B. Count the numbers of faces, edges, and vertices of a prism.
C. Find the perimeters of polygons.
D. Find the areas of polygons.
E. Find the circumferences and areas of circles.
F. Find the surface areas of rectangular prisms.
G. Find the volumes of rectangular prisms, pyramids, cylinders, or cones.
H. Apply computational skills in real-life situations.
I. Solve a problem by writing a simpler problem.

Chapter 13 RATIONAL NUMBERS
A. Compare the absolute values of rational numbers.
B. Add and subtract integers.
C. Add and subtract rational numbers.
D. Multiply and divide integers.
E. Multiply and divide rational numbers.
F. Write a standard numeral in scientific notation and vice versa.
G. Name points in a coordinate plane.
H. Apply computational skills in real-life situations.
I. Solve a problem by selecting an appropriate strategy.

Chapter 14 EQUATIONS
A. Simplify numerical expressions.
B. Evaluate algebraic expressions.
C. Write algebraic expressions and sentences.
D. Graph the solution to an equation or inequality.
E. Solve equations.
F. Graph solutions to equations with two variables.
G. Apply computational skills in real-life situations.

To: The Teachers

From: The Authors

Re: Philosophy for Practical Mathematics: Skills and Concepts

Most of your students have probably not enjoyed their study of mathematics very much over the years, nor have they been able to relate it to their current or future lives. Yet we know how essential it is that these students achieve a certain level of understanding about mathematics and are able to use that understanding to solve everyday, real-life problems.

Practical Mathematics: Skills and Concepts was written to emphasize the relevance of mathematics to everyday life. Students will review basic mathematics skills and learn how they apply to consumer and career situations.

For students who find learning mathematics difficult, this book will assist them in the following ways:

—*Basic skills* are broken down into subskills that are clearly explained in carefully worked-out examples.

—*Conceptual understanding* is fostered through the use of discussion questions designed to have the students explain their reasoning rather than just give a rote answer.

—*Common student errors* are quickly identified in Checkpoint exercises that appear in many lessons on computational skills.

—*Problem-solving situations* provide opportunities for students to make conjectures, try various methods, verify results, and learn from errors as well as from successes.

—Activities appear throughout the book to help students acquire greater proficiency with the *calculator*, and show them the need to use *estimation* to check the reasonableness of their answers.

Overall, in our lessons and exercises, we have tried to show students the usefulness and importance of mathematical skills and logical reasoning in surviving in the real world. We hope, therefore, that this book will help them better bridge the gap between what they learn in school and what they do outside of school now and in the future.

Marguerite M. Fredrick

Robert D. Postman

Steven J. Leinwand

Laurence R. Wantuck

Chapter 1 Numeration

OBJECTIVE

- Test the prerequisite skills needed to learn the concepts developed in Chapter 1.

USING THE TEST

The Pre-Skills Test is designed to diagnose students' strengths and weaknesses on prerequisite skills necessary to study the mathematics in Chapter 1.

Assign the Pre-Skills Test. Allow students to work together in pairs or small groups. Group members should help those who demonstrate a misunderstanding of a concept or a weakness in a skill.

The following table correlates the items on the Pre-Skills Test with the prerequisite skill and the lesson(s) in the chapter for which it is needed.

Item(s)	Prerequisite skills	Lesson(s)
1–5	Identify points on a number line.	1.2, 1.8
6–11	Complete counting sequences.	1.2, 1.4
12–14	Write numbers given word names.	1.2, 1.3
15–17	Write numbers in expanded form.	1.2, 1.3
18–22	Round numbers to the nearest 10.	1.8
23–50	Recall basic facts.	1.9

You may wish to use the Skills Inventory Test available in the Teacher's Resource Package to establish groups for three levels of instruction or to determine the students' readiness to begin the test.

SUPPLEMENTARY MATERIAL

TRP Skills Inventory Test, pp. 1–8

Pre-Skills Test

Use the number line to identify the point using a letter.

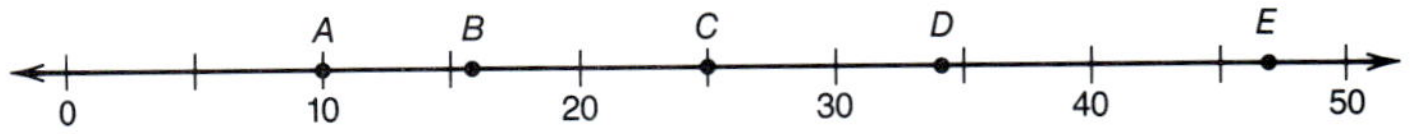

1. 10 *A* **2.** 25 *C* **3.** 34 *D* **4.** 16 *B* **5.** 47 *E*

Complete the sequence.

6. 25, 30, 35, 40, ■, ■, ■ 45, 50, 55

7. 46, 56, 66, ■, ■, ■ 76, 86, 96

8. 100, 200, 300, ■, ■, ■ 400, 500, 600

9. 2,000, 4,000, 6,000, ■, ■ 8,000; 10,000

10. 0.1, 1.1, 2.1, ■, ■, ■ 3.1, 4.1, 5.1

11. 5.4, 5.3, ■, ■, 5.0 5.2, 5.1

Write the number.

12. eleven 11 **13.** twenty-one 21 **14.** thirty-three 33

Complete.

15. 15 = 1 ten ■ ones 5 **16.** 27 = ■ tens 7 ones 2 **17.** 47 = 4 tens ■ ones 7

Round to the nearest 10. Use the number line.

18. 8 10 **19.** 22 20 **20.** 29 30 **21.** 41 40 **22.** 46 50

Find the answer.

23. 7 + 2 = 9 **24.** 9 + 5 = 14 **25.** 6 + 7 = 13 **26.** 8 + 4 = 12 **27.** 7 + 7 = 14 **28.** 9 + 6 = 15 **29.** 5 + 4 = 9

30. 12 − 6 = 6 **31.** 14 − 9 = 5 **32.** 15 − 7 = 8 **33.** 13 − 4 = 9 **34.** 10 − 8 = 2 **35.** 16 − 7 = 9 **36.** 17 − 8 = 9

37. 4 × 3 = 12 **38.** 5 × 5 = 25 **39.** 8 × 3 = 24 **40.** 7 × 6 = 42 **41.** 5 × 9 = 45 **42.** 8 × 7 = 56 **43.** 6 × 4 = 24

44. 6)54 9 **45.** 8)72 9 **46.** 7)56 8 **47.** 9)81 9 **48.** 6)42 7 **49.** 5)40 8 **50.** 3)24 8

Chapter 1

NUMERATION

We use numbers in many different ways. What would happen if all numbers disappeared one day?

OBJECTIVE

- Formulate problems using text and data.

USING THE CHAPTER OPENER

Discuss students' answers to the question on the page. Some problems that might arise include the inability to tell time, to deliver mail based on addresses, to price items for sale, to issue licenses or credit cards.

Ask students what they could substitute for numbers and guide them in considering the practical aspects of their suggestions. If students suggest some form of symbols to replace 0 to 9, discuss the fact that the numerals 0 to 9 are symbols for numbers; and in this problem situation, the numbers as well as the symbols have disappeared.

COOPERATIVE LEARNING

Write the following on the chalkboard.

NEEDED: System to replace missing numbers. Must be efficient and easy to learn.

Assign groups of students the task of devising a new numeration system. Explain that it need not be a base-10 system. Then have each group show examples of their new system applied to those areas they feel are most important.

As an extension, the groups could also show how their system might eliminate the use of numbers in some areas. For example, temperature could be reported as hot, warm, or cold rather than in degrees.

Allow students time to study each display, and then conduct a class discussion of the merits and pitfalls of each to determine if there is one system the students feel works best.

SUPPLEMENTARY MATERIALS

TRP Group Projects, p. 1

Objective

- Classify numbers according to their use as a count, a measure, a location, or a code.

Teaching the Lesson

INTRODUCTION Begin the discussion by pointing out that numbers are used in different ways. Have students suggest ways in which numbers appear in their everyday life. Examples might include: telephone numbers, street addresses, zip codes, dates, temperatures, and prices.

INSTRUCTION Use Example 1 to lead students to see how numbers answer different questions.

Example 2 shows how numbers are used on a check. Use the table above Example 2 to show how a number's classification is determined by which question it answers.

The text offers the general question "How much?" to assist in classifying a number as a measure. You may wish to discuss other more specific questions that would help classify a number as a unit of measure. Examples might include: How big? How long? How tall? How heavy? How wide?

FOR DISCUSSION This question is designed to lead students to discover that the context in which a number is used will determine its classification.

Sample answers for this discussion question are:

count—10 students

location—10th St., July 10th

measure—10 in.

code—10 FMZ

Accept any reasonable answers offered by the students that demonstrate their understanding of this concept.

1.1 NUMBERS EVERYWHERE

We use numbers in many different ways. Does 45 Pearl Street mean there are 45 houses on Pearl Street? Is Highway 95 longer than Route 4?

Numbers answer the questions, "how many," "how much," "where," "when," and "which one."

EXAMPLE 1 Read the following story. Does the number tell "how many," "how much," "where," "when," or "which one"?

The Grade 9 class trip to Washington, D.C. cost $500 a student. On September 18th, 37 students toured 1600 Pennsylvania Avenue.

THINK: 9 tells "which one."
18th tells "when."
1600 tells "where."
$500 tells "how much."
37 tells "how many."

A number can be classified as a count, a measure, a location, or a code.

Use	Question	Classification
17 albums	How many?	Count
88 km, $2.15	How much?	Measure
5th Avenue, Oct. 12, 1492	Where?/When?	Location
555-7676, ITX 5871	Which one?	Code

EXAMPLE 2 Classify each number on the check below.

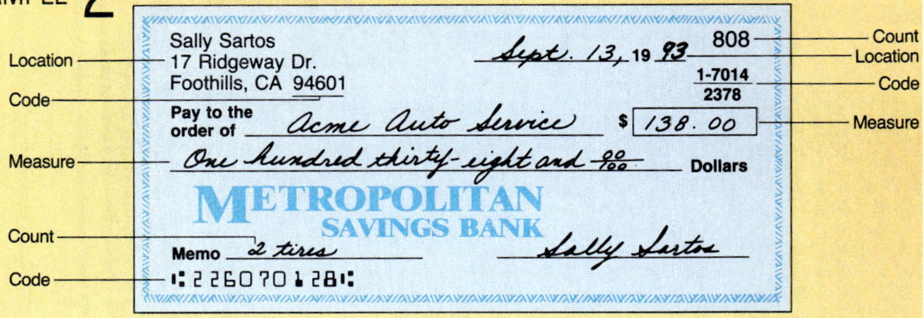

FOR DISCUSSION See TE side column.

Explain how 10 could be used as a count, a measure, a location, or a code.

PRACTICE EXERCISES

1. How much?; 2. Measure; 3. Which one?; 4. Code; 5. How many?; 6. Count; 7. When?; 8. Location; 9. How much?; 10. Measure; 11. Where?; 12. Location; 13. Which one?; 14. Code; 15. When?; 16. Location; 17. How much?; 18. Measure; 19. Where?; 20. Location

Complete the table. Does the number tell "how many," "how much," "where," "when," or "which one"? Then, classify the number.

Use	Question	Classification
9 kg	• 1. ■	• 2. ■
409-1467	3. ■	4. ■
15 books	5. ■	6. ■
June 30, 1989	7. ■	8. ■
$38.99	9. ■	10. ■

Use	Question	Classification
12th Street	11. ■	12. ■
Model number	13. ■	14. ■
Birth date	15. ■	16. ■
Height	17. ■	18. ■
Latitude	19. ■	20. ■

Use the bill to classify the number.

• **21.** 284-4916-75 • **22.** 35
23. 02874394 **24.** 5.67
25. 9-27-89 **26.** 3
27. 12459-3801 **28.** 12.68
29. 67912358 **30.** 1

Quality Stores

Acct. No. 284-4916-75
Martha Rumple
35 Ranchero Drive
Los Pedros, NM 12459-3801

Date	Reference	Quantity	Amount
9-13-89	02874394	3	16.23
9-27-89	67912358	1	5.67
9-29-89	03491726	4	12.68

21. code; 22. location; 23. code; 24. measure; 25. location; 26. count; 27. location; 28. measure; 29. code; 30. count

Solve.

31. Class 10a took a trip to the museum on October 3rd. The bus cost $5 and admission to the museum was $2. Which number is a code? 10a

32. The historical site at 125 Main St. was built in 1738. It is open from 10 to 3 every day. Admission is $1.50. Which numbers are locations?
125, 1738, 10, 3

TIME OUT

Numbers represented by dots as a geometric figure are called **figurate numbers.** Here are the first four triangular numbers.
See page 477 for answers.

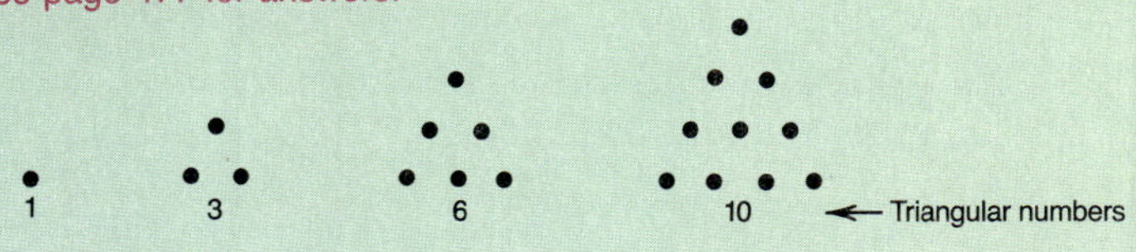

1. Arrange dots to form the first four square numbers.

2. Arrange dots to form the first four rectangular numbers.

TIME OUT Explain that *figurate numbers* are named by the figure or shape they form. Review with students the characteristics of triangles, squares, and rectangles. Demonstrate the use of dots to form and find triangular numbers. Lead students to see how triangular numbers can be expressed as the sum of consecutive numbers.

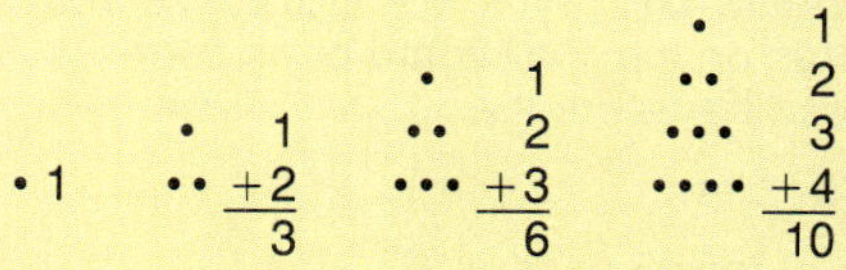

ASSIGNMENTS

Level 1	1–10, 21–31
Level 2	1–20, Odd 21–31, 32, TO
Level 3	11–32, TO

FOLLOW-UP ACTIVITY

APPLICATION Have students look for numbers on the sides of a cereal box, the sports page of a newspaper, or a supermarket advertisement. Have them cut out a section that shows all four classifications of numbers. Have students make a chart to show how to classify the numbers.

SUPPLEMENTARY MATERIALS

TRP Practice, p. 1

TRP Reteaching, p. 1

TRP Transparency 1

OBJECTIVES

- Identify the place and value of a digit in whole numbers.
- Write word names for numbers and vice versa.

TEACHING THE LESSON

WARM-UP Write the following exercises on the chalkboard or an overhead transparency.

Complete.

1. 2 tens 3 ones = ■ (23)
2. 7 tens 0 ones = ■ (70)
3. 2 hundreds 3 tens 5 ones = ■ (235)
4. 3 hundreds 0 tens 2 ones = ■ (302)
5. 9 hundreds 0 tens 0 ones = ■ (900)

INTRODUCTION Write 4671804 on the chalkboard. Ask students to read this number. Note the difficulty without using commas. Review the procedure of placing commas every three digits from right to left.

INSTRUCTION As students examine the place-value chart, emphasize that each digit has a place value 10 times the value of the place to the right.

As you discuss Example 1, make sure students understand that to find the value of a digit, they must first identify its place.

In Example 2, point out that the way we read a number is commonly referred to as its short word name. Make sure students understand how the period names are used in reading a number. Elicit from them that the ones period is the only period not named when a number is read.

As you discuss Example 3, carefully note the use of zeros as placeholders in the hundred-thousands and ten-thousands place.

1.2 PLACE VALUE: WHOLE NUMBERS

Our number system is made up of 10 digits: 0, 1, 2, 3, 4, 5, 6, 7, 8, and 9.

To write numbers greater than 9, a place-value system has been developed. Study the following place-value chart. Notice that each digit has a value 10 times greater than the value of the place to its right.

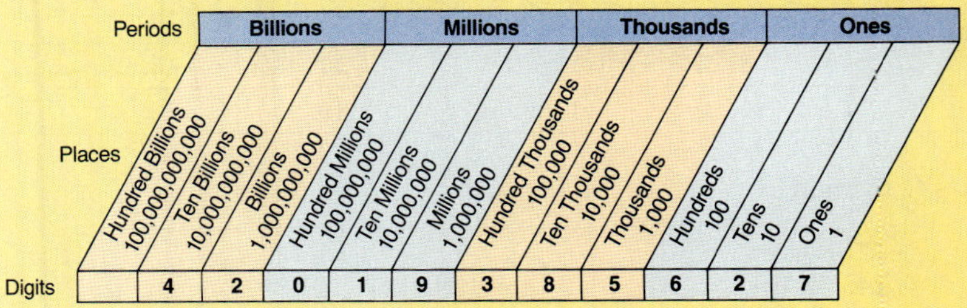

You can find the value of a digit in a number by multiplying the digit by its place value.

EXAMPLE 1 Find the place and the value of 4 in the number 42,019,385,627.

THINK: In the chart above, the 4 is in the ten-billions place.

DIGIT	×	PLACE VALUE	=	VALUE
4	×	10,000,000,000	=	40,000,000,000

You can read and write word names for numbers using the period names.

EXAMPLE 2 Read 42,019,385,627. Then write the word name.

READ: 42 billion, 19 million, 385 thousand, 627

WORD NAME: forty-two billion, nineteen million, three hundred eighty-five thousand, six hundred twenty-seven

To write a number from a word name, start with the digit in the greatest place. Then write the digit for each place to the right. Write zeros as placeholders if necessary. Use commas to separate the periods.

EXAMPLE 3 Write the number for two million, six thousand, three hundred twenty-nine.

THINK:	Millions	Hundred thousands	Ten thousands	Thousands	Hundreds	Tens	Ones
WRITE:	2	0	0	6	3	2	9

NUMBER: 2,006,329

PRACTICE EXERCISES

See Extra Practice, page 426.

Find the place and the value of the underlined digit. See page 477 for answers to Exercises 1–36.

• **1.** 341	• **2.** 625	**3.** 2,847	**4.** 9,564
5. 32,419	**6.** 78,627	**7.** 80,439	**8.** 93,205
9. 321,604	**10.** 459,328	**11.** 845,169	**12.** 896,437
13. 3,002,158	**14.** 7,325,893	**15.** 70,138,625	**16.** 113,864,018
17. 2,589,411,007	**18.** 39,425,687,990	**19.** 739,435,689,101	**20.** 963,004,211,357

Write the word name.

• **21.** 175	• **22.** 358	**23.** 4,568	**24.** 8,604
25. 23,961	**26.** 48,543	**27.** 67,801	**28.** 73,496
29. 325,406	**30.** 462,001	**31.** 580,102	**32.** 706,005
33. 1,890,000	**34.** 2,340,010	**35.** 3,400,009	**36.** 8,060,123

Write the number.

• **37.** 336 thousand, 154 336,154

• **38.** 893 thousand, 201 893,201

39. 4 million, 4 thousand, 4 4,004,004

40. 7 million, 16 thousand, 89 7,016,089

41. 5 billion, 28 thousand, 50 5,000,028,050

42. 85 billion, 347 85,000,000,347

43. forty-seven thousand, two hundred six 47,206

44. two hundred twenty-five thousand, sixty 225,060

45. sixty-five million, forty-six thousand, one hundred fifty 65,046,150

46. thirty-four billion, nine hundred five million, sixty-two 34,905,000,062

47. Copy this blank check. Use the information provided to fill it out.

Amount: $109.58

Date: 9-15-93

Check number: 562

Memo: Tune-up

Pay to: Gregg's Garage

48. Design your own check. Then fill out one of your checks for a down payment on a car.

ASSIGNMENTS

Level 1	1–12, 21–28, 37–40, 43–44, 47, 48
Level 2	1–16, Odd 21–35, 37–48
Level 3	13–20, 29–48

FOLLOW-UP ACTIVITY

ENRICHMENT Provide students with the names of the next eight periods after the billions period. Write the following on the chalkboard or an overhead transparency.

Word name	Number	Power of 10
1 hundred	100	10^2
1 thousand	1,000	10^3
1 million	1,000,000	10^6
1 billion	■	■ (10^9)
1 trillion	■	■ (10^{12})
1 quadrillion	■	■ (10^{15})
1 quintillion	■	■ (10^{18})
1 sextillion	■	■ (10^{21})
1 septillion	■	■ (10^{24})
1 octillion	■	■ (10^{27})
1 nentillion	■	■ (10^{30})
1 decitillion	■	■ (10^{33})

(Check each answer to be sure that the number of zeros matches the corresponding power of 10.)

Have the students write the number and the power of 10 associated with each word name. You may wish to have them work together to make a place-value chart on poster board, similar to the one on page 4, through the decitillions period.

SUPPLEMENTARY MATERIALS

TRP Practice, p. 2

TRP Reteaching, p. 2

TRP Lesson Aids, pp. 1, 2

TRP Transparency 1

OBJECTIVES

- Identify the place and value of a digit in decimals.
- Write word names for decimals and vice versa.

TEACHING THE LESSON

WARM-UP Write the following exercises on the chalkboard or an overhead transparency.

Complete.

1. 6 tenths = ■ (0.6)
2. 8 tenths = ■ (0.8)
3. 51 hundredths = ■ (0.51)
4. 93 hundredths = ■ (0.93)
5. 7 hundredths = ■ (0.07)

INTRODUCTION Have a volunteer draw a place-value chart on the chalkboard that shows the hundreds place through the hundredths place. Encourage students to look for similarities on both sides of the decimal point. You may wish to use the following diagram to illustrate the relationship between the places to the right and left of the ones place.

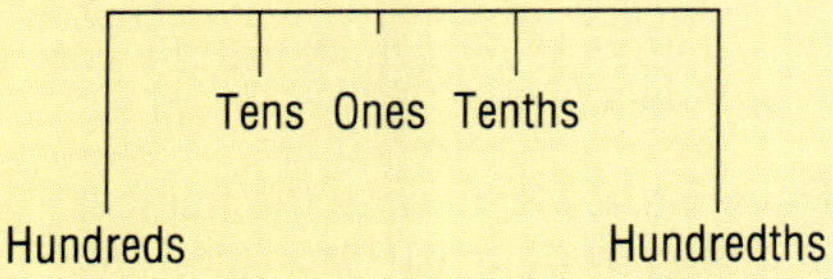

Have the students extend the place-value chart through millionths using this method.

INSTRUCTION As you discuss Example 1, point out that the procedure for finding the value of a digit in a decimal is the same as that for whole numbers. First identify the place of the digit, and then multiply the digit by its place value.

In Example 2, show the necessity of using the hyphen in the place-value names. You can illustrate this using 0.0900 (nine hundred ten-thousandths).

1.3 PLACE VALUE: DECIMALS

You can extend the place-value chart to write decimals less than one.

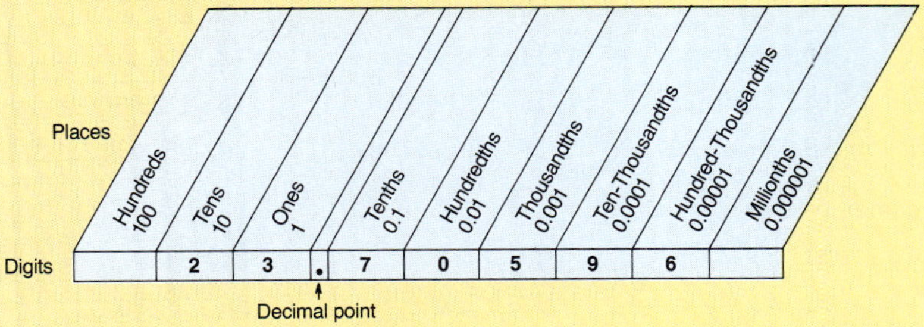

You can find the value of a digit in a decimal by multiplying the digit by its place value.

EXAMPLE 1 Find the place and the value of 5 in the number 23.70596.

THINK: In the chart above, the 5 is in the thousandths place.

DIGIT	×	PLACE VALUE	=	VALUE
5	×	0.001	=	0.005

The decimal point that appears between the ones and tenths place is read as *"and."*

EXAMPLE 2 Read 23.70596. Then write the word name.

READ: 23 *and* 70,596 hundred-thousandths

WORD NAME: twenty-three and seventy thousand, five hundred ninety-six hundred-thousandths

Sometimes zeros are needed as place holders when writing decimals.

EXAMPLE 3 Write the number for seventy thousand and thirty-seven hundred-thousandths.

THINK: Hundred-thousandths means 5 decimal places.

WRITE: 70,000.00037

FOR DISCUSSION

See TE side column.

You have two 5-digit decimals. The middle digit in each number is the same. Do the middle digits have the same value? Use examples to explain your answer.

PRACTICE EXERCISES

See Extra Practice, page 426.

Find the place and the value of the underlined digit. See page 477 for answers to Exercises 1–36.

•**1.** 23.1	•**2.** 45.36	**3.** 61.47	**4.** 93.85
5. 1.046	**6.** 8.193	**7.** 14.987	**8.** 63.795
9. 5.0036	**10.** 17.8934	**11.** 59.7012	**12.** 125.4987
13. 6.39872	**14.** 10.10106	**15.** 18.76304	**16.** 78.04921
17. 0.529716	**18.** 2.940372	**19.** 11.693724	**20.** 41.100846

Write the word name.

•**21.** 0.5	•**22.** 1.8	**23.** 16.9	**24.** 59.6
25. 0.04	**26.** 0.29	**27.** 1.51	**28.** 2.89
29. 0.007	**30.** 0.039	**31.** 0.241	**32.** 1.372
33. 0.0002	**34.** 0.0017	**35.** 0.0356	**36.** 0.1245

Write the number.

•**37.** 4 and 27 hundredths 4.27

•**38.** 13 and 63 hundredths 13.63

39. 195 thousandths 0.195

40. 66 thousandths 0.066

41. 12 and 249 thousandths 12.249

42. 49 and 7 thousandths 49.007

43. 1,259 ten-thousandths 0.1259

44. 48 ten-thousandths 0.0048

45. 83 and 3,459 ten-thousandths 83.3459

46. 56 and 23 ten-thousandths 56.0023

47. 129 hundred-thousandths 0.00129

48. 315 millionths 0.000315

49. 1,238 billionths 0.000001238

50. 342 and 25 billionths 342.000000025

51. four and five thousand three ten-thousandths 4.5003

52. six thousand and sixty-six hundred-thousandths 6,000.00066

Solve.

•**53.** A car's odometer reads 2,105.75. Write the word name for this number.
Two thousand one hundred five and seventy-five hundredths

54. The manual for your new car says, "Bring the car in for a check at five thousand miles." What would your odometer show for this number? 5,000

55. Classify the following numbers:

Power Steering Fluid

Part No.: 1050017
Size: 32 oz
Usage: Power steering

1050017—Code 32—measure

56. The manual states that your tires should be rotated every 15,000 miles. Your odometer reads:

1 4 5 8 7 . 6

Are you overdue in rotating your tires?
No

As you discuss Example 3, ask students to identify the decimal part of the word name. (It follows *and*.)

FOR DISCUSSION Students should explain that the number of digits in a number does not affect the value as much as the placement of the decimal point. For example, in the numbers 37,924 and 37.924 the middle digit, 9, has very different values.

ASSIGNMENTS

Level 1	1–12, 21–28, 37–42, 53–54
Level 2	1–16, Odd 21–35, 37–46, 53–56
Level 3	13–20, 29–36, 43–56

FOLLOW-UP ACTIVITY

APPLICATION Have students work in groups of three. Have each group write a short story of 200 words or less that incorporates 10 decimals in the text. Some topics may be: sports, construction, cars. Have the groups exchange papers and read the stories aloud. You may want to have the students write the numbers as a volunteer reads the story.

SUPPLEMENTARY MATERIALS

TRP Practice, p. 3

TRP Reteaching, p. 3

TRP Lesson Aids, p. 1

TRP Transparency 2

OBJECTIVES

- Compare whole numbers and decimals.
- Order whole numbers and decimals.

TEACHING THE LESSON

WARM-UP Write the following exercises on the chalkboard or an overhead transparency.

For each pair of numbers, name the greatest place in which there is a different digit.

1. 8,870 / 8,780 **2.** 10.006 / 14.689 **3.** 637,912 / 635,912

4. 45.701 / 45.713 **5.** 0.0056 / 0.0038 **6.** 31.09613 / 31.09615

(**1.** hundreds **2.** ones **3.** thousands **4.** hundredths **5.** thousandths **6.** hundred-thousandths)

INTRODUCTION Present the following words: bigger, larger, taller, heavier, lighter, wider, and so on. Ask students what all these words have in common. They should note that all are used for comparing. Discuss the words and symbols used in mathematics to compare numbers: =, <, and >.

INSTRUCTION Example 1 shows how to compare whole numbers that have an equal number of places. Remind students how easy it is to compare whole numbers with an unequal number of digits—the number with more places is always greater.

Use Example 2 to extend this method of comparing numbers with decimal numbers less than 1. Use the following diagram to show why adding zeros to a decimal number will not change its value.

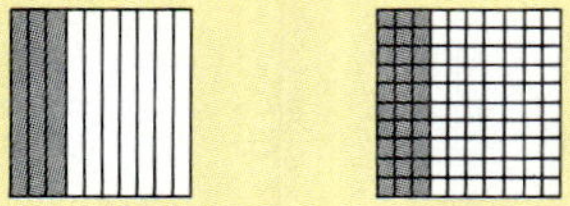

0.3 = 0.30

1.4 COMPARING AND ORDERING NUMBERS

The air distance between New York and Honolulu is 4,964 mi. Is this distance less than (<), equal to (=), or greater than (>) the air distance between New York and Istanbul?

Air Distance (in miles)

Cities	Moscow	New York	Paris
Hong Kong	4,439	8,054	5,985
Honolulu	7,037	4,964	7,438
Istanbul	1,091	4,975	1,400

To solve, you need to compare the air distances. To compare numbers, start at the left and move right until the digits in the same place are different.

EXAMPLE 1 Compare: 4,964 ● 4,975. Write =, <, or >.

THINK: The digits are the same in the thousands and hundreds places. The digits differ in the tens place.

4,964
4,975

Since 6 < 7; 4,964 < 4,975.
↑ **is less than**

The air distance between New York and Honolulu is less than the air distance between New York and Istanbul.

When comparing decimals, you sometimes have to write additional zeros.

EXAMPLE 2 Compare:

a. 0.293 ● 0.2

THINK: 0.2 = 0.200

0.293
0.200

Since 9 > 0, 0.293 > 0.2.
↑ **is greater than**

b. 0.293 ● 0.2936

THINK: 0.293 = 0.2930

0.2930
0.2936

Since 0 < 6, 0.293 < 0.2936.
↑ **is less than**

Three or more numbers can be arranged in order by repeated comparisons.

EXAMPLE 3 Arrange 748.25, 749.12, 748.04 in order from greatest to least.

THINK: The digits are the same in the hundreds and tens places. The digits differ in the ones place.

748.25
749.12
748.04

Since 9 is the greatest, 749.12 is the greatest.

Now compare: 748.25 / 748.04 Since 2 > 0, 748.25 > 748.04.

748.04 is the least.

The numbers in order from greatest to least are 749.12, 748.25, 748.04.

PRACTICE EXERCISES

See Extra Practice, page 426.

Compare. Write =, <, or >.

1. 342 ● 915 <
2. 1,000 ● 999 >
3. 5,638 ● 5,836 <
4. 27,825 ● 27,826 <
5. 493,926 ● 943,962 <
6. 540,500 ● 540,050 >
7. 67,989,192 ● 67,990,192 <
8. 78,340,982 ● 78,340,892 >
9. 0.02 ● 0.2 <
10. 0.9 ● 0.90 =
11. 649.81 ● 650 <
12. 70.5 ● 7.05 >
13. 96.31 ● 936.1 <
14. 1.902 ● 1.934 <
15. 0.71 ● 0.671 >
16. 12.012 ● 12,012 <
17. 10,056.95 ● 10,060.05 <
18. 580.632 ● 581.002 <
19. 25,000.465 ● 25,000 >
20. 1.3842 ● 1.38421 <
21. 35.7004 ● 35.70040 =
22. 823,918.67205 ● 823,918.6725 <
23. 9.084153 ● 9.085143 <
24. 29,621,000.93 ● 29,621,001.03 <
25. 439,708.0023 ● 439,780.0203 <
26. 37,294.5400 ● 37,294.54 =
27. 852,001.3205 ● 852,001.32005 >
28. 300,001,094.56 ● 300,001,094.553 >

Choose the number that is least.

29. **a.** 4,521 **b.** 541 **c.** 1,425 b
30. **a.** 6,739 **b.** 6.739 **c.** 67.39 b
31. **a.** 5.0095 **b.** 5.0059 **c.** 5.00095 c
32. **a.** 10.6803 **b.** 10.8603 **c.** 10.8630 a

33. 9,742; 7,492; 4,279; 2,972
34. 0.863; 0.6308; 0.3086; 0.0863
35. 42,915; 9.730; 4.2915; 0.937
36. 208; 82.67; 78.625; 60.28
37. 36,081; 36,180; 38,016; 306,810
38. 0.527; 0.57; 0.7; 20.75
39. 3,428.8065; 3,428.8605; 3,428,856; 3,428,865

Arrange in order from greatest to least.

33. 4,279; 2,972; 9,742; 7,492
34. 0.0863; 0.863; 0.6308; 0.3086
35. 42,915; 0.937; 4.2915; 9.730
36. 78.625; 82.67; 208; 60.28

Arrange in order from least to greatest.

37. 36,081; 36,180; 306,810; 38,016
38. 0.527; 0.57; 20.75; 0.7
39. 3,428,865; 3,428.8605; 3,428.8065; 3,428,856

Use the table at the top of page 8 to solve.

40. Which is greater, the distance between Moscow and Istanbul or the distance between Paris and Istanbul?
Paris and Istanbul
41. Which is less, the distance between Paris and Hong Kong or the distance between New York and Istanbul?
New York and Istanbul
42. Would you classify the numbers in the table as counts, measures, locations, or codes? Measures
43. The greatest distance is between which two cities?
New York and Hong Kong

In Example 3, note that the ordering of a group of numbers is accomplished through repeated comparisons of pairs of numbers. Provide additional practice by having students arrange the following numbers in order from least to greatest.

45,970 45,709 47,905 45,907

(45,709; 47,905; 45,907; 45,970)

ASSIGNMENTS

Level 1	1–16, 29–36, 40–42
Level 2	1–20, 29–42
Level 3	13–28, 33–43

FOLLOW-UP ACTIVITY

COOPERATIVE LEARNING
Have students work in groups of four. Each group makes a set of 52 number cards out of poster board. Assign each group a range, such as 4-digit numbers, for writing the set of numbers.

Have students shuffle the cards and place the deck face down. Distribute all the cards equally. Each student, in turn, lays down a card. The four cards are compared and the card with the greatest number wins. The student holding the winning card collects the other players' cards. Continue until all the cards have been played. The winner is the student with the most cards.

SUPPLEMENTARY MATERIALS

TRP Practice, p. 4

TRP Reteaching, p. 4

TRP Lesson Aids, p. 3

TRP Transparency 2

OBJECTIVE

- Solve problems that involve reading electric meters.

TEACHING THE LESSON

WARM-UP Write the following exercises on the chalkboard or an overhead transparency.

Find the value of the underlined digit.

1. 362 **2.** 673 **3.** 2,589

4. 14,576 **5.** 39,208 **6.** 70,699

(**1.** 60 **2.** 600 **3.** 2,000 **4.** 10,000 **5.** 30,000 **6.** 70,000)

INTRODUCTION Ask the students to write the number that is 1 more than 19; 39; 79. (20; 40; 80) Ask them what digit is in the ones place of each of their answers. (0)

You may wish to use an abacus to review the concept that in our place-value system the greatest digit that can be shown in any place is 9. When we want to show 1 more than 9, we increase the digit at the left by 1, and we write a 0 in the given place.

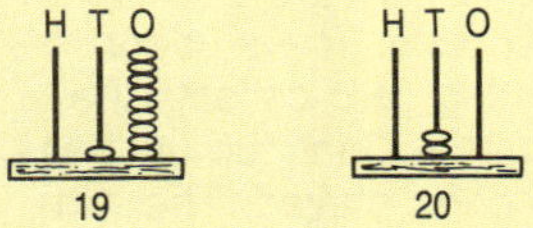

INSTRUCTION In the first illustration, point out that the arrows indicate the direction in which the dials move. Also, point out that the ones dial makes a complete revolution (from 0 to 0) before the tens dial moves one full unit; the tens dial makes a complete revolution (from 0 back to 0) before the hundreds dial moves one full unit, and so on.

After discussing the first example, you may wish to present a situation where the pointer is between 0 and 1. In this case, the student should read the 0.

PROBLEM *Solving* APPLICATION

1.5 CONSUMER: READING METERS

Electric energy is measured in **kilowatt-hours (kWh)**. One kilowatt is 1,000 **watts (W)**. When an appliance uses 1,000 W in an hour, 1 kWh of energy is used.

The more electrical energy that is used, the greater the cost. Meters record the amount of electricity used.

Read the dials from left to right. If the pointer is between two numbers, read the lower number. When the pointer is between 9 and 0, read 9.

Find the number of kilowatt-hours represented on this electric meter.

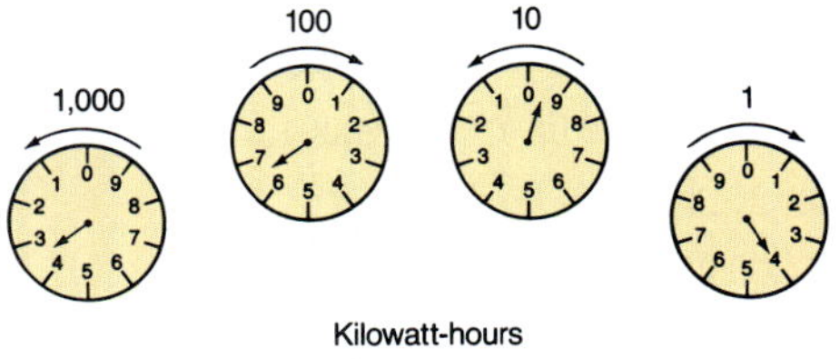

Thousands and tens dials move counterclockwise. Hundreds and ones dials move clockwise.

1. Read the thousands dial.
   THINK: The pointer is between 3 and 4. Read 3.
2. Read the hundreds dial.
   THINK: The pointer is between 6 and 7. Read 6.
3. Read the tens dial.
   THINK: The pointer is between 9 and 0. Read 9.
4. Read the ones dial.
   THINK: The pointer is at 4. Read 4.

So, the reading on the electric meter is 3,694 kWh.

FOR DISCUSSION

See TE side column.

If the pointer on a dial is between 9 and 0, why would you read 9 and not 0?

Write the number of kilowatt-hours represented on the electric meter.

• **1.** 5,927

2. 8,139

3. 3,795

4. 6,002

Use the electric meter at the right for Exercises 5–8.

January 31 Reading

• **5.** Write the number that is 10 more than the January 31 reading.
2,749

6. Write the number that is 100 more than the January 31 reading.
2,839

7. Write the number that is 1,000 more than the January 31 reading.
3,739

8. Write the number that is 1 more than the January 31 reading.
2,740

This electric meter has five dials. The dial at the left is for tens of thousands of kilowatts. Use this meter for Exercises 9–15.

September 15 Reading

• **9.** Write the number of kilowatt-hours represented in the September 15 reading.
23,416

• **10.** Write the number that is 20 more than the September 15 reading.
23,436

11. Write the number that is 500 more than the September 15 reading.
23,916

12. Write the number that is 20,000 more than the September 15 reading.
43,416

13. Write the number that is 6,000 more than the September 15 reading.
29,416

14. What is the greatest number that can be represented on this meter?
99,999

15. Draw a meter with five dials. Then draw pointers to represent 70,806.
Check students' drawings.

FOR DISCUSSION Elicit from students that when the pointer is between 9 and 0, the 0 represents the larger number. That is, it represents 10 on that dial. Thus, the smaller number, 9, is read.

ASSIGNMENTS

Level 1	1–13
Level 2	1–15
Level 3	Even 2–14, 15

FOLLOW-UP ACTIVITY

ENRICHMENT Have the students keep a weekly record of the reading on the electric meters in their homes or apartment buildings. Have them record the data in a table.

After a specified number of weeks, have students calculate the weekly use of electric energy. You may also ask students to compare their tables of data and discuss the reasons why one family may use a significantly greater amount of electric energy than another family.

SUPPLEMENTARY MATERIALS

TRP Practice, p. 5

TRP Transparency 3

OBJECTIVE

- Solve problems that involve reading and interpreting data in postage tables.

TEACHING THE LESSON

WARM-UP Write the following exercises on the chalkboard or an overhead transparency.

One pound is 16 oz. Tell whether each weight is more than 3 lb, but less than 4 lb.

1. 2 lb 12 oz (no)

2. 3 lb 6 oz (yes)

3. 3 lb 15 oz (yes)

4. 0.5 lb (no)

INTRODUCTION Encourage students to discuss experiences they have had mailing large packages to friends or relatives. Have them recall the procedure that the postal clerk followed to determine the cost of mailing a package.

INSTRUCTION Be sure that students understand that the first table is used in determining the cost of mailing, first class, packages or letters with weights that do not exceed 12 oz. The second table is used for determining the cost of mailing, first class, packages or letters with weights that exceed 12 oz.

Point out that only a portion of the second table is provided in the text. A postal clerk has the complete table available for reference.

ASSIGNMENTS

Level 1	1–13
Level 2	1–14
Level 3	Odd 1–11, 13–14

SUPPLEMENTARY MATERIALS

TRP Practice, p. 6

TRP Transparency 3

PROBLEM *Solving* APPLICATION

1.6 CAREER: POSTAL CLERK

Nicholas Tutela is a postal clerk. The ability to read tables is important in his job. The tables below show the rates for mailing letters or packages first class. Notice that if a package weighs more than 5 lb, the cost depends on the number of zones it will travel as well as on its weight.

What is the cost of shipping a 2 lb 8 oz package to Zone 6?

THINK: 2 lb 8 oz is over 2 lb, but does not exceed 3 lb.

Find the entry across row 3 and under Zone 6 of the second table.

The cost is $4.10.

FIRST CLASS

LETTER RATES:
1st ounce . . . . 29¢
Each additional ounce . . . . 23¢

For Pieces Not Exceeding (oz.)	*The Rate is*	For Pieces Not Exceeding (oz.)	*The Rate is*
1	$0.29	7	$1.67
2	0.52	8	1.90
3	0.75	9	2.13
4	0.98	10	2.36
5	1.21	11	2.59
6	1.44	12	2.82

FOR PIECES OVER 12 OUNCES SEE PRIORITY MAIL RATES

PRIORITY MAIL

Weight up to and not exceeding-pound(s)	Rate					
	Local zones 1, 2, and 3	Zone 4	Zone 5	Zone 6	Zone 7	Zone 8
1	$2.90	$2.90	$2.90	$2.90	$ 2.90	$ 2.90
2	2.90	2.90	2.90	2.90	2.90	2.90
3	4.10	4.10	4.10	4.10	4.10	4.10
4	4.65	4.65	4.65	4.65	4.65	4.65
5	5.45	5.45	5.45	5.45	5.45	5.45
6	5.55	5.75	6.10	6.85	7.65	8.60
7	5.70	6.10	6.70	7.55	8.50	9.65
8	5.90	6.50	7.30	8.30	9.40	10.70
9	6.10	7.00	7.95	9.05	10.25	11.75
10	6.35	7.55	8.55	9.80	11.15	12.80

Find the cost of mailing, first class, a package with the given weight.

1. 3 oz $0.75 **2.** 8 oz $1.90

3. 10 oz $2.36 **4.** 15 oz $2.90

5. 1 lb 3 oz $2.90 **6.** 1 lb 8 oz $2.90

Find the cost of mailing the package.

7. 2 lb 6 oz, Zone 3 $4.10

8. 2 lb 6 oz, Zone 5 $4.10

9. 3 lb 4 oz, Zone 4 $4.65

10. 3 lb 4 oz, Zone 8 $4.65

11. 4.5 lb, Zone 2 $5.45

12. 4.5 lb, Zone 7 $5.45

Solve.

13. A package weighs 6.25 lb. What is the least amount and the greatest amount that the first-class postage might be? $5.70; $9.65

14. For which package is the first-class postage greater?

Package A: 7.5 lb, Zone 6
Package B: 8.2 lb, Zone 5 Package A

MIDCHAPTER REVIEW

Use the store receipt to classify each number. [2]

1. 1 measure
2. 9/8/93 location
3. 123 location
4. 4 count
5. (303) 555-3535 code
6. 9.00 measure
7. 80739 code
8. 2 count
9. 19.90 measure
10. CR26 code

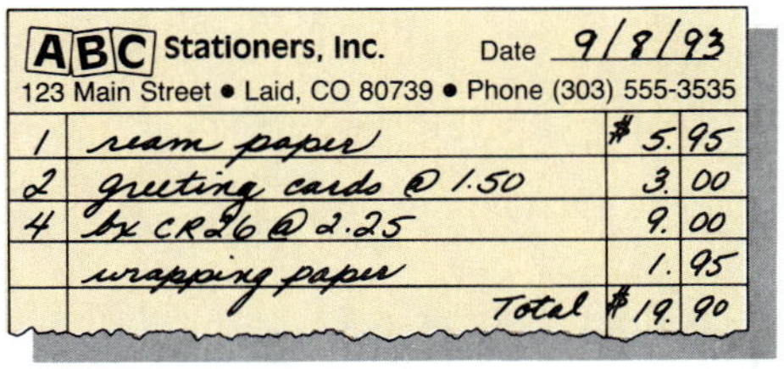

See page 477 for answers to Exercises 11–22.

Find the place and the value of the underlined digit. [4, 6]

11. 4,925
12. 67,819
13. 342,896
14. 7,105,264
15. 16.9
16. 38.72
17. 5.0575
18. 89.13706

Write the word name for the number. [4, 6]

19. 426
20. 279,300
21. 10.562
22. 0.0129

Write the number. [4, 6]

23. 205 thousand, 506 205,506
24. four hundred twenty-nine billion, eighty-four thousand, seventy-five 429,000,084,075
25. ten and one hundred seventy-seven thousandths 10.177
26. forty-five and thirty-two hundred-thousandths 45.00032

Compare. Write =, <, or >. [8]

27. 627 ● 617 >
28. 6,892 ● 6,829 >
29. 3,000 ● 30,000 <
30. 152 ● 15.2 >
31. 26.9 ● 26.900 =
32. 8.05 ● 8.005 >
33. 9.5624 ● 9.5723 <
34. 3.70956 ● 4.71 <
35. 325.81922 ● 325.81923 <

Arrange in order from least to greatest. [8]

36. 9,528; 5,928; 9,285; 9,852
5,928; 9,285; 9,528; 9,852
37. 562; 9.532; 0.565; 0.562
0.562; 0.565; 9.532; 562

Solve. [10]

38. Write the number of kilowatt-hours represented on the electric meter. 1,541

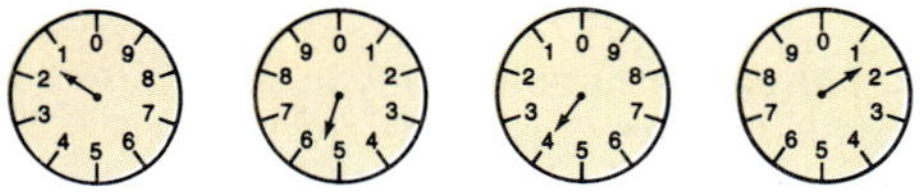

OBJECTIVE

- Evaluate student progress.

USING THE REVIEW

This page provides a means for informally evaluating students' understanding of the skills and concepts covered so far in this chapter.

Have the students look at the page to familiarize themselves with the various question formats that are presented. Discuss any questions that they may have. Then ask them to complete the page independently.

In addition to grading them individually, you may wish to review the answers to the questions collectively with the students.

Page references appear in brackets. They refer to pages on which a particular skill was introduced.

Before continuing on to the topics found in the remainder of the chapter, you may wish to have students review any skills or concepts in which they have demonstrated weakness.

OBJECTIVE

- Identify needed information to solve problems.

TEACHING THE LESSON

WARM-UP Write the following exercises on the chalkboard or an overhead transparency.

Compare. Use =, <, >.

1. 80 ● 85 (<)

2. 54 ● 45 (>)

3. 100 ● 100 (=)

4. 709 ● 790 (<)

5. 4,050 ● 4,500 (<)

6. 9,860 ● 9,680 (>)

INTRODUCTION Present the following situations to the class. Have students discuss the information needed to answer each question.

1. Who is taller, Jane or her cousin, Bill? (Jane's height, Bill's height)

2. How long does it take Jim to walk to school? (the time he leaves home, the time he arrives at school)

Guide students to realize that in everyday life, sometimes the first step in solving a problem is to decide what information is needed.

INSTRUCTION As you discuss the situation, you may want to draw a diagram on the chalkboard. Label the height of both waterfalls and the width of the Horseshoe Falls.

Use the question in B to show that sometimes there is not enough information to solve a problem. Discuss ways of obtaining the missing information, such as consulting an encyclopedia or calling the tourist bureau in Niagara Falls.

PROBLEM Solving STRATEGY

1.7 IDENTIFYING NEEDED INFORMATION

Situation:

Niagara Falls consists of two waterfalls, the Horseshoe Falls and the American Falls. The Horseshoe Falls is about 173 ft high and 2,600 ft wide at its widest point. The American Falls is about 182 ft high.

Strategy:

Identify only the information you need to solve a problem. Sometimes you will not have enough information.

Applying the Strategy:

A. Which falls is higher?

THINK: The Horseshoe Falls is 173 ft high.

The American Falls is 182 ft high.

Compare the heights. 173 < 182

The American Falls is higher.

B. Which falls is wider?

THINK: The Horseshoe Falls is 2,600 ft wide.

The width of the American Falls is not given.

This problem cannot be solved.

PRACTICE EXERCISES

See Extra Practice, p. 427.

Which sentence describes the problem?

• **1.** The ledge of the Horseshoe Falls wears away at the rate of about 3 in. a year. The ledge of the American Falls erodes more slowly. How much does the ledge of the American Falls erode each year?

a. Not enough information is given to solve the problem.

b. More information than you need to solve the problem is given. a

• **2.** The Horseshoe Falls is about 2,600 ft wide at its widest point. The American Falls is about 182 ft high and 1,000 ft wide at its widest point. Which falls is the widest?

a. Not enough information is given to solve the problem.

b. More information than you need to solve the problem is given. b

Solve if possible. If not, identify the needed information.

3. The Niagara River plunges into a gorge. The gorge consists of 120 ft of layers of different stone and 80 ft of limestone. Which part of the gorge is thicker, the layers of different stone or the limestone?
Layers of different stone

4. Niagara Falls was originally at Lewiston, New York, but it gradually moved about 7 mi back upstream toward Lake Erie. How long ago was Niagara Falls at Lewiston, New York?
Needed: The rate at which the falls moved upstream.

5. There are four observation towers so that tourists can view the falls. One tower is 282 ft high. Another tower is 500 ft high. Which of the four towers is the highest?
Needed: Which of the four towers is the highest?

6. Niagara Falls Park was established in 1885. The park covers 430 acres. Queen Victoria Park was established in 1886. It covers 196 acres. Which park was larger when it was established?
Needed: Size of parks when established.

7. Rock slides have changed the appearance of Niagara Falls. In 1931, about 80,000 tons of rock fell from the American Falls. Then years later, about 185,000 tons of rock fell. In which year did a greater amount of rock fall?
Needed: The year in which 185,000 tons of rock fell.

8. Three sightseeing boats took tourists to the base of the American Falls twice in 1 day. Boat A took 186 tourists. Boat B took 168 tourists. Boat C took 196 tourists. List the boats in order from the greatest number of tourists on board to the least number of tourists on board.
C, A, B

ASSIGNMENTS

Level 1	1–6
Level 2	1–8
Level 3	1–8

FOLLOW-UP ACTIVITY

SITUATIONAL PROBLEM SOLVING Your family is planning a 3-day trip to Niagara Falls. You are in charge of planning the details. As you organize all aspects of the trip, consider these problems:

- How will you pay for the trip?
- How will you travel to Niagara Falls?
- Where will you stay?
- What will you see and do?
- Where will you eat?

SUPPLEMENTARY MATERIALS

TRP Practice, p. 7

TRP Applications, p. 1

TRP Transparency 4

OBJECTIVE

- Round whole numbers and decimals.

TEACHING THE LESSON

WARM-UP Write the following exercises on the chalkboard or an overhead transparency.

Name the place of the underlined digit.

1. 43,861 **2.** 92.56

3. 84,760,015 **4.** 25.604

(**1.** hundreds **2.** tenths **3.** hundred thousands **4.** thousandths)

INTRODUCTION Read the following newspaper headline: "90,000 Attend Super Bowl." Discuss the reasons for using approximate numbers in everyday life.

INSTRUCTION Discuss the introductory situation. Then use the number line for additional examples. Write a list of numbers less than 4,000 on the chalkboard. Have students use the number line to determine the nearest thousand. Ask students to decide whether the nearest thousand is higher or lower than the number, and record the results in a table like the one below.

Number	Lower thousand	Higher thousand
1520		2000
1460	1000	

Direct students' attention to the rules for rounding. Have them use the rules to verify their results on the number line.

Use Example 1 to reinforce the rules for rounding. Emphasize that when working with whole numbers, all digits to the right of the rounded place change to zero.

1.8 ROUNDING NUMBERS

Video Views rented 3,972 movies during the summer season. About how many movies did they rent?

You can find *about how many* by rounding.

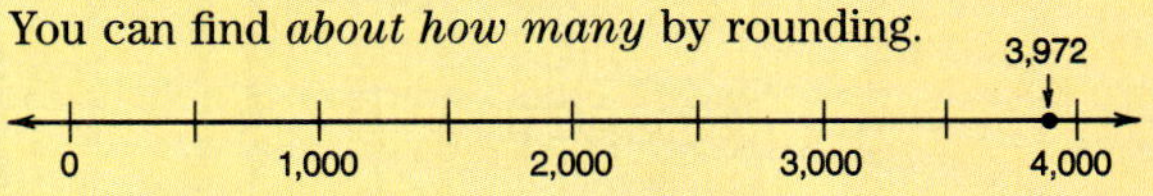

3,972 is closer to 4,000 than to 3,000. So, about 4,000 movies were rented.

> When you round to a given place, look at the digit to the right of that place. If the digit is
>
> Less than 5 → round down   5 or greater → round up

EXAMPLE 1 Round 67,438 to the nearest hundred.

1. Find the hundreds place. — 67,438 (↑ under 4)
   THINK: 3 is to the right. 3 < 5, so round down.
2. To round down, keep the digit in the hundreds place. Change the digits to the right to zero. — 67,400

67,438 rounded to the nearest hundred is 67,400.

When rounding decimals, you may have to drop some digits.

EXAMPLE 2 Round 5.291 to the nearest tenth.

1. Find the tenths place. — 5.291 (↑ under 2)
   THINK: 9 is to the right. 9 > 5, so round up.
2. To round up, increase the digit in the tenths place by 1. Drop the digits to the right. — 5.3

5.291 rounded to the nearest tenth is 5.3.

When working with money, sometimes you have to round to the nearest cent.

EXAMPLE 3 The tax on an item is $2.195. Round the tax to the nearest cent.

1. Find the cents. — $2.195 (↑ under 9)
   THINK: 5 is to the right. 5 = 5, so round up.
2. To round up, increase the cents digit by 1. Drop the digit to the right. — $2.20

The tax is $2.20.

PRACTICE EXERCISES

See Extra Practice, page 427.
See page 477 for answers to Exercises 1–32.

Round to the nearest hundred and to the nearest thousand.

• **1.** 2,782	• **2.** 3,984	**3.** 5,283	**4.** 6,096
5. 8,599	**6.** 17,851	**7.** 63,800	**8.** 74,210
9. 89,954	**10.** 93,150	**11.** 354,167	**12.** 489,649
13. 666,111	**14.** 899,521	**15.** 942,235	**16.** 999,999

Round to the nearest tenth and to the nearest hundredth.

• **17.** 0.0193	• **18.** 0.077	**19.** 1.549	**20.** 4.051
21. 16.951	**22.** 39.945	**23.** 85.563	**24.** 0.045
25. 0.4811	**26.** 2.1673	**27.** 11.1598	**28.** 55.1231
29. 0.35432	**30.** 5.12345	**31.** 23.66666	**32.** 48.98326

Round to the nearest dollar.

• **33.** \$5.15 \$5.00	• **34.** \$39.50 \$40.00	**35.** \$65.39 \$65.00	**36.** \$483.98 \$484.00
37. \$8.358 \$8.00	**38.** \$17.163 \$17.00	**39.** \$32.843 \$33.00	**40.** \$133.518 \$134.00

Round to the nearest cent.

• **41.** \$0.858 \$0.86	• **42.** \$1.899 \$1.90	**43.** \$16.455 \$16.46	**44.** \$24.296 \$24.30
45. \$0.1462 \$0.15	**46.** \$8.3586 \$8.36	**47.** \$50.8192 \$50.82	**48.** \$138.1955 \$138.20

Round to the underlined place.

• **49.** 3,586 4,000	• **50.** 49.527 49.5	**51.** 87,841 87,800	**52.** 24.09 24.1
53. 64.938 64.94	**54.** \$36.865 \$36.87	**55.** 601,498 601,500	**56.** \$16.324 \$16.00

Solve.

• **57.** Tax on a video tape came to \$0.3594. What is the tax to the nearest cent? \$0.36

58. Video View made \$1,484.54 last week. This week it made \$1,470.18. During which week did it make more money? Last week

59. Brian spends about \$8.97 a month to rent movies. Last year he spent \$107.64. How much did he spend last year to the nearest dollar? \$108.00

60. A number was rounded to 150,000. Give three possible numbers that could have been rounded to 150,000. Accept any reasonable answers.

Example 2 shows how to apply the rules for rounding to decimals. Point out that zeros do not need to be written as placeholders when rounding to a decimal place.

Use Example 3 to show that money amounts should have two digits to the right of the decimal point. Discuss how rounding up the digit 9 will affect the rounded place and the place to the left. Have students round other numbers with successive 9s.

COMMON ERROR

When rounding numbers, students frequently round to the leading digit. For example, in Exercise 6, they would round 17,851 to 20,000 instead of to 18,000. To assist these students, have them underline the place to which the number is to be rounded.

ASSIGNMENTS

Level 1	1–12, 17–24, 33–36, 41–44, 57–58
Level 2	1–12, Odd 17–55, 57–59
Level 3	9–16, Even 18–56, 57–60

FOLLOW-UP ACTIVITY

MENTAL MATH Divide the class into two teams. Write a number on the chalkboard. Call out a place to which the number can be rounded. Have a student from the first team write the rounded number on the chalkboard. Call out another place to which the number can be rounded and have a student from the other team write the rounded number on the chalkboard. Award a point for each correct answer. The first team to score 10 points wins.

SUPPLEMENTARY MATERIALS

TRP Practice, p. 8

TRP Reteaching, p. 5

TRP Transparency 4

OBJECTIVES

- Write repeated multiplications in exponent form.
- Write powers of 10 in exponent form.
- Write numbers given in exponent form as products.

TEACHING THE LESSON

WARM-UP Write the following exercises on the chalkboard or an overhead transparency.

Find the product.

1. 7×7 (49)
2. $1 \times 1 \times 1$ (1)
3. 3×3 (9)
4. 9×9 (81)
5. 10×10 (100)
6. $2 \times 2 \times 2$ (8)

INTRODUCTION Review the term *factor.* Point out that in each Warm-Up exercise, the factors are identical. Ask students to identify the Warm-Up exercises having two identical factors. (Exercises 1, 3, 4, 5) Indicate that these factors are squared. Then have students identify the exercises having three identical factors. (Exercises 2, 6) Indicate that these factors are cubed.

INSTRUCTION Make sure students understand that the term *repeated* implies the same number and the same operation. Write examples of exponential notation on the chalkboard. Call on students to identify the base and the exponent and read each expression.

Use Examples 1 and 2 to reinforce the meaning of exponential notation and what is meant by a number squared, a number cubed, and a number to any power.

Have students look for patterns in the chart of powers of 10. Discuss the relationship between the exponent in the power of 10 and the number of zeros in the product. Use Example 3 to reinforce the generalization.

1.9 EXPONENT FORM

A short way to show repeated multiplication is **exponent** form. For example, 3^2 is a number written in exponent form. The exponent, 2, tells how many times 3 is used as a factor; 3 is called the **base.**

Base → $3^2 = 3 \times 3 = 9$

↑ **Exponent** ↑ **Product**

3^2 is read as three to the second power, or three squared.

EXAMPLE 1 Write $2 \times 2 \times 2$ in exponent form.

$2 \times 2 \times 2$ **THINK:** The base is 2.
2 is used as a factor 3 times.
The exponent is 3.

$2 \times 2 \times 2 = 2^3$

2^3 is read as two to the third power, or two cubed.

To find the product of a number in exponent form, first write it as a repeated multiplication.

EXAMPLE 2 Write the product of 2^4.

2^4 **THINK:** 2 is used as a factor 4 times.

$2^4 = 2 \times 2 \times 2 \times 2 = 16$

Study the table below. Do you see a pattern?

Repeated Factors	Product	Exponent Form
10×10	100	10^2
$10 \times 10 \times 10$	1,000	10^3
$10 \times 10 \times 10 \times 10$	10,000	10^4
$10 \times 10 \times 10 \times 10 \times 10$	100,000	10^5
$10 \times 10 \times 10 \times 10 \times 10 \times 10$	1,000,000	10^6

Notice that when 10 is raised to any power, the exponent tells the number of zeros in the product.

EXAMPLE 3 a. Write 1,000,000 in exponent form.

1,000,000 **THINK:** 6 zeros
$1{,}000{,}000 = 10^6$

b. Write the product of 10^8.

10^8 **THINK:** 8 zeros
$10^8 = 100{,}000{,}000$

18 CHAPTER 1

FOR DISCUSSION See TE side column.

Look at the patterns. What do you notice?

Pattern 1: $3^1 = 3$ $5^1 = 5$ $7^1 = 7$ $8^1 = 8$

Pattern 2: $2^0 = 1$ $4^0 = 1$ $6^0 = 1$ $9^0 = 1$

PRACTICE EXERCISES

See Extra Practice, page 427.

Write in exponent form.

• **1.** 2 squared 2^2 • **2.** 7 cubed 7^3 **3.** 9 to the third power 9^3

4. 10 to the fifth power 10^5 **5.** 12 to the sixth power 12^6 **6.** 18 to the second power 18^2

7. $2 \times 2 \times 2 \times 2 \times 2$ 2^5 **8.** $3 \times 3 \times 3$ 3^3 **9.** 4×4 4^2

10. $8 \times 8 \times 8 \times 8$ 8^4 **11.** $12 \times 12 \times 12$ 12^3 **12.** $13 \times 13 \times 13 \times 13$ 13^4

Write as a product.

• **13.** 2^2 4 • **14.** 7^0 1 **15.** 3^2 9 **16.** 12^0 1 **17.** 3^3 27 **18.** 6^2 36 **19.** 1^3 1

Write in exponent form.

• **20.** 100 10^2 **21.** 10,000 10^4 **22.** 10,000,000 10^7 **23.** 1,000,000,000 10^9

Write as a product. 25. 1,000 26. 100,000 27. 100 29. 10,000

• **24.** 10^0 1 • **25.** 10^3 **26.** 10^5 **27.** 10^2 **28.** 1^8 1 **29.** 10^4 **30.** 10^1 10

Solve.

31. X-rays have a frequency of more than 10,000,000,000 vibrations per second. Write this number in exponent form. 10^{10}

32. The diameter of Venus is 7,700 mi. The diameter of Earth is 7,926 mi. Which number is greater? 7,926

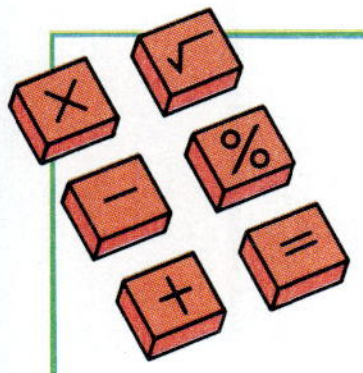

CALCULATOR

Most calculators have an x^2 key and $\sqrt{\ }$ key. The x^2 key is called the **square** key. The $\sqrt{\ }$ is called the **square root** key. Square and square root are called inverse operations. This means one undoes the other. For example, enter 6, then press the square key. What does the display show? Now press the square root key. What does the display show? Now repeat the procedure for 8. What is the definition of square root? Guess the square root of the number. Use the calculator to check your answer.

1. 4 2 **2.** 9 3 **3.** 16 4 **4.** 25 5 **5.** 49 7

FOR DISCUSSION Students should explain that the exponents 1 and 0 have special properties. Any number raised to the first power is the number itself. Any number raised to the zero power is 1.

CALCULATOR You may wish to have students familiarize themselves with the square key on their calculators before completing this activity. Have students square numbers 1–10 using their calculator, and write the results in a table. (See below.)

Then have the students complete the Calculator activity. Encourage them to extend the table to include a third column, "Square root." Have them complete the table before doing Exercises 1–5.

ASSIGNMENTS

Level 1	1–9, Even 14–30, 31
Level 2	1–12, Odd 13–31, 32, CA
Level 3	10–32, CA

FOLLOW-UP ACTIVITY

CALCULATOR Have students use their calculators to make a table similar to the following for numbers 1 to 15.

Number	Squared	Cubed	Square root
1	1	1	1
2	4	8	1.414
3	9	27	1.332
4	16	64	2
5	25	125	2.236
6	36	216	2.449
7	49	343	2.646
8	64	512	2.828
9	81	729	3
10	100	1,000	3.162
11	121	1,331	3.317
12	144	1,728	3.464
13	169	2,197	3.606
14	196	2,744	3.742
15	225	3,375	3.873

SUPPLEMENTARY MATERIALS

TRP Practice, p. 9

TRP Reteaching, p. 6

TRP Transparency 5

OBJECTIVE

- Solve problems that involve reading and interpreting data from a menu.

TEACHING THE LESSON

WARM-UP Read the following exercises aloud.

Round to the nearest dollar.

1. \$3.19 (\$3.00) **2.** \$5.92 (\$6.00)

3. \$7.53 (\$8.00) **4.** \$9.45 (\$9.00)

5. \$6.25 (\$6.00) **6.** \$4.50 (\$5.00)

INTRODUCTION Ask the students to recall an occasion when they ate a meal in a restaurant. Were there any unusual words or phrases on the menu? Did the menu make it clear what the cost of each item was? Encourage students to discuss their experiences with menus and restaurants.

INSTRUCTION Point out that people often read signs, displays, or menus quickly, which can sometimes lead to difficulties. Emphasize the importance of reading menus carefully.

Discuss the meaning of the three key phrases used in this lesson:

Entrée

A la carte

Complete dinner

Before assigning the exercises, be sure that the students understand the difference between a la carte and a complete dinner.

FOR DISCUSSION Students should realize that there are additional costs associated with the dinner. For example, in most states there is a sales tax that is added to the bill. Also, people are expected to leave a tip for the waiter or waitress.

PROBLEM Solving APPLICATION

1.10 CONSUMER: UNDERSTANDING MENUS

You and a friend decide to eat at the Hearthstone Restaurant. The waiter gives you a menu. There are some words and phrases you must understand. **A la carte** indicates a separate price for each item not bought as a part of a complete dinner. An **entrée** is a main dish.

HEARTHSTONE RESTAURANT

MENU

Dinners include entrée, side order, beverage, and dessert.

ENTRÉE

	A la Carte	Complete Dinner
Filet of sole	\$ 9.95	\$11.50
Salmon steak	\$10.95	\$12.50
Spaghetti and meatballs	\$ 6.25	\$ 7.75
Roast chicken	\$ 8.75	\$10.25
Chopped sirloin	\$ 7.50	\$ 8.95

SIDE ORDERS

French fried potatoes	\$0.85	*Cole slaw*	\$0.75
Vegetable	\$0.85	*Salad*	\$1.10

DESSERTS

Fruit and cheese	\$1.15	*Jello*	\$0.85
Yogurt	\$1.10	*Rice pudding*	\$0.95

BEVERAGES

Milk	\$0.75	*Herbal tea*	\$0.85
Lemonade	\$0.95	*Orange juice*	\$0.90

FOR DISCUSSION See TE side column.

Why would the total cost for the filet of sole complete dinner be more than \$11.50?

Use the Hearthstone Restaurant menu on page 20. Write the cost.

• **1.** A complete dinner with filet of sole, salad, milk, and jello $11.50

2. Roast chicken if a person does not order a complete dinner $8.75

3. A complete dinner with chopped sirloin, vegetable, herbal tea, and yogurt $8.95

Solve.

• **4.** In the a la carte column for entrées, how many items cost less than $9.00? 3 items

5. In the a la carte column for entrées, how many items cost more than $10.00? 1 item

6. In the complete dinner column, how many items cost less than $9.00? 2 items

7. In the complete dinner column, how many items cost more than $11.00? 2 items

8. Which of the a la carte entrées is the least expensive? spaghetti and meatballs

9. Which dessert is the most expensive? fruit and cheese

In Exercises 10–15, give the answer to the nearest dollar.

• **10.** Salmon steak ordered a la carte $11.00

11. Spaghetti and meatballs ordered a la carte $6.00

12. A complete dinner with filet of sole as the entrée $12.00

13. Yogurt ordered a la carte $1.00

14. If you order a la carte, about how much more does roast chicken cost than chopped sirloin? $1.00

15. If you order a complete dinner, about how much more does filet of sole cost than spaghetti and meatballs? $4.00

ASSIGNMENTS

Level 1	1–11
Level 2	1–14
Level 3	Odd 1–15

FOLLOW-UP ACTIVITY

COOPERATIVE LEARNING

Have groups of three or four students work together to develop and organize a menu for an "Ethnic Food Fair" that might be held to raise money for various school activities. After they develop the menu, ask students to write several problems based on the data given in the menu. Have students solve the problems.

SUPPLEMENTARY MATERIALS

TRP Practice, p. 10

TRP Transparency 5

OBJECTIVES

- Review vocabulary.
- Practice key chapter concepts and skills.

USING THE REVIEW

The Chapter Review is designed to help students prepare for taking the Chapter Test. The first section focuses on vocabulary. It requires that students select a word(s) to complete statements. The second section presents practice exercises of key mathematical skills. Under each directive there is a sample exercise with the answer.

Each item on the review is referenced to the page on which the topic is taught in the Pupil's Edition. You may wish to have students refer to these pages to help review any concepts or skills they have not yet mastered.

It is suggested that students work in small-sized heterogeneous cooperative learning groups. Some cooperative learning methods that may be used are as follows:

1. After each student has independently completed the entire Chapter Review, a discussion should follow within each group about the solutions to the practice exercises.
2. The group can complete the entire Chapter Review by working together to discuss the sample exercises and then to answer the practice exercises.

End the lesson with an entire class discussion in which any questions brought up in group discussions are presented and answered.

CHAPTER REVIEW

Vocabulary Choose the letter of the word(s) that completes each statement.

1. A number that tells "where" is classified as a ■. [2] c
a. count **b.** measure **c.** location **d.** code

2. You can find the value of a digit by multiplying the digit by its ■. [4] c
a. word name **b.** period **c.** place value **d.** numeral

3. 5^3 is read as five to the third ■. [18] a
a. power **b.** base **c.** exponent **d.** factor

Skills Find the value of the underlined digit. [4, 6]

78,$\underline{2}$47 *200*	**4.** 9$\underline{2}$4,518 20,000	**5.** $\underline{4}$,806,231 4,000,000	**6.** 23,$\underline{6}$27,856 600,000
7. 0.5$\underline{6}$7 0.06	**8.** 0.034$\underline{2}$ 0.0002	**9.** 3.25$\underline{6}$7 0.006	**10.** 14.003$\underline{5}$6 0.0005

Compare. Write =, <, or >. [8]

765 ● 76 >	**11.** 5,408 ● 4,379 >	**12.** 23,456 ● 23,546 <	**13.** 80,060 ● 80,006 >
14. 0.56 ● 0.68 <	**15.** 4.57 ● 4.569 >	**16.** 0.129 ● 0.1290 =	**17.** 34.876 ● 34.872 >

Arrange in order from least to greatest. [8]

19. 68,999; 74,876; 86,098
20. 4,899; 4,978; 5,098; 5,456

567; 470; 322 *322; 470; 567*

18. 2,908; 3,541; 3,154 2,908; 3,154; 3,541

19. 74,876; 68,999; 86,098

20. 5,098; 5,456; 4,978; 4,899

21. 0.456; 0.57; 0.392
0.392; 0.456; 0.57

22. 3.9; 3.29; 2.98; 2.098
2.098; 2.98; 3.29; 3.9

Round the number to the underlined place. [16]

$\underline{3}$4,876 *30,000*	**23.** 70$\underline{9}$,543 710,000	**24.** 45,$\underline{8}$76 45,900	**25.** 97$\underline{6}$,568 977,000
26. 0.$\underline{4}$56 0.5	**27.** 3.4$\underline{6}$8 3.47	**28.** 2.6$\underline{7}$19 2.67	**29.** 15.$\underline{6}$666 15.7

Write in exponent form. [18]

4 cubed *4^3*	**30.** 3 to the ninth power 3^9	**31.** $8 \times 8 \times 8 \times 8$ 8^4
32. 5 squared 5^2	**33.** 6 to the fourth power 6^4	**34.** $7 \times 7 \times 7 \times 7 \times 7$ 7^5

Chapter Test

Use the ad to classify each number.

1. 2 count **2.** 1993 location **3.** 12 count

4. 6810 code **5.** 555-1020 code **6.** 6 p.m. location

> **For Sale:**
> 2 tires for a 1993 Comet
> 12-speed bike, Model 6810
> Call 555-1020 after 6 p.m.

Write the word name for the number. See page 477 for answers to Exercises 7–10.

7. 12,489 **8.** 240,007 **9.** 5.841 **10.** 0.0487

Write the number.

11. twelve billion, nine thousand, five hundred two 12,000,009,502

12. nineteen and one hundred forty-one ten-thousandths 19.0141

Find the value of the underlined digit.

13. $\underline{2}$89,650 200,000 **14.** 2$\underline{7}$8,598,643 70,000,000 **15.** 1,01$\underline{5}$,734 5,000 **16.** $\underline{8}$4,073,219 80,000,000

17. 4.92$\underline{9}$ 0.009 **18.** 2.094$\underline{4}$06 0.0004 **19.** 577.$\underline{7}$73 0.7 **20.** 7.03678$\underline{5}$ 0.00008

Compare. Write =, <, or >.

21. 6,739 ● 6,719 > **22.** 58,326 ● 85,362 < **23.** 4.0056 ● 4.0506 <

24. 0.0048 ● 0.00480 = **25.** 123,995 ● 123,895 > **26.** 90.937 ● 90.9371 <

Arrange in order from least to greatest.

27. 48.321; 48.32; 40.8832
40.8832; 48.32; 48.321

28. 586.17; 508.617; 586.107; 561.87
508.617; 561.87; 586.107; 586.17

Round 87,915.0649 to the nearest:

29. thousand. 88,000 **30.** tenth. 87,915.1 **31.** hundredth. 87,915.06 **32.** hundred. 87,900

Write in exponent form.

33. $7 \times 7 \times 7$ 7^3 **34.** 3 squared 3^2 **35.** $4 \times 4 \times 4 \times 4$ 4^4 **36.** 14 cubed 14^3

Write as a product.

37. 2^3 8 **38.** 10^{10} 10,000,000,000 **39.** 1^{46} 1 **40.** 8 squared 64

Solve if possible. If not, identify the needed information.

41. At Magic Mountain park in California, the Sky Tower is 385 ft high. At Six Flags Over Georgia, the parachute jump is 225 ft high. Which is taller?
Sky Tower

42. The October 15 reading of an electric meter was 41,347 kWh. Write the number that is 500 kWh more.
41,847 kWh

Objective

- Evaluate achievement of the chapter objectives.

Using the Test

The Chapter Test may be used as a posttest to evaluate student achievement. However, you may wish to use the Chapter Posttest offered in the Teacher's Resource Package or to design your own chapter test. If this page is not used as a test, you may wish to assign it as additional review or practice.

The test items are correlated to the chapter objectives in the table below.

Chapter objectives	Test items
A. Identify the uses of numbers.	1–6
B. Write word names for numbers and vice versa.	7–12
C. Identify the values of digits in whole numbers and decimals.	13–20
D. Compare and order whole numbers and decimals.	21–28
E. Round whole numbers and decimals.	29–32
F. Write numbers in exponent form and vice versa.	33–40
G. Identify needed information in solving problems.	41
H. Apply computational skills in real-life situations.	42

Supplementary Materials

TRP Ch. 1 Posttest Form A, pp. 1–2

TRP Ch. 1 Posttest Form B, pp. 1–2

Objectives

- Write standard numerals for Egyptian and Roman numerals.
- Write Egyptian and Roman numerals for standard numerals.

Using the Page

Before referring to the text, have students give examples of other number systems used in the past such as the Egyptian and Roman systems. Point out that computation was not always easy in these number systems, and so computational devices, such as the abacus, were used.

Write the symbols for the Egyptian and Roman number systems on the chalkboard. Guide students to realize that the Egyptian system is similar to ours because it is based on powers of 10. Point out that both systems use repetition of symbols to write numbers. For example, 200 is written as  in the Egyptian system and CC in the Roman system. Students should note that the Egyptian system uses only addition to represent numbers while the Roman system uses both addition and subtraction.

Students often encounter difficulty with the grouping techniques used in the Roman system. Illustrate the four examples on the chalkboard and emphasize the grouping technique for MCMLXXIV (1,974) and MMXLIX (2,049). You may want to use the following numerals to provide more practice with the grouping technique.

CXIV (114)	CDXLIV (444)
MCMLIX (1,959)	MMMXCVIII (3,098)

Enrichment ANCIENT NUMBER SYSTEMS

The Egyptian number system is based on the number 10. Each symbol stands for a different power of 10.

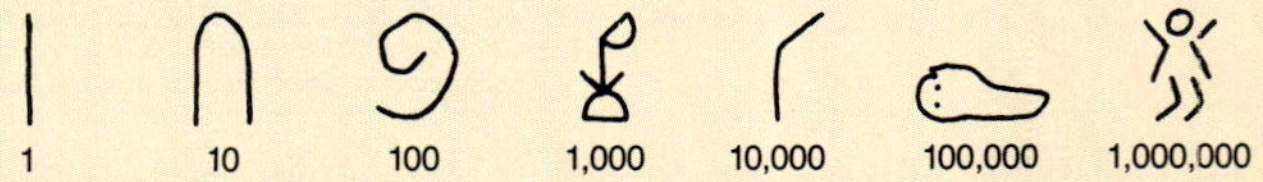

The Egyptian system uses addition. The value of a number is found by adding the values of the symbols.

Example

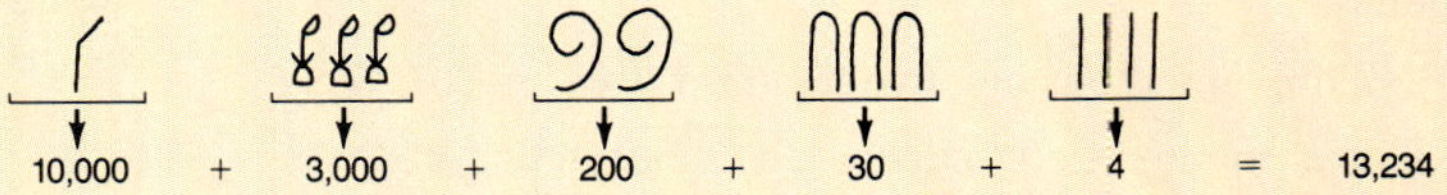

The Roman number system uses the following symbols.

I	V	X	L	C	D	M
1	5	10	50	100	500	1,000

This system uses addition and subtraction. If a symbol has a value greater than or equal to the value of the symbol to the right, add. If a symbol has a value less than the value of the symbol to the right, subtract.

Examples M CM LXX IV: 1,000 + 900 + 70 + 4 = 1,974   MM XL IX: 2,000 + 40 + 9 = 2,049

Write our numbers.

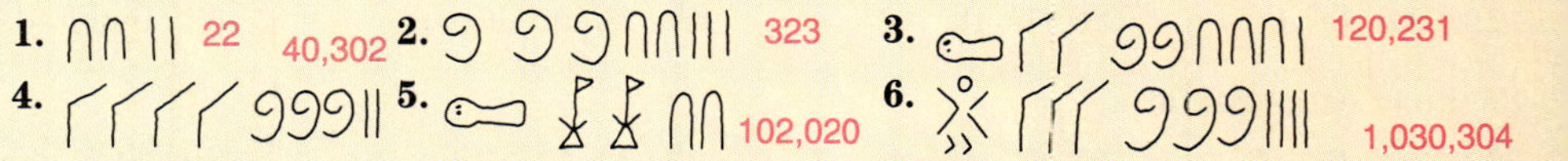

Write Egyptian numbers. See page 477 for answers to Exercises 7–11.

7. 25 **8.** 46 **9.** 131 **10.** 158 **11.** 1,254

Write our numbers.

12. L X X V I 76 **13.** C C L X V 265 **14.** D C C C X X X I I 832

15. X C V 95 **16.** C D L I V 454 **17.** M C M L I X 1,959

Write Roman numbers.

18. 13 XIII **19.** 38 XXXVIII **20.** 156 CLVI **21.** 509 DIX **22.** 944 CMXLIV

24 CHAPTER 1

CALCULATOR BASIC OPERATIONS ON A CALCULATOR

A calculator can save you time and increase the accuracy of your computations. The four operations are performed as shown below.

Operation	Calculator Entry	Display
Add: 6,936 + 7,098	[6] [9] [3] [6] [+] [7] [0] [9] [8] [=]	14,034.
Subtract: 32.715 − 9.846	[3] [2] [.] [7] [1] [5] [−] [9] [.] [8] [4] [6] [=]	22.869
Multiply: 542 × 1,714	[5] [4] [2] [×] [1] [7] [1] [4] [=]	928,988.
Divide: 3.66 ÷ 0.48	[3] [.] [6] [6] [÷] [.] [4] [8] [=]	7.625

Several operations can be performed in sequence. You do not have to use the = key after each operation.

Compute: 729 × 38 ÷ 18

Operation	Calculator Entry	Display
1. Multiply.	[7] [2] [9] [×] [3] [8] [÷] ← Get ready to divide.	27,702.
2. Divide.	[1] [8] [=]	1,539.

If a number is entered incorrectly, the **CE (Clear Entry)** key can be used to make the necessary correction.

Correct, then complete the following computation on a calculator.

Add: 13.95 + 4.78 (Error: 47.8)

		Display
Calculator entry:	[1] [3] [.] [9] [5] [+] [4] [7] [.] [8]	47.8
1. Press the CE key to clear 47.8 from the calculator.		0.
2. Complete the computation. Enter:	[4] [.] [7] [8] [=]	18.73

Use a calculator to compute. Use the Clear Entry key if you enter a number incorrectly.

1. 35,248 + 9,675 44,923

2. 8,200 − 7,957 243

3. 97 × 964 93,508

4. 5,280 ÷ 176 30

5. 4.738 + 19.824 24.562

6. 342.15 − 18.76 323.39

7. 0.034 × 4,673 158.882

8. 193.2 ÷ 34.5 5.6

9. 478 + 9,732 + 5,286 15,496

10. 378.4 − 59.7 + 18.9 337.6

11. 36 × 207 × 188 1,400,976

12. 36.3 × 48 ÷ 7.2 242

OBJECTIVE

- Perform the four basic operations on a calculator.

USING THE PAGE

Point out that when a decimal less than 1 is entered in a calculator, it is not necessary to enter a zero to the left of the decimal point. Also, caution students to read displayed numbers carefully. A calculator does not show commas separating the periods in a number, and substantial errors in magnitude occur if the number of digits in a display is not observed carefully.

In this lesson, some of the exercises involve more than one operation. These exercises have been chosen so that operations may be performed in the order given. Nonscientific calculators are not programmed to do certain operations in the mathematically correct order: multiplication and division from left to right, then addition and subtraction. For example, if the numbers and the operations in the expression 4 + 6 × 2 are entered in the given order, a nonscientific calculator will give an incorrect answer. This type of calculator performs the operations in the order in which they are entered.

$4 + 6 \times 2 = 10 \times 2 = 20$ Incorrect answer

The correct answer is given by a scientific calculator, which performs the multiplication first.

$4 + 6 \times 2 = 4 + 12 = 16$ Correct answer

Whenever computation is not the object of a lesson, students should be encouraged to use a calculator.

On page 425, more will be said about the correct use of scientific and nonscientific calculators to do computations in which the order of operations affects the answers.

OBJECTIVE

- Test the prerequisite skills needed to learn the concepts developed in Chapter 2.

USING THE TEST

The Pre-Skills Test is designed to diagnose students' strengths and weaknesses on prerequisite skills necessary to study the mathematics in Chapter 2.

Assign the Pre-Skills Test. Allow the students to work together in pairs or small groups. Group members should help those who demonstrate a misunderstanding of a concept or a weakness in a skill.

Some items in the test are referenced to the pages on which the topics are taught in the Pupil's Edition. You may wish to have students refer to these pages for review.

The following table correlates the items on the Pre-Skills Test with the prerequisite skill and the lesson(s) in the chapter for which it is needed.

Item(s)	Prerequisite skill	Lesson(s)
1–4	Round whole numbers.	2.1, 2.6
5–8	Write numbers in expanded form.	2.2, 2.3
9–20	Add whole numbers.	2.2
21–32	Subtract whole numbers.	2.3
33–44	Multiply whole numbers.	2.5–2.8
45–54	Divide whole numbers.	2.5, 2.6, 2.10–2.12

Pre-Skills TEST

Round to the nearest hundred and to the nearest thousand. [16]

1. 4,839 4,800; 5,000	**2.** 5,999 6,000; 6,000	**3.** 16,573 16,600; 17,000	**4.** 238,421 238,400; 238,0

Complete. [4]

5. 56 = ■ tens 6 ones 5 **6.** 87 = 8 tens ■ ones 7

7. 748 = ■ hundreds 4 tens 8 ones 7

8. 650 = 6 hundreds ■ tens 0 ones 5

Add.

9. 5 + 8 = 13	**10.** 8 + 6 = 14	**11.** 6 + 4 = 10	**12.** 3 + 8 = 11	**13.** 8 + 7 = 15	**14.** 7 + 9 = 16
15. 30 + 60 = 90	**16.** 40 + 50 = 90	**17.** 800 + 800 = 1,600	**18.** 1 + 4 + 3 = 8	**19.** 80 + 70 + 20 = 170	**20.** 400 + 300 + 500 = 1,200

Subtract.

21. 17 − 9 = 8	**22.** 11 − 6 = 5	**23.** 15 − 7 = 8	**24.** 9 − 3 = 6	**25.** 14 − 7 = 7	**26.** 15 − 6 = 9
27. 80 − 30 = 50	**28.** 90 − 40 = 50	**29.** 800 − 700 = 100	**30.** 140 − 70 = 70	**31.** 120 − 50 = 70	**32.** 160 − 80 = 80

Multiply.

33. 7 × 3 = 21	**34.** 9 × 4 = 36	**35.** 5 × 8 = 40	**36.** 2 × 6 = 12	**37.** 4 × 8 = 32	**38.** 5 × 3 = 15
39. 6 × 7 = 42	**40.** 8 × 7 = 56	**41.** 9 × 3 = 27	**42.** 7 × 9 = 63	**43.** 6 × 6 = 36	**44.** 9 × 8 = 72

Divide.

45. 63 ÷ 9 7	**46.** 28 ÷ 7 4	**47.** 54 ÷ 6 9	**48.** 72 ÷ 8 9	**49.** 49 ÷ 7 7
50. 27 ÷ 5 5 R2	**51.** 37 ÷ 4 9 R1	**52.** 60 ÷ 8 7 R4	**53.** 56 ÷ 6 9 R2	**54.** 53 ÷ 9 5 R8

Chapter 2 WHOLE NUMBER COMPUTATION

Community centers offer a variety of activities and services for people of all ages. What would the ideal community center provide for your community?

OBJECTIVE

- Formulate problems using text and data.

USING THE CHAPTER OPENER

Discuss students' answers to the question on the page. The nature of your community will influence students' ideas, but guide them to recognize that teenagers aren't the only ones who need activities or services. Some types of community center programs are day care (for the elderly as well as children), sports, clubs, psychological counseling, suicide hot lines, and Alcoholics Anonymous. List all suggestions on the chalkboard.

COOPERATIVE LEARNING

Write the following on the chalkboard.

New Community Center Room
Size: 50 ft × 25 ft
Budget for staff: $35,000 per year
Furniture will be donated

Separate the class into groups for the project. Explain that each group is to put together a proposal for the use of this new community center room. You may want to have the groups research community needs if time permits, or they can use their own judgment based on the earlier discussions.

The groups are to choose the services and activities for their proposal, develop a schedule for use of the room (taking into consideration the possibility of using room dividers to make smaller rooms), and apportion the funds for each program's staff and equipment or supplies.

If students find that their budget doesn't cover all their programs, have them cut back their plans. Then encourage them to devise a plan for attracting volunteers using the cut-back information to support their efforts.

SUPPLEMENTARY MATERIALS

TRP Group Projects, p. 2

OBJECTIVES

- Estimate sums and differences by rounding.
- Estimate sums and differences by front-end estimation.

TEACHING THE LESSON

WARM-UP Write the following exercises on the chalkboard or an overhead transparency.

Round the number to its greatest place.

1. 485 **2.** 739 **3.** 8,042

4. 3,516 **5.** 16,951 **6.** 623,806

(**1.** 500 **2.** 700 **3.** 8,000 **4.** 4,000 **5.** 20,000 **6.** 600,000)

INTRODUCTION Discuss with students reasons for using an estimated sum or difference. Student responses may include: to find out about how much is needed, to determine if an answer is reasonable, to determine if you have enough money to buy a group of items.

INSTRUCTION As you discuss Example 1, point out that all addends have the same number of places. Therefore, all three addends are rounded to the thousands place. However, in Example 2, the numbers do not have the same number of places. Each number must be rounded to the greatest place. Therefore, 11,157 is rounded to 10,000 and 8,301 is rounded to 8,000.

Use Example 3 to establish how front-end estimation is a quick way to estimate. Point out that only the digits in the greatest place are added or subtracted. Example 3a shows three addends, but only the digits in the thousands place are added. Guide students in seeing why the actual sum has to be greater than the estimate. In Example 3b, students subtract only the digits in the ten-thousands place. Since the subtrahend does not have a digit in the ten-thousands place, zero is subtracted from 7.

2.1 ESTIMATING SUMS AND DIFFERENCES

For a 3-day bike-a-thon, the Big Wheels Cycle Club raised \$2,492, \$3,096, and \$2,987. About how much money did they raise in all?

To find *about how much,* use estimation.

You can estimate a sum or difference by rounding each number to its greatest place. Then, add or subtract.

EXAMPLE 1 Estimate: \$2,492 + \$3,096 + \$2,987

\$2,492 + \$3,096 + \$2,987
↓ ↓ ↓
\$2,000 + \$3,000 + \$3,000 = \$8,000

So, the Cycle Club raised about \$8,000.

Sometimes you must round the numbers to different places.

EXAMPLE 2 Estimate: 11,157 − 8,301

1. Round each number to the greatest place.
2. Subtract.

11,157 → 10,000
− 8,301 → 8,000
2,000

The estimated difference is 2,000.

You can also estimate using **front-end estimation.** Add or subtract only the digits in the greatest place.

EXAMPLE 3 Estimate:

a. 5,479 + 6,890 + 239

THINK: The greatest place is thousands. Add only the digits in the thousands place.

5 + 6 = 11

5,479
6,890
+ 239
11,000

b. 78,152 − 6,583

THINK: The greatest place is ten thousands. Subtract only the digit in the ten-thousands place.

7 − 0 = 7

78,152
− 6,583
70,000

FOR DISCUSSION See TE side column.

Why will front-end estimation for addition always be an underestimation?

PRACTICE EXERCISES

See Extra Practice, page 428.

5. 400,000 **10.** 320,000 **15.** 1,300,000
20. 260,000 **25.** 300,000 **30.** 50,000

Estimate by rounding.

• **1.** 739 + 421 1,100	**2.** 1,583 + 6,291 8,000	**3.** 2,986 + 1,123 4,000	**4.** 43,023 + 29,436 70,000	**5.** 183,426 + 158,559
6. 468 + 37 540	**7.** 2,804 + 657 3,700	**8.** 6,239 + 118 6,100	**9.** 11,219 + 2,658 13,000	**10.** 287,999 + 17,823
11. 862 942 +871 2,700	**12.** 3,682 4,286 +9,870 18,000	**13.** 4,765 5,821 +8,601 20,000	**14.** 11,006 29,132 +32,446 70,000	**15.** 666,231 476,003 +123,459
16. 361 802 + 54 1,250	**17.** 4,362 399 + 213 4,600	**18.** 5,673 4,986 + 211 11,200	**19.** 29,014 981 + 3,025 34,000	**20.** 28,697 186,760 + 34,876
• **21.** 627 − 374 200	**22.** 9,428 − 5,037 4,000	**23.** 6,829 − 2,268 5,000	**24.** 57,342 − 19,820 40,000	**25.** 379,005 − 124,198
26. 109 − 67 30	**27.** 3,711 − 994 3,000	**28.** 5,468 − 967 4,000	**29.** 12,287 − 7,594 2,000	**30.** 107,812 − 45,485

Estimate by using front-end estimation.

35. 700,000 **40.** 600,000 **45.** 1,500,000
50. 200,000 **55.** 200,000 **60.** 700,000

• **31.** 739 + 421 1,100	**32.** 6,291 + 1,583 7,000	**33.** 6,823 + 3,104 9,000	**34.** 43,123 + 12,999 50,000	**35.** 586,036 + 228,451
36. 836 + 22 800	**37.** 4,223 + 349 4,000	**38.** 5,982 + 411 5,000	**39.** 34,097 + 1,789 30,000	**40.** 607,098 + 12,569
41. 316 250 +123 600	**42.** 4,361 2,102 +1,254 7,000	**43.** 5,786 1,102 +2,006 8,000	**44.** 21,009 33,436 +42,985 90,000	**45.** 613,890 506,231 +438,506
46. 482 41 +512 900	**47.** 1,056 2,213 + 872 3,000	**48.** 912 3,069 +6,730 9,000	**49.** 31,986 82,411 + 6,121 110,000	**50.** 212,765 17,645 + 5,680
• **51.** 356 − 127 200	**52.** 6,436 − 4,672 2,000	**53.** 5,208 − 1,341 4,000	**54.** 42,118 − 34,882 10,000	**55.** 379,005 − 124,198
56. 984 − 26 900	**57.** 5,434 − 267 5,000	**58.** 9,896 − 398 9,000	**59.** 26,402 − 3,857 20,000	**60.** 704,012 − 13,564

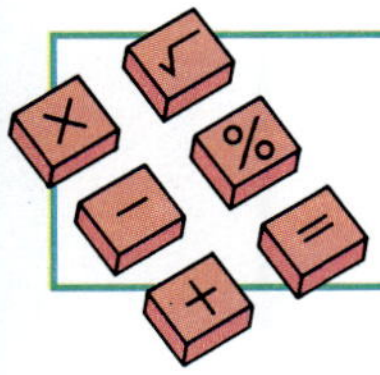

CALCULATOR

See page 477 for answers.

Use a calculator to find the exact answers to Exercises 6 through 25. Compare these answers to your estimates. Were they close?

FOR DISCUSSION Students should explain that front-end estimation always drops part or all of a number. Therefore, an addition answer is always an underestimate.

CALCULATOR This activity is designed to emphasize a major purpose of estimation, that of checking the reasonableness of an exact answer.

ASSIGNMENTS

Level 1	Odd 1–55, 56–59, CA
Level 2	Even 2–54, 56–60, CA
Level 3	6–24, Odd 31–55, 56–60, CA

FOLLOW-UP ACTIVITY

APPLICATION Present each student with the task of spending $250 during a shopping spree. Provide copies of newspaper ads, catalogs, or prepared price lists. Be sure to include items of interest to your students, for example, clothes and audio equipment. Instruct the students to use only estimates in determining what to buy. The total amount of the purchases should be close to, but not exceed, $250.

SUPPLEMENTARY MATERIALS

TRP Practice, p. 11

TRP Reteaching, p. 7

TRP Transparency 6

OBJECTIVES

- Add whole numbers.
- Estimate and check the reasonableness of answers.

TEACHING THE LESSON

WARM-UP Read the following exercises aloud. Instruct students to write their answers on a piece of paper.

1. 7 + 4 (11) **2.** 5 + 9 (14)

3. 2 + 9 (11) **4.** 6 + 8 (14)

5. 3 + 7 (10) **6.** 8 + 9 (17)

INTRODUCTION Use place-value manipulatives to work through 25 + 47. Demonstrate the regrouping.

$$\begin{array}{r} {\scriptstyle 1} \\ 25 \\ +47 \\ \hline 72 \end{array}$$

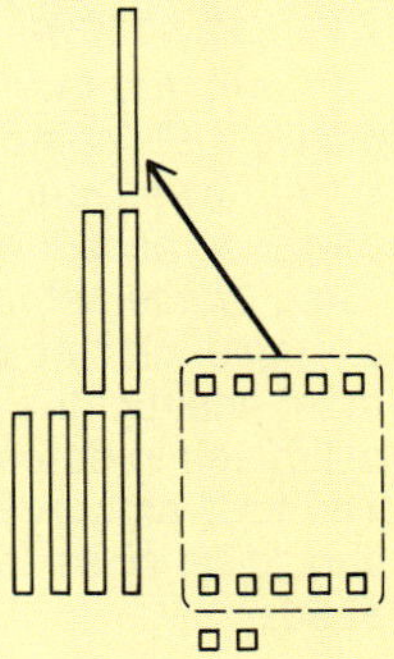

INSTRUCTION Use Example 1 to present the addition algorithm. Stress the alignment of digits in columns, and the regrouping. Make sure students understand that 13 ones are 1 ten 3 ones and that 19 tens are 1 hundred 9 tens. Discuss **addend** and **sum**.

The estimate in Example 2 is found by rounding. Some students may estimate using front-end estimation. The estimate would be 150,000.

As you discuss Example 3, note how the Commutative and Associative Properties justify the check of adding in reverse order.

FOR DISCUSSION Students should explain that each of the three addends is less than 1,000. The sum must be less than 3,000. Therefore, 4,087 is incorrect.

2.2 ADDING

There are 840 miles of coastline in California, 296 miles in Oregon, and 157 miles in Washington. How many miles of coastline are there in all?

To find the total number of miles, add the three distances.

EXAMPLE 1 Add: 840 + 296 + 157

1. Add the ones. Regroup.

$$\begin{array}{r} {\scriptstyle 1} \\ 840 \\ 296 \\ +157 \\ \hline 3 \end{array}$$

2. Add the tens. Regroup.

$$\begin{array}{r} {\scriptstyle 11} \\ 840 \\ 296 \\ +157 \\ \hline 93 \end{array}$$

3. Add the hundreds. Regroup.

$$\begin{array}{rl} {\scriptstyle 11} & \\ 840 & \leftarrow \\ 296 & \leftarrow \textbf{Addends} \\ +157 & \leftarrow \\ \hline 1{,}293 & \leftarrow \textbf{Sum} \end{array}$$

So, there are 1,293 miles of coastline.

To determine if an exact answer is reasonable, compare the answer to an estimate. If they are close, the answer is reasonable.

EXAMPLE 2 Add: 87,950 and 71,285. Is your answer reasonable?

1. Estimate.

$$\begin{array}{rcr} 87{,}950 & \rightarrow & 90{,}000 \\ +71{,}285 & \rightarrow & +70{,}000 \\ \hline & & 160{,}000 \end{array}$$

2. Add.

$$\begin{array}{r} 87{,}950 \\ +71{,}285 \\ \hline 159{,}235 \end{array}$$

The estimate, 160,000, and the exact answer, 159,235, are close. The answer is reasonable.

To check your addition, you can add in the reverse order.

EXAMPLE 3 Add: 4,935 + 138,861 + 72,049 + 6,113. Check your answer.

1. Add down.

$$\begin{array}{r} 4{,}935 \\ 138{,}861 \\ 72{,}049 \\ +\quad 6{,}113 \\ \hline 221{,}958 \end{array} \downarrow$$

2. Check up.

$$\begin{array}{r} 4{,}935 \\ 138{,}861 \\ 72{,}049 \\ +\quad 6{,}113 \\ \hline 221{,}958 \end{array} \uparrow$$

FOR DISCUSSION See TE side column.

Explain how you can tell just by looking that the given sum is incorrect: 976 + 854 + 792 = 4,087.

PRACTICE EXERCISES See Extra Practice, page 428.

Estimate by rounding. Which exact answer is reasonable?

• **1.** 673 + 985 + 261	b	**a.** 119	**b.** 1,919	**c.** 1,119
2. 506 + 742 + 328	a	**a.** 1,576	**b.** 2,076	**c.** 14,476
3. 5,291 + 8,702 + 4,095	c	**a.** 29,088	**b.** 26,088	**c.** 18,088
4. 7,936 + 298 + 1,031	a	**a.** 9,265	**b.** 14,265	**c.** 22,265

Add. Check your answer.

• **5.** 24 +86 110	• **6.** 19 +32 51	**7.** 75 +48 123	**8.** 94 + 7 101	**9.** 89 + 3 92
10. 372 +549 921	**11.** 503 +467 970	**12.** 728 +360 1,088	**13.** 831 + 69 900	**14.** 967 + 64 1,031
15. 3,942 +5,607 9,549	**16.** 9,314 +6,812 16,126	**17.** 6,982 +5,436 12,418	**18.** 7,873 + 951 8,824	**19.** 8,324 + 807 9,131
20. 87,024 +13,142 100,166	**21.** 75,087 +49,932 125,019	**22.** 83,012 +14,624 97,636	**23.** 4,863 +94,057 98,920	**24.** 90,816 + 7,643 98,459
25. 854,017 +513,946 1,367,963	**26.** 760,912 +841,536 1,602,448	**27.** 982,413 +116,908 1,099,321	**28.** 359,806 + 67,238 427,044	**29.** 613,501 + 39,489 652,990
30. 86 57 +23 166	**31.** 139 243 +478 860	**32.** 4,236 5,113 +8,982 18,331	**33.** 15,982 26,875 +33,483 76,340	**34.** 383,051 127,987 +431,856 942,894
35. 107 936 641 +285 1,969	**36.** 2,076 8,341 6,593 +1,845 18,855	**37.** 516 4,389 63 +9,482 14,450	**38.** 92,037 6,139 804,625 + 39 902,840	**39.** 7,256 511,833 76,032 +568,912 1,164,033

First estimate by rounding. Then find the exact answer.

• **40.** 684 + 9,705 + 734,281 710,700; 744, 670 **41.** 17,827 + 3,496 + 28,092 53,000; 49,415

42. 28,045 + 396 + 7,198 37,400; 35,639 **43.** 7,831 + 64,189 + 24,815 88,000; 96,835

Solve. Use the table.

• **44.** Estimate the length of the total Pacific coastline. R: 8,000 mi; FE: 5,000 mi

45. Find the actual length of the total Pacific coastline. 7,623 mi

46. Rank the states by the length of coastline from longest to shortest.
AL, CA, HA, OR, WA

PACIFIC COASTLINE

State	Length (mi)
California	840
Oregon	296
Washington	157
Alaska	5,580
Hawaii	750

COMMON ERROR

Students may forget to regroup and write a sum incorrectly. For example, when adding 79 + 56, they may write:

$$\begin{array}{r} 79 \\ +56 \\ \hline 1{,}215 \end{array}$$

Work with these students in small groups with manipulative materials. Have them show regrouping ones and regrouping tens.

ASSIGNMENTS

Level 1	1–24, 44, 45
Level 2	1–4, Odd 5–43, 44–46
Level 3	Even 16–42, 44–46

FOLLOW-UP ACTIVITY

ENRICHMENT Have students use addition to find palindromes. A **palindrome** is a number, word, or phrase that is read the same backward and forward. Have students choose a 2-digit number and follow these steps.

1. Write the number.
2. Reverse the digits and write the new number under the original number.
3. Add.
4. Use the sum and repeat steps 2 and 3 until you find a palindrome.

$$\begin{array}{r} 68 \\ +86 \\ \hline 154 \\ +451 \\ \hline 605 \\ +506 \\ \hline 1{,}111 \end{array}$$

SUPPLEMENTARY MATERIALS

TRP Practice, p. 12

TRP Reteaching, p. 8

TRP Transparency 6

OBJECTIVES

- Subtract whole numbers.
- Estimate and check the reasonableness of answers.

TEACHING THE LESSON

WARM-UP Read the following exercises aloud. Instruct students to write their answers on a piece of paper.

1. 5 − 3 (2) **2.** 9 − 6 (3)

3. 12 − 5 (7) **4.** 13 − 8 (5)

5. 14 − 5 (9) **6.** 10 − 4 (6)

INTRODUCTION Use place-value manipulatives to work through 34 − 18. Demonstrate the regrouping of 1 ten to 10.

```
 2 14
  3̸4̸
− 18
  16
```

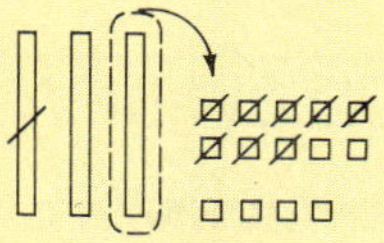

INSTRUCTION Use Example 1 to present the subtraction algorithm. Make sure students not only understand the regrouping process but know when to regroup.

As you discuss Example 2, point out that the first step is to regroup the hundreds. Since most students have difficulty subtracting across zeros, have a volunteer explain every step of the regrouping process.

Example 3 uses estimation by rounding to check the reasonableness of an exact answer. Some students may estimate using front-end estimation; the estimate is also 7,000.

Use Example 4 to show how to check subtraction. For the check to be effective, stress that students must add carefully.

2.3 SUBTRACTING

The ROBO Inc. car manufacturing plant made 5,682 cars. The records show 729 cars were not assembled by robots. How many cars were assembled by robots?

To find the difference, subtract 729 from 5,682.

EXAMPLE 1 Subtract: 5,682 − 729

1. Regroup. Subtract the ones.	2. Subtract the tens.	3. Regroup. Subtract the hundreds.	4. Subtract the thousands.
7 12 5,6 8̸ 2̸ − 729 3	7 12 5,6 8̸ 2̸ − 729 53	4 16 7 12 5̸,6̸ 8̸ 2̸ − 729 953	4 16 7 12 5̸,6̸ 8̸ 2̸ − 729 4,953

So, 4,953 cars were assembled by robots.

When you regroup across zeros, you have to regroup more than once.

EXAMPLE 2 Subtract: 600 − 187

1. Regroup.	2. Regroup.	3. Subtract.
THINK: 6 hundreds = 5 hundreds, 10 tens	**THINK: 10 tens = 9 tens, 10 ones**	
5 10 6̸ 0̸ 0 −187	9 5 1̸0̸ 10 6̸ 0̸ 0̸ −187	9 5 1̸0̸ 10 6̸ 0̸ 0̸ −187 413

So, 600 − 187 = 413.

Remember, you can determine if an exact answer is reasonable by comparing the answer to an estimate.

EXAMPLE 3 Subtract: 9,693 − 2,874. Is your answer reasonable?

1. Estimate.	2. Subtract.
9,693 → 10,000 −2,874 → − 3,000 7,000	9,693 −2,874 6,819

Since 7,000 and 6,819 are close, the exact answer is reasonable.

You can use addition to check subtraction.

EXAMPLE 4 Subtract: 82,603 − 16,571. Check your answer.

1. Subtract.

```
  82,603
− 16,571
  66,032
```

2. Check.

```
  66,032
+ 16,571
  82,603
```

CHECKPOINT

Write the letter of the correct answer.

Subtract.

1. 50,000 − 19,381 d **a.** 41,729 **b.** 31,711 **c.** 49,381 **d.** 30,619
2. 16,428 − 7,523 a **a.** 8,905 **b.** 9,105 **c.** 9,945 **d.** 8,945

PRACTICE EXERCISES

See Extra Practice, page 429.

Estimate by rounding. Which exact answer is reasonable?

• 1. 912 − 234 b **a.** 368 **b.** 678 **c.** 858
2. 5,709 − 2,457 b **a.** 2,252 **b.** 3,252 **c.** 5,252
3. 8,178 − 605 b **a.** 3,573 **b.** 7,573 **c.** 8,573
4. 19,936 − 6,292 c **a.** 8,644 **b.** 9,464 **c.** 13,644

Subtract. Check your answer.

• 5. 97 − 47 50
• 6. 43 − 29 14
7. 51 − 33 18
8. 82 − 7 75
9. 70 − 9 61
10. 837 − 245 592
11. 653 − 491 162
12. 924 − 828 96
13. 501 − 76 425
14. 900 − 45 855
15. 3,419 − 2,652 767
16. 8,738 − 2,804 5,934
17. 3,651 − 1,836 1,815
18. 5,017 − 849 4,168
19. 3,802 − 956 2,846
20. 56,957 − 34,817 22,140
21. 26,815 − 13,906 12,909
22. 71,657 − 65,738 5,919
23. 90,508 − 6,591 83,917
24. 34,090 − 904 33,186
25. 452,311 − 109,533 342,778
26. 611,092 − 272,041 339,051
27. 923,187 − 867,059 56,128
28. 260,900 − 59,654 201,246
29. 501,087 − 43,852 457,235

First estimate by rounding. Then find the exact answer. **32.** 30,000; 26,562 **35.** 8,600; 8,687

• **30.** 611 − 421 200; 190
• **31.** 7,986 − 2,814 5,000; 5,172
32. 38,999 − 12,437
33. 383 − 42 360; 341
34. 8,822 − 396 8,600; 8,426
35. 9,123 − 436
36. 36,123 − 8,041 32,000; 28,082
37. 59,001 − 4,124 56,000; 54,877
38. 683,421 − 36,923 660,000; 646,498

CHECKPOINT Distribute four index cards to each student on which the answer choices a–d appear. After allowing students a few minutes, ask each to raise the letter that indicates their answer choice. Scan the cards and identify those students who need additional attention before proceeding to the Practice Exercises. The incorrect answer choices include common errors.

TIME OUT Have students draw the scales on graph paper. If the scales are precise, students will find locating the answer easier.

COMMON ERROR

When subtracting, some students subtract the lesser digit from the greater digit regardless of placement. For example, in item 1 of Checkpoint they would select c. Work with these students in small groups, using manipulative materials.

ASSIGNMENTS

Level 1	Odd 1–37, 39–52
Level 2	1–4, Odd 5–37, 39–54, TO
Level 3	Even 16–50, 52–54, TO

FOLLOW-UP ACTIVITY

ENRICHMENT Write the following example on the chalkboard or an overhead transparency.

```
 99,999
− 9,999
 86,997
```

Tell the students that the answer is correct. However, there are four nines (three in 99,999 and one in 9,999) that should be sixes. Ask students to write the example correctly.
(96,696 − 9,699 = 86,997)

SUPPLEMENTARY MATERIALS

TRP Practice, p. 13

TRP Reteaching, p. 9

TRP Transparency 7

MIXED REVIEW

Find the value of the underlined digit.

• **39.** 765,034 700,000
• **40.** 18,345,097 8,000,000
41. 674,091,254 90,000
42. 2.00865 0.0006
43. 0.671294 0.001
44. 19.385014 0.000004

Compare. Write =, >, <.

• **45.** 9,457 ● 9,449 >
• **46.** 74,653 ● 74,660 <
47. 201,567 ● 199,567 >
48. 8.43 ● 8.430 =
49. 2.0056 ● 2.056 <
50. 243.9 ● 243.8876 >

Solve.

• **51.** Robo's goal for the year is to build 25,000 cars. To date, 9,476 cars have been built. How many more cars are needed to reach the goal? 15,524 cars

52. The Bion car plant produces 12,493 cars. All but 2,874 are made by robots. How many more cars are assembled by robots than by people? 9,619 cars

53. Gross sales of the Jetstar from September to November were 6 thousand, 104 thousand, and 83 thousand. Write the number for the total gross sales. 193,000 sales

54. The Zap model car has 3,194 moving parts. The Zing model has 23,842, and the Zoom model has 9,375. How many moving parts are there in all three models? 36,411 parts

TIME OUT

You can make an interesting computing device called a **nomograph**.

1. On a piece of graph paper, draw three horizontal lines as shown. Label them *A*, *B*, and *C*.
2. Mark the scales as shown.

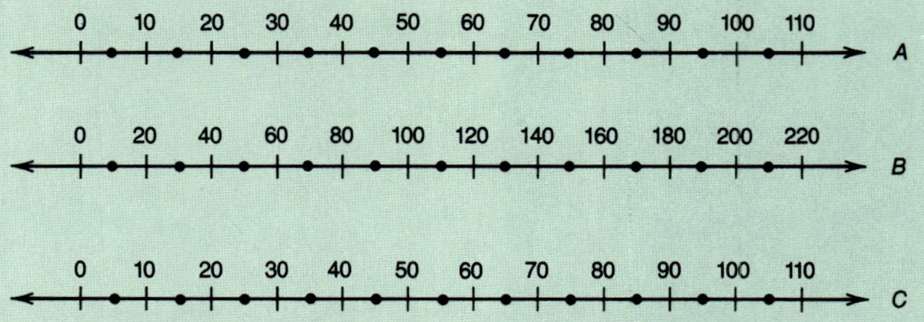

To find 70 − 25, place a straightedge so that it crosses *A* at 25 and *B* at 70. The answer is where it crosses *C* at 45.

Use your nomograph to find the difference.

1. 80 − 15 65 **2.** 110 − 5 105 **3.** 180 − 70 110 **4.** 125 − 40 85

PROBLEM *Solving* APPLICATION

2.4 CONSUMER: MAKING CHANGE

In daily life people often need to count money or make change. Change is generally given using the least number of coins and bills. Start with the coin of least value. Add on until the original amount of money given is reached.

Suppose you work in a stationery shop. A customer buys an item that costs 63 cents. He hands you a dollar bill. Count out the change and identify the amount of money he receives.

Give: 2 pennies	**THINK:** 63 + 2 = 65	Say: 65 cents
Give: 1 dime	**THINK:** 65 + 10 = 75	Say: 75 cents
Give: 1 quarter	**THINK:** 75 + 25 = 100	Say: 1 dollar (100 cents)

So, the customer receives 2 pennies, 1 dime, and 1 quarter.
The total amount of money he receives is $0.37.

FOR DISCUSSION

See TE side column.

Sometimes you may not have all types of coins. Suppose you did not have quarters. How could you have given the change in the situation above?

List the least amount of coins or coins and bills you would use to make change for the purchase. Then identify the total amount of change the customer would receive.

See page 477 for answers to Exercises 1–8.

Cost of Item	Money Paid
• **1.** 79 cents	$1 bill
• **2.** 17 cents	$1 bill
3. $0.28	$1 bill
4. $0.66	$1 bill
5. $1.24	two $1 bills
6. $1.53	two $1 bills
7. $3.91	$5 bill
8. $7.53	$20 bill

Solve.

9. Mrs. Carson buys an item for $0.32. She hands the cashier a $1 bill. The cashier has only pennies and dimes to use as change. List the least amount of coins and the total amount she would receive. 6 d, 8 p; $0.68

10. In problem 9, suppose the cashier had only pennies, nickels, quarters, and half-dollars to use as change. List Mrs. Carson's change in the fewest coins possible. 1 hd, 3 n, 3 p

OBJECTIVE

- Solve problems that involve making change.

TEACHING THE LESSON

WARM-UP Read the following exercises aloud. Have students add mentally to find the sum.

1. 53 + 2 (55)	**4.** 75 + 25 (100)
2. 85 + 5 (90)	**5.** 57 + 3 (60)
3. 40 + 10 (50)	**6.** 86 + 4 (90)

INSTRUCTION Review the value of the different coins: penny, nickel, dime, quarter, half-dollar. You may wish to use coins to demonstrate the adding-on technique described in the first example.

Discuss the fact that some stores have cash registers that show the amount of change needed. Then the clerk decides on the denominations of bills and coins to use in making change. Elicit from students that the clerk usually begins with the largest denomination of bills or coins.

FOR DISCUSSION Students should realize that the change for this situation could be given in various ways.

For example: 2 pennies, 1 nickel, 3 dimes; 2 pennies, 3 nickels, 2 dimes; 2 pennies, 7 nickels.

ASSIGNMENTS

Level 1	1–6, 9
Level 2	1–10
Level 3	3–10

SUPPLEMENTARY MATERIALS

TRP Practice, p. 14

TRP Transparency 7

OBJECTIVES

- Multiply by multiples of powers of 10.
- Divide by multiples of powers of 10.

TEACHING THE LESSON

WARM-UP Write the following exercises on the chalkboard or an overhead transparency.

Write the number.

1. 7 thousands (7,000)
2. 8 hundreds (800)
3. 6 tens (60)
4. 4 ten thousands (40,000)
5. 9 hundreds (900)
6. 3 thousands (3,000)

INTRODUCTION Use place-value manipulatives to represent 2 × 300. Make a set of 3 hundreds. 2 × 300 means 2 sets, each containing 3 hundreds, or 600.

INSTRUCTION As you study the multiplication patterns, have the students discover the relationship between the zeros in the factors and the zeros in the product.

You may also want to present Example 2a vertically. Note, however, how the steps of the procedure are reversed. Write the zeros first. Then multiply the lead digits. As you discuss Example 2b, some students may think that there is an extra zero in the answer. Explain that when you multiply the lead digits, 8 and 5, the product is 40. This accounts for the "extra" zero.

You may also want to present Example 3a in standard form. Emphasize that you are dividing both the dividend and divisor by 100 to simplify the problem.

$$8\not{0}\not{0}\overline{)720{,}0\not{0}\not{0}} = 900$$

2.5 MULTIPLYING AND DIVIDING: POWERS OF 10

It is estimated that Americans receive 5,000 tons of advertising mail each day. How many tons do they receive in 5 days?

To find how many tens, multiply.

EXAMPLE 1 Multiply: 5 × 5,000

THINK: 5 × 5 thousands = 25 thousands
WRITE: 5 × 5,000 = 25,000

So, Americans receive 25,000 tons of advertising mail in 5 days.

Discover a shortcut for multiplying multiples of powers of 10.

9 × 10 = 90	2 × 30 = 60	70 × 40 = 2,800
9 × 100 = 900	2 × 300 = 600	70 × 400 = 28,000
9 × 1,000 = 9,000	2 × 3,000 = 6,000	70 × 4,000 = 280,000

To multiply multiples of powers of 10:

1. Multiply the nonzero digits.
2. Write as many zeros as there are zeros in each factor.

EXAMPLE 2 Multiply:

a. 5,000 × 30

THINK: 5 × 3 = 15

So, 5,000 × 30 = 150,000.

b. 8,000 × 500

THINK: 8 × 5 = 40

So, 8,000 × 500 = 4,000,000.

Patterns also show a shortcut for dividing multiples of powers of 10.

80 ÷ 8 = 10	70 ÷ 10 = 7	270 ÷ 30 = 9
800 ÷ 8 = 100	700 ÷ 10 = 70	2,700 ÷ 30 = 90
8,000 ÷ 8 = 1,000	7,000 ÷ 10 = 700	27,000 ÷ 30 = 900

To divide multiples of powers of 10:

1. Cross out the same number of zeros in each of the numbers.
2. Divide the result.

EXAMPLE 3 Divide:

a. 720,000 ÷ 800

THINK: $\frac{720{,}0\not{0}\not{0}}{8\not{0}\not{0}}$ ← **divided by**

72 ÷ 8 = 9
7,200 ÷ 8 = 900

So, 720,000 ÷ 800 = 900.

b. 400,000 ÷ 8,000

THINK: $\frac{400{,}\not{0}\not{0}\not{0}}{8{,}\not{0}\not{0}\not{0}}$

40 ÷ 8 = 5
400 ÷ 8 = 50

So, 400,000 ÷ 8,000 = 50.

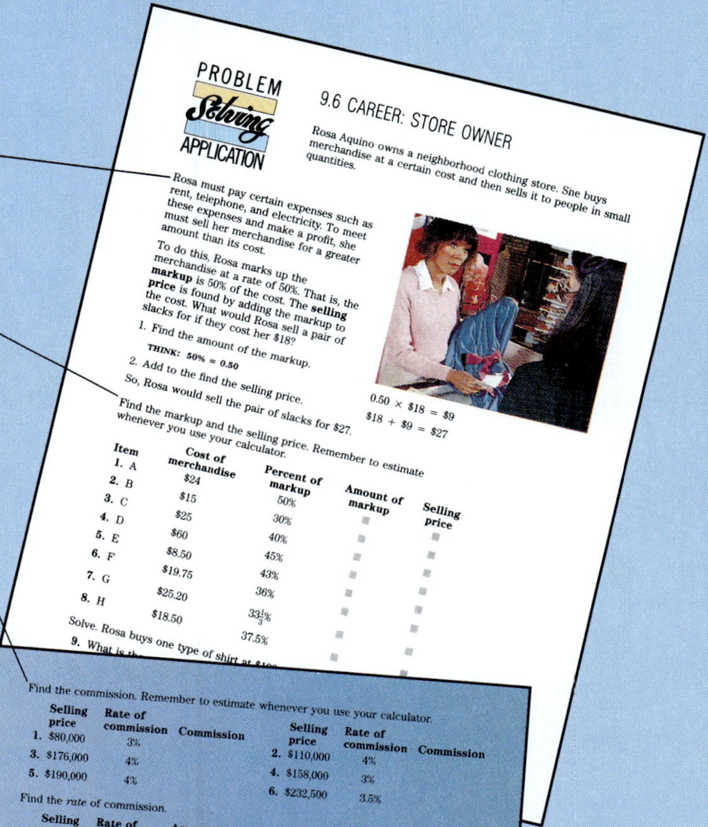

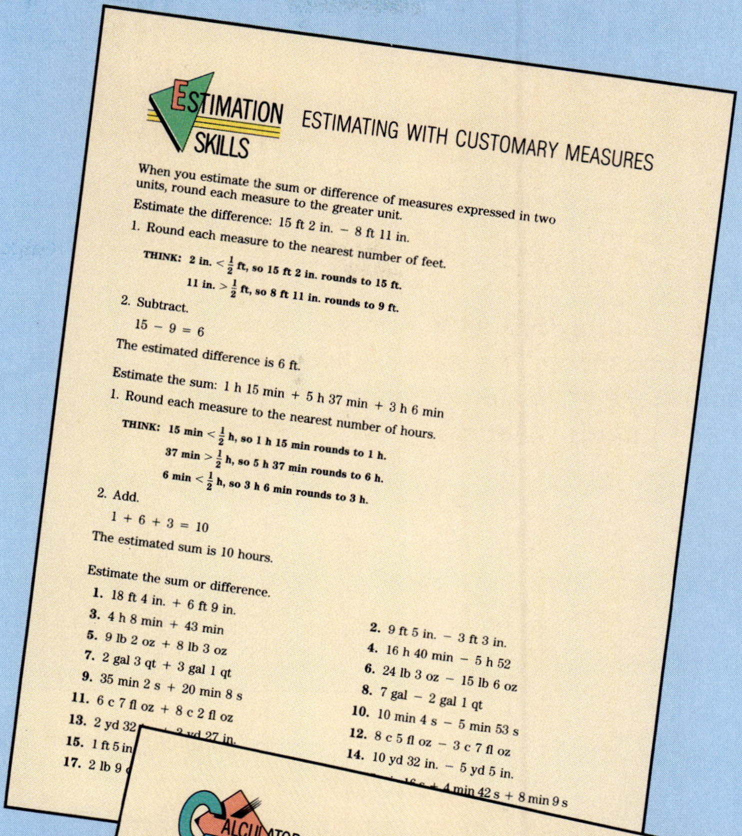

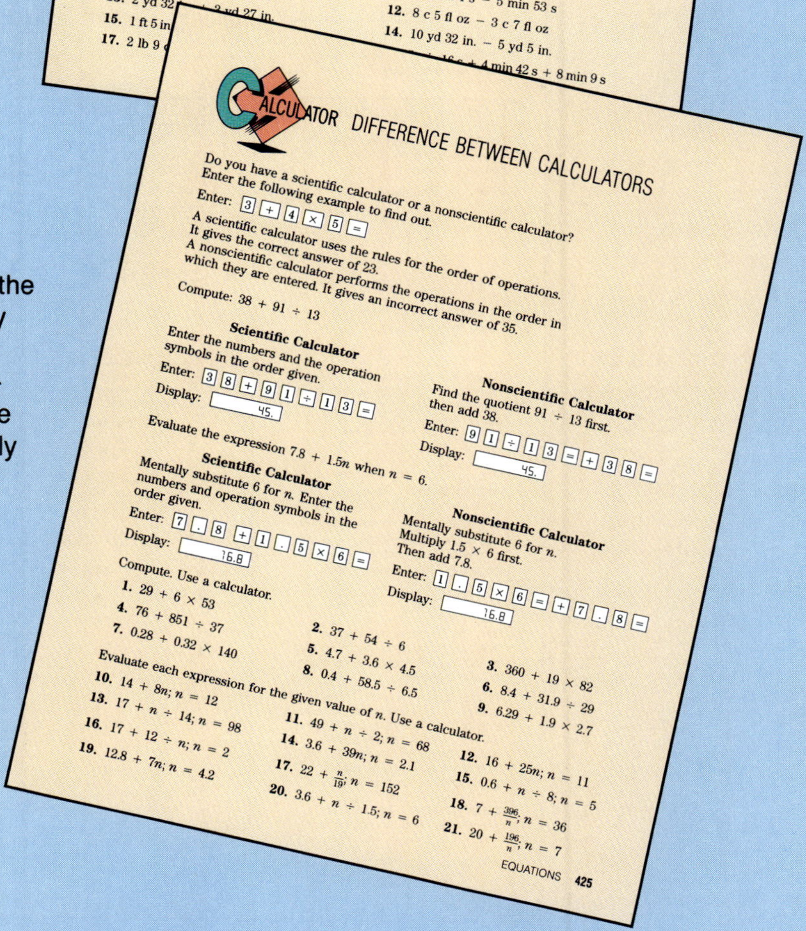

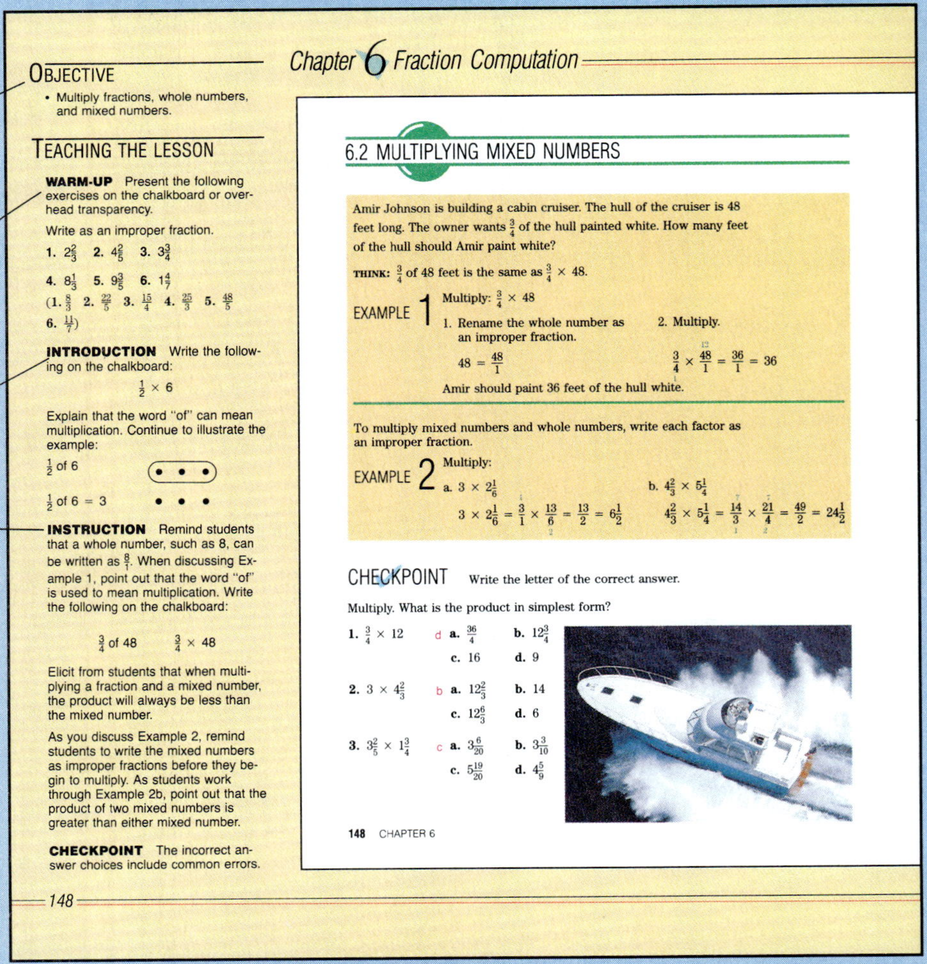

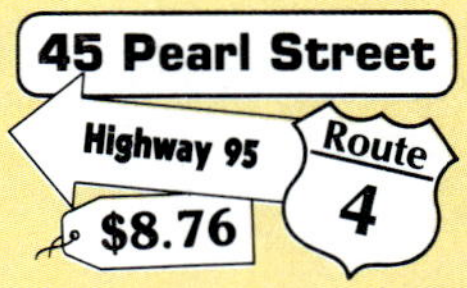

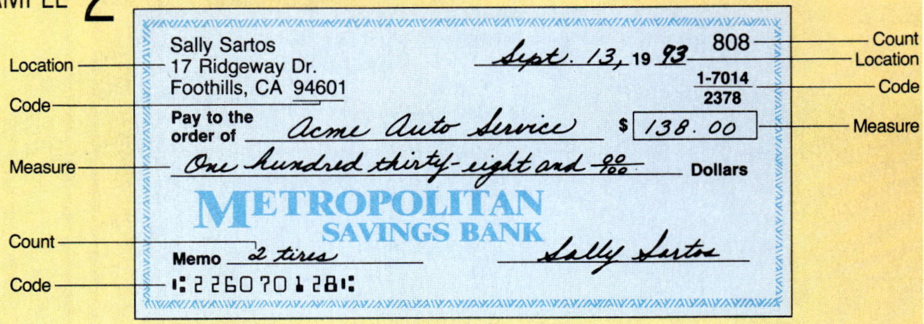

PRACTICE EXERCISES

See Extra Practice, page 426.

Find the place and the value of the underlined digit. See page 477 for answers to Exercises 1–36.

• **1.** $23.\underline{1}$	• **2.** $45.3\underline{6}$	**3.** $61.\underline{4}7$	**4.** $93.8\underline{5}$
5. $1.0\underline{4}6$	**6.** $8.19\underline{3}$	**7.** $14.\underline{9}87$	**8.** $63.79\underline{5}$
9. $5.00\underline{3}6$	**10.** $17.\underline{8}934$	**11.** $59.701\underline{2}$	**12.** $125.49\underline{8}7$
13. $6.3987\underline{2}$	**14.** $10.\underline{1}0106$	**15.** $18.76\underline{3}04$	**16.** $78.0\underline{4}921$
17. $0.529\underline{7}16$	**18.** $2.940\underline{3}72$	**19.** $11.6937\underline{2}4$	**20.** $41.10084\underline{6}$

Write the word name.

• **21.** 0.5	• **22.** 1.8	**23.** 16.9	**24.** 59.6
25. 0.04	**26.** 0.29	**27.** 1.51	**28.** 2.89
29. 0.007	**30.** 0.039	**31.** 0.241	**32.** 1.372
33. 0.0002	**34.** 0.0017	**35.** 0.0356	**36.** 0.1245

Write the number.

• **37.** 4 and 27 hundredths 4.27

• **38.** 13 and 63 hundredths 13.63

39. 195 thousandths 0.195

40. 66 thousandths 0.066

41. 12 and 249 thousandths 12.249

42. 49 and 7 thousandths 49.007

43. 1,259 ten-thousandths 0.1259

44. 48 ten-thousandths 0.0048

45. 83 and 3,459 ten-thousandths 83.3459

46. 56 and 23 ten-thousandths 56.0023

47. 129 hundred-thousandths 0.00129

48. 315 millionths 0.000315

49. 1,238 billionths 0.000001238

50. 342 and 25 billionths 342.000000025

51. four and five thousand three ten-thousandths 4.5003

52. six thousand and sixty-six hundred-thousandths 6,000.00066

Solve.

• **53.** A car's odometer reads 2,105.75. Write the word name for this number.
Two thousand one hundred five and seventy-five hundredths

54. The manual for your new car says, "Bring the car in for a check at five thousand miles." What would your odometer show for this number? 5,000

55. Classify the following numbers:

Power Steering Fluid

Part No.: 1050017
Size: 32 oz
Usage: Power steering

1050017—Code   32—measure

56. The manual states that your tires should be rotated every 15,000 miles. Your odometer reads:

1 4 5 8 7 . 6

Are you overdue in rotating your tires?
No

As you discuss Example 3, ask students to identify the decimal part of the word name. (It follows *and*.)

FOR DISCUSSION Students should explain that the number of digits in a number does not affect the value as much as the placement of the decimal point. For example, in the numbers 37,924 and 37.924 the middle digit, 9, has very different values.

ASSIGNMENTS

Level 1	1–12, 21–28, 37–42, 53–54
Level 2	1–16, Odd 21–35, 37–46, 53–56
Level 3	13–20, 29–36, 43–56

FOLLOW-UP ACTIVITY

APPLICATION Have students work in groups of three. Have each group write a short story of 200 words or less that incorporates 10 decimals in the text. Some topics may be: sports, construction, cars. Have the groups exchange papers and read the stories aloud. You may want to have the students write the numbers as a volunteer reads the story.

SUPPLEMENTARY MATERIALS

TRP Practice, p. 3

TRP Reteaching, p. 3

TRP Lesson Aids, p. 1

TRP Transparency 2

MIDCHAPTER REVIEW

Use the store receipt to classify each number. [2]

1. 1 measure **2.** 9/8/93 location
3. 123 location **4.** 4 count
5. (303) 555-3535 code **6.** 9.00 measure
7. 80739 code **8.** 2 count
9. 19.90 measure **10.** CR26 code

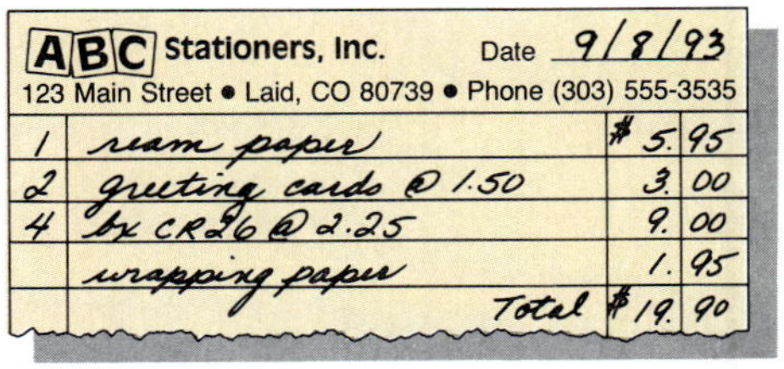

See page 477 for answers to Exercises 11–22.

Find the place and the value of the underlined digit. [4, 6]

11. 4,$\underline{9}$25 **12.** 6$\underline{7}$,819 **13.** $\underline{3}$42,896 **14.** $\underline{7}$,105,264
15. 16.$\underline{9}$ **16.** 38.7$\underline{2}$ **17.** 5.0$\underline{5}$75 **18.** 89.1370$\underline{6}$

Write the word name for the number. [4, 6]

19. 426 **20.** 279,300 **21.** 10.562 **22.** 0.0129

Write the number. [4, 6]

23. 205 thousand, 506 205,506
24. four hundred twenty-nine billion, eighty-four thousand, seventy-five 429,000,084,075
25. ten and one hundred seventy-seven thousandths 10.177
26. forty-five and thirty-two hundred-thousandths 45.00032

Compare. Write =, <, or >. [8]

27. 627 ● 617 > **28.** 6,892 ● 6,829 > **29.** 3,000 ● 30,000 <
30. 152 ● 15.2 > **31.** 26.9 ● 26.900 = **32.** 8.05 ● 8.005 >
33. 9.5624 ● 9.5723 < **34.** 3.70956 ● 4.71 < **35.** 325.81922 ● 325.81923 <

Arrange in order from least to greatest. [8]

36. 9,528; 5,928; 9,285; 9,852
5,928; 9,285; 9,528; 9,852

37. 562; 9.532; 0.565; 0.562
0.562; 0.565; 9.532; 562

Solve. [10]

38. Write the number of kilowatt-hours represented on the electric meter. 1,541

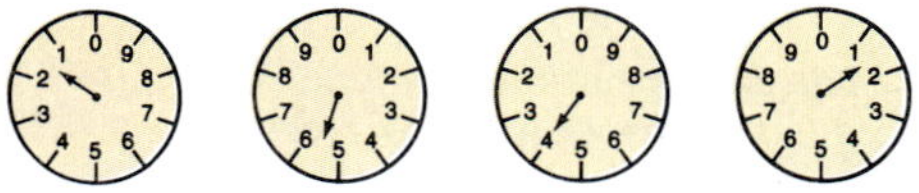

OBJECTIVE

- Evaluate student progress.

USING THE REVIEW

This page provides a means for informally evaluating students' understanding of the skills and concepts covered so far in this chapter.

Have the students look at the page to familiarize themselves with the various question formats that are presented. Discuss any questions that they may have. Then ask them to complete the page independently.

In addition to grading them individually, you may wish to review the answers to the questions collectively with the students.

Page references appear in brackets. They refer to pages on which a particular skill was introduced.

Before continuing on to the topics found in the remainder of the chapter, you may wish to have students review any skills or concepts in which they have demonstrated weakness.

OBJECTIVES

- Review vocabulary.
- Practice key chapter concepts and skills.

USING THE REVIEW

The Chapter Review is designed to help students prepare for taking the Chapter Test. The first section focuses on vocabulary. It requires that students select a word(s) to complete statements. The second section presents practice exercises of key mathematical skills. Under each directive there is a sample exercise with the answer.

Each item on the review is referenced to the page on which the topic is taught in the Pupil's Edition. You may wish to have students refer to these pages to help review any concepts or skills they have not yet mastered.

It is suggested that students work in small-sized heterogeneous cooperative learning groups. Some cooperative learning methods that may be used are as follows:

1. After each student has independently completed the entire Chapter Review, a discussion should follow within each group about the solutions to the practice exercises.
2. The group can complete the entire Chapter Review by working together to discuss the sample exercises and then to answer the practice exercises.

End the lesson with an entire class discussion in which any questions brought up in group discussions are presented and answered.

CHAPTER REVIEW

Vocabulary Choose the letter of the word(s) that completes each statement.

1. A number that tells "where" is classified as a ■. [2] c
a. count **b.** measure **c.** location **d.** code

2. You can find the value of a digit by multiplying the digit by its ■. [4] c
a. word name **b.** period **c.** place value **d.** numeral

3. 5^3 is read as five to the third ■. [18] a
a. power **b.** base **c.** exponent **d.** factor

Skills Find the value of the underlined digit. [4, 6]

78,247 *200*

4. 924,518 20,000 **5.** 4,806,231 4,000,000 **6.** 23,627,856 600,000

7. 0.567 0.06 **8.** 0.0342 0.0002 **9.** 3.2567 0.006 **10.** 14.00356 0.0005

Compare. Write =, <, or >. [8]

765 ● 76 *>*

11. 5,408 ● 4,379 > **12.** 23,456 ● 23,546 < **13.** 80,060 ● 80,006 >

14. 0.56 ● 0.68 < **15.** 4.57 ● 4.569 > **16.** 0.129 ● 0.1290 = **17.** 34.876 ● 34.872 >

Arrange in order from least to greatest. [8]

567; 470; 322 *322; 470; 567*

18. 2,908; 3,541; 3,154 2,908; 3,154; 3,541

19. 74,876; 68,999; 86,098 **19.** 68,999; 74,876; 86,098

20. 5,098; 5,456; 4,978; 4,899 **20.** 4,899; 4,978; 5,098; 5,456

21. 0.456; 0.57; 0.392 0.392; 0.456; 0.57

22. 3.9; 3.29; 2.98; 2.098 2.098; 2.98; 3.29; 3.9

Round the number to the underlined place. [16]

34,876 *30,000*

23. 709,543 710,000 **24.** 45,876 45,900 **25.** 976,568 977,000

26. 0.456 0.5 **27.** 3.468 3.47 **28.** 2.6719 2.67 **29.** 15.6666 15.7

Write in exponent form. [18]

4 cubed *4^3*

30. 3 to the ninth power 3^9 **31.** $8 \times 8 \times 8 \times 8$ 8^4

32. 5 squared 5^2 **33.** 6 to the fourth power 6^4 **34.** $7 \times 7 \times 7 \times 7 \times 7$ 7^5

OBJECTIVES

- Write standard numerals for Egyptian and Roman numerals.
- Write Egyptian and Roman numerals for standard numerals.

USING THE PAGE

Before referring to the text, have students give examples of other number systems used in the past such as the Egyptian and Roman systems. Point out that computation was not always easy in these number systems, and so computational devices, such as the abacus, were used.

Write the symbols for the Egyptian and Roman number systems on the chalkboard. Guide students to realize that the Egyptian system is similar to ours because it is based on powers of 10. Point out that both systems use repetition of symbols to write numbers. For example, 200 is written as 9 9 in the Egyptian system and CC in the Roman system. Students should note that the Egyptian system uses only addition to represent numbers while the Roman system uses both addition and subtraction.

Students often encounter difficulty with the grouping techniques used in the Roman system. Illustrate the four examples on the chalkboard and emphasize the grouping technique for MCMLXXIV (1,974) and MMXLIX (2,049). You may want to use the following numerals to provide more practice with the grouping technique.

CXIV (114)	CDXLIV (444)
MCMLIX (1,959)	MMMXCVIII (3,098)

ENRICHMENT ANCIENT NUMBER SYSTEMS

The Egyptian number system is based on the number 10. Each symbol stands for a different power of 10.

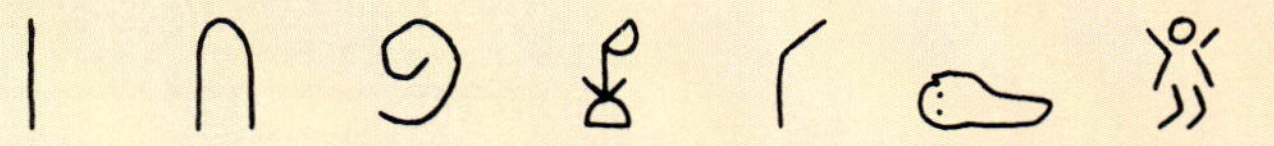

1 10 100 1,000 10,000 100,000 1,000,000

The Egyptian system uses addition. The value of a number is found by adding the values of the symbols.

Example

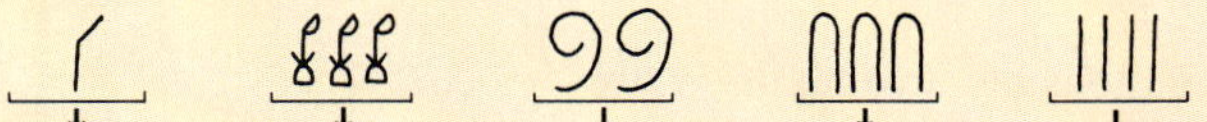

10,000 + 3,000 + 200 + 30 + 4 = 13,234

The Roman number system uses the following symbols.

I	V	X	L	C	D	M
1	5	10	50	100	500	1,000

This system uses addition and subtraction. If a symbol has a value greater than or equal to the value of the symbol to the right, add. If a symbol has a value less than the value of the symbol to the right, subtract.

Examples M CM LXX IV: 1,000 + 900 + 70 + 4 = 1,974 MM XL IX: 2,000 + 40 + 9 = 2,049

Write our numbers.

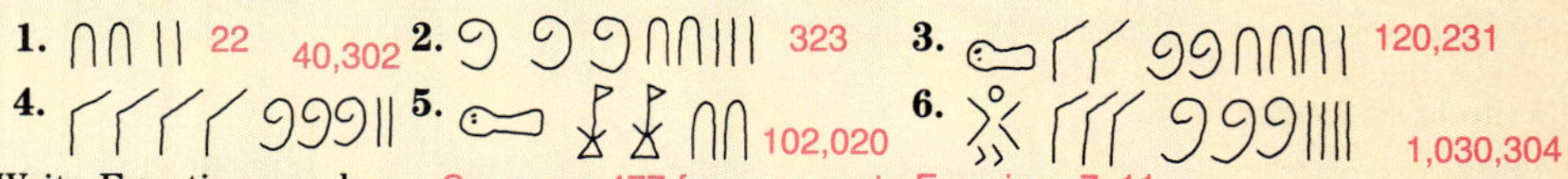

Write Egyptian numbers. See page 477 for answers to Exercises 7–11.

7. 25 **8.** 46 **9.** 131 **10.** 158 **11.** 1,254

Write our numbers.

12. L X X V I 76 **13.** C C L X V 265 **14.** D C C C X X X I I 832
15. X C V 95 **16.** C D L I V 454 **17.** M C M L I X 1,959

Write Roman numbers.

18. 13 XIII **19.** 38 XXXVIII **20.** 156 CLVI **21.** 509 DIX **22.** 944 CMXLIV

OBJECTIVES

- Estimate sums and differences by rounding.
- Estimate sums and differences by front-end estimation.

TEACHING THE LESSON

WARM-UP Write the following exercises on the chalkboard or an overhead transparency.

Round the number to its greatest place.

1. 485 **2.** 739 **3.** 8,042

4. 3,516 **5.** 16,951 **6.** 623,806

(**1.** 500 **2.** 700 **3.** 8,000 **4.** 4,000 **5.** 20,000 **6.** 600,000)

INTRODUCTION Discuss with students reasons for using an estimated sum or difference. Student responses may include: to find out about how much is needed, to determine if an answer is reasonable, to determine if you have enough money to buy a group of items.

INSTRUCTION As you discuss Example 1, point out that all addends have the same number of places. Therefore, all three addends are rounded to the thousands place. However, in Example 2, the numbers do not have the same number of places. Each number must be rounded to the greatest place. Therefore, 11,157 is rounded to 10,000 and 8,301 is rounded to 8,000.

Use Example 3 to establish how front-end estimation is a quick way to estimate. Point out that only the digits in the greatest place are added or subtracted. Example 3a shows three addends, but only the digits in the thousands place are added. Guide students in seeing why the actual sum has to be greater than the estimate. In Example 3b, students subtract only the digits in the ten-thousands place. Since the subtrahend does not have a digit in the ten-thousands place, zero is subtracted from 7.

2.1 ESTIMATING SUMS AND DIFFERENCES

For a 3-day bike-a-thon, the Big Wheels Cycle Club raised $2,492, $3,096, and $2,987. About how much money did they raise in all?

To find *about how much,* use estimation.

You can estimate a sum or difference by rounding each number to its greatest place. Then, add or subtract.

EXAMPLE 1 Estimate: $2,492 + $3,096 + $2,987

$2,492 + $3,096 + $2,987
↓ ↓ ↓
$2,000 + $3,000 + $3,000 = $8,000

So, the Cycle Club raised about $8,000.

Sometimes you must round the numbers to different places.

EXAMPLE 2 Estimate: 11,157 − 8,301

1. Round each number to the greatest place.
2. Subtract.

$$\begin{array}{rcr} 11{,}157 & \rightarrow & 10{,}000 \\ -\ 8{,}301 & \rightarrow & 8{,}000 \\ \hline & & 2{,}000 \end{array}$$

The estimated difference is 2,000.

You can also estimate using **front-end estimation.** Add or subtract only the digits in the greatest place.

EXAMPLE 3 Estimate:

a. 5,479 + 6,890 + 239

THINK: The greatest place is thousands. Add only the digits in the thousands place.

5 + 6 = 11

$$\begin{array}{r} 5{,}479 \\ 6{,}890 \\ +\ \ 239 \\ \hline 11{,}000 \end{array}$$

b. 78,152 − 6,583

THINK: The greatest place is ten thousands. Subtract only the digit in the ten-thousands place.

7 − 0 = 7

$$\begin{array}{r} 78{,}152 \\ -\ 6{,}583 \\ \hline 70{,}000 \end{array}$$

FOR DISCUSSION See TE side column.

Why will front-end estimation for addition always be an underestimation?

See Extra Practice, page 428.

5. 400,000 **10.** 320,000 **15.** 1,300,000
20. 260,000 **25.** 300,000 **30.** 50,000

PRACTICE EXERCISES

Estimate by rounding.

• **1.** 739 +421 1,100	**2.** 1,583 +6,291 8,000	**3.** 2,986 +1,123 4,000	**4.** 43,023 +29,436 70,000	**5.** 183,426 +158,559
6. 468 + 37 540	**7.** 2,804 + 657 3,700	**8.** 6,239 + 118 6,100	**9.** 11,219 + 2,658 13,000	**10.** 287,999 + 17,823
11. 862 942 +871 2,700	**12.** 3,682 4,286 +9,870 18,000	**13.** 4,765 5,821 +8,601 20,000	**14.** 11,006 29,132 +32,446 70,000	**15.** 666,231 476,003 +123,459
16. 361 802 + 54 1,250	**17.** 4,362 399 + 213 4,600	**18.** 5,673 4,986 + 211 11,200	**19.** 29,014 981 + 3,025 34,000	**20.** 28,697 186,760 + 34,876
• **21.** 627 −374 200	**22.** 9,428 −5,037 4,000	**23.** 6,829 −2,268 5,000	**24.** 57,342 −19,820 40,000	**25.** 379,005 −124,198
26. 109 − 67 30	**27.** 3,711 − 994 3,000	**28.** 5,468 − 967 4,000	**29.** 12,287 − 7,594 2,000	**30.** 107,812 − 45,485

35. 700,000 **40.** 600,000 **45.** 1,500,000
50. 200,000 **55.** 200,000 **60.** 700,000

Estimate by using front-end estimation.

• **31.** 739 +421 1,100	**32.** 6,291 +1,583 7,000	**33.** 6,823 +3,104 9,000	**34.** 43,123 +12,999 50,000	**35.** 586,036 +228,451
36. 836 + 22 800	**37.** 4,223 + 349 4,000	**38.** 5,982 + 411 5,000	**39.** 34,097 + 1,789 30,000	**40.** 607,098 + 12,569
41. 316 250 +123 600	**42.** 4,361 2,102 +1,254 7,000	**43.** 5,786 1,102 +2,006 8,000	**44.** 21,009 33,436 +42,985 90,000	**45.** 613,890 506,231 +438,506
46. 482 41 +512 900	**47.** 1,056 2,213 + 872 3,000	**48.** 912 3,069 +6,730 9,000	**49.** 31,986 82,411 + 6,121 110,000	**50.** 212,765 17,645 + 5,680
• **51.** 356 −127 200	**52.** 6,436 −4,672 2,000	**53.** 5,208 −1,341 4,000	**54.** 42,118 −34,882 10,000	**55.** 379,005 −124,198
56. 984 − 26 900	**57.** 5,434 − 267 5,000	**58.** 9,896 − 398 9,000	**59.** 26,402 − 3,857 20,000	**60.** 704,012 − 13,564

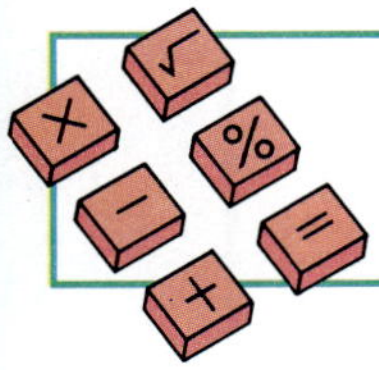

CALCULATOR

See page 477 for answers.

Use a calculator to find the exact answers to Exercises 6 through 25. Compare these answers to your estimates. Were they close?

FOR DISCUSSION Students should explain that front-end estimation always drops part or all of a number. Therefore, an addition answer is always an underestimate.

CALCULATOR This activity is designed to emphasize a major purpose of estimation, that of checking the reasonableness of an exact answer.

ASSIGNMENTS

Level 1	Odd 1–55, 56–59, CA
Level 2	Even 2–54, 56–60, CA
Level 3	6–24, Odd 31–55, 56–60, CA

FOLLOW-UP ACTIVITY

APPLICATION Present each student with the task of spending $250 during a shopping spree. Provide copies of newspaper ads, catalogs, or prepared price lists. Be sure to include items of interest to your students, for example, clothes and audio equipment. Instruct the students to use only estimates in determining what to buy. The total amount of the purchases should be close to, but not exceed, $250.

SUPPLEMENTARY MATERIALS

TRP Practice, p. 11

TRP Reteaching, p. 7

TRP Transparency 6

OBJECTIVES

- Add whole numbers.
- Estimate and check the reasonableness of answers.

TEACHING THE LESSON

WARM-UP Read the following exercises aloud. Instruct students to write their answers on a piece of paper.

1. 7 + 4 (11) **2.** 5 + 9 (14)

3. 2 + 9 (11) **4.** 6 + 8 (14)

5. 3 + 7 (10) **6.** 8 + 9 (17)

INTRODUCTION Use place-value manipulatives to work through 25 + 47. Demonstrate the regrouping.

$$\begin{array}{r} {}^{1} \\ 25 \\ +47 \\ \hline 72 \end{array}$$

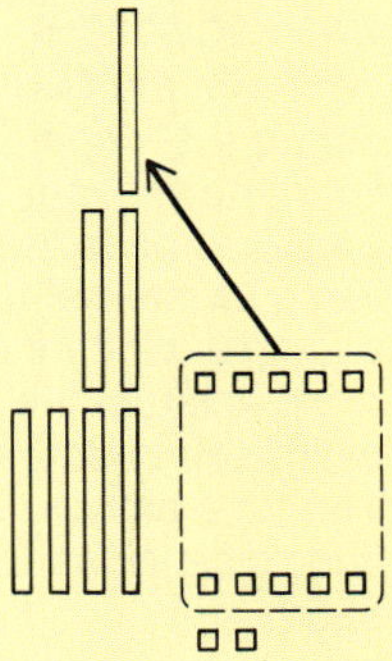

INSTRUCTION Use Example 1 to present the addition algorithm. Stress the alignment of digits in columns, and the regrouping. Make sure students understand that 13 ones are 1 ten 3 ones and that 19 tens are 1 hundred 9 tens. Discuss **addend** and **sum**.

The estimate in Example 2 is found by rounding. Some students may estimate using front-end estimation. The estimate would be 150,000.

As you discuss Example 3, note how the Commutative and Associative Properties justify the check of adding in reverse order.

FOR DISCUSSION Students should explain that each of the three addends is less than 1,000. The sum must be less than 3,000. Therefore, 4,087 is incorrect.

2.2 ADDING

There are 840 miles of coastline in California, 296 miles in Oregon, and 157 miles in Washington. How many miles of coastline are there in all?

To find the total number of miles, add the three distances.

EXAMPLE 1 Add: 840 + 296 + 157

1. Add the ones. Regroup.

$$\begin{array}{r} {}^{1} \\ 840 \\ 296 \\ +157 \\ \hline 3 \end{array}$$

2. Add the tens. Regroup.

$$\begin{array}{r} {}^{1\,1} \\ 840 \\ 296 \\ +157 \\ \hline 93 \end{array}$$

3. Add the hundreds. Regroup.

$$\begin{array}{rl} {}^{1\,1} & \\ 840 & \leftarrow \\ 296 & \leftarrow \text{Addends} \\ +157 & \leftarrow \\ \hline 1{,}293 & \leftarrow \text{Sum} \end{array}$$

So, there are 1,293 miles of coastline.

To determine if an exact answer is reasonable, compare the answer to an estimate. If they are close, the answer is reasonable.

EXAMPLE 2 Add: 87,950 and 71,285. Is your answer reasonable?

1. Estimate.

$$\begin{array}{rcr} 87{,}950 & \rightarrow & 90{,}000 \\ +71{,}285 & \rightarrow & +70{,}000 \\ \hline & & 160{,}000 \end{array}$$

2. Add.

$$\begin{array}{r} 87{,}950 \\ +71{,}285 \\ \hline 159{,}235 \end{array}$$

The estimate, 160,000, and the exact answer, 159,235, are close. The answer is reasonable.

To check your addition, you can add in the reverse order.

EXAMPLE 3 Add: 4,935 + 138,861 + 72,049 + 6,113. Check your answer.

1. Add down.

$$\begin{array}{r} 4{,}935 \\ 138{,}861 \\ 72{,}049 \\ +\ \ 6{,}113 \\ \hline 221{,}958 \end{array}\ \downarrow$$

2. Check up.

$$\begin{array}{r} 4{,}935 \\ 138{,}861 \\ 72{,}049 \\ +\ \ 6{,}113 \\ \hline 221{,}958 \end{array}\ \uparrow$$

FOR DISCUSSION See TE side column.

Explain how you can tell just by looking that the given sum is incorrect: 976 + 854 + 792 = 4,087.

OBJECTIVES

- Subtract whole numbers.
- Estimate and check the reasonableness of answers.

TEACHING THE LESSON

WARM-UP Read the following exercises aloud. Instruct students to write their answers on a piece of paper.

1. 5 − 3 (2) **2.** 9 − 6 (3)

3. 12 − 5 (7) **4.** 13 − 8 (5)

5. 14 − 5 (9) **6.** 10 − 4 (6)

INTRODUCTION Use place-value manipulatives to work through 34 − 18. Demonstrate the regrouping of 1 ten to 10.

```
 2 14
  3̸4̸
 −18
  16
```

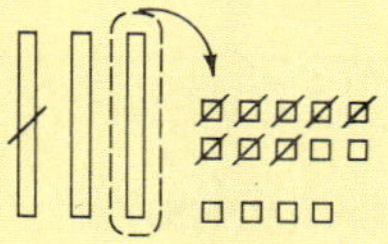

INSTRUCTION Use Example 1 to present the subtraction algorithm. Make sure students not only understand the regrouping process but know when to regroup.

As you discuss Example 2, point out that the first step is to regroup the hundreds. Since most students have difficulty subtracting across zeros, have a volunteer explain every step of the regrouping process.

Example 3 uses estimation by rounding to check the reasonableness of an exact answer. Some students may estimate using front-end estimation; the estimate is also 7,000.

Use Example 4 to show how to check subtraction. For the check to be effective, stress that students must add carefully.

2.3 SUBTRACTING

The ROBO Inc. car manufacturing plant made 5,682 cars. The records show 729 cars were not assembled by robots. How many cars were assembled by robots?

To find the difference, subtract 729 from 5,682.

EXAMPLE 1 Subtract: 5,682 − 729

1. Regroup. Subtract the ones.	2. Subtract the tens.	3. Regroup. Subtract the hundreds.	4. Subtract the thousands.
7 12 5,682 − 729 3	7 12 5,682 − 729 53	4 16 7 12 5,682 − 729 953	4 16 7 12 5,682 − 729 4,953

So, 4,953 cars were assembled by robots.

When you regroup across zeros, you have to regroup more than once.

EXAMPLE 2 Subtract: 600 − 187

1. Regroup.	2. Regroup.	3. Subtract.
THINK: 6 hundreds = 5 hundreds, 10 tens	**THINK: 10 tens = 9 tens, 10 ones**	
5 10 600 −187	9 5 10 10 600 −187	9 5 10 10 600 −187 413

So, 600 − 187 = 413.

Remember, you can determine if an exact answer is reasonable by comparing the answer to an estimate.

EXAMPLE 3 Subtract: 9,693 − 2,874. Is your answer reasonable?

1. Estimate.	2. Subtract.
9,693 → 10,000 −2,874 → − 3,000 7,000	9,693 −2,874 6,819

Since 7,000 and 6,819 are close, the exact answer is reasonable.

You can use addition to check subtraction.

EXAMPLE 4 Subtract: 82,603 − 16,571. Check your answer.

1. Subtract.

$$\begin{array}{r} 82{,}603 \\ -\,16{,}571 \\ \hline 66{,}032 \end{array}$$

2. Check.

$$\begin{array}{r} 66{,}032 \\ +\,16{,}571 \\ \hline 82{,}603 \end{array}$$

CHECKPOINT

Write the letter of the correct answer.

Subtract.

1. 50,000 − 19,381 d **a.** 41,729 **b.** 31,711 **c.** 49,381 **d.** 30,619
2. 16,428 − 7,523 a **a.** 8,905 **b.** 9,105 **c.** 9,945 **d.** 8,945

PRACTICE EXERCISES

See Extra Practice, page 429.

Estimate by rounding. Which exact answer is reasonable?

• 1. 912 − 234 b **a.** 368 **b.** 678 **c.** 858
2. 5,709 − 2,457 b **a.** 2,252 **b.** 3,252 **c.** 5,252
3. 8,178 − 605 b **a.** 3,573 **b.** 7,573 **c.** 8,573
4. 19,936 − 6,292 c **a.** 8,644 **b.** 9,464 **c.** 13,644

Subtract. Check your answer.

• 5. 97 − 47 50
• 6. 43 − 29 14
7. 51 − 33 18
8. 82 − 7 75
9. 70 − 9 61
10. 837 − 245 592
11. 653 − 491 162
12. 924 − 828 96
13. 501 − 76 425
14. 900 − 45 855
15. 3,419 − 2,652 767
16. 8,738 − 2,804 5,934
17. 3,651 − 1,836 1,815
18. 5,017 − 849 4,168
19. 3,802 − 956 2,846
20. 56,957 − 34,817 22,140
21. 26,815 − 13,906 12,909
22. 71,657 − 65,738 5,919
23. 90,508 − 6,591 83,917
24. 34,090 − 904 33,186
25. 452,311 − 109,533 342,778
26. 611,092 − 272,041 339,051
27. 923,187 − 867,059 56,128
28. 260,900 − 59,654 201,246
29. 501,087 − 43,852 457,235

First estimate by rounding. Then find the exact answer. **32.** 30,000; 26,562 **35.** 8,600; 8,687

• 30. 611 − 421 200; 190
• 31. 7,986 − 2,814 5,000; 5,172
32. 38,999 − 12,437
33. 383 − 42 360; 341
34. 8,822 − 396 8,600; 8,426
35. 9,123 − 436
36. 36,123 − 8,041 32,000; 28,082
37. 59,001 − 4,124 56,000; 54,877
38. 683,421 − 36,923 660,000; 646,498

CHECKPOINT Distribute four index cards to each student on which the answer choices a–d appear. After allowing students a few minutes, ask each to raise the letter that indicates their answer choice. Scan the cards and identify those students who need additional attention before proceeding to the Practice Exercises. The incorrect answer choices include common errors.

TIME OUT Have students draw the scales on graph paper. If the scales are precise, students will find locating the answer easier.

COMMON ERROR

When subtracting, some students subtract the lesser digit from the greater digit regardless of placement. For example, in item 1 of Checkpoint they would select c. Work with these students in small groups, using manipulative materials.

ASSIGNMENTS

Level 1	Odd 1–37, 39–52
Level 2	1–4, Odd 5–37, 39–54, TO
Level 3	Even 16–50, 52–54, TO

FOLLOW-UP ACTIVITY

ENRICHMENT Write the following example on the chalkboard or an overhead transparency.

$$\begin{array}{r} 99{,}999 \\ -\,9{,}999 \\ \hline 86{,}997 \end{array}$$

Tell the students that the answer is correct. However, there are four nines (three in 99,999 and one in 9,999) that should be sixes. Ask students to write the example correctly.
(96,696 − 9,699 = 86,997)

SUPPLEMENTARY MATERIALS

TRP Practice, p. 13

TRP Reteaching, p. 9

TRP Transparency 7

MIXED REVIEW

Find the value of the underlined digit.

•**39.** $\underline{7}65{,}034$ 700,000
•**40.** $1\underline{8}{,}345{,}097$ 8,000,000
41. $674{,}0\underline{9}1{,}254$ 90,000
42. $2.008\underline{6}5$ 0.0006
43. $0.67\underline{1}294$ 0.001
44. $19.38501\underline{4}$ 0.000004

Compare. Write =, >, <.

•**45.** 9,457 ● 9,449 >
•**46.** 74,653 ● 74,660 <
47. 201,567 ● 199,567 >
48. 8.43 ● 8.430 =
49. 2.0056 ● 2.056 <
50. 243.9 ● 243.8876 >

Solve.

•**51.** Robo's goal for the year is to build 25,000 cars. To date, 9,476 cars have been built. How many more cars are needed to reach the goal? 15,524 cars

52. The Bion car plant produces 12,493 cars. All but 2,874 are made by robots. How many more cars are assembled by robots than by people? 9,619 cars

53. Gross sales of the Jetstar from September to November were 6 thousand, 104 thousand, and 83 thousand. Write the number for the total gross sales. 193,000 sales

54. The Zap model car has 3,194 moving parts. The Zing model has 23,842, and the Zoom model has 9,375. How many moving parts are there in all three models? 36,411 parts

TIME OUT

You can make an interesting computing device called a **nomograph**.

1. On a piece of graph paper, draw three horizontal lines as shown. Label them *A*, *B*, and *C*.
2. Mark the scales as shown.

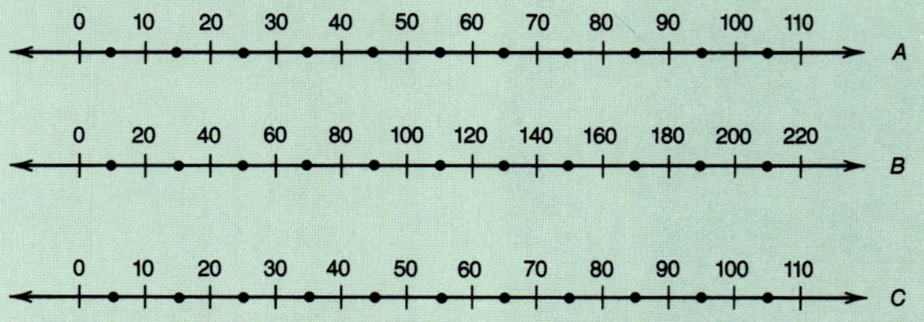

To find 70 − 25, place a straightedge so that it crosses *A* at 25 and *B* at 70. The answer is where it crosses *C* at 45.

Use your nomograph to find the difference.

1. 80 − 15 65 **2.** 110 − 5 105 **3.** 180 − 70 110 **4.** 125 − 40 85

CHECKPOINT

Write the letter of the correct answer.

Multiply or divide.

1. 769 × 1,000 — b — **a.** 7,690,000 **b.** 769,000 **c.** 76,900
2. 20 × 50,000 — a — **a.** 1,000,000 **b.** 100,000 **c.** 10,000
3. 3,600 ÷ 40 — b — **a.** 9 **b.** 90 **c.** 900
4. 3,000,000 ÷ 600 — c — **a.** 50 **b.** 500 **c.** 5,000

PRACTICE EXERCISES

See Extra Practice, page 429.

14. 30,000,000 **17.** 1,500,000,000

Multiply.

• **1.** 20 × 10 200 • **2.** 70 × 60 4,200 **3.** 800 × 80 64,000 **4.** 600 × 20 12,000

5. 900 × 100 90,000 **6.** 600 × 800 480,000 **7.** 400 × 700 280,000 **8.** 800 × 900 720,000

9. 6,000 × 700 4,200,000 **10.** 800 × 5,000 4,000,000 **11.** 400 × 600 240,000

12. 2,000 × 9,000 18,000,000 **13.** 4,000 × 6,000 24,000,000 **14.** 5,000 × 6,000

15. 5,000 × 70,000 350,000,000 **16.** 4,596 × 10,000 45,960,000 **17.** 3,000 × 500,000

18. 4,387 × 100,000 438,700,000 **19.** 70,000 × 400,000 28,000,000,000 **20.** 300 × 6,000,000 1,800,000,000

Divide.

• **21.** 800 ÷ 10 80 • **22.** 600 ÷ 30 20 **23.** 1,200 ÷ 20 60 **24.** 2,100 ÷ 70 30

25. 54,000 ÷ 90 600 **26.** 72,000 ÷ 80 900 **27.** 120,000 ÷ 40 3,000 **28.** 150,000 ÷ 50 3,000

29. 7,000 ÷ 100 70 **30.** 9,000 ÷ 300 30 **31.** 15,000 ÷ 500 30

32. 28,000 ÷ 7,000 4 **33.** 420,000 ÷ 6,000 70 **34.** 300,000 ÷ 5,000 60

35. 270,000 ÷ 90,000 3 **36.** 930,000 ÷ 10,000 93 **37.** 200,000 ÷ 40,000 5

38. 3,690,000 ÷ 10,000 369 **39.** 4,000,000 ÷ 50,000 80 **40.** 6,300,000 ÷ 90,000 70

Solve.

41. Advertising time for a television special cost $140,000 a minute. How much did it cost for 4 min? $560,000

42. A poll showed that 2,000 teenagers in each of 5,000 cities chose a product because of an ad. How many teenagers chose the product? 10,000,000 teenagers

TIME OUT

As a promotion, a local record store ran a contest. The contest was to guess how many pennies there are in a stack that is 1 yd high. Jenna guessed 100 pennies, Frank guessed 1,000 pennies, and Ben guessed 10,000 pennies. They knew that 20 pennies in a stack is about 1 in. high. Whose guess was closest to the actual answer? Frank's

CHECKPOINT The incorrect answer choices include common errors students make.

TIME OUT You may want to prominently display a yardstick in the classroom to remind students that there are 36 in. in 1 yd. Discuss how students solved the problem. Some students may have multiplied 20 and 36, while others may have used estimation.

COMMON ERROR

When multiplying by multiples of powers of 10, some students may forget to put the zero in the product when the product of the lead digits contains a zero. For example, in Checkpoint 2 they would select choice b. Have these students write the exercise in vertical form, first writing the zeros and then multiplying the lead digits.

ASSIGNMENTS

Level 1	Odd 1–11, Odd 21–31, 41–42
Level 2	Even 2–40, 41–42, TO
Level 3	12–20, 29–40, 41–42, TO

FOLLOW-UP ACTIVITY

MENTAL MATH Read the following exercises aloud. Ask for two volunteers. Have one find the product using a calculator while the other finds the product mentally.

60 × 200 (12,000)
30 × 400 (12,000)
200 × 400 (80,000)
500 × 1,000 (500,000)
5 × 90 (450)
600 × 2,000 (1,200,000)

Award a point to the student who is first in writing a correct answer. Use this activity to demonstrate that there are times when computing mentally is faster than using the calculator.

SUPPLEMENTARY MATERIALS

TRP Practice, p. 15
TRP Reteaching, p. 10
TRP Transparency 8

OBJECTIVES

- Estimate products by rounding.
- Estimate quotients by rounding and using compatible numbers.

TEACHING THE LESSON

WARM-UP Write the following exercises on the chalkboard or an overhead transparency.

Round the number to its greatest place.

1. 82 **2.** 506

3. 4,781 **4.** 5,328

5. 46,209 **6.** 489,081

(**1.** 80 **2.** 500 **3.** 5,000 **4.** 5,000 **5.** 50,000 **6.** 500,000)

INTRODUCTION Discuss situations where an estimate of a product or a quotient is needed. Student responses may include: to estimate cost when more than one of an item is bought or to estimate quantity when distributing items among a group of people.

INSTRUCTION As you discuss Example 1, point out that it answers both questions posed by the introductory problem. Remind students of the rules for multiplying with multiples of powers of 10. Point out that in Example 1a, a 1-digit factor is not rounded. In Example 1b students should realize that an "extra" zero appears because the product of the lead digits 5 × 4 is 20.

Use Example 2 to show students when rounding results in a simple division. Point out that in Example 2a, the 1-digit divisor is not rounded. Emphasize that if the same method is used in Example 3, the result would not be a simple division.

Use Example 3 to introduce the concept of rounding to compatible numbers. Show the students that in Example 3a, the 1-digit divisor is not rounded. The dividend is rounded to a compatible number.

2.6 ESTIMATING PRODUCTS AND QUOTIENTS

An animal shelter takes care of 374 stray animals a week. About how many animals does it care for in 8 weeks? in a year (52 weeks)?

To find *about how many,* use estimation.

You can estimate a product by rounding each number to its greatest place. Then multiply.

EXAMPLE 1 Estimate:

a. 8 × 374

$$\begin{array}{r} 374 \\ \times\ 8 \end{array} \rightarrow \begin{array}{r} 400 \\ \times\ 8 \\ \hline 3{,}200 \end{array}$$

About 3,200 animals are cared for in 8 weeks.

b. 52 × 374

$$\begin{array}{r} 374 \\ \times\ 52 \end{array} \rightarrow \begin{array}{r} 400 \\ \times\ 50 \\ \hline 20{,}000 \end{array}$$

About 20,000 animals are cared for in a year.

You can also estimate a quotient by rounding each number to its greatest place. Then divide.

EXAMPLE 2 Estimate:

a. 773 ÷ 8

773 ÷ 8
↓ ↓
800 ÷ 8 = 100

The estimated quotient is 100.

b. 56,214 ÷ 32

56,214 ÷ 32
↓ ↓
60,000 ÷ 30 = 2,000

The estimated quotient is 2,000.

Another way you can estimate a quotient is by using **compatible numbers,** numbers that divide evenly.

EXAMPLE 3 Estimate:

a. 196 ÷ 7

THINK: 21 ÷ 7 = 3

196 ÷ 7
↓ ↓
210 ÷ 7 = 30

b. 1,102 ÷ 58

THINK: 12 ÷ 6 = 2

1,102 ÷ 58
↓ ↓
1,200 ÷ 60 = 20

FOR DISCUSSION See TE side column.

Estimate 743 ÷ 8 by rounding each number to its greatest place, and then by using compatible numbers. Which method was easier? Explain.

PRACTICE EXERCISES

See Extra Practice, page 429.
31. 5,000 **32.** 10,000 **33.** 10,000
34. 10,000 **35.** 10,000

Estimate by rounding.

• **1.** 49 × 7 — 350	**2.** 56 × 8 — 480	**3.** 792 × 9 — 7,200	**4.** 539 × 6 — 3,000	**5.** 816 × 7 — 5,600
6. 2,134 × 4 — 8,000	**7.** 6,982 × 6 — 42,000	**8.** 7,316 × 9 — 63,000	**9.** 8,012 × 5 — 40,000	**10.** 9,986 × 7 — 70,000
11. 94 × 43 — 3,600	**12.** 76 × 58 — 4,800	**13.** 84 × 17 — 1,600	**14.** 67 × 32 — 2,100	**15.** 73 × 41 — 2,800
16. 390 × 69 — 28,000	**17.** 496 × 23 — 10,000	**18.** 695 × 48 — 35,000	**19.** 2,394 × 78 — 160,000	**20.** 6,221 × 11 — 60,000
• **21.** 5)235 — 40	**22.** 7)749 — 100	**23.** 8)408 — 50	**24.** 6)594 — 100	**25.** 5)215 — 40
26. 3)5,925 — 2,000	**27.** 4)7,604 — 2,000	**28.** 5)2,195 — 400	**29.** 6)3,120 — 500	**30.** 7)7,385 — 1,000
31. 8)37,904	**32.** 9)94,986	**33.** 4)39,126	**34.** 3)28,666	**35.** 4)36,436
36. 19)834 — 40	**37.** 48)986 — 20	**38.** 27)592 — 20	**39.** 39)19,426 — 500	**40.** 23)59,611 — 3,000

Estimate using compatible numbers.

• **41.** 352 ÷ 8 — 40	**42.** 527 ÷ 9 — 60	**43.** 652 ÷ 7 — 90	**44.** 133 ÷ 4 — 30
45. 4,123 ÷ 6 — 700	**46.** 8,236 ÷ 9 — 900	• **47.** 5,802 ÷ 83 — 70	**48.** 1,391 ÷ 22 — 60
49. 3,597 ÷ 41 — 90	**50.** 4,814 ÷ 87 — 50	**51.** 30,752 ÷ 43 — 800	**52.** 29,456 ÷ 39 — 700

Solve.

53. A kennel uses 387 cases of dog food a month. There are 24 cans in each case. About how many cans does the kennel use? 8,000 cans

54. Last year, the number of registered German shepherds dropped from 65,073 to 59,450. How many fewer registrations were there last year? 5,623 registrations

55. In 1983, 64,389 Labrador retrievers were registered in 936 kennel clubs. About the same number were registered in each club. About how many were registered in each club? 70 Labradors

56. In a recent year, there were 1,085,248 registrations in the American Kennel Club. Round the number of registrations to the nearest million. 1,000,000 registrations

CALCULATOR

See page 477 for answers.

Use a calculator to find the exact answers to Exercises 21 through 40. Compare these answers to your estimates. Were they close?

In Example 3b, the dividend and the 2-digit divisor are both rounded. The divisor is rounded to its greatest place. The dividend is rounded so that it is a compatible number, a multiple of the divisor.

FOR DISCUSSION Students should explain that rounding yields 700 ÷ 8, numbers which are difficult to divide. Rounding to compatible numbers yields 720 ÷ 8.

CALCULATOR This activity is designed to help students see the range of answers that result when students estimate by rounding.

ASSIGNMENTS

Level 1	Odd 1–15, Odd 21–29, Odd 41–51, 53–54
Level 2	Even 2–40, Even 42–52, 53–56, CA
Level 3	6–20, 31–40, 45–56, CA

FOLLOW-UP ACTIVITY

COOPERATIVE LEARNING Form heterogeneous groups of about five students. Have each group find the exact answers to Exercises 11–14 using a calculator. Then have them compare the exact answers and their estimates in order to answer the following questions.

1. When both numbers are rounded up, how does the estimated answer compare with the actual answer? (The estimated answer is greater. See Exercise 12.)
2. When both numbers are rounded down, how does the estimated answer compare with the actual answer? (The estimated answer is lesser. See Exercise 11.)

SUPPLEMENTARY MATERIALS

TRP Practice, p. 16

TRP Reteaching, p. 11

TRP Transparency 8

OBJECTIVES

- Multiply by 1-digit and 2-digit numbers.
- Estimate and check the reasonableness of answers.

TEACHING THE LESSON

WARM-UP Read the following exercises aloud. Instruct students to write their answers on a piece of paper.

1. 5×7 (35) **2.** 7×8 (56)
3. 9×8 (72) **4.** 9×0 (0)
5. 4×3 (12) **6.** 8×6 (48)

INTRODUCTION Use place-value manipulatives to work through 3×24. Demonstrate the regrouping of 12 ones as 1 ten 2 ones.

$$\begin{array}{r} {\scriptstyle 1} \\ 24 \\ \times 3 \\ \hline 72 \end{array}$$

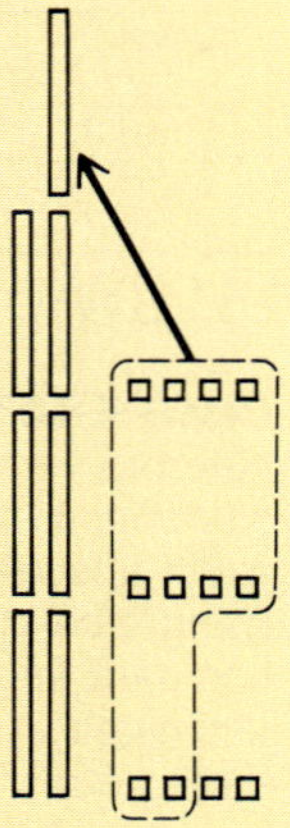

INSTRUCTION As you discuss Example 1, which presents the multiplication algorithm with a 1-digit multiplier, make sure students know how to regroup. Review the terms *factor* and *product.*

Example 2 extends the algorithm to 2-digit multipliers. Encourage students to regroup mentally to avoid careless errors as each partial product is determined. Elicit from students that they write a zero in the second partial product because they are actually multiplying by 30.

Use Example 3 to reinforce the use of estimation as a means for determining the reasonableness of an answer.

2.7 MULTIPLYING BY 1- AND 2-DIGIT NUMBERS

The school cafeteria sells an average of 2,049 sandwiches each week. How many sandwiches does it sell in a month (4 weeks)?

To find the number of sandwiches, multiply 2,049 by 4.

EXAMPLE 1 Multiply: $4 \times 2{,}049$

1. Multiply the ones. Regroup.
2. Multiply the tens. Regroup.
3. Multiply the hundreds.
4. Multiply the thousands.

$$\begin{array}{r} {\scriptstyle 3} \\ 2{,}049 \\ \times \quad 4 \\ \hline 6 \end{array} \qquad \begin{array}{r} {\scriptstyle 1\,3} \\ 2{,}049 \\ \times \quad 4 \\ \hline 96 \end{array} \qquad \begin{array}{r} {\scriptstyle 1\,3} \\ 2{,}049 \\ \times \quad 4 \\ \hline 196 \end{array} \qquad \begin{array}{r} {\scriptstyle 1\,3} \\ 2{,}049 \\ \times \quad 4 \\ \hline 8{,}196 \end{array}$$

(2,049 and 4: **Factors**; 8,196: **Product**)

So, it sells an average of 8,196 sandwiches in a month.

You can use the same steps to multiply greater numbers.

EXAMPLE 2 Multiply: 38×97

1. Multiply by the ones.
2. Multiply by the tens.
3. Add the partial products.

$$\begin{array}{r} 97 \\ \times 38 \\ \hline 776 \end{array} \leftarrow \mathbf{8 \times 97} \qquad \begin{array}{r} 97 \\ \times 38 \\ \hline 776 \\ 2\,910 \end{array} \leftarrow \mathbf{30 \times 97} \qquad \begin{array}{r} 97 \\ \times 38 \\ \hline 776 \\ 2\,910 \\ \hline 3{,}686 \end{array}$$

So, $38 \times 97 = 3{,}686$.

To determine if an exact answer is reasonable, compare the answer with an estimate.

EXAMPLE 3 Multiply: 29×814. Is your answer reasonable?

1. Estimate.
2. Multiply.

$$\begin{array}{r} 814 \rightarrow \\ \times \;\; 29 \rightarrow \\ \hline \end{array} \begin{array}{r} 800 \\ \times \;\; 30 \\ \hline 24{,}000 \end{array} \qquad \begin{array}{r} 814 \\ \times \;\; 29 \\ \hline 7\,326 \\ 16\,280 \\ \hline 23{,}606 \end{array}$$

Since the estimate, 24,000, and the exact answer, 23,606, are close, the answer is reasonable.

PRACTICE EXERCISES See Extra Practice, page 430.

Estimate by rounding. Which answer is reasonable?

• **1.** 6 × 241	b	**a.** 146	**b.** 1,446	**c.** 2,446
2. 5 × 3,096	c	**a.** 150,480	**b.** 25,480	**c.** 15,480
3. 42 × 78	b	**a.** 1,876	**b.** 3,276	**c.** 6,876
4. 71 × 409	a	**a.** 29,039	**b.** 19,309	**c.** 2,930

Multiply.

• **5.** 13 × 6 78 **6.** 72 × 4 288 **7.** 86 × 5 430 **8.** 51 × 6 306 **9.** 97 × 9 873

10. 142 × 6 852 **11.** 217 × 7 1,519 **12.** 344 × 8 2,752 **13.** 563 × 9 5,067 **14.** 893 × 9 8,037

15. 1,236 × 5 6,180 **16.** 3,542 × 3 10,626 **17.** 4,615 × 2 9,230 **18.** 2,538 × 4 10,152 **19.** 5,173 × 6 31,038

20. 6,094 × 7 42,658 **21.** 5,607 × 9 50,463 **22.** 3,940 × 6 23,640 **23.** 7,815 × 8 62,520 **24.** 6,009 × 7 42,063

25. 16,423 × 4 65,692 **26.** 29,415 × 5 147,075 **27.** 36,017 × 6 216,102 **28.** 41,009 × 8 328,072 **29.** 67,149 × 9 604,341

• **30.** 83 × 27 2,241 **31.** 56 × 19 1,064 **32.** 92 × 53 4,876 **33.** 78 × 16 1,248 **34.** 65 × 38 2,470

35. 367 × 25 9,175 **36.** 290 × 56 16,240 **37.** 648 × 32 20,736 **38.** 815 × 24 19,560 **39.** 509 × 47 23,923

40. 1,123 × 28 31,444 **41.** 2,983 × 32 95,456 **42.** 4,786 × 41 196,226 **43.** 5,980 × 56 334,880 **44.** 7,004 × 82 574,328

• **45.** 3 × 95 285 **46.** 7 × 908 6,356 **47.** 9 × 5,602 50,418 **48.** 7 × 8,432 59,024

• **49.** 37 × 52 1,924 **50.** 60 × 318 19,080 **51.** 32 × 806 25,792 **52.** 28 × 629 17,612

53. 41 × 1,456 59,696 **54.** 53 × 2,004 106,212 **55.** 64 × 6,559 419,776 **56.** 80 × 6,117 489,360

First estimate by rounding. Then find the exact answer.

• **57.** 2,308 × 5 10,000; 11,540 **58.** 4,879 × 8 40,000; 39,032 • **59.** 38 × 57 2,400; 2,166 **60.** 82 × 32 2,400; 2,624 **61.** 216 × 74 14,000; 15,984

Solve.

62. 279 students each raised $46 for a community fund raiser. How much money did they raise? $12,834

63. Class enrollments are 987, 893, 905, and 936. The school gym seats 3,500. Can all the classes fit in the gym at the same time? No

ASSIGNMENTS

Level 1	1–4, Odd 5–51, 57, 58, 62, 63
Level 2	1–4, Even 6–60, 62, 63
Level 3	25–44, Odd 45–61, 62, 63

FOLLOW-UP ACTIVITY

CALCULATOR Have students use a calculator to explore number patterns. Ask students to continue and describe the following number pattern:

1 × 9 + 2 = 11
12 × 9 + 3 = 111
123 × 9 + 4 = 1,111
1,234 × 9 + 5 = 11,111

SUPPLEMENTARY MATERIALS

TRP Practice, p. 17

TRP Reteaching, p. 12

TRP Transparency 9

OBJECTIVE

- Multiply by 3-digit numbers.

TEACHING THE LESSON

WARM-UP Write the following exercises on the chalkboard or an overhead transparency.

Multiply.

1. 6 × 249 (1,494)

2. 4 × 356 (1,424)

3. 5 × 6,370 (31,850)

4. 3 × 4,253 (12,759)

INTRODUCTION Review multiplication by 1-digit multipliers. Use place-value manipulatives to work through 2 × 219. Demonstrate the regrouping.

```
  1
 219
 ×2
 438
```

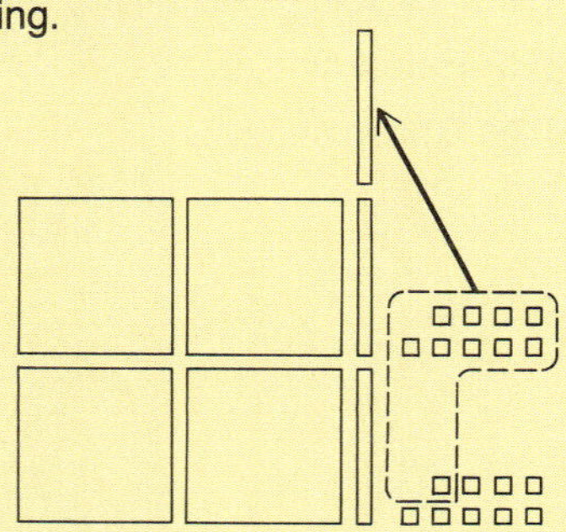

INSTRUCTION As you discuss Example 1, stress the importance of writing the partial products in columns. Encourage students to write zeros in the second and third partial products for tens and hundreds. Remind students to write the dollar sign in the product.

When you discuss Example 2, you may wish to present the following alternative method for multiplying 803 × 4,906. Lead the students to conclude that writing a row of zeros as the second partial product is unnecessary.

```
    4,906
    × 803
   14 718
   00 000   ←Unnecessary
3 924 800
3,939,518
```

2.8 MULTIPLYING BY 3-DIGIT NUMBERS

Dan's Discount Stores have VCRs on sale for $298. In 1 week, 367 VCRs were sold. What was the total amount of sales?

To find the total amount of sales, multiply $298 by 367.

EXAMPLE 1 Multiply: 367 × $298

1. Multiply.

```
    $298
   × 367
   2 086 ←  7 × 298
  17 880 ←  60 × 298
  89 400 ←  300 × 298
```

2. Add the partial products.

```
    $298
   × 367
   2 086
  17 880
  89 400
$109,366
```

So, the total amount of sales was $109,366.

Work carefully when the bottom factor has one or more zeros.

EXAMPLE 2 Multiply: 803 × 4,906

```
    4,906
   ×  803
   14 718 ←  3 × 4,906
3 924 800 ←  800 × 4,906
3,939,518
```

It is often helpful to choose the number you will use as the bottom factor. Try to make the multiplication easier.

EXAMPLE 3 Multiply:

a. 839 × 2,004

THINK: 2,004 has two nonzero digits.

```
      839
   ×2,004
    3 356 ← Two partial products
1 678 000 ←
1,681,356
```

b. 419 × 737

THINK: In 737, there are two 7s.

```
    419
   ×737
  2 933 ← Same nonzero
 12 570   digits in the
293 300 ← partial products
308,803
```

FOR DISCUSSION See TE side column.

Look at the following example: 892 × 635 = 56,640. Is the answer reasonable? Explain.

PRACTICE EXERCISES

See Extra Practice, page 430.

See page 477 for answers to Exercises 1–64.

Multiply.

• **1.** 729 × 132 **2.** 563 × 245 **3.** 612 × 328 **4.** 435 × 517 **5.** 347 × 193

6. 951 × 376 **7.** 816 × 439 **8.** 712 × 635 **9.** 967 × 249 **10.** 658 × 427

11. 5,416 × 325 **12.** 4,176 × 193 **13.** 3,284 × 267 **14.** 5,318 × 241 **15.** 4,935 × 564

16. 3,618 × 127 **17.** 4,589 × 236 **18.** 5,912 × 348 **19.** 6,142 × 459 **20.** 7,196 × 517

• **21.** 495 × 703 **22.** 360 × 580 **23.** 704 × 600 **24.** 527 × 203 **25.** 803 × 450

26. 1,639 × 200 **27.** 3,258 × 460 **28.** 7,503 × 305 **29.** 6,479 × 506 **30.** 7,857 × 308

31. 218 × 677 **32.** 241 × 303 **33.** 167 × 414 **34.** 479 × 808 **35.** 825 × 626

36. 3,007 × 213 **37.** 4,256 × 848 **38.** 7,700 × 541 **39.** 4,114 × 623 **40.** 6,006 × 589

• **41.** 288 × 673 **42.** 764 × 222 **43.** 345 × 7,000 **44.** 6,060 × 539

45. 707 × 123 **46.** 163 × 528 **47.** 404 × 209 **48.** 127 × 5,004

49. 8,006 × 412 **50.** 919 × 246 **51.** 4,182 × 936 **52.** 636 × 8,923

53. 321 × 865 **54.** 159 × 827 **55.** 803 × 507 **56.** 821 × 936

57. 5,006 × 127 **58.** 962 × 141 **59.** 6,431 × 322 **60.** 239 × 9,002

61. 8,623 × 428 **62.** 5,002 × 128 **63.** 7,514 × 424 **64.** 165 × 8,008

Use the table to solve.

Find the total amount of sales for:

• **65.** televisions. $65,403

66. stereos. $60,507

67. computers. $42,704

68. How many items were sold in all? 807 items

69. What was the total amount of sales for all the items? $277,980

DAN'S DISCOUNT SALES

Item/Price	Number Sold	Total Sales
VCR/$298	367	$109,366
TV/$507	129	■
Stereo/$249	243	■
Computer/$628	68	■
Total	■	■

Example 3 is designed to show students the advantages of using the commutative property in selecting the bottom factors.

FOR DISCUSSION Students should estimate to find out if the answer is reasonable. If they estimate the answer by rounding:

892 × 635
↓ ↓
900 × 600 = 540,000

The estimate is far greater than the given answer. Therefore, the given answer cannot be correct.

COMMON ERROR

Students may place the partial products in columns incorrectly. Have students use grid paper to line up the digits.

ASSIGNMENTS

Level 1	1–10, 21–25, 31–35, 41–52, 65–69
Level 2	11–52, 65–69
Level 3	31–40, 53–69

FOLLOW-UP ACTIVITY

APPLICATION Have students make a chart of the total sales for each of the following items in Bill's Bargain Stores: 483 calculators at $15 each, 167 telephones at $39 each, 72 typewriters at $297 each, and 253 copiers at $148. Have students find the total number of items sold and the total sales for all the items.

SUPPLEMENTARY MATERIALS

TRP Practice, p. 18

TRP Reteaching, p. 13

TRP Transparency 9

OBJECTIVE

- Evaluate student progress.

USING THE REVIEW

This page provides a means for informally evaluating students' understanding of the skills and concepts covered so far in this chapter.

Have the students look at the page to familiarize themselves with the various question formats that are presented. Discuss any questions that they may have. Then ask them to complete the page independently.

In addition to grading them individually, you may wish to review the answers to the questions collectively with the students.

Page references appear in brackets. They refer to pages on which a particular skill was introduced.

Before continuing on to the topics found in the remainder of the chapter, you may wish to have students review any skills or concepts in which they have demonstrated weakness.

MIDCHAPTER REVIEW

Estimate by rounding. [28]

1. 842 + 396 = 1,200	**2.** 3,719 + 5,478 = 9,000	**3.** 623 + 2,534 + 6,195 = 9,600	**4.** 4,857 − 3,149 = 2,000	**5.** 56,493 − 7,346 = 53,000

Estimate using front-end estimation. [28]

6. 893 − 346 = 500	**7.** 9,863 − 1,347 = 8,000	**8.** 6,982 − 783 = 6,000	**9.** 59,007 − 45,113 = 10,000	**10.** 623,456 − 54,897 = 600,000

Add or subtract. Check your answer. [30, 32]

11. 536 + 342 = 878	**12.** 819 + 675 = 1,494	**13.** 5,093 + 9,847 = 14,940	**14.** 26,534 + 37,628 = 64,162	**15.** 629,845 + 356,971 = 986,816
16. 729 + 15 + 648 = 1,392	**17.** 935 + 194 + 382 = 1,511	**18.** 5,497 + 2,386 + 4,713 + 8,565 = 21,161	**19.** 98 + 8,567 + 481 + 6,005 = 15,151	**20.** 725,347 + 52,485 + 600,721 + 8,340 = 1,386,893
21. 786 − 357 = 429	**22.** 629 − 35 = 594	**23.** 8,596 − 5,738 = 2,858	**24.** 5,774 − 865 = 4,909	**25.** 7,000 − 3,829 = 3,171
26. 3,000 − 537 = 2,463	**27.** 94,617 − 28,369 = 66,248	**28.** 93,648 − 74,956 = 18,692	**29.** 529,400 − 132,675 = 396,725	**30.** 600,000 − 250,785 = 349,215

Multiply. [36]

31. 20 × 10 200 **32.** 300 × 40 12,000 **33.** 500 × 30 15,000

34. 600 × 30 18,000 **35.** 500 × 600 300,000 **36.** 400 × 6,000 2,400,000

Divide. [36]

37. 600 ÷ 20 30 **38.** 900 ÷ 30 30 **39.** 8,000 ÷ 40 200

40. 50,000 ÷ 50 1,000 **41.** 6,000 ÷ 300 20 **42.** 20,000 ÷ 400 50

Solve. [35]

43. Joe Moser bought an album for $7.98. He gave the cashier a $10 bill. List the least amount of coins and bills he would receive as change.
2 pennies, two $1 bills

44. Ellen Kamp bought a record for $1.79. She gave the clerk two $1 bills. List the least amount of coins she would receive as change.
2 dimes, 1 penny

PROBLEM Solving APPLICATION

2.9 CAREER: NURSE

Luveen Smith is a nurse in a hospital. She helps people who are sick. She also helps people avoid sickness through immunization programs. Often in her work she must keep records of the people she has attended to and helped.

This table shows the number of people immunized for various diseases at three hospitals. Use the table to complete Exercises 1–10. Remember to estimate whenever you use your calculator.

Types of immunization	Number of people immunized			Types of immunization	Number of people immunized		
	Hospital A	Hospital B	Hospital C		Hospital A	Hospital B	Hospital C
Diphtheria	805	427	608	Polio	1,674	923	1,527
Flu	3,349	1,103	1,649	Rubella	248	189	256
Measles	405	389	426	Smallpox	637	420	489
Mumps	297	187	288	Tuberculin	479	342	461

What was the total number of people immunized for:

• **1.** diphtheria? 1,840

2. polio? 4,124

How many more people were immunized for measles than for mumps:

• **3.** at Hospital A? 108

4. at Hospital C? 138

What was the total number of immunizations for:

5. rubella and smallpox? 2,239

6. mumps and measles? 1,992

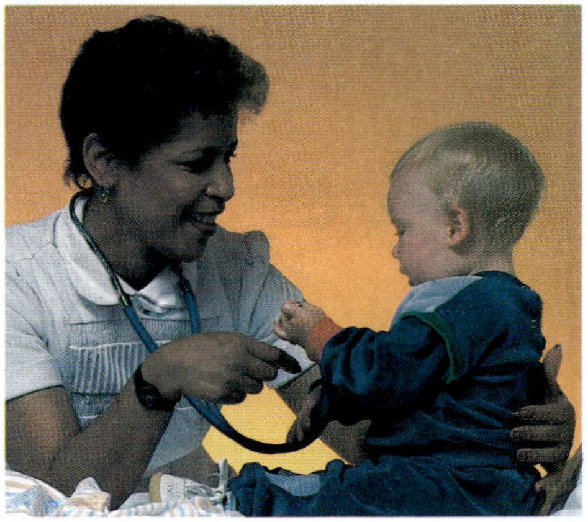

Solve.

7. Which two hospitals immunized the greatest number of people? A, C

8. Which two hospitals immunized the least number of people? B, C

9. The number of diphtheria immunizations at Hospital A was about how many times as great as the number immunized for diphtheria at Hospital B? two times

10. The number of flu immunizations at Hospital A was about how many times as great as the number of flu immunizations at Hospital B? three times

OBJECTIVE

- Solve problems that involve computation with whole numbers.

TEACHING THE LESSON

WARM-UP Write the following exercises on the chalkboard or an overhead transparency.

Use these numbers to answer the questions.

2,785 2,801 997

1. Which number is greatest?
2. What is the sum of the numbers?
3. How much greater is the greatest number than the least number?

(**1.** 2,801 **2.** 6,583 **3.** 1,804)

INSTRUCTION As students examine the table, make sure they understand two facts: Each row shows the number of immunizations of a specific type at the three hospitals. Each column shows the number of different types of immunizations at one of the hospitals.

Point out that in Problems 5 and 6 the students must locate the two appropriate rows and find the sum of the six numbers. In Problems 9 and 10, students should estimate to find the answer.

ASSIGNMENTS

Level 1	1–6, 9–10
Level 2	1–10
Level 3	Even 2–6, 7–10

SUPPLEMENTARY MATERIALS

TRP Practice, p. 19

TRP Transparency 10

OBJECTIVE

- Divide by 1-digit numbers.

TEACHING THE LESSON

WARM-UP Read the following exercises aloud. Instruct students to write their answers on a piece of paper.

1. 45 ÷ 5 (9)	**2.** 36 ÷ 9 (4)
3. 42 ÷ 7 (6)	**4.** 72 ÷ 8 (9)
5. 63 ÷ 7 (9)	**6.** 48 ÷ 6 (8)

INTRODUCTION Use place-value manipulatives to work through 78 ÷ 3. Have students divide 6 tens into 3 groups of 2 tens. Then demonstrate regrouping 1 ten as 10 ones. Have students complete the division by dividing 18 ones into 3 groups of 6 ones.

```
  26
3)78
```

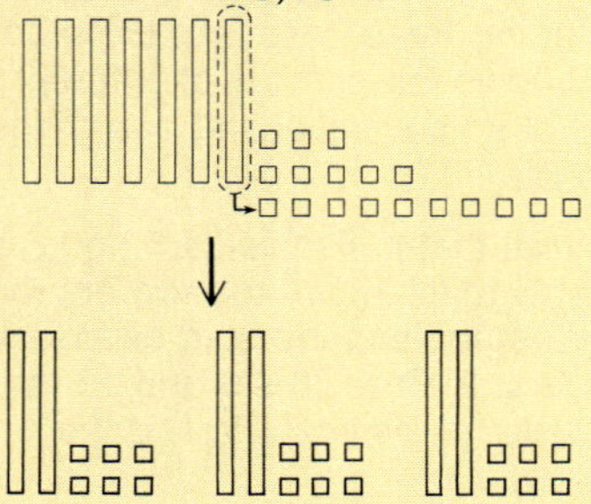

INSTRUCTION Use Example 1 to present the division algorithm. Review the terms *divisor, dividend, quotient,* and *remainder.* Stress the importance of determining where the quotient should begin.

Use Example 2 to show students how zero is used as a place holder in the quotient.

Use Example 3 to show how each part of the division example is used in the check. Make sure the students add the remainder to the product of the quotient and divisor.

2.10 DIVIDING BY A 1-DIGIT DIVISOR

The Computer Corner stores had a sale on software. The warehouse had 317 software packages to distribute equally to 5 different stores. How many software packages did each store receive? How many packages were left?

To find the number of packages delivered to each store, divide 317 by 5.

EXAMPLE 1

Divide: 5)317

1. Determine the first place that can be divided.

   5)317 **THINK:** 5 > 3 Can't divide the hundreds.
   5 < 31 Divide the tens.

2. Divide the tens.

```
   6
5)317
  30
   1
```

3. Divide the ones.
   Write the remainder.

```
   63 R2
5)317
  30
   17
   15
    2   ← Remainder
```

So, each store received 63 software packages and 2 were left over.

Sometimes when you divide, you need a zero placeholder in the quotient.

EXAMPLE 2

Divide: 3)2,418

1. Determine the first place that can be divided.

   3)2,418 **THINK:** 3 > 2 Can't divide the thousands.
   3 < 24 Divide the hundreds.

2. Divide the hundreds.

```
    8
3)2,418
  2 4
    0
```

3. Divide the tens.

```
    80
3)2,418
  2 4
    01
```

4. Divide the ones.

```
    806
3)2,418
  2 4
    018
     18
      0
```

See Extra Practice, page 430.

You can use multiplication to check division.

EXAMPLE 3 Divide: 3,412 ÷ 7. Check your answer.

1. Divide.

```
Quotient →     487 R3 ← Remainder
Divisor  →  7)3,412   ← Dividend
              2 8
                61
                56
                 52
                 49
                  3
```

2. Check.

```
  487 ← Quotient
×   7 ← Divisor
3,409
+   3 ← Remainder
3,412 ← Dividend
```

PRACTICE EXERCISES

41. 3,165 **42.** 3,051 **43.** 9,324 **44.** 7,092 **45.** 7,605
46. 9,581 **47.** 8,171 **48.** 4,252 R3 **49.** 18,071 **50.** 8,305
51. 4,125 **52.** 11,213 **53.** 17,156 **54.** 5,165 R2 **55.** 6,127 R1

Divide.

• **1.** 6)824 137 R2	**2.** 4)729 182 R1	**3.** 9)459 51	**4.** 5)830 166	**5.** 8)548 68 R4
6. 7)636 90 R6	**7.** 3)180 60	**8.** 5)354 70 R4	**9.** 9)918 102	**10.** 6)653 108 R5
• **11.** 8)842 105 R2	**12.** 5)249 49 R4	**13.** 7)161 23	**14.** 6)590 98 R2	**15.** 9)837 93
16. 3)630 210	**17.** 5)700 140	**18.** 8)440 55	**19.** 7)500 71 R3	**20.** 6)200 33 R2
21. 3)1,296 432	**22.** 7)3,584 512	**23.** 4)8,092 2,023	**24.** 6)3,726 621	**25.** 8)7,632 954
26. 5)4,520 904	**27.** 9)6,395 710 R5	**28.** 7)2,429 347	**29.** 8)1,523 190 R3	**30.** 3)8,427 2,809
31. 7)2,358 336 R6	**32.** 4)7,012 1,753	**33.** 6)3,288 548	**34.** 8)5,028 628 R4	**35.** 9)6,327 703
36. 4)8,400 2,100	**37.** 5)6,500 1,300	**38.** 8)6,000 750	**39.** 5)3,600 720	**40.** 7)4,000 571 R3
41. 5)15,825	**42.** 7)21,357	**43.** 4)37,296	**44.** 9)63,828	**45.** 6)45,630
46. 6)57,486	**47.** 8)65,368	**48.** 9)38,271	**49.** 5)90,355	**50.** 7)58,135
51. 6)24,750	**52.** 7)78,491	**53.** 4)68,624	**54.** 7)36,157	**55.** 8)49,017

Divide.

• **56.** 372 ÷ 6 62	• **57.** 896 ÷ 4 224	**58.** 215 ÷ 7 30 R5	**59.** 567 ÷ 3 189
60. 500 ÷ 4 125	**61.** 200 ÷ 8 25	**62.** 600 ÷ 7 85 R5	**63.** 900 ÷ 5 180
64. 4,028 ÷ 4 1,007	**65.** 3,692 ÷ 4 923	**66.** 5,348 ÷ 7 764	**67.** 27,229 ÷ 9 3,025 R4

Solve.

68. There are 8,750 sheets of computer paper divided evenly into 7 boxes. How many sheets are in each box? 1,250 sheets

69. The computer work center sells for $149. If each piece is bought separately, it costs $174. How much is saved by buying the complete center? $25

COMMON ERROR

Students may omit zero as a placeholder in the quotient or write it in the wrong place.

```
    57 R3      or      570 R3
6)3,045             6)3,045
  3 0                 3 0
    045                 045
     42                  42
      3                   3
```

To avoid this error, have students show the multiplication by zero.

ASSIGNMENTS

Level 1	Odd 1–45, 56–59, 68, 69
Level 2	Even 2–46, 60–69
Level 3	35–69

FOLLOW-UP ACTIVITY

ENRICHMENT Long division is a process of making estimates of the quotient and further refining these estimates. The following pyramid method of long division may help students understand the process more clearly. The name is derived from the "pyramid" that is formed by the estimates.

Divide: 7)4,863

```
      4
     90
    600 R5 = 694 R5
7)4,863
  4 200
    663
    630
     33
     28
      5
```

Have students use the pyramid method to divide in Exercises 21–25.

SUPPLEMENTARY MATERIALS

TRP Practice, p. 20

TRP Reteaching, p. 14

TRP Transparency 10

OBJECTIVES

- Divide by 2-digit numbers.
- Estimate and check the reasonableness of answers.

TEACHING THE LESSON

WARM-UP Write the following exercises on the chalkboard or an overhead transparency.

Compare. Write =, <, or >.

1. 83 ● 78	**2.** 15 ● 21
3. 62 ● 59	**4.** 298 ● 300
5. 634 ● 635	**6.** 987 ● 978

(**1.** > **2.** < **3.** > **4.** < **5.** < **6.** >)

INTRODUCTION Have students use manipulatives to work through 143 ÷ 13. Provide them with 1 hundred, 4 tens, 3 ones. Guide them as they change the 1 hundred to 10 tens. Then have them separate the blocks into groups of 1 ten 3 ones. Guide them as they change tens to ones. Have them count the number of groups to find the quotient. (11)

INSTRUCTION Use Example 1 to extend the division algorithm to a 2-digit divisor. Review how to determine the number of digits in the quotient shown in Step 1. In Step 2, have a student explain why 5 was chosen as the estimate. In Step 3, remind students that they must check their estimate by comparing 10 with the divisor, 21.

Use Example 2 to show students that sometimes it is necessary to correct an estimate. In Step 3, stress the process of comparing 345 with 310. Have these students say "345 is greater than 310. Can't subtract. Correct the estimate. Try 4."

Use Example 3 to show students how to use estimation as a means for determining the reasonableness of an answer. Have students note how the estimate was made by rounding to compatible numbers.

2.11 DIVIDING BY A 2-DIGIT DIVISOR

Sarah O'Brien has 115 record albums in her collection. She can put 21 albums in a case for display. How many cases can she fill? How many albums will be left over?

To find how many cases, divide 115 by 21.

EXAMPLE 1

Divide: 21)115

1. Determine the first place that can be divided.

21)115 **THINK:** 21 > 1 Can't divide the hundreds.
21 > 11 Can't divide the tens.
21 < 115 Divide the ones.

2. Estimate the first digit. **THINK:** 21)115 or 2)11 Try 5.

3. Divide the ones.
Write the remainder.

```
      5 R10
21)115        } Remainder
   105
    10
```

Sarah can fill 5 display cases, and 10 albums will be left over.

Sometimes your estimate of the first digit of a quotient may be too large.

EXAMPLE 2

Divide: 69)3,105

1. Determine the first place that can be divided.

69)3,105 **THINK:** 69 > 3 Can't divide the thousands.
69 > 31 Can't divide the hundreds.
69 < 310 Divide the tens.

2. Estimate the first digit. **THINK:** 69)310 or 6)31 Try 5.

3. Divide the tens.

```
      5
69)3,105
   3 45    345 > 310
           5 is too large.
           Try 4.

      4
69)3,105
   2 76
     34
```

4. Divide the ones.

```
      45
69)3,105
   2 76
     345
     345
       0
```

You can estimate to determine if your answer is reasonable.

EXAMPLE 3 Divide: 34)17,238. Is your answer reasonable?

1. Estimate using compatible numbers.

   34)17,238 → 30)18,000, quotient 600

2. Divide.

```
      507
34)17,238
   17 0
      238
      238
        0
```

Since the estimate, 600, and the exact answer, 507, are close, the answer is reasonable.

PRACTICE EXERCISES See Extra Practice, page 431.

Divide.

• 1. 32)268 8 R12	2. 51)397 7 R40	3. 71)651 9 R12	4. 32)109 3 R13	5. 76)542 7 R10
• 6. 60)415 6 R55	7. 58)406 7	8. 64)510 7 R62	9. 43)297 6 R39	10. 24)127 5 R7
11. 97)291 3	12. 34)284 8 R12	13. 42)336 8	14. 98)600 6 R12	15. 86)430 5
16. 23)4,600 200	17. 40)8,805 220 R5	18. 67)3,618 54	19. 18)9,342 519	20. 56)1,960 35
21. 19)1,292 68	22. 75)6,000 80	23. 96)2,692 28 R4	24. 27)3,726 138	25. 81)5,675 70 R5
26. 68)1,872 27 R36	27. 35)2,253 64 R13	28. 79)3,950 50	29. 90)5,130 57	30. 26)9,724 374
31. 30)67,020 2,234	32. 25)15,050 602	33. 54)16,524 306	34. 16)77,000 4,812 R8	35. 82)29,274 357
36. 47)33,135 705	37. 91)57,057 627	38. 60)93,960 1,566	39. 74)38,702 523	40. 36)84,996 2,361
41. 18)73,836 4,102	42. 26)29,199 1,123 R1	43. 36)20,989 583 R1	44. 34)30,334 892 R6	45. 54)35,208 652

First estimate by using compatible numbers. Then find the exact answer.

• 46. 459 ÷ 51 10; 9 • 47. 837 ÷ 27 30; 31 48. 4,760 ÷ 68 70; 70 49. 1,890 ÷ 54 40; 35

50. 57,792 ÷ 48 1,000; 1,204 51. 59,328 ÷ 72 800; 824 52. 498 ÷ 93 5; 5 R33 53. 6,408 ÷ 92 70, 69 R60

Solve.

54. Delta Jazz sold 43,884 albums of their new hit in 12 weeks. On an average, how many did they sell each week? 3,657 albums

55. Buzz Recording Studio made $96,252 from 26 records sold or played last year. On an average, how much did it make from each record? $3,702

COMMON ERROR

Students may forget to insert a zero as a placeholder in the quotient. Encourage students to check their answers.

ASSIGNMENTS

Level 1	Odd 1–29, 54
Level 2	Even 2–40, 46–49, 54–55
Level 3	31–55

FOLLOW-UP ACTIVITY

CALCULATOR If students use a calculator to divide, they may get a decimal answer. This is often no help in determining if the remainder is correct. Have the students use exercises which have remainders to practice the following steps for finding the remainder.

Use a calculator to divide 2,725 ÷ 16. Express the remainder as a whole number.

1. Divide to find the quotient.

   2,725 ÷ 16 = 170.3125

2. Multiply the whole number part of the quotient by the divisor.

   16 × 170 = 2,720

3. Subtract this product from the dividend. The result is the remainder.

   2,725 − 2,720 = 5

4. Check the answer, 170 R5.

   Quotient × divisor + remainder = dividend

   170 × 16 = 2,720 + 5 = 2,725

SUPPLEMENTARY MATERIALS

TRP Practice, p. 21

TRP Reteaching, p. 15

TRP Transparency 11

OBJECTIVE

- Divide by 3-digit numbers.

TEACHING THE LESSON

WARM-UP Write the following exercises on the chalkboard or an overhead transparency.

Estimate by rounding to compatible numbers.

1. 38)1,569 **2.** 62)4,793

3. 28)1,816 **4.** 72)41,013

5. 91)81,946

(**1.** 40 **2.** 80 **3.** 60 **4.** 600 **5.** 900)

INTRODUCTION Review the division process. Help the students outline the steps.

1. Determine the number of digits in the quotient.
2. Estimate the first digit in the quotient.
3. Divide.
4. Check.

INSTRUCTION Use Example 1 to point out to students that dividing with a 3-digit divisor involves the same process as dividing with a 2-digit divisor. Encourage the students to check their answers.

You may wish to present the following example to students who demonstrate difficulty in adjusting their estimates. Divide: 62,347 ÷ 378. Note that 3)6 suggests that the first digit is a 2. However, since $2 \times 378 = 756$ and $756 > 623$, students should see the need to lower the estimate to 1.

Use Example 2 to stress the necessity of the end-zeros as placeholders in the quotient.

FOR DISCUSSION Students should explain that the greatest possible remainder is 466 if the divisor is 467.

2.12 DIVIDING BY A 3-DIGIT DIVISOR

The Small Car Parts Company makes 3,500 spark plugs an hour.
A box of spark plugs contains 432 spark plugs.
How many boxes will 3,500 spark plugs fill?
How many spark plugs will be left over?

You can divide to find the number of boxes.

EXAMPLE 1 Divide: 432)3,500. Check your answer.

1. Determine the first place that can be divided.

432)3,500 **THINK:**

$432 > 3$	Can't divide the thousands.
$432 > 35$	Can't divide the hundreds.
$432 > 350$	Can't divide the tens.
$432 < 3,500$	Divide the ones.

2. Estimate the first digit. **THINK:** 432)3,500 or 4)35 Try 8.

3. Divide the ones.

```
       8R44
432)3,500
    3 456
       44
```

4. Check.

```
   432
 ×   8
 3,456
+   44
 3,500
```

So, 8 boxes will be filled, and 44 spark plugs will be left over.

Sometimes you will need a zero placeholder in the quotient.

EXAMPLE 2 Divide: 318)95,405

1. Determine the first place that can be divided.

318)95,405 **THINK:**

$318 > 9$	Can't divide the ten thousands.
$318 > 95$	Can't divide the thousands.
$318 < 954$	Divide the hundreds.

2. Estimate the first digit. **THINK:** 318)954 or 3)9 Try 3.

3. Divide the hundreds.

```
        3
318)95,405
    95 4
       0
```

4. Divide the tens.

```
        30
318)95,405
    95 4
       00
```

5. Divide the ones.

```
        300R5
318)95,405
    95 4
       005
         0
         5
```

FOR DISCUSSION
See TE side column.

When dividing by 467, what is the greatest possible remainder? Explain.

PRACTICE EXERCISES
See Extra Practice, page 431.

Divide.

• **1.** 9,804 ÷ 258 38	**2.** 8,480 ÷ 200 42 R80	**3.** 9,216 ÷ 156 59 R12	
4. 5,440 ÷ 136 40	**5.** 5,042 ÷ 168 30 R2	**6.** 33,748 ÷ 418 80 R308	
7. 25,599 ÷ 179 143 R2	**8.** 28,341 ÷ 308 92 R5	**9.** 25,761 ÷ 515 50 R11	
10. 341,440 ÷ 231 1,478 R22	**11.** 505,891 ÷ 569 889 R50	**12.** 119,179 ÷ 587 203 R18	
• **13.** 450)9,008 20 R8	**14.** 256)8,960 35	**15.** 652)5,868 9	**16.** 328)9,840 30
17. 176)8,096 46	**18.** 563)8,000 14 R118	**19.** 298)7,158 24 R6	**20.** 765)6,200 8 R80
21. 592)4,144 7	**22.** 397)5,955 15	**23.** 418)5,435 13 R1	**24.** 205)8,200 40
25. 896)8,095 9 R31	**26.** 143)26,169 183	**27.** 326)16,300 50	**28.** 228)17,784 78
29. 135)40,500 300	**30.** 600)90,000 150	**31.** 159)16,695 105	**32.** 252)32,870 130 R110
33. 739)72,422 98	**34.** 946)61,500 65 R10	**35.** 479)99,632 208	**36.** 857)68,600 80 R40
37. 118)107,262 909	**38.** 468)131,508 281	**39.** 645)598,560 928	**40.** 113)469,628 4,156

MIXED REVIEW

Find the answer.

• **41.** 675 + 387 1,062	• **42.** 1,209 − 97 1,112	**43.** 20 × 600 12,000	**44.** 9 × 132 1,188
45. 400,000 ÷ 50 8,000	**46.** 6,400 ÷ 80 80	**47.** 9)2,720 302 R2	**48.** 7)1,854 264 R6

Solve.

• **49.** 480 fan belts are packed in a box. How many boxes will 45,600 fan belts fill? 95 boxes

50. Gloria works a 35-hour week. If she earns $420 a week, how much does she earn an hour? $12

51. Rosella's car averages 31 miles per gallon (mi/gal). How many gallons of gas will be needed for a trip of 713 mi? 23 gal

52. Adam's gas tank holds 20 gal of gas. His car averages 28 mi/gal. How far can he travel if his gas tank is full? 560 mi

TIME OUT If it takes 8 min to cut a log into 3 pieces, how long will it take to cut a log into 4 pieces? (*Hint:* How many cuts are needed to make 3 pieces?) 12 min

TIME OUT Encourage students who are having difficulty to draw a diagram. This should help students see that 2 cuts are needed to make 3 pieces.

8 min ÷ 2 cuts = 4 min per cut

Since you need 3 cuts to make 4 pieces, you need 3 × 4 or 12 min.

COMMON ERROR

Students may omit the end-zeros as placeholders in the quotient. Encourage these students to always check their answers.

ASSIGNMENTS

Level 1	1–6, Odd 13–23, 41–52
Level 2	1–9, Even 14–40, 41–52, TO
Level 3	7–12, 29–40, 49–52, TO

FOLLOW-UP ACTIVITY

REINFORCEMENT Write the following on the chalkboard:

```
             15                  51 R834
1. 361)37,905       2. 942)50,868
       36 1                47 10
        1 805               17 76
        1 805                9 42
            0                8 34
```

Tell students that there are errors in each example. Have students identify the errors and rework each example correctly. (**1.** The zero in the quotient is missing. **2.** The initial error is that 4,710 is inaccurately subtracted from 5,086.)

SUPPLEMENTARY MATERIALS

TRP Practice, p. 22

TRP Reteaching, p. 16

TRP Transparency 11

OBJECTIVE

- Solve multistep problems by writing subproblems.

TEACHING THE LESSON

WARM-UP Write the following exercises on the chalkboard or an overhead transparency.

Compute.

1. 7 + 86 + 25 (118)
2. 172 + 265 (437)
3. 500 − 268 (232)
4. 704 − 569 (135)
5. 8 × 124 (992)
6. 4 × 85 (340)
7. 726 ÷ 6 (121)
8. 304 ÷ 4 (76)

INTRODUCTION Read the following problems aloud. Have students explain which operation is used to solve them.

1. There are 100 students going on a field trip. Each bus holds 25 students. How many buses are needed? (divide)
2. Mrs. Brooks took a group of 25 students on the morning museum tour. She took 36 students on the afternoon tour. How many students did she take altogether? (add)
3. A museum guide had 6 packages of tour maps. There are 12 maps in each package. How many maps were there altogether? (multiply)

INSTRUCTION As you discuss the situation, refer to the calorie table. Explain that a calorie is a unit of measure of the energy value of foods.

Point out that there is only one question asked in the problem. However, it cannot be answered without first finding the total number of calories in Ted's breakfast outlined in Subproblem 1.

PROBLEM *Solving* STRATEGY

2.13 WRITING SUBPROBLEMS

Situation:

Ted Cohen wanted to have a 300-calorie (cal) breakfast. He ate 1 cup of oatmeal, an orange, and a glass of skimmed milk. How many fewer calories did he have than he wanted?

Strategy:

Find subproblems you can solve to help you arrive at a final solution.

Applying the Strategy:

Subproblem 1: How many calories were in Ted's breakfast? 134 + 60 + 90 = 284 cal

THINK: Oatmeal: 134 cal Orange: 60 cal
Skimmed milk: 90 cal

Subproblem 2: How many fewer calories did he have than he wanted? 300 − 284 = 16 cal

CALORIC COUNT OF SOME BASIC FOODS

Milk group	Meat group	Bread-cereal group	Vegetable-fruit group
1 cup whole milk 165	1 hard-boiled egg 75	1 slice whole wheat bread 55	1 medium orange 60
1 cup skimmed milk 90	1 serving tuna fish 172	1 cup brown rice 648	1 medium apple 72
1 slice Swiss cheese 105	1 small piece chicken 220	1 cup oatmeal 134	1 serving broccoli 45
1 cup plain yogurt 125	1 cup beans 242	1 small roll 122	1 medium baked potato 95

PRACTICE EXERCISES See Extra Practice, page 431.

Use the calorie table on page 52 to solve.

A casserole serves six people. It has 160 cal from the vegetable-fruit group, 648 cal from the bread-cereal group, and 680 cal from the meat group.

• **1.** What is the total number of calories in this casserole? 1,488 cal

• **2.** How many calories are in each of six servings? 248 cal

Ted made a casserole out of two cups of brown rice, four slices of Swiss cheese, and one serving of tuna fish.

3. How many calories were in two cups of brown rice? 1,296 cal

4. How many calories were in four slices of Swiss cheese? 420 cal

5. What was the total number of calories in the casserole? 1,888 cal

Solve the problem by writing subproblems. Use the calorie table.

6. Ted wanted to have a 500-cal lunch. He ate one small piece of chicken, a roll, and a cup of whole milk. How many more calories were in the lunch than he wanted? 7 cal

7. Ted is preparing dinner for his friends. He bought one bag of potatoes for $1.95, one bunch of broccoli for $0.89, and one package of chicken for $4.80. How much change did Ted receive from a $10 bill? $2.36

8. Amy had one cup of baked beans and a slice of whole wheat bread for lunch. Ted ate 1 cup of oatmeal and an apple for lunch. How many fewer calories did Ted's lunch contain? 91 cal

9. Ted wants to prepare a 380-cal lunch. He chose two slices of whole wheat bread, two slices of Swiss cheese, and one piece of fruit. Should he choose an apple or an orange? An orange

Reinforce the idea that in order to solve multistep problems, students should ask these questions:

1. What information is given?
2. What other information is needed to answer the question?
3. How can the information be found?

ASSIGNMENTS

Level 1	1–6
Level 2	1–8
Level 3	1–9

FOLLOW-UP ACTIVITY

SITUATIONAL PROBLEM SOLVING Your class is assigned to plan the school cafeteria lunch menu for 5 days. Each meal must be nutritionally balanced. As you do your research, consider these questions:

1. What is a nutritionally balanced meal?
2. About how many calories should each meal contain for students your age?
3. How many servings from each food group should be included in a meal?
4. How practical is it to prepare your meals for large numbers of students?

SUPPLEMENTARY MATERIALS

TRP Practice, p. 23

TRP Applications, p. 2

TRP Transparency 12

OBJECTIVE

- Decide, given relevant information, which extracurricular activity to pursue.

TEACHING THE LESSON

WARM-UP Write the following exercises on the chalkboard or an overhead transparency.

Complete the work schedule.

	Mon.	Tue.	Wed.	Thurs.	Fri.
3–4 PM					
4–5 PM					
5–6 PM	Peter		Peter		

Peter: M, W, 5–6 PM

Jane: M, W, F, 3–5 PM

Mark: Th, 3–6 PM; F, 5–6 PM

Lucy: T, 4–6 PM

INTRODUCTION Present the problem situation to the class. Discuss extracurricular activities with them, and ask students what factors they think are important when selecting such an activity. List the factors on the chalkboard.

INSTRUCTION Have students read the descriptions of the four activities and discuss the notes on each and how they could be interpreted.

Direct attention to the decision-making factors. Ask the students if there are any factors they would add to the list from those written on the chalkboard.

Have students complete Exercises 1–9 in the table in order to compare the four clubs. Ask them to write conclusions that can be drawn from the data and discuss the appropriateness of these conclusions and possible misinterpretations.

DECISION MAKING

2.14 CHOOSING AN ACTIVITY

(Open-ended problem solving)

Many high school students participate in extracurricular activities. Some of them have a real interest in the sport or club. Others get involved because their friends are.

PROBLEM

Lisa is trying to choose an after-school activity. She cut out the following from the school bulletin and added some notes of her own. She has numbered each activity in order of her interest in it.

BAND
Practice: Wed., Fri., 3:00–5:00PM
Play at all games, local parades, and rallies. Schedule varies.
Must audition for membership and submit a letter of recommendation from student's music teacher.

1. I can read music and play the flute. Maybe my brother's music teacher will write a letter for me. SCOTT is in the band!

DANCE CLUB
Rehearsals: Mon., Thurs., 3:00–5:00PM
Performances: Last Friday of the month, 8:00PM
Must audition for membership.

2. I've studied dance, I need the exercise, I know many of the people in the club.

FUTURE TEACHERS' CLUB
Meetings: Thurs., 3:00PM
For all students planning to enter the field of teaching.

3. That's me.
I wonder what they do at meetings??

SCIENCE CLUB
Meetings: Wed., 3:00PM
For all students interested in Science.
Compete in Science Fairs!
Must have at least a B average in Science.

4. Science is my best subject.
This might be fun, but it could be a lot of work.

DECISION-MAKING FACTORS

Interest   Qualifications   Schedule

DECISION-MAKING COMPARISONS

Complete the table.

Some answers may vary. Accept reasonable answers based on students' rationales.

	Band	Dance Club	Future Teachers' Club	Science Club
Interest	Very interested	• 1. ■	• 2. ■	3. ■
Qualifications	4. ■	5. ■	An interest in teaching	6. ■
Schedule	7. ■	8. ■	9. ■	Wed., 3 P.M.

1. Interested 2. Somewhat interested 3. Hesitant 4. Audition, letter 5. Audition 6. B average 7. Wed., Fri., 3–5 P.M. plus events 8. Mon., Thurs., 3–5 P.M. Last Fri./mo, 8 P.M. 9. Thurs., 3 P.M.

MAKING THE DECISIONS

• **10.** If interest were the only criteria, which activity would Lisa choose? Band

• **11.** Based on her qualifications, which groups would most likely accept Lisa as a member? Dance Club, Future Teachers' Club, Science Club

12. Lisa forgot to consider her part-time job. If she works Tuesdays and Wednesdays from 3:00–6:00 P.M. and Saturdays from noon–8:00 P.M., which activities can she fit into her schedule? Dance Club and Future Teachers' Club

13. If Lisa's friend Gary is willing to trade his Friday afternoon work hours for Lisa's Wednesday hours, how does this change her options? She can add Science Club.

14. If Lisa is serious about being a dancer, what should she consider before deciding to work on Fridays? She might be too tired to do her best dancing.

15. Which activities would be most helpful to Lisa in her future career? Future Teachers' Club, Science Club

16. If Lisa's part-time job is office work, which activity would be best for her health? Dance Club

17. Which activities would you recommend to Lisa? Why? Answers will vary.

Assign Exercises 10–18 and have students work in small groups. Encourage them to explain and defend their answers and decisions as necessary. Continually reinforce the fact that individual preferences and situations strongly influence decisions of this type. Observe the groups as they work and interact as necessary to draw nonparticipating students into the discussion.

ASSIGNMENTS

Level 1	1–17
Level 2	1–17
Level 3	1–17

FOLLOW-UP ACTIVITY

APPLICATION Provide students with a list of the extracurricular activities available in your school. Explain that their assignment will be to choose an activity; therefore, they are to begin by gathering as much information as they can about the activities that interest them. Once they have their data, they are to make up a list of factors that apply to the situation. Finally, they are to choose an activity and give the reasons for their decision.

SUPPLEMENTARY MATERIALS

TRP Practice, p. 24

TRP Transparency 12

Objectives

- Review vocabulary.
- Practice key chapter concepts and skills.

Using the Review

The Chapter Review is designed to help students prepare for taking the Chapter Test. The first section focuses on vocabulary. It requires that students select words to complete statements. The second section presents practice exercises of key mathematical skills. Under each directive there is a sample exercise with an answer.

Each item on the review is referenced to the page on which the topic is taught in the Pupil's Edition. You may wish to have students refer to these pages to help review any concepts or skills they have not yet mastered.

It is suggested that students work in small-sized heterogeneous cooperative learning groups. Some cooperative learning methods that may be used are as follows:

1. After each student has independently completed the entire Chapter Review, a discussion should follow within each group about the solutions to the practice exercises.
2. The group can complete the entire Chapter Review by working together to discuss the sample exercises and then to answer the practice exercises.

End the lesson with an entire class discussion in which any questions brought up in group discussions are presented and answered.

Chapter Review

Vocabulary Choose the letter of the word(s) that completes the statement.

1. The answer in addition is called the ■. [30] b
   a. addend **b.** sum **c.** estimate **d.** product
2. Compatible numbers are numbers that ■. [38] b
   a. end in zero **b.** divide evenly **c.** are rounded **d.** are great
3. The product of two whole numbers is greater than either ■. [40] c
   a. addend **b.** sum **c.** factor **d.** quotient

Skills Estimate by rounding. [28, 38]

3,796 + 169 *4,200*	**4.** 567 + 345 900	**5.** 1,237 + 4,813 6,000	**6.** 176,482 + 563,201 800,000
7. 821 − 436 400	**8.** 3,243 − 1,726 1,000	**9.** 36,265 − 27,401 10,000	**10.** 471,673 − 323,851 200,000
11. 46 × 9 450	**12.** 681 × 4 2,800	**13.** 721 × 15 14,000	**14.** 4,396 × 21 80,000
15. 329 ÷ 2 150	**16.** 4,385 ÷ 5 800	**17.** 905 ÷ 33 30	**18.** 53,744 ÷ 46 1,000

Estimate by using compatible numbers. [38]

71)22,104 *300*	**19.** 6)410 70	**20.** 8)376 50	**21.** 7)692 100
22. 40)3,331 80	**23.** 62)4,624 80	**24.** 36)28,341 700	**25.** 24)74,116 3,000

Add or subtract. [30, 32]

456 + 870 = *1,326*	**26.** 62 + 43 105	**27.** 94 − 63 31	**28.** 642 − 314 328
29. 843 + 19 862	**30.** 43,201 + 842 44,043	**31.** 7,699 − 4,021 3,678	**32.** 90,468 − 37,491 52,977

Multiply or divide. [40, 42, 46, 48, 50]

357 × 19 = *6,783*	**33.** 32 × 9 288	**34.** 5)524 104 R4	**35.** 43)788 18 R14
36. 40 × 900 36,000	**37.** 5,000 × 80 400,000	**38.** 60)5,400 90	**39.** 900)270,000 300
40. 784 × 21 16,464	**41.** 140 × 683 95,620	**42.** 87)5,682 65 R27	**43.** 190)36,401 191 R111

CHAPTER TEST

Estimate by rounding.

1. 705 + 364 — 1,100

2. 4,235 + 2,563 + 337 — 7,300

3. 8,317 − 2,852 — 5,000

4. 9,604 − 715 — 9,300

5. 2,364 × 7 — 14,000

6. 4,864 × 15 — 100,000

Estimate by using front-end estimation.

7. 2,426 + 951 — 2,000

8. 38,499 + 11,987 — 40,000

9. 25,012 − 1,683 — 20,000

Estimate by using compatible numbers.

10. 9)7,352 — 800

11. 19)6,514 — 300

12. 58)17,589 — 300

Add.

13. 9,643 + 5,717 — 15,360

14. 43,869 + 69,454 — 113,323

15. 1,264 + 38 + 815 — 2,117

16. 56,729 + 1,845 + 66,083 — 124,657

Subtract.

17. 5,044 − 2,381 — 2,663

18. 7,000 − 965 — 6,035

19. 60,236 − 28,675 — 31,561

20. 21,812 − 9,023 — 12,789

Multiply.

21. 50 × 300 — 15,000

22. 70 × 700 — 49,000

23. 800 × 400 — 320,000

24. 900 × 1,200 — 1,080,000

25. 6,432 × 8 — 51,456

26. 847 × 56 — 47,432

27. 4,573 × 99 — 452,727

28. 805 × 614 — 494,270

Divide.

29. 6,300 ÷ 70 — 90

30. 9,600 ÷ 80 — 120

31. 2,400 ÷ 800 — 3

32. 4,200 ÷ 600 — 7

33. 8)3,784 — 473

34. 9)5,042 — 560.2222

35. 29)8,816 — 304

36. 152)4,560 — 30

Solve.

37. On Monday, Maria sold 12 novels, 11 mysteries, and 4 cookbooks. On Tuesday, she sold 14 novels, 6 cookbooks, and 3 mysteries. On which day did she sell more books?
Monday

38. Bob Tompkins bought a paperback mystery for $4.95. He gave the clerk a $20 bill. List the least amount of coins and bills he would receive as change.
one $10-bill, one $5-bill, and one nickel

OBJECTIVE

- Evaluate achievement of the chapter objectives.

USING THE TEST

The Chapter Test may be used as a posttest to evaluate student achievement. However, you may wish to use the Chapter Posttest offered in the Teacher's Resource Package or to design your own chapter test. If this page is not used as a test, you may wish to assign it as additional review or practice.

The test items are correlated to the chapter objectives in the table below.

Chapter objectives	Test items
A. Estimate by rounding.	1–6
B. Estimate by using front-end estimation.	7–9
C. Estimate by using compatible numbers.	10–12
D. Add or subtract whole numbers.	13–20
E. Multiply or divide whole numbers.	21–36
F. Write subproblems to solve problems.	37
G. Apply computational skills in real-life situations.	38

SUPPLEMENTARY MATERIALS

TRP Ch. 2 Posttest Form A, pp. 1–2

TRP Ch. 2 Posttest Form B, pp. 1–2

OBJECTIVES

- To determine whether a number is divisible by 2, 5, or 10 by using the ones-digit rule.
- To determine whether a number is divisible by 3 or 9 by using the digit-sum rule.

USING THE PAGE

Write 2)912 on the chalkboard. Have a volunteer complete the example. Point out that since the answer is 456, with no remainder, we say 912 is divisible by 2. Have another volunteer complete 5)912. Since the answer is 182 R2, 912 is not divisible by 5.

Point out that there are tests for divisibility that show, without dividing, whether one number is divisible by another.

Guide students in discovering the rule for divisibility by 2. Write:

Divisible by 2	Not divisible by 2
24, 76, 250, 5,352, 9,868	11, 45, 879, 903 2,081, 7,569

Ask how the two groups differ. (Those that are divisible by 2 are even numbers.)

Repeat with similar examples to guide students in discovering the tests for divisibility by 5 and by 10.

Since the tests for divisibility by 3 and by 9 are less obvious, use the following examples to demonstrate them:

3)564   9)76,851

Students will be interested to learn that they can use the rules for divisibility by 2, 3, 5, 9, 10 to test for divisibility by 6; 15; 18; 20; 27; 30; 45; 50; 54; 60; 90; 100; 135; 150; 180; 270; 450; 900; 1,350; and 2,700. This is done by combining the rules for the factors of the divisors listed above. For example, a number is divisible by 18 if it is divisible by 2 and by 9.

ENRICHMENT DIVISIBILITY

When you say that 15 **is divisible by** 3, it means that when you divide 15 by 3, the remainder is 0. You can use the following rules to determine if a number is divisible by 2, 3, 5, 9, or 10.

A number is divisible by:

2 if the ones digit is even.
3 if the sum of the digits is divisible by 3.
5 if the ones digit is 0 or 5.
9 if the sum of the digits is divisible by 9.
10 if the ones digit is 0.

Example Is 336 divisible by 2, 3, 5, 9, 10, or none of these?

1. Look at the ones digit.

THINK: The ones digit is 6, an even number.

336 is divisible by 2,
but not divisible by 5 or 10.

2. Look at the sum of the digits.

3 + 3 + 6 = 12

THINK: 12 is divisible by 3, but not divisible by 9.

336 is divisible by 3,
but not divisible by 9.

In some cases you can combine the rules to make up other divisibility rules. For example, a number is divisible by 6 if it is divisible by 2 and 3.

Example Is 252 divisible by 6?

1. Look at the ones digit.

THINK: The ones digit is 2, an even number.

252 is divisible by 2.

2. Look at the sum of the digits.

2 + 5 + 2 = 9

THINK: 9 is divisible by 3.

252 is divisible by 3.

So, 252 is divisible by 6.

Is the number divisible by 2, 3, 5, 9, 10, or none of these?

1. 42 2, 3 **2.** 60 2, 3, 5, 10 **3.** 91 none **4.** 105 3, 5 **5.** 162 2, 3, 9
6. 301 none **7.** 350 2, 5, 10 **8.** 450 2, 3, 5, 9, 10 **9.** 1,611 3, 9 **10.** 3,627 3, 9

Is the number divisible by 6, 15, or neither of these?

11. 60 6, 15 **12.** 72 6 **13.** 109 neither **14.** 135 15 **15.** 1,422 6
16. 2,463 neither **17.** 4,500 6, 15 **18.** 6,750 15 **19.** 13,500 6, 15 **20.** 2,088 6

ESTIMATION SKILLS — ADJUSTING FRONT-END ESTIMATES

You can usually adjust an estimated sum in order to bring it closer to the actual sum. To adjust a front-end estimate, consider the digits in the next-to-greatest place.

Estimate by using front-end estimation: 8,216,450 + 3,735,420

1. Estimate.

THINK: The greatest place is millions. Add the digits in the millions place.

8,216,450
+3,735,420
11,000,000 ← **First estimate**

2. Adjust the estimate.

THINK: Add the digits in the hundred-thousands place.

8,|216,450
+3,|735,420 } about 1 million
12,000,000 ← **Adjusted estimate**

Estimate by using front-end estimation: 9,135 + 616 + 5,315 + 1,478

1. Estimate.

THINK: The greatest place is thousands. Add the digits in the thousands place.

9,135
616
5,315
+1,478
15,000 ← **First estimate**

2. Adjust the estimate.

THINK: Add the digits in the hundreds place.

9,|135
|616 } about 1 thousand
5,|315
+1,|478 } about $\frac{1}{2}$ thousand
16,500 ← **Adjusted estimate**

Estimate by using front-end estimation. Then adjust your estimate.

1. 865 + 632 — 1,500
2. 1,256 + 5,317 — 6,500
3. 35,408 + 25,320 — 60,000
4. 424,375 + 91,429 — 500,000
5. 9,433,698 + 8,052,728 — 17,500,000
6. 942 + 628 + 136 — 1,700
7. 3,848 + 1,204 + 462 — 5,500
8. 12,426 + 2,342 + 1,068 — 15,000
9. 738,420 + 903,725 + 367,416 — 2,000,000
10. 5,802,624 + 1,629,283 + 8,541,739 — 16,000,000
11. 6,281 + 846 + 1,592 + 3,468 — 12,000
12. 38,604 + 20,318 + 52,125 + 65,277 — 175,000
13. 11,764 + 21,068 + 4,283 + 3,265 — 40,000
14. 880,425 + 691,076 + 316,351 + 822,075 — 2,700,000
15. 236,914 + 874,003 + 556,947 + 463,081 — 2,100,000
16. 92,143 + 205,526 + 9,187 + 312,416 — 600,000
17. 2,438,007 + 1,204,318 + 938,409 + 1,072,361 — 5,500,000
18. 1,946,146 + 2,038,308 + 457,162 + 615,519 — 5,000,000
19. 7,249,196 + 85,156 + 714,039 + 2,445,136 — 10,500,000
20. 46,193 + 12,468,148 + 5,156,749 + 34,476,374 — 50,000,000

OBJECTIVE

- Adjust estimated sums found by using the front-end method.

USING THE PAGE

As you go over the examples, emphasize the use of the numbers $\frac{1}{2}$ and 1 in the adjustment of front-end estimates. In the first example, we think: 2 hundred thousand + 7 hundred thousand is about 1 million. In the second example: 1 hundred + 6 hundred + 3 hundred is 1 thousand, and 4 hundred is about $\frac{1}{2}$ thousand.

You may wish to have students compare the adjusted estimates in the examples with estimates obtained by rounding each number to its greatest place.

Example 1		Example 2
		9,000
		600
8,000,000		5,000
+4,000,000		+1,000
12,000,000	← Estimates by rounding →	15,600

In the first example, either method produces the estimate 12,000,000, and this is very close to the actual sum, 11,951,870. In the second example, it happens that the actual sum, 16,544, is closer to the adjusted estimate shown on the pupil's page than it is to the estimate found by rounding.

Help students to understand that there are many methods of estimation. In choosing an appropriate method, consideration should be given to the ease of calculation and to the degree of accuracy required. For example, a simple front-end estimate may be adequate if only a rough idea of a sum is needed. If a closer estimate is desired, the student may want to estimate by rounding or may choose to make an adjusted front-end estimate.

Objective

- Test the prerequisite skills needed to learn the concepts developed in Chapter 3.

Using the Test

The Pre-Skills Test is designed to diagnose students' strengths and weaknesses on prerequisite skills necessary to study the mathematics in Chapter 3.

Assign the Pre-Skills Test. Allow the students to work together in pairs or small groups. Group members should help those who demonstrate a misunderstanding of a concept or a weakness in a skill.

The items in the test are referenced to the pages on which the topics are taught in the Pupil's Edition. You may wish to have students refer to these pages for review.

The following table correlates the items on the Pre-Skills Test with the prerequisite skill and the lesson(s) in the chapter for which it is needed.

Item(s)	Prerequisite skill	Lesson(s)
1–8	Round to a given decimal place.	3.1, 3.2, 3.7, 3.10, 3.11
9–19	Estimate with whole numbers.	3.1, 3.2, 3.7, 3.10, 3.11
20–35	Add or subtract whole numbers.	3.1, 3.2
36–41	Multiply and divide whole numbers by powers of 10.	3.4
42–45	Multiply whole numbers.	3.7
46–48	Divide whole numbers.	3.10, 3.11

Chapter 3 Decimal Computation

Pre Skills Test

Round to the given place. [16]

Number	Tenths	Thousandths	Ones	Hundreds
27.0938	**1.** ■ 27.1	**2.** ■ 27.094	**3.** ■ 27	**4.** ■ 0
0.85032	**5.** ■ 0.9	**6.** ■ 0.850	**7.** ■ 1	**8.** ■ 0

Estimate by rounding. [28, 38]

9. 428 + 369 — 800
10. 23,109 + 42,063 — 60,000
11. 3,491 − 807 — 2,200
12. 73,804 − 9,531 — 60,000

13. 32 × 76 — 2,400
14. 692 × 841 — 560,000
15. 42)375 — 10
16. 645)29,871 — 50

Estimate using compatible numbers. [38]

17. 3)27,014 — 9,000
18. 86)47,301 — 500
19. 57)326,085 — 5,000

Add or subtract. [30, 32]

20. 459 + 456 — 915
21. 896 + 37 — 933
22. 787 − 89 — 698
23. 923 + 787 — 1,710

24. 5,296 + 743 — 6,039
25. 74,802 − 5,064 — 69,738
26. 88,843 + 10,529 — 99,372
27. 19,042 − 12,894 — 6,148

28. 23,003 + 34,991 — 57,994
29. 50,339 − 5,892 — 44,447
30. 17,859 + 2,789 — 20,648
31. 45,006 − 4,869 — 40,137

32. 19,004 − 8,994 — 10,010
33. 99,018 − 39,896 — 59,122
34. 165,565 + 13,006 — 178,571
35. 644,887 − 78,998 — 565,889

Multiply or divide. [36, 40, 42, 46, 48, 50]

36. 62 × 100 — 6,200
37. 3,401 × 1,000 — 3,401,000
38. 732 × 10,000 — 7,320,000

39. 2,300 ÷ 100 — 23
40. 100,000 ÷ 100 — 1,000
41. 8,400 ÷ 10,000 — 0.84

42. 27 × 83 — 2,241
43. 192 × 68 — 13,056
44. 731 × 596 — 435,676
45. 4,085 × 637 — 2,602,145

46. 8)4,232 — 529
47. 49)34,398 — 702
48. 301)807,425 — 2,682 R143

Chapter 3
DECIMAL COMPUTATION

People often hold fund-raising events for special purchases. How could you and your classmates raise enough money to buy a personal computer for your class?

OBJECTIVE

- Formulate problems using text and data.

USING THE CHAPTER OPENER

Discuss fund-raising events in general and then move on to the question presented on the page. Fund-raising takes many forms ranging from garage sales and flea markets to $1,000-a-plate charity dinners and celebrity tennis matches. Guide the students in focusing on events they could realistically sponsor. These might include a car wash, a walk-a-thon, a spaghetti dinner, or a baby-sitting service. They should consider their talents and abilities as a class and also how much free time they have available. Another important consideration is the kinds of events people in your community would support.

COOPERATIVE LEARNING

Write the following on the chalkboard.

COMPUTER FUND

Cost: $2,249 (or fill in the best price the class found)

PTA will pay half if students raise the money for the balance.

Assign groups of students the task of devising a plan for a way the class can raise $1,125 (or half the price of the computer). They are to consider the points discussed earlier before making their decision. Instruct them to include all pertinent information in their plan—schedules, prices or fees, work assignments, publicity, and so on.

When the projects are presented to the class, ask them to choose the best plan based on originality as well as thoroughness.

SUPPLEMENTARY MATERIALS

TRP Group Projects, p. 3

OBJECTIVES

- Add decimals.
- Check the reasonableness of answers using estimation.

TEACHING THE LESSON

WARM-UP Write the following exercises on the chalkboard or on an overhead transparency.

Find the sum.

1. $65 + 98$
2. $381 + 647$
3. $25{,}089 + 642$
4. $36 + 1{,}029 + 842$
5. $954 + 43 + 3{,}578$

(**1.** 163 **2.** 1,028 **3.** 25,731 **4.** 1,907 **5.** 4,575)

INTRODUCTION Point out to students that adding decimals is a skill used every day by people who handle money. Ask students to name situations when they need to add money, such as buying lunch in the cafeteria or buying food at the grocery store. Ask students how they add money amounts. Use examples such as $5.61 + $3.27 or $16.09 + $3.25 to establish that decimal points must be aligned.

INSTRUCTION Use Example 1 to show the importance of lining up the decimal points. Have students write the decimal point in the sum before actually adding as a safeguard. The addition is then done as with whole numbers, and students should use the commutative property to check their answers.

Explain to students that writing extra zeros at the end of a decimal number does not change its value. In Example 2, 4.06 and 4.060 or 17.9 and 17.900 are equivalent decimals. In addition, having the same number of decimal places to the right of the decimal point is not necessary, but it is helpful.

3.1 ADDING DECIMALS

In New York City, the average rainfall for October is 3.4 in. In November, it is 4.1 in. On the average, how much rain falls in New York during these 2 months?

Add 3.4 and 4.1 to find the average rainfall for these months.

EXAMPLE 1 Add: 3.4 + 4.1

1. Line up the decimal points.

$$\begin{array}{r} 3.4 \\ +4.1 \\ \hline . \end{array}$$

2. Add as with whole numbers.

$$\begin{array}{r} 3.4 \\ +4.1 \\ \hline 7.5 \end{array}$$

So, the average rainfall for October and November is 7.5 in.

Sometimes it is helpful to use equivalent decimals so that each addend has the same number of places.

EXAMPLE 2 Add: 4.06 + 17.9 + 35.082

1. Line up the decimal points.

$$\begin{array}{r} 4.06 \\ 17.9 \\ +35.082 \\ \hline . \end{array}$$

2. Add.

THINK: 4.06 = 4.060
17.9 = 17.900

$$\begin{array}{r} 4.060 \\ 17.900 \\ +35.082 \\ \hline 57.042 \end{array}$$

As with whole numbers, you can estimate to determine if a sum is reasonable.

EXAMPLE 3 Add: 0.84 + 0.7 + 29.082. Is your answer reasonable?

1. Estimate.

$$\begin{array}{rcr} 0.84 & \rightarrow & 0.8 \\ 0.7 & \rightarrow & 0.7 \\ +29.082 & \rightarrow & 30.0 \\ \hline & & 31.5 \end{array}$$

2. Add.

$$\begin{array}{r} 0.840 \\ 0.700 \\ +29.082 \\ \hline 30.622 \end{array}$$

Since 31.5 and 30.622 are close, the answer is reasonable.

CHECKPOINT

Write the letter of the correct answer.

Add.

1. 0.6 + 3 + 45.08 a **a.** 48.68 **b.** 54.08 **c.** 45.98 **d.** 45.17
2. $64.29 + $203.75 a **a.** $268.04 **b.** $26.80 **c.** $2680.04 **d.** $2.68

PRACTICE EXERCISES See Extra Practice, page 432.

Add.

1. 49.6 + 27.2 = 76.8	2. 83.1 + 51.9 = 135.0	3. 70.59 + 18.64 = 89.23	4. 0.832 + 0.513 = 1.345	5. 0.0052 + 0.9614 = 0.9666
• 6. 72.083 + 19.5 = 91.583	7. 47.6 + 82.054 = 129.654	8. 6.0738 + 5.1 = 11.1738	9. 7.602 + 1.8573 = 9.4593	10. 38.0006 + 97 = 135.0006
11. 0.82 + 6.1 = 6.92	12. 3.4 + 17.092 = 20.492	13. 54 + 3.019 = 57.019	14. 680.07 + 4.529 = 684.599	15. $493.00 + 86.87 = $579.87
16. $6.83, 4.51, + 0.39 = $11.73	17. 3.71, 0.48, + 9.06 = 13.25	18. $42.63, 78, + 51.85 = $172.48	19. 63.07, 92.48, + 14.65 = 170.20	20. 13.784, 4.063, + 0.905 = 18.752

21. 0.0019 + 0.0263 0.0282 22. 3.024 + 17.8 + 5 25.824 23. 65.3 + 0.94 + 2.071 68.311

24. 48.079 + 0.846 48.925 25. $4.63 + $0.93 $5.56 26. 0.901 + 0.099 1

First estimate by rounding, and then find the exact answer.

•27. 806.09 + 27.614 — 830; 833.704	•28. 798.14 + 12.80 — $810; 810.94	29. 8.005 + 5.7319 — 14; 13.7369	30. 0.9216 + 0.4738 — 1.4; 1.3954	31. 38.06 + 8.0807 — 50; 46.1407
32. 25.1 + 13.84 — 40; 38.94	33. 108.65 + 329.57 — 400; 438.22	34. $64.87 + 19.63 — $80; $84.50	35. 0.83 + 0.417 — 1.2; 1.247	36. 0.059 + 0.4026 — 0.46; 0.4616

Solve for n.

•37. 76.4 + 3.05 + 28 = n 107.45 38. 0.05 + 0.839 + 7 = n 7.889

39. 0.505 + 0.04 + 0.055 = n 0.6 40. 6.29 + 18 + 0.075 = n 24.365

Use the table for Exercises 41–44.
Solve.

YEARLY TOTAL PRECIPITATION (in.)

	Bismarck	Boston	Chicago	Denver	Reno
Rain	14.77	50.24	34	16.49	4.28
Snow	72.5	44.6	38.7	65.73	16.5

•41. During the year, did more rain or snow fall altogether in the cities? Snow

42. Which city had more rain than snow? Boston; 5.64 in.

43. Estimate the total precipitation for each city by rounding. 80 in.; 90 in.; 70 in.; 90 in.; 20 in.

44. Round the snowfall for each city to the nearest inch. About how many feet of snow fell in each city? 6 ft; 4 ft; 3 ft; 5 ft; 1 ft

Use Example 3 to show estimation as a check of the reasonableness of the answer. Remind students to round to the greatest place value and write extra zeros to the right of the decimal point.

CHECKPOINT Distribute to each student four index cards on which the answer choices a–d appear. After allowing students a few minutes, ask each to raise the letter that indicates their answer choice. Scan the cards and identify those students who need additional attention before proceeding to the Practice Exercises. The incorrect answer choices include common errors.

COMMON ERROR

Some students will line up digits regardless of the decimal point. For example, in Item 1 of Checkpoint, they would select *b*. You may want those students to turn lined paper sideways so that they have distinct columns for each place value and a separate one for the decimal point.

ASSIGNMENTS

Level 1	1–20, 27–31, 41, 42
Level 2	6–15, Even 16–36, 41–44
Level 3	Odd 11–35, 37–44

FOLLOW-UP ACTIVITY

APPLICATION Provide students with ads from grocery stores. Tell students to plan a shopping trip, buying food items needed to cook dinner for their families. Stress that nutritious foods from the four basic food groups should be included. Have students find the total cost of their shopping trips.

SUPPLEMENTARY MATERIALS

TRP Practice, p. 25

TRP Reteaching, p. 17

TRP Transparency 13

OBJECTIVES

- Subtract decimals.
- Check the reasonableness of answers using estimation.

TEACHING THE LESSON

WARM-UP Write the following exercises on the chalkboard or on an overhead transparency.

Find the difference.

1. $53 - 28$ **2.** $705 - 361$

3. $4{,}295 - 367$ **4.** $8{,}000 - 5{,}492$

(**1.** 25 **2.** 344 **3.** 3,928 **4.** 2,508)

INTRODUCTION Ask students to name situations when they need to subtract money, such as finding differences in prices or making change. Elicit from students how they find the difference in money amounts. Include examples such as \$12.34 − \$7.89, \$1.84 − \$0.69 and \$5.00 − \$2.79 to establish aligning decimal points and writing extra zeros.

INSTRUCTION Use Example 1 to stress lining up decimal points in subtraction of decimals. When students align the decimal points, have them write a decimal point in the difference before subtracting. Subtraction is done as with whole numbers. Encourage students to check their answers with addition.

Example 2 shows the necessity of having the same number of decimal places to the right of the decimal in both the minuend and the subtrahend. You may want to review the regrouping process across zeros.

Have students estimate by rounding in Example 3 to check the reasonableness of their answer. This process is particularly helpful in assessing alignment errors.

3.2 SUBTRACTING DECIMALS

During the 1972 Olympics, Monika Zehrt from East Germany ran the 400-m dash in 51.08 seconds. In 1984, Valerie Brisco-Hooks of the United States finished it in 48.83 seconds. How much faster was the 1984 record?

Subtract 48.83 from 51.08 to find the difference between the winning times.

EXAMPLE 1 Subtract: 51.08 − 48.83

1. Line up the decimal points.

$$\begin{array}{r} 51.08 \\ -\,48.83 \\ \hline . \end{array}$$

2. Subtract as with whole numbers.

$$\begin{array}{r} 51.08 \\ -\,48.83 \\ \hline 2.25 \end{array}$$

So, the 1984 record was 2.25 seconds faster.

Sometimes it is helpful to use equivalent decimals to subtract.

EXAMPLE 2 Subtract: 23 − 7.94

1. Line up the decimal points.

$$\begin{array}{r} 23. \\ -\,7.94 \\ \hline . \end{array}$$

2. Subtract.

THINK: 23 = 23.00

$$\begin{array}{r} 23.00 \\ -\,7.94 \\ \hline 15.06 \end{array}$$

As with whole numbers. you can estimate the difference to determine if your answer is reasonable.

EXAMPLE 3 Subtract: 0.7916 − 0.027. Is your answer reasonable?

1. Estimate.

$$\begin{array}{rcr} 0.7916 & \rightarrow & 0.80 \\ -\,0.027 & \rightarrow & \underline{0.03} \\ & & 0.77 \end{array}$$

2. Subtract.

$$\begin{array}{r} 0.7916 \\ -\,0.0270 \\ \hline 0.7646 \end{array}$$

Since 0.77 and 0.7646 are close, the answer is reasonable.

FOR DISCUSSION See TE side column.

Is 5.25 a reasonable answer for 18.67 − 0.1342? Explain.

PRACTICE EXERCISES See Extra Practice, p. 432.

Subtract.

• **1.** 3.9 − 1.6 = 2.3	**2.** 73.8 − 26.2 = 47.6	**3.** 63.18 − 47.09 = 16.09	**4.** 17.21 − 12.59 = 4.62	**5.** 5.989 − 0.849 = 5.140
• **6.** 28.67 − 17.041 = 11.629	**7.** 71.8 − 19.638 = 52.162	**8.** 92 − 56.082 = 35.918	**9.** 6.38 − 4.7094 = 1.6706	**10.** 0.6834 − 0.21 = 0.4734
11. 973.061 − 85.042 = 888.019	**12.** 5.2897 − 0.8245 = 4.4652	**13.** 5.0004 − 2.917 = 2.0834	**14.** 2 − 0.87 = 1.13	**15.** 37 − 0.235 = 36.765

16. 0.9007 − 0.5819 0.3188 **17.** 7.003 − 4.08 2.923 **18.** 13 − 8.49 4.51 **19.** 0.63 − 0.327 0.303

20. \$24.10 − \$6.00 \$18.10 **21.** \$12.24 − \$7.45 \$4.79 **22.** 84.3 − 7.62 76.68 **23.** 16 − 4.7026 11.2974

First estimate by rounding, then find the exact answer.

• **24.** 7 − 3.43 4; 3.57	• **25.** 0.8061 − 0.782 0; 0.0241	**26.** 64.2 − 7.928 52; 56.272	**27.** \$143.00 − 5.16 \$135; \$137.84	**28.** 492.06 − 8.527 480; 483.533

Solve for n.

• **29.** $12.864 - 5.017 = n$ 7.847 • **30.** $6.731 - 4.98 = n$ 1.751 **31.** $4.805 - 0.3906 = n$ 4.4144

32. $n - 5.83 = 16$ 21.83 **33.** $0.1064 - n = 0.0089$ 0.0975 **34.** $n - 2.048 = 7$ 9.048

Use the table to the right for Exercises 35–37.

• **35.** What is the difference between men's and women's 200-m times? 1.62 s

36. What is the difference between men's and women's 400-m times? 4.31 s

37. Write the men's and women's times for the 800-m event in minutes and seconds. Men's: 1 min 41.73 s; women's: 1 min 53.28 s

WORLD TRACK RECORDS (in seconds)

	100 m	200 m	400 m	800 m
Men's time	9.86	19.72	43.29	101.73
Women's time	10.49	21.34	47.60	113.28

TIME OUT

During a field day competition, one of the events was the Ladder Lag. The goal was to get to the twelfth step of a ladder in the shortest time. The "lag" was that for every four steps you climbed, you had to go back three steps.

Alex took 2 seconds to climb up four steps and lag back three steps. How long did it take Alex to get to the twelfth step? More than 16 s, less than 18 s

FOR DISCUSSION Students should explain that an estimate of only the whole numbers will quickly show 20 − 0, or 20. So, 5.25 cannot be a reasonable answer.

TIME OUT You may wish to have the students draw a diagram of a ladder with 12 steps to help them solve the problem.

COMMON ERROR

Some students will not write the necessary extra zeros in the minuend and just bring down the extra digits in the subtrahend.

$$\begin{array}{r} 82.6 \\ -34.715 \\ \hline 47.915 \end{array}$$

Have these students write extra zeros and circle them in red before they begin to subtract to avoid this error.

ASSIGNMENTS

Level 1	Even 2–28, 35–37
Level 2	Odd 1–33, 35–37, TO
Level 3	Odd 11–27, 29–37, TO

FOLLOW-UP ACTIVITY

ENRICHMENT Have students use a current almanac to find the Olympic records set for the 400-m dash in 1972, 1976, 1980, 1984, 1988, and 1992. Have them find the difference in times for each consecutive year.

SUPPLEMENTARY MATERIALS

TRP Practice, p. 26

TRP Reteaching, p. 18

TRP Transparency 13

OBJECTIVE

- Solve problems that involve finding gross earnings, deductions, and net pay.

TEACHING THE LESSON

WARM-UP Write the following exercises on the chalkboard or an overhead transparency.

Find the answer.

1. $6.65 + 1.25
2. $92.50 − 11.80
3. $15.28 + 14.90 + 11.90
4. $156.00 − 27.45
5. $275.65 − 187.92
6. $2.78 + 13.55 + 75.19

(**1.** $7.90 **2.** $80.70 **3.** $42.08 **4.** $128.55 **5.** $87.73 **6.** $91.52)

INTRODUCTION Ask the students whether any of them (or other members of their families) have jobs for which they are paid by check. Discuss the following questions: Were the checks written for the total amount that they had earned? Were some deductions taken from their total earnings? If so, what types of deductions?

INSTRUCTION As you discuss Helga Schmidt's paycheck stub, be sure the students understand the distinction between gross earnings and net pay. Emphasize that net pay is the amount that a person receives after all the deductions have been made.

Use the second example to prepare students for the problems they will encounter in Problems 5–10.

You may wish to have students use calculators to help them solve the problems. Notice that Problem 10 introduces another important type of deduction, the deduction for retirement. Problems 11 and 12 involve estimation. In Problem 12, students must realize that net pay + deductions = gross earnings.

PROBLEM Solving APPLICATION

3.3 CONSUMER: NET PAY

Helga Schmidt works for the Omega Tax Company. Each week she is paid by check. The stub of her paycheck gives information about her earnings and deductions.

Earnings			
Period Ending	Regular Pay	Overtime	Gross Earnings
5_14_93	280.00	42.00	322.00

Deductions				
F.I.C.A.	Federal Withholding Tax	State Withholding Tax	Medical Insurance	Total Deductions
24.18	40.00	6.44	2.56	73.18
				Net Pay 248.82

The **gross earnings,** or **gross pay,** is the total amount of money earned from regular pay and overtime pay.

The **deductions** are all the various amounts subtracted for taxes and other purposes. F.I.C.A. indicates Social Security taxes.

The **net pay** is the amount that Helga receives after the deductions are subtracted.

As the stub of the paycheck shows, Helga's gross earnings are $322.00. (Dollar signs are omitted on the stub.) There are four types of deductions, and the total deductions are $73.18. Helga's net pay is $248.82.

$322.00 − $73.18 = $248.82

Len Borden works part-time in a factory. His gross earnings are $225.00 per week. The deductions from his gross earnings are shown below. What is his net pay?

F.I.C.A.	Federal Withholding Tax	State Withholding Tax	Medical Insurance
16.80	25.00	4.50	1.15

1. Add to find the total of the deductions. $16.80 + $25.00 + $4.50 + $1.15 = $47.45
2. Subtract $47.45 from the gross earnings to find the net pay. $225.00 − $47.45 = $177.55

So, Len Borden's net pay is $177.55 per week.

Use the stub from Helga's paycheck (page 66) to answer Exercises 1–4.

• **1.** What is Helga's regular pay? $280.00

2. How much greater are Helga's gross earnings than her regular pay? $42

3. Which of the four types of deductions is the greatest amount? Federal Withholding Tax

4. How much greater is the deduction for F.I.C.A. than the deduction for state withholding tax? $17.74

Solve. Remember to estimate whenever you use your calculator.

• **5.** Tina Adams works part-time. Her gross earnings are $185.50 per week. Total deductions are $40.81. What is her net pay? $144.69

6. Raul Cortez has a full-time job. His gross earnings are $395.00 per week. Total deductions are $94.86. What is his net pay? $300.14

7. One week Carl Bennett earns $385.00 in regular pay and $52.00 in overtime pay. Total deductions that week are $113.62. What is his net pay? $323.38

8. One week Jill Donnelly earns $412.00 in regular pay and $83.00 in overtime pay. Total deductions that week are $130.48. What is her net pay? $364.52

9. Paul Herson works part-time. His gross earnings are $255.00. The deductions are shown below. Find the total deductions and his net pay. $55.52; $199.48

F.I.C.A.	Federal Withholding Tax	State Withholding Tax	Medical Insurance
19.15	30.00	5.10	1.27

10. Anna Zagat works full-time. Her gross earnings are $473.00. The deductions are shown below. What is her net pay? $323.10

F.I.C.A.	Federal Withholding Tax	State Withholding Tax	Medical Insurance	Retirement
35.52	76.00	9.46	4.12	24.80

Estimate.

11. Suppose you work full-time and your gross earnings are $519.00 per week. Total deductions from your gross earnings are $124.78. About how much is your net pay? $400.00

12. Your friend says that his net pay is $387.62 per week. He also says that the total deductions are $131.12. About how much are his gross earnings? $500.00

ASSIGNMENTS

Level 1	1–9
Level 2	1–12
Level 3	3–12

FOLLOW-UP ACTIVITY

MENTAL MATH Have students find the missing amounts in the following exercises.

	Gross earnings	Total deductions	Net pay
1.	$400	$100	$ ■ (300)
2.	450	110	■ (340)
3.	280	70	■ (210)
4.	560	■ (120)	440
5.	690	■ (170)	520
6.	1,000	■ (250)	750

SUPPLEMENTARY MATERIALS

TRP Practice, p. 27

TRP Transparency 14

Objective

- Multiply or divide decimals by powers of 10.

Teaching the Lesson

WARM-UP Write the following exercises on the chalkboard or on an overhead transparency.

Multiply or divide.

1. 30 × 40 **2.** 600 × 200

3. 2,000 × 200 **4.** 9,000 × 100

5. 80 ÷ 20 **6.** 600 ÷ 300

7. 2,000 ÷ 100 **8.** 1,500 ÷ 500

(**1.** 1,200 **2.** 120,000 **3.** 400,000 **4.** 900,000 **5.** 4 **6.** 2 **7.** 20 **8.** 3)

INTRODUCTION Write the following money amounts on the chalkboard: \$9.20, \$92.00, \$920.00. Elicit from students that as the decimal point moves to the right, the value increases. Then write the following sequence on the chalkboard: \$637.00, \$63.70, \$6.37. Lead students to discover that as the decimal point moves left, the value decreases.

INSTRUCTION Use Example 1 to introduce multiplication of a decimal number by a power of 10. Help students see that stating the shortcut as moving the decimal point to the right is true for both whole and decimal numbers. Help students to see that as they multiply by 10, 100, or 1,000, the product is greater.

Example 2 shows that in division of a decimal number by a power of 10, the decimal point is moved to the left. Elicit from students that as they divide by powers of 10, the quotient becomes lesser.

3.4 MULTIPLYING AND DIVIDING BY POWERS OF 10

The cafeteria at the Widgit Corporation sells yogurt in servings of 3.25 oz. How many ounces would be needed for 10 servings? 100 servings? 1,000 servings?

To multiply a decimal by a power of 10, move the decimal point as many places to the right as there are zeros in the power of 10.

EXAMPLE 1 Multiply: 3.25 by 10, by 100, and by 1,000.

10 × 3.25 = 3.2 5 → 32.5

THINK: One zero, so move the decimal point one place to the right.

100 × 3.25 = 3.2 5 → 325

THINK: Two zeros, so move the decimal point two places to the right.

1,000 × 3.25 = 3.2 5 0 → 3,250

THINK: Three zeros, so move the decimal point three places to the right. Write a zero in the product to place the decimal point correctly.

The cafeteria needs 32.5 oz of yogurt for 10 people, 325 oz for 100 people, and 3,250 oz for 1,000 people.

To divide a decimal by a power of 10, move the decimal point as many places to the left as there are zeros in the power of 10.

EXAMPLE 2 Divide: 61.7 by 10, by 100, and by 1,000.

61.7 ÷ 10 → 6 1. 7 → 6.17

THINK: One zero, so move the decimal point one place to the left.

61.7 ÷ 100 → 0 6 1. 7 → 0.617

THINK: Two zeros, so move the decimal point two places to the left.

61.7 ÷ 1,000 → 0 6 1. 7 → 0.0617

THINK: Three zeros, so move the decimal point three places to the left. Write a zero in the quotient to place the decimal point correctly.

CHECKPOINT

Write the letter of the correct answer.

Multiply or divide.

1. 5.68 × 100	b	**a.** 0.0568	**b.** 568	**c.** 5,680	**d.** 56,800
2. 46.3 × 10,000	c	**a.** 0.00463	**b.** 4,630	**c.** 463,000	**d.** 4,630,000
3. 795 ÷ 100	b	**a.** 0.00795	**b.** 7.95	**c.** 79.50	**d.** 79,500
4. 1.04 ÷ 1,000	b	**a.** 0.000104	**b.** 0.00104	**c.** 1,040	**d.** 104,000

PRACTICE EXERCISES

See Extra Practice, page 432.

1. 504 2. 72 3. 797 4. 65.2
5. 4,280 6. 3,670 7. 589,000 8. 8.4
9. 627.3 10. 64,200 11. 730 12. 4,009
13. 890,000 14. 3,460 15. 580 16. 6.1
17. 49,010 18. 81,429 19. 5,600 20. 57,200

Multiply.

• **1.** 50.4 × 10	• **2.** 7.2 × 10	**3.** 79.7 × 10	**4.** 0.652 × 100
5. 4.280 × 1,000	**6.** 3.67 × 1,000	**7.** 58.9 × 10,000	**8.** 0.0084 × 1,000
9. 6.273 × 100	**10.** 6.42 × 10,000	**11.** 7.3 × 100	**12.** 4.009 × 1,000
13. 89 × 10,000	**14.** 3.46 × 1,000	**15.** 5.8 × 100	**16.** 0.61 × 10
17. 49.01 × 1,000	**18.** 814.29 × 100	**19.** 0.56 × 10,000	**20.** 57.2 × 1,000

21. 0.55 22. 0.9 23. 0.4317 24. 73.4
25. 0.0652 26. 0.00029 27. 0.0864 28. 4.307
29. 5.8629 30. 0.027 31. 7.92 32. 0.3065

Divide.

• **21.** 5.5 ÷ 10	• **22.** 9 ÷ 10	**23.** 43.17 ÷ 100	**24.** 734 ÷ 10
25. 6.52 ÷ 100	**26.** 0.029 ÷ 100	**27.** 86.4 ÷ 1,000	**28.** 43.07 ÷ 10
29. 586.29 ÷ 100	**30.** 2.7 ÷ 100	**31.** 79.2 ÷ 10	**32.** 30.65 ÷ 100
33. 69.32 ÷ 10	**34.** 851 ÷ 100	**35.** 203.1 ÷ 100	**36.** 14 ÷ 1,000
37. 4.83 ÷ 1,000	**38.** 10.5 ÷ 10,000	**39.** 0.39 ÷ 10,000	**40.** 0.065 ÷ 100

33. 6.932 34. 8.51 35. 2.031 36. 0.014
37. 0.00483 38. 0.00105 39. 0.000039 40. 0.00065

Complete the table.

Number	×100	÷1,000	×10,000	÷100,000
6.002	• **41.** ▩	• **42.** ▩	**43.** ▩	**44.** ▩
2.58	**45.** ▩	**46.** ▩	**47.** ▩	**48.** ▩
0.037	**49.** ▩	**50.** ▩	**51.** ▩	**52.** ▩

41. 600.2 42. 0.006002 43. 60,020 44. 0.00006002
45. 258 46. 0.00258 47. 25,800 48. 0.0000258
49. 3.7 50. 0.000037 51. 370 52. 0.00000037

MIXED REVIEW

Estimate by rounding.

• **53.** 9,468 + 7,512 17,000
54. 4,614 − 1,638 3,000
55. 7,142 + 8,813 + 849 16,800
56. 4 × 623 2,400
57. 29 × 802 24,000
58. 2,846 ÷ 59 50

Solve.

• **59.** Ms. Sloan, the cafeteria manager, paid $0.23 each for 100 grapefruit. What was the total cost? $23.00

60. Ms. Sloan paid $22.50 for 15 lb of prepared tuna salad. What was the cost per pound? $1.50

61. Lucy Davis spent $3.25 for a sandwich, $2.85 for salad, and $0.95 for a carton of juice. How much did she spend altogether? $7.05

62. Ms. Sloan needs 80 apples. Each apple costs $0.35. If she buys 100 apples, the cost drops to $0.25 each. Should she buy 100 apples? Explain. Yes; she will save $3.00.

CHECKPOINT The incorrect answer choices include common errors students make.

COMMON ERROR

Some students will move the decimal point in the opposite direction. For example, in Item 1 of Checkpoint, they would choose a. Encourage these students to estimate (6 × 100) to guarantee a reasonable answer.

ASSIGNMENTS

Level 1	Odd 1–61
Level 2	Even 2–58, 59–62
Level 3	13–20, 33–40, 53–62

FOLLOW-UP ACTIVITY

REINFORCEMENT Have students write the numbers 1, 2, 3, and 4 on four different sheets of paper, and a large × or ÷ on two other sheets. Hold up a card with a number written on it, and then another card with that number multiplied or divided by 10, 100, 1,000, or 10,000. Have the students hold up 1, 2, 3, or 4 to show how many zeros were written or crossed out, and an × or ÷ to show which operation was used. For example, for 67 and 67,000, students would hold up 3 and ×.

SUPPLEMENTARY MATERIALS

TRP Practice, p. 28

TRP Reteaching, p. 19

TRP Transparency 14

OBJECTIVE

- Solve problems by guessing and checking.

TEACHING THE LESSON

WARM-UP Write the following exercises on the chalkboard or an overhead transparency.

Multiply.

1. 10 × \$0.25 **2.** 9 × \$0.10

3. 14 × \$0.10 **4.** 15 × \$0.25

5. 13 × \$0.05 **6.** 12 × \$0.12

(**1.** \$2.50 **2.** \$0.90 **3.** \$1.40 **4.** \$3.75 **5.** \$0.65 **6.** \$1.44)

INTRODUCTION Pose this question to the class. "I am thinking of a number between 1 and 50. What number am I thinking of?" Have students try to guess the number. After each guess say "higher" or "lower." Continue until the correct number is guessed. Discuss the fact that by narrowing the range within which the number lies, the number of guesses is minimized.

INSTRUCTION After discussing the situation, point out that sometimes there is no obvious way to compute a solution to a problem. A guess is a quick and easy way to explore the problem. Use the information obtained on the first guess as a basis for making subsequent guesses.

You may wish to have quarters and dimes available for demonstration purposes. Encourage students to check their work using a calculator.

PROBLEM *Solving* STRATEGY

3.5 GUESSING AND CHECKING

Situation:

The Eloni family belongs to a record and tape club. Mr. Eloni collected \$7.20 from Bryan, who took quarters and dimes from his bank to pay for the tapes he ordered. Bryan gave his father 33 coins. How many of each coin did Bryan give his father?

Strategy:

Guessing and checking can often help you to solve a problem.

Applying the Strategy:

Guess 1	**Check**	
20 quarters 13 dimes	20 × \$0.25 = \$5.00 13 × \$0.10 = +1.30 \$6.30	THINK: \$6.30 < \$7.20 Make another guess. Try more quarters to raise the total.
Guess 2	**Check**	
28 quarters 5 dimes	28 × \$0.25 = \$7.00 5 × \$0.10 = +0.50 \$7.50	THINK: \$7.50 > \$7.20 Make another guess. Try fewer quarters to lower the total.
Guess 3	**Check**	
26 quarters 7 dimes	26 × \$0.25 = \$6.50 7 × \$0.10 = +0.70 \$7.20	Bryan gave his father 26 quarters and 7 dimes.

PRACTICE EXERCISES See Extra Practice, page 433.

Check each guess. Then choose the correct guess.

• **1.** The Eloni family ordered some records and tapes. They ordered four more tapes than records. If they ordered 14 items this month, how many were records? b
   a. Guess: 3 records
   b. Guess: 5 records
   c. Guess: 6 records

• **2.** Bryan paid his father for some tapes. Mr. Eloni gave his son $1.90 in quarters and nickels for change. He gave him 18 coins. How many coins of each kind were there? c
   a. Guess: 3 quarters, 15 nickels
   b. Guess: 4 quarters, 14 nickels
   c. Guess: 5 quarters, 13 nickels

Solve the problem by using the guess-and-check method.

3. Dee Eloni gave her mother $6.90 in quarters and dimes to pay for a record she ordered. Dee gave her 30 coins. How many of each coin did Dee give her mother? 26 quarters, 4 dimes

4. In September, the Eloni family ordered records and tapes. They ordered five more tapes than records. If they ordered 17 items, how many were records? 6 records

5. During the past 3 years, the Eloni family bought 108 new tapes. If they bought 15 tapes more each year than the year before, how many tapes did the Elonis buy last year? 51 tapes

6. During a special sale, tapes sell for $6 each and records for $4 each. The Eloni family spent $60 in all for tapes and records. They bought the same number of tapes as records. How many of each item did they buy? 6 tapes, 6 records

7. In October, the Elonis ordered 23 records and tapes. There were seven fewer tapes than records. How many of each did the Elonis order? 15 records, 8 tapes

8. Ben's Record Store ordered records and tapes. They ordered 584 more tapes than records. If they ordered 2,200 items, how many were records? 808 records

ASSIGNMENTS

Level 1	1–2, Odd 3–7
Level 2	1–7
Level 3	3–8

FOLLOW-UP ACTIVITY

SITUATIONAL PROBLEM SOLVING Present students with the following problem: Your family is considering joining a record and tape club where records and tapes are sent on a monthly basis. You are in charge of researching the different clubs, and choosing the club to join. What questions would you want answered before you made your decision?

SUPPLEMENTARY MATERIALS

TRP Practice, p. 29

TRP Applications, p. 3

TRP Transparency 15

OBJECTIVE

- Evaluate student progress.

USING THE REVIEW

This page provides a means for informally evaluating students' understanding of the skills and concepts covered so far in this chapter.

Have the students look at the page to familiarize themselves with the various question formats that are presented. Discuss any questions that they may have. Then ask them to complete the page independently.

In addition to grading them individually, you may wish to review the answers to the questions collectively with the students.

Page references appear in brackets. They refer to pages on which a particular skill was introduced.

Before continuing on to the topics found in the remainder of the chapter, you may wish to have students review any skills or concepts in which they have demonstrated weakness.

MIDCHAPTER REVIEW

Add. [62]

1. 36.04 +17.95 53.99	**2.** 5.693 +8.042 13.735	**3.** 0.617 +0.028 0.645	**4.** 0.0053 +0.0271 0.0324	**5.** 0.054 +0.738 0.792
6. 6.39 +18.6 24.99	**7.** 11.2 + 4.835 16.035	**8.** 0.0935 +0.008 0.1015	**9.** 0.7605 +0.09 0.8505	**10.** 52.4 + 0.736 53.136
11. 4.62 0.095 +57.8 62.515	**12.** 63 8.401 + 0.5 71.901	**13.** 6.903 0.72 +0.006 7.629	**14.** 136.5 0.837 + 25.06 162.397	**15.** 0.037 72.9 + 4.85 77.787

16. 38.16 + 5.29 43.45 **17.** 5.06 + 0.984 6.044 **18.** 149.6 + 0.0037 149.6037

19. 58.64 + 9.87 + 126.34 194.85 **20.** 4 + 0.867 + 15.24 20.107 **21.** 0.0407 + 3 + 9.5 12.5407

Subtract. [64]

22. 738.1 −596.8 141.3	**23.** 42.09 −23.86 18.23	**24.** 6.431 −2.948 3.483	**25.** 0.807 −0.594 0.213	**26.** 9.801 −2.572 7.229
27. 300 − 62.37 237.63	**28.** 0.72 −0.493 0.227	**29.** 0.0829 −0.043 0.0399	**30.** 0.71 −0.6208 0.0892	**31.** 4.8 −1.756 3.044

32. 86.54 − 27.36 59.18 **33.** 395.06 − 184.97 210.09 **34.** 703.01 − 96.28 606.73

35. 26.07 − 8.351 17.719 **36.** 19 − 0.8065 18.1935 **37.** 0.6 − 0.5792 0.0208

Multiply or divide. [68]

38. 8.365 × 10,000 83,650 **39.** 0.693 × 1,000 693 **40.** 0.00562 × 100 0.562

41. 231.09 × 1,000 231,090 **42.** 9.72 × 10,000 97,200 **43.** 0.000093 × 100,000 9.3

44. 75.69 ÷ 1,000 0.07569 **45.** 0.463 ÷ 100 0.00463 **46.** 607.094 ÷ 10,000 0.0607094

47. 531 ÷ 10,000 0.0531 **48.** 8.67 ÷ 1,000 0.00867 **49.** 0.00234 ÷ 100,000 0.0000000234

Solve for n. [68]

50. $6.37 \times 100 = n$ 637 **51.** $0.80547 \times 10{,}000 = n$ 8,054.7 **52.** $75.69 \div 100 = n$ 0.7569

Solve. [66]

53. Marilyn Holmes works in a bakery. Her gross earnings are $325.50 per week. Total deductions are $97.75. What is her net pay? $227.75

54. John Jacobson earns $430.25 in regular pay. He earns $37.00 in overtime pay. Total deductions that week are $175.50. What is his net pay? $291.75

PROBLEM
Solving
APPLICATION

3.6 CONSUMER: CHECKING ACCOUNTS

If you have a checking account, it is important to keep an accurate record of your balance. The **balance** is the amount of money that is in the account.

When you write a check, subtract that amount to find the new balance. When you make a deposit, add that amount to find the new balance.

JANE CALDWELL CHECKBOOK RECORD

Check Number	Date	Description of Transaction	Amount of Check	Amount of Deposit	Balance	
					250	00
1	11/6	Ted's Market	34.75		34	75
					215	25
	11/8	Deposit		95.87	95	87
					311	12

The checkbook record shows a beginning balance of $250.00.

On 11/6, Jane wrote a check for $34.75 to Ted's Market. The new balance was $215.25.

On 11/8, Jane deposited $95.87. The new balance was $311.12.

Find each new balance after the given transaction.

Check Number	Date	Description of Transaction	Amount of Check	Amount of Deposit	Balance	
					375	00
	1/29	Deposit		50.85		
					• 1. ■	
1	2/3	John Forbes	55.25			
					• 2. ■	
2	2/12	Book Mart	19.75			
					3. ■	
	2/15	Deposit		50.85		
					4. ■	
3	2/21	Pat Simms, Inc.	165.32			
					5. ■	

1. 425.85 **2.** 370.60 **3.** 350.85 **4.** 401.70 **5.** 236.38

Solve.

6. On 3/1, Sue opened a checking account with $425.00. On 3/8, she deposited $65.38 and wrote checks for $27.60 and $120.95. On 3/22, she deposited $75.32. What was Sue's new balance? $417.15

OBJECTIVE

- Solve problems that involve finding the new balance in a checking account.

TEACHING THE LESSON

WARM-UP Write the following exercises on the chalkboard or an overhead transparency.

Find the sum or difference.

1. $32.15 + 16.15 **2.** $76.20 − 14.08

3. $83.76 + 12.46 **4.** $95.00 − 12.85

5. $126.78 + 79.23 **6.** $156.01 − 87.54

(**1.** $48.30 **2.** $62.12 **3.** $96.22 **4.** $82.15 **5.** $206.01 **6.** $68.47)

INSTRUCTION As you examine Jane Caldwell's checking account, be sure that students understand these points: When Jane writes a check, the result is a decrease in the amount of money in the account. When she makes a deposit, the result is an increase in the amount in the account.

Point out that in writing in a checkbook record, dollar signs and decimal points are omitted since there is usually a vertical line that separates the spaces for dollars and cents.

You may wish to have students use calculators in working out the solution to Problem 6.

ASSIGNMENTS

Level 1	1–5
Level 2	1–6
Level 3	1–6

SUPPLEMENTARY MATERIALS

TRP Practice, p. 30

TRP Lesson Aids, p. 2

TRP Transparency 15

OBJECTIVES

- Multiply decimals.
- Check the reasonableness of answers using estimation.

TEACHING THE LESSON

WARM-UP Write the following exercises on the chalkboard or an overhead transparency.

Estimate. Then find the exact answer.

1. 283 × 14 **2.** 784 × 69 **3.** 5,213 × 752

(**1.** 3,000; 3,962 **2.** 56,000; 54,096 **3.** 4,000,000; 3,920,176)

INTRODUCTION Write the following addition sentence on the chalkboard: 1.4 + 1.4 + 1.4 + 1.4 = ? Ask students to find the sum. (5.6) Then ask them to rewrite the problem as a multiplication sentence. (4 × 1.4 = 5.6)

INSTRUCTION Example 1 establishes the rule for correctly placing the decimal point in the product. Stress that the numbers of decimal places in the factors should be added.

Inserting zero placeholders is illustrated in Example 2. Emphasize that the zero(s) must always be placed in front of the product. Zeros at the end of a decimal number do not change its value. (0.38 ≠ 0.0038 but 0.38 = 0.3800)

Use Example 3 to reinforce the importance and usefulness of estimating the product. Point out that each factor is rounded to its greatest nonzero place.

3.7 MULTIPLYING DECIMALS

Tom Rosicki's car goes 24.4 miles on a gallon of gasoline.
When his fuel tank is full, it holds 18 gallons.
How many miles can his car travel on a full tank of gas?

Multiply 24.4 by 18 to find the total number of miles.

EXAMPLE 1 Multiply: 18 × 24.4

1. Multiply as with whole numbers.

```
  244
×  18
 1952
 244
 4392
```

2. Count the decimal places in the factors.

THINK: The product has the same number of decimal places as the factors.

```
  24.4 ← 1 decimal place
×   18 ← 0 decimal places
 439.2 ← 1 decimal place
```

So, Tom's car can travel 439.2 miles on a full tank of gasoline.

Sometimes extra zeros are needed as placeholders.

EXAMPLE 2 Multiply: 0.0092 × 0.53

1. Multiply as with whole numbers.

```
  00092
×   053
    276
   4600
   4876
```

2. Count the decimal places in the factors. Use zeros as place holders.

```
  0.0092   ← 4 decimal places
×   0.53   ← 2 decimal places
0.004876   ← 6 decimal places
```

So, 0.092 × 0.53 = 0.004876.

Estimating your answer first can help you determine the reasonableness of your answer.

EXAMPLE 3 Multiply: $7.32 × 0.08. Check the reasonableness of your answer.

1. Estimate.

```
$ 7.32 → $   7  ← 0 decimal places
× 0.08 →  0.08  ← 2 decimal places
         $ 0.56 ← 2 decimal places
```

2. Multiply.

```
$   7.32 ← 2 decimal places
  × 0.08 ← 2 decimal places
 $0.5856 ← 4 decimal places
```

Since $0.56 is close to $0.5856, the answer is reasonable.
When the product represents money, round to the nearest cent.

So, $7.32 × 0.08 = $0.5856 —**rounds to**→ $0.59.

PRACTICE EXERCISES See Extra Practice, page 433.

Multiply.

• **1.** 5.6×7.9 44.24	• **2.** 1.04×32 33.28	**3.** 42×3.7 155.4	**4.** 28.6×5.9 168.74	**5.** 3.69×0.14 0.5166
6. 43.28×0.016 0.69248	**7.** 2.73×0.004 0.01092	**8.** 0.0084×0.7 0.00588	**9.** 0.0034×0.071 0.0002414	**10.** 0.237×0.008 0.001896
11. \$52.68 × 8.37 \$440.93	**12.** \$27.02 × 0.04 \$1.08	**13.** \$0.24 × 0.83 \$0.20	**14.** \$35.70 × 0.09 \$3.21	**15.** \$80.92 × 3.06 \$247.62
16. 0.62×0.38 0.2356	**17.** 94.2×0.083 7.8186	**18.** 76.1×15 1,141.5	**19.** 306×8.4 2,570.4	**20.** \$93.56 × 37 \$3,461.72
21. 1.005×7.3 7.3365	**22.** 62.19×0.008 0.49752	**23.** 8.007×0.119 0.952833	**24.** 3.921×0.018 0.070578	**25.** 85×0.8 68
26. 63×0.4 25.2	**27.** 418×0.5 209	**28.** 12.3×8.6 105.78	**29.** 1.085×0.036 0.03906	**30.** 140×5.8 812

First estimate by rounding, and then find the exact answer.

• **31.** 792×0.6 480; 475.2	• **32.** 6.95×0.18 1.4; 1.251	**33.** 52×1.3 50; 67.6	**34.** 4.09×3.7 16; 15.133	**35.** \$60.08 × 0.51 \$30; \$30.64
36. 418×0.5 200; 209	**37.** 603×2.9 1,800; 1,748.7	**38.** 87×3.4 270; 295.8	**39.** 83.2×0.74 56; 61.568	**40.** 0.7083×0.045 0.035; 0.0318735

41. 7.1×0.008 0.056; 0.0568 **42.** 0.92×0.005 0.0045; 0.0046 **43.** 5.9×0.26 1.8; 1.534

Solve for n.

• **44.** $3.08 \times 0.65 = n$ 2.002 **45.** $0.52 \times \$81.30 = n$ \$42.28 **46.** $(0.407 - 0.111) \times 0.28 = n$ 0.08288

Solve.

• **47.** In 1940, a reort showed that consumers spent \$16.6 billion on food. In 1980, a report showed that they spent 18.25 times that amount. How much money was spent on food in 1980? \$302.95 billion

48. In 1920, a Census report showed that the population per square mile in Illinois was 115.7 people. In 1980, it was reported to be 205.3 people. What was the increase in population per square mile? 89.6 people per sq mi

49. In 1950, a government report showed that \$21.7 billion was spent on housing. Write that number in standard form. \$21,700,000,000

50. A report showed that a car could travel 25.4 mi on a gallon of gasoline. How far could a car go on 20 gal of gas? 508 mi

COMMON ERROR

Some students may count decimal places from the left instead of the right. Encourage these students to use estimation to check the reasonableness of the answer.

ASSIGNMENTS

Level 1	Odd 1–43, 47, 48
Level 2	Even 2–44, 47–50
Level 3	Odd 17–43, 44–50

FOLLOW-UP ACTIVITY

ESTIMATION Have students explore estimation by looking in the Practice Exercises to find factors that are close to 1 or $\frac{1}{2}$. Have students only focus on exercises where the second factor is a whole number. Help students to use this estimation "shortcut." For example, 0.53×792 is about $\frac{1}{2} \times 800 = 400$.

SUPPLEMENTARY MATERIALS

TRP Practice, p. 31

TRP Reteaching, p. 20

TRP Transparency 16

OBJECTIVE

- Solve multistep problems that involve finding the cost of electricity.

TEACHING THE LESSON

WARM-UP Write the following exercises on the chalkboard or an overhead transparency.

Find the product.

1. 6 × 365 (2,190)

2. 0.25 × 1,240 (310)

3. 1.125 × 0.13 (0.14625)

4. 1,460 × $0.03 ($43.80)

5. 2,190 × $0.025 ($54.75)

INTRODUCTION Discuss why electrical bills are likely to be higher during the summer months. Ask which appliances are used more during the summer.

INSTRUCTION Be sure students understand the basic plan: First, find the total number of hours the appliance is used. Next, find the cost per hour for using the appliance. Finally, multiply the total number of hours by the cost per hour.

In Step 2, point out that the notation $0.0315 indicates 3.15¢, or about 3¢.

You may wish to have students use calculators to help them solve the problems in this lesson. Notice that Problem 7 requires the students to find the total number of minutes the toaster is used in 60 days. Then they express this time as hours and find the cost.

ASSIGNMENTS

Level 1	1–6
Level 2	1–8
Level 3	3–8

SUPPLEMENTARY MATERIALS

TRP Practice, p. 32
TRP Transparency 16

PROBLEM Solving APPLICATION

3.8 CONSUMER: COST OF ELECTRICITY

The amount of electrical power used by appliances varies. The table below shows the approximate power used in an hour by electrical appliances. If the cost of electricity per kilowatt-hour is known, as well as the number of hours an appliance is used, the electrical costs can be computed.

Appliance	Power used (kW)
Air conditioner	1.250
Clothes dryer	3.750
Radio	0.125
Refrigerator	0.250
Television set	0.225
Toaster	0.775
Vacuum cleaner	0.550

The television set is used an average of 4 hours per day. If the cost of electricity is $0.14 per kilowatt-hour, what is the cost of electricity for the television set for a year (365 days)?

1. Multiply to find the number of hours the set is used per year.

HOURS USED PER DAY	×	NUMBER OF DAYS	=	HOURS USED PER YEAR
4	×	365	=	1,460

2. Multiply to find the cost of using the set for 1 hour.

POWER IN KILOWATTS	×	COST PER KILOWATT-HOUR	=	COST OF USE PER HOUR
0.225	×	$0.14	=	$0.0315

3. Multiply the number of hours by the cost per hour.

1,460 × $0.0315 = $45.99

So, the cost of electricity for the television set is $45.99 per year.

Use the table to find the cost. Assume the cost of electricity is $0.14 per kilowatt-hour. One year is 365 days, or 52 weeks. Remember to estimate whenever you use your calculator.

• **1.** The air conditioner is used 8 hours per day for 60 days. $84.00

2. The vacuum cleaner is used an average of 2 hours per week for 1 year. $8.01

3. The refrigerator is used 24 hours per day for 1 year. $306.60

4. The radio is used 6 hours per day for 1 year. $38.33

Use $0.16 as the cost of electricity per kilowatt-hour in Exercises 5–8.

• **5.** The clothes dryer is used an average of 4 hours per week for 1 year. $124.80

6. The air conditioner is used 9 hours per day for 60 days. What is the cost? $108.00

7. The toaster is used 10 minutes a day for 60 days. What is the cost? $1.24

8. Which appliance costs about 60 cents an hour to use? clothes dryer

PROBLEM Solving APPLICATION

3.9 CAREER: LAB TECHNICIAN

Carol Hormel is a lab technician. She performs experiments, makes measurements, and records the results of the experiments.

The table below shows the amount of three compounds produced in a series of five experiments.

AMOUNT OF COMPOUND PRODUCED (LITERS)

Experiment number	Compound A	Compound B	Compound C
101	0.342	0.175	0.675
102	0.287	0.148	0.551
103	0.625	0.332	1.209
104	0.536	1.078	0.261
105	0.858	1.736	0.403

Solve. Remember to estimate whenever you use your calculator.

How many liters of the three compounds were produced:

• **1.** in Experiment 101? 1.192

2. in Experiment 103? 2.166

In Experiment 102, how much more was produced of:

• **3.** Compound A than Compound B? 0.139 L

4. Compound C than Compound A? 0.264 L

List the amounts in order from greatest to least.

5. The compounds produced in Experiment 103 C, A, B

6. The compounds produced in Experiment 105 B, A, C

7. Did Experiment 101 produce more than five times as much Compound C as Compound B? no

8. Did Experiment 104 produce more than four times as much Compound B as Compound C? yes

9. Which experiments produced about twice as much Compound A as Compound B? Experiments 101, 102, 103

10. Which experiments produced about twice as much Compound B as Compound A? Experiments 104, 105

OBJECTIVE

• Solve problems that involve computation with decimals.

TEACHING THE LESSON

WARM-UP Write the following exercises on the chalkboard or an overhead transparency.

Find the sum or difference.

1. 0.256 + 0.312 (0.568)

2. 0.763 − 0.241 (0.522)

3. 0.457 + 0.809 (1.266)

4. 0.732 − 0.489 (0.243)

5. 0.428 + 0.329 + 0.278 (1.035)

INSTRUCTION You may wish to show the students a container that holds a liter as well as a container that holds 0.250 L.

Before assigning the problems, be sure the students understand that each row of the table shows the amounts of the three compounds produced in a given experiment. Each column of the table shows the amounts of a given compound produced in the five experiments.

You may wish to have students use a calculator to help them solve Problems 1–4. Notice that Problems 9 and 10 involve estimation.

ASSIGNMENTS

Level 1	1–8
Level 2	1–10
Level 3	1, 3, 5–10

SUPPLEMENTARY MATERIALS

TRP Practice, p. 33

TRP Transparency 17

OBJECTIVE

- Divide a decimal by a whole number.

TEACHING THE LESSON

WARM-UP Write the following exercises on the chalkboard or an overhead transparency.

Divide.

1. $6\overline{)180}$ **2.** $8\overline{)5,608}$

3. $72\overline{)1,296}$ **4.** $93\overline{)2,332}$

(**1.** 30 **2.** 701 **3.** 18 **4.** 25 R7)

INTRODUCTION Ask students how much each person would receive if \$4.80 was divided among two people, four people, six people, and eight people. Write the division examples and answers on the chalkboard. Compare the location of the decimal point in each dividend and quotient.

$\begin{array}{r} \$2.40 \\ 2\overline{)\$4.80} \end{array}$ $\begin{array}{r} \$1.20 \\ 4\overline{)\$4.80} \end{array}$

$\begin{array}{r} \$0.80 \\ 6\overline{)\$4.80} \end{array}$ $\begin{array}{r} \$0.60 \\ 8\overline{)\$4.80} \end{array}$

INSTRUCTION Use Example 1 to establish the rule for decimal quotients. Make sure students realize that the division algorithm is unchanged.

Example 2 shows that once the decimal point is in the quotient, there must be a digit over every corresponding digit in the dividend. Make sure students understand that the 0 before the 2 in the quotient is necessary.

Use Example 3 to show the necessity of writing extra zeros in the dividend.

3.10 DECIMAL QUOTIENTS

Video Vibes magazine had an introductory offer of \$33.75 for nine issues. What was the cost of a single issue?

To find the cost per issue, divide \$33.75 by 9.

EXAMPLE 1 Divide: \$33.75 ÷ 9

1. Write a decimal point in the quotient directly above the decimal point in the dividend.

$$\begin{array}{r} . \\ 9\overline{)\$33.75} \end{array}$$

2. Divide as with whole numbers.

$$\begin{array}{r} \$\ 3.75 \\ 9\overline{)\$33.75} \\ \underline{27} \\ 6\,7 \\ \underline{6\,3} \\ 4\,5 \\ \underline{4\,5} \\ 0 \end{array}$$

So, each issue costs \$3.75.

Estimating a quotient first is a good way to check the reasonableness of your answer. Sometimes you will need to insert zeros in the quotient as placeholders.

EXAMPLE 2 Divide: 0.3924 ÷ 18. Check the reasonableness of your answer.

1. Estimate by using compatible numbers.
2. Divide.

Zero placeholder

$$18\overline{)0.3924} \rightarrow \begin{array}{r} .0200 \\ 20\overline{)0.4000} \end{array}$$

$$\begin{array}{r} .0218 \\ 18\overline{)0.3924} \\ \underline{36} \\ 32 \\ \underline{18} \\ 144 \\ \underline{144} \\ 0 \end{array}$$

Since 0.0218 is close to 0.0200, the answer is reasonable.

Sometimes you need to write one or more zeros in the dividend to complete the division.

EXAMPLE 3 Divide:

a. 13.4 ÷ 4

```
    3.35
4)13.40
  12
   14
   12
    20
    20
     0
```

b. 3 ÷ 8

```
   0.375
8)3.000
  24
   60
   56
    40
    40
     0
```

PRACTICE EXERCISES

See Extra Practice, page 433.

37. 6; 6.2 38. 1; 0.91 39. 0.07; 0.06 40. 4; 3.74
41. 0.004; 0.0037 42. 0.5; 0.57 43. 0.07; 0.073 44. 0.006; 0.0061

Divide.

• 1. 4)27.32 6.83	2. 6)$45.06 $7.51	3. 35)634.55 18.13	4. 14)44.24 3.16
• 5. 5)8 1.6	6. 4)37 9.25	7. 8)7 0.875	8. 12)9 0.75
9. 5)0.0095 0.0019	10. 6)0.0042 0.0007	11. 2)0.48 0.24	12. 12)0.0132 0.0011
13. 7)0.63 0.09	14. 11)0.264 0.024	15. 94)3.478 0.037	16. 7)$21.42 $3.06
17. 5)$43.05 $8.61	18. 8)0.0592 0.0074	19. 3)23.46 7.82	20. 67)348.4 5.2
21. 49)4.067 0.083	22. 38)36.48 0.96	23. 3)$12.39 $4.13	24. 7)142.8 20.4
25. 4)5.308 1.327	26. 7)1.057 0.151	27. 9)48.6027 5.4003	28. 8)0.2448 0.0306
29. 4)54 13.5	30. 5)23 4.6	31. 8)123 15.375	32. 12)117 9.75
33. 17)1.02 0.06	34. 36)619.2 17.2	35. 32)$26.88 $0.84	36. 23)4.37 0.19

First estimate using compatible numbers. Then find the exact quotient.

•37. 173.6 ÷ 28	•38. 69.16 ÷ 76	39. 3.54 ÷ 59	40. 168.3 ÷ 45
41. 0.2368 ÷ 64	42. 52.44 ÷ 92	43. 3.504 ÷ 48	44. 0.1159 ÷ 19
45. 0.2257 ÷ 37 0.005; 0.0061	46. 130.208 ÷ 26 5; 5.008	47. 1.6274 ÷ 79 0.02; 0.0206	48. 48.6567 ÷ 81 0.6; 0.6007

Solve.

49. Daisy Comics prints 2.44 million comic books every month. The same number of comics are sent to each of eight distribution centers. How many comics are sent to each center?
305,000 comic books

50. In 1985, Daisy Comics had 3.6 million readers. By 1993, the number of readers will increase by 2.02 million. What will the readership be in 1993?
5.62 million readers

COMMON ERROR

Some students will forget to write zero placeholders in the quotient. To avoid this error, have these students circle every digit to the right of the decimal point in the dividend. Stress that there must be that many digits to the right of the decimal point in the quotient.

ASSIGNMENTS

Level 1	1–16, Odd 17–47, 49, 50
Level 2	Even 2–50
Level 3	Odd 17–47, 49, 50

FOLLOW-UP ACTIVITY

ESTIMATION Provide each student with a worksheet of division examples as follows:

Example	Estimate	Actual	Off
8)63.92			
5)0.9165			

Have students write an estimate of the answer. Next, have them find the actual quotient using a calculator. Then have them subtract to see how far off their estimate was.

Frequent use of this type of activity will help to establish good number sense, and hone estimating skills.

SUPPLEMENTARY MATERIALS

TRP Practice, p. 34

TRP Reteaching, p. 21

TRP Transparency 17

OBJECTIVE

- Divide decimals.

TEACHING THE LESSON

WARM-UP Write the following exercises on the chalkboard or an overhead transparency.

Multiply or divide.

1. 700 × 6	**2.** 40 × 50
3. 4,000 × 80	**4.** 600 × 60
5. 800 ÷ 20	**6.** 100 ÷ 50
7. 7,000 ÷ 10	**8.** 9,000 ÷ 900

(**1.** 4,200 **2.** 2,000 **3.** 320,000 **4.** 36,000 **5.** 40 **6.** 2 **7.** 700 **8.** 10)

INTRODUCTION Write the following division examples on the chalkboard.

$6\overline{)54}$ = 9 $60\overline{)540}$ = 9 $600\overline{)5{,}400}$ = 9

Discuss with students why the quotients are the same in each example. (Both the divisor and the dividend were multiplied by the same number—first 10, then 100.)

INSTRUCTION Use Example 1 to establish the rule for dividing decimals. Stress that the decimal point must be moved in both the dividend and the divisor. Elicit from students how to determine how many places to move the decimal point (as many places as there are digits to the right of the decimal point).

3.11 DIVIDING BY DECIMALS

In a laboratory experiment, the temperature of a solution being heated rose 1° every 0.5 min. How many degrees did the temperature rise after 22.5 min?

To find the number of degrees, divide 22.5 by 0.5.

When the divisor is a decimal, multiply both divisor and dividend by the same power of 10 to make the divisor a whole number.

EXAMPLE 1

Divide: 22.5 ÷ 0.5

1. Make the divisor a whole number.

THINK: Multiply the divisor and dividend by 10.

$0.5\overline{)22.5}$

2. Divide.

```
       45.
05.)225.
     20
      25
      25
       0
```

So, the temperature rose 45°.

Sometimes you need to round the quotient.

EXAMPLE 2

Find 49.308 ÷ 7.05 to the nearest hundredth.

1. Make the divisor a whole number.

THINK: Multiply the divisor and dividend by 100.

$7.05\overline{)49.308}$

2. Divide to thousandths. Round to the nearest hundredth.

```
          6.994 → 6.99
705.)4,930.800
     4,230
       700 8
       634 5
        66 30
        63 45
         2 850
         2 820
            30
```

So, 49.308 ÷ 7.05 to the nearest hundredth is 6.99.

To estimate with a decimal divisor, first multiply to make the divisor a whole number.

EXAMPLE 3 Estimate: 0.73)0.2628

1. Make the divisor a whole number.

0.73)0.2628 → 73.)26.28

2. Estimate by using compatible numbers.

0.4
70)28.00

CHECKPOINT

Write the letter of the correct answer.

Divide.

1. 0.9)0.225	b	**a.** 0.025	**b.** 0.25	**c.** 2.5	**d.** 25
2. 3.6)164.52	c	**a.** 0.457	**b.** 4.57	**c.** 45.7	**d.** 457
3. 6.47)$17.83	b	**a.** $0.28	**b.** $2.76	**c.** $27.60	**d.** $276
4. 0.0089)0.658	c	**a.** 0.739	**b.** 7.39	**c.** 73.9	**d.** 739

PRACTICE EXERCISES

See Extra Practice, page 434.

Name the power of 10 needed to make the divisor a whole number. Write 10; 100; 1,000; or 10,000.

• **1.** 83.6)4.92 10 • **2.** 5.39)87.4 100 **3.** 0.0072)5.68 10,000 **4.** 9.063)2.7 1,000

5. 0.376)98.506 1,000 **6.** 0.843)52 1,000 **7.** 4.0815)28.6 10,000 **8.** 2.861)7.69 1,000

Divide.

• **9.** 0.9)1.8 2 • **10.** 0.4)4.8032 12.008 **11.** 0.3)16.38 54.6 **12.** 0.5)5.95 11.9

13. 0.06)4.6728 77.88 **14.** 0.08)0.06952 0.869 **15.** 0.07)63.504 907.2 **16.** 0.02)38.6 1,930

17. 0.67)0.00201 0.003 **18.** 0.21)0.147 0.7 **19.** 2.3)14.03 6.1 **20.** 5.7)23.94 4.2

21. 6.8)0.0238 0.0035 **22.** 1.9)6.232 3.28 **23.** 0.037)0.2405 6.5 **24.** 7.2)22.176 3.08

25. 0.205)1.23 6 **26.** 0.714)22.848 32 **27.** 0.832)79.872 96 **28.** 0.603)4.2813 7.1

29. 4.83)3.5259 0.73 **30.** 0.510)1621.8 3,180 **31.** 18.5)5.6425 0.305 **32.** 3.59)3.1233 0.87

Divide. Round to the nearest hundredth.

33. 0.94)3.50432 3.73 **34.** 0.041)0.37187 9.07 **35.** 0.58)0.05568 0.10

36. 0.025)0.014575 0.58 **37.** 0.307)12.28921 40.03 **38.** 51.9)1074.33 20.70

Example 2 illustrates how to divide to a specific place value. Be sure students understand that they must divide one place further than asked for in order to round back.

Use Example 3 to emphasize that when estimating, first make the divisor a whole number before determining which compatible numbers to use. Point out that compatible numbers are used for estimation.

CHECKPOINT The incorrect answer choices include common errors students make.

TIME OUT Suggest that students work backward to reach the *finishing number.* Have students use a calculator to check their results.

COMMON ERROR

Some students may forget to move the decimal point in the dividend. For example, in Checkpoint 1 they would select choice a. You may wish to have these students write the division examples as follows:

8.63)3.7401
× 100 × 100
863)374.01

ASSIGNMENTS

Level 1	Odd 1–73, 75, 76, TO
Level 2	Even 2–76, 77, 77, TO
Level 3	Odd 21–65, 75–78, TO

FOLLOW-UP ACTIVITY

MENTAL MATH Have students use shortcuts for dividing certain decimals. Help students to discover the patterns.

To divide by 0.25, multiply by 4.

125 ÷ 0.25 = ■

THINK: 125 × 4 = 500
So, 125 ÷ 0.25 = 500.

To divide by 0.5, simply multiply by 2.

78 ÷ 0.5 = ■

THINK: 78 × 2 = 156
So, 78 ÷ 0.5 = 156.

Have students use the shortcuts to find the quotient.

1. 0.25)750 **2.** 0.25)1,000

3. 0.5)980 **4.** 0.5)1,364

(**1.** 3,000 **2.** 4,000 **3.** 1,960 **4.** 2,728)

SUPPLEMENTARY MATERIALS

TRP Practice, p. 35

TRP Reteaching, p. 22

TRP Transparency 18

First estimate each quotient by using compatible numbers, and then find the exact answer.

•**39.** 0.007)51.94	•**40.** 0.04)1.638	**41.** 0.8)3.07	**42.** 0.005)0.04998
43. 4.2)97.44	**44.** 0.15)3.75	**45.** 6.1)14.64	**46.** 0.038)1.7005
47. 53.8)28.245	**48.** 0.704)2.64	**49.** 3.05)65.88	**50.** 0.619)0.3095
51. 0.02)8.36	**52.** 0.06)0.0582	**53.** 0.008)0.00424	**54.** 0.04)0.29
55. 8.4)5.04	**56.** 0.76)0.2204	**57.** 6.11)2.444	**58.** 0.473)0.2838

39. 7,000; 7,420 **40.** 40; 40.95 **41.** 4; 3.8375 **42.** 10; 9.996
43. 20; 23.2 **44.** 20; 25 **45.** 2; 2.4 **46.** 40; 44.75
47. 0.6; 0.525 **48.** 4; 3.75 **49.** 20; 21.6 **50.** 0.5; 0.5
51. 400; 418 **52.** 1; 0.97 **53.** 0.5; 0.53 **54.** 7; 7.25
55. 0.6; 0.6 **56.** 0.3; 0.29 **57.** 0.4; 0.4 **58.** 0.5; 0.6

MIXED REVIEW

Find the answer.

•**59.** 4,899 + 6,559 = 11,458	•**60.** 5,238 + 9,784 = 15,022	**61.** 56,896 + 36,998 + 7,596 = 101,490	**62.** 101,237 + 987,786 + 89,867 = 1,178,890
63. 9,697 − 899 = 8,798	**64.** 78,213 − 8,989 = 69,224	**65.** 87,657 − 78,986 = 8,671	**66.** 785,433 − 699,850 = 85,583
67. 679 × 76 = 51,604	**68.** 188 × 86 = 16,168	**69.** 997 × 877 = 874,369	**70.** 5,785 × 749 = 4,332,965
71. 0.096 − 0.078 0.018	**72.** 45.998 + 4.31 50.308	**73.** 894.03 × 100 89,403	**74.** 73.906 ÷ 1,000 0.073906

Solve.

•**75.** Jan spends 4.9 hours a week working in the lab. She has logged 189.63 hours. How many weeks has she worked? 38.7 weeks

76. A box of test tubes weighs 12.8 lb. A shipment of test tubes arrives that weighs 307.2 lb. How many boxes of test tubes are inside? 24 boxes

77. Bart typed reports at 57.4 words per minute. By using a word processing program, his speed increased to 61.2 words per minute. What was the increase in words per minute? 3.8 words per minute

78. A printer in the lab prints 210 characters per second. It is in use from 7:30 A.M. until 4:30 P.M. How many characters does it print in a day? 6,804,000 characters

TIME OUT

Complete the following puzzle to find the "starting number."

Starting number **Finishing number**

■ + 2.4 = ■ ÷ 3 = 7.2 19.2; 21.6

PROBLEM Solving APPLICATION

3.12 CONSUMER: FINDING MILES PER GALLON

It is useful for a car owner to know how many miles a car can travel per gallon of gasoline. A car will usually travel more **miles per gallon (mpg)** in highway driving than in city driving.

In highway driving, a car travels 567 mi on 18 gal of gasoline. How many miles does it travel per gallon?

THINK: **NUMBER OF MILES** ÷ **NUMBER OF GALLONS USED** = **MPG**

567 ÷ 18 = 31.5

So, the car travels 31.5 mpg in highway driving.

Sometimes, the answer is rounded to the nearest whole number. In the situation above, the car travels about 32 mpg.

Complete. Remember to estimate whenever you use your calculator.

Number of miles (highway)	Number of gallons used	mpg	Number of miles (city)	Number of gallons used	mpg
• **1.** 520	20	■	• **2.** 280	16	■
3. 513	18	■	**4.** 270	20	■
5. 272	10	■	**6.** 129	8.6	■
7. 288	9.6	■	**8.** 183	12.2	■

Solve. **1.** 26; **2.** 17.5; **3.** 28.5; **4.** 13.5; **5.** 27.2; **6.** 15; **7.** 30; **8.** 15

• **9.** In city driving, Ricardo's car travels 189 mi on 14 gal of gasoline. How many miles does his car travel per gallon? 13.5 mpg

10. Ricardo drove 346 mi on the highway and used 16.4 gal of gasoline. What is the mpg? 21.1 mpg

11. A Sunburst KX travels 26 mpg in city driving. How far will it travel on 11.5 gal of gasoline? 299 mi

12. The Trotter travels 37 mpg on the highway and 28 mpg in the city. How much farther will it go on 12.5 gal on the highway than in the city? 112.5 mi

OBJECTIVE

- Solve problems that involve finding the number of miles per gallon (mpg) for a car.

TEACHING THE LESSON

WARM-UP Write the following exercises on the chalkboard or an overhead transparency.

Find the quotient. Divide to one decimal place if necessary.

1. 480 ÷ 30 (16)
2. 345 ÷ 15 (23)
3. 392 ÷ 16 (24.5)
4. 114 ÷ 7.6 (15)
5. 328 ÷ 10 (32.8)
6. 226 ÷ 11.3 (20)

INSTRUCTION Write *mpg* on the chalkboard. Ask the students to tell you what the abbreviation means. As you discuss the example, point out that miles per gallon is sometimes referred to as the fuel economy of a car. The greater the number of miles per gallon, the less the amount of fuel needed for a given trip.

You may wish to have students use calculators to help them solve the problems in this lesson. Notice that in Problem 11, the students must multiply to find the answer. Problem 12 is a multistep problem that requires subtraction and multiplication.

ASSIGNMENTS

Level 1	1–10
Level 2	1–12
Level 3	Even 2–12

SUPPLEMENTARY MATERIALS

TRP Practice, p. 36

TRP Transparency 18

OBJECTIVE

- Decide, given relevant information, where to shop for various items.

TEACHING THE LESSON

WARM-UP Write the following exercises on the chalkboard or an overhead transparency.

Compare.

1. \$159 ● \$239
2. \$647 ● \$674
3. \$24 ● \$24
4. \$393 ● \$339
5. \$281 ● \$182
6. \$444 ● \$3,333
7. \$46 ● \$60
8. \$7,848 ● \$7,839

(**1.** $<$ **2.** $<$ **3.** $=$ **4.** $>$ **5.** $>$ **6.** $<$ **7.** $<$ **8.** $>$)

INTRODUCTION Present the problem situation to the class. Discuss shopping in terms of choosing where to buy, and ask students what factors they think are important when selecting a store. List the factors on the chalkboard.

INSTRUCTION Have students study the data presented for each store. You may want to discuss the differences between shopping in town, at a large spectacular mall, and at a small mall.

Direct attention to the decision-making factors. Ask the students if there are any factors they would add to the list from those written on the chalkboard.

Have students complete Exercises 1–12 in the table in order to compare the three stores. Ask them to write conclusions that can be drawn from the data and discuss the appropriateness of these conclusions and any possible misinterpretations.

Assign Exercises 13–25 and have students work in small groups. Encourage them to explain and defend their answers as necessary. Remind them that their own preferences and situations will often influence their answers. Observe the groups as they work and interact as necessary to draw nonparticipating students into the discussion.

3.13 WHERE TO SHOP

DECISION MAKING

(Open-ended problem solving)

Most people have several stores to choose from when they go shopping. Choosing the right store can save time and money, but there are other benefits to be gained from shopping wisely.

PROBLEM

Carla and Tony have just moved from Ohio to Texas. They each need a new suit for the warmer climate that they can wear on job interviews. They also want to replace some of the items they sold before they moved. They made a list and checked department store ads for prices. They also made some notes about the stores and their locations.

	Downtown Department Store	Exclusive Department Store	Discount Department Store
Tony:			
Suit	\$ 159 – \$ 239	\$ 265 and up	\$75
Tie	\$ 12 – \$ 29	\$ 19 – \$ 79	2 for \$15
desklamp	\$ 39	\$ 39	not available
19" Television	\$ 297	\$ 499	\$ 249
Carla			
Suit	\$100 – \$250	\$265 – \$500	\$50
briefcase	\$150	\$275	not available
clock radio	\$ 79	\$ 79	\$ 13
Calculator	\$ 19	\$ 49	\$ 5
Location	2 mi away	30 mi away at Marvelous Mall	10 mi away at Mini-Mall
Hours	Mon.-Wed. 10 am - 6 pm Thurs. – Sat. 10 am - 9 pm	7 days 10 am – 9 pm	Mon.- Sat. 9 am - 9 pm
Parking	none	3,000 Cars	100 Cars
Public Transportation	yes	none	none
Number of stores in area	30	150	10
Selection	average	extensive	limited
Quality	Well-Known moderate-priced brands, some better labels	Well-known brands, designer fashions, latest trends	Some brand names, mostly store labels
Other	Free delivery, free alternations on all clothing	Free delivery, free alterations on men's suits	Free delivery for furniture and major appliances, no alterations

DECISION-MAKING FACTORS

Cost | Selection | Convenience | Quality

DECISION-MAKING COMPARISONS

Some answers may vary. Accept reasonable answers based on students' rationales.

Compare the three stores by completing the table.

	Downtown	Exclusive	Discount	
Cost	•1. ■	•2. ■	Inexpensive	Moderate; Expensive
Selection	Average	3. ■	4. ■	Extensive; Limited
Convenience				
Location	5. ■	Inconvenient	6. ■	Convenient; Inconvenient
Parking	7. ■	Extensive	8. ■	None; Limited
Shopping hours	9. ■	10. ■	Convenient	Convenient; Convenient
Quality	11. ■	Excellent	12. ■	Good; Fair

MAKING THE DECISIONS

Where should Tony and Carla shop:

•**13.** if cost were the only factor? Discount

•**14.** if selection were the only factor? Exclusive

15. if convenience were the only factor? Downtown or Exclusive

16. if quality were the only factor? Exclusive

17. How much could Carla save on a suit by shopping at Discount rather than Exclusive? $215–$450

18. If Carla is looking for a suit that is very fashionable, where should she shop? Exclusive

19. Tony wants a suit that will wear well and look good. If his total budget is $500, where should he shop for the suit? Downtown

20. If Carla and Tony didn't have a car, where would they have to shop? Downtown

21. If the television set is the same brand and model at all three stores, where should Tony buy it? Why? Discount; price difference is worth the trip to the store.

22. Would it make sense to buy the desk lamp and clock radio at Downtown or at Exclusive? Why? Downtown; same price, more convenient if there is a problem after the purchase.

23. The briefcases vary in price and quality. What other factors should Carla consider? Answers will vary. Examples: size, durability, style

24. Why do people, even if they can't afford to shop there, visit large malls with expensive stores? Answers will vary. Examples: free entertainment, pleasant surroundings, see latest fashions and trends.

25. If you were Tony or Carla, where would you buy each of the items on his or her list? Answers will vary.

ASSIGNMENTS

Level 1	1–25
Level 2	1–25
Level 3	1–25

FOLLOW-UP ACTIVITY

APPLICATION Ask students to collect newspaper ads and catalogs from various stores. Gather as much material as possible and keep it together for all students to use.

Explain to the students that they will be assigned the task of "shopping" for one complete outfit (excluding underwear). Their budget will be $100. Encourage them to keep in mind the factors they considered in the lesson and to add any others they choose to form their own list. After they select the clothing and the stores, have them present their selections and give the reasons for their decisions. If necessary, remind students of the convenience factor as it relates to the location of the stores they chose or the number of stores they used to put together their outfit. Of course, if these students have a lot of free time, convenience may not even be a factor.

As an extension, you may want to reassign the activity with a budget of $1,000 for one outfit.

SUPPLEMENTARY MATERIALS

TRP Practice, p. 37

TRP Applications, p. 4

TRP Transparency 19

OBJECTIVES

- Review vocabulary.
- Practice key chapter concepts and skills.

USING THE REVIEW

The Chapter Review is designed to help students prepare for taking the Chapter Test. The first section focuses on vocabulary. It requires that students select words to complete statements. The second section presents practice exercises of key mathematical skills. Under each directive there is a sample exercise with the answer.

Each item on the review is referenced to the page on which the topic is taught in the Pupil's Edition. You may wish to have students refer to these pages to help review any concepts or skills they have not yet mastered.

It is suggested that students work in small-sized heterogeneous cooperative learning groups. Some cooperative learning methods that may be used are as follows:

1. After each student has independently completed the entire Chapter Review, a discussion should follow within each group about the solutions to the practice exercises.
2. The group can complete the entire Chapter Review by working together to discuss the sample exercises and then to answer the practice exercises.

End the lesson with an entire class discussion in which any questions brought up in group discussions are presented and answered.

CHAPTER REVIEW

Vocabulary Choose the letter of the word that completes the statement.

1. When multiplying decimals, the number of decimal places in the product is equal to the number of decimal places in the ■. [74] b
   a. addends **b.** factors **c.** quotient **d.** sum
2. The first step in dividing by a decimal is to move the decimal point in the ■. [80] a
   a. divisor **b.** quotient **c.** dividend **d.** remainder

Skills Add. [62]

Sample: 2.87 + 6.453 = *9.323*

3. 48.66 + 20.05 — 68.71 **4.** 0.86 + 3.99 — 4.85 **5.** $17.84 + 3.08 + 9.02 — $29.94 **6.** 6.76 + 0.0025 + 100.3 — 107.0625

7. 0.632 + 1.987 2.619 **8.** 8.362 + 0.0119 8.3739 **9.** 0.3437 + 3.3347 3.6784

Subtract. [64]

Sample: 73.090 − 38.145 = *34.945*

10. 39.32 − 32.46 — 6.86 **11.** 48.06 − 7.333 — 40.727 **12.** 86.4 − 0.003 — 86.397 **13.** 12.345 − 11.9834 — 0.3616

14. 1.0934 − 0.679 0.4144 **15.** 20.004 − 11.441 8.563 **16.** 1.0348 − 0.9697 0.0651

Multiply or divide by powers of 10. [68] **17.** 934.7 **18.** 742,000 **19.** 14,368.1 **20.** 1.6743 **21.** 4.5098 **22.** 8.99763

Sample: 82.5 × 100 = *8,250*; 77.82 ÷ 10 = *7.782*

17. 9.347 × 100 **18.** 74.2 × 10,000 **19.** 14.3681 × 1,000
20. 1,674.3 ÷ 1,000 **21.** 45,098 ÷ 10,000 **22.** 899,763 ÷ 100,000

Multiply. [74]

Sample: 8.76 × 0.08 = *0.7008*

23. 9.89 × 0.32 — 3.1648 **24.** 14.64 × 1.1 — 16.104 **25.** 128.3 × 0.89 — 114.187 **26.** 100.06 × 1.066 — 106.66396

27. 1,660.2 × 0.75 1,245.15 **28.** 14,000.3 × 3.03 42,420.909 **29.** 0.6897 × 10.3 7.10391

Divide. [78, 80]

Sample: 8.2)36.9 = *4.5*

30. 118.4 ÷ 4 29.6 **31.** 108.8 ÷ 8 13.6 **32.** 91.5 ÷ 5 18.3

33. 0.476)0.119 0.25 **34.** 7.32)4.8312 0.66 **35.** 1.03)$15.45 $15.00 **36.** 2.002)10.8108 5.4

Chapter Test

Estimate by rounding to the greatest place.

1. 16.83 + 5.07 = 25

2. 304.87 + 69.14 = 370

3. 48.05 − 26.93 = 20

4. 300.62 − 46.38 = 250

5. 42.6 × 2.1 80

6. 0.409 × 2.006 0.8

7. 4.68)19.83 4

8. 0.003)0.91 300

Compute.

9. 27.65 + 18.09 = 45.74

10. 2.09 + 86.5 = 88.59

11. 7.395 − 2.086 = 5.309

12. 93.7 − 6.082 = 87.618

13. 75.06 + 3.94 79

14. 8.052 + 13.09 21.142

15. 4.0086 + 0.029 4.0376

16. 403.8 − 56.7 347.1

17. \$400 − \$67.29 \$332.71

18. 238 − 0.923 237.077

19. 63.74 × 1,000 63,740

20. 0.809 × 10,000 8,090

21. 5.2064 × 10,000,000 52,064,000

22. 18.951 ÷ 100 0.18951

23. 0.031 ÷ 1,000 0.000031

24. 620 ÷ 100,000 0.0062

25. 8.3 × 7 = 58.1

26. 95 × 0.6 = 57

27. 4.7 × 1.5 = 7.05

28. 3.0617 × 0.005 = 0.0153085

29. 64.7 × 1.8 = 116.46

30. 0.0049 × 6.3 = 0.03087

31. 0.0671 × 0.48 = 0.032208

32. 5.009 × 0.082 = 0.410738

33. 0.08)0.376 4.7

34. 6.3)15.12 2.4

35. 0.47)0.02397 0.051

36. 0.051)0.0001836 0.0036

Divide. Round to the nearest hundredth.

37. 5.109 ÷ 0.004 1277.25

38. 72.08 ÷ 0.19 379.37

39. 0.01362 ÷ 0.54 0.03

Solve.

40. Juan collects baseball and football cards. He has 44 more baseball cards than football cards. If he has 112 cards, how many are football cards? 34

41. Geno scored 38 more points at basketball than Bill. Together they scored 96 points. How many points did Bill score? 29 points

42. Sarah drove 378 mi on the highway. She used 14 gal of gasoline. What was the mpg? 27 mpg

43. Victoria's motorbike gets 48 mpg. How many miles can she travel on 3.8 gal of gasoline? 182.4 mi

Objective

- Evaluate achievement of the chapter objectives.

Using the Test

The Chapter Test may be used as a posttest to evaluate student achievement. However, you may wish to use the Chapter Posttest offered in the Teacher's Resource Package or to design your own chapter test. If this page is not used as a test, you may wish to assign it as additional review or practice.

The test items are correlated to the chapter objectives in the table below.

Chapter objectives	Test items
A. Estimate with decimals.	1–8
B. Add or subtract decimals.	9–18
C. Multiply or divide decimals by powers of 10.	19–24
D. Multiply or divide decimals.	25–39
E. Guess and check to solve problems.	40–41
F. Apply computational skills in real-life situations.	42–43

Supplementary Materials

TRP Ch. 3 Posttest Form A, pp. 1–2

TRP Ch. 3 Posttest Form B, pp. 1–2

OBJECTIVES

- Review and maintain previously taught concepts and skills.
- Practice taking tests in a multiple-choice format.

USING THE REVIEW

Assign the Cumulative Review to all students. Provide students with an answer sheet to record their answers.

Each Cumulative Review gives students an opportunity to practice taking tests that are written in multiple-choice format.

Provide appropriate remedial help for students having difficulty with any of the skills and concepts on these pages.

CUMULATIVE REVIEW

Choose the letter of the correct answer.

1. What is the standard numeral for 50,000 + 400 + 9? c

a. 549 **b.** 5,409 **c.** 50,409 **d.** none of these

2. What is the standard numeral for three hundred nine ten-thousandths? d

a. 309.0001 **b.** 0.0390 **c.** 0.3090 **d.** none of these

3. What is 4.5068 rounded to the nearest hundredth? c

a. 4.507 **b.** 4.506 **c.** 4.51 **d.** none of these

4. Which of the following decimals is greater than 0.0784? c

a. 0.00999 **b.** 0.0613 **c.** 0.08 **d.** none of these

5. What is the total amount of money earned on regular and overtime pay? b

a. net pay **b.** gross pay **c.** deduction **d.** none of these

Select the best estimated answer.

6. 43.6 − 2.79	c	**a.** 15	**b.** 31	**c.** 37	**d.** 45
7. 9.6 × 50.17	b	**a.** 540	**b.** 500	**c.** 400	**d.** 60
8. 22.5 + 9.45 + 10.6	b	**a.** 30	**b.** 40	**c.** 110	**d.** 120
9. 3.53 ÷ 0.09	c	**a.** 4	**b.** 5	**c.** 40	**d.** 50

Compute.

10. 45 + 219 + 804 a

a. 1,068 **b.** 1,078 **c.** 2,068 **d.** none of these

11. 45,098 − 3,789 c

a. 42,209 **b.** 42,309 **c.** 41,309 **d.** none of these

12. 43.6 × 5.7 d

a. 24.852 **b.** 237.42 **c.** 228.42 **d.** none of these

13. $6.7\overline{)13.668}$ b

a. 0.024 **b.** 2.04 **c.** 2.4 **d.** none of these

14. 80,000 ÷ 100 b

a. 80 **b.** 800 **c.** 8,000 **d.** none of these

Solve.

15. A train goes from Miami to Jacksonville, a distance of 652 km; from Jacksonville to Savannah, 225 km; and from Savannah to Richmond, 805 km. What is the total distance by train from Miami to Richmond? b

a. 2,452 km **b.** 1,682 km **c.** 982 km **d.** none of these

16. A stockholder owns 244.3 shares of stock in a mutual fund. If each share is worth $2.15, what is the value of her stock to the nearest cent? a

a. $525.25 **b.** $425.25 **c.** $200.25 **d.** none of these

17. Pearl's gross earnings for the month are $1,575.00. A total of $439.63 for taxes is deducted. An additional $6.32 is deducted for insurance. What is Pearl's net pay? b

a. $1,135.37 **b.** $1,129.05 **c.** $2,020.95 **d.** none of these

18. Just off the coast of California is an undersea mountain called San Juan Seamount. How far below the surface of the water is the top of the mountain if the ocean floor depth is 3,658 m and the mountain rises 3,105 m? d

a. 7,863 m **b.** 3,553 m **c.** 653 m **d.** none of these

19. At $1.78 a dozen, how many dozen oranges can you buy for $8.90? d

a. 10 dozen **b.** 8 dozen **c.** 6 dozen **d.** none of these

20. A new apartment complex has 542 units. Each unit is furnished with a refrigerator that costs $329. What is the total cost of the refrigerators? a

a. $178,318 **b.** $202,680 **c.** $168,308 **d.** none of these

THINKING ABOUT MATH See TE side column.

1. Why will front-end estimation for addition always result in an underestimate?

2. If you wanted to estimate the quotient of 7.45 ÷ 2.9, what compatible numbers would you use?

3. Look at the following number sentence: 976 + 876 + 776 = 4,087. Explain how you can tell by looking that the sum is incorrect.

4. The pointer on an electric meter is between 9 and 0. Which number would you read? Why?

THINKING ABOUT MATH

The questions in this section can be used in discussions with individual students or with small groups. Guide students in conveying their ideas clearly and precisely.

Listed below are expected student answers. However, accept any reasonable answer.

1. In front-end estimation, only the lead digits are added. Because the digits that follow are not taken into account, the estimate will always be less than the actual sum.
2. Use the compatible numbers 6 and 3. 6 ÷ 3 = 2; 7.45 ÷ 2.9 = 2.57.
3. The sum is incorrect because there are only three addends, each of which is less than 1,000. So, the sum is less than 3,000.
4. 9 is the final number on an electric meter, and 0 begins the cycle. When the pointer is between two numbers, the lower number is read.

SUPPLEMENTARY MATERIALS

TRP Cum. Test, Ch. 1–3, pp. 1–4

OBJECTIVE

- Test the prerequisite skills needed to learn the concepts developed in Chapter 4.

USING THE TEST

The Pre-Skills Test is designed to diagnose students' strengths and weaknesses on prerequisite skills necessary to study the mathematics in Chapter 4.

Assign the Pre-Skills Test. Allow students to work together in pairs or small groups. Group members should help those who demonstrate a misunderstanding of a concept or a weakness in a skill.

The items in the test are referenced to the pages on which the topics are taught in the Pupil's Edition. You may wish to have students refer to these pages for review.

The following table correlates the items on the Pre-Skills Test with the prerequisite skill and the lesson(s) in the chapter for which it is needed.

Item(s)	Prerequisite skill	Lesson(s)
1–4	Add whole and decimal numbers.	4.1, 4.3, 4.5
5–12	Subtract whole and decimal numbers.	4.3, 4.4, 4.5
13–20	Multiply whole and decimal numbers.	4.5, 4.11
21–28	Divide whole and decimal numbers.	4.5
29–30	Order numbers.	4.1, 4.4
31–42	Round numbers.	4.5, 4.8, 4.9, 4.11

Pre-Skills Test

Add or subtract. [30, 32, 62, 64]

1. 7 6 8 5 3 +6 35	**2.** 3.2 4.9 2.7 3.5 3.1 +4.2 21.6	**3.** 382 387 376 372 383 +385 2,285	**4.** 72.1 71.9 72.6 72.2 71.9 +72.3 433
5. 48 − 6 42	**6.** 861 − 29 832	**7.** 540 −176 364	**8.** 9,863 −2,794 7,069
9. 8,645 −1,295 7,350	**10.** 80,000 −69,257 10,743	**11.** 60.1 −27.8 32.3	**12.** 39.6 −15.9 23.7

Multiply or divide. [36, 40, 46, 68, 78]

13. 37 ×10 370	**14.** 2.7 × 10 27	**15.** 8.75 × 10 87.5	**16.** 14.87 × 10 148.7
17. 54 × 3 162	**18.** 74 × 7 518	**19.** 17 ×15 255	**20.** 36 ×21 756
21. 10)6,900 690	**22.** 10)581 58.1	**23.** 10)5.1 0.51	**24.** 10)4.83 0.483
25. 6)486 81	**26.** 8)2,056 257	**27.** 7)48.3 6.9	**28.** 4)35.16 8.79

Order from least to greatest. [8]

29. 421, 514, 96, 105, 557
96, 105, 421, 514, 557

30. 15.47, 59.65, 58.7, 454.5
15.47, 58.7, 59.65, 454.5

Round to the nearest hundred. [16]

31. 705 700 **32.** 428 400 **33.** 547 500 **34.** 8,492 8,500

Round to the nearest thousand. [16]

35. 6,027 6,000 **36.** 2,490 2,000 **37.** 8,664 9,000 **38.** 4,583 5,000

Round to the nearest hundred thousand. [16]

39. 457,658 500,000 **40.** 963,872 1,000,000 **41.** 5,672,541 5,700,000 **42.** 28,678,994 28,700,000

OBJECTIVE

- Formulate problems using text and data.

USING THE CHAPTER OPENER

Begin the discussion of the question by allowing students to speculate on the answer, encouraging them to support their answers with some explanation. For example, a student may say that more people are going to college because there are more algebra and geometry courses given than general mathematics. Lead them to see that this is speculation and they are assuming things that may not be true. Ask how they could get a more accurate answer to the question. When someone suggests taking a survey, discuss ways in which this could be done.

COOPERATIVE LEARNING

Assign groups for the project and explain that they are to find the answer to the question using some type of survey. Each group is to decide how many students to survey at each grade level based on the total school population. Then they should write their question or questions on a form they design and gather the information. They should find that there will be students with plans other than the two given. Therefore, they should make provisions for recording all other answers in some way.

When the surveys have been completed, each group is to organize the data in a manner of their choosing. In doing this they will have to determine how best to present the "other" answers to the question. For example, if one or two students name a particular goal, it probably should be listed as "other."

SUPPLEMENTARY MATERIALS

TRP Group Projects, p. 4

OBJECTIVE

- Construct frequency tables and histograms.

TEACHING THE LESSON

WARM-UP Write the following sets of data on the chalkboard or an overhead transparency.

Order from least to greatest.

1. 20; 26; 23; 21; 29; 27; 25; 28

2. 105; 100; 108; 107; 102; 104

3. 43; 27; 35; 29; 38; 41; 34; 32

4. 12.6; 8.2; 17.5; 8.7; 12.4; 17.9

(**1.** 20; 21; 23; 25; 26; 27; 28; 29
2. 100; 102; 104; 105; 107; 108
3. 27; 29; 32; 34; 35; 38; 41; 43
4. 8.2; 8.7; 12.4; 12.6; 17.5; 17.9)

INTRODUCTION On the chalkboard or an overhead transparency, draw a table like the one shown below. List all the letters from A–Z.

Letter	Tally	Frequency
A		
B		
C		
⋮		

Make a tally for the first letter of each student's name. Count the marks and record the frequency.

INSTRUCTION Use Example 1 to establish the steps for making a frequency table. Elicit from students the reason for ordering the data (to simplify interpretation). Stress that tally marks are an efficient way to keep count. The / across a group of |||| easily identifies a group of 5. Encourage students to check that the frequency total is the same as the number of data items.

4.1 COLLECTING DATA

Clark Watson has a part-time job at the West Side Recreation Center. His boss wants to know the ages of the teenagers in the center's judo class. Clark records the ages of everyone in the class. Here is the information, or **data.**

AGES OF STUDENTS IN THE JUDO CLASS

14	15	17	15	16
13	15	14	17	18
14	16	15	16	13
17	18	14	15	16

Clark organizes the data into a **frequency table** to make it easier to understand.

EXAMPLE 1 Make a frequency table.

1. Arrange the data in descending order.
2. Make a tally for each age.
3. Count the tallies. Write the frequencies.
4. Total all the frequencies.

AGES: JUDO CLASS

Age	Tally	Frequency
18	II	2
17	III	3
16	IIII	4
15	~~IIII~~	5
14	IIII	4
13	II	2
Total		20

When the data contains many different numbers, group the data into equal intervals.

EXAMPLE 2 Make a grouped-frequency table for the data.

1. Group the data in intervals.
2. Total the frequencies for each interval.
3. Total all the frequencies.

Age	Frequency
17–18	5
15–16	9
13–14	6
Total	20

You can show the data from a grouped-frequency table in a **histogram.**

EXAMPLE 3

NUMBER OF DAYS BETWEEN VISITS TO THE CENTER

Number of days	0–4	5–9	10–14	15–19
Frequency	3	8	6	4

Make a histogram for the data.

1. Draw a horizontal axis and a vertical axis. Label the horizontal axis "Number of Days." Label the vertical axis "Frequency."
2. Mark the lowest value of each interval on the horizontal axis.
3. Make a bar as wide as the interval and as high as the frequency.
4. Write a title for the histogram.

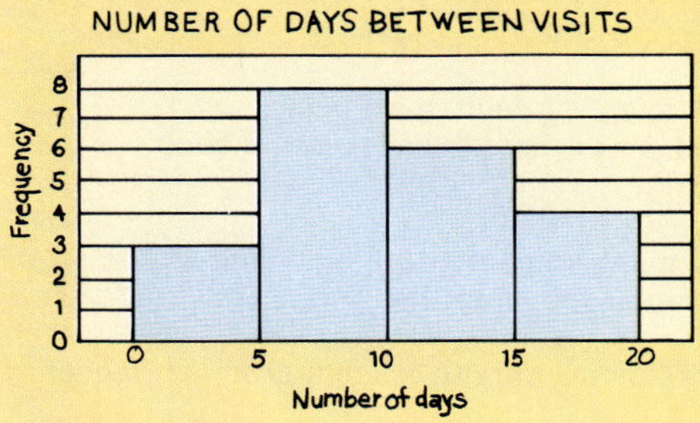

PRACTICE EXERCISES

Use the frequency table at the right to answer Exercises 1–4.

DISTANCE FROM THE RECREATION CENTER

Number of blocks	Tally	Frequency
5	卌 卌 卌 III	18
4	卌 卌 III	13
3	卌 卌 卌 IIII	19
2	卌 卌 卌 卌 II	22
1	卌 卌 卌 I	16

• **1.** How many members live two blocks from the recreation center? 22

• **2.** How many members live four or more blocks from the center? 31

3. How many blocks from the center does the greatest number of members live? 2

4. How many members were included in the survey? 88

Use the following data to make a frequency table.

5. Clark took a survey of 15 teenagers to see how many times a month they visited the recreation center.
Here is the data:

15, 13, 12, 14, 11, 16, 15, 12,
10, 14, 17, 15, 16, 11, 13

Check students' tables.

Example 2 shows how to group data for a frequency table. Discuss how to set up equal intervals.

Example 3 explains how to use grouped-frequency data to make a histogram. Check that students make the bar width correctly (must have equal distance on either side of the middle number). Stress that adjacent bars must touch. Make sure students understand how the last value of 20 on the horizontal scale was determined. Note that it is the lowest value of the next interval, 20–24.

ASSIGNMENTS

Level 1	1–11, 13–27
Level 2	1–11, 13–27
Level 3	1–12, Odd 13–27

FOLLOW-UP ACTIVITY

ENRICHMENT Demonstrate the stem-and-leaf plot as another way of displaying the following data:

28, 32, 28, 23, 49, 21, 35, 47,
36, 24, 32, 37, 44, 28, 41, 35

List the tens digits in ascending order in a vertical column. This forms the stem.

2
3
4

Draw a vertical line. Then write each ones digit across from its corresponding tens digit. These digits form the leaves.

2	8, 8, 3, 1, 4, 8
3	2, 5, 6, 2, 7, 5
4	9, 7, 4, 1

Arrange the leaves in ascending order for ease of interpretation.

2	1, 3, 4, 8, 8, 8
3	2, 2, 5, 5, 6, 7
4	1, 4, 7, 9

Have students complete Exercise 10 using a stem-leaf plot. Discuss the advantages of this method.

SUPPLEMENTARY MATERIALS

TRP Practice, p. 38

TRP Reteaching, p. 23

TRP Transparency 19

Use the grouped-frequency table at the right to answer Exercises 6–9.

MEMBERSHIP: RECREATION CENTER

Number of years	Frequency
0–1	37
2–3	58
4–5	52
6–7	21
8–9	3

• **6.** For how many years has the greatest number of teenagers belonged to the center? 2–3

7. For how many years has the fewest number of teenagers belonged to the center? 8–9

8. How many students have been members of the center for less than 2 years? 37

9. How many students have belonged to the center for more than 3 years? 76

Use the following data to make a grouped-frequency table.

10. Clark asked 30 students how many dances they attended in one year at the West Side Recreation Center.
Here are the results:

15, 20, 18, 12, 10, 17, 24, 21, 18, 8, 12, 24, 18, 16, 9,
15, 21, 24, 13, 17, 22, 7, 19, 23, 16, 14, 10, 8, 9, 22

Check students' tables.

11. Use the following data to make a histogram.

AGE OF ATHLETES: WEST SIDE RECREATION CENTER

Age	5–7	8–10	11–13	14–16	17–19
Frequency	12	18	27	36	24

Check students' histograms.

12. The West Side Recreation Club held a raffle to raise money. Fifteen club members sold raffle tickets. Here is the number of tickets each member sold. Use the data to make a histogram.

4, 8, 14, 26, 7, 9, 16, 30, 2, 24, 15, 5, 28, 11, 44

Check students' histograms.

MIXED REVIEW

Multiply or divide.

• **13.** 28×1.3 36.4

• **14.** 4.7×2.6 12.22

15. 3.01×0.02 0.0602

16. 0.003×0.09 0.00027

17. 83.04×0.75 62.28

18. $5\overline{)23.4}$ 4.68

19. $0.48\overline{)0.8736}$ 1.82

20. $8\overline{)6}$ 0.75

21. $6.7\overline{)2.3316}$ 0.348

22. $0.26\overline{)185.12}$ 712

23. 1.7×0.55 0.935

24. $0.9\overline{)3.87}$ 4.3

25. $0.53\overline{)0.02756}$ 0.052

26. 843×0.4 337.2

27. $4.8\overline{)1.728}$ 0.36

PROBLEM Solving APPLICATION

4.2 CONSUMER: FINDING THE PRICE OF A NEW CAR

The prices of new cars can vary. There is a base price which is the dealer's price for the car without any optional equipment. If a car includes extra equipment, or if you want certain types of optional equipment, there are extra charges for the equipment.

OPTIONAL EQUIPMENT

Air conditioning (manual)	$795	Electronic instrument cluster	$357
Air conditioning (deluxe)	$956	Power door locks	$195
Rear window defroster	$149	Illuminated entry system	$ 87
Automatic transmission	$482	Extended range fuel tank	$ 54

The table above lists the prices of some optional equipment. You are considering buying a car that has a base price of $9,245. Automatic transmission and power steering/brakes are standard equipment in the car. You want this optional equipment: air-conditioning (manual), rear window defroster, and power door locks. What is the price of the car?

THINK: Add the base price and the prices for the optional equipment.

$9,245 + $795 + $149 + $195 = $10,384

So, the price of your car with the optional equipment is $10,384.

Find the price of the car with the optional equipment.
Remember to estimate whenever you use your calculator.

	Car	Base Price	Optional Equipment
1.	Panther	$10,850	Air-conditioning (deluxe), rear window defroster, extended range fuel tank $12,009
2.	Eagle	$11,930	Electronic instrument cluster, extended range fuel tank $12,341
3.	Dynamo	$ 9,770	Air-conditioning (deluxe), electronic instrument cluster, power door locks, illuminated entry system $11,365
4.	TMX	$ 5,860	Automatic transmission, air-conditioning (manual), rear window defroster $7,286
5.	Meteor	$12,580	Electronic instrument cluster, rear window defroster, illuminated entry system $13,173

OBJECTIVE

- Solve problems that involve reading and interpreting data from a price list.

TEACHING THE LESSON

WARM-UP Write the following exercises on the chalkboard or an overhead transparency.

$875 $429 $942 $859 $635

1. Which amount above is greatest in value? ($942)
2. Which amount is least in value? ($429)
3. Add the three greatest amounts. ($2,676)

INTRODUCTION You may wish to display a large magazine or newspaper advertisement for a new car. Discuss the advertised base price with the students. Will a buyer actually pay just the base price? Encourage students to give their opinions.

INSTRUCTION As you discuss the introductory example, have the students refer to the price list to locate each of the three specified optional equipment items and identify the price of that item.

You may wish to have students use calculators to help them solve the problems in this lesson.

ASSIGNMENTS

Level 1	1–4
Level 2	1–5
Level 3	1–5

SUPPLEMENTARY MATERIALS

TRP Practice, p. 39

TRP Applications, p. 5

TRP Transparency 20

OBJECTIVE

- Read and interpret data from bar and line graphs.

TEACHING THE LESSON

INTRODUCTION Present the following on the chalkboard or an overhead transparency.

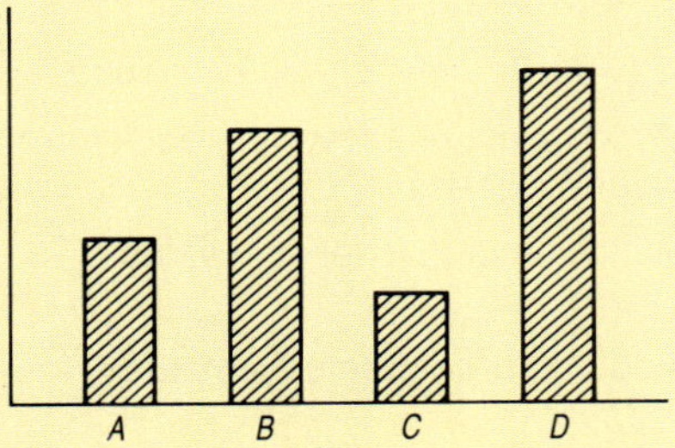

Ask students what types of questions can be answered using the graph (questions of comparison). Ask what information is needed to make the graph more useful (title, scale).

INSTRUCTION Point out that in Example 1 all the components of a bar graph are present, enabling students to read it. Discuss the visual ease of making comparisons on a bar graph.

In Example 2, make sure that students understand how to use the different keys on a double-bar graph. Emphasize that color and shading are the two most common ways used to differentiate the two groups being compared. Point out that the first bar always represents the 6–11 age group.

In Example 3, stress that the jagged section of the vertical scale simply means that the numbers from 0 to 350 are not written. Students may wonder why this graph is called a broken-line graph when there are no "breaks" in the line. The term is used to distinguish it from a straight line.

4.3 READING AND INTERPRETING DATA

You can see comparisons on a **bar graph.**

EXAMPLE 1 The horizontal bar graph at the right shows the costs of some breakfast foods. Which food is the most expensive?

To find the most expensive food, look for the longest bar.

Cereal costs the most.

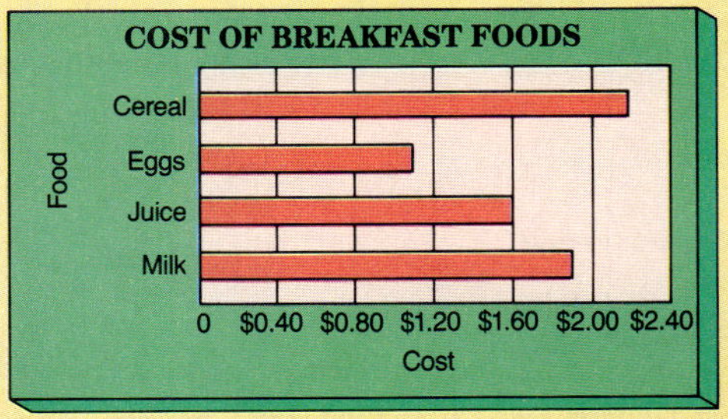

You can compare two sets of data on a **double-bar graph.**

EXAMPLE 2 The vertical double-bar graph at the right shows the favorite brands of cereal for two different age groups. Which cereal is equally popular with both groups?

To find the cereal that is equally liked by both groups, look for bars that have the same height.

Corice is equally liked by both age groups.

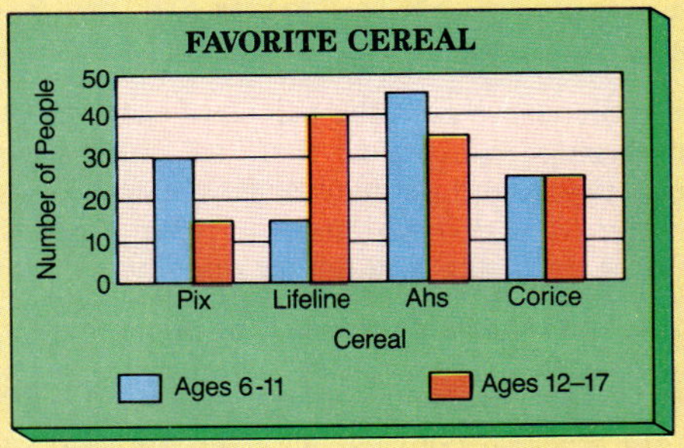

You can see changes over time on a **broken-line graph.**

EXAMPLE 3 The broken-line graph at the right shows egg production in Maryland from 1980 to 1984. Did egg production increase or decrease between 1980 and 1981?

Look at the line segment between 1980 and 1981. The line segment rises.

Egg production increased between 1980 and 1981.

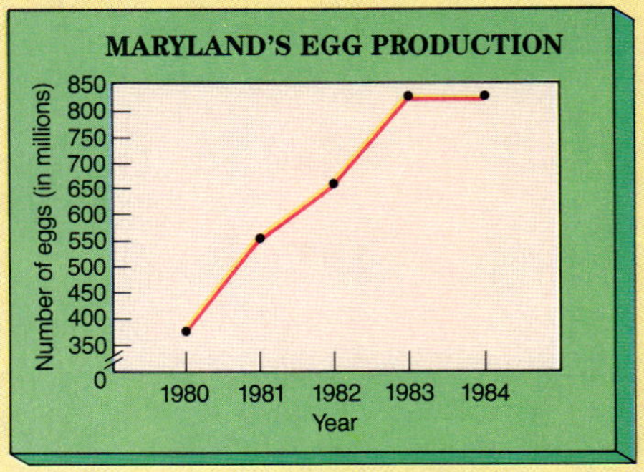

FOR DISCUSSION See TE side column.

What are the advantages of showing data on a graph instead of a table? Think of a situation in which showing data on a graph would be more effective than using a table.

PRACTICE EXERCISES See Extra Practice, page 434.

Use the graph in Example 1 to answer Exercises 1–8.

• **1.** Which food costs the least? eggs

• **2.** Which is more expensive, juice or milk? milk

3. About how much does a box of cereal cost? $2.20

4. About how much more does milk cost than eggs? $0.80

5. If you buy cereal and eggs, about how much would you spend? $3.30

6. Which item costs about twice the price of eggs? cereal

7. How much would all the items cost? $6.80

8. If you pay for all the items with a $10 bill, how much change would you receive? $3.20

Use the graph in Example 2 to answer Exercises 9–18.

• **9.** How many people in the 6–11 age group like Pix? 30

• **10.** How many people in the 12–17 age group like Ahs? 35

11. Which age group prefers Lifeline? 12–17

12. About how many more 6- to 11-year-olds prefer Ahs than 12- to 17-year-olds? 10

13. How many people chose Corice as their favorite cereal? 50

14. About how many more 12- to 17-year-olds prefer Lifeline to Pix? 25

15. How many more people liked Lifeline than Corice? 5

16. How many people chose Ahs as their favorite cereal? 80

17. How many more 6- to 11-year-olds prefer Ahs to Pix? 15

18. How many people are included in the survey? 230

Use the graph in Example 3 to answer Exercises 19–20.

19. Did egg production increase or decrease between 1982 and 1983? increased

20. What happened to egg production between 1983 and 1984? no change

21. Use an almanac to find out if egg production in Maryland increased, decreased, or stayed the same after 1984. increased

FOR DISCUSSION Students should explain that a graph provides visual facility in reading and interpreting data. Magazines, papers, and presentations are some situations in which graphs are used to make a visual impact.

ASSIGNMENTS

Level 1	1–6, 9–16, 19–20
Level 2	1–20
Level 3	Even 2–8, 9–21

FOLLOW-UP ACTIVITY

COOPERATIVE LEARNING Separate the class into groups of 3–5 students. Provide each group with a bar graph and a broken-line graph from magazines or newspapers. Have each group write three or four questions based on each graph. On a separate sheet, have them provide the answers. Have the groups exchange papers and answer the questions.

SUPPLEMENTARY MATERIALS

TRP Practice, p. 40

TRP Reteaching, p. 24

TRP Transparency 20

OBJECTIVE

- Find the mean.

TEACHING THE LESSON

WARM-UP Write the following sets of data on the chalkboard or an overhead transparency.

Add, then divide. Round your answer to the nearest tenth.

1. $(6 + 4 + 5) \div 3$

2. $(7 + 13 + 6 + 3) \div 4$

3. $(5.6 + 2.75 + 3.65 + 1.5) \div 4$

4. $(9 + 2.2 + 7.685 + 0.79 + 1) \div 5$

(**1.** 5 **2.** 7.3 **3.** 3.4 **4.** 4.1)

INTRODUCTION Discuss the definition of *average.* As an adjective, it means normal or typical. As a noun, it means a number that typifies a set of numbers. An arithmetic mean (usually referred to as simply *mean*) is one "average" for a set of numbers. It is found the same way an arithmetic average is found. *Average* usually refers to the mean but the median and the mode are also averages.

INSTRUCTION Remind students that the procedure for finding the mean is the same as that for finding the average. Use Example 1 to review the process.

Example 2 reviews computation with decimals and involves rounding the answer to a specific place.

FOR DISCUSSION Students should explain that with an average breakdown of six times a month, it would seem wise to invest in a new milkshake machine.

4.4 MEAN

Randy Johnson manages a fast-food restaurant called Smiley's. The head office wants to know the average length of employment at Randy's restaurant. Randy compares the starting date and ending date for every employee who left during the past 6 months. Here is the number of months that each of those employees worked at Smiley's.

6, 14, 8, 3, 12, 9, 15, 18, 14

To find the average length of employment, Randy finds the mean of this data. The **mean** of a set of data is the same as the average.

EXAMPLE 1 Find the mean of the number of months worked by the former employees.

1. Add. $6 + 14 + 8 + 3 + 12 + 9 + 15 + 18 + 14 = 99$

2. Divide by the number of addends. $\frac{99}{9} = 11$

So, the average length of employment at Randy's restaurant was 11 months.

When computing with decimals, you may need to round the mean to the nearest tenth.

EXAMPLE 2 Find the mean to the nearest tenth: 8.8, 17.6, 11.4, 18.9, 7.3, 10.7

1. Add. $8.8 + 17.6 + 11.4 + 18.9 + 7.3 + 10.7 = 74.7$

2. Divide by the number of addends. $\frac{74.7}{6} = 12.45$

3. Round the answer to the nearest tenth. 12.45 rounds to 12.5.

The mean is 12.5 to the nearest tenth.

FOR DISCUSSION

See TE side column.

Find the average number of times that the milk shake machine has broken down during the past six months.

Jan.—4 breakdowns; Feb.—1 breakdown;
Mar.—4 breakdowns; Apr.—6 breakdowns;
May—9 breakdowns; June—13 breakdowns

If you were the manager of Smiley's, would you buy a new milkshake machine? Explain.

PRACTICE EXERCISES See Extra Practice, page 434.

Find the mean. Round to the nearest tenth.

• **1.** 4, 6, 8, 10 7

• **2.** 19, 15, 26, 38, 32 26

3. 63, 72, 58, 67, 41, 77 63

4. 476, 938, 217, 557 547

5. 702, 149, 678, 350, 836 543

6. 4, 795; 2,618; 997; 1,008; 6,242 3,132

7. 3.5, 4.1, 6.7, 7.9 5.6

8. 24.6, 57.3, 44.1, 39.8, 64.5 46.1

9. 1.8, 6.2, 9.3, 4.5, 10.2, 7.7, 3.6 6.2

10. 5.46, 12.7, 17.36, 8.4, 9.03 10.6

Find the missing data.

11. Data: 10, 6, 9, 14, ■ 11
Mean: 10

12. Data: 27.3, 14.6, 32.7, ■ 24.2
Mean: 24.7

MIXED REVIEW

Add or subtract.

• **13.** 4.56 0.02 + 114.3 118.88	• **14.** 0.006 17.1 + 5.27 22.376	**15.** 92.5 0.007 + 4.93 97.437	**16.** 467 3.21 18.6 + 0.005 488.815	**17.** 9.26 15.3 0.004 + 891.5 916.064
18. 72.6 − 4.8 67.8	**19.** 9.03 − 0.7 8.33	**20.** 45.1 − 28.63 16.47	**21.** 884.76 − 235.08 649.68	**22.** 9.02 − 0.005 9.015
23. 4.26 19.073 + 0.8 24.133	**24.** 71.3 − 68.51 2.79	**25.** 19.48 − 0.807 18.673	**26.** 5.3 + 0.857 6.157	**27.** 80 − 5.27 74.73

Solve.

Randy asks 20 customers how often they visit the restaurant each month. Here are the results.

2, 5, 1, 3, 0, 4, 8, 6, 5, 2,
1, 9, 2, 4, 8, 7, 3, 0, 5, 5

28. Make a grouped-frequency table for this data. Check students' tables.

29. Find the mean number of times these customers visit the restaurant in 1 month. 4

TIME OUT Take a survey of 25 people to find out how many times they go to a fast-food restaurant in a month. Find the mean. Compare it with those of your classmates. Are the results about the same?

TIME OUT Have students discuss the results of their surveys. Elicit possible reasons for the differences in their results.

COMMON ERROR

Students may make careless errors in adding or dividing. Encourage them to double-check their computation.

ASSIGNMENTS

Level 1	1–9, Odd 13–27, TO
Level 2	Even 2–28, TO
Level 3	7–12, Odd 13–27, 28, 29, TO

FOLLOW-UP ACTIVITY

CALCULATOR From 9:00 A.M. to 9:15 A.M., Smiley's has 16 customers who make purchases in the following amounts:

$1.59, $2.21, $0.89, $3.44,
$4.00, $2.61, $3.27, $1.49,
$5.29, $2.12, $2.47, $3.06,
$1.75, $2.57, $0.89, $3.19

Have the students use their calculators to find the average amount spent by the customers. Remind them to round their answers to the nearest cent. ($2.55)

SUPPLEMENTARY MATERIALS

TRP Practice, p. 41

TRP Reteaching, p. 25

TRP Transparency 21

OBJECTIVE

- Find the range, median, and mode.

TEACHING THE LESSON

WARM-UP Write the following exercises on the chalkboard or an overhead transparency.

Find the difference.

1. 83 − 21 **2.** 94 − 26

3. 72 − 37

Order from least to greatest.

4. 29; 16; 24; 28; 19; 21; 18

5. 53; 68; 65; 62; 57; 52; 64

6. 12; 19; 34; 22; 13; 27; 26

(**1.** 62 **2.** 68 **3.** 35
4. 16; 18; 19; 21; 24; 28; 29
5. 52; 53; 57; 62; 64; 65; 68
6. 12; 13; 19; 22; 26; 27; 34)

INTRODUCTION Have students study the following sets of data and the corresponding means.

Data	Mean
3, 24, 6, 8, 9, 8, 7, 7	9
13, 38, 31, 92, 35, 37	41
5, 44, 17, 96, 23, 79	44

Ask if the mean is really a representative number for the data. Elicit that the spread between the least and the greatest numbers is very great. Therefore, the mean is not representative of the data.

INSTRUCTION Use Example 1 to introduce *range* as a measure of variability. Help students to realize that when there is a large range for a small data set, the mean is often misleading. The range is helpful to determine how much the data varies. It is sometimes written using the actual number, e.g., 94–21 (read "from 94 to 21").

4.5 RANGE, MEDIAN, AND MODE

Animals have been clocked for maximum speeds. This table shows the speeds of some animals. The speeds vary from 12 mph (miles per hour) to 70 mph.

Animal	Speed (mph)
Cheetah	70
Lion	50
Cat	30
Bear	30
Squirrel	12

The **range** of a set of data shows how much the data varies. It is the difference between the greatest number and the least number.

70 − 12 = 58 The range for this set of data is 58 mph.

EXAMPLE 1 Find the range: 45, 79, 21, 86, 55, 94

94 ← **Greatest number**
−21 ← **Least number**
73 ← **Range**

The range for this set of data is 73.

The **median** gives the middle value when the numbers are listed in order.
The median of the data given in the table is 30 mph.

Sometimes a set of data has two middle numbers.
To find the median, find the average of the middle numbers.

EXAMPLE 2 Find the median: 26, 15, 48, 53, 74, 39

1. List the numbers in order from greatest to least. 74, 53, 48, 39, 26, 15
2. Find the average of the middle numbers. $\frac{48 + 39}{2} = \frac{87}{2} = 43.5$

The median of this set of data is 43.5.

The **mode** is the value that occurs most often.
The mode of the data given in the table is 30 mph.

A set of data may have no mode, or it may have more than one mode.

EXAMPLE 3 Find the mode:

a. 46, 3, 57, 89, 2, 17, 101, 33
THINK: Every value appears only once.
This set of data has no mode.

b. 19.4, 89.6, 75.34, 19.4, 67.85, 0.01, 89.6, 35.8
THINK: The values 19.4 and 89.6 each appear twice.
This set of data has two modes: 19.4 and 89.6.

FOR DISCUSSION

See TE side column.

Here are the speeds at which a racehorse ran 1 mile on 5 different days.

Day	1	2	3	4	5
mph	35	31	28	35	35

Which gives a better idea of the speed at which the horse usually runs, the mean or the mode? Explain.

PRACTICE EXERCISES

1. 9 2. 6 3. 6 4. 6 5. 31 6. 22 7. none 8. 21.2 9. 123 10. 108 11. 73 12. 112 13. 11 14. 9 15. 9 16. 8.3 17. 4.9 18. 6 19. 3.7; 7.5 20. 6 21. 37.6 22. 32.9 23. none 24. 35.9 25. 524.3 26. 475 27. 475 28. 421.5

Find the statistical measures for the set of data. See Extra Practice, page 435.
Round the answer to the nearest tenth.

Data	Range	Median	Mode	Mean
4, 6, 10, 11, 6, 2, 6, 3	•1. ■	•2. ■	•3. ■	•4. ■
22, 18, 4, 27, 35	5. ■	6. ■	7. ■	8. ■
108, 73, 196, 110, 73	9. ■	10. ■	11. ■	12. ■
12, 9, 3, 11, 9, 13, 2, 10, 9, 5	13. ■	14. ■	15. ■	16. ■
5, 3.7, 8.6, 7.5, 3.7, 7.5, 6	17. ■	18. ■	19. ■	20. ■
17, 54.6, 28.2, 47, 32.9	21. ■	22. ■	23. ■	24. ■
589.7, 400, 475, 65.4, 475, 523.9	25. ■	26. ■	27. ■	28. ■

Solve.

29. Scientists measured the speed at which six sharks swam in the ocean.

Shark	1	2	3	4	5	6
mph	37	32	36	34	39	38

Find the mean. 36 mph

30. At the World Championship Tortoise Races, the six fastest turtles moved at the following speeds.

0.2 mph, 0.3 mph, 0.4 mph,
0.5 mph, 0.46 mph, 0.34 mph

Find the median. 0.37 mph

31. Greyhounds are racing dogs. Here are the speeds, in miles per hour, at which 12 greyhounds finished a 1-mi race:

38, 35, 38, 36, 37, 38, 34, 37, 33, 36, 39, 32

Make a frequency table to show this data. Check students' tables.

The *mean,* the *median,* and the *mode* are measures of central tendency. They each describe an average or central number of the set. How the measure will be used determines which one to choose.

As you work through Examples 2 and 3, discuss situations in which those measures might be useful. For example, the *median* is useful in describing prices and salaries since the extremes don't affect it. *Mode* is useful in reordering since it gives the most frequently needed item. Stress that while mean, median, and mode all represent the data, they can differ and, in some cases, misrepresent the data.

FOR DISCUSSION Students should explain that the mean is 32.8 and the mode is 35. The mode represents the horse's speed 60% of the time; therefore, it is more accurate.

ASSIGNMENTS

Level 1	1–16, 29, 30
Level 2	9–31
Level 3	17–31

FOLLOW-UP ACTIVITY

APPLICATION Have the students keep a record of the daily high and low temperatures for 1 week. Then have them find the range, mean, and median for each set of data.

SUPPLEMENTARY MATERIALS

TRP Practice, p. 42

TRP Reteaching, p. 26

TRP Transparency 21

OBJECTIVE

- Solve logical reasoning problems by making a table.

TEACHING THE LESSON

WARM-UP Read the following clues to the class.

Clue A: *X* is an even number.
Clue B: *X* is divisible by 3.
Clue C: *X* is less than 10.

Then ask the following questions:

1. Could *X* be 8? (No, because 8 is not divisible by 3.)
2. Could *X* be 6? (Yes, because 6 would satisfy the conditions in all 3 clues.)

INTRODUCTION Present these facts to the class:

1. Some cats have long hair.
2. Al owns a cat.

Discuss which of the following conclusions could be drawn from the given facts.

a. Al's cat must have long hair. (no)

b. Al's cat might have long hair. (yes)

Use this example to point out that many of the problems we solve in daily life do not involve numbers. Introduce the idea that logical reasoning helps us solve problems like these by enabling us to rule out choices that do not fit given conditions.

INSTRUCTION After discussing the situation, use the chalkboard to show how to set up a table for organizing the given facts. Emphasize that since each person belongs to only one club, there will be only one *yes* in each column and row of the table.

PROBLEM *Solving* STRATEGY

4.6 MAKING A TABLE

Situation:

Bill, Tom, Mary, and Sue each belong to one of the following after-school clubs: stamp, music, tennis, and hiking. Mary and Tom tried out for the tennis club, but did not make it. Sue does not like to participate in sports activities. Sue and Mary went to hear the music club perform. Which club does each person belong to?

Strategy:

Making a table can help you solve some problems.

Applying the Strategy:

Clue 1: Mary and Tom tried out for the tennis club but did not make it. Write *No* to show that Mary and Tom are not in the tennis club.

	Clubs			
	Stamp	Music	Tennis	Hiking
Bill				
Tom			No	
Mary			No	
Sue				

Clue 2: Sue does not like to participate in sports. Write *No* to show that Sue is not in the tennis or hiking clubs. Write *Yes* to show that Bill belongs to the tennis club.

	Clubs			
	Stamp	Music	Tennis	Hiking
Bill	No	No	Yes	No
Tom			No	
Mary			No	
Sue			No	No

Clue 3: Sue and Mary went to hear the music club perform. Write *No* to show that Sue and Mary are not in the music club. Why do you know that:

- Sue is in the stamp club?
- Mary is in the hiking club?
- Tom is in the music club?

	Clubs			
	Stamp	Music	Tennis	Hiking
Bill	No	No	Yes	No
Tom	No	Yes	No	No
Mary	No	No	No	Yes
Sue	Yes	No	No	No

PRACTICE EXERCISES

Read the problem. Then decide which statement is true.

• 1. Jamie collects stamps either from foreign countries or from the United States. He does not collect stamps from foreign countries. a
   a. Jamie collects U.S. stamps.
   b. Jamie collects stamps from foreign countries.

• 2. Some hikers like to camp. Amelia is a camper. b
   a. Amelia is a hiker.
   b. Amelia may or may not be a hiker.

Solve by making a table.

3. Eli, Bert, and June exercise regularly. One jogs, one swims, and one plays golf. The jogger and Bert work together. Eli and the jogger went to see the golfer play in a tournament. Which kind of exercise does each person do? Eli swims; Bert golfs; June jogs

4. Willy, Julia, Anne, and Sean all have carpentry as their hobby. They won the first four prizes in a carpentry contest. Sean won third prize. Julia did not win second prize. If Ann won first prize, which prize did Willy win? Second prize

5. Jason, Eve, Ben, and Kerry belong to a cooking club. They each baked different muffins: blueberry, banana, apple, and bran. Kerry needed to wash some berries for her muffins. Eve needed to remove the seeds from her fruit. Ben's muffins did not have any fruit. Which kind of muffins did each person bake? Jason—banana; Eve—apple; Ben—bran; Kerry—blueberry

6. Mr. Moore, Mrs. Donato, and Mr. Alen are the teachers for these clubs: photography, science, and acting. Mr. Moore and the photography club teacher are cousins. The science club teacher and Mr. Alen have lunch together. Mr. Alen is not the photography club teacher. What does each person teach? Alen—acting; Donato—photography; Moore—science

As you use Clues 1 and 2 to complete more of the table, stress that the "yes" and "no" answers are not guesses but logical conclusions drawn from the given facts. Stress the importance of completing a row or column with *no's* once the *yes* has been filled in.

After discussing Clue 3 and writing *no* to show that Sue and Mary are not in the music club, ask students how they decided that Sue must be in the stamp club. (The space for the stamp club is the only space left in Sue's row, and the other three spaces are all marked *no*). Use a similar approach to explain the answers to the last two questions.

ASSIGNMENTS

Level 1	1–4
Level 2	1–5
Level 3	3–6

FOLLOW-UP ACTIVITY

SITUATIONAL PROBLEM SOLVING Your school is considering having after-school clubs. Your class is in charge of surveying the students in your school to see which clubs interest them. You will use the results of this survey to develop a plan to present to the principal. What questions would you want answered by students before you developed a plan to present?

SUPPLEMENTARY MATERIALS

TRP Practice, p. 43

TRP Transparency 22

TRP Applications, p. 6

OBJECTIVE

- Evaluate student progress.

USING THE REVIEW

This page provides a means for informally evaluating students' understanding of the skills and concepts covered so far in this chapter.

Have the students look at the page to familiarize themselves with the various question formats that are presented. Discuss any questions that they may have. Then ask them to complete the page independently.

In addition to grading them individually, you may wish to review the answers to the questions collectively with the students.

Page references appear in brackets. They refer to pages on which a particular skill was introduced.

Before continuing on to the topics found in the remainder of the chapter, you may wish to have students review any skills or concepts in which they have demonstrated weakness.

MIDCHAPTER REVIEW

Use the following data to make a frequency distribution table. [92]

1. Number of children in some families: Check students' tables.

2, 3, 2, 1, 5, 2, 6, 3, 1, 1, 2, 4,
3, 5, 2, 1, 8, 2, 4, 4, 3, 1, 1, 6, 5

Use the data below to complete the grouped-frequency table. [92]

74, 133, 4, 127, 20, 30,
5, 147, 100, 27, 142, 121,
141, 125, 67, 150, 130, 68,
140, 132, 76, 86, 42, 55,
59, 67, 102, 9, 124, 95

Interval	Frequency
0–30	**2.** ■ 6
31–60	**3.** ■ 3
61–90	**4.** ■ 6
91–120	**5.** ■ 3
121–150	**6.** ■ 12

Use the graph at the right to answer the questions. [96]

7. Which station is the most popular? WJBN

8. Which stations are preferred by more than 40,000 people? WLQZ, WKNY, WJBN, WCLH

9. How many more people chose WJBN than WLQZ? 25,000

10. How many chose WGPX and WCLH? 85,000

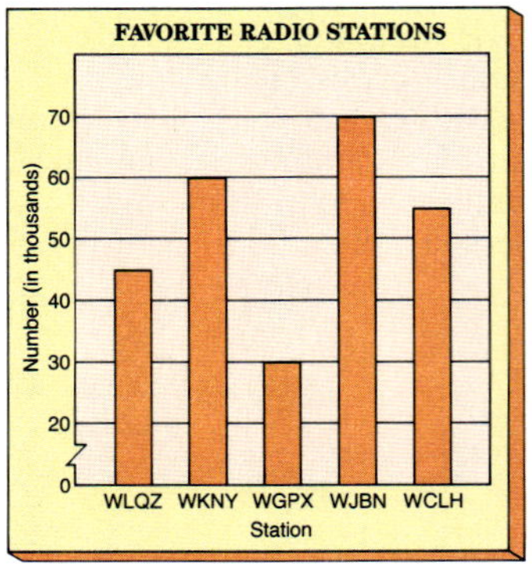

Find the statistical measures for each set of data. [98, 100]

	Range	Mean	Median	Mode
20, 15, 18, 22, 10, 12, 16, 17, 19, 21	**11.** ■ 12	**12.** ■ 17	**13.** ■ 17.5	**14.** ■ none
9, 11, 10, 8, 13, 12, 11, 6, 7, 9	**15.** ■ 7	**16.** ■ 9.6	**17.** ■ 9.5	**18.** ■ 9, 11

Solve. [95]

19. A 1993 Towne Coupe costs $18,735. Cruise control is optional for $934. What is the total price of the car? $19,669

104 CHAPTER 4

PROBLEM Solving APPLICATION

4.7 CAREER: ECOLOGIST

Anton Kluzewski is an ecologist. He is interested in the environment and the ways in which we can reduce the dangers of pollution. At present he is working for an environmental protection agency.

The map below gives information about the distribution of hazardous waste sites in a recent year.

Hazardous Waste Sites (Final and Proposed) — Number of Sites on National Priority List

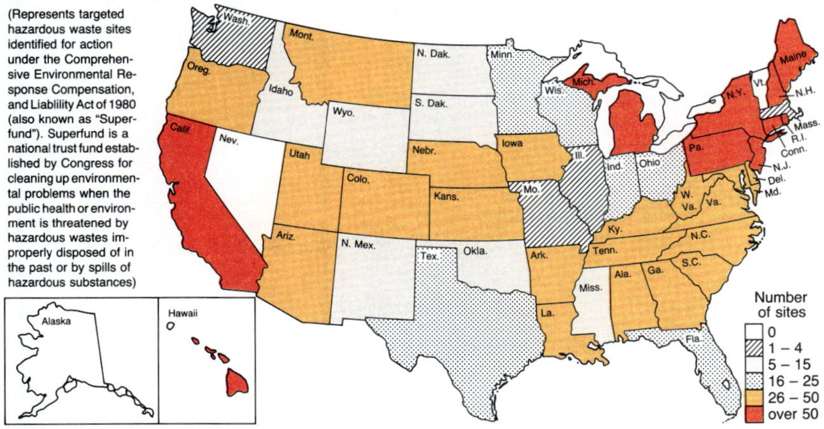

How many sites are in:

• **1.** Texas? 16-25

• **2.** California? over 50

3. Nevada? 0

4. Illinois? 1–4

Which states have:

5. over 50 of the sites?
CA, NY, PA, MI, CT, HI, NJ, VT, ME, RI, NH

6. 26–50 of the sites?
TX, FL, MN, IN, OH, WI

Which state has more of the sites?

7. Nevada or Utah? Utah

8. Georgia or Florida? Georgia

9. South Dakota or Montana? Montana

10. Minnesota or Arkansas? Arkansas

Solve.

11. Are there more states that have sites than have 16–25 sites? yes

12. Make a tally that shows the number of states that have these numbers of sites: 0, 1–4, 5–15, 16–25, 26–50, over 50.
Check students' tallies.

OBJECTIVE

- Solve problems that involve reading and interpreting data from a map.

TEACHING THE LESSON

WARM-UP Display a large map of the United States and have the students complete these exercises.

List:

1. two states that border on the Pacific Ocean.
2. three states that border on the Atlantic Ocean.
3. three states that border on Canada.

INTRODUCTION Ask students what they have read or heard about pollution and hazardous waste. Let students share relevant facts and opinions.

INSTRUCTION Point out that the map in the text is used as a graphic device for presenting data. Discuss the key shown below the map. Students should realize that the darker the shading, the greater the number of hazardous waste sites.

Notice that Problem 12 requires students to make a tally using the same categories that are used in the key for the map.

ASSIGNMENTS

Level 1	1–10
Level 2	1–12
Level 3	Even 2–12

SUPPLEMENTARY MATERIALS

TRP Practice, p. 44

TRP Transparency 22

OBJECTIVE

- Construct bar graphs.

TEACHING THE LESSON

WARM-UP Write the following exercises on the chalkboard or an overhead transparency.

Write the first five multiples of each number.

1. 5 **2.** 20 **3.** 25 **4.** 50

(**1.** 5; 10; 15; 20; 25 **2.** 20; 40; 60; 80; 100 **3.** 25; 50; 75; 100; 125 **4.** 50; 100; 150; 200; 250)

INTRODUCTION Pass out copies of two or more bar graphs from a newspaper or magazine. Ask students questions about the components of a bar graph. For example, What is the title of the graph? What is represented on the horizontal axis? The vertical axis? What is the scale?

INSTRUCTION Use the Introduction as a lead-in to making a bar graph. Students may need help in deciding on a scale. The largest number on the scale must be greater than the largest number in the data set. Rounding will help determine the span of the numbers. The intervals must be large enough to show the data clearly but small enough to fit the graph in the allotted space.

A simple technique to determine a scale is to choose an interval and divide it into the largest number to see how many scale markings are needed. In Example 1, if intervals of 20 were chosen, it would be $\frac{100}{20} = 5$. If intervals of 10 were chosen, it would become $\frac{100}{10} = 10$. This is a quick and easy way to determine the space needed for the scale axis.

4.8 BAR GRAPHS

A **bar graph** helps you compare data quickly and easily. Each bar has a title, a horizontal axis, and a vertical axis. A scale on one of the axes helps you find the number represented by the bar.

EXAMPLE 1 The table shows the sales of computer games in 1 year at the ZAP Computer Stores. Show this data on a vertical bar graph.

1. Round the data to a convenient place to determine a scale.
2. Show the scale on the vertical axis. Label the scale.
3. Label the horizontal axis.
4. Draw a bar to show the sales of each game.
5. Give the graph a title.

TOP-SELLING COMPUTER GAMES

Game	Copies sold (in thousands)	
	Actual	Rounded
York	78	80
Zeon	57	60
Hotshot	52	50
Radar	38	40

A survey was taken to find out what accessories customers are buying for their computers.

EXAMPLE 2 Make a horizontal bar graph to show this data.

1. Round the data to the nearest million. Choose a scale.

COMPUTER ACCESSORIES

Item	Units sold (in millions)	
	Actual	Rounded
Floppy disk	8.118	8
Hard disk	1.836	2
Impact printer	4.429	4
Modem	2.052	2

2. Show and label the scale on the horizontal axis.
3. Label the vertical axis.
4. Draw the bars.
5. Write a title for the graph.

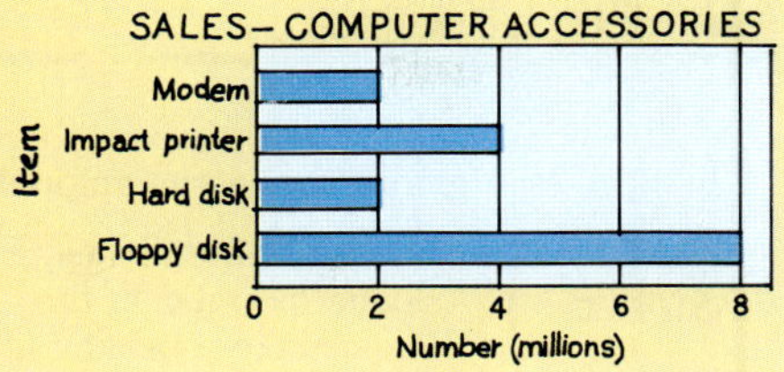

PRACTICE EXERCISES

• **1.** Make a vertical bar graph to show the numbers of magazine subscriptions sold. Check students' graphs.

SUBSCRIPTIONS SOLD—COMPUTER MAGAZINES

Magazine	Subscriptions sold
Munch	10,980
PC Life	9,870
PRINT!	6,400
T+	12,460

2. The ZAP Computer Stores asked their customers how they used their computers most often. Make a horizontal bar graph to show the primary uses of the computer. Check students' graphs.

PRIMARY USES—PERSONAL COMPUTERS

Function	Number
Database	4,902
Games	2,487
Spreadsheet	10,053
Word processing	12,685

3. The advertising manager of SOLDOUT Computer Company wants to include a graph of software sales in his monthly report. Use the scales below to make two vertical bar graphs. Which graph gives a better representation of the data?

Scale 1: 0 to 40,000; intervals of 10,000
Scale 2: 30,000 to 40,000; intervals of 2,000

SOFTWARE SALES—AUGUST

Software	Number sold
WRITE RIGHT	30,956
FIGURE-OUT	39,493
INFO FAST	31,611
JAZZ IV	35,082

Check students' graphs; Graph 2

4. A store sells four different brands of computer diskettes. They want to sell only two brands. Make two horizontal bar graphs using the scales below to show the sales for each brand. Which graph gives a better representation of the data?

Scale 1: 0 to 5,000 with intervals of 1,000
Scale 2: 500 to 4,500 with intervals of 500

BRANDS OF DISKETTES

Brand	Sales
4KB	3,826
COPY-IT	634
GUARDIAN	2,529
DISK III	996

Check students' graphs; Graph 2

Explain that a good technique in graphing is to start the scale at 0. A graph can be misleading if it starts at another number.

Use the scale to determine the height of the bars. The bars should be uniform in width and evenly spaced. The width of the bars and the amount of space between them may depend on the available space.

Use Example 2 to show that the type of bar graph (in this case horizontal) determines on which axis the scale is shown.

ASSIGNMENTS

Level 1	1–2
Level 2	1–3
Level 3	1–4

FOLLOW-UP ACTIVITY

ENRICHMENT Have students make double-bar graphs. A double-bar graph is used to compare two sets of data on the same graph. A key is needed to identify each bar. Stress that the same set of data must appear first each time.

SUPPLEMENTARY MATERIALS

TRP Practice, p. 45
TRP Reteaching, p. 29
TRP Lesson Aids, p. 4
TRP Transparency 23

OBJECTIVE

- Construct line graphs.

TEACHING THE LESSON

WARM-UP Write the following exercises on the chalkboard or an overhead transparency.

Round each number to the underlined place.

1. 73 **2.** 239 **3.** 45

4. 562 **5.** 3,095 **6.** 9,501

(**1.** 70 **2.** 200 **3.** 50 **4.** 600 **5.** 3,000 **6.** 10,000)

INTRODUCTION Present the following on the chalkboard.

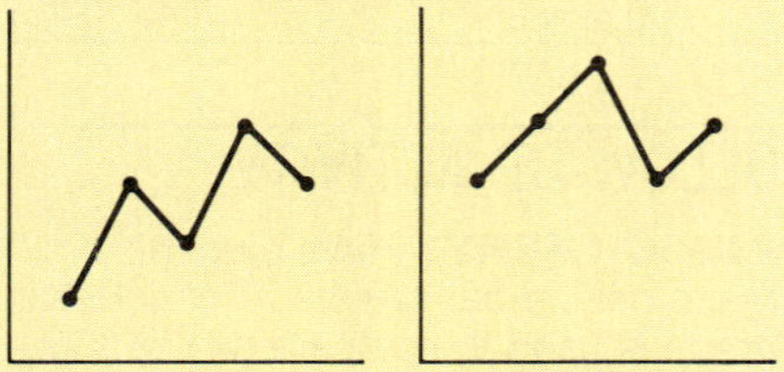

A B

Tell students that A and B each represent a stock's growth over the same period of time. Ask if they can tell which one is worth more. (No, because the scales are missing.)

INSTRUCTION Use the Introduction to lead into the components of a broken-line graph. Review the process of determining a scale:

$$\frac{\text{Greatest data value}}{\text{Interval}} = \text{scale markings}$$

Emphasize that in a broken-line graph, the horizontal axis is usually used for time, the vertical axis for the scale.

4.9 LINE GRAPHS

A **broken-line graph** shows the amount of change over time.

EXAMPLE 1 Home satellite dishes became popular in the 1980s. Make a broken-line graph to show the sales from 1981 to 1985.

1. Round the data to a convenient place to determine a scale.
2. Draw and label the horizontal and vertical axes.
3. Locate the point for each piece of data.
4. Draw line segments to connect the points in order.
5. Write a title for the graph.

SALES—HOME SATELLITE DISHES

Year	Number sold (in thousands)	
	Actual	Rounded
1981	32	30
1982	44	40
1983	53	50
1984	114	110
1985	209	210

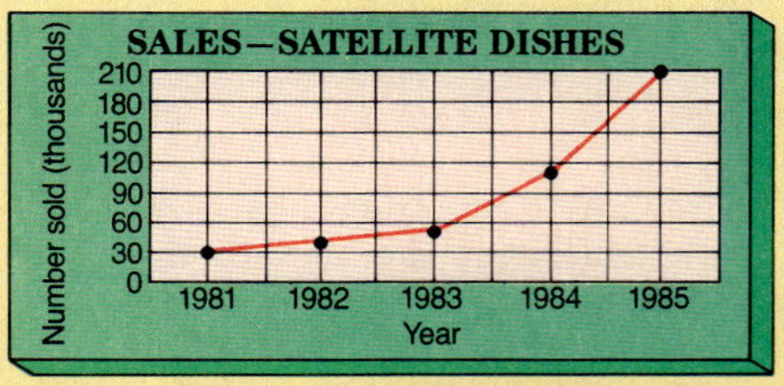

Sometimes you need to break the scale to fit the data on the graph.

EXAMPLE 2 Make a broken-line graph to show the sales of video-cassette recorders for the period shown.

1. Round the data to a convenient place to determine the scale.
2. Draw and label the scale on the vertical axis. Use a jagged line to show the break in the scale from 0 to 2,000.
3. Draw and label the horizontal axis.
4. Locate and connect the points.
5. Write a title for the graph.

SALES—VIDEO-CASSETTE RECORDERS

Month	Number sold	
	Actual	Rounded
October	2,058	2,100
November	2,404	2,400
December	2,348	2,300

Scale: 2,000 to 2,500; intervals of 100

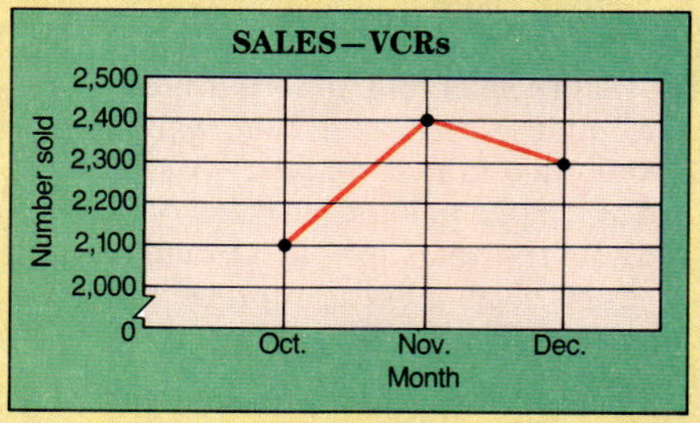

FOR DISCUSSION See TE side column.

Would a bar graph or a broken-line graph be better to show each of the following sets of data? Explain.

Favorite colors of different age groups
Hourly temperatures during a 24-h period

PRACTICE EXERCISES

Use each set of data to make a broken-line graph.

• **1.** Jarene kept a record of the number of calls she received on her answering machine in 1 week. Check students' graphs.

DAILY CALLS: ANSWERING MACHINE

Day	Mon.	Tues.	Wed.	Thurs.	Fri.	Sat.	Sun.
Number of calls	3	2	6	8	5	10	2

2. The manager of the rock band "STAR" tracked the sales of the group's latest music video. Check students' graphs.

MONTHLY SALES— "FIREBALL" VIDEO MUSIC

Month	July	Aug.	Sept.	Oct.	Nov.	Dec.
Number sold	7,834	12,642	25,869	46,008	43,683	37,304

3. An airline company has gathered information about the number of mobile telephones installed in airplanes. Check students' graphs.

MOBILE TELEPHONE INSTALLATIONS

Year	1981	1982	1983	1984	1985	1986
Number installed	24	15	65	89	100	895

4. For the past 3 years, Green Ridge High School has made a videotape of the highlights of the school year in addition to the yearbook. Make a double broken-line graph to compare the sales of the video and the yearbook over the past 3 years. Use - - - - - to show the sales of the video. Use ——— to show the sales of the yearbook. Check students' graphs.

YEARBOOK/VIDEO SALES

Year	Yearbook	Video
1987	363	52
1988	287	101
1989	167	198

Help the students plot the points. Instruct them to go across the horizontal axis and up the vertical axis to plot a point. This approach will lay the foundation for the "across-and-up" method of plotting (x, y) points on a coordinate grid. Stress that the points must be connected in order.

Use Example 2 to explain the jagged line on the vertical axis. The sensible graphing technique of starting the scale at 0 is used. Space constraints necessitate using 2,000 on the first scale marking.

FOR DISCUSSION Bar graphs are better for comparisons (favorite colors). Broken-line graphs are better for a change over a period of time (hourly temperatures).

ASSIGNMENTS

Level 1	1–2
Level 2	1–3
Level 3	1–4

FOLLOW-UP ACTIVITY

APPLICATION Have the students choose a stock from the stock exchange listings in the newspaper. Then have them graph the stock's closing price for 5 days.

SUPPLEMENTARY MATERIALS

TRP Practice, p. 46

TRP Reteaching, p. 28

TRP Lesson Aids, p. 4

TRP Transparency 23

OBJECTIVE

- Solve problems that involve using a mail-order catalog.

TEACHING THE LESSON

WARM-UP Write the following exercises on the chalkboard or an overhead transparency.

Find the answer.

1. 4 × \$12.50 (\$50.00)
2. 6 × \$7.95 (\$47.70)
3. \$14.50 + \$27.65 (\$42.15)
4. \$18.60 + \$11.50 + \$15.90 (\$46.00)

INTRODUCTION Ask students whether they or members of their family have ever ordered items from a catalog. Have students tell what types of items were ordered, how long it took to receive the items, and other details. Point out that often items ordered from a catalog are less expensive than similar items purchased in a store.

Elicit from students that a mail-order company does not need to arrange and display its stock in an expensive setting, and can reduce prices accordingly.

INSTRUCTION As you discuss the introductory example, point out that the M. M. Green Company does not charge for regular shipment by mail. Make sure students understand that many mail-order companies do charge for delivery, and the customer must add the appropriate amount to the cost of merchandise.

Point out that a common error mail-order customers make is to forget to multiply the amount times the unit price. Often customers simply enter the unit price under the total amount.

PROBLEM Solving APPLICATION

4.10 CONSUMER: ORDERING FROM A CATALOG

Mail-order shopping is a large business in the United States. Often, catalogs provide a wide selection of different items. When ordering by mail, the consumer must give the correct information on the order form.

In this order, compute the cost of the shirts, the cost of the socks, and the total cost of the two types of items.

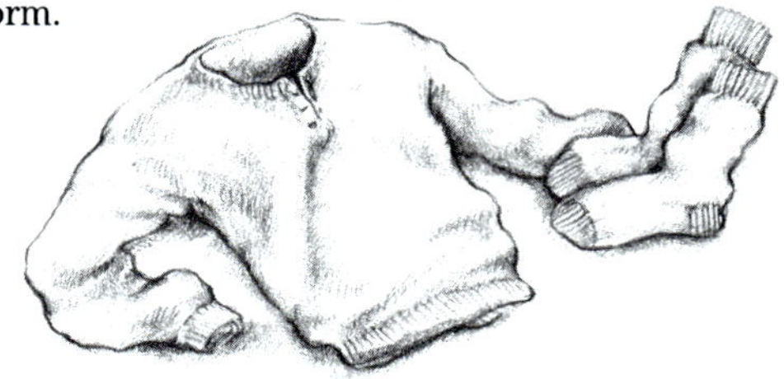

STOCK NO.	ITEM	COLOR NO.	SIZE	HOW MANY	UNIT PRICE	TOTAL AMOUNT
3899RR	Shirts	94	M	3	\$22.50	■
1076RR	Socks	42	M	5 pair	\$ 3.95	■
					TOTAL AMOUNT FOR ALL ITEMS	■
				Regular shipping and handling FREE within U.S.	Paid by M.M. Green, Inc.	

	NUMBER BOUGHT × UNIT PRICE = COST OF THE ITEMS
1. Multiply to find the cost of the shirts.	3 × \$22.50 = \$67.50
2. Multiply to find the cost of the socks.	5 × \$ 3.95 = \$19.75
3. Add to find the total cost.	\$67.50 + \$19.75 = \$87.25

So, the total cost of the two types of items is \$87.25.

FOR DISCUSSION

See TE side column.

What services may customers want that M. M. Green, Inc. would charge extra for?

Use the information in this order form to answer Exercises 1–10. Remember to estimate whenever you use your calculator.

STOCK NO.	ITEM	COLOR NO.	SIZE	HOW MANY	UNIT PRICE	TOTAL AMOUNT
2759RR	Solid shirt	45	M	4	$18.50	■
2760RR	Striped shirt	43	M	2	$20.50	■
3064NN	Shorts	42	32	1 pair	$31.25	■
1076RR	Socks	44	M	3 pair	$ 3.95	■
2853RR	Sweater	43	M	2	$31.95	■

What is the cost of the:

• **1.** solid shirts? $74.00

• **2.** striped shirts? $41.00

3. shorts? $31.25

4. socks? $11.85

5. sweaters? $63.90

6. What is the total cost of all the items? $222.00

Find the total cost of the order.

• **7.** 3 shirts, stock no. 2760RR; 4 sweaters, stock no. 2853RR $189.30

8. 6 pairs of socks, stock no. 1076RR; 2 shirts, stock no. 2759RR; 2 pairs of shorts, stock no. 3064NN $123.20

Estimate the answer.

• **9.** About how much is the cost of 5 striped shirts, stock no. 2760RR? $100

10. Jan has $24 to spend on socks, stock no. 1076RR. About how many pairs of socks can she buy? about 6 pairs

Solve.

11. Five types of slacks cost these amounts per pair: $27.95; $23.95; $25.50; $23.95; $32.50. What is the range of the amounts? $8.55

12. Which is the greatest—the mean, the median, or the mode of the costs of the five types of slacks? (See Exercise 11.) the mean

Note that Problems 1–5 provide step-by-step calculations that lead to the answer to Problem 6. Estimation is used in Problems 9 and 10. You may wish to have students use calculators to solve Problems 7 and 8 in this lesson.

FOR DISCUSSION Students should point out that customers may want extra services such as faster delivery service, alterations, or gift wrapping.

ASSIGNMENTS

Level 1	1–8, 11
Level 2	1–11
Level 3	6–12

FOLLOW-UP ACTIVITY

APPLICATION Provide students with several different types of mail-order catalogs. Tell students that they each have $100 to spend on merchandise. Have the students construct order forms, using catalogs of their choice, to show the items they would order.

SUPPLEMENTARY MATERIALS

TRP Practice, p. 47

TRP Applications, p. 7

TRP Transparency 24

OBJECTIVE

- Construct pictographs.

TEACHING THE LESSON

WARM-UP Write the following exercises on the chalkboard or an overhead transparency.

Multiply each number by 10,000.

1. 5 **2.** 8 **3.** 10 **4.** 6

(**1.** 50,000 **2.** 80,000 **3.** 100,000 **4.** 60,000)

INTRODUCTION Ask students if they can complete the quote: A picture is worth ■. (a thousand words) (Ask students why this might be true. Explain that a pictograph uses symbols to represent data.)

INSTRUCTION Discuss whether a pictograph is more accurate or less accurate than a bar graph or a broken-line graph. Elicit from students how accurate they think a pictograph can be. (less accurate since the symbol stands for rounded amounts)

Use Example 1 to explain how to determine the number of robots. Ask students if they think exactly 55,000 robots were used. Help them see that this number is an estimate.

Example 2 shows how to take actual figures and translate them into symbols. Explain the importance of choosing a symbol and a key that will provide as much accuracy as possible. A symmetrical symbol works well. Make sure that students round the data to determine a suitable key. Show students that rounding to other than the nearest 500,000 would have resulted in either misrepresentation (nearest million would have made 4,346,000 → 4,000,000) or having to represent awkward parts of the whole. (Nearest hundred thousand would have resulted in 1.2, 4.3, 5.2, and 2.9.)

4.11 PICTOGRAPHS

A **pictograph** uses a picture or a symbol to represent the data. A **key** shows how many each symbol represents.

EXAMPLE 1 How many robots are used in the computer industry?

To find how many robots are used, count the symbols.

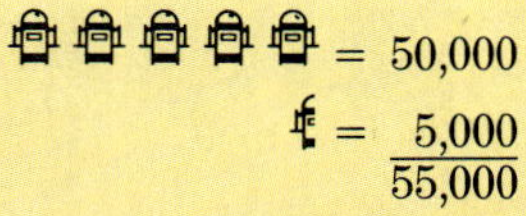

ROBOTS USED IN INDUSTRY

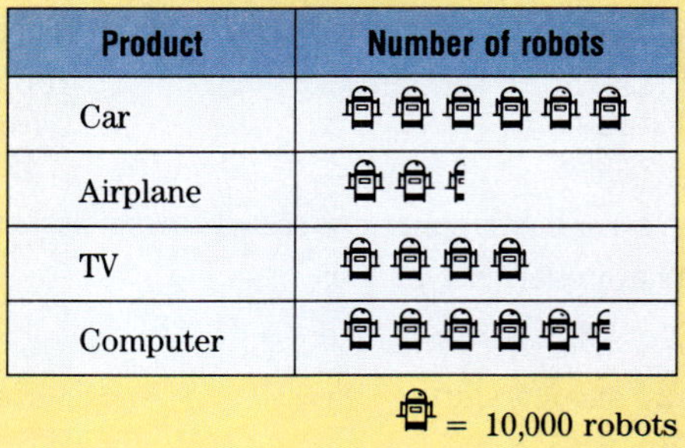

So, 55,000 robots are used in the computer industry.

When you make a pictograph, choose an appropriate symbol for the data.

EXAMPLE 2 Make a pictograph to show the number of employees in each U.S. industry.

EMPLOYEES IN U.S. INDUSTRIES—1980

Industry	Number of employees
Mining	1,207,000
Construction	4,346,000
Finance	5,160,000
Government	2,866,000

1. Round the numbers to an appropriate place.
2. Choose a symbol. Decide how many it represents.

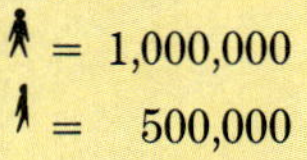

3. Determine the number of symbols needed.

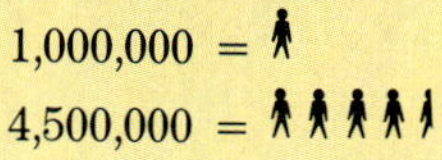

4. Label the vertical axis.
5. Draw the symbols for each number.
6. Write a title for the graph.

EMPLOYEES IN U.S. INDUSTRIES—1980

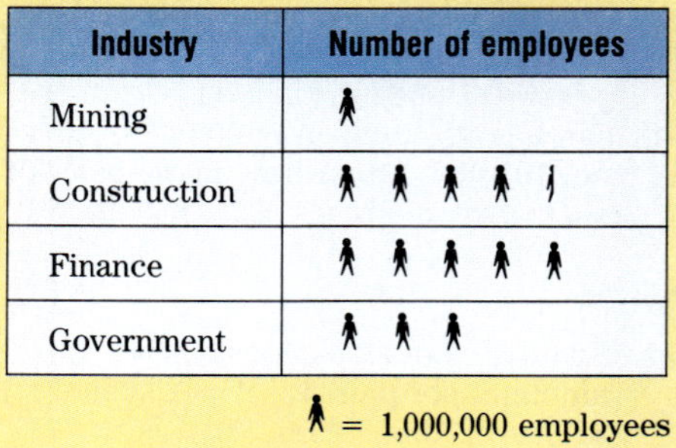

PRACTICE EXERCISES

Use the pictograph to answer Exercises 1–6.

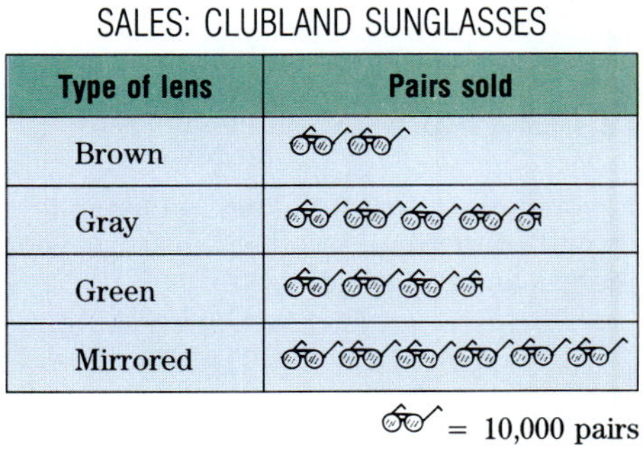

• **1.** How many pairs of sunglasses does each symbol represent? 10,000

• **2.** Which lens had sales of 35,000? green

3. How many pairs of mirrored lenses were sold? 60,000

4. Which lenses had sales of more than 40,000 pairs? gray, mirrored

5. How many more pairs of gray lenses were sold than green lenses? 10,000

6. Which type of lens had sales three times as high as another? mirrored, 3 times as many as brown

Use the given data to make a pictograph. Check students' graphs.

• **7.** WORLD SHARK RECORDS

Type	Weight (lb)
Blue	437
Hammerhead	991
White	2,664
Mako	1,080
Tiger	1,780

8. U.S. PUBLIC LIBRARIES

City	Volumes
Erie, PA	420,355
Tucson, AZ	760,000
Austin, TX	817,338
Tampa, FL	1,324,262
Baltimore, MD	1,892,972

9. WORLD CORN PRODUCTION

Country	Tons
Italy	6,820,000
Brazil	21,098,000
Mexico	12,215,000
France	9,100,000

10. LEADING MERCHANT FLEETS

Country	Number of ships
Liberia	2,019
Greece	2,454
Japan	1,712
Panama	3,290

TIME OUT

Find the numbers in this set of data:

Clues:
My mean is 18.4.
My mode is 18.
My median is 18.
My range is 8.
My greatest number is 22.
14, 18, 18, 20, 22

To determine how many symbols are needed, divide each rounded number in the data set by the number each symbol represents:

$$\frac{4{,}500{,}000}{1{,}000{,}000} = 4.5$$

Students should see 4.5 as four 1,000,000-symbols and 0.5 or $\frac{1}{2}$ as half of a 1,000,000 symbol.

TIME OUT You may want to give students who are having difficulty an additional hint: there are five numbers in the set of data. Have students volunteer the answer.

ASSIGNMENTS

Level 1	1–8
Level 2	1–9, TO
Level 3	1–10, TO

FOLLOW-UP ACTIVITY

COOPERATIVE LEARNING Provide students with back issues of newspapers and magazines. Have them go through each one to find pictographs. Working in groups of four or five students, have them cut out all the pictographs they can find. After mounting them on construction paper, students should make a list of how pictographs are used, the most popular items pictographs represent, etc.

SUPPLEMENTARY MATERIALS

TRP Practice, p. 48

TRP Reteaching, p. 29

TRP Transparency 24

OBJECTIVE

- Solve problems that involve finding and comparing rental costs.

TEACHING THE LESSON

WARM-UP Write the following exercises on the chalkboard or an overhead transparency.

Find the answer.

1. 12 × 355 (4,260)
2. 295 + 3,540 (3,835)
3. 3,720 − 3,480 (240)
4. $276 + $49.75 ($325.75)
5. 12 × $42.85 ($514.20)

INTRODUCTION Ask students if any of their older brothers or sisters or friends live in their own apartments. Have volunteers tell whether finding the apartment was easy or difficult, and whether the costs of living in an apartment were less than or greater than anticipated.

INSTRUCTION As you discuss the introductory problem, be sure the students understand what a security deposit is. It is an amount that is not applied to a person's rent while the person is living in the apartment. Rather, the building owner holds it to guard against various contingencies, such as damage to the apartment by the tenant. If the tenant does significant damage to the apartment, the security deposit may be forfeited, or appropriate deductions may be made from the deposit.

Before students begin the problems, be sure they read and understand the information given in the signs on page 114. In Problem 13, students must subtract and divide to find the monthly rental. You may wish to have students use calculators to help them solve the problems in this lesson.

PROBLEM Solving APPLICATION

4.12 CONSUMER: RENTING APARTMENTS

Statistics show that most young adults live in rental apartments.

When renting an apartment, you need to understand all the costs that are involved. The monthly rent is not your only cost. Usually, you also need to pay a security deposit. The security deposit may be equal to 1 or 2 months' rent.

In some apartment buildings, a charge for gas and electricity is included as part of the rent. In other buildings, you must pay the companies for the gas and electricity used.

WASHINGTON APARTMENTS	
Studio	$275/month
Two rooms	$355/month
Three rooms	$440/month
Four rooms (two bedrooms)	$538/month
Gas and electricity not included.	
Security deposit: 1 month's rent.	

JEFFERSON MANOR	
Studio	$316/month
Two rooms	$404/month
Three rooms	$498/month
Four rooms (two bedrooms)	$602/month
Gas and electricity included.	
Security deposit: 1 month's rent.	

You decide to rent a studio apartment in Washington Apartments. How much money will you pay the owner in the first year?

THINK: In the first year, you will need to pay the security deposit plus 12 months' rent.

1. Multiply to find the cost of 12 months' rent. 12 × $275 = $3,300
2. Add to find the total amount.
   THINK: The security deposit is equal to 1 month's rent, or $275. $275 + $3,300 = $3,575

So, you will pay the owner $3,575 in the first year.

Use the signs on page 114 to answer the following exercises. Remember to estimate whenever you use your calculator.

• **1.** What is the monthly rent for a two-room apartment in Washington Apartments? $355

• **2.** What is the monthly rent for a two-room apartment in Jefferson Manor? $404

3. How much more is the rent for a two-room apartment in Jefferson Manor than in Washington Apartments? $49

4. Does the rent for a two-room apartment in Jefferson Manor include anything that is not included in the rent for a two-room apartment in Washington Apartments? If so, what? gas and electricity

5. Ann Ramirez has a two-room apartment in Washington Apartments. Her gas and electricity bills average about $53/month. Are her monthly expenses for rent, gas, and electricity less than they would be if she had a two-room apartment in Jefferson Manor? no

6. How much money does Ann Ramirez pay for rent, gas, and electricity in 6 months? $2,448

Find the total amount paid the owner in the first year for the following apartments.

• **7.** A studio apartment in Jefferson Manor $4,108

8. A two-room apartment in Washington Apartments $4,615

9. A three-room apartment in Jefferson Manor $6,474

10. A four-room apartment in Washington Apartments $6,994

Solve.

• **11.** Jack O'Hara rents a two-room apartment in Jefferson Manor. He stays for 8 months. When he moves, the security deposit is returned to him. How much rent does he pay in the 8 months? $3,232

12. Della Truro rents a three-room apartment in Washington Apartments. Her gas and electricity bills average about $62.50 per month. How much money does she pay for rent, gas, and electricity in 10 months? $5.025

13. In 6 months, Glen Forbes pays a total of $2,562 for rent, gas, and electricity. Of this amount, $432 is paid for gas and electricity. What is the monthly rental for his apartment. In which building does he live? $355/month for two rooms in Washington Apartments

ASSIGNMENTS

Level 1	1–11
Level 2	1–13
Level 3	Odd 1–13

FOLLOW-UP ACTIVITY

ENRICHMENT Distribute various newspaper ads that show apartment rentals. Have students write problems based on finding the costs of yearly rentals (including the security deposits) for various apartments. Then ask students to solve each other's problems.

SUPPLEMENTARY MATERIALS

TRP Practice, p. 49

TRP Transparency 25

OBJECTIVES

- Review vocabulary.
- Practice key chapter concepts and skills.

USING THE REVIEW

The Chapter Review is designed to help students prepare for taking the Chapter Test. The first section focuses on vocabulary. It requires that students select words to complete statements. The second section presents practice exercises of key mathematical skills. Under each directive there is a sample exercise with the answer.

Each item on the review is referenced to the page on which the topic is taught in the Pupil's Edition. You may wish to have students refer to these pages to help review any concepts or skills they have not yet mastered.

It is suggested that students work in small-sized heterogeneous cooperative learning groups. Some cooperative learning methods that may be used are as follows:

1. After each student has independently completed the entire Chapter Review, a discussion should follow within each group about the solutions to the practice exercises.
2. The group can complete the entire Chapter Review by working together to discuss the sample exercises and then to answer the practice exercises.

End the lesson with an entire class discussion in which any questions brought up in group discussions were presented and answered.

CHAPTER REVIEW

Vocabulary Write the letter of the correct answer.

1. Which of the following is not a measure of central tendency? [100] b
a. mean **b.** histogram **c.** median **d.** mode

2. What is the difference between the greatest number and the least number in a set of data? [100] c
a. mode **b.** mean **c.** range **d.** frequency table

Skills Find the statistical measures for the set of data. [98, 100]

Data: 55, 76, 67, 47, 55
Range: *29* Median: *55* Mode: *55* Mean: *60*

3. 33.2 **4.** 35.5 **5.** 35.5 **6.** 31.7 **7.** 673 **8.** 318 **9.** none **10.** 353.5 **11.** 6,522 **12.** 2,974 **13.** none **14.** 3,870.5

	Range	Median	Mode	Mean
Data: 21.7, 35.5, 49.5, 16.3, 35.5	**3.** ■	**4.** ■	**5.** ■	**6.** ■
Data: 108, 93, 426, 766, 518, 210	**7.** ■	**8.** ■	**9.** ■	**10.** ■
Data: 1,506; 8,028; 4,441; 1,507	**11.** ■	**12.** ■	**13.** ■	**14.** ■

Use the frequency table to the right to answer Exercises 15–17. [92]

How many walkers participated? *98*

MILES COMPLETED IN WALK-A-THON

Miles	Walkers
0–5	16
6–10	25
11–15	39
16–20	18

15. How many people completed more than 10 miles? 57

16. How many people walked less than 11 miles? 41

17. How many miles did the least number of people complete? 0–5 miles

18. The table at the right shows the number of listeners at WJJK during a 5-month period. Use the data to make a broken-line graph. [108]
Check students' graphs.

LISTENING AUDIENCE—WJJK

Month	Number of listeners
January	53,000
February	54,000
March	52,500
April	55,000
May	53,500

CHAPTER TEST

Use the following data to make a grouped-frequency table.

1. Number of library books checked out per person per year:

48, 10, 89, 63, 21, 15, 41, 10, 32, 18, 48, 3, 19, 25, 55
Check students' tables.

Find the statistical measures for the set of data in Exercise 1. Round to the nearest hundredth.

2. Range ■ 86 **3.** Median ■ 25 **4.** Mode ■ 10, 48 **5.** Mean ■ 33.13

Use the bar graph at the right to answer Exercises 6 and 7.

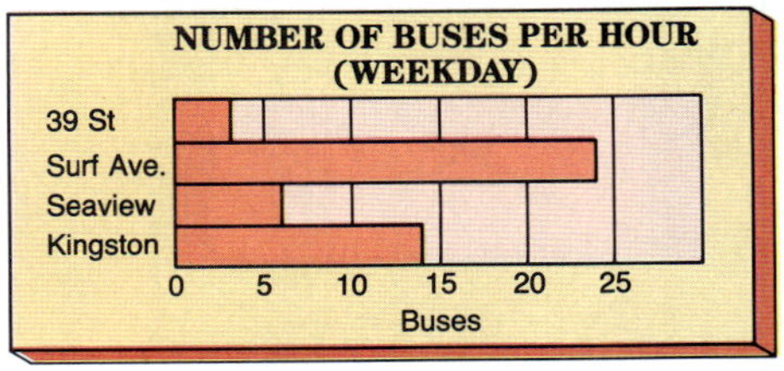

6. Which route has the most buses?
Surf Ave.

7. About how many more buses run on the Seaview route than on the 39 St. route? 2

Make a broken-line graph by using the following data.

8. The Tidal Waves have released a new album. Here are the sales for the first 4 weeks. Check students' graphs.

Week 1: 189,764 Week 3: 500,794
Week 2: 385,618 Week 4: 412,308

Solve.

9. Clara, Leo, Jim, and Heather each belong to one of the following after-school clubs: photography, music, arts and crafts, and drama. Which club does each person belong to? Use the following clues to find out. Leo—photography; Clara—drama; Jim—crafts; Heather—music

Clue 1: Leo and Jim do not like to perform.
Clue 2: Jim and Heather do not own cameras.
Clue 3: Heather watched a performance of the drama club.

Solve.

10. The Kunderas rent an apartment for $815 per month. The security deposit is 1 month's rent. Gas and electricity are furnished. What will the Kunderas pay for their apartment in the first year? $10,595

11. A Sidewider sells for $9,349. Power steering/power brakes are standard equipment. Air-conditioning costs an extra $855. What is the total cost of the Sidewider? $10,204

OBJECTIVE

- Evaluate achievement of the chapter objectives.

USING THE TEST

The Chapter Test may be used as a posttest to evaluate student achievement. However, you may wish to use the Chapter Posttest offered in the Teacher's Resource Package or to design your own chapter test. If this page is not used as a test, you may wish to assign it as additional review or practice.

The test items are correlated to the chapter objectives in the table below.

Chapter objectives	Test items
A. Make a grouped-frequency table.	1
B. Find statistical measures.	2–5
C. Read and interpret data from a graph.	6–7
D. Make a broken-line graph.	8
E. Make a table to solve problems.	9
F. Apply computational skills to real-life situations.	10–11

SUPPLEMENTARY MATERIALS

TRP Ch. 4 Posttest Form A, pp. 1–2

TRP Ch. 4 Posttest Form B, pp. 1–2

OBJECTIVE

- Use a scattergram to interpret data.

USING THE PAGE

Before referring to the lesson, ask students the following questions:

- What do you think the relationship is between the number of hours you study for a test and your test grade?
- What do you think the relationship is between the time you spend taking a test and the number of errors you make?

Discuss the graphs on the lesson page. Point out that *correlation* is the relationship between two variables or events. There can be a *positive correlation* (as one variable increases so does the other), or a *negative correlation* (as one variable increases the other decreases). Sometimes there is no correlation, that is, there is no relationship between the variables or events.

Emphasize that the location of a particular point is not important in this type of graph. Instead, the type of correlation is determined by how the points cluster.

As a follow-up activity, take a class poll to determine if there is any correlation between the distance a student lives from school and the likelihood that he or she will participate in after-school activities. Discuss the accuracy of the conclusions. Ask if there are other factors that would influence a student's decision to participate in after-school activities.

ENRICHMENT SCATTERGRAMS

A **scattergram** is a graph that shows whether a correlation exists between two sets of data.

Simone graphed the two scattergrams below. The graph on the left shows the number of hours she studied for each test and her grade for each test. The graph on the right shows the total time each student spent working on a test and the total number of each student's errors.

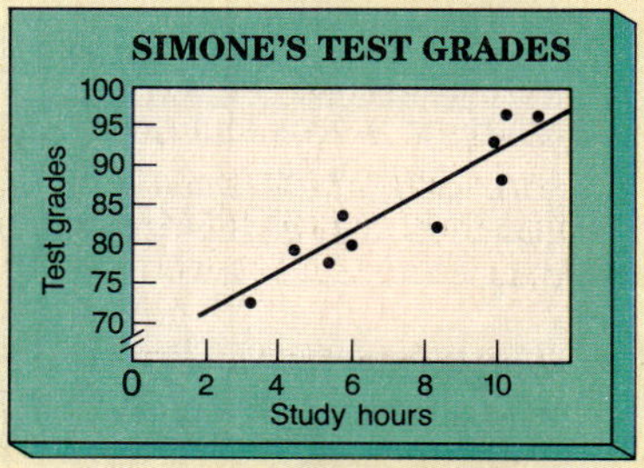

Notice that the line slants upward. This graph shows a **positive correlation** between test grades and study hours. This means Simone's test grades increased as her study hours increased.

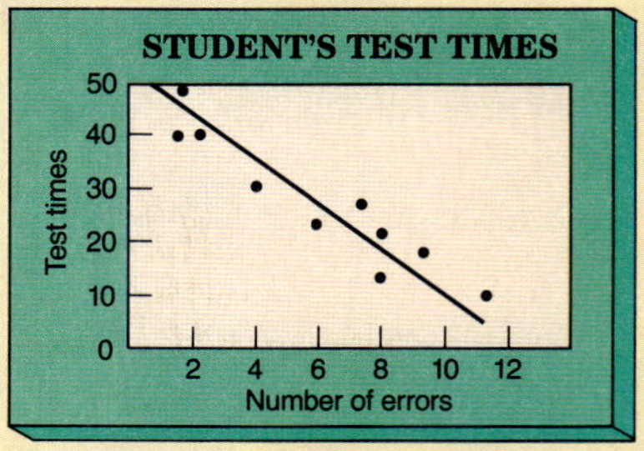

Notice that the line slants downward. This graph shows a **negative correlation** between test time and number of errors. This means their number of errors decreased as test time increased.

If a straight line cannot be drawn close to the points, there is **no correlation** between the data.

Examples Tell which kind of correlation you would probably see in each scattergram.

A scattergram that shows speed versus distance.

THINK: As speed increases, distance increases.

So, this scattergram would show a positive correlation.

A scattergram that shows speed versus fuel economy.

THINK: As speed increases, fuel economy decreases.

So, this scattergram would show a negative correlation.

Tell which kind of correlation you would probably see in the scattergram.

1. A scattergram that shows speed versus stopping distance positive

2. A scattergram that shows a person's weight versus salary no correlation

3. A scattergram that shows practice hours versus number of base hits positive

4. A scattergram that shows a baby's age in months versus number of sleeping hours negative

MENTAL MATH CLUSTERING

Sometimes several addends are "clustered" around each other. That is, they are all close to the same value. In such cases, you can multiply to find an estimate. Using the estimate, you add and subtract mentally to find the actual sum.

Add mentally: 95 + 87 + 96

1. Estimate by multiplying.

   THINK: Each of the 3 numbers is about 90.

   3 × 90 = 270 ⟵ **Estimated sum**

2. Adjust the estimate.

   THINK: 95 is 5 more than 90. 5 more than 270 is 275.
   87 is 3 less than 90. 3 less than 275 is 272.
   96 is 6 more than 90. 6 more than 272 is 278.

The sum is 278.

Add mentally: 62 + 58 + 59 + 56 + 61

1. Estimate by multiplying.

   THINK: Each of the 5 numbers is about 60.

   5 × 60 = 300 ⟵ **Estimated sum**

2. Adjust the estimate.

   THINK: 62 is 2 more than 60. 2 more than 300 is 302.
   58 is 2 less than 60. 2 less than 302 is 300.
   59 is 1 less than 60. 1 less than 300 is 299.
   56 is 4 less than 60. 4 less than 299 is 295.
   61 is 1 more than 60. 1 more than 295 is 296.

The sum is 296.

Find the sum mentally. Use the clustering method.

1. 22 + 25 + 16 63
2. 75 + 80 + 83 238
3. 21 + 19 + 16 + 24 80
4. 35 + 44 + 42 + 38 159
5. 27 + 25 + 33 + 30 115
6. 85 + 85 + 79 + 81 330
7. 51 + 60 + 64 + 54 + 69 298
8. 69 + 71 + 78 + 65 + 62 345
9. 56 + 41 + 45 + 51 + 55 248
10. 104 + 98 + 102 + 95 + 96 495
11. 90 + 98 + 88 + 92 + 85 453
12. 13 + 24 + 17 + 23 + 27 + 22 126

OBJECTIVE

- Add several numbers mentally, using the clustering method.

USING THE PAGE

You may wish to point out that an estimated sum can be adjusted more easily if there are pairs of addends that would result in the addition and subtraction of the same number. To illustrate this shortcut, consider the steps in Example 2 in which the estimate is adjusted for 62 and for 58.

62 is 2 more than 60.

2 more than **300** is 302.

58 is 2 less than 60.

2 less than 302 is **300**.

Addition and subtraction are inverse operations. Therefore, since 2 is added to 300 and then 2 is subtracted, you are back to 300 again. You do not need to consider 62 and 58 when you adjust the estimate.

Similarly, you do not need to consider 59 and 61, since 59 is 1 less than 60 and 61 is 1 more than 60. To adjust the estimated sum of 300, you need to use only the remaining addend, 56.

THINK: 56 is 4 less than 60.
4 less than 300 is 296.

The sum is 296.

At first, some students may find it difficult to use the clustering technique. Point out that methods of mental computation, as well as methods of estimation, become truly helpful only when their use is almost automatic. This takes practice! Be on the lookout for opportunities to use mental arithmetic and estimates in the classroom. Also, encourage students to use these techniques whenever they can in their daily lives.

OBJECTIVE

- Test the prerequisite skills needed to learn the concepts developed in Chapter 5.

USING THE TEST

The Pre-Skills Test is designed to diagnose students' strengths and weaknesses on prerequisite skills necessary to study the mathematics in Chapter 5.

Assign the Pre-Skills Test. Allow students to work together in pairs or small groups. Group members should help those who demonstrate a misunderstanding of a concept or a weakness in a skill.

The following table correlates the items on the Pre-Skills Test with the prerequisite skill and the lesson(s) in the chapter for which it is needed.

Item(s)	Prerequisite skill	Lesson
1–4	Identify points on a number line.	5.1
5–8	Order whole numbers.	5.4
9–12	Complete counting sequences.	5.2
13–18	Find missing factors.	5.2
19–24	Multiply, then add.	5.5
25–32	Divide whole numbers.	5.7
33–40	Divide, round answer to nearest hundredth.	5.7

Pre-Skills Test

Use the number line to identify each point. [8]

1. A 86 **2.** B 97 **3.** C 69 **4.** D 54

Arrange in order from greatest to least. [8]

5. 45, 76, 28, 66, 70 76, 70, 66, 45, 28

6. 54, 60, 98, 65, 56 98, 65, 60, 56, 54

7. 508, 58, 85, 580, 805 805, 580, 508, 85, 58

8. 917, 179, 197, 719, 791 917, 791, 719, 197, 179

Complete the sequence.

9. 4, 8, 12, ■, ■, ■, ■ 16, 20, 24, 28

10. 7, 14, 21, ■, ■, ■, ■ 28, 35, 42, 49

11. 12, 24, 36, ■, ■, ■, ■ 48, 60, 72, 84

12. 15, 30, 45, ■, ■, ■, ■ 60, 75, 90, 105

Complete.

13. 1 × ■ = 12 12
2 × ■ = 12 6
3 × ■ = 12 4

14. 1 × ■ = 15 15
3 × ■ = 15 5

15. 1 × ■ = 18 18
2 × ■ = 18 9
3 × ■ = 18 6

16. 1 × ■ = 27 27
3 × ■ = 27 9

17. 1 × ■ = 30 30
2 × ■ = 30 15
3 × ■ = 30 10
5 × ■ = 30 6

18. 1 × ■ = 42 42
2 × ■ = 42 21
3 × ■ = 42 14
6 × ■ = 42 7

Compute.

19. $(6 \times 7) + 2$ 44 **20.** $(4 \times 5) + 3$ 23 **21.** $(3 \times 8) + 6$ 30

22. $(4 \times 10) + 7$ 47 **23.** $(5 \times 11) + 1$ 56 **24.** $(6 \times 12) + 8$ 80

Divide. Name the quotient and remainder. [46, 48]

25. $9 \div 4$ 2 R1 **26.** $15 \div 6$ 2 R3 **27.** $25 \div 8$ 3 R1 **28.** $40 \div 7$ 5 R5

29. $20 \div 10$ 2 **30.** $37 \div 15$ 2 R7 **31.** $47 \div 12$ 3 R11 **32.** $50 \div 16$ 3 R2

Divide. Round the answer to the nearest hundredth. [78]

33. $9\overline{)5}$ 0.56 **34.** $8\overline{)7}$ 0.88 **35.** $6\overline{)8}$ 1.33 **36.** $7\overline{)23}$ 3.29

37. $24\overline{)13}$ 0.54 **38.** $12\overline{)7}$ 0.58 **39.** $15\overline{)34}$ 2.27 **40.** $24\overline{)55}$ 2.29

Chapter 5

FRACTIONS

Teamwork is necessary in baseball and many other sports. In what situations can working as a team be useful?

OBJECTIVE

- Formulate problems using text and data.

USING THE CHAPTER OPENER

Discuss the question on the page, making a list of situations on the chalkboard. Students might suggest various team sports; operating a business such as a store, restaurant, or factory; professional groups such as doctors or lawyers, even the teaching staff of a high school. Students may not realize, however, that there are actually few jobs that can be performed truly independently. Ask for some examples. They might suggest the following, but point out how these usually require other people.

tennis pro (trainer, coach, staff)

doctor (nurse, receptionist, accountant, lab to run tests, etc.)

Truly individual occupations include artists, independent vendors, housekeepers, consultants and so on.

COOPERATIVE LEARNING

Write the following on the chalkboard or substitute whatever sport is most popular in your area.

New softball league being formed

We need: teams, sponsors, equipment, uniforms, and more

Assign groups of students the task of organizing an imaginary softball team. They are to devise a practical plan for each aspect of the situation. As plans progress, allow the class time to analyze each and determine what is missing, which is most feasible, and which is the most original. Each overall plan should include a plan to attract a coach and talented team members, as well as sponsors to pay for uniforms and equipment and a budget for uniforms, equipment, and supplies.

SUPPLEMENTARY MATERIALS

TRP Group Projects, p. 5

OBJECTIVES

- Identify fractions as part of a whole or part of a set.
- Identify fractions on a number line.

TEACHING THE LESSON

WARM-UP Present the following exercises on the chalkboard or on an overhead transparency.

Write the number of each point.

1. *A* **2.** *B* **3.** *C* **4.** *D*

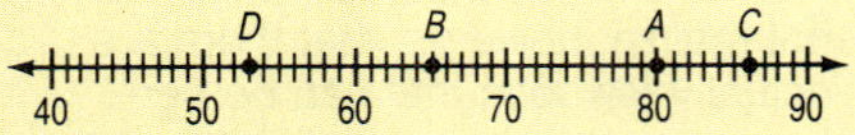

(**1.** 80 **2.** 65 **3.** 86 **4.** 53)

INTRODUCTION You may wish to inform students that fractions were used as early as 2000 B.C. by the Babylonians and Egyptians. Later, they were used by the Greeks and Romans. However, they were not written as they are today. The use of a numerator and denominator, separated by a fraction bar, was introduced in Europe by the Arabs around the twelfth century.

INSTRUCTION In discussing Example 1, be sure to emphasize that the pizza is divided into 8 **equal** parts. Ask students what fractions would be represented by 1, 3, and 7 parts of the pizza. $\left(\frac{1}{8}, \frac{3}{8}, \frac{7}{8}\right)$

In discussing Example 2, note that the process is the same as in Example 1. Emphasize that the 9 circles may be thought of as one whole. Thus, each circle will represent $\frac{1}{9}$ of the set.

You may wish to relate Example 3 to the distance from school to a home 1 mi away. Let 0 be the location of the school and 1 the location of the home.

FOR DISCUSSION Students should explain that since the denominator represents the total number of parts into which a whole number is divided, it must be at least 1. Therefore, it cannot be zero.

5.1 INTRODUCTION TO FRACTIONS

You can write a fraction to name part of a whole.

The top number, the **numerator,** identifies the number of equal parts being considered.

The bottom number, the **denominator,** identifies the total number of equal parts.

EXAMPLE 1 A pizza was sliced into 8 equal pieces. Jim ate 3 slices. What fraction of the pie did Jim eat?

THINK: **Numerator →** 3 **← Slices eaten**
Denominator → 8 **← Total number of slices**

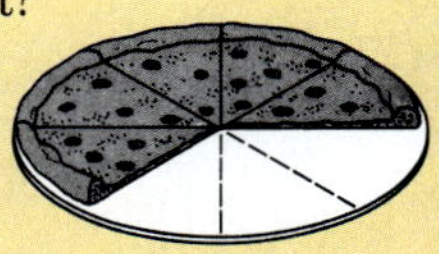

Jim ate $\frac{3}{8}$ of the pie.

You can write a fraction to name part of a group of objects.

EXAMPLE 2 What fraction of the circles are red? What fraction of the circles are blue?

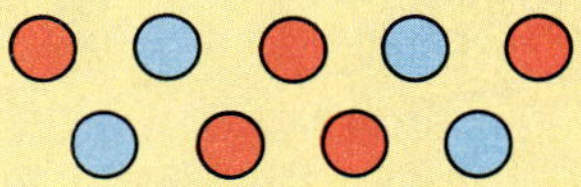

5 **← Number of red circles** 9 **← Total number of circles**	4 **← Number of blue circles** 9 **← Total number of circles**
$\frac{5}{9}$ of the circles are red.	$\frac{4}{9}$ of the circles are blue.

You can also write a fraction to identify a point on a number line.

EXAMPLE 3 What is point *A* on the number line?

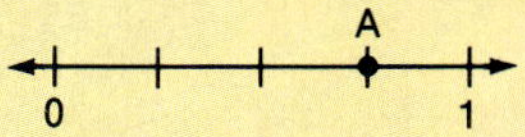

THINK: The distance from 0 to 1 is divided into 4 equal parts. The distance from 0 to *A* is 3 of these parts.

Point *A* is $\frac{3}{4}$ on the number line.

FOR DISCUSSION See TE side column.

Can the denominator of a fraction ever be zero? Explain. (*Hint:* What is the least number of parts that you could divide an object into?)

PRACTICE EXERCISES See Extra Practice, page 435.

Write a fraction for the shaded part.

• 1. 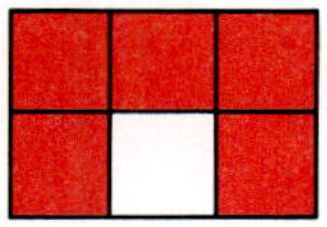 $\frac{5}{6}$

• 2. 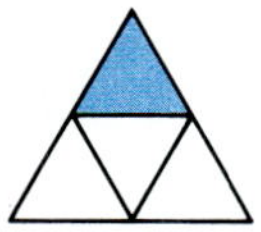 $\frac{1}{4}$

3. 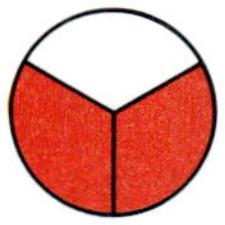 $\frac{2}{3}$

4.  $\frac{9}{16}$

5. 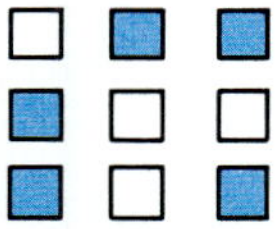 $\frac{5}{9}$

6. 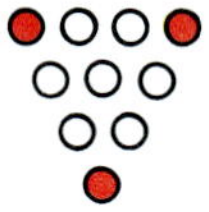 $\frac{3}{10}$

7. 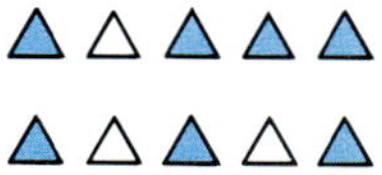 $\frac{7}{10}$

8. 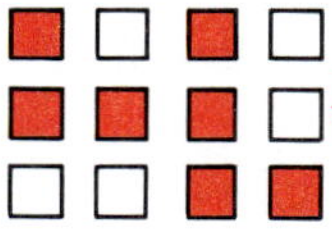 $\frac{7}{12}$

9.  $\frac{3}{4}$

10. 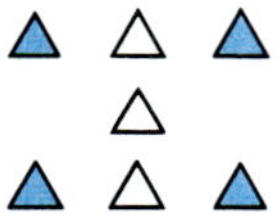 $\frac{4}{7}$

11.  $\frac{9}{14}$

12. 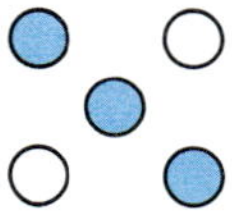 $\frac{3}{5}$

Write a fraction for the point.

• 13. 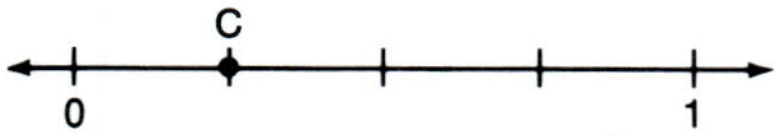

$\frac{1}{4}$

• 14. 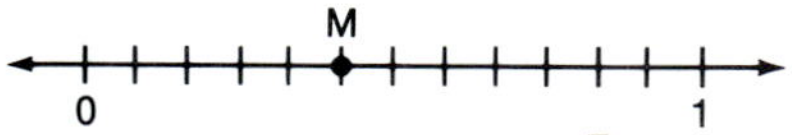

$\frac{5}{12}$

15. 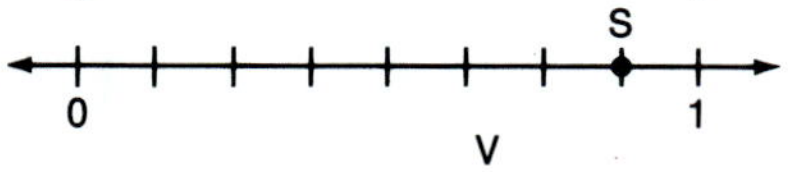

$\frac{7}{8}$

16. 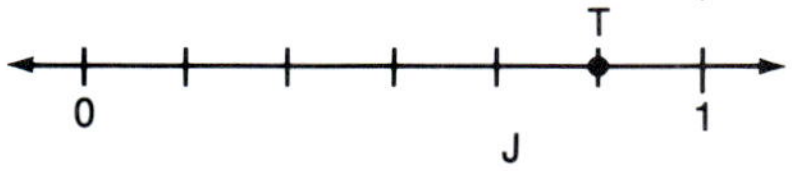

$\frac{5}{6}$

17. 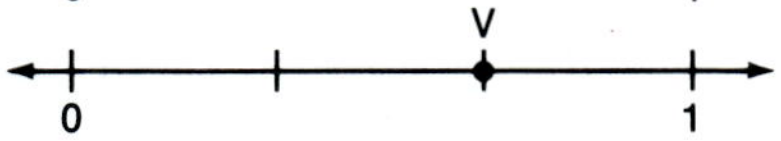

$\frac{2}{3}$

18. 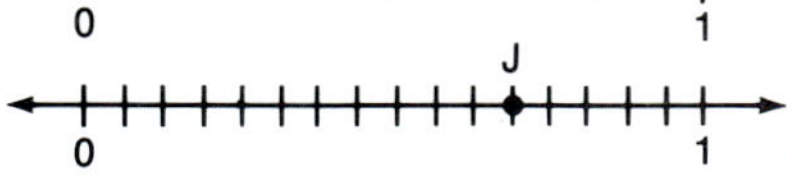

$\frac{11}{16}$

Solve.

• **19.** There are six compartments in a freezer. If five of the compartments are filled, what fraction of the freezer is filled? $\frac{5}{6}$

20. A pizza costs \$6.99. Each extra topping costs \$0.98. How much will it cost to buy a pizza with meatballs, peppers, and onions? \$9.93

21. James ordered a pizza that was divided into 10 equal slices. Three of the slices were topped with mushrooms. What fraction of the pizza did not have mushrooms? $\frac{7}{10}$

22. Ellen has nine friends at a party. She plans to serve each friend three slices of pizza. If each pizza has eight slices, how many pizzas does she need? 4 pizzas

TIME OUT The world's largest pizza was baked in 1978 at the Oma Pizza Restaurant in Glen Falls, New York. It weighed 18,664 pounds and was cut into 60,318 slices. About how many years would it take one very hungry person to eat the entire pie if that person eats three slices every day? about 55 years

TIME OUT A two-step calculation is necessary to solve the problem.

1. $3 \times 365 = 1{,}095$
2. $60{,}318 \div 1{,}095 \approx 55$

COMMON ERROR

In determining the number of equal parts into which the number line has been divided, some students may count the number of points on the line rather than the number of jumps. In Example 3, these students will say that the line is divided into 5 equal parts, and point *A* is 4 from 0. Have the students mark the jumps on the number line to identify the number of equal parts.

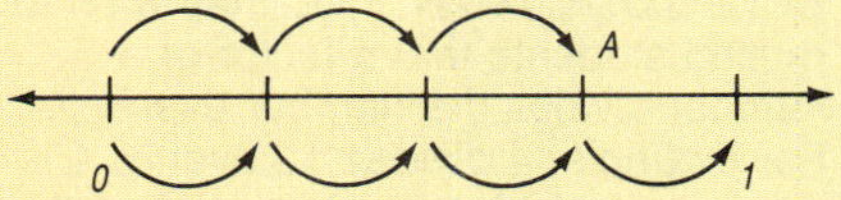

ASSIGNMENTS

Level 1	Odd 1–17, 19, 20
Level 2	Even 2–22, TO
Level 3	Even 6–18, 19–22, TO

FOLLOW-UP ACTIVITY

ENRICHMENT Have students name the following fractions.

1. The denominator is 5 and the numerator is three less than the denominator. $\left(\frac{2}{5}\right)$
2. The numerator is 7 and the denominator is twice as large as the numerator. $\left(\frac{7}{14}\right)$
3. The sum of the numerator and denominator is 5 and their product is 6. $\left(\frac{2}{3} \text{ or } \frac{3}{2}\right)$

SUPPLEMENTARY MATERIALS

TRP Practice, p. 50

TRP Reteaching, p. 30

TRP Lesson Aids, p. 3

TRP Transparency 25

OBJECTIVES

- Find the GCF of a pair of numbers.
- Find the LCM of a pair of numbers.

TEACHING THE LESSON

WARM-UP Present the following exercises on the chalkboard or on an overhead transparency.

Complete these sequences.

1. 3, 6, 9, ■, ■, ■ (12, 15, 18)
2. 6, 12, 18, ■, ■, ■ (24, 30, 36)
3. 4, 8, 12, ■, ■, ■ (16, 20, 24)
4. 8, 16, 24, ■, ■, ■ (32, 40, 48)

INTRODUCTION You may wish to remind students that a factor of a number divides the number evenly. For example, 4 divides 12 evenly. 4 is a factor of 12. Ask students to find the other factors of 12.

INSTRUCTION Discuss the common factors of 12 and 18. Since 6 is the greatest common factor of 12 and 18, it is the largest number that divides both of them evenly.

For Example 1 you may wish to point out that the set of factors common to 16 and 24 turns out to be the set of factors of 8 (the GCF of 16 and 24).

In discussing Example 2, point out that the multiples of 3 can be obtained by counting by threes; the multiples of 4 are determined by counting by fours.

CHECKPOINT Distribute to each student four index cards on which the answer choices a–d appear. After allowing students a few minutes, ask each to raise the letter that indicates their answer choice. Scan the cards and identify those students who need additional attention before proceeding to the Practice Exercises. The incorrect answer choices include common errors.

5.2 GCF AND LCM

To find the factors of 16, list all the numbers which when multiplied give you 16.

$16 = 1 \times 16 \qquad 16 = 2 \times 8 \qquad 16 = 4 \times 4$

So, the factors of 16 are: 1, 2, 4, 8, 16.

Some numbers have common factors. The greatest of these common factors is called the **greatest common factor, GCF.**

EXAMPLE 1

Find the greatest common factor (GCF) of 16 and 24.

1. List all the factors of each number. 16: 1, 2, 4, 8, 16; 24: 1, 2, 3, 4, 6, 8, 12, 24
2. Identify the common factors. 1, 2, 4, 8
3. Choose the greatest common factor. 8

So, 8 is the greatest common factor (GCF) of 16 and 24.

The product of any number and 3 is a **multiple** of 3. Here are some multiples of 3.

$0 \times 3 = 0$	$3 \times 3 = 9$	$6 \times 3 = 18$
$1 \times 3 = 3$	$4 \times 3 = 12$	$7 \times 3 = 21$
$2 \times 3 = 6$	$5 \times 3 = 15$	$8 \times 3 = 24$

You can write the multiples of 3 as: 0, 3, 6, 9, 12, 15, 18, 21, 24, ... (The three dots mean *and so on.*)

Some numbers have common multiples. The least of these common multiples, other than 0, is called the **least common multiple, LCM.**

EXAMPLE 2

Find the least common multiple (LCM) of 3 and 4.

1. List some multiples of each number. 3: 0, 3, 6, 9, 12, 15, 18, 21, 24, . . . 4: 0, 4, 8, 12, 16, 20, 24, . . .
2. Identify the common multiples. 0, 12, 24, . . .
3. Choose the least common multiple, other than 0. 12

So, 12 is the least common multiple (LCM) of 3 and 4.

CHECKPOINT

Write the letter of the correct answer.

1. Find the greatest common factor (GCF) of 24 and 32. b
   a. 4 **b.** 8 **c.** 12 **d.** 96
2. Find the least common multiple (LCM) of 24 and 32. d
   a. 8 **b.** 48 **c.** 64 **d.** 96

PRACTICE EXERCISES See Extra Practice, page 435.

Use the table to determine the greatest common factor (GCF).

Number	Factors	Number	Factors
6	1, 2, 3, 6	16	1, 2, 4, 8, 16
8	1, 2, 4, 8	18	1, 2, 3, 6, 9, 18
9	1, 3, 9	24	1, 2, 3, 4, 6, 8, 12, 24
12	1, 2, 3, 4, 6, 12	36	1, 2, 3, 4, 6, 9, 12, 18, 36

• **1.** 6, 8 2 • **2.** 8, 12 4 **3.** 9, 18 9 **4.** 6, 16 2 **5.** 8, 18 2
6. 9, 24 3 **7.** 12, 6 6 **8.** 18, 24 6 **9.** 12, 18 6 **10.** 24, 36 12

Find the greatest common factor (GCF).

•**11.** 4, 6 2 •**12.** 9, 12 3 **13.** 5, 7 1 **14.** 4, 12 4 **15.** 10, 15 5
16. 12, 14 2 **17.** 18, 30 6 **18.** 12, 20 4 **19.** 15, 21 3 **20.** 14, 35 7
21. 11, 21 1 **22.** 24, 32 8 **23.** 18, 27 9 **24.** 24, 42 6 **25.** 27, 45 9
26. 56, 72 8 **27.** 36, 54 9 **28.** 42, 60 6 **29.** 15, 75 15 **30.** 56, 64 8

Use the table to determine the least common multiple (LCM).

Number	Multiples	Number	Multiples
2	0, 2, 4, 6, 8, 10, 12, 14, . . .	8	0, 8, 16, 24, 32, 40, 48, 56, . . .
3	0, 3, 6, 9, 12, 15, 18, 21, . . .	9	0, 9, 18, 27, 36, 45, 54, 63, . . .
4	0, 4, 8, 12, 16, 20, 24, 28, . . .	12	0, 12, 24, 36, 48, 60, 72, 84, . . .
5	0, 5, 10, 15, 20, 25, 30, 35, . . .	16	0, 16, 32, 48, 64, 80, 96, 112, . . .

•**31.** 2, 3 6 •**32.** 2, 4 4 **33.** 3, 5 15 **34.** 3, 9 9 **35.** 2, 5 10
36. 4, 5 20 **37.** 9, 12 36 **38.** 8, 12 24 **39.** 8, 16 16 **40.** 12, 16 48

Find the least common multiple (LCM).

•**41.** 4, 6 12 •**42.** 3, 15 15 **43.** 4, 3 12 **44.** 6, 9 18 **45.** 9, 15 45
46. 14, 21 42 **47.** 12, 15 60 **48.** 9, 12 36 **49.** 24, 8 24 **50.** 5, 6 30
51. 12, 18 36 **52.** 27, 18 54 **53.** 15, 20 60 **54.** 12, 16 48 **55.** 10, 12 60
56. 24, 36 72 **57.** 20, 25 100 **58.** 28, 21 84 **59.** 15, 18 90 **60.** 64, 48 192

COMMON ERROR

Students may often confuse the GCF with the LCM. For example, in item 1 under Checkpoint they would select d. It may be helpful if you equate "multiple" with "more" and note that a multiple of a number must be equal to or greater than the number. A factor of a number is its divisor, and thus must be less than or equal to the number.

ASSIGNMENTS

Level 1	1–20, 31–50
Level 2	Odd 1–59
Level 3	21–30, 51–60

FOLLOW-UP ACTIVITY

ENRICHMENT Show how to find the prime factorization of a number. Begin by reviewing the definitions of a prime number and a composite number. A **prime number** is a number whose factors are 1 and itself. A **composite number** is a number that has more than two factors.

Students may draw a factor tree to find the prime factorization of a composite number.

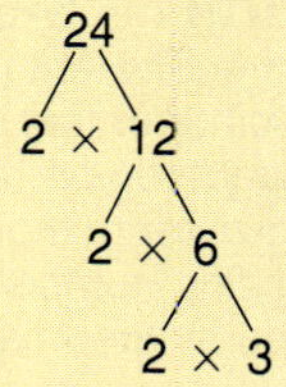

$24 = 2 \times 2 \times 2 \times 3$

Have students draw factor trees to find the prime factorizations of 18, 40, and 60.

SUPPLEMENTARY MATERIALS

TRP Practice, p. 51

TRP Reteaching, p. 31

TRP Transparency 26

OBJECTIVES

- Write a fraction in lowest terms.
- Find the missing term for two equivalent fractions.
- Determine if two fractions are equivalent.

TEACHING THE LESSON

WARM-UP Present the following exercises on the chalkboard or on an overhead transparency.

Find the GCF for each pair of numbers.

1. 3 and 6 (3) **2.** 8 and 12 (4)

3. 4 and 14 (2) **4.** 3 and 9 (3)

5. 24 and 64 (8) **6.** 45 and 90 (45)

INTRODUCTION Have students draw rectangular regions 1 cm by 6 cm. Guide them in dividing one region into 6 equal parts, and another region into 12 equal parts. Have them shade three parts in the first region, and six parts in the second region. Have them write a fraction to represent the shaded parts in each region. Students may recognize that these are equivalent fractions.

INSTRUCTION In discussing the illustrations of equivalent fractions in this lesson, point out that although each subsequent region has been divided into a greater number of parts, the corresponding number of parts that have been shaded has also increased.

Emphasize that you can multiply or divide the numerator and denominator by any nonzero number to form an equivalent fraction.

Note in Example 1 that by using the GCF as the divisor, you can always reduce to lowest terms in one step.

In discussing Example 2, point out that when going from a smaller denominator to a larger one, multiply; when going from a larger denominator to a smaller one, divide.

5.3 EQUIVALENT FRACTIONS

Equivalent fractions name the same amount using different terms.
$\frac{2}{3}$, $\frac{4}{6}$, and $\frac{8}{12}$ are equivalent fractions.

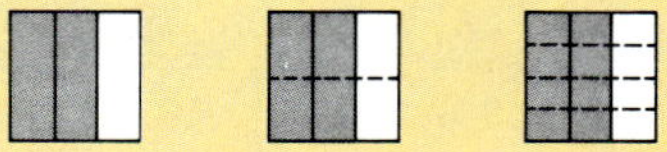

$\frac{2}{3} = \frac{4}{6} = \frac{8}{12}$

You can find an equivalent fraction by multiplying or dividing the numerator and denominator of a fraction by the same nonzero number.

To find fractions equivalent to $\frac{4}{6}$, multiply or divide the numerator and denominator by 2.

Multiplying by 2

$\frac{4}{6} = \frac{4 \times 2}{6 \times 2} = \frac{8}{12}$

Dividing by 2

$\frac{4}{6} = \frac{4 \div 2}{6 \div 2} = \frac{2}{3}$

$\frac{2}{3}$ is a fraction in lowest terms. The greatest common factor (GCF) of the numerator and denominator is 1. You can write a fraction in lowest terms by dividing the numerator and denominator by the GCF.

EXAMPLE 1 Write $\frac{10}{25}$ in lowest terms.

THINK: GCF of 10 and 25 is 5. Divide by 5. $\frac{10 \div 5}{25 \div 5} = \frac{2}{5}$

So, $\frac{10}{25}$ in lowest terms is $\frac{2}{5}$.

Sometimes you need to find the missing term of two equivalent fractions.

EXAMPLE 2 Complete: $\frac{3}{4} = \frac{\blacksquare}{12}$

THINK: **4 × ■ = 12**
4 × 3 = 12

Multiply the numerator by 3.

$\frac{3}{4} = \frac{9}{12}$ (×3 numerator, ×3 denominator)

So, 9 is the missing term.

You can compare cross products to determine if two fractions are equivalent.

EXAMPLE 3 Using cross products, determine if the fractions are equivalent.

a. $\frac{3}{5}$ and $\frac{9}{12}$

$\frac{3}{5} \times \frac{9}{12}$ **THINK:** **3 × 12 = 36**
9 × 5 = 45
36 ≠ 45

So, $\frac{3}{5}$ and $\frac{9}{12}$ are not equivalent.

b. $\frac{4}{8}$ and $\frac{7}{14}$

$\frac{4}{8} \times \frac{7}{14}$ **THINK:** **4 × 14 = 56**
7 × 8 = 56
56 = 56

So, $\frac{4}{8}$ and $\frac{7}{14}$ are equivalent.

FOR DISCUSSION See TE side column.

Is it possible for a fraction to be in lowest terms if the numerator and denominator are both even numbers? Explain.

PRACTICE EXERCISES See Extra Practice, page 436.

Write the fraction in lowest terms.

•1. $\frac{3}{9}$ $\frac{1}{3}$ •2. $\frac{9}{18}$ $\frac{1}{2}$ 3. $\frac{8}{24}$ $\frac{1}{3}$ 4. $\frac{7}{21}$ $\frac{1}{3}$ 5. $\frac{8}{12}$ $\frac{2}{3}$ 6. $\frac{4}{14}$ $\frac{2}{7}$ 7. $\frac{12}{15}$ $\frac{4}{5}$ 8. $\frac{6}{8}$ $\frac{3}{4}$

9. $\frac{12}{18}$ $\frac{2}{3}$ 10. $\frac{10}{15}$ $\frac{2}{3}$ 11. $\frac{15}{18}$ $\frac{5}{6}$ 12. $\frac{18}{24}$ $\frac{3}{4}$ 13. $\frac{12}{16}$ $\frac{3}{4}$ 14. $\frac{9}{24}$ $\frac{3}{8}$ 15. $\frac{14}{21}$ $\frac{2}{3}$ 16. $\frac{9}{15}$ $\frac{3}{5}$

17. $\frac{12}{21}$ $\frac{4}{7}$ 18. $\frac{20}{35}$ $\frac{4}{7}$ 19. $\frac{18}{21}$ $\frac{6}{7}$ 20. $\frac{15}{20}$ $\frac{3}{4}$ 21. $\frac{16}{24}$ $\frac{2}{3}$ 22. $\frac{15}{21}$ $\frac{5}{7}$ 23. $\frac{21}{28}$ $\frac{3}{4}$ 24. $\frac{20}{24}$ $\frac{5}{6}$

25. $\frac{12}{36}$ $\frac{1}{3}$ 26. $\frac{24}{36}$ $\frac{2}{3}$ 27. $\frac{24}{27}$ $\frac{8}{9}$ 28. $\frac{15}{24}$ $\frac{5}{8}$ 29. $\frac{16}{20}$ $\frac{4}{5}$ 30. $\frac{12}{28}$ $\frac{3}{7}$ 31. $\frac{36}{45}$ $\frac{4}{5}$ 32. $\frac{50}{75}$ $\frac{2}{3}$

Complete.

•33. $\frac{1}{3} = \frac{■}{9}$ 3 •34. $\frac{3}{5} = \frac{■}{10}$ 6 35. $\frac{2}{3} = \frac{■}{6}$ 4 36. $\frac{4}{7} = \frac{■}{21}$ 12 37. $\frac{5}{6} = \frac{■}{18}$ 15

38. $\frac{3}{4} = \frac{■}{16}$ 12 39. $\frac{7}{8} = \frac{■}{24}$ 21 40. $\frac{8}{9} = \frac{■}{27}$ 24 41. $\frac{6}{9} = \frac{12}{■}$ 18 42. $\frac{5}{7} = \frac{35}{■}$ 49

43. $\frac{12}{15} = \frac{■}{5}$ 4 44. $\frac{9}{18} = \frac{■}{6}$ 3 45. $\frac{16}{48} = \frac{■}{12}$ 4 46. $\frac{24}{36} = \frac{■}{9}$ 6 47. $\frac{18}{24} = \frac{■}{8}$ 6

48. $\frac{20}{32} = \frac{■}{16}$ 10 49. $\frac{16}{40} = \frac{■}{10}$ 4 50. $\frac{30}{54} = \frac{■}{18}$ 10 51. $\frac{12}{42} = \frac{4}{■}$ 14 52. $\frac{24}{48} = \frac{3}{■}$ 6

Using cross products, determine if the fractions are equivalent. Write $=$ or $\neq$.

•53. $\frac{2}{3}$ ● $\frac{6}{9}$ $=$ •54. $\frac{3}{4}$ ● $\frac{9}{16}$ $\neq$ 55. $\frac{1}{5}$ ● $\frac{4}{20}$ $=$ 56. $\frac{5}{6}$ ● $\frac{10}{12}$ $=$ 57. $\frac{8}{15}$ ● $\frac{4}{5}$ $\neq$

58. $\frac{9}{12}$ ● $\frac{15}{20}$ $=$ 59. $\frac{12}{20}$ ● $\frac{15}{25}$ $=$ 60. $\frac{8}{12}$ ● $\frac{12}{18}$ $=$ 61. $\frac{10}{16}$ ● $\frac{18}{24}$ $\neq$ 62. $\frac{12}{21}$ ● $\frac{20}{35}$ $=$

Solve. Use the picture at the right.

•63. Alice Tompkins bought several jars of baby food. What fraction of the jars are filled with pears? Give two equivalent fractions. $\frac{2}{8}$, $\frac{1}{4}$

64. What fraction of the jars are filled with vegetables? Give three equivalent fractions. $\frac{4}{8}$, $\frac{2}{4}$, $\frac{1}{2}$

In Example 3a, identify the numerators of the two fractions when the denominators are both 60, 5 × 12. Elicit from students what the cross products identify in Example 3b (the numerators of the two fractions when the denominators are both 112, 8 × 14).

FOR DISCUSSION Students should explain that since the numerator and denominator are even, they have a common factor of 2. Therefore, the fractions are not in lowest terms.

COMMON ERROR

In writing equivalent fractions, some students will forget to multiply or divide both the numerator and denominator by the same number. Require these students to show all the steps outlined at the beginning of the lesson.

ASSIGNMENTS

Level 1	1–32, Odd 33–61, 63
Level 2	Even 2–62, 63, 64
Level 3	25–42, 53–64

FOLLOW-UP ACTIVITY

ENRICHMENT Show how to use prime factorization to simplify fractions.

Example: Write $\frac{27}{45}$ in lowest terms.

Write the prime factorization of both terms, and cancel common factors.

$$\frac{27}{45} = \frac{\not{3} \times \not{3} \times 3}{\not{3} \times \not{3} \times 5} = \frac{3}{5}$$

Have students use prime factorization to simplify these fractions.

1. $\frac{32}{56}$ $\left(\frac{4}{7}\right)$ 2. $\frac{27}{36}$ $\left(\frac{3}{4}\right)$ 3. $\frac{30}{54}$ $\left(\frac{5}{9}\right)$

4. $\frac{42}{48}$ $\left(\frac{7}{8}\right)$ 5. $\frac{36}{45}$ $\left(\frac{4}{5}\right)$ 6. $\frac{60}{72}$ $\left(\frac{5}{6}\right)$

SUPPLEMENTARY MATERIALS

TRP Practice, p. 52

TRP Reteaching, p. 32

TRP Transparency 26

OBJECTIVE

- Compare and order fractions.

TEACHING THE LESSON

WARM-UP Present the following exercises on the chalkboard.

Find the LCM.

1. 4 and 16 **2.** 6 and 9

3. 12 and 15 **4.** 2, 3, and 5

5. 3, 7, and 9

(**1.** 16 **2.** 18 **3.** 60 **4.** 30 **5.** 63)

INTRODUCTION Draw the following diagram on the chalkboard or overhead transparency.

Ask students to identify what fractional part of each figure is shaded. Then ask which fraction is greater by determining which figure has more shaded regions.

Write $\frac{5}{8} > \frac{3}{8}$.

INSTRUCTION Use Example 1 to reinforce the concept that comparison of the numerators of two fractions with the same denominator will determine which fraction is greater.

As you discuss Example 2, ask students what common denominator was selected and why. Note that 24 is the LCM of 3 and 8. Demonstrate by listing the multiples of 3 and 8.

Multiples of 3: 3, 6, 9, 12, 15, 18, 21, 24

Multiples of 8: 8, 16, 24

Make sure students understand how to find each cross product in Example 3. Remind them that the cross products represent the numerators when a common denominator is established by multiplying the existing denominators.

5.4 COMPARING FRACTIONS

Bill and Carol were asked to paint school murals. Their murals were to be the same size. After a week, Bill had completed $\frac{9}{16}$ of his mural, and Carol had completed $\frac{5}{16}$ of her mural. Who had completed a greater portion of their mural?

$\frac{9}{16}$

$\frac{5}{16}$

You can compare two fractions that have the same denominator by comparing their numerators.

EXAMPLE 1 Compare: $\frac{9}{16}$ and $\frac{5}{16}$

THINK: Since $9 > 5$, $\frac{9}{16} > \frac{5}{16}$.

Bill painted a greater portion of his mural.

You can compare two fractions that have different denominators.

EXAMPLE 2 Compare: $\frac{3}{8}$ and $\frac{2}{3}$

1. Write equivalent fractions using a common denominator. $\frac{3}{8} = \frac{3 \times 3}{8 \times 3} = \frac{9}{24}$ $\frac{2}{3} = \frac{2 \times 8}{3 \times 8} = \frac{16}{24}$

   THINK: The LCM of 8 and 3 is 24. Use 24 as the denominator.

2. Compare the numerators. Since $9 < 16$, $\frac{9}{24} < \frac{16}{24}$.

So, $\frac{3}{8} < \frac{2}{3}$.

Another way you can compare fractions is by comparing cross products.

EXAMPLE 3 Compare: $\frac{7}{9}$ and $\frac{5}{8}$

1. Find the first cross product. $\frac{7}{9} \searrow \frac{5}{8}$ $7 \times 8 = 56$
2. Find the second cross product. $\frac{7}{9} \swarrow \frac{5}{8}$ $5 \times 9 = 45$
3. Compare cross products. Since $56 > 45$, $\frac{7}{9} > \frac{5}{8}$.

So, $\frac{7}{9} > \frac{5}{8}$.

You can order three or more fractions by comparing fractions two at a time.

EXAMPLE 4 Order $\frac{2}{5}$, $\frac{1}{4}$, and $\frac{5}{8}$ from greatest to least.

1. Write equivalent fractions using a common denominator. $\frac{2}{5} = \frac{16}{40}$ $\frac{1}{4} = \frac{10}{40}$ $\frac{5}{8} = \frac{25}{40}$
2. Compare. $\frac{16}{40} > \frac{10}{40}$ $\frac{25}{40} > \frac{16}{40}$
3. List the equivalent fractions in order from greatest to least. $\frac{25}{40}, \frac{16}{40}, \frac{10}{40}$
4. List the original fractions in order from greatest to least. $\frac{5}{8}, \frac{2}{5}, \frac{1}{4}$

FOR DISCUSSION See TE side column.

Look at the following set of fractions: $\frac{1}{4}, \frac{1}{8}, \frac{1}{16}$.
Can you identify the greatest fraction by just looking? Explain.

PRACTICE EXERCISES See Extra Practice, page 436.

Compare using equivalent fractions. Write =, <, or >.

• 1. $\frac{3}{5}$ ● $\frac{4}{5}$ <	• 2. $\frac{7}{9}$ ● $\frac{15}{9}$ <	3. $\frac{9}{10}$ ● $\frac{3}{10}$ >	4. $\frac{16}{25}$ ● $\frac{24}{25}$ <	5. $\frac{11}{12}$ ● $\frac{3}{12}$ >
6. $\frac{4}{6}$ ● $\frac{2}{3}$ =	7. $\frac{3}{8}$ ● $\frac{1}{4}$ >	8. $\frac{7}{10}$ ● $\frac{3}{5}$ >	9. $\frac{5}{6}$ ● $\frac{11}{12}$ <	10. $\frac{3}{5}$ ● $\frac{9}{15}$ =
11. $\frac{1}{8}$ ● $\frac{1}{3}$ <	12. $\frac{1}{4}$ ● $\frac{1}{5}$ >	13. $\frac{2}{3}$ ● $\frac{1}{5}$ >	14. $\frac{1}{2}$ ● $\frac{3}{7}$ >	15. $\frac{4}{5}$ ● $\frac{7}{8}$ <
16. $\frac{4}{9}$ ● $\frac{5}{6}$ <	17. $\frac{2}{9}$ ● $\frac{4}{5}$ <	18. $\frac{5}{7}$ ● $\frac{2}{3}$ >	19. $\frac{2}{9}$ ● $\frac{3}{8}$ <	20. $\frac{9}{10}$ ● $\frac{3}{4}$ >

Compare using cross products. Write =, <, or >.

• 21. $\frac{4}{7}$ ● $\frac{5}{8}$ <	• 22. $\frac{7}{8}$ ● $\frac{8}{9}$ <	23. $\frac{3}{5}$ ● $\frac{12}{20}$ =	24. $\frac{8}{16}$ ● $\frac{5}{10}$ =	25. $\frac{4}{9}$ ● $\frac{3}{8}$ >
26. $\frac{5}{7}$ ● $\frac{8}{11}$ <	27. $\frac{9}{12}$ ● $\frac{3}{4}$ =	28. $\frac{2}{3}$ ● $\frac{17}{25}$ <	29. $\frac{7}{9}$ ● $\frac{10}{12}$ <	30. $\frac{11}{15}$ ● $\frac{7}{10}$ >
31. $\frac{9}{10}$ ● $\frac{17}{20}$ >	32. $\frac{7}{10}$ ● $\frac{30}{100}$ >	33. $\frac{2}{11}$ ● $\frac{3}{10}$ <	34. $\frac{11}{20}$ ● $\frac{17}{30}$ <	35. $\frac{15}{16}$ ● $\frac{4}{5}$ >

You may also wish to establish the following relationship between cross products and fractions:

If first cross product > second cross product, then first fraction > second fraction.

If first cross product < second cross product, then first fraction < second fraction.

If first cross product = second cross product, then first fraction = second fraction.

Step 1 under Example 4 may need a detailed look. First have students determine the LCM of 4, 5, and 8 by listing. Then ask students to name the factor by which the numerator and denominator of each fraction was multiplied.

$\left(\frac{2 \times 8}{5 \times 8} = \frac{16}{40} \quad \frac{1 \times 10}{4 \times 10} = \frac{10}{40} \quad \frac{5 \times 5}{8 \times 5} = \frac{25}{40}\right)$

FOR DISCUSSION Students should explain that if the numerator stays constant, then a fraction decreases in value as the denominator increases. This can easily be demonstrated by taking a piece of paper and continually folding it. Students can then observe that as the number of regions increases, the size of each region decreases.

COMMON ERROR

When ordering fractions, many students will ignore the denominators and simply select the fraction with the greater numerator. Ask students why we cannot compare numerators when comparing fractions with different denominators. Elicit that since the denominators are different, the pieces being compared are unequal in size.

Work individually with students, using fraction strips to show that the same strip can be folded into halves, fourths, eighths, and sixteenths. Reinforce the concept that the regions are different in size.

ASSIGNMENTS

Level 1	1–30, Even 36–50, 52–70
Level 2	Odd 1–51, 52–73
Level 3	16–20, 31–51, Odd 53–67, 68–73

FOLLOW-UP ACTIVITY

ESTIMATION Begin developing a number sense of the relative sizes of fractions as compared with 0, $\frac{1}{2}$, and 1.

Present the following fractions:

$\frac{6}{7}, \frac{1}{10}, \frac{8}{9}, \frac{5}{6}, \frac{7}{12}, \frac{3}{7}, \frac{3}{16}, \frac{5}{9}, \frac{4}{5}$

Ask students how they could identify the fractions that are close to 1 in value. Elicit that the numerator and denominator should be about the same. Using that guideline, select the fractions. $\left(\frac{6}{7}, \frac{8}{9}, \frac{5}{6}, \frac{4}{5}\right)$

Similarly, develop the idea that when the denominator is much greater than the numerator, the fraction is close to 0. When the denominator is about twice the size of the numerator, the fraction is close to $\frac{1}{2}$.

(Close to 0: $\frac{1}{10}, \frac{3}{16}$; close to $\frac{1}{2}$: $\frac{7}{12}, \frac{5}{9}$.)

SUPPLEMENTARY MATERIALS

TRP Practice, p. 53

TRP Reteaching, p. 33

TRP Lesson Aids, p. 4

TRP Transparency 27

Arrange in order from greatest to least.

•**36.** $\frac{3}{8}, \frac{7}{8}, \frac{5}{8}$ $\frac{7}{8}, \frac{5}{8}, \frac{3}{8}$

•**37.** $\frac{7}{10}, \frac{4}{10}, \frac{8}{10}$ $\frac{8}{10}, \frac{7}{10}, \frac{4}{10}$

38. $\frac{1}{3}, \frac{5}{6}, \frac{7}{12}$ $\frac{5}{6}, \frac{7}{12}, \frac{1}{3}$

39. $\frac{2}{5}, \frac{9}{15}, \frac{1}{3}$ $\frac{9}{15}, \frac{2}{5}, \frac{1}{3}$

40. $\frac{1}{2}, \frac{3}{8}, \frac{5}{6}$ $\frac{5}{6}, \frac{1}{2}, \frac{3}{8}$

41. $\frac{7}{8}, \frac{3}{5}, \frac{9}{10}$ $\frac{9}{10}, \frac{7}{8}, \frac{3}{5}$

42. $\frac{3}{6}, \frac{3}{4}, \frac{3}{8}, \frac{3}{12}$ $\frac{3}{4}, \frac{3}{6}, \frac{3}{8}, \frac{3}{12}$

43. $\frac{1}{2}, \frac{7}{9}, \frac{2}{3}, \frac{5}{6}$ $\frac{5}{6}, \frac{7}{9}, \frac{2}{3}, \frac{1}{2}$

Arrange in order from least to greatest.

•**44.** $\frac{1}{6}, \frac{0}{6}, \frac{5}{6}$ $\frac{0}{6}, \frac{1}{6}, \frac{5}{6}$

•**45.** $\frac{17}{24}, \frac{9}{24}, \frac{11}{24}$ $\frac{9}{24}, \frac{11}{24}, \frac{17}{24}$

46. $\frac{1}{2}, \frac{3}{5}, \frac{7}{10}$ $\frac{1}{2}, \frac{3}{5}, \frac{7}{10}$

47. $\frac{2}{3}, \frac{7}{8}, \frac{11}{24}$ $\frac{11}{24}, \frac{2}{3}, \frac{7}{8}$

48. $\frac{1}{3}, \frac{2}{5}, \frac{1}{6}$ $\frac{1}{6}, \frac{1}{3}, \frac{2}{5}$

49. $\frac{5}{6}, \frac{10}{11}, \frac{2}{3}$ $\frac{2}{3}, \frac{5}{6}, \frac{10}{11}$

50. $\frac{5}{6}, \frac{5}{9}, \frac{5}{8}, \frac{5}{12}$ $\frac{5}{12}, \frac{5}{9}, \frac{5}{8}, \frac{5}{6}$

51. $\frac{1}{2}, \frac{3}{4}, \frac{4}{5}, \frac{7}{12}$ $\frac{1}{2}, \frac{7}{12}, \frac{3}{4}, \frac{4}{5}$

MIXED REVIEW

Find the answer.

•**52.** $4{,}329 + 6{,}820$ 11,149

•**53.** $617 + 4{,}832 + 9{,}590$ 15,039

54. $81.6 + 21.9$ 103.5

55. $36.25 + 18.95$ 55.2

56. $8{,}639 - 259$ 8,380

57. $16{,}007 - 9{,}625$ 6,382

58. $29.8 - 16.9$ 12.9

59. $84.63 - 29.87$ 54.76

60. 500×300 150,000

61. 436×28 12,208

62. 16.8×24 403.2

63. 36.83×4.9 180.467

64. $72{,}000 \div 90$ 800

65. $3{,}264 \div 8$ 408

66. $88.2 \div 18$ 4.9

67. $22.31 \div 2.3$ 9.7

Solve.

•**68.** On Monday, Carol used $\frac{1}{3}$ can of red paint and $\frac{1}{2}$ can of blue paint. Did Carol use more red paint or blue paint? blue paint

•**69.** Nails at least $\frac{3}{4}$ inches long are needed for a construction project. Can nails that are $\frac{5}{8}$ inches in length be used? No

70. Bill used a can of paint for every 210 square feet of the mural. If the mural covers 472.5 square feet, how many cans of paint did Bill use? Write your answer as a decimal. 2.25 cans

71. Of 75 students surveyed, 65 approved of the school murals. What fraction of the students approved of the murals? Write the answer in lowest terms. $\frac{13}{15}$

72. To frame each mural, Sam used 90.25 feet of wood and 26 nails. How much wood does Sam need for two murals? 180.5 ft

73. Bill was asked to paint a mural for the town. He was paid \$8.50 an hour. If the mural took him 40.5 hours, how much money did Bill earn? \$344.25

MIDCHAPTER REVIEW

Write a fraction for the shaded part. [122]

1. 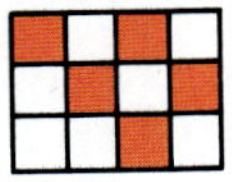 $\frac{5}{12}$

2. 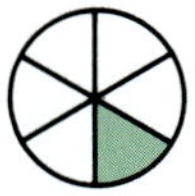 $\frac{1}{6}$

3. 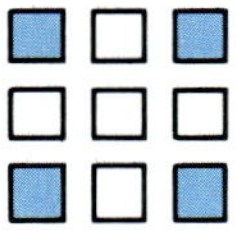 $\frac{4}{9}$

4. 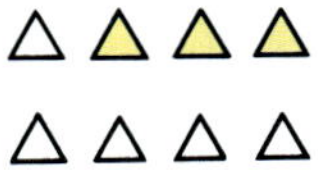 $\frac{3}{8}$

Write a fraction for the point. [122]

5. 

$\frac{1}{3}$

6. 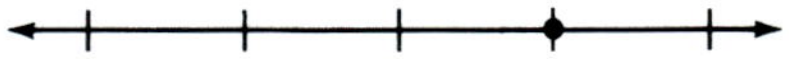

$\frac{3}{4}$

7. 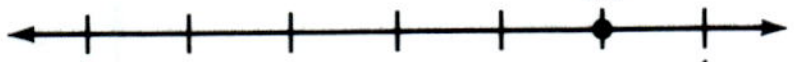

$\frac{5}{6}$

8. 

$\frac{7}{12}$

Find the greatest common factor. [124]

9. 7 and 14 7 **10.** 6 and 8 2 **11.** 14 and 16 2 **12.** 16, 20, and 28 4

Find the least common multiple. [124]

13. 2 and 4 4 **14.** 6 and 9 18 **15.** 8 and 18 72 **16.** 5, 9, and 15 45

Write the fraction in lowest terms. [126]

17. $\frac{3}{9}$ $\frac{1}{3}$ **18.** $\frac{7}{21}$ $\frac{1}{3}$ **19.** $\frac{8}{36}$ $\frac{2}{9}$ **20.** $\frac{10}{12}$ $\frac{5}{6}$ **21.** $\frac{21}{30}$ $\frac{7}{10}$ **22.** $\frac{54}{81}$ $\frac{2}{3}$

Complete. [126]

23. $\frac{1}{4} = \frac{■}{12}$ 3 **24.** $\frac{1}{3} = \frac{3}{■}$ 9 **25.** $\frac{9}{15} = \frac{■}{5}$ 3 **26.** $\frac{35}{63} = \frac{■}{9}$ 5

Arrange in order from greatest to least. [128]

27. $\frac{3}{10}, \frac{7}{10}, \frac{1}{10}$ $\frac{7}{10}, \frac{3}{10}, \frac{1}{10}$

28. $\frac{5}{12}, \frac{5}{6}, \frac{5}{9}$ $\frac{5}{6}, \frac{5}{9}, \frac{5}{12}$

29. $\frac{3}{4}, \frac{1}{22}, \frac{5}{8}$ $\frac{3}{4}, \frac{5}{8}, \frac{1}{22}$

30. $\frac{1}{4}, \frac{7}{16}, \frac{3}{8}, \frac{1}{2}$ $\frac{1}{2}, \frac{7}{16}, \frac{3}{8}, \frac{1}{4}$

Solve. [122, 128]

31. Susan earned $48 last weekend. She spent $8 on an album. What fraction of her earnings did she spend on the album? Write the answer in simplest form. $\frac{1}{6}$

32. Joseph collects sheet music. $\frac{1}{4}$ of his collection is pop, $\frac{5}{12}$ is jazz, and $\frac{1}{3}$ is rock. Which type of music makes up the largest part of his collection? jazz

OBJECTIVE

- Evaluate student progress.

USING THE REVIEW

This page provides a means for informally evaluating students' understanding of the skills and concepts covered so far in this chapter.

Have the students look at the page to familiarize themselves with the various question formats that are presented. Discuss any questions that they may have. Then ask them to complete the page independently.

In addition to grading them individually, you may wish to review the answers to the questions collectively with the students.

Page references appear in brackets. They refer to pages on which a particular skill was introduced.

Before continuing on to the topics found in the remainder of the chapter, you may wish to have students review any skills or concepts in which they have demonstrated weakness.

OBJECTIVES

- Write a mixed number as an improper fraction.
- Write an improper fraction as a whole or mixed number.

TEACHING THE LESSON

WARM-UP Present the following exercises on the chalkboard or on an overhead transparency.

Compute.

1. $3 \times 4 + 6$ (18)
2. $7 \times 4 + 3$ (31)
3. $5 \times 7 + 4$ (39)
4. $6 \times 6 + 5$ (41)
5. $4 \times 9 + 1$ (37)

INTRODUCTION Draw the following illustrations on the chalkboard. Ask students if they can think of a way to write a number that represents each illustration.

1. 

2. 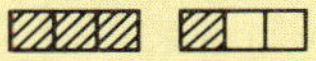

3. 

(1. $2\frac{3}{4}$ 2. $1\frac{1}{3}$ 3. $4\frac{1}{2}$)

INSTRUCTION In discussing Example 1 point out that multiplying the whole number times the denominator of the fraction identifies how many fourths are in the whole number part of the mixed number. Adding the numerator to the product gives us the total number of fourths in the mixed number.

In discussing Example 2, draw a diagram to illustrate the conversion of the improper fraction to a mixed number.

$$\frac{9}{5}$$

$$\frac{5}{5} + \frac{4}{5}$$

$$1 + \frac{4}{5} = 1\frac{4}{5}$$

5.5 MIXED NUMBERS

A **mixed number** consists of a whole number and a fraction.

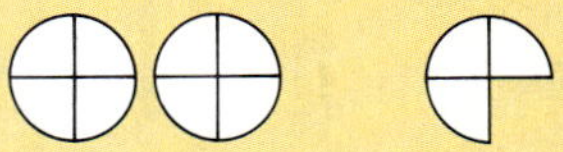

$2 + \frac{3}{4} = 2\frac{3}{4}$

You can write a mixed number as a fraction.

EXAMPLE 1 Write $2\frac{3}{4}$ as a fraction.

1. Multiply the whole number and the denominator. Add the numerator. $2\frac{3}{4}$ **THINK: $4 \times 2 = 8$, $8 + 3 = 11$**
2. Write the sum over the denominator. $\frac{11}{4}$

So, $2\frac{3}{4} = \frac{11}{4}$.

Fractions with numerators equal to or greater than the denominators are called **improper fractions.** You can write an improper fraction as a mixed number.

EXAMPLE 2 Write $\frac{28}{8}$ as a mixed number in simplest form.

1. Divide the numerator by the denominator. Write the remainder as a fraction.

$$\begin{array}{r} 3 \\ 8\overline{)28} \\ \underline{24} \\ 4 \end{array}$$

$3\frac{4}{8}$ **THINK: 4 is the remainder. 8 is the divisor.**

2. Write the fraction in lowest terms. $3\frac{4}{8} = 3\frac{1}{2}$ ← **In simplest form**

So, $\frac{28}{8} = 3\frac{1}{2}$.

CHECKPOINT

Write the letter of the correct answer.

Complete. Choose the answer written in simplest form.

1. $3\frac{2}{3} = ■$	d	**a.** $\frac{18}{3}$	**b.** $\frac{9}{3}$	**c.** $\frac{15}{3}$	**d.** $\frac{11}{3}$
2. $4\frac{1}{8} = ■$	a	**a.** $\frac{33}{8}$	**b.** $\frac{8}{8}$	**c.** $\frac{12}{8}$	**d.** $\frac{3}{18}$
3. $\frac{13}{5} = ■$	c	**a.** $2\frac{1}{5}$	**b.** $3\frac{1}{5}$	**c.** $2\frac{3}{5}$	**d.** $5\frac{3}{5}$
4. $\frac{16}{6} = ■$	c	**a.** $2\frac{4}{6}$	**b.** $3\frac{3}{4}$	**c.** $2\frac{2}{3}$	**d.** $3\frac{1}{2}$

Make sure students understand that a fraction is in simplest form when it is in lowest terms. A mixed number is in simplest form when its fractional part is less than 1 and in lowest terms.

CHECKPOINT The incorrect answer choices include common errors students make.

PRACTICE EXERCISES

See Extra Practice, page 436.

Write as an improper fraction.

• **1.** $2\frac{1}{3}$ $\frac{7}{3}$ • **2.** $1\frac{2}{3}$ $\frac{5}{3}$ **3.** $3\frac{1}{2}$ $\frac{7}{2}$ **4.** $1\frac{3}{4}$ $\frac{7}{4}$ **5.** $2\frac{3}{5}$ $\frac{13}{5}$ **6.** $1\frac{5}{8}$ $\frac{13}{8}$ **7.** $2\frac{4}{7}$ $\frac{18}{7}$ **8.** $5\frac{2}{5}$ $\frac{27}{5}$

9. $7\frac{4}{5}$ $\frac{39}{5}$ **10.** $4\frac{3}{7}$ $\frac{31}{7}$ **11.** $6\frac{5}{6}$ $\frac{41}{6}$ **12.** $5\frac{2}{9}$ $\frac{47}{9}$ **13.** $3\frac{2}{3}$ $\frac{11}{3}$ **14.** $2\frac{3}{4}$ $\frac{11}{4}$ **15.** $4\frac{3}{8}$ $\frac{35}{8}$ **16.** $3\frac{4}{5}$ $\frac{19}{5}$

17. $5\frac{1}{3}$ $\frac{16}{3}$ **18.** $2\frac{4}{9}$ $\frac{22}{9}$ **19.** $2\frac{3}{8}$ $\frac{19}{8}$ **20.** $3\frac{5}{7}$ $\frac{26}{7}$ **21.** $5\frac{7}{9}$ $\frac{52}{9}$ **22.** $3\frac{5}{8}$ $\frac{29}{8}$ **23.** $4\frac{2}{3}$ $\frac{14}{3}$ **24.** $3\frac{3}{8}$ $\frac{27}{8}$

25. $6\frac{3}{4}$ $\frac{27}{4}$ **26.** $2\frac{5}{6}$ $\frac{17}{6}$ **27.** $4\frac{7}{8}$ $\frac{39}{8}$ **28.** $5\frac{3}{8}$ $\frac{43}{8}$ **29.** $17\frac{2}{3}$ $\frac{53}{3}$ **30.** $13\frac{5}{9}$ $\frac{122}{9}$ **31.** $15\frac{6}{7}$ $\frac{111}{7}$ **32.** $16\frac{2}{3}$ $\frac{50}{3}$

Write as a whole number or mixed number in simplest form.

• **33.** $\frac{3}{2}$ $1\frac{1}{2}$ • **34.** $\frac{4}{3}$ $1\frac{1}{3}$ **35.** $\frac{6}{2}$ 3 **36.** $\frac{5}{3}$ $1\frac{2}{3}$ **37.** $\frac{7}{4}$ $1\frac{3}{4}$ **38.** $\frac{8}{3}$ $2\frac{2}{3}$ **39.** $\frac{6}{5}$ $1\frac{1}{5}$ **40.** $\frac{10}{4}$ $2\frac{1}{2}$

41. $\frac{7}{2}$ $3\frac{1}{2}$ **42.** $\frac{11}{6}$ $1\frac{5}{6}$ **43.** $\frac{12}{5}$ $2\frac{2}{5}$ **44.** $\frac{16}{8}$ 2 **45.** $\frac{15}{9}$ $1\frac{2}{3}$ **46.** $\frac{14}{3}$ $4\frac{2}{3}$ **47.** $\frac{20}{9}$ $2\frac{2}{9}$ **48.** $\frac{46}{8}$ $5\frac{3}{4}$

49. $\frac{36}{10}$ $3\frac{3}{5}$ **50.** $\frac{85}{7}$ $12\frac{1}{7}$ **51.** $\frac{63}{9}$ 7 **52.** $\frac{25}{8}$ $3\frac{1}{8}$ **53.** $\frac{32}{6}$ $5\frac{1}{3}$ **54.** $\frac{18}{5}$ $3\frac{3}{5}$ **55.** $\frac{28}{8}$ $3\frac{1}{2}$ **56.** $\frac{39}{9}$ $4\frac{1}{3}$

57. $\frac{17}{3}$ $5\frac{2}{3}$ **58.** $\frac{21}{7}$ 3 **59.** $\frac{30}{4}$ $7\frac{1}{2}$ **60.** $\frac{27}{6}$ $4\frac{1}{2}$ **61.** $\frac{45}{8}$ $5\frac{5}{8}$ **62.** $\frac{39}{5}$ $7\frac{4}{5}$ **63.** $\frac{22}{8}$ $2\frac{3}{4}$ **64.** $\frac{38}{3}$ $12\frac{2}{3}$

MIXED REVIEW

Here are Stephanie's science grades in the second semester.

Test 1: 85 Test 2: 88 Test 3: 92 Test 4: 78 Test 5: 92

Compute.

• **65.** Range = ■ 14 **66.** Median = ■ 88 **67.** Mode = ■ 92 **68.** Mean = ■ 87

Use the list of ingredients to solve.

• **69.** Jon has a $\frac{1}{4}$ cup for measuring. How many $\frac{1}{4}$ cups of wheat flour does he need? 7

70. How many $\frac{1}{4}$ cups of honey does he need? 6

71. Which is greater, the amount of water or the amount of honey? honey

HONEY OATMEAL BREAD

1 c oats	$1\frac{3}{4}$ c sifted wheat flour
1 stick butter	1 t baking soda
$1\frac{1}{4}$ c boiling water	$\frac{3}{4}$ t salt
$1\frac{1}{2}$ c honey	1 t cinnamon
1 t vanilla	$\frac{1}{4}$ t nutmeg
2 eggs	

COMMON ERROR

When writing mixed numbers as fractions, some students will multiply the whole number by the numerator and then add the denominator to the product. For example, in item 2 under Checkpoint they would select c. Students making these errors should be encouraged to draw diagrams to illustrate the correct process.

ASSIGNMENTS

Level 1	Even 2–64, 65–71
Level 2	Odd 1–63, 65–71
Level 3	17–32, 49–71

FOLLOW-UP ACTIVITY

APPLICATION Present students with the task of finding mixed numbers in the world around them. Have them make a poster of the practical applications of mixed numbers. These applications may include recipes, stock market reports, house building, decorating. Have students write a summary of the ways that mixed numbers are used, and give an example of how mixed numbers are used in each situation.

SUPPLEMENTARY MATERIALS

TRP Practice, p. 54

TRP Reteaching, p. 34

TRP Transparency 27

OBJECTIVE

- Interpret the quotient and remainder when solving division problems.

TEACHING THE LESSON

WARM-UP Write the following exercises on the chalkboard or on an overhead transparency.

Divide.

1. $7\overline{)92}$ (13 R1)
2. $12\overline{)39}$ (3 R3)
3. $30\overline{)462}$ (15 R12)
4. $45\overline{)905}$ (20 R5)

INTRODUCTION Pose this situation to the class. "There are 44 cars waiting to take the ferry across the river. The ferry can only hold 8 cars at one time." Ask the following questions:

1. If the ferry will only cross the river when it is full, how many trips will it make? (5 trips)
2. If all the cars must cross the river, how many trips will the ferry make? (6 trips)

Elicit that in the first question, the remainder is not important. In the second question, the remainder is important. A sixth trip must be made even though the ferry will not be full.

INSTRUCTION After discussing the situation, explain that four different questions requiring the given division problem will be asked. As you discuss A, stress that you only are concerned with cars that are full, so the remainder is not important.

In B, stress that if all 372 people went on the ride, the last car would have been partially filled with 12 people.

In C, point out that the size of the remainder is not important. If there was a remainder of any size at all, the seventh car would be needed.

PROBLEM *Solving* STRATEGY

5.6 INTERPRETING REMAINDERS

Situation:

The first and largest Ferris Wheel was built in Chicago in 1893. It had 36 cars. Each car held 60 people. There were 372 people waiting to go on the Ferris Wheel.

Strategy:

The same division example can be used to answer different questions about a situation. Make sure that you answer the question that is asked.

Applying the Strategy:

This division example can be used to answer different questions about seating people on the Ferris Wheel.

$$\begin{array}{r} 6\text{ R}12 \\ 60\overline{)372} \\ \underline{360} \\ 12 \end{array}$$

Notice the different answers to each of the questions below.

A. How many cars would have been filled by 372 people?
Use only the quotient. — 6 cars would have been filled.

B. How many people would have been in the last car?
Use only the remainder. — 12 people would have been in the last car.

C. How many cars would have been needed to hold 372 people?
Raise the quotient by 1. — **THINK:** 6 full cars + 1 car with 12 people.
7 cars would have been needed.

D. What fraction of the last car would have been filled?
Use a fractional remainder. — **THINK:** **People in last car →** / **People in full car →** $\frac{12}{60} = \frac{1}{5}$

$\frac{1}{5}$ of the last car would have been filled.

PRACTICE EXERCISES See Extra Practice, page 437.

Write the letter of the correct answer.

• **1.** The Whirly Bird ride holds 25 people. If 185 people are taking a ride, how many people will go on the last trip? b
a. Use only the quotient.
b. Use only the remainder.
c. Raise the quotient by 1.

• **2.** There are 39 people who want to ride the Safari Train. If 9 people can ride in each car, how many cars are needed? c
a. Use only the quotient.
b. Use only the remainder.
c. Raise the quotient by 1.

Solve.

Rose sold sweatshirts at an amusement park. She packed 20 sweatshirts in each carton. She had 185 sweatshirts to pack.

3. How many cartons did Rose fill? 9 cartons

4. How many sweatshirts did she put into the last partially filled carton? 5 sweatshirts

5. How many cartons did Rose use to pack 185 sweatshirts? 10 cartons

6. What fraction of the last carton was filled? $\frac{1}{4}$

Jeff sold lemonade at an amusement park. He made 1 qt (32 oz) at a time. He served the lemonade in 6-oz cups.

7. How many cups could Jeff fill with 1 qt of lemonade? 5 cups

8. How many ounces of lemonade were poured into the last cup? 2 oz

9. How many cups were needed to serve the quart of lemonade? 6 cups

10. What fraction of the last cup was filled? $\frac{1}{3}$

11. One paper cup is needed for each of 115 people at the Snack Bar. If a package contains 20 cups, how many packages are needed? 6 packages

12. Bob has $20 to spend on tickets for rides at the amusement park. How many tickets can he buy if each ticket costs $3? 6 tickets

In D, point out that the size of the remainder is important because writing a fraction requires knowing the number of people in a filled car as well as the number of people in the partially filled car. Encourage students to reread the problem and consider whether their answer correctly responds to the question.

ASSIGNMENTS

Level 1	1–8
Level 2	1–12
Level 3	3–12

FOLLOW-UP ACTIVITY

SITUATIONAL PROBLEM SOLVING You are in charge of organizing the sports events for your school's field day. Decide how to schedule the events. As you organize all aspects of the field day, consider these questions.

1. How many students will participate?
2. What events should you include?
3. Where will the events take place?
4. Will there be awards?

SUPPLEMENTARY MATERIALS

TRP Practice, p. 55

TRP Transparency 28

TRP Applications, p. 8

OBJECTIVES

- Write a decimal as a fraction or mixed number.
- Write a fraction or a mixed number as a terminating or repeating decimal.

TEACHING THE LESSON

WARM-UP Present the following exercises on the chalkboard or on an overhead transparency.

Divide.

1. $4\overline{)5}$ (1.25) **2.** $4\overline{)1}$ (0.25)
3. $10\overline{)8}$ (0.8) **4.** $4\overline{)3}$ (0.75)
5. $14\overline{)7}$ (0.5) **6.** $100\overline{)9}$ (0.09)

INTRODUCTION Ask students to think of situations in which a fraction is expressed as a decimal, such as recording the time of a race, computer programming, or using a calculator.

INSTRUCTION In discussing Example 1, point out that in a decimal, the denominator is not written. It is understood to be a power of 10 such as 10, 100, 1,000, and so on. Emphasize that in converting a decimal to a fraction, students need to identify the numerator and denominator.

The number of digits to the right of the decimal point tell which power of 10 to use for the denominator. The digits of the decimal identify the numerator.

Before discussing Examples 2 and 3, make sure that students understand the division required for changing a fraction to a decimal. Point out that because the division comes to an end in Examples 2 and 3, these decimals are called *terminating* decimals.

You may wish to have students continue the division in Example 4 beyond the thousandths place. At some point, stop and elicit from students that this division will never come to an end. Note that this is an example of a *nonterminating* decimal.

5.7 FRACTIONS AND DECIMALS

You can write a decimal as a fraction or a mixed number.

EXAMPLE 1 Write as a fraction or mixed number in simplest form.

a. 0.08

THINK: 0.08 means 8 hundredths

$0.08 = \frac{8}{100} = \frac{2}{25}$

b. 2.3

THINK: 2.3 means 2 and 3 tenths

$2.3 = 2\frac{3}{10}$

You can write a fraction as a decimal by dividing the numerator by the denominator.

EXAMPLE 2 Write $\frac{3}{4}$ as a decimal.

$$\frac{3}{4} \rightarrow \begin{array}{r} 0.75 \\ 4\overline{)3.00} \\ \underline{2\,8} \\ 20 \\ \underline{20} \\ 0 \end{array}$$

THINK: Additional zeros are needed to complete the division.

So, $\frac{3}{4} = 0.75$.

You can also write a mixed number as a decimal.

EXAMPLE 3 Write $2\frac{1}{8}$ as a decimal.

1. Write the fraction as a decimal. $\frac{1}{8} \rightarrow 8\overline{)1.000}$ = 0.125
2. Add the whole number. $2 + 0.125 = 2.125$

So, $2\frac{1}{8} = 2.125$.

Sometimes when you divide to change a fraction to a decimal, the remainder will never be zero. These are called **nonterminating decimals.**

EXAMPLE 4 Write $\frac{2}{9}$ as a decimal.

$$\begin{array}{r} 0.222\ldots \\ 9\overline{)2.000} \\ \underline{1\,8} \\ 20 \\ \underline{18} \\ 20 \\ \underline{18} \\ 2 \end{array}$$

Notice that the remainder will never be 0. This answer can be written in two ways:

$\frac{2}{9} = 0.22$ when rounded to the nearest hundredth.

$\frac{2}{9} = 0.\overline{2}$ which means the digit, 2, repeats.

CHECKPOINT

Write the letter of the correct answer.

1. What is 2.82 as a mixed number in simplest form? a

a. $2\frac{41}{50}$ **b.** $\frac{41}{50}$ **c.** $\frac{282}{100}$ **d.** $2\frac{82}{100}$

2. What is $\frac{3}{5}$ as a decimal? c

a. 1.6 **b.** 0.06 **c.** 0.6 **d.** 0.006

3. What is $\frac{3}{7}$ as a decimal rounded to the nearest hundredth? d

a. 0.043 **b.** 4.28 **c.** 42.85 **d.** 0.43

PRACTICE EXERCISES

See Extra Practice, page 437.

Write as a fraction or mixed number in simplest form.

• **1.** 0.3 $\frac{3}{10}$ **2.** 0.6 $\frac{3}{5}$ **3.** 0.5 $\frac{1}{2}$ **4.** 0.8 $\frac{4}{5}$ **5.** 0.9 $\frac{9}{10}$ • **6.** 2.7 $2\frac{7}{10}$

7. 3.4 $3\frac{2}{5}$ **8.** 4.2 $4\frac{1}{5}$ **9.** 0.12 $\frac{3}{25}$ **10.** 0.25 $\frac{1}{4}$ **11.** 0.35 $\frac{7}{20}$ **12.** 0.56 $\frac{14}{25}$

13. 0.75 $\frac{3}{4}$ **14.** 2.47 $2\frac{47}{100}$ **15.** 8.05 $8\frac{1}{20}$ **16.** 3.75 $3\frac{3}{4}$ **17.** 0.221 $\frac{221}{1000}$ **18.** 0.792 $\frac{99}{125}$

19. 0.125 $\frac{1}{8}$ **20.** 0.075 $\frac{3}{40}$ **21.** 0.625 $\frac{5}{8}$ **22.** 4.375 $4\frac{3}{8}$ **23.** 5.025 $5\frac{1}{40}$ **24.** 2.085 $2\frac{17}{200}$

Write the fraction or mixed number as a decimal. **34.** 2.25; **35.** 5.9; **36.** 3.2; **37.** 8.75; **38.** 6.8; **39.** 11.1; **40.** 6.02

• **25.** $\frac{1}{2}$ 0.5 **26.** $\frac{3}{4}$ 0.75 **27.** $\frac{4}{5}$ 0.8 **28.** $\frac{1}{4}$ 0.25 **29.** $\frac{2}{5}$ 0.4 **30.** $\frac{7}{10}$ 0.7 **31.** $\frac{1}{5}$ 0.2 **32.** $\frac{1}{20}$ 0.05

• **33.** $1\frac{3}{5}$ 1.6 **34.** $2\frac{1}{4}$ **35.** $5\frac{9}{10}$ **36.** $3\frac{1}{5}$ **37.** $8\frac{3}{4}$ **38.** $6\frac{4}{5}$ **39.** $11\frac{1}{10}$ **40.** $6\frac{1}{50}$

41. $\frac{1}{8}$ 0.125 **42.** $\frac{3}{8}$ 0.375 **43.** $2\frac{7}{8}$ 2.875 **44.** $\frac{6}{15}$ 0.4 **45.** $9\frac{5}{8}$ 9.625 **46.** $\frac{5}{16}$ 0.3125 **47.** $4\frac{1}{16}$ 4.0625 **48.** $10\frac{5}{16}$ 10.3125

Write the fraction or mixed number as a decimal. Round to the nearest hundredth.

• **49.** $\frac{1}{3}$ 0.33 • **50.** $\frac{5}{6}$ 0.83 **51.** $\frac{2}{7}$ 0.29 **52.** $\frac{4}{9}$ 0.44 **53.** $4\frac{3}{14}$ 4.21 **54.** $7\frac{5}{9}$ 7.56 **55.** $6\frac{5}{12}$ 6.42 **56.** $9\frac{7}{15}$ 9.47

Write the fraction or mixed number as a decimal. Use the bar notation.

• **57.** $\frac{1}{3}$ $0.\overline{3}$ • **58.** $\frac{7}{9}$ $0.\overline{7}$ **59.** $\frac{6}{11}$ $0.\overline{54}$ **60.** $\frac{2}{3}$ $0.\overline{6}$ **61.** $2\frac{2}{9}$ $2.\overline{2}$ **62.** $4\frac{6}{11}$ $4.\overline{54}$ **63.** $5\frac{1}{3}$ $5.\overline{3}$ **64.** $7\frac{4}{9}$ $7.\overline{4}$

Solve.

65. Ellen Travers wants to make hamburgers for her party. She needs $2\frac{1}{2}$ pounds of meat. She buys 2.75 pounds. Does she have enough hamburger? yes

66. Ellen is making fruit salad for the party. She buys $2\frac{1}{2}$ pounds of bananas and 2.25 pounds of grapes. Does she buy more bananas or more grapes? bananas

CHECKPOINT The incorrect answer choices include common errors students make.

COMMON ERROR

In converting a fraction to a decimal, students will sometimes divide the numerator into the denominator. For example, in item 2 under Checkpoint, they would select choice a. If students check the reasonableness of their answers, it will help eliminate this error. Since $\frac{3}{5} < 1$ and $1.\overline{6} > 1$, $1.\overline{6}$ is not a reasonable answer.

ASSIGNMENTS

Level 1	Odd 1–63, 65–66
Level 2	Even 2–64, 65–66
Level 3	13–24, Odd 25–63, 64–66

FOLLOW-UP ACTIVITY

ESTIMATION Remind students that if a fraction is less than 1, then its decimal equivalent must be less than 1. Conversely, if a mixed number is greater than 1, its decimal equivalent must be greater than 1. By quick inspection they can then determine whether the following conversions are reasonable.

1. $\frac{5}{6} = 1.2$ (no)

2. $1\frac{4}{5} = 1.8$ (yes)

3. $2\frac{1}{2} = 2.5$ (yes)

4. $\frac{11}{20} = 0.55$ (yes)

5. $3\frac{1}{3} = 0.6$ (no)

6. $\frac{37}{40} = 1.16$ (no)

SUPPLEMENTARY MATERIALS

TRP Practice, p. 56

TRP Reteaching, p. 35

TRP Transparency 28

OBJECTIVE

- Solve problems that involve reading and interpreting data from a circle graph.

TEACHING THE LESSON

WARM-UP Give the students the following exercises orally.

Which is greater?

1. $\frac{1}{4}$ or $\frac{3}{4}$ $(\frac{3}{4})$ **2.** $\frac{1}{2}$ or $\frac{1}{8}$ $(\frac{1}{2})$

3. $\frac{2}{5}$ or $\frac{2}{15}$ $(\frac{2}{5})$ **4.** $\frac{2}{3}$ or $\frac{5}{6}$ $(\frac{5}{6})$

INTRODUCTION Ask students how they decide to spend their allowances or their earnings from part-time jobs. Ask if anyone has used a budget to determine how the money should be spent. Discuss the advantages and disadvantages of a budget.

INSTRUCTION As you discuss the opening paragraph, you may wish to point out that nearly all companies, departments of government, and so on, use budgets to help plan how money will be spent.

Problems 1–4 involve reading the circle graph, and Problems 5–9 involve comparing fractions. In Problems 10–12, students must use their knowledge of equivalent fractions and decimals to find the appropriate items on the graph.

ASSIGNMENTS

Level 1	Odd 1–13
Level 2	1–13
Level 3	Even 2–12, 13–15

SUPPLEMENTARY MATERIALS

TRP Practice, p. 57

TRP Applications, p. 9

TRP Transparency 29

PROBLEM Solving APPLICATION

5.8 CONSUMER: BUDGETING

A **budget** is a plan for using money. The Williams family decided to control their expenses by making and following a budget. This circle graph shows how they decided to allot their family income after taxes.

THE WILLIAMS' BUDGET

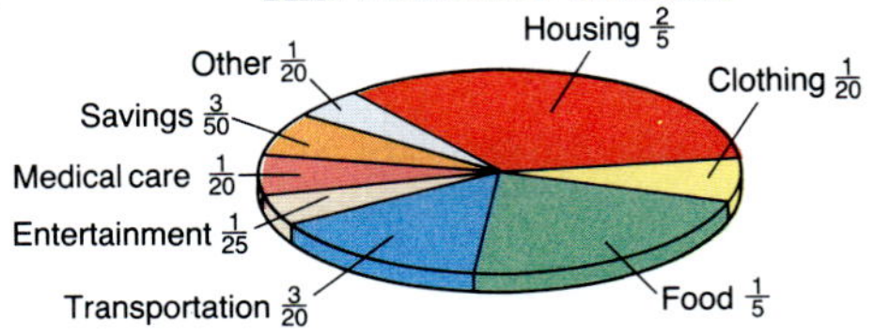

Use the circle graph to answer Exercises 1–12.

What part of the budget is allotted to:

• **1.** housing? $\frac{2}{5}$ **2.** food? $\frac{1}{5}$

3. clothing? $\frac{1}{20}$ **4.** savings? $\frac{3}{50}$

Did they allot more money to:

• **5.** transportation or clothing? transportation **6.** entertainment or medical care? medical care

7. food or clothing? food **8.** savings or entertainment? savings

9. For which type of expense did they allot the greatest part of their budget? housing

Which expense represents: **10.** housing **11.** entertainment **12.** transportation

• **10.** $\frac{40}{100}$ of the budget? **11.** $\frac{4}{100}$ of the budget? **12.** 0.15 of the budget?

Solve. **14.** food, $\frac{4}{25}$; housing, $\frac{3}{10}$; clothing, $\frac{3}{50}$; savings, $\frac{1}{8}$; transportation, $\frac{1}{5}$; entertainment, $\frac{1}{20}$; medical care, $\frac{1}{20}$; other, $\frac{11}{200}$

• **13.** The Bartner family made a budget. They allotted their income as follows: food, 0.16; housing, 0.30; clothing, 0.06; savings, 0.125; transportation, 0.20; entertainment, 0.05; medical care, 0.05; other, 0.055. For which types of expense did they allot more than 0.15? food, housing, transportation

14. Rewrite the Bartners' budget using fractions in lowest terms.

15. Suppose you wanted to arrange the Bartner budget items in order from greatest to least. Do you think it would be easier to use decimals or fractions? Why? decimals; fractions may require changing to a common denominator

PROBLEM Solving APPLICATION

5.9 CAREER: ADVERTISING EXECUTIVE

Michelle Morales works for a small advertising agency. She is the Account Executive for a car manufacturer. It is her job to make and present an advertising campaign budget for the new Aurora sports car.

Michelle used a **divided-bar graph** in her presentation to the client.

Aurora Advertising Campaign

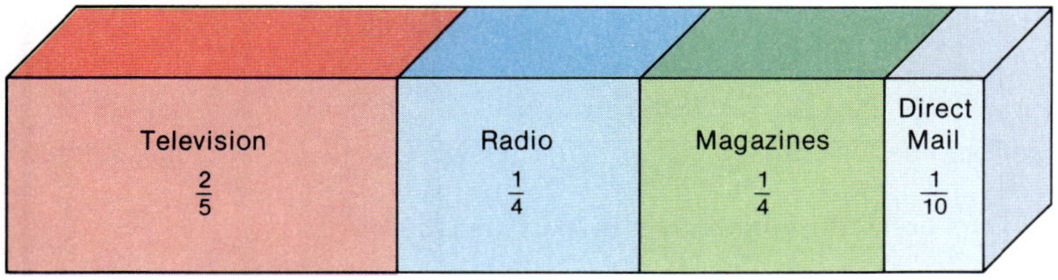

Use the divided-bar graph to answer Exercises 1–10.

What part of the advertising budget is devoted to:

• **1.** radio? $\frac{1}{4}$ • **2.** television? $\frac{2}{5}$ **3.** magazines? $\frac{1}{4}$ **4.** direct mail? $\frac{1}{10}$

5. Which type of advertising received the smallest part of the budget? direct mail

6. Which types of advertising were allotted equal parts of the budget? radio and magazines

Solve.

• **7.** Use graph paper to draw a rectangle 20 units long and 5 units wide. Show the Aurora graph using decimals instead of fractions. Check students' graphs

8. What is the sum of the decimals in the graph you drew for Exercise 7? 1

9. The part of the budget allotted for television is how many times as great as the part allotted for direct mail? (Use the decimals in your graph.) 4 times

10. Suppose a total of $8,500,000 will be used for the Aurora campaign. How much money will be used for each of the four types of advertising? TV, $3,400,000; radio, $2,125,000; magazines, $2,125,000; direct mail, $850,000

11. Why do you think Michelle has chosen to spend more money on television advertising than on any other type of advertising?

12. Which types of magazines would be best for advertising a new sports car? Why? For Exercises 11 and 12, accept all reasonable answers. See TE side column.

OBJECTIVE

• Solve problems that involve reading and interpreting data from a divided-bar graph.

TEACHING THE LESSON

WARM-UP Write the following exercises on the chalkboard or an overhead transparency.

Write the decimal.

1. $\frac{1}{10}$ = ▩ (0.1) **2.** $\frac{3}{4}$ = ▩ (0.75)

3. $\frac{4}{5}$ = ▩ (0.8) **4.** $\frac{9}{20}$ = ▩ (0.45)

INTRODUCTION Point out that businesses, like families, use budgets regularly. List the expenses that the following small businesses are likely to incur.

- florist
- gas station
- fast-food restaurant

Guide students to realize that all businesses have expenses for overhead (rent, telephone, heat, electricity), salaries, and insurance. Then list the other expenses that are unique to the businesses cited above.

INSTRUCTION Point out that a divided-bar graph, like a circle graph, shows parts of a whole. This divided-bar graph is similar to the Williams family's circle graph in that each graph shows parts of a budget.

Notice that in Problems 7–10, the students make use of decimals; these problems help maintain previously learned concepts and skills.

ASSIGNMENTS

Level 1	1–8
Level 2	1–12
Level 3	Odd 1–9, 11–12

SUPPLEMENTARY MATERIALS

TRP Practice, p. 58

TRP Transparency 29

OBJECTIVES

- Review vocabulary.
- Practice key chapter concepts and skills.

USING THE REVIEW

The Chapter Review is designed to help students prepare for taking the Chapter Test. The first section focuses on vocabulary. It requires that students select a word(s) to complete statements. The second section presents practice exercises of key mathematical skills. Under each directive there is a sample exercise with the answer.

Each item on the review is referenced to the page on which the topic is taught in the Pupil's Edition. You may wish to have students refer to these pages to help review any concepts or skills they have not yet mastered.

It is suggested that students work in small-sized heterogeneous cooperative learning groups. Some cooperative learning methods that may be used are as follows:

1. After each student has independently completed the entire Chapter Review, a discussion should follow within each group about the solutions to the practice exercises.
2. The group can complete the entire Chapter Review by working together to discuss the sample exercises and then to answer the practice exercises.

End the lesson with an entire class discussion in which any questions brought up in group discussions are presented and answered.

CHAPTER REVIEW

Vocabulary. Choose the letter of the word(s) that completes each statement.

1. The ■ of the fraction identifies the total number of equal parts. [122] b

a. numerator **b.** denominator **c.** mixed number **d.** multiple

2. Fractions that name the same amount using different terms are ■. [126] c

a. nonterminating **b.** common **c.** equivalent **d.** cross products

3. A whole number and a fraction is called a ■. [132] a

a. mixed number **b.** GCF **c.** LCM **d.** multiple

Skills. Find the greatest common factor (GCF). [124]

4, 10 *2*

4. 3, 9 3 **5.** 15, 25 5 **6.** 12, 20 4 **7.** 18, 30 6

Find the least common multiple (LCM). [124]

4, 5 *20*

8. 2, 6 6 **9.** 6, 10 30 **10.** 8, 12 24 **11.** 10, 25 50

Write the fraction in lowest terms. [126]

$\frac{6}{9}$ $\frac{2}{3}$

12. $\frac{12}{15}$ $\frac{4}{5}$ **13.** $\frac{14}{18}$ $\frac{7}{9}$ **14.** $\frac{15}{21}$ $\frac{5}{7}$ **15.** $\frac{25}{35}$ $\frac{5}{7}$

Complete. [126]

$\frac{9}{12} = \frac{■}{4}$ *3*

16. $\frac{1}{4} = \frac{■}{36}$ 9 **17.** $\frac{2}{3} = \frac{■}{12}$ 8 **18.** $\frac{20}{24} = \frac{■}{6}$ 5 **19.** $\frac{12}{16} = \frac{3}{■}$ 4

Compare. Write =, <, or >. [128]

$\frac{2}{3}$ ● $\frac{3}{4}$ <

20. $\frac{1}{2}$ ● $\frac{1}{3}$ > **21.** $\frac{3}{4}$ ● $\frac{12}{16}$ = **22.** $\frac{3}{7}$ ● $\frac{4}{5}$ < **23.** $\frac{7}{8}$ ● $\frac{2}{3}$ >

Write as an improper fraction. [132]

$2\frac{3}{4}$ $\frac{11}{3}$

24. $1\frac{2}{3}$ $\frac{5}{3}$ **25.** $5\frac{2}{7}$ $\frac{37}{7}$ **26.** $3\frac{5}{6}$ $\frac{23}{6}$ **27.** $8\frac{3}{10}$ $\frac{83}{10}$

Write the fraction or mixed number as a decimal. [136]

$1\frac{1}{2}$ *1.5*

28. $\frac{1}{4}$ 0.25 **29.** $\frac{3}{5}$ 0.6 **30.** $1\frac{1}{10}$ 1.1 **31.** $2\frac{4}{5}$ 2.8 **32.** $3\frac{1}{8}$ 3.12

CHAPTER TEST

Write a fraction for the shaded part.

1. 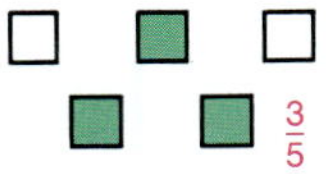 $\frac{3}{5}$

2. 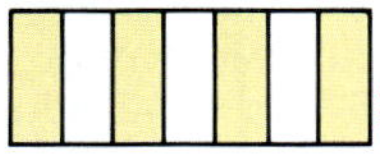 $\frac{4}{7}$

Write a fraction for the point.

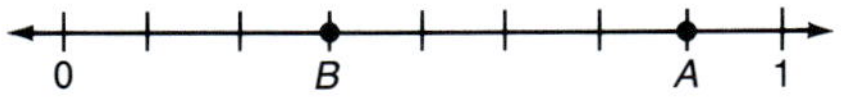

3. A $\frac{7}{8}$

4. B $\frac{3}{8}$

Find the greatest common factor (GCF).

5. 6 and 15 3

6. 6, 12, and 24 6

Find the least common multiple (LCM).

7. 4 and 10 20

8. 2, 5, and 8 40

Write the fraction in lowest terms.

9. $\frac{15}{40}$ $\frac{3}{8}$

10. $\frac{9}{12}$ $\frac{3}{4}$

Complete.

11. $\frac{1}{5} = \frac{■}{15}$ 3

12. $\frac{9}{21} = \frac{■}{7}$ 3

Compare. Write >, <, or =.

13. $\frac{5}{8}$ ● $\frac{3}{8}$ >

14. $\frac{3}{4}$ ● $\frac{9}{12}$ =

15. $\frac{1}{5}$ ● $\frac{1}{4}$ <

16. $\frac{2}{3}$ ● $\frac{5}{8}$ >

Write as an improper fraction.

17. $1\frac{1}{4}$ $\frac{5}{4}$

18. $3\frac{2}{3}$ $\frac{11}{3}$

19. $6\frac{4}{9}$ $\frac{58}{9}$

Write as a whole number or mixed number.

20. $\frac{5}{3}$ $1\frac{2}{3}$

21. $\frac{15}{6}$ $2\frac{1}{2}$

22. $\frac{21}{7}$ 3

Write as a fraction or mixed number in simplest form.

23. 0.8 $\frac{4}{5}$

24. 0.24 $\frac{6}{25}$

25. 5.3 $5\frac{3}{10}$

26. 10.45 $10\frac{9}{20}$

Write as a decimal. Round to the nearest hundredth when necessary.

27. $\frac{1}{4}$ 0.25

28. $\frac{9}{20}$ 0.45

29. $2\frac{1}{2}$ 2.5

30. $5\frac{5}{6}$ 5.83

Solve.

31. Larry has $38. How many T-shirts can he buy if each T-shirt costs $8?
4 tee shirts

32. Hamburger buns come 8 to a package. If there are 36 hamburgers, how many packages are needed?
5 packages

33. What fraction of Jim's expenses is spent on books? $\frac{1}{8}$

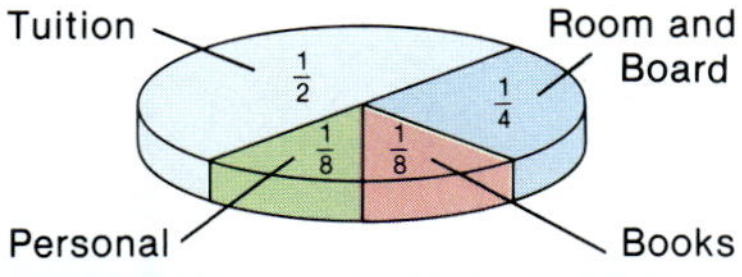

34. Will more money be spent on advertising on TV or radio? TV

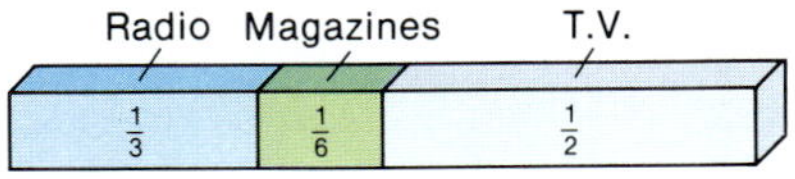

OBJECTIVE

- Evaluate achievement of the chapter objectives.

USING THE TEST

The Chapter Test may be used as a posttest to evaluate student achievement. However, you may wish to use the Chapter Posttest offered in the Teacher's Resource Package or to design your own chapter test. If this page is not used as a test, you may wish to assign it as additional review or practice.

The test items are correlated to the chapter objectives in the table below.

Chapter objectives	Test items
A. Write fractions.	1–4
B. Find the GCF or LCM.	5–8
C. Write fractions in lowest terms.	9–12
D. Compare fractions.	13–16
E. Write improper fractions as whole numbers, or mixed numbers and vice versa.	17–22
F. Write decimals as fractions or mixed numbers.	23–26
G. Write fractions or mixed numbers as decimals.	27–30
H. Interpret remainders to solve problems.	31–32
I. Apply computational skills in real-life situations.	33–34

SUPPLEMENTARY MATERIALS

TRP Ch. 5 Posttest Form A, pp. 1–2

TRP Ch. 5 Posttest Form B, pp. 1–2

OBJECTIVE

- Classify fractions by whether their decimal equivalents are terminating or nonterminating.

USING THE PAGE

The examples show a shortcut to determine whether the decimal equivalent of a fraction will be terminating or nonterminating. The shortcut works because a terminating decimal can be written as a fraction with a denominator of a power of 10 such as 10, 100, or 1,000.

Since the prime factorization of any power of 10 is 2^x, 5^y, or $2^x \times 5^y$, all the factors of the denominator must be 2 or 5 for the decimal to be terminating. For example, the prime factorization of 1,000 is $2^3 \times 5^3$. Guide students in finding the prime factorizations for the following denominators and in identifying which decimal equivalents will terminate.

$\frac{4}{15}$ (3×5) $\quad$ $\frac{12}{50}$ (2×5^2)

$\frac{36}{75}$ (3×5^2) $\quad$ $\frac{57}{100}$ $(2^2 \times 5^2)$

After students have completed the exercises, encourage them to use their calculators to find the decimal equivalent of each fraction and to check the accuracy of their predictions.

ENRICHMENT TERMINATING DECIMALS

All fractions can be written as **terminating** or **nonterminating decimals.** 0.435 is a terminating decimal. 0.434343... is a nonterminating decimal. A fraction written in simplest form will be a terminating decimal only if all the prime factors of the denominator are 2 or 5.

Examples Will $\frac{28}{50}$ be a terminating decimal?

1. Write $\frac{28}{50}$ in simplest form. $\frac{28}{50} = \frac{14}{25}$
2. Factor the denominator.

   $25 = 5 \times 5$

   THINK: 5 is the only factor.

So, $\frac{28}{50}$ will terminate.

Will $\frac{10}{24}$ be a terminating decimal?

1. Write $\frac{10}{24}$ in simplest form. $\frac{10}{24} = \frac{5}{12}$
2. Factor the denominator.

   $12 = 4 \times 3 = 2 \times 2 \times 3$

   THINK: 3 is a prime factor.

So, $\frac{10}{24}$ will not terminate.

Will the decimal terminate? Write *yes* or *no*.

1. $\frac{1}{2}$ yes	**2.** $\frac{2}{10}$ yes	**3.** $\frac{14}{18}$ no	**4.** $\frac{6}{20}$ yes
5. $\frac{33}{45}$ no	**6.** $\frac{13}{16}$ yes	**7.** $\frac{18}{24}$ yes	**8.** $\frac{9}{20}$ yes
9. $\frac{4}{25}$ yes	**10.** $\frac{3}{125}$ yes	**11.** $\frac{9}{21}$ no	**12.** $\frac{6}{8}$ yes
13. $\frac{3}{32}$ yes	**14.** $\frac{6}{27}$ no	**15.** $\frac{5}{75}$ no	**16.** $\frac{121}{625}$ yes
17. $\frac{3}{36}$ no	**18.** $\frac{8}{128}$ yes	**19.** $\frac{12}{15}$ yes	**20.** $\frac{10}{65}$ no

CALCULATOR FRACTIONS ON A CALCULATOR

You have learned that a fraction can be written as a decimal by dividing the numerator by the denominator.

$$\frac{1}{4} \rightarrow 4\overline{)1.00}\ \ 0.25$$

You can also use a calculator to change a fraction or mixed number to a decimal.

Change $\frac{1}{3}$ to a decimal.

Procedure	Calculator Entry	Calculator Display
Divide the numerator by the denominator.	1 ÷ 3 =	0.33333333

$\frac{1}{3} = 0.333$ ← **Repeating decimal rounded to the nearest thousandth**

Change $5\frac{3}{8}$ to a decimal.

Procedure	Calculator Entry	Calculator Display
Change the fraction to a decimal.	3 ÷ 8 =	0.375
Add the whole number.	+ 5 =	5.375

$5\frac{3}{8} = 5.375$

Use a calculator to change the fraction or mixed number to a decimal. When necessary, round your answer to the nearest thousandth.

1. $\frac{5}{8}$ 0.625 **2.** $\frac{7}{20}$ 0.35 **3.** $\frac{13}{25}$ 0.52 **4.** $\frac{3}{7}$ 0.429 **5.** $\frac{5}{6}$ 0.833

6. $4\frac{3}{5}$ 4.6 **7.** $8\frac{18}{25}$ 8.72 **8.** $7\frac{11}{16}$ 7.688 **9.** $6\frac{1}{6}$ 6.167 **10.** $8\frac{2}{3}$ 8.667

Use a calculator to change the fraction to a decimal.

11. $\frac{1}{11}$ 0.091 **12.** $\frac{2}{11}$ 0.182 **13.** $\frac{3}{11}$ 0.273 **14.** $\frac{4}{11}$ 0.364 **15.** $\frac{5}{11}$ 0.455

What patterns do you see? Use the pattern to write a decimal for the fraction.

16. $\frac{6}{11}$ 0.545 **17.** $\frac{7}{11}$ 0.636 **18.** $\frac{8}{11}$ 0.727 **19.** $\frac{9}{11}$ 0.818 **20.** $\frac{10}{11}$ 0.909

OBJECTIVE

- Change a fraction or mixed number to a decimal using a calculator.

USING THE PAGE

Assist students in using a calculator to change a fraction to a decimal. Remind students that they divide the numerator by the denominator. Repeating decimals show up on the calculator display as repeating digits. Point out that the calculator display only shows decimals to 9 digits. In reality these digits do not come to an end. For our purposes, however, students may round these decimals to the nearest thousandth.

You may wish to do the exercises with the class or assign them to a group or to individual students for independent work.

You may wish to extend this activity by having students use the calculator to find decimal patterns for the following fractions.

$\frac{1}{9}\ \frac{2}{9}\ \frac{3}{9}\ \frac{4}{9}\ \frac{5}{9}\ \frac{6}{9}\ \frac{7}{9}\ \frac{8}{9}$

$\frac{1}{9} = 0.1111111$

$\frac{2}{9} = 0.2222222$

$\frac{3}{9} = 0.3333333$

$\frac{4}{9} = 0.4444444$

$\frac{5}{9} = 0.5555555$

$\frac{6}{9} = 0.6666666$

$\frac{7}{9} = 0.7777777$

$\frac{8}{9} = 0.8888888$

OBJECTIVE

- Test the prerequisite skills needed to learn the concepts developed in Chapter 6.

USING THE TEST

The Pre-Skills Test is designed to diagnose students' strengths and weaknesses on prerequisite skills necessary to study the mathematics in Chapter 6.

Assign the Pre-Skills Test. Allow the students to work together in pairs or in small groups. Group members should help those who demonstrate a misunderstanding of a concept or a weakness in a skill.

The following table correlates the items on the Pre-Skills Test with the prerequisite skill and the lesson(s) in the chapter for which it is needed.

Item(s)	Prerequisite skill	Lesson(s)
1–8	Find the GCF.	6.1
9–16	Write fractions in simplest form.	6.1–6.4, 6.7–6.10
17–26	Write whole and mixed numbers as improper fractions.	6.2, 6.4
27–36	Write improper fractions as mixed numbers.	6.2, 6.4
37–41	Find the LCM.	6.7–6.10
42–51	Compare fractions.	6.12

Pre-Skills Test

Find the greatest common factor (GCF). [124]

1. 8, 6 2 **2.** 16, 24 8 **3.** 21, 81 3 **4.** 54, 18 18

5. 6, 16 2 **6.** 12, 16 4 **7.** 11, 19 1 **8.** 18, 45 9

Write the fraction in lowest terms. [126]

9. $\frac{6}{8}$ $\frac{3}{4}$ **10.** $\frac{3}{9}$ $\frac{1}{3}$ **11.** $\frac{4}{10}$ $\frac{2}{5}$ **12.** $\frac{8}{12}$ $\frac{2}{3}$

13. $\frac{16}{24}$ $\frac{2}{3}$ **14.** $\frac{7}{21}$ $\frac{1}{3}$ **15.** $\frac{6}{15}$ $\frac{2}{5}$ **16.** $\frac{12}{21}$ $\frac{4}{7}$

Write as an improper fraction. [132]

17. 3 $\frac{3}{1}$ **18.** $4\frac{1}{5}$ $\frac{21}{5}$ **19.** $7\frac{5}{8}$ $\frac{61}{8}$ **20.** $5\frac{1}{6}$ $\frac{31}{6}$ **21.** $8\frac{4}{9}$ $\frac{76}{9}$

22. $4\frac{1}{9}$ $\frac{37}{9}$ **23.** $6\frac{2}{3}$ $\frac{20}{3}$ **24.** $2\frac{7}{8}$ $\frac{23}{8}$ **25.** $10\frac{1}{2}$ $\frac{21}{2}$ **26.** $9\frac{2}{3}$ $\frac{29}{3}$

Write as a mixed number in simplest form. [132]

27. $\frac{5}{3}$ $1\frac{2}{3}$ **28.** $\frac{8}{5}$ $1\frac{3}{5}$ **29.** $\frac{23}{3}$ $7\frac{2}{3}$ **30.** $\frac{13}{2}$ $6\frac{1}{2}$ **31.** $\frac{19}{5}$ $3\frac{4}{5}$

32. $\frac{9}{4}$ $2\frac{1}{4}$ **33.** $\frac{33}{6}$ $5\frac{1}{2}$ **34.** $\frac{20}{2}$ 10 **35.** $\frac{26}{8}$ $3\frac{1}{4}$ **36.** $\frac{15}{4}$ $3\frac{3}{4}$

Find the least common multiple (LCM) for the pair of numbers. [124]

37. 3, 4 12 **38.** 6, 8 24 **39.** 9, 6 18 **40.** 12, 8 24 **41.** 12, 16 48

Compare. Use <, >, or =. [128]

42. $\frac{4}{5}$ ● $\frac{3}{5}$ > **43.** $\frac{1}{2}$ ● $\frac{1}{3}$ > **44.** $\frac{7}{10}$ ● $\frac{7}{8}$ < **45.** $\frac{5}{12}$ ● $\frac{7}{12}$ < **46.** $\frac{31}{40}$ ● $\frac{31}{60}$ >

47. $\frac{3}{8}$ ● $\frac{1}{2}$ < **48.** $\frac{2}{4}$ ● $\frac{1}{2}$ = **49.** $\frac{5}{6}$ ● $\frac{1}{2}$ > **50.** $\frac{7}{12}$ ● $\frac{1}{2}$ > **51.** $\frac{7}{16}$ ● $\frac{1}{2}$ <

Chapter 6
FRACTION COMPUTATION

Often people have to follow the same time schedule each day. What are the advantages or disadvantages?

OBJECTIVE

- Formulate problems using text and data.

USING THE CHAPTER OPENER

During the discussion of the question, develop a list of advantages and disadvantages. Some advantages might include helping to assure an efficient use of time or a way for busy people to keep track of everything. Some disadvantages could be restricting activities or not being able to do things on the spur of the moment.

COOPERATIVE LEARNING

Write the following on the chalkboard.

PART-TIME JOBS AVAILABLE
Sales help wanted—Sam's Clothing
M–Th 3:00–6:00 P.M.
F 3:00–9:00 P.M. $3.75/hour
Sat 9:00 A.M.–6:00 P.M.

Burger Palace now hiring
Up to $4.00 per hour + free lunch
Sat. and Sun. 7:00 A.M. to 3:00 P.M.

Pizza Masters
Delivery person—bicycle provided
Thurs., Fri. 6:00–9:00 P.M.
Sat., Sun. 2:00–10:00 P.M.
$4.25/hour

Discuss the concept of job sharing, explaining that in some companies part-time employees may share a single job. This necessitates the need for schedules. Explain to the groups that they are to select one of the three jobs listed for a job-sharing situation. They are to base their choice on an agreement they reach. Note that they could choose to split the job into unequal parts if one student wants to work more hours than another.

Have them present their selection and why they chose it, a schedule for the hours each will work, and how much each will earn per week.

SUPPLEMENTARY MATERIALS

TRP Group Projects, p. 6

OBJECTIVE

- Multiply fractions.

TEACHING THE LESSON

WARM-UP Present the following exercises on the chalkboard or overhead transparency.

Find the GCF.

1. 9, 12 **2.** 15, 10 **3.** 7, 17
4. 10, 36 **5.** 12, 39 **6.** 24, 88

(**1.** 3 **2.** 5 **3.** 1 **4.** 2 **5.** 3 **6.** 8)

INTRODUCTION Illustrate the product of $\frac{1}{2} \times \frac{2}{3}$ by using geometric regions. Show $\frac{1}{2}$ by using an overhead transparency. Then show $\frac{2}{3}$. Next, superimpose the $\frac{1}{2}$ over the $\frac{2}{3}$. The number of parts that are crosshatched show the numerator of the product. The denominator is shown by the number of parts into which the rectangle is divided.

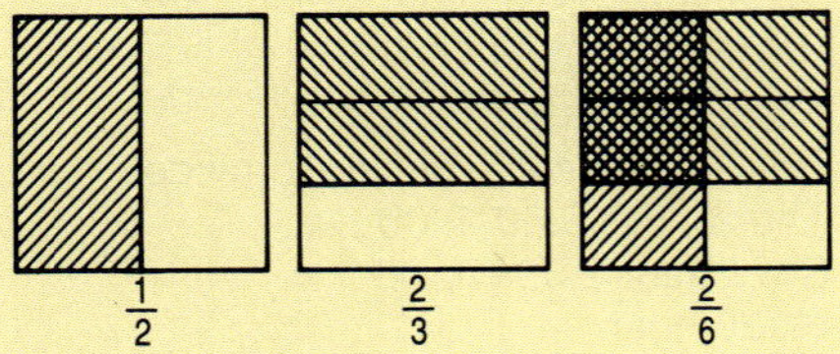

INSTRUCTION In discussing Example 1, remind students that a fraction is in lowest terms when the GCF of the numerator and denominator is 1. Point out that since the GCF of 2 and 15 is 1, the product is already in lowest terms.

In discussing Example 2, have students compare the two factors and the product. Elicit that the product of two proper fractions will always be less than either fraction.

For Example 3, point out that if all the common factors are eliminated before multiplying, then the product will always be in lowest terms.

6.1 MULTIPLYING FRACTIONS

Frank Torelli saves $\frac{1}{5}$ of his income. He deposits $\frac{2}{3}$ of his savings into a savings account. He buys stock with the rest of his savings. What part of his total earnings does Frank deposit into his savings account?

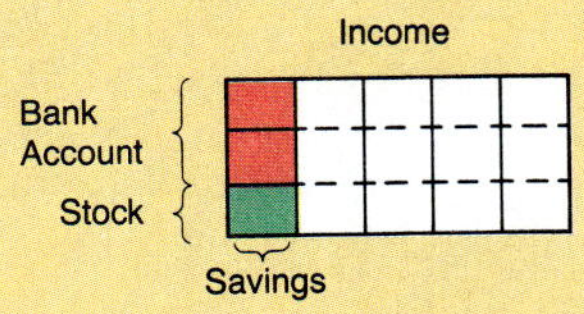

Multiply $\frac{1}{5}$ by $\frac{2}{3}$ to find the part of his total earnings that Frank puts into his savings account.

EXAMPLE 1 Multiply: $\frac{2}{3} \times \frac{1}{5}$

1. Multiply the numerators.

$$\frac{2}{3} \times \frac{1}{5} = 2$$

2. Multiply the denominators.

$$\frac{2}{3} \times \frac{1}{5} = \frac{2}{15}$$

Frank deposits $\frac{2}{15}$ of his income into a savings account.

Sometimes the answer needs to be written in simplest form.

EXAMPLE 2 Multiply: $\frac{3}{4} \times \frac{8}{9}$ Write the answer in simplest form.

1. Multiply.

$$\frac{3}{4} \times \frac{8}{9} = \frac{24}{36}$$

2. Write the product in simplest form.

$$\frac{24}{36} = \frac{24 \div 12}{36 \div 12} = \frac{2}{3}$$

So, $\frac{3}{4} \times \frac{8}{9} = \frac{2}{3}$.

When numerators and denominators have common factors, you can use a shortcut to multiply.

EXAMPLE 3 Multiply: $\frac{4}{9} \times \frac{5}{8} \times \frac{3}{7}$

1. Find common factors. Simplify the fractions.

**THINK: 4 is the GCF of 4 and 8.
3 is the GCF of 3 and 9.**

$$\frac{\overset{1}{\cancel{4}}}{\underset{3}{\cancel{9}}} \times \frac{5}{\underset{2}{\cancel{8}}} \times \frac{\overset{1}{\cancel{3}}}{7}$$

2. Multiply.

$$\frac{\overset{1}{\cancel{4}}}{\underset{3}{\cancel{9}}} \times \frac{5}{\underset{2}{\cancel{8}}} \times \frac{\overset{1}{\cancel{3}}}{7} = \frac{5}{42}$$

CHECKPOINT

Write the letter of the correct answer.

Multiply. What is the product in simplest form?

1. $\frac{3}{8} \times \frac{5}{6}$ c **a.** $\frac{8}{14}$ **b.** $\frac{15}{48}$ **c.** $\frac{5}{16}$ **d.** $\frac{35}{86}$

2. $\frac{3}{8} \times \frac{1}{2} \times \frac{4}{9}$ d **a.** $\frac{8}{19}$ **b.** $\frac{12}{144}$ **c.** $\frac{1}{24}$ **d.** $\frac{1}{12}$

PRACTICE EXERCISES

See Extra Practice, page 438.

Multiply. Write the product in simplest form.

• **1.** $\frac{1}{3} \times \frac{1}{2}$ $\frac{1}{6}$ **2.** $\frac{1}{4} \times \frac{1}{2}$ $\frac{1}{8}$ **3.** $\frac{2}{3} \times \frac{1}{5}$ $\frac{2}{15}$ **4.** $\frac{1}{6} \times \frac{5}{8}$ $\frac{5}{48}$ **5.** $\frac{5}{9} \times \frac{1}{2}$ $\frac{5}{18}$ **6.** $\frac{1}{7} \times \frac{5}{6}$ $\frac{5}{42}$

7. $\frac{2}{3} \times \frac{2}{5}$ $\frac{4}{15}$ **8.** $\frac{3}{4} \times \frac{3}{4}$ $\frac{9}{16}$ **9.** $\frac{3}{4} \times \frac{5}{7}$ $\frac{15}{28}$ **10.** $\frac{5}{9} \times \frac{2}{3}$ $\frac{10}{27}$ **11.** $\frac{2}{3} \times \frac{4}{5}$ $\frac{8}{15}$ **12.** $\frac{2}{5} \times \frac{3}{7}$ $\frac{6}{35}$

• **13.** $\frac{3}{4} \times \frac{4}{5}$ $\frac{3}{5}$ **14.** $\frac{2}{3} \times \frac{5}{6}$ $\frac{5}{9}$ **15.** $\frac{3}{4} \times \frac{3}{9}$ $\frac{1}{4}$ **16.** $\frac{2}{7} \times \frac{3}{8}$ $\frac{3}{28}$ **17.** $\frac{3}{8} \times \frac{4}{6}$ $\frac{1}{4}$ **18.** $\frac{5}{6} \times \frac{3}{5}$ $\frac{1}{2}$

19. $\frac{5}{12} \times \frac{2}{3}$ $\frac{5}{18}$ **20.** $\frac{3}{4} \times \frac{7}{10}$ $\frac{21}{40}$ **21.** $\frac{9}{15} \times \frac{5}{8}$ $\frac{3}{8}$ **22.** $\frac{2}{3} \times \frac{6}{13}$ $\frac{4}{13}$ **23.** $\frac{8}{11} \times \frac{3}{8}$ $\frac{3}{11}$ **24.** $\frac{4}{5} \times \frac{3}{8}$ $\frac{3}{10}$

25. $\frac{4}{15} \times \frac{5}{12}$ $\frac{1}{9}$ **26.** $\frac{3}{14} \times \frac{7}{12}$ $\frac{1}{8}$ **27.** $\frac{9}{10} \times \frac{10}{11}$ $\frac{9}{11}$ **28.** $\frac{11}{20} \times \frac{5}{22}$ $\frac{1}{8}$ **29.** $\frac{10}{21} \times \frac{14}{25}$ $\frac{4}{15}$ **30.** $\frac{5}{6} \times \frac{3}{10}$ $\frac{1}{4}$

31. $\frac{1}{4} \times \frac{8}{9}$ $\frac{2}{9}$ **32.** $\frac{4}{5} \times \frac{2}{3}$ $\frac{8}{15}$ **33.** $\frac{3}{8} \times \frac{1}{6}$ $\frac{1}{16}$ **34.** $\frac{7}{12} \times \frac{3}{14}$ $\frac{1}{8}$ **35.** $\frac{1}{8} \times \frac{1}{4}$ $\frac{1}{32}$ **36.** $\frac{9}{16} \times \frac{2}{3}$ $\frac{3}{8}$

Multiply. Use the shortcut when possible.

• **37.** $\frac{1}{3} \times \frac{2}{5} \times \frac{1}{2}$ $\frac{1}{15}$ • **38.** $\frac{2}{3} \times \frac{3}{4} \times \frac{1}{2}$ $\frac{1}{4}$ **39.** $\frac{1}{3} \times \frac{1}{8} \times \frac{4}{5}$ $\frac{1}{30}$ **40.** $\frac{3}{4} \times \frac{2}{3} \times \frac{3}{5}$ $\frac{3}{10}$

41. $\frac{4}{5} \times \frac{5}{8} \times \frac{1}{2}$ $\frac{1}{4}$ **42.** $\frac{6}{9} \times \frac{3}{4} \times \frac{3}{4}$ $\frac{3}{8}$ **43.** $\frac{5}{24} \times \frac{8}{15} \times \frac{9}{10}$ $\frac{1}{10}$ **44.** $\frac{10}{21} \times \frac{6}{15} \times \frac{7}{8}$ $\frac{1}{6}$

MIXED REVIEW

Add or subtract.

• **45.** $347 + 68$ 415 **46.** $1,980 + 3,268$ 5,248 **47.** $825,971 + 3,009$ 828,980 **48.** $61,803 + 2,999$ 64,802 **49.** $572 + 46$ 618

• **50.** $47 - 28$ 19 **51.** $852 - 690$ 162 **52.** $78,542 - 12,855$ 65,687 **53.** $623,401 - 9,426$ 613,975 **54.** $937 - 486$ 451

Solve.

55. Yoko spends $\frac{1}{8}$ of her total income on entertainment. Of that amount, she spends $\frac{2}{5}$ on movies. What part of her total income does she spend on movies? $\frac{1}{20}$

56. Erik deducted $\frac{1}{4}$ of his income for taxes. He received $\frac{1}{3}$ of his total deductions as a tax refund. What part of his total income did he receive as a refund? $\frac{1}{12}$

CHECKPOINT Distribute to each student four index cards on which the answer choices a–d appear. After allowing students a few minutes, ask them to raise the card that indicates their answer choice. Scan the cards and identify those students who need additional attention before proceeding to the Practice Exercises. The incorrect answer choices include common errors.

COMMON ERROR

When multiplying fractions, some students will add rather than multiply. For example, in item 1 under Checkpoint, these students will select choice a. To prevent this error, have these students circle the operation sign before they begin to work each exercise.

ASSIGNMENTS

Level 1	Odd 1–55
Level 2	Even 2–54, 55, 56
Level 3	25–44, Odd 45–53, 55–56

FOLLOW-UP ACTIVITY

ESTIMATION Explain how to use estimation to multiply fractions. Provide the following guidelines: If a fraction is less then $\frac{1}{2}$, round up to $\frac{1}{2}$. If a fraction is greater than or equal to $\frac{1}{2}$, round up to 1.

Have students estimate by rounding.

1. $\frac{3}{5} \times \frac{2}{7}$ **2.** $\frac{3}{8} \times \frac{1}{4}$ **3.** $\frac{2}{3} \times \frac{1}{3}$

4. $\frac{4}{5} \times \frac{1}{2}$ **5.** $\frac{3}{4} \times \frac{8}{9}$ **6.** $\frac{5}{6} \times \frac{3}{4}$

(**1.** $\frac{1}{2}$ **2.** $\frac{1}{4}$ **3.** $\frac{1}{2}$ **4.** 1 **5.** 1 **6.** 1)

SUPPLEMENTARY MATERIALS

TRP Practice, p. 59

TRP Reteaching, p. 36

TRP Transparency 30

OBJECTIVE

- Multiply fractions, whole numbers, and mixed numbers.

TEACHING THE LESSON

WARM-UP Present the following exercises on the chalkboard or overhead transparency.

Write as an improper fraction.

1. $2\frac{2}{3}$ **2.** $4\frac{2}{5}$ **3.** $3\frac{3}{4}$

4. $8\frac{1}{3}$ **5.** $9\frac{3}{5}$ **6.** $1\frac{4}{7}$

(**1.** $\frac{8}{3}$ **2.** $\frac{22}{5}$ **3.** $\frac{15}{4}$ **4.** $\frac{25}{3}$ **5.** $\frac{48}{5}$ **6.** $\frac{11}{7}$)

INTRODUCTION Write the following on the chalkboard:

$$\frac{1}{2} \times 6$$

Explain that the word "of" can mean multiplication. Continue to illustrate the example:

$\frac{1}{2}$ of 6 ● ● ●

$\frac{1}{2}$ of 6 = 3 ● ● ●

INSTRUCTION Remind students that a whole number, such as 8, can be written as $\frac{8}{1}$. When discussing Example 1, point out that the word "of" is used to mean multiplication. Write the following on the chalkboard:

$\frac{3}{4}$ of 48 $\frac{3}{4} \times 48$

Elicit from students that when multiplying a fraction and a mixed number, the product will always be less than the mixed number.

As you discuss Example 2, remind students to write the mixed numbers as improper fractions before they begin to multiply. As students work through Example 2b, point out that the product of two mixed numbers is greater than either mixed number.

CHECKPOINT The incorrect answer choices include common errors.

6.2 MULTIPLYING MIXED NUMBERS

Amir Johnson is building a cabin cruiser. The hull of the cruiser is 48 feet long. The owner wants $\frac{3}{4}$ of the hull painted white. How many feet of the hull should Amir paint white?

THINK: $\frac{3}{4}$ of 48 feet is the same as $\frac{3}{4} \times 48$.

EXAMPLE 1 Multiply: $\frac{3}{4} \times 48$

1. Rename the whole number as an improper fraction.

$$48 = \frac{48}{1}$$

2. Multiply.

$$\frac{3}{4} \times \frac{48}{1} = \frac{36}{1} = 36$$

Amir should paint 36 feet of the hull white.

To multiply mixed numbers and whole numbers, write each factor as an improper fraction.

EXAMPLE 2 Multiply:

a. $3 \times 2\frac{1}{6}$

$$3 \times 2\frac{1}{6} = \frac{3}{1} \times \frac{13}{6} = \frac{13}{2} = 6\frac{1}{2}$$

b. $4\frac{2}{3} \times 5\frac{1}{4}$

$$4\frac{2}{3} \times 5\frac{1}{4} = \frac{14}{3} \times \frac{21}{4} = \frac{49}{2} = 24\frac{1}{2}$$

CHECKPOINT

Write the letter of the correct answer.

Multiply. What is the product in simplest form?

1. $\frac{3}{4} \times 12$ d **a.** $\frac{36}{4}$ **b.** $12\frac{3}{4}$ **c.** 16 **d.** 9

2. $3 \times 4\frac{2}{3}$ b **a.** $12\frac{2}{3}$ **b.** 14 **c.** $12\frac{6}{3}$ **d.** 6

3. $3\frac{2}{5} \times 1\frac{3}{4}$ c **a.** $3\frac{6}{20}$ **b.** $3\frac{3}{10}$ **c.** $5\frac{19}{20}$ **d.** $4\frac{5}{9}$

PRACTICE EXERCISES

See Extra Practice, page 438.

Multiply. Write the product in simplest form.

- • **1.** $2 \times \frac{1}{2}$ 1
- **2.** $3 \times \frac{2}{3}$ 2
- **3.** $5 \times \frac{3}{5}$ 3
- **4.** $\frac{1}{2} \times 6$ 3
- **5.** $\frac{2}{3} \times 6$ 4
- **6.** $8 \times \frac{3}{4}$ 6
- **7.** $\frac{1}{3} \times 9$ 3
- **8.** $12 \times \frac{5}{6}$ 10
- **9.** $5 \times \frac{2}{3}$ $3\frac{1}{3}$
- **10.** $\frac{4}{5} \times 3$ $2\frac{2}{5}$
- • **11.** $2 \times 1\frac{1}{2}$ 3
- **12.** $3 \times 2\frac{1}{3}$ 7
- **13.** $2 \times 3\frac{1}{2}$ 7
- **14.** $4 \times 3\frac{1}{4}$ 13
- **15.** $3 \times 2\frac{2}{3}$ 8
- **16.** $1\frac{3}{4} \times 8$ 14
- **17.** $2 \times 2\frac{5}{8}$ $5\frac{1}{4}$
- **18.** $3\frac{1}{6} \times 24$ 76
- **19.** $3\frac{7}{20} \times 5$ $16\frac{3}{4}$
- **20.** $4 \times 2\frac{1}{12}$ $8\frac{1}{3}$
- **21.** $\frac{7}{8} \times 2\frac{1}{2}$ $2\frac{3}{16}$
- **22.** $\frac{3}{5} \times 4\frac{1}{3}$ $2\frac{3}{5}$
- **23.** $7\frac{1}{4} \times \frac{8}{9}$ $6\frac{4}{9}$
- **24.** $6\frac{2}{3} \times \frac{4}{5}$ $5\frac{1}{3}$
- **25.** $\frac{5}{6} \times 5\frac{1}{3}$ $4\frac{4}{9}$
- • **26.** $1\frac{1}{5} \times 2\frac{1}{2}$ 3
- **27.** $2\frac{2}{3} \times 1\frac{1}{3}$ $3\frac{5}{9}$
- **28.** $2\frac{1}{4} \times 2\frac{2}{3}$ 6
- **29.** $1\frac{3}{5} \times 2\frac{1}{2}$ 4
- **30.** $2\frac{2}{3} \times 3\frac{1}{2}$ $9\frac{1}{3}$
- **31.** $4\frac{3}{4} \times 2\frac{1}{3}$ $11\frac{1}{12}$
- **32.** $3\frac{3}{4} \times 1\frac{3}{4}$ $6\frac{9}{16}$
- **33.** $4\frac{2}{3} \times 1\frac{2}{7}$ 6
- **34.** $4\frac{1}{2} \times 2\frac{2}{3}$ 12
- **35.** $1\frac{1}{3} \times 3\frac{2}{7}$ $4\frac{8}{21}$
- **36.** $2\frac{1}{4} \times 3$ $6\frac{3}{4}$
- **37.** $3\frac{2}{3} \times 3\frac{3}{4}$ $13\frac{3}{4}$
- **38.** $\frac{2}{5} \times 3\frac{1}{2}$ $1\frac{2}{5}$
- **39.** $5\frac{1}{3} \times 2\frac{3}{8}$ $12\frac{2}{3}$
- **40.** $\frac{1}{2} \times 9\frac{2}{3}$ $4\frac{5}{6}$
- **41.** $5\frac{1}{5} \times 6\frac{1}{4}$ $32\frac{1}{2}$
- **42.** $7\frac{2}{3} \times 2$ $15\frac{1}{3}$
- **43.** $\frac{1}{3} \times 3\frac{2}{4}$ $1\frac{1}{6}$
- **44.** $6\frac{2}{5} \times 10$ 64
- **45.** $10\frac{1}{2} \times \frac{4}{7}$ 6
- **46.** $\frac{2}{5} \times \frac{3}{8} \times 10$ $1\frac{1}{2}$
- **47.** $2\frac{1}{2} \times \frac{1}{3} \times \frac{1}{4}$ $\frac{5}{24}$
- **48.** $3\frac{2}{3} \times \frac{1}{4} \times 3$ $2\frac{3}{4}$
- **49.** $\frac{2}{3} \times \frac{1}{5} \times 2\frac{3}{4}$ $\frac{11}{30}$
- **50.** $1\frac{1}{2} \times 1\frac{1}{4} \times 1\frac{1}{3}$ $2\frac{1}{2}$
- **51.** $4\frac{2}{5} \times 8\frac{1}{3} \times 2\frac{1}{4}$ $82\frac{1}{2}$
- **52.** $5\frac{2}{5} \times 7\frac{1}{3} \times 6\frac{1}{11}$ $241\frac{1}{5}$
- **53.** $9\frac{3}{5} \times 6\frac{1}{8} \times 2\frac{4}{7}$ $151\frac{1}{5}$
- **54.** $\left(2 + 1\frac{1}{2}\right) \times \frac{4}{7}$ 2
- **55.** $\left(1\frac{1}{4} + 3\frac{3}{8}\right) \times \frac{1}{2}$ $2\frac{5}{16}$
- **56.** $\left(\frac{3}{4} + \frac{7}{8}\right) \times \frac{2}{3}$ $1\frac{1}{12}$
- **57.** $\left(\frac{2}{3} + 1\frac{1}{2}\right) \times \frac{3}{4}$ $1\frac{5}{8}$

Solve.

• **58.** Maria's boat has a maximum speed of 36 miles per hour. Its cruising speed is $\frac{3}{4}$ of its maximum speed. What is the boat's cruising speed? 27 mph

59. Latoya earns $18.00 an hour as a carpenter building boats. One week she works 42.5 hours. How much money does she earn that week? $765.00

60. Cesar's budget allows him to spend $78,000 on lumber. He has spent $45,963. How much more can he spend without going over budget? $32,037

61. A ship can travel at a rate of $18\frac{3}{4}$ miles per hour. How far can it travel in $3\frac{1}{5}$ hours? 60 mi

TIME OUT A ship is sailing from New York City to Southhampton, England. The total distance is about 3,600 miles. The ship has traveled $\frac{1}{2}$ the distance to the halfway point. How far has the ship traveled? 900 mi

TIME OUT It may be helpful for students to draw a number line to illustrate that "$\frac{1}{2}$ the distance to the halfway point" is $\frac{1}{4}$."

COMMON ERROR

When multiplying mixed numbers, some students will multiply whole numbers and fractions instead of writing the mixed numbers as improper fractions. For example, in item 3 of Checkpoint, these students will select a. Have these students follow this procedure before they begin to multiply:

$$3\frac{2}{5} = \frac{(5 \times 3) + 2}{5} = \frac{17}{5}$$

$$1\frac{3}{4} = \frac{(1 \times 4) + 3}{4} = \frac{7}{4}$$

ASSIGNMENTS

Level 1	Odd 1–45, 58–59
Level 2	Even 2–48, 58–61, TO
Level 3	31–61

FOLLOW-UP ACTIVITY

APPLICATION Present the following recipe. Direct students to rewrite the recipe, to make $2\frac{1}{2}$ times the amount.

White Bread	
2 pkgs. dry yeast	2 tbsp salt
$\frac{1}{2}$ cup water	2 cups milk
$\frac{1}{3}$ cup sugar	$1\frac{1}{2}$ cups water
$\frac{1}{3}$ cup shortening	10 cups flour

SUPPLEMENTARY MATERIALS

TRP Practice, p. 60

TRP Reteaching, p. 37

TRP Transparency 30

OBJECTIVES

- Find the reciprocal of a number.
- Divide fractions and whole numbers.

TEACHING THE LESSON

WARM-UP Present the following exercises on the chalkboard or overhead transparency.

Multiply. Write the product in simplest form.

1. $\frac{3}{4} \times \frac{4}{3}$ **2.** $\frac{7}{2} \times \frac{2}{7}$ **3.** $8 \times \frac{1}{8}$

4. $\frac{1}{9} \times 9$ **5.** $1\frac{2}{3} \times \frac{3}{5}$ **6.** $\frac{5}{17} \times 3\frac{2}{5}$

(All products are 1.)

INTRODUCTION Ask students to name situations when they may need to divide fractions, such as following sewing patterns, modifying recipes, or computing distances.

INSTRUCTION In discussing the introductory problem, elicit from students that the quotient will be more than 12. Have students count the number of sections in the illustration to verify the quotient.

In Examples 1 and 2, emphasize that the first step is to find the reciprocal of the divisor. Extend the instruction by having students write the reciprocal of:

$14(\frac{1}{14})$ $\frac{2}{3}(\frac{3}{2})$ $7(\frac{1}{7})$ $\frac{8}{5}(\frac{5}{8})$

CHECKPOINT The incorrect answer choices include common errors.

COMMON ERROR

Some students use the reciprocal of the dividend when dividing fractions. For example, in item 2 of Checkpoint, these students will select d. Remind students to use the reciprocal of the divisor, the number following the division symbol. Have students identify the divisor in Exercises 25-30.

6.3 DIVIDING FRACTIONS

If the product of two numbers equals 1, the numbers are called **reciprocals.**

$\frac{3}{4} \times \frac{4}{3} = \frac{12}{12} = 1$ — **Reciprocals**

$6 \times \frac{1}{6} = \frac{6}{1} \times \frac{1}{6} = \frac{6}{6} = 1$ — **Reciprocals**

$2\frac{2}{3} \times \frac{3}{8} = \frac{8}{3} \times \frac{3}{8} = \frac{24}{24} = 1$ — **Reciprocals**

To find the reciprocal of a number, write the number as a fraction and reverse the numerator and denominator.

At the Snapper Creek Scout Camp, the scouts make their own tent stakes. How many stakes $\frac{3}{4}$ ft long can be made out of a piece of lumber that is 12 ft long?

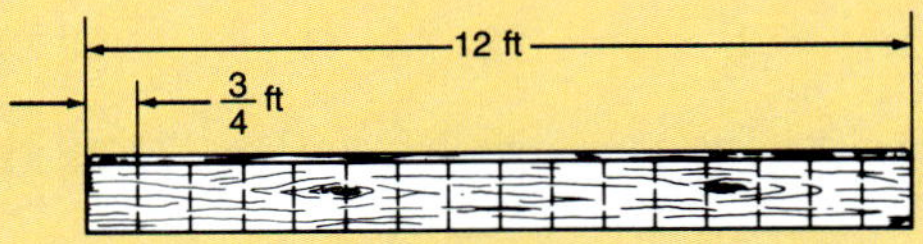

Divide 12 by $\frac{3}{4}$ to find the number of stakes that can be made.

To divide fractions, multiply the dividend by the reciprocal of the divisor.

EXAMPLE 1 Divide: $12 \div \frac{3}{4}$

1. Find the reciprocal of the divisor. The reciprocal of $\frac{3}{4}$ is $\frac{4}{3}$.
2. Multiply. $\frac{12}{1} \times \frac{4}{3} = \frac{\overset{4}{\cancel{12}}}{1} \times \frac{4}{\underset{1}{\cancel{3}}} = \frac{16}{1} = 16$

You can make 16 stakes that are $\frac{3}{4}$ ft long from a 12-ft board.

When the divisor is a whole number, follow the same procedure as when dividing fractions.

EXAMPLE 2 Divide: $\frac{2}{3} \div 8$

1. Find the reciprocal of the divisor. The reciprocal of 8 is $\frac{1}{8}$.
2. Multiply. $\frac{2}{3} \times \frac{1}{8} = \frac{\overset{1}{\cancel{2}}}{3} \times \frac{1}{\underset{4}{\cancel{8}}} = \frac{1}{12}$

CHECKPOINT

Write the letter of the correct answer.

Divide. What is the quotient in simplest form?

1. $\frac{1}{2} \div \frac{2}{3}$ a **a.** $\frac{3}{4}$ **b.** $1\frac{1}{3}$ **c.** $\frac{1}{6}$ **d.** $\frac{1}{3}$

2. $\frac{8}{9} \div 4$ c **a.** $\frac{8}{36}$ **b.** $\frac{4}{9}$ **c.** $\frac{2}{9}$ **d.** $4\frac{1}{2}$

PRACTICE EXERCISES

See Extra Practice, page 438.

Find the reciprocal of each number.

• **1.** 3 $\frac{1}{3}$ • **2.** 7 $\frac{1}{7}$ **3.** 15 $\frac{1}{15}$ **4.** 36 $\frac{1}{36}$ **5.** $\frac{1}{4}$ $\frac{4}{1}$ **6.** $\frac{1}{7}$ $\frac{7}{1}$ **7.** $\frac{1}{12}$ $\frac{12}{1}$ **8.** $\frac{1}{40}$ $\frac{40}{1}$

9. $\frac{3}{5}$ $\frac{5}{3}$ **10.** $\frac{4}{9}$ $\frac{9}{4}$ **11.** $\frac{2}{7}$ $\frac{7}{2}$ **12.** $\frac{8}{3}$ $\frac{3}{8}$ **13.** $\frac{3}{11}$ $\frac{11}{3}$ **14.** $\frac{6}{5}$ $\frac{5}{6}$ **15.** $\frac{15}{12}$ $\frac{12}{15}$ **16.** $\frac{25}{3}$ $\frac{3}{25}$

17. $1\frac{1}{3}$ $\frac{3}{4}$ **18.** $2\frac{3}{4}$ $\frac{4}{11}$ **19.** $4\frac{2}{5}$ $\frac{5}{22}$ **20.** $3\frac{1}{6}$ $\frac{6}{19}$ **21.** $5\frac{2}{7}$ $\frac{7}{37}$ **22.** $6\frac{3}{8}$ $\frac{8}{51}$ **23.** $10\frac{1}{2}$ $\frac{2}{21}$ **24.** $12\frac{5}{6}$ $\frac{6}{77}$

Divide. Write the quotient in simplest form.

• **25.** $\frac{2}{3} \div \frac{3}{4}$ $\frac{8}{9}$ **26.** $\frac{1}{2} \div \frac{2}{3}$ $\frac{3}{4}$ **27.** $\frac{3}{8} \div \frac{1}{2}$ $\frac{3}{4}$ **28.** $\frac{3}{4} \div \frac{5}{6}$ $\frac{9}{10}$ **29.** $\frac{2}{9} \div \frac{2}{3}$ $\frac{1}{3}$ **30.** $\frac{1}{8} \div \frac{5}{6}$ $\frac{3}{20}$

31. $\frac{4}{5} \div \frac{5}{6}$ $\frac{24}{25}$ **32.** $\frac{2}{7} \div \frac{7}{8}$ $\frac{16}{49}$ **33.** $\frac{1}{5} \div \frac{3}{4}$ $\frac{4}{15}$ **34.** $\frac{5}{6} \div \frac{8}{9}$ $\frac{15}{16}$ **35.** $\frac{3}{8} \div \frac{6}{7}$ $\frac{7}{16}$ **36.** $\frac{1}{4} \div \frac{7}{8}$ $\frac{2}{7}$

37. $\frac{3}{4} \div \frac{2}{3}$ $1\frac{1}{8}$ **38.** $\frac{1}{2} \div \frac{1}{4}$ 2 **39.** $\frac{4}{5} \div \frac{3}{4}$ $1\frac{1}{15}$ **40.** $\frac{5}{8} \div \frac{5}{9}$ $1\frac{1}{8}$ **41.** $\frac{8}{9} \div \frac{2}{3}$ $1\frac{1}{3}$ **42.** $\frac{4}{7} \div \frac{2}{5}$ $1\frac{3}{7}$

• **43.** $\frac{3}{4} \div 3$ $\frac{1}{4}$ **44.** $\frac{2}{3} \div 5$ $\frac{2}{15}$ **45.** $\frac{2}{5} \div 4$ $\frac{1}{10}$ **46.** $\frac{3}{8} \div 9$ $\frac{1}{24}$ **47.** $\frac{8}{9} \div 12$ $\frac{2}{27}$ **48.** $\frac{5}{6} \div 10$ $\frac{1}{12}$

49. $2 \div \frac{4}{5}$ $2\frac{1}{2}$ **50.** $4 \div \frac{2}{3}$ 6 **51.** $8 \div \frac{8}{9}$ 9 **52.** $10 \div \frac{5}{6}$ 12 **53.** $14 \div \frac{4}{7}$ $24\frac{1}{2}$ **54.** $24 \div \frac{3}{8}$ 64

Solve.

• **55.** A hiking trail is $\frac{3}{4}$ mile long. It is marked by stakes $\frac{1}{8}$ mile apart. How many stakes mark the trail? (*Hint:* There is no marker at the beginning of the trail.) 6 stakes

56. It takes 8 stakes to put up 1 tent. How many stakes are needed for 6 tents? 48 stakes

57. Each scout is served two $\frac{1}{4}$-lb hamburgers at lunch. How many pounds of ground beef are needed to make hamburgers for 24 scouts? 12 lb

58. The scouts wanted to cook frog legs, but they only had $1\frac{1}{2}$ pounds of frog legs. Rewrite the recipe, showing the ingredients they need.
$1\frac{1}{2}$ lb frog legs; $\frac{1}{4}$ cup bread crumbs; $\frac{1}{8}$ cup Parmesan cheese; $\frac{1}{2}$ teaspoon salt; $\frac{1}{8}$ cup flour; 1 egg; $\frac{1}{8}$ cup cooking oil; 1 tablespoon parsley

FRIED FROG LEGS

3 lb frog legs	2 tablespoons parsley
$\frac{1}{2}$ cup bread crumbs	$\frac{1}{4}$ cup cooking oil
1 teaspoon salt	2 eggs
$\frac{1}{4}$ cup flour	
$\frac{1}{4}$ cup Parmesan cheese	

ASSIGNMENTS

Level 1	Odd 1–53, 55–56
Level 2	Even 2–54, 55–58
Level 3	Odd 17–53, 55–58

FOLLOW-UP ACTIVITY

ESTIMATION Have students experiment with fractions, mixed numbers and whole numbers to complete these sentences.

1. If the dividend is a mixed number and the divisor is a proper fraction, the quotient will be ■ than the dividend. (greater)
2. If the dividend is a fraction and the divisor is a whole number, the quotient will be ■ than the dividend. (less)
3. If the dividend is a fraction and the divisor is a mixed number, the quotient will be ■ than the dividend. (less)

Then have students identify which of the following division sentences are incorrect, based on these generalizations.

1. $6 \div \frac{4}{5} = 7\frac{1}{2}$ **2.** $2\frac{1}{3} \div \frac{1}{5} = \frac{3}{35}$

3. $\frac{1}{2} \div 4\frac{1}{2} = \frac{1}{9}$ **4.** $8 \div \frac{2}{3} = \frac{1}{12}$

(2 and 4 are incorrect)

SUPPLEMENTARY MATERIALS

TRP Practice, p. 61

TRP Reteaching, p. 38

TRP Transparency 31

OBJECTIVE

- Divide fractions, whole numbers, and mixed numbers.

TEACHING THE LESSON

WARM-UP Present the following exercises on the chalkboard or overhead transparency.

Find the reciprocal.

1. $\frac{3}{5}$ **2.** 15 **3.** $1\frac{3}{4}$

4. $4\frac{2}{3}$ **5.** $2\frac{4}{5}$ **6.** $15\frac{5}{6}$

(**1.** $\frac{5}{3}$ **2.** $\frac{1}{15}$ **3.** $\frac{4}{7}$ **4.** $\frac{3}{14}$ **5.** $\frac{5}{14}$ **6.** $\frac{6}{95}$)

INTRODUCTION Bring in a copy of the stock market page from the newspaper. Elicit from students that all stock prices are listed using fractions and mixed numbers. Ask the students how $\$2\frac{1}{4}$ would be written using dollars and cents notation.

INSTRUCTION In discussing Example 1, have students estimate whether the quotient should be greater than or less than 1. They should respond that the quotient should be close to 36.

For Example 2, students should respond that since they are dividing a fraction by approximately the number 3, the quotient will be less than 1.

In Example 3, since students are dividing a smaller number by a larger number, they should respond that the quotient will be less than 1.

CHECKPOINT The incorrect answer choices include common errors.

CALCULATOR This activity demonstrates how to multiply fractions and mixed numbers on a calculator.

6.4 DIVIDING MIXED NUMBERS

The price of one share of stock in the New Age Electronics Company is $\$2\frac{1}{4}$. Cathy Willis has \$72 to invest. How many shares of stock can she buy?

Divide 72 by $2\frac{1}{4}$ to find how many shares of stock Cathy can buy.

To divide by a mixed number, first rename the mixed number as an improper fraction. Then divide as you would with fractions.

EXAMPLE 1 Divide: $72 \div 2\frac{1}{4}$

$$72 \div 2\frac{1}{4} = \frac{72}{1} \div \frac{9}{4} = \frac{\overset{8}{\cancel{72}}}{1} \times \frac{4}{\underset{1}{\cancel{9}}} = 32$$

Cathy can buy 32 shares of stock in the New Age Electronics Company.

When the dividend is a fraction, divide as you would with fractions.

EXAMPLE 2 Divide: $\frac{5}{6} \div 3\frac{1}{3}$

$$\frac{5}{6} \div 3\frac{1}{3} = \frac{5}{6} \div \frac{10}{3} = \frac{\overset{1}{\cancel{5}}}{\underset{2}{\cancel{6}}} \times \frac{\overset{1}{\cancel{3}}}{\underset{2}{\cancel{10}}} = \frac{1}{4}$$

Remember to rename both mixed numbers as fractions before dividing.

EXAMPLE 3 Divide: $1\frac{1}{2} \div 3\frac{3}{4}$

$$1\frac{1}{2} \div 3\frac{3}{4} = \frac{3}{2} \div \frac{15}{4} = \frac{\overset{1}{\cancel{3}}}{\underset{1}{\cancel{2}}} \times \frac{\overset{2}{\cancel{4}}}{\underset{5}{\cancel{15}}} = \frac{2}{5}$$

CHECKPOINT

Write the letter of the correct answer.

Divide. What is the quotient in the simplest form?

1. $6 \div 1\frac{1}{2}$	c	**a.** $\frac{12}{3}$	**b.** 9	**c.** 4	**d.** $\frac{1}{4}$
2. $2\frac{2}{3} \div 1\frac{1}{4}$	a	**a.** $2\frac{2}{15}$	**b.** $\frac{15}{32}$	**c.** $\frac{2}{15}$	**d.** $1\frac{13}{15}$
3. $\frac{4}{7} \div 3\frac{1}{2}$	d	**a.** 2	**b.** $6\frac{1}{8}$	**c.** $\frac{49}{8}$	**d.** $\frac{8}{49}$
4. $4\frac{1}{5} \div \frac{3}{8}$	b	**a.** $\frac{15}{168}$	**b.** $11\frac{1}{5}$	**c.** $\frac{63}{40}$	**d.** $1\frac{23}{40}$

PRACTICE EXERCISES See Extra Practice, page 439.

Divide. Write the quotient in simplest form.

• **1.** $5 \div 1\frac{2}{3}$ 3	**2.** $3 \div 2\frac{3}{4}$ $1\frac{1}{11}$	**3.** $10 \div 3\frac{1}{3}$ 3	**4.** $15 \div 2\frac{1}{7}$ 7	**5.** $8 \div 2\frac{2}{5}$ $3\frac{1}{3}$
• **6.** $2\frac{1}{2} \div 5$ $\frac{1}{2}$	**7.** $1\frac{2}{3} \div 10$ $\frac{1}{6}$	**8.** $4\frac{3}{4} \div 3$ $1\frac{7}{12}$	**9.** $5\frac{3}{5} \div 2$ $2\frac{4}{5}$	**10.** $3\frac{3}{8} \div 3$ $1\frac{1}{8}$
• **11.** $\frac{2}{3} \div 1\frac{1}{2}$ $\frac{4}{9}$	**12.** $\frac{4}{9} \div 2\frac{2}{3}$ $\frac{1}{6}$	**13.** $\frac{2}{5} \div 4\frac{1}{2}$ $\frac{4}{45}$	**14.** $\frac{3}{4} \div 2\frac{1}{4}$ $\frac{1}{3}$	**15.** $\frac{7}{10} \div 4\frac{1}{5}$ $\frac{1}{6}$
16. $2\frac{2}{5} \div \frac{3}{5}$ 4	**17.** $2\frac{1}{4} \div \frac{2}{3}$ $3\frac{3}{8}$	**18.** $4\frac{3}{4} \div \frac{5}{6}$ $5\frac{7}{10}$	**19.** $2\frac{2}{5} \div \frac{3}{10}$ 8	**20.** $2\frac{5}{6} \div \frac{3}{4}$ $3\frac{7}{9}$
• **21.** $1\frac{3}{4} \div 2\frac{1}{3}$ $\frac{3}{4}$	**22.** $1\frac{2}{3} \div 2\frac{1}{2}$ $\frac{2}{3}$	**23.** $2\frac{2}{3} \div 1\frac{3}{5}$ $1\frac{2}{3}$	**24.** $3\frac{3}{5} \div 4\frac{1}{4}$ $\frac{72}{85}$	**25.** $4\frac{1}{2} \div 2\frac{5}{8}$ $1\frac{5}{7}$
26. $3\frac{3}{4} \div 4\frac{2}{3}$ $\frac{45}{56}$	**27.** $2\frac{3}{4} \div 2\frac{4}{5}$ $\frac{55}{56}$	**28.** $4\frac{2}{3} \div 5\frac{1}{6}$ $\frac{28}{31}$	**29.** $6\frac{1}{8} \div 1\frac{3}{4}$ $3\frac{1}{2}$	**30.** $1\frac{2}{3} \div 2\frac{1}{12}$ $\frac{4}{5}$
31. $5\frac{1}{3} \div \frac{3}{4}$ $7\frac{1}{9}$	**32.** $3 \div 4\frac{2}{3}$ $\frac{9}{14}$	**33.** $12\frac{1}{4} \div 1\frac{2}{5}$ $8\frac{3}{4}$	**34.** $10\frac{5}{8} \div 15\frac{5}{6}$ $\frac{51}{76}$	**35.** $8\frac{2}{3} \div 4$ $2\frac{1}{6}$

Solve.

36. The price of Fast Lane Records stock is $\$5\frac{3}{4}$ per share. How many shares can be purchased with \$230?
40 shares

37. 250 shares of Trans Air stock were purchased for \$1,200. What was the price per share? \$4.80

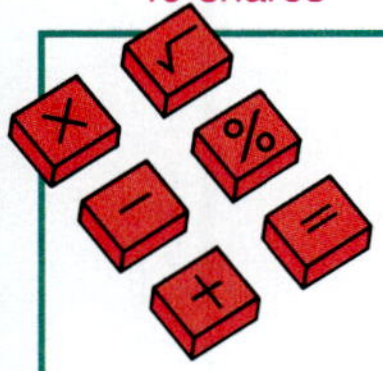

CALCULATOR

1. \$22.02; **2.** \$106.04; **3.** \$26.24
4. \$29.77; **5.** \$184.07

You can use what you know about changing a fraction to a decimal to help you multiply fractions and mixed numbers on a calculator.

Ground beef—\$1.79 per lb   Amount—$24\frac{3}{4}$ lb   Cost—■

1. Change $24\frac{3}{4}$ to a decimal.

$24\frac{3}{4} = 24.75$

2. Use your calculator to multiply. Round to the nearest cent.

$24.75 \times \$1.79 = \$44.3025 \rightarrow \$44.31$

Copy and complete the table. Use your calculator to find the cost of each type of meat and the total cost.

	Chicken	Steak	Turkey	Bacon	
Number of pounds	$18\frac{1}{2}$	$21\frac{1}{4}$	$26\frac{1}{2}$	$15\frac{3}{4}$	
Cost per pound	\$1.19	\$4.99	\$0.99	\$1.89	Total cost
Cost	**1.** ■	**2.** ■	**3.** ■	**4.** ■	**5.** ■

COMMON ERROR

Some students will forget to multiply by the reciprocal after first renaming mixed numbers as improper fractions. For example, in item 1 of Checkpoint, these students would select b. Have these students write out each step to reinforce multiplying by the reciprocal.

ASSIGNMENTS

Level 1	Odd 1–35, 36
Level 2	Even 2–34, 36–37, CA
Level 3	21–37, CA

FOLLOW-UP ACTIVITY

APPLICATION Have students bring in newspapers that list prices from the New York Stock Exchange. Ask them to select the stocks they would purchase for \$1,000 and record the number of shares of each stock. At the end of each week, have them compute the value of their stock and determine their net gain or loss.

SUPPLEMENTARY MATERIALS

TRP Practice, p. 62

TRP Reteaching, p. 39

TRP Transparency 31

OBJECTIVE

- Solve problems by working backward.

TEACHING THE LESSON

WARM-UP Write the following exercises on the chalkboard or an overhead transparency.

Find the missing number.

1. 45 + ■ = 57 (12)
2. ■ − 17 = 15 (32)
3. 2 × ■ = 24 (12)
4. ■ − 9 = 15 (24)

INTRODUCTION Present the following situation: You need to decide when to leave home in the morning to get to work on time. How can you decide? Elicit from students that they must work backward from the time work starts and subtract the amount of time it takes to reach work.

INSTRUCTION After reading the situation, review the given facts. Ask the following questions:

1. How many weeks of work are dealt with in the problem? (2)
2. How are the number of hours Alonso worked described? (as half as many as last week)
3. Are you given the number of hours he worked last week? (no)
4. What do you need to find? (the number of hours Alonso worked last week)

As you discuss each step of the solution, emphasize the concept that multiplying by $\frac{1}{2}$ is the same as dividing by 2. Encourage students to check answers by substituting back into the original number sentence.

PROBLEM Solving STRATEGY

6.5 WORKING BACKWARD

Situation:

Alonso Ortega has a part-time job at Ernie's Garden Center. This week he worked $\frac{1}{2}$ as many hours as he did last week. He worked 10 hours this week. How many hours did he work last week?

Strategy:

Working backward can help you find a solution to some problems.

Applying the Strategy:

THINK: **$\frac{1}{2}$ of the number of hours worked last week = the number of hours worked this week**

$\frac{1}{2}$ of ■ hours = 10 hours

$$10 \div \frac{1}{2} = \frac{10}{1} \times \frac{2}{1} = \frac{20}{1} = 20$$

Alonso worked 20 hours last week.

FOR DISCUSSION

See TE side column.

Blanca deposits her paycheck in a savings account. She has $285 in her account. Last week she deposited $100. Two weeks ago she withdrew $40. How much was in Blanca's account before she withdrew the money?

1. What facts are you given?
2. Explain how you can work backward to solve the problem.
3. What is the solution?

PRACTICE EXERCISES See Extra Practice, page 439.

Solve the problem by working backward.

• 1. During the first week of October, Steve worked $\frac{1}{3}$ as many hours as he did the second week of October. He worked 9 hours the first week.

   a. How many times as many hours did Steve work during the second week as compared with the first week? 3

   b. How many hours did he work the second week of October? 27

• 2. A customer owes Ernie's Garden Center \$75. Last week she charged a \$20 plant to her account. Three weeks ago she paid \$30 of her bill.

   a. How much was the customer's bill before she charged the \$20 plant? \$55

   b. How much was the customer's bill before she paid \$30? \$85

Steve made this sale sign for a weekend special. Find the original price.

3. Begonia \$4

4. Geranium \$3.50

5. Fern \$8.50

SALE: 1/2 OFF REGULAR PRICE		
Begonias	—	Now \$2.00 each
Geraniums	—	Now \$1.75 each
Ferns	—	Now \$4.25 each

6. Donna works in the greenhouse. This week she worked $\frac{2}{3}$ as many hours as she did last week. She worked 36 hours this week. How many hours did she work last week? 54 h

7. Ernie has \$3,000 in his bank account. Yesterday he deposited \$500. Last week he withdrew \$1,000. How much was in Ernie's account before he withdrew the money? \$3,500

8. In the garden shop, Fred bought twice as many plants as Juanita. Juanita bought three fewer plants than Ann. Ann bought five plants. How many plants did each person buy? Fred—4 plants; Juanita—2 plants; Ann—5 plants.

9. Betty wants radishes by July 12. The seed package states that radishes take 24 days to grow. Steve looks at the calendar and tells Betty there are 30 days in June. What is the last day that Betty can plant radish seeds? June 18

FOR DISCUSSION

1. Students should explain that Blanca now has \$285 in a savings account. Last week she added \$100 to it. Two weeks ago she withdrew \$40.
2. Students should explain that they must subtract \$100 from \$285 to find the amount of money Blanca had in her account before she deposited \$100 (\$185). They could add \$40 to \$185 to find the amount of money Blanca had two weeks before she withdrew \$40.
3. The solution is \$225.

   ■ + \$100 − \$40 = \$285

ASSIGNMENTS

Level 1	1–6
Level 2	1–8
Level 3	3–9

FOLLOW-UP ACTIVITY

SITUATIONAL PROBLEM SOLVING Have students work in small groups. Have them consider the following problem: You are considering taking a part-time job at the local garden center. Before you talk to the manager, you want to prepare a list of questions to find out as much as you can about the job. Devise a list of questions that would help you to find out the necessary information.

SUPPLEMENTARY MATERIALS

TRP Practice, p. 63

TRP Applications, p. 10

TRP Transparency 32

OBJECTIVE

- Evaluate student progress.

USING THE REVIEW

This page provides a means for informally evaluating students' understanding of the skills and concepts covered so far in this chapter.

Have the students look at the page to familiarize themselves with the various question formats that are presented. Discuss any questions that they may have. Then ask them to complete the page independently.

In addition to grading them individually, you may wish to review the answers to the questions collectively with the students.

Page references appear in brackets. They refer to pages on which a particular skill was introduced.

Before continuing on to the topics found in the remainder of the chapter, you may wish to have students review any skills or concepts in which they have demonstrated weakness.

MIDCHAPTER REVIEW

Multiply. Write the product in simplest form. [146, 148]

1. $\frac{1}{2} \times \frac{1}{4}$ $\frac{1}{8}$	**2.** $\frac{1}{8} \times \frac{1}{3}$ $\frac{1}{24}$	**3.** $\frac{1}{6} \times \frac{1}{5}$ $\frac{1}{30}$	**4.** $\frac{2}{3} \times \frac{1}{5}$ $\frac{2}{15}$	**5.** $\frac{5}{8} \times \frac{1}{2}$ $\frac{5}{16}$
6. $\frac{4}{5} \times \frac{1}{2}$ $\frac{2}{5}$	**7.** $\frac{2}{3} \times \frac{2}{5}$ $\frac{4}{15}$	**8.** $\frac{7}{8} \times \frac{3}{4}$ $\frac{21}{32}$	**9.** $\frac{5}{12} \times \frac{2}{3}$ $\frac{5}{18}$	**10.** $\frac{3}{5} \times \frac{3}{8}$ $\frac{9}{40}$
11. $\frac{1}{6} \times \frac{2}{7}$ $\frac{1}{21}$	**12.** $\frac{3}{4} \times \frac{2}{9}$ $\frac{1}{6}$	**13.** $\frac{5}{8} \times \frac{3}{10}$ $\frac{3}{16}$	**14.** $\frac{5}{6} \times \frac{3}{5}$ $\frac{1}{2}$	**15.** $\frac{7}{8} \times \frac{4}{9}$ $\frac{7}{18}$
16. $6 \times \frac{2}{3}$ 4	**17.** $5 \times \frac{7}{10}$ $3\frac{1}{2}$	**18.** $\frac{3}{4} \times 12$ 9	**19.** $\frac{5}{6} \times 7$ $5\frac{5}{6}$	**20.** $15 \times \frac{4}{5}$ 12
21. $\frac{1}{2} \times 3\frac{1}{4}$ $1\frac{5}{8}$	**22.** $\frac{2}{3} \times 6\frac{1}{2}$ $4\frac{1}{3}$	**23.** $3\frac{3}{4} \times \frac{5}{8}$ $2\frac{11}{32}$	**24.** $4\frac{2}{3} \times \frac{3}{8}$ $1\frac{3}{4}$	**25.** $\frac{3}{5} \times 8\frac{1}{2}$ $5\frac{1}{10}$
26. $1\frac{1}{3} \times 2\frac{1}{2}$ $3\frac{1}{3}$	**27.** $3\frac{1}{4} \times 5\frac{1}{2}$ $17\frac{7}{8}$	**28.** $4\frac{2}{5} \times 3\frac{1}{6}$ $13\frac{14}{15}$	**29.** $8\frac{3}{8} \times 1\frac{3}{4}$ $14\frac{21}{32}$	**30.** $2\frac{1}{6} \times 3\frac{5}{12}$ $7\frac{29}{72}$

Find the reciprocal.

31. $\frac{2}{3}$ $\frac{3}{2}$	**32.** $\frac{4}{9}$ $\frac{9}{4}$	**33.** $\frac{7}{15}$ $\frac{15}{7}$	**34.** 6 $\frac{1}{6}$	**35.** 18 $\frac{1}{18}$
36. $2\frac{1}{5}$ $\frac{5}{11}$	**37.** $3\frac{1}{7}$ $\frac{7}{22}$	**38.** $10\frac{1}{2}$ $\frac{2}{21}$	**39.** $6\frac{3}{4}$ $\frac{4}{27}$	**40.** $12\frac{4}{5}$ $\frac{5}{64}$

Divide. Write the quotient in simplest form. [150, 152]

41. $\frac{2}{3} \div \frac{5}{6}$ $\frac{4}{5}$	**42.** $\frac{2}{3} \div \frac{4}{5}$ $\frac{5}{6}$	**43.** $\frac{3}{4} \div \frac{3}{8}$ 2	**44.** $\frac{5}{6} \div \frac{4}{9}$ $1\frac{7}{8}$	**45.** $\frac{6}{7} \div \frac{3}{4}$ $1\frac{1}{7}$
46. $6 \div \frac{2}{5}$ 15	**47.** $\frac{7}{8} \div 3$ $\frac{7}{24}$	**48.** $9 \div \frac{1}{2}$ 18	**49.** $10 \div \frac{7}{10}$ $14\frac{2}{7}$	**50.** $\frac{5}{6} \div 10$ $\frac{1}{12}$
51. $2\frac{3}{4} \div 6$ $\frac{11}{24}$	**52.** $3\frac{2}{3} \div 4$ $\frac{11}{12}$	**53.** $5\frac{3}{5} \div 7$ $\frac{4}{5}$	**54.** $8 \div 2\frac{2}{3}$ 3	**55.** $12 \div 1\frac{1}{5}$ 10
56. $2\frac{1}{2} \div \frac{1}{4}$ 10	**57.** $1\frac{3}{5} \div \frac{7}{10}$ $2\frac{2}{7}$	**58.** $\frac{5}{6} \div 4\frac{1}{3}$ $\frac{5}{26}$	**59.** $\frac{7}{12} \div 2\frac{1}{4}$ $\frac{7}{27}$	**60.** $\frac{5}{9} \div 4\frac{2}{3}$ $\frac{5}{42}$
61. $1\frac{1}{3} \div 2\frac{1}{6}$ $\frac{8}{13}$	**62.** $3\frac{1}{2} \div 1\frac{3}{4}$ 2	**63.** $3\frac{2}{5} \div 1\frac{5}{6}$ $1\frac{47}{55}$	**64.** $4\frac{1}{2} \div 5\frac{2}{3}$ $\frac{27}{34}$	**65.** $2\frac{7}{8} \div 3\frac{1}{4}$ $\frac{23}{26}$

Solve by working backward. [154]

66. Isabel bought twice as many albums as Sam. Sam bought 3 more albums than Becky. Becky bought 2 albums. How many albums did each person buy? Becky—2 albums, Sam—5 albums, Isabel—10 albums

67. Jack works at Disco Record Store. This week he worked $\frac{1}{2}$ as many hours as he did last week. Last week he worked 14 hours. How many hours did he work this week? 7 hours

PROBLEM *Solving* APPLICATION

6.6 CAREER: CARPENTER

Kristin Leonard is a carpenter. In her work, she must be able to make careful measurements. Often, she needs to use fractions and mixed numbers.

A board 9 ft long is cut into sections of the given length. How many sections are there?

• **1.** Sections $\frac{3}{4}$ ft long 12

2. Sections $1\frac{1}{2}$ ft long 6

A board $7\frac{1}{2}$ ft long is cut into equal parts. How long is each part?

• **3.** 6 equal parts $1\frac{1}{4}$ feet

4. 10 equal parts $\frac{3}{4}$ foot

Each drawing represents a board marked off into equal parts. Find the total length of the board.

5.

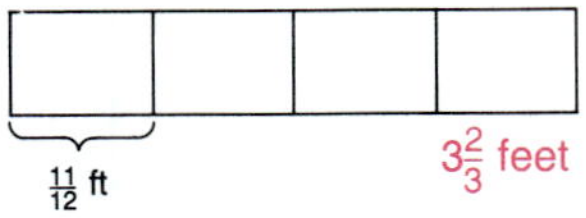

$3\frac{2}{3}$ feet

6.

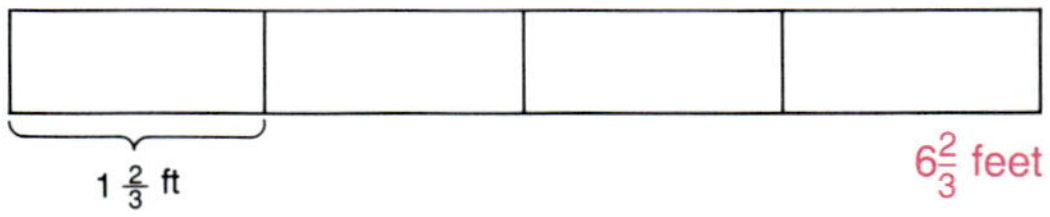

$6\frac{2}{3}$ feet

Write the measurement using a fraction or a mixed number in lowest terms.

• **7.** 0.75 yd $\frac{3}{4}$ yd **8.** 2.5 in. $2\frac{1}{2}$ in. **9.** 1.125 in. $1\frac{1}{8}$ in. **10.** 0.9375 in. $\frac{15}{16}$ in.

Solve.

11. A board of lumber called a "2-by-4" is actually $1\frac{5}{8}$ in. thick and $3\frac{5}{8}$ in. wide. Kristen stacks four of these boards as shown at the right. What is the total height of the stack? $6\frac{1}{2}$ in.

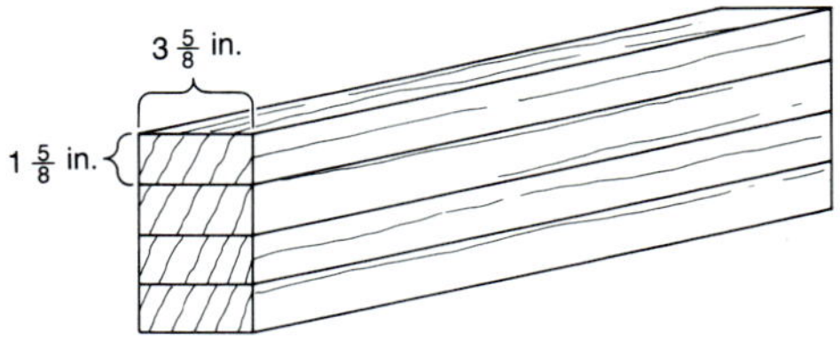

12. Kristen had a board $11\frac{1}{2}$ ft long. She sawed it in half and used one piece. Then she sawed the remaining piece in half and used one of the pieces. How long was the remaining piece? $2\frac{7}{8}$ ft

OBJECTIVE

- Solve problems that involve computation with fractions and decimals.

TEACHING THE LESSON

WARM-UP Write the following exercises on the chalkboard or an overhead transparency.

Solve.

1. $3 \times \frac{11}{12}$ $(2\frac{3}{4})$ **2.** $4 \times 1\frac{1}{2}$ (6)

3. $6 \times 1\frac{3}{4}$ $(10\frac{1}{2})$ **4.** $6 \div \frac{1}{2}$ (12)

INTRODUCTION Discuss the various tasks that carpenters encounter in their work. Make a list and remember to include tasks such as estimating how much a job will cost. Discuss how a knowledge of mathematics is helpful in dealing with each task.

INSTRUCTION Since computing with fractions and mixed numbers is often difficult for some students, you may wish to suggest that students check their answers using inverse operations. For example, Problems 1–4 are solved by division, and the answers can be checked by multiplication. Problems 5 and 6 are solved by multiplication, and the answers can be checked by division.

Problems 7–10 require the students to use their knowledge of the relationship between decimals and fractions.

In Problem 11, students must use the drawing to determine the necessary data, and then multiply.

ASSIGNMENTS

Level 1	1–8
Level 2	1–12
Level 3	Even 2–12

SUPPLEMENTARY MATERIALS

TRP Practice, p. 64

TRP Transparency 32

OBJECTIVE

- Add fractions.

TEACHING THE LESSON

WARM-UP Present the following exercises on the chalkboard or overhead transparency.

Find the LCM for each pair of numbers.

1. 3, 8 **2.** 6, 8 **3.** 9, 3

4. 24, 16 **5.** 24, 36 **6.** 18, 27

(**1.** 24 **2.** 24 **3.** 9 **4.** 48 **5.** 72 **6.** 54)

INTRODUCTION Have students discuss how they spend the money they earn. You may wish to have them identify fixed expenses, such as school lunches, and incidental expenses, such as entertainment. Point out that living expenses such as housing and food take a large part of most budgets.

INSTRUCTION Discuss the illustration showing $\frac{3}{8} + \frac{1}{8}$. Be sure that students understand that each part of the region represents $\frac{1}{8}$ of the total budget.

In discussing Example 1, remind students that a fraction is in simplest form if it is written in lowest terms.

In discussing Example 2, you may wish to provide students with the following illustration.

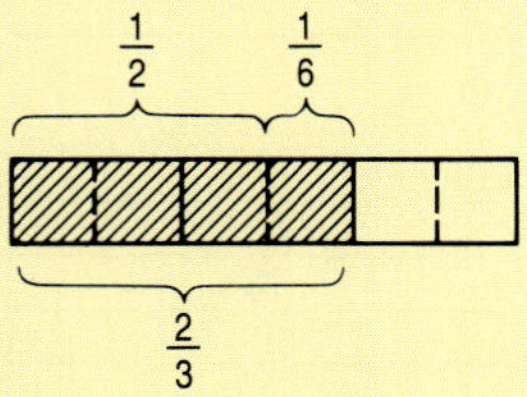

6.7 ADDING FRACTIONS

Gerry Thomas spends $\frac{3}{8}$ of her income on rent and $\frac{1}{8}$ of her income on utilities. What part of her income is spent on rent and utilities?

You can add $\frac{3}{8}$ and $\frac{1}{8}$ to find the answer.

EXAMPLE 1 Add: $\frac{3}{8} + \frac{1}{8}$

1. Add the numerators.
2. Write the sum over the denominator.
3. Write the answer in simplest form.

$$\begin{array}{r} \frac{3}{8} \\ +\frac{1}{8} \\ \hline 4 \end{array} \qquad \begin{array}{r} \frac{3}{8} \\ +\frac{1}{8} \\ \hline \frac{4}{8} \end{array} \qquad \begin{array}{r} \frac{3}{8} \\ +\frac{1}{8} \\ \hline \frac{4}{8} = \frac{1}{2} \end{array}$$

Gerry spends $\frac{1}{2}$ of her income on rent and utilities.

Sometimes you need to add fractions that have different denominators.

EXAMPLE 2 Add: $\frac{1}{2} + \frac{1}{6}$

1. Write equivalent fractions with a common denominator.
2. Add.
3. Write the answer in simplest form.

$$\begin{array}{r} \frac{1}{2} = \frac{3}{6} \\ +\frac{1}{6} = \frac{1}{6} \\ \hline \end{array} \qquad \begin{array}{r} \frac{1}{2} = \frac{3}{6} \\ +\frac{1}{6} = \frac{1}{6} \\ \hline \frac{4}{6} \end{array} \qquad \begin{array}{r} \frac{1}{2} = \frac{3}{6} \\ +\frac{1}{6} = \frac{1}{6} \\ \hline \frac{4}{6} = \frac{2}{3} \end{array}$$

When the sum is greater than 1, rename the sum as a mixed number.

EXAMPLE 3 Add:

a. $$\begin{array}{r} \frac{2}{3} \\ +\frac{5}{8} \\ \hline \end{array} \qquad \begin{array}{r} \frac{2}{3} = \frac{16}{24} \\ +\frac{5}{8} = \frac{15}{24} \\ \hline \frac{31}{24} = 1\frac{7}{24} \end{array}$$

b. $$\begin{array}{r} \frac{1}{3} \\ \frac{1}{4} \\ +\frac{5}{6} \\ \hline \end{array} \qquad \begin{array}{r} \frac{1}{3} = \frac{4}{12} \\ \frac{1}{4} = \frac{3}{12} \\ +\frac{5}{6} = \frac{10}{12} \\ \hline \frac{17}{12} = 1\frac{5}{12} \end{array}$$

CHECKPOINT

Write the letter of the correct answer.

Add. What is the sum in simplest form?

1. $\frac{1}{8} + \frac{3}{4}$ d **a.** $\frac{3}{32}$ **b.** $\frac{4}{8}$ **c.** $\frac{4}{12}$ **d.** $\frac{7}{8}$

2. $\frac{4}{5} + \frac{2}{3}$ c **a.** $\frac{6}{15}$ **b.** $\frac{8}{15}$ **c.** $1\frac{7}{15}$ **d.** $\frac{6}{8}$

PRACTICE EXERCISES

See Extra Practice, page 439.

Add. Write the sum in simplest form.

• **1.** $\frac{1}{5} + \frac{2}{5}$ $\frac{3}{5}$	**2.** $\frac{3}{7} + \frac{2}{7}$ $\frac{5}{7}$	**3.** $\frac{5}{9} + \frac{2}{9}$ $\frac{7}{9}$	**4.** $\frac{2}{5} + \frac{3}{5}$ 1	**5.** $\frac{3}{8} + \frac{5}{8}$ 1
6. $\frac{3}{4} + \frac{3}{4}$ $1\frac{1}{2}$	**7.** $\frac{5}{6} + \frac{2}{6}$ $1\frac{1}{6}$	**8.** $\frac{3}{5} + \frac{4}{5}$ $1\frac{2}{5}$	**9.** $\frac{5}{12} + \frac{9}{12}$ $1\frac{1}{6}$	**10.** $\frac{4}{10} + \frac{8}{10}$ $1\frac{1}{5}$
• **11.** $\frac{1}{2} + \frac{2}{5}$ $\frac{9}{10}$	**12.** $\frac{2}{3} + \frac{1}{4}$ $\frac{11}{12}$	**13.** $\frac{2}{5} + \frac{1}{3}$ $\frac{11}{15}$	**14.** $\frac{1}{3} + \frac{3}{5}$ $\frac{14}{15}$	**15.** $\frac{2}{3} + \frac{2}{9}$ $\frac{8}{9}$
• **16.** $\frac{3}{4} + \frac{2}{3}$ $1\frac{5}{12}$	**17.** $\frac{1}{3} + \frac{6}{9}$ 1	**18.** $\frac{3}{4} + \frac{3}{8}$ $1\frac{1}{8}$	**19.** $\frac{2}{5} + \frac{3}{4}$ $1\frac{3}{20}$	**20.** $\frac{3}{4} + \frac{4}{5}$ $1\frac{11}{20}$
21. $\frac{1}{8} + \frac{2}{5}$ $\frac{21}{40}$	**22.** $\frac{2}{9} + \frac{4}{9}$ $\frac{2}{3}$	**23.** $\frac{1}{4} + \frac{8}{9}$ $1\frac{5}{36}$	**24.** $\frac{3}{4} + \frac{1}{6}$ $\frac{11}{12}$	**25.** $\frac{5}{6} + \frac{3}{8}$ $1\frac{5}{24}$
26. $\frac{3}{4} + \frac{5}{12}$ $1\frac{1}{6}$	**27.** $\frac{3}{16} + \frac{3}{8}$ $\frac{9}{16}$	**28.** $\frac{7}{16} + \frac{3}{16}$ $\frac{5}{8}$	**29.** $\frac{3}{10} + \frac{5}{8}$ $\frac{37}{40}$	**30.** $\frac{1}{6} + \frac{4}{15}$ $\frac{13}{30}$

31. $\frac{1}{2} + \frac{1}{2} + \frac{1}{3}$ $1\frac{1}{3}$ **32.** $\frac{1}{8} + \frac{1}{3} + \frac{1}{4}$ $\frac{17}{24}$ **33.** $\frac{1}{2} + \frac{1}{4} + \frac{5}{6}$ $1\frac{7}{12}$ **34.** $\frac{2}{5} + \frac{1}{2} + \frac{3}{4}$ $1\frac{13}{20}$

35. $\frac{9}{24} + \frac{3}{8}$ $\frac{3}{4}$ **36.** $\frac{1}{2} + \frac{3}{7}$ $\frac{13}{14}$ **37.** $\frac{3}{4} + \frac{7}{10}$ $1\frac{9}{20}$ **38.** $\frac{3}{8} + \frac{7}{8}$ $1\frac{1}{4}$ **39.** $\frac{1}{4} + \frac{3}{5}$ $\frac{17}{20}$ **40.** $\frac{3}{5} + \frac{5}{6}$ $1\frac{13}{30}$

• **41.** $\frac{1}{3} + \frac{2}{3} + \frac{1}{4}$ $1\frac{1}{4}$ **42.** $\frac{1}{2} + \frac{1}{8} + \frac{3}{4}$ $1\frac{3}{8}$ **43.** $\frac{4}{9} + \frac{1}{6} + \frac{2}{3}$ $1\frac{5}{18}$ **44.** $\frac{1}{8} + \frac{1}{6} + \frac{1}{4}$ $\frac{13}{24}$ **45.** $\frac{3}{5} + \frac{1}{4} + \frac{5}{8}$ $1\frac{19}{40}$ **46.** $\frac{2}{3} + \frac{3}{8} + \frac{1}{6}$ $1\frac{5}{24}$

Solve.

• **47.** Akira spends $\frac{3}{8}$ of her income on rent and $\frac{1}{12}$ of her income on utilities. What part of her income is spent on housing? $\frac{11}{24}$

48. Sally spends $\frac{1}{5}$ of her income on food. If she earns \$450 per week, how much money does she spend each week on food? \$90

49. Carol spends $\frac{1}{3}$ of her pay for food, $\frac{1}{4}$ for rent, and $\frac{1}{8}$ for entertainment. The rest she saves. What part of her pay does Carol spend? $\frac{17}{24}$

50. Kyle earns \$2,500 per month. He spends \$250 per month on entertainment. What part of his salary does he spend on entertainment? $\frac{1}{10}$

As you discuss Example 3b, point out the need to determine a common denominator for the three fractions.

CHECKPOINT The incorrect answer choices include common errors.

COMMON ERROR

Some students will make the error of adding the numerators and then the denominators when adding fractions. For example, in item 1 of Checkpoint, these students will choose c. Have these students work with fraction strips to show the addition. Elicit that the denominator does not change when adding the numerators.

ASSIGNMENTS

Level 1	1–25, 35–40, 47, 48
Level 2	Even 2–46, 47–50
Level 3	Odd 21–45, 47–50

FOLLOW-UP ACTIVITY

ENRICHMENT Some students may be interested in a shortcut for adding fractions. Provide students with the following examples on the chalkboard or overhead transparency.

$$\frac{2}{5} + \frac{3}{7} = \frac{(2 \times 7) + (3 \times 5)}{5 \times 7} = \frac{14 + 15}{35} = \frac{29}{35}$$

$$\frac{2}{3} + \frac{1}{8} = \frac{(2 \times 8) + (1 \times 3)}{3 \times 8} = \frac{16 + 3}{24} = \frac{19}{24}$$

Have students use this method to work Exercises 11–15.

SUPPLEMENTARY MATERIALS

TRP Practice, p. 65

TRP Reteaching, p. 40

TRP Transparency 33

OBJECTIVE

- Add fractions, whole numbers, and mixed numbers.

TEACHING THE LESSON

WARM-UP Present the following exercises on the chalkboard or overhead transparency.

Write as a mixed number.

1. $\frac{7}{6}$ **2.** $\frac{9}{8}$ **3.** $\frac{19}{12}$

4. $\frac{25}{24}$ **5.** $\frac{23}{20}$ **6.** $\frac{41}{24}$

(**1.** $1\frac{1}{6}$ **2.** $1\frac{1}{8}$ **3.** $1\frac{7}{12}$ **4.** $1\frac{1}{24}$ **5.** $1\frac{3}{20}$ **6.** $1\frac{17}{24}$)

INTRODUCTION You may wish to show and discuss various types of road maps with your students. Discuss how to read the mileage between various cities and highway intersections.

INSTRUCTION In discussing Example 1, note that the denominators are the same. Thus, they need only to add the numerators and write the sum over the common denominator. Remind the students that to simplify mixed numbers means to write the fraction part in lowest terms.

For Examples 2 and 3, make sure that students understand that the fractions are added first with the sum being renamed as a mixed number. Then, the whole number part must be added to the whole number parts of the two mixed numbers.

6.8 ADDING MIXED NUMBERS

Nicole Williams rides her bicycle $8\frac{1}{10}$ miles from Ellisville to Norton, then $3\frac{3}{10}$ miles from Norton to Millstown. How far does she ride?

To find the total distance, add $8\frac{1}{10}$ miles and $3\frac{3}{10}$ miles.

EXAMPLE 1 Add: $8\frac{1}{10} + 3\frac{3}{10}$

1. Add the fractions.
$$\begin{array}{r} 8\frac{1}{10} \\ +3\frac{3}{10} \\ \hline \frac{4}{10} \end{array}$$
2. Add the whole numbers.
$$\begin{array}{r} 8\frac{1}{10} \\ +3\frac{3}{10} \\ \hline 11\frac{4}{10} \end{array}$$
3. Write the answer in simplest form.
$$\begin{array}{r} 8\frac{1}{10} \\ +3\frac{3}{10} \\ \hline 11\frac{4}{10} = 11\frac{2}{5} \end{array}$$

The distance from Ellisville to Millstown is $11\frac{2}{5}$ miles.

Sometimes you need to rename the sum of the fractions.

EXAMPLE 2 Add: $2\frac{3}{8} + 4\frac{7}{8}$

1. Add the fractions. Rename.
$$\begin{array}{r} 2\frac{3}{8} \\ +4\frac{7}{8} \\ \hline \frac{10}{8} = 1\frac{2}{8} \end{array}$$
2. Add the whole numbers.
$$\begin{array}{r} \overset{1}{2}\frac{3}{8} \\ +4\frac{7}{8} \\ \hline 7\frac{2}{8} \end{array}$$
3. Write the answer in simplest form.
$$\begin{array}{r} 2\frac{3}{8} \\ +4\frac{7}{8} \\ \hline 7\frac{2}{8} = 7\frac{1}{4} \end{array}$$

Sometimes the fractions have different denominators.

EXAMPLE 3 Add: $1\frac{3}{5} + 2\frac{1}{10} + 4\frac{3}{4}$

1. Write equivalent fractions with a common denominator.
$$\begin{array}{rcl} 1\frac{3}{5} & = & 1\frac{12}{20} \\ 2\frac{1}{10} & = & 2\frac{2}{20} \\ +4\frac{3}{4} & = & 4\frac{15}{20} \end{array}$$
2. Add the fractions. Rename.
$$\begin{array}{rcl} 1\frac{3}{5} & = & 1\frac{12}{20} \\ 2\frac{1}{10} & = & 2\frac{2}{20} \\ +4\frac{3}{4} & = & 4\frac{15}{20} \\ \hline & & \frac{29}{20} = 1\frac{9}{20} \end{array}$$
3. Add the whole numbers.
$$\begin{array}{rcl} 1\frac{3}{5} & = & \overset{1}{1}\frac{12}{20} \\ 2\frac{1}{10} & = & 2\frac{2}{20} \\ +4\frac{3}{4} & = & 4\frac{15}{20} \\ \hline & & 8\frac{9}{20} \end{array}$$

PRACTICE EXERCISES See Extra Practice, page 439.

Add. Write the sum in simplest form.

- • **1.** $5\frac{1}{8} + 3\frac{3}{8}$ $8\frac{1}{2}$ **2.** $2\frac{2}{5} + 4\frac{2}{5}$ $6\frac{4}{5}$ **3.** $4\frac{3}{7} + 2\frac{2}{7}$ $6\frac{5}{7}$ **4.** $8\frac{5}{9} + 7\frac{4}{9}$ 16 **5.** $9\frac{3}{4} + 3\frac{3}{4}$ $13\frac{1}{2}$
- • **6.** $4\frac{3}{8} + 3\frac{1}{4}$ $7\frac{5}{8}$ **7.** $6\frac{3}{4} + 7\frac{1}{6}$ $13\frac{11}{12}$ **8.** $5\frac{2}{3} + 6\frac{2}{9}$ $11\frac{8}{9}$ **9.** $9\frac{2}{3} + 3\frac{5}{6}$ $13\frac{1}{2}$ **10.** $7\frac{3}{4} + 7\frac{5}{8}$ $15\frac{3}{8}$
- **11.** $7\frac{3}{4} + 4\frac{2}{3}$ $12\frac{5}{12}$ **12.** $9\frac{2}{3} + 2\frac{3}{5}$ $12\frac{4}{15}$ **13.** $8\frac{1}{2} + 8\frac{2}{3}$ $17\frac{1}{6}$ **14.** $6\frac{3}{4} + 3\frac{5}{6}$ $10\frac{7}{12}$ **15.** $9\frac{2}{3} + 9\frac{4}{5}$ $19\frac{7}{15}$
- **16.** $12\frac{3}{4} + 4\frac{3}{8}$ $17\frac{1}{8}$ **17.** $3\frac{11}{12} + 1\frac{1}{8}$ $5\frac{1}{24}$ **18.** $23\frac{1}{2} + 5\frac{1}{6}$ $28\frac{2}{3}$ **19.** $38\frac{2}{3} + 2\frac{4}{15}$ $40\frac{14}{15}$ **20.** $42\frac{7}{12} + 31\frac{5}{16}$ $73\frac{43}{48}$
- **21.** $3 + 2\frac{7}{8}$ $5\frac{7}{8}$ **22.** $9\frac{3}{5} + 6$ $15\frac{3}{5}$ **23.** $18\frac{1}{6} + 7$ $25\frac{1}{6}$ **24.** $13 + 12\frac{2}{3}$ $25\frac{2}{3}$ **25.** $15\frac{4}{5} + 18$ $33\frac{4}{5}$
- **26.** $5\frac{1}{6} + \frac{1}{6}$ $5\frac{1}{3}$ **27.** $6\frac{2}{5} + 4\frac{1}{5}$ $10\frac{3}{5}$ **28.** $4\frac{1}{2} + 6\frac{1}{3}$ $10\frac{5}{6}$ **29.** $\frac{1}{4} + 8\frac{2}{3}$ $8\frac{11}{12}$ **30.** $5 + 6\frac{1}{2}$ $11\frac{1}{2}$
- **31.** $7\frac{2}{5} + \frac{3}{4}$ $8\frac{3}{20}$ **32.** $6\frac{3}{10} + 5\frac{4}{10}$ $11\frac{7}{10}$ **33.** $8 + 6\frac{4}{5}$ $14\frac{4}{5}$ **34.** $5\frac{5}{6} + 2\frac{2}{5}$ $8\frac{7}{30}$ **35.** $9\frac{5}{12} + 7\frac{7}{12}$ 17
- • **36.** $4\frac{1}{4} + 1\frac{3}{4} + 8\frac{3}{4}$ $14\frac{3}{4}$ **37.** $5\frac{1}{2} + 4\frac{1}{3} + 3\frac{3}{4}$ $13\frac{7}{12}$ **38.** $6\frac{1}{12} + 7\frac{2}{3} + 18\frac{3}{4}$ $32\frac{1}{2}$ **39.** $\frac{5}{9} + 18 + 7\frac{1}{3}$ $25\frac{8}{9}$ **40.** $12\frac{3}{8} + 54 + \frac{7}{12}$ $66\frac{23}{24}$

Solve. Use the map at the right.

• **41.** Susan lives $3\frac{1}{2}$ miles north of Millstown. Bill lives $2\frac{2}{5}$ miles west of Millstown. How far does Susan live from Bill? $5\frac{9}{10}$ mi

42. To drive from Norton to Sunset Mt., you must pass through Millstown. How far is Sunset Mt. from Norton? $17\frac{3}{10}$ mi

43. The distance from Norton to Brownsville is $1\frac{1}{3}$ times the distance from Norton to Ellisville. How far is Brownsville from Norton? $10\frac{4}{5}$ mi

44. Larry jogged from Millstown to Sunset Mt. in $1\frac{3}{4}$ hours. What was his average rate of speed? 8 mph

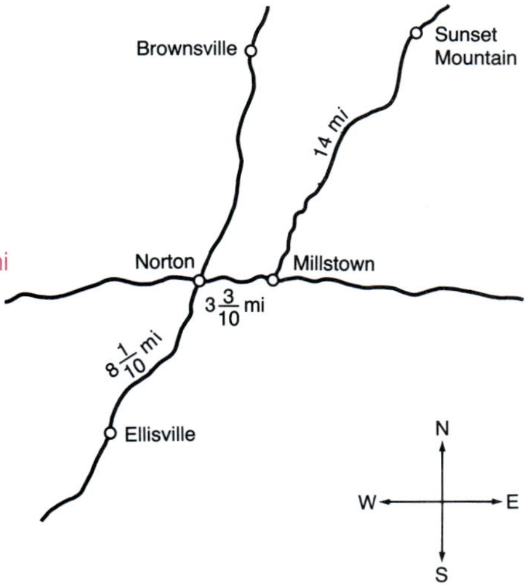

ASSIGNMENTS

Level 1	Odd 1–39, 41–42
Level 2	Even 2–44
Level 3	Odd 21–39, 41–44

FOLLOW-UP ACTIVITY

COOPERATIVE LEARNING
Have students work in small groups. Ask them to bring in road maps of your state to class. Select two cities and determine the shortest mileage route between the cities. You may want to discuss the fact that the shortest route may not always be the fastest one.

Have each group plan a trip between two cities. Instruct them to calculate the estimated time of arrival based on an average speed of 40 mph. Discuss ways of determining the cost of gasoline needed for the trip.

SUPPLEMENTARY MATERIALS

TRP Practice, p. 66

TRP Reteaching, p. 41

TRP Transparency 33

OBJECTIVE

- Subtract fractions.

TEACHING THE LESSON

WARM-UP Present the following exercises on the chalkboard or overhead transparency.

Write equivalent fractions with the same denominator.

1. $\frac{5}{6}, \frac{2}{3}$ **2.** $\frac{1}{2}, \frac{3}{8}$ **3.** $\frac{2}{3}, \frac{5}{8}$

4. $\frac{2}{3}, \frac{3}{5}$ **5.** $\frac{3}{4}, \frac{2}{3}$ **6.** $\frac{19}{21}, \frac{2}{3}$

(**1.** $\frac{5}{6}, \frac{4}{6}$ **2.** $\frac{4}{8}, \frac{3}{8}$ **3.** $\frac{16}{24}, \frac{15}{24}$ **4.** $\frac{10}{15}, \frac{9}{15}$ **5.** $\frac{9}{12}, \frac{8}{12}$ **6.** $\frac{19}{21}, \frac{14}{21}$)

INTRODUCTION Discuss with students whether people usually get paid by the hour or by the job for doing lawn work. Discuss the advantages and disadvantages of each method of payment.

INSTRUCTION Discuss the illustration that shows $\frac{3}{4} - \frac{1}{4}$. Elicit from the students that the total time for cutting the lawn plus the time for raking the lawn should equal the total time.

In discussing Example 2, remind students that they must always find a common denominator before they begin to subtract fractions with different denominators.

CHECKPOINT The incorrect answer choices include common errors.

COMMON ERROR

When subtracting fractions, some students will make the error of subtracting the numerators and then the denominators. In item 3 of Checkpoint, these students would choose a. Have these students work with fraction strips to show the subtraction.

6.9 SUBTRACTING FRACTIONS

Tom Miller worked in his yard for $\frac{3}{4}$ of an hour. He spent $\frac{1}{4}$ of an hour cutting the lawn and the rest of the time raking leaves. How much time did Tom spend raking leaves?

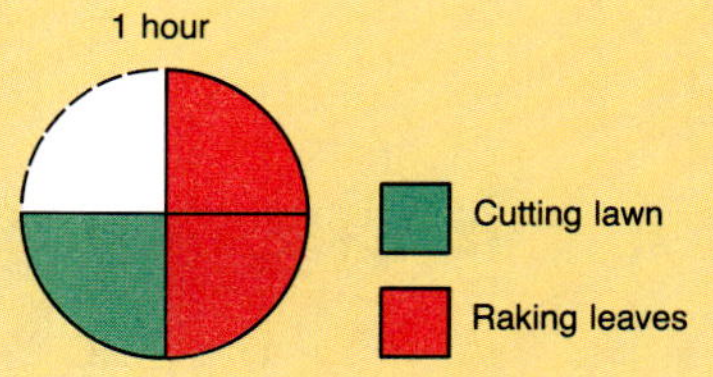

THINK: You know the total and one part. You can subtract $\frac{1}{4}$ from $\frac{3}{4}$ to find the missing part.

EXAMPLE 1 Subtract: $\frac{3}{4} - \frac{1}{4}$

1. Subtract the numerators. $\frac{3}{4} - \frac{1}{4} = 2$
2. Write the difference over the denominator. $\frac{3}{4} - \frac{1}{4} = \frac{2}{4}$
3. Write the answer in simplest form. $\frac{2}{4} = \frac{1}{2}$

Tom spent $\frac{1}{2}$ of an hour raking leaves.

Sometimes you need to subtract fractions with different denominators.

EXAMPLE 2 Subtract: $\frac{11}{15} - \frac{2}{5}$

1. Write equivalent fractions with a common denominator.
$$\frac{11}{15} = \frac{11}{15}, \quad -\frac{2}{5} = \frac{6}{15}$$
2. Subtract.
$$\frac{11}{15} = \frac{11}{15}, \quad -\frac{2}{5} = \frac{6}{15}, \quad \frac{5}{15}$$
3. Write the answer in simplest form.
$$\frac{5}{15} = \frac{1}{3}$$

CHECKPOINT

Write the letter of the correct answer.

Subtract. What is the difference in simplest form?

1. $\frac{5}{6} - \frac{1}{6}$ c **a.** $\frac{4}{6}$ **b.** $\frac{4}{12}$ **c.** $\frac{2}{3}$ **d.** $\frac{1}{3}$

2. $\frac{7}{9} - \frac{1}{3}$ d **a.** $\frac{6}{27}$ **b.** $\frac{6}{6}$ **c.** $\frac{2}{3}$ **d.** $\frac{4}{9}$

3. $\frac{5}{7} - \frac{2}{3}$ d **a.** $\frac{3}{4}$ **b.** $\frac{3}{21}$ **c.** $\frac{3}{10}$ **d.** $\frac{1}{21}$

4. $\frac{3}{5} - \frac{2}{9}$ a **a.** $\frac{17}{45}$ **b.** $\frac{17}{35}$ **c.** $\frac{11}{45}$ **d.** $\frac{1}{5}$

PRACTICE EXERCISES

See Extra Practice, page 440.

Subtract. Write the difference in simplest form.

• 1. $\frac{2}{3} - \frac{1}{3}$ $\frac{1}{3}$
2. $\frac{4}{5} - \frac{2}{5}$ $\frac{2}{5}$
3. $\frac{7}{8} - \frac{4}{8}$ $\frac{3}{8}$
4. $\frac{4}{6} - \frac{3}{6}$ $\frac{1}{6}$
5. $\frac{5}{7} - \frac{2}{7}$ $\frac{3}{7}$
6. $\frac{7}{9} - \frac{5}{9}$ $\frac{2}{9}$
7. $\frac{5}{8} - \frac{3}{8}$ $\frac{1}{4}$
8. $\frac{7}{12} - \frac{5}{12}$ $\frac{1}{6}$
9. $\frac{9}{16} - \frac{3}{16}$ $\frac{3}{8}$
10. $\frac{17}{24} - \frac{11}{24}$ $\frac{1}{4}$

• 11. $\frac{3}{4} - \frac{1}{2}$ $\frac{1}{4}$
12. $\frac{5}{6} - \frac{2}{3}$ $\frac{1}{6}$
13. $\frac{3}{4} - \frac{3}{8}$ $\frac{3}{8}$
14. $\frac{2}{3} - \frac{2}{9}$ $\frac{4}{9}$
15. $\frac{1}{2} - \frac{3}{8}$ $\frac{1}{8}$
16. $\frac{7}{8} - \frac{1}{4}$ $\frac{5}{8}$
17. $\frac{11}{12} - \frac{2}{3}$ $\frac{1}{4}$
18. $\frac{13}{16} - \frac{3}{4}$ $\frac{1}{16}$
19. $\frac{7}{12} - \frac{1}{4}$ $\frac{1}{3}$
20. $\frac{9}{10} - \frac{3}{5}$ $\frac{3}{10}$
21. $\frac{2}{3} - \frac{1}{2}$ $\frac{1}{6}$
22. $\frac{2}{3} - \frac{2}{5}$ $\frac{4}{15}$
23. $\frac{3}{4} - \frac{2}{3}$ $\frac{1}{12}$
24. $\frac{1}{2} - \frac{2}{5}$ $\frac{1}{10}$
25. $\frac{2}{3} - \frac{5}{8}$ $\frac{1}{24}$
26. $\frac{2}{3} - \frac{3}{8}$ $\frac{7}{24}$
27. $\frac{5}{6} - \frac{3}{8}$ $\frac{11}{24}$
28. $\frac{8}{9} - \frac{5}{6}$ $\frac{1}{18}$
29. $\frac{4}{5} - \frac{4}{7}$ $\frac{8}{35}$
30. $\frac{5}{9} - \frac{1}{2}$ $\frac{1}{18}$

• 31. $\frac{11}{12} - \frac{3}{12}$ $\frac{2}{3}$
32. $\frac{2}{3} - \frac{2}{15}$ $\frac{8}{15}$
33. $\frac{7}{10} - \frac{7}{12}$ $\frac{7}{60}$
34. $\frac{19}{21} - \frac{4}{7}$ $\frac{1}{3}$
35. $\frac{9}{10} - \frac{3}{10}$ $\frac{3}{5}$
36. $\frac{5}{12} - \frac{3}{14}$ $\frac{17}{84}$
37. $\frac{11}{12} - \frac{5}{18}$ $\frac{23}{36}$
38. $\frac{11}{18} - \frac{5}{9}$ $\frac{1}{18}$
39. $\frac{15}{16} - \frac{13}{16}$ $\frac{1}{8}$
40. $\frac{13}{24} - \frac{12}{36}$ $\frac{5}{24}$

MIXED REVIEW

Multiply or divide.

• 41. 23×9 207
• 42. 187×40 7,480
43. $7,301 \times 568$ 4,146,968
44. $24,664 \times 100$ 2,466,400
45. $32,412 \times 372$ 12,057,264
46. $3\overline{)204}$ 68
47. $33\overline{)1,914}$ 58
48. $100\overline{)5,869}$ 58.69
49. $541\overline{)8,926.5}$ 16.5
50. $89\overline{)3,675.7}$ 41.3

Solve.

• 51. Gerry mows the lawn in $\frac{2}{3}$ of an hour. Matt mows the lawn in $\frac{3}{4}$ of an hour. How much longer does it take Matt to mow the lawn? 5 min

52. It takes Frank $\frac{5}{6}$ of an hour to mow and edge the lawn. He can mow the lawn in $\frac{2}{3}$ of an hour. How long does it take him to edge the lawn? 10 min

53. Esther charges \$5.25 per hour for doing lawn work. How much would she charge for $3\frac{2}{3}$ hours? \$19.25

54. Fran earned \$13.75 for working $2\frac{1}{2}$ hours mowing lawns. What was her average hourly rate of pay? \$5.50 per hour

ASSIGNMENTS

Level 1	Odd 41–50, 51–52
Level 2	Even 2–50, 51–54
Level 3	21–40, Odd 41–54

FOLLOW-UP ACTIVITY

ENRICHMENT Draw the following circle graph on the chalkboard.

TIME COLLEGE FRESHMEN SPEND STUDYING

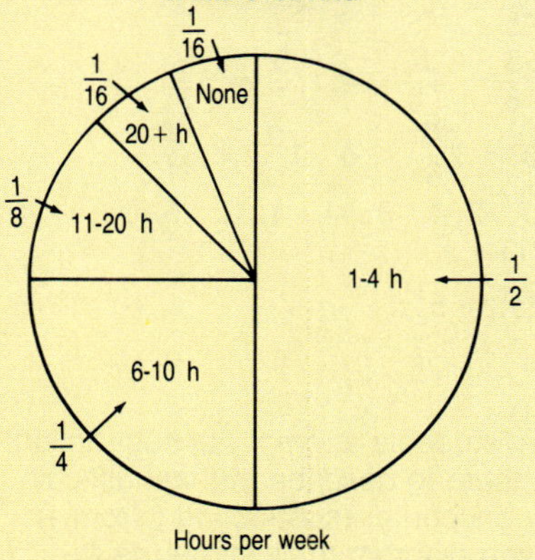

Have students write questions based on comparing the information in the graph. Then have the students solve each other's problems.

SUPPLEMENTARY MATERIALS

TRP Practice, p. 67

TRP Reteaching, p. 42

TRP Transparency 34

OBJECTIVE

- Subtract fractions, whole numbers, and mixed numbers.

TEACHING THE LESSON

WARM-UP Present the following exercises on the chalkboard or overhead transparency.

Complete.

1. $3\frac{2}{5} = 2\frac{■}{5}$ 2. $4\frac{1}{3} = 3\frac{■}{3}$
3. $5\frac{3}{8} = 4\frac{■}{8}$ 4. $7\frac{5}{7} = 6\frac{■}{7}$
5. $8 = 7\frac{■}{6}$ 6. $9\frac{9}{12} = 8\frac{■}{12}$

(**1.** 7 **2.** 4 **3.** 11 **4.** 12 **5.** 6 **6.** 21)

Write: $5\frac{1}{4}$
$-2\frac{3}{4}$

Have students attempt the subtraction. Ask them to describe the first difficulty they encounter (subtracting $\frac{3}{4}$ from $\frac{1}{4}$). Draw a diagram to illustrate the regrouping involved.

INSTRUCTION In discussing Example 1, note that the two denominators are the same. Point out that they first subtract the fractions, and then the whole numbers.

In Example 2, make sure they understand that they cannot subtract $\frac{3}{4}$ from $\frac{1}{4}$. Thus, they will need to regroup before subtracting.

For Example 3, make sure the students understand why $8\frac{3}{12} = 7\frac{15}{12}$. The following illustration may help them understand the process.

$$8\frac{3}{12} = 8 + \frac{3}{12} = 7 + 1 + \frac{3}{12}$$
$$= 7 + \frac{12}{12} + \frac{3}{12}$$
$$= 7 + \frac{15}{12} = 7\frac{15}{12}$$

CHECKPOINT The incorrect answer choices include common errors.

6.10 SUBTRACTING MIXED NUMBERS

A mutton snapper weighed $9\frac{7}{8}$ pounds. A red grouper weighed $5\frac{3}{8}$ pounds. How much more did the mutton snapper weigh?

THINK: You are comparing weights. Subtract to find how much more the mutton snapper weighs.

EXAMPLE 1 Subtract: $9\frac{7}{8} - 5\frac{3}{8}$

1. Subtract the fractions.
$$\begin{array}{r} 9\frac{7}{8} \\ -5\frac{3}{8} \\ \hline \frac{4}{8} \end{array}$$

2. Subtract the whole numbers.
$$\begin{array}{r} 9\frac{7}{8} \\ -5\frac{3}{8} \\ \hline 4\frac{4}{8} \end{array}$$

3. Write the answer in simplest form.
$$4\frac{4}{8} = 4\frac{1}{2}$$

The mutton snapper weighs $4\frac{1}{2}$ lb more than the red grouper.

Sometimes you need to rename when subtracting mixed numbers.

EXAMPLE 2 Subtract: $4\frac{1}{4} - 2\frac{3}{4}$

1. Rename. Subtract the fractions.
$$\begin{array}{rcr} 4\frac{1}{4} & = & 3\frac{5}{4} \\ -2\frac{3}{4} & = & 2\frac{3}{4} \\ \hline & & \frac{2}{4} \end{array}$$

2. Subtract the whole numbers.
$$\begin{array}{rcr} 4\frac{1}{4} & = & 3\frac{5}{4} \\ -2\frac{3}{4} & = & 2\frac{3}{4} \\ \hline & & 1\frac{2}{4} \end{array}$$

3. Write the answer in simplest form.
$$1\frac{2}{4} = 1\frac{1}{2}$$

Sometimes you need to rename twice when subtracting mixed numbers.

EXAMPLE 3 Subtract: $8\frac{1}{4} - 5\frac{5}{6}$

1. Write equivalent fractions with a common denominator.
$$\begin{array}{rcr} 8\frac{1}{4} & = & 8\frac{3}{12} \\ -5\frac{5}{6} & = & 5\frac{10}{12} \end{array}$$

2. Rename if necessary. Subtract.
$$\begin{array}{rcr} 8\frac{3}{12} & = & 7\frac{15}{12} \\ -5\frac{10}{12} & = & 5\frac{10}{12} \\ \hline & & 2\frac{5}{12} \end{array}$$

CHECKPOINT

Write the letter of the correct answer.

Subtract. What is the difference in simplest form?

1. $8\frac{3}{4} - 5\frac{1}{4}$ d **a.** $3\frac{2}{4}$ **b.** $2\frac{1}{2}$ **c.** $2\frac{2}{4}$ **d.** $3\frac{1}{2}$
2. $7\frac{11}{12} - 3\frac{3}{4}$ a **a.** $4\frac{1}{6}$ **b.** $4\frac{8}{12}$ **c.** $4\frac{2}{12}$ **d.** $4\frac{8}{8}$
3. $4\frac{1}{3} - 1\frac{2}{5}$ d **a.** $3\frac{1}{2}$ **b.** $3\frac{1}{15}$ **c.** $\frac{14}{15}$ **d.** $2\frac{14}{15}$

PRACTICE EXERCISES

See Extra Practice, page 440.

Subtract. Write the difference in simplest form.

• **1.** $4\frac{2}{3} - 2\frac{1}{3}$ $2\frac{1}{3}$ **2.** $6\frac{4}{5} - 3\frac{2}{5}$ $3\frac{2}{5}$ **3.** $5\frac{5}{8} - 1\frac{3}{8}$ $4\frac{1}{4}$ **4.** $7\frac{5}{6} - 4\frac{2}{6}$ $3\frac{1}{2}$ **5.** $9\frac{6}{7} - 6\frac{2}{7}$ $3\frac{4}{7}$

• **6.** $6\frac{7}{8} - 3\frac{1}{4}$ $3\frac{5}{8}$ **7.** $18\frac{5}{6} - 5\frac{2}{3}$ $13\frac{1}{6}$ **8.** $13\frac{7}{9} - 11\frac{1}{6}$ $2\frac{11}{18}$ **9.** $5\frac{3}{4} - 1\frac{2}{3}$ $4\frac{1}{12}$ **10.** $11\frac{2}{3} - 9\frac{1}{2}$ $2\frac{1}{6}$

11. $9\frac{2}{3} - 6\frac{1}{5}$ $3\frac{7}{15}$ **12.** $10\frac{11}{12} - 3\frac{5}{8}$ $7\frac{7}{24}$ **13.** $8\frac{13}{16} - 2\frac{1}{4}$ $6\frac{9}{16}$ **14.** $7\frac{4}{5} - 2\frac{1}{4}$ $5\frac{11}{20}$ **15.** $16\frac{8}{9} - 7\frac{2}{3}$ $9\frac{2}{9}$

•**16.** $8\frac{3}{4} - 3\frac{5}{6}$ $4\frac{11}{12}$ **17.** $10\frac{1}{3} - 3\frac{1}{2}$ $6\frac{5}{6}$ **18.** $9\frac{2}{3} - 8\frac{5}{6}$ $\frac{5}{6}$ **19.** $13\frac{3}{8} - 1\frac{3}{4}$ $11\frac{5}{8}$ **20.** $24\frac{2}{9} - 17\frac{5}{12}$ $6\frac{29}{36}$

21. $7\frac{1}{4} - 3\frac{1}{2}$ $3\frac{3}{4}$ **22.** $8\frac{2}{3} - 1\frac{4}{5}$ $6\frac{13}{15}$ **23.** $10\frac{1}{6} - 4\frac{5}{8}$ $5\frac{13}{24}$ **24.** $15\frac{5}{12} - 6\frac{1}{2}$ $8\frac{11}{12}$ **25.** $11\frac{1}{6} - 7\frac{3}{5}$ $3\frac{17}{30}$

•**26.** $9\frac{3}{4} - 6\frac{1}{2}$ $3\frac{1}{4}$ **27.** $7\frac{5}{12} - \frac{5}{6}$ $6\frac{7}{12}$ **28.** $12\frac{8}{9} - 6\frac{5}{9}$ $6\frac{1}{3}$ **29.** $4\frac{2}{3} - 3$ $1\frac{2}{3}$ **30.** $8\frac{4}{5} - \frac{3}{5}$ $8\frac{1}{5}$

31. $23\frac{11}{16} - 14\frac{3}{4}$ $8\frac{15}{16}$ **32.** $37\frac{3}{8} - \frac{7}{12}$ $36\frac{19}{24}$ **33.** $4\frac{3}{8} - 4$ $\frac{3}{8}$ **34.** $17\frac{1}{8} - 6\frac{4}{5}$ $10\frac{13}{40}$ **35.** $8 - 3\frac{2}{5}$ $4\frac{3}{5}$

Write +, −, ×, or ÷ to make the sentence true.

36. $2\frac{1}{4}$ ● $3\frac{1}{3} = 7\frac{1}{2}$ × **37.** $6\frac{3}{8}$ ● $5\frac{1}{6} = 11\frac{13}{24}$ + **38.** $5\frac{1}{2}$ ● $2\frac{3}{4} = 2$ ÷ **39.** $9\frac{1}{6}$ ● $4\frac{5}{8} = 4\frac{13}{24}$ −

MIXED REVIEW

Add or subtract.

•**40.** 6.21 + 4.7 = 10.91 **41.** 43.344 + 5.03 = 48.374 **42.** 17.4 + 2.28 = 19.68 **43.** 72.29 + 6.15 = 78.44 **44.** 83.156 + 41.97 = 125.126

45. 6 − 4.2 1.8 **46.** 87.3 + 104.39 191.69 **47.** 28.3 − 12.5 15.8 **48.** 9 + 47.6 56.6

TIME OUT Students may find it helpful to make a list to help them organize the information in the problem.

COMMON ERROR

When subtracting fractions and mixed numbers, some students will forget to write the fraction in simplest form. For example, in item 1 of Checkpoint, these students will choose a. Have these students review the definition of simplest form. Then provide them with a list of fractions. Have them identify the fractions that are written in simplest form.

ASSIGNMENTS

Level 1	1–15, 26–35, 40–50
Level 2	Odd 1–47, 49–52, TO
Level 3	Even 16–34, 36–39, 49–54, TO

FOLLOW-UP ACTIVITY

APPLICATION Ask students to bring in newspaper and magazine articles and/or advertisements in which mixed numbers are used. Discuss the situations in which mixed numbers are commonly found. Have each student make up a problem-solving situation using the mixed numbers found in the newspapers and magazines.

Have volunteers write their problems on the chalkboard for the class to solve.

SUPPLEMENTARY MATERIALS

TRP Practice, p. 68

TRP Reteaching, p. 43

TRP Transparency 34

Solve.

• **49.** The world record for a green sunfish is $2\frac{1}{8}$ lb. The record for red-ear sunfish is $4\frac{1}{2}$ lb. How much heavier is the red-ear sunfish? $2\frac{3}{8}$ lb

• **50.** Camila caught a bluegill weighing $3\frac{7}{8}$ lb. The world record for a bluegill is $4\frac{3}{4}$ lb. How much less than the world record did Camila's fish weigh? $\frac{7}{8}$ lb

51. The world record for a white marlin is $181\frac{7}{8}$ lb. Larry caught a white marlin that weighed $\frac{4}{5}$ of the record weight. How much did Larry's marlin weigh? $145\frac{1}{2}$ lb

52. Eric caught 3 mutton snappers that weighed $4\frac{1}{2}$ lb, $6\frac{2}{3}$ lb, and $5\frac{3}{4}$ lb. He sold the fish for $3 per pound. How much money did he earn? $50.75

53. Joyce caught four perch during the tournament. Their weights were: 1 lb; $1\frac{1}{4}$ lb; $1\frac{1}{2}$ lb; $1\frac{3}{4}$ lb. Find the average weight of the fish. $1\frac{3}{8}$ lb

54. List the lengths of these fish in order from the longest to the shortest. 2 ft; $1\frac{5}{6}$ ft; $1\frac{3}{4}$ ft; 20 in.

Contestant	Kevin	Sarah	John	Glenda
Length of pike	2 ft	20 in.	$1\frac{3}{4}$ ft	$1\frac{5}{6}$ ft

TIME OUT

You are the bookkeeper at Daisy's Diner.
Every week the employees give you a time sheet.
This week they played a joke.
They gave their times in a puzzle.

Use the clues below to match each employee with the number of hours he or she worked.

Employee	Lonnie	Donna	Bob	Steve	Ken
Hours	$42\frac{1}{4}$	46	$39\frac{1}{2}$	$36\frac{1}{4}$	$38\frac{1}{2}$

- Lonnie worked $2\frac{3}{4}$ fewer hours than Donna.
- Donna worked more hours than Steve but fewer than Ken.
- Bob worked $7\frac{1}{4}$ hours a day for 5 days.
- If you divide the number of hours that Steve worked by 7, the answer would be $5\frac{1}{2}$.
- Ken worked $7\frac{1}{2}$ hours more than Steve.

Bob—$36\frac{1}{4}$;
Steve—$38\frac{1}{2}$;
Lonnie—$39\frac{1}{2}$;
Donna—$42\frac{1}{4}$;
Ken—46

PROBLEM *Solving* APPLICATION

6.11 CONSUMER: DETERMINING CAB FARES

You enter a taxicab in Ocean City and see this sign. You know that your destination is about 1 mi away. How much will your fare be?

OCEAN CITY RATES
75¢ first $\frac{1}{9}$ mi
20¢ each additional $\frac{1}{9}$ mi

1. What is the cost of the first $\frac{1}{9}$ mi? $0.75
2. How much longer is the trip?

   THINK: 1 mi = $\frac{9}{9}$ mi $\frac{9}{9} - \frac{1}{9} = \frac{8}{9}$
3. What is the cost of the $\frac{8}{9}$ mi? $8 \times \$0.20 = \$1.60$
4. What is the total cost? $\$0.75 + \$1.60 = \$2.35$

So, the total fare for your trip is $2.35.

Use the sign above to answer Exercises 1–4.
Remember to estimate whenever you use your calculator.

What is the cost of a trip of:

• **1.** $1\frac{4}{9}$ mi? $3.15 **2.** 2 mi? $4.15 **3.** $4\frac{1}{9}$ mi? $7.95 **4.** $3\frac{2}{3}$ mi? $7.15

Taxicab rates for New London are posted at the right. Use these rates to answer Exercises 5–12.

NEW LONDON RATES
85¢ first $\frac{1}{5}$ mi
35¢ each additional $\frac{1}{5}$ mi

What is the cost of a trip of:

• **5.** 1 mi? $2.25

6. 2 mi? $4.00

7. 3 mi? $5.75

8. 5 mi? $9.25

9. 10 mi? $18.00

10. Does a trip of 2 mi cost twice as much as a trip of 1 mi? no

11. Does a trip of 1 mi cost exactly $\frac{1}{3}$ as much as a trip of 3 mi? no

12. What amount will the meter show for a trip of $4\frac{1}{2}$ mi? Explain. $8.20. The meter will show the price of $4\frac{2}{5}$ mi.

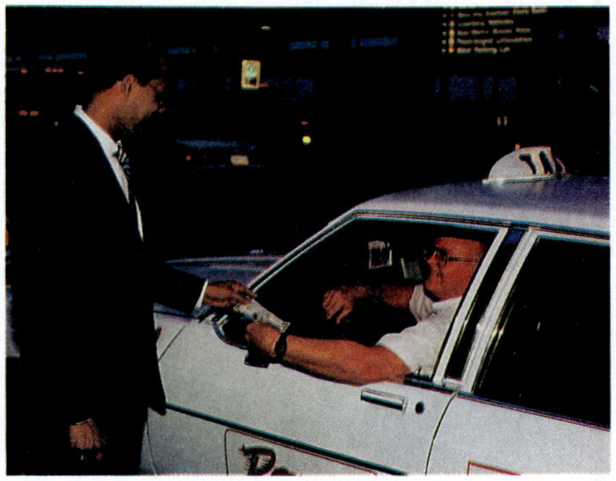

OBJECTIVE

• Solve problems that involve using data from a cab rate sign.

TEACHING THE LESSON

WARM-UP Write the following exercises on the chalkboard or an overhead transparency.

1. $1 = \frac{\blacksquare}{8}$ (8) **4.** $2 = \frac{\blacksquare}{9}$ (18)

2. $1 = \frac{\blacksquare}{5}$ (5) **5.** $1\frac{5}{8} = \frac{\blacksquare}{8}$ (13)

3. $2 = \frac{\blacksquare}{10}$ (20) **6.** $2\frac{3}{5} = \frac{\blacksquare}{5}$ (13)

INSTRUCTION Discuss the introductory example carefully, since it involves a multistep problem. Make sure the students understand each step:

Find the cost of the first $\frac{1}{9}$ mi.

Find the cost of the remaining $\frac{8}{9}$ mi.

Then add the two costs.

Problems 1–4 are similar to the introductory problem. However, students must realize that in each problem they should begin by expressing the given distance as a fraction, and then proceed to calculate the cost.

Problems 10 and 11 can help students realize that a short cab ride costs more per mile than a longer cab ride.

You may wish to have students use calculators to help them solve the problems in this lesson.

ASSIGNMENTS

Level 1	1–9
Level 2	1–11
Level 3	1–12

SUPPLEMENTARY MATERIALS

TRP Practice, p. 69

TRP Transparency 35

OBJECTIVES

- Estimate sums and differences of mixed numbers.
- Estimate products and quotients of mixed numbers.

TEACHING THE LESSON

WARM-UP Present the following exercises on the chalkboard or overhead transparency.

Compare. Use >, <, or =.

1. $\frac{2}{3}$ ● $\frac{1}{2}$ **2.** $\frac{3}{5}$ ● $\frac{1}{2}$ **3.** $\frac{2}{4}$ ● $\frac{1}{2}$

4. $\frac{5}{6}$ ● $\frac{1}{2}$ **5.** $\frac{3}{8}$ ● $\frac{1}{2}$ **6.** $\frac{5}{9}$ ● $\frac{1}{2}$

(**1.** > **2.** > **3.** = **4.** > **5.** < **6.** >)

INTRODUCTION Discuss situations in which we round mixed numbers to the nearest whole number. For example, the height of a person is usually rounded to the nearest inch. A person's weight is normally rounded to the nearest pound. Have students give other examples where mixed numbers are rounded to the nearest whole number.

INSTRUCTION In discussing the rounding of mixed numbers, it will be helpful for students to determine whether a fraction is greater or less than $\frac{1}{2}$. Present this simple rule: Double the numerator. If this number is less than the denominator, the fraction is less than $\frac{1}{2}$. If the number is more than the denominator, the fraction is more than $\frac{1}{2}$.

For Example 2, you may wish to have students draw number lines to verify which whole numbers are closest to the mixed numbers.

6.12 ESTIMATING WITH MIXED NUMBERS

Ray Harper is a tailor for the Main Event Circus. He is taking an inventory of his supply of fabrics. Here are two situations in which he estimates with mixed numbers.

One roll has $6\frac{3}{4}$ yd of blue satin. Another roll has $8\frac{1}{3}$ yd of blue satin. About how many yards of blue satin are in stock?

Ray needs $24\frac{1}{2}$ yd of beaded material. He has $10\frac{1}{6}$ yd. About how many more yards should he order?

EXAMPLE 1 Estimate:

a. $6\frac{3}{4} + 8\frac{1}{3}$

1. Round. $6\frac{3}{4} \to 7$, $+8\frac{1}{3} \to 8$
2. Add. 15

Ray has about 15 yd of blue satin in stock.

b. $24\frac{1}{2} - 10\frac{1}{6}$

1. Round. $24\frac{1}{2} \to 25$, $-10\frac{1}{6} \to 10$
2. Subtract. 15

Ray should order about 15 yd of beaded material.

Use the same method to estimate products and quotients.

EXAMPLE 2 Estimate:

a. $7\frac{3}{4} \times 8\frac{1}{8}$

1. Round. $7\frac{3}{4} \times 8\frac{1}{8}$ → 8×8
2. Multiply. $8 \times 8 = 64$

$7\frac{3}{4} \times 8\frac{1}{8}$ is about 64.

b. $5\frac{2}{3} \div 9\frac{3}{7}$

1. Round. $5\frac{2}{3} \div 9\frac{3}{7}$ → $6 \div 9$
2. Divide. $6 \div 9 = \frac{6}{9} = \frac{2}{3}$

$5\frac{2}{3} \div 9\frac{3}{7}$ is about $\frac{2}{3}$.

FOR DISCUSSION

See TE side column.

How will the estimated sum compare with the actual answer:

1. If you round both mixed numbers up?

2. If you round both mixed numbers down?

3. If you round one mixed number up and the other down?

PRACTICE EXERCISES See Extra Practice, page 440.

Round to the nearest whole number.

• **1.** $4\frac{1}{2}$ 5 **2.** $6\frac{1}{3}$ 6 **3.** $9\frac{3}{4}$ 10 **4.** $3\frac{3}{5}$ 4 **5.** $8\frac{3}{7}$ 8

6. $13\frac{1}{4}$ 13 **7.** $23\frac{1}{2}$ 24 **8.** $71\frac{4}{5}$ 72 **9.** $65\frac{6}{11}$ 66 **10.** $81\frac{6}{13}$ 81

Estimate.

• **11.** $3\frac{1}{4} + 4\frac{2}{3}$ 8 **12.** $8\frac{1}{7} + 3\frac{1}{3}$ 11 **13.** $9\frac{5}{8} + 4\frac{3}{4}$ 15 **14.** $6\frac{2}{3} + 5\frac{1}{6}$ 12 **15.** $7\frac{5}{9} + 4\frac{2}{3}$ 13

• **16.** $3\frac{1}{3} - 2\frac{1}{4}$ 1 **17.** $4\frac{1}{2} - 1\frac{2}{3}$ 3 **18.** $5\frac{7}{8} - 1\frac{2}{9}$ 5 **19.** $6\frac{1}{3} - 4\frac{2}{5}$ 2 **20.** $8\frac{1}{2} - 2\frac{3}{8}$ 7

• **21.** $5\frac{1}{5} \times 1\frac{3}{4}$ 10 **22.** $6\frac{1}{2} \times \frac{7}{8}$ 7 **23.** $2\frac{3}{4} \times 9\frac{2}{3}$ 30 **24.** $4\frac{3}{5} \times 6\frac{5}{6}$ 35 **25.** $3\frac{1}{2} \times 6\frac{1}{8}$ 24

• **26.** $21\frac{1}{3} \div 6\frac{2}{3}$ 3 **27.** $5\frac{3}{4} \div \frac{7}{8}$ 6 **28.** $5\frac{3}{4} \div 2\frac{3}{4}$ 2 **29.** $4\frac{3}{8} \div 2\frac{1}{8}$ 2 **30.** $15\frac{5}{6} \div 3\frac{3}{4}$ 4

31. $4\frac{1}{5} - 2\frac{1}{6}$ 2 **32.** $12\frac{1}{3} \times 3\frac{1}{5}$ 36 **33.** $14\frac{2}{3} \div 5\frac{1}{4}$ 3 **34.** $23\frac{4}{5} + 3\frac{1}{6}$ 27 **35.** $5\frac{5}{6} \times 21\frac{3}{8}$ 126

36. $9\frac{3}{8} \div 2\frac{9}{16}$ 3 **37.** $9\frac{3}{4} + 14\frac{3}{8}$ 24 **38.** $18\frac{2}{5} - 15\frac{5}{6}$ 2 **39.** $8\frac{3}{7} - 7\frac{5}{6}$ 0 **40.** $8\frac{1}{2} \times 31\frac{2}{3}$ 288

RAY HARPER'S FILE OF ANIMAL SIZES

Animal	Neck size	Sleeve length (ft)	Waist size (ft)
Ollie the Ostrich	$12\frac{3}{4}$ in.	—	—
Sparkle the Monkey	$8\frac{1}{2}$ in.	3	$2\frac{1}{2}$
Teddy the Tiger	$3\frac{1}{4}$ ft	$1\frac{1}{12}$	$4\frac{2}{3}$
Minnie the Elephant	$8\frac{3}{4}$ ft	$2\frac{5}{6}$	$15\frac{7}{12}$

Solve. Use the data from the table to estimate the answer.

• **41.** About how much larger is Minnie's waist size than Sparkle's? 13 ft

42. About how many times larger is Minnie's neck size than Ollie's? 9 times

43. Ray is making a new costume for Teddy. The sleeves will be blue satin. Estimate the amount of blue satin that Ray needs. (*Remember:* A tiger has four legs.) 4 ft

44. Ray wants to make 6 bow ties for Ollie. Each bow tie needs $1\frac{1}{2}$ ft of fabric. He has 6 ft of fabric. How many bow ties can he make? How much more fabric does he need to make all 6 bow ties? 4 bow ties; 3 ft

FOR DISCUSSION Students should explain that if both mixed numbers are rounded up, the estimated sum will be greater than the actual answer. If both mixed numbers are rounded down, the estimate will be less than the actual answer. If one mixed number is rounded up and the other is rounded down, the estimate will be very close to the actual answer.

ASSIGNMENTS

Level 1	Odd 1–43
Level 2	Even 2–44
Level 3	Odd 11–39, 41–44

FOLLOW-UP ACTIVITY

ENRICHMENT Have students choose the operation sign to complete the exercise.

		Estimated answer	
1.	$1\frac{1}{2}$ ● $2\frac{2}{3}$ ⟶	5	(+)
2.	$4\frac{4}{5}$ ● $2\frac{7}{8}$ ⟶	2	(−)
3.	$1\frac{3}{4}$ ● $\frac{1}{2}$ ⟶	4	(÷)
4.	$3\frac{1}{3}$ ● 4 ⟶	12	(×)

SUPPLEMENTARY MATERIALS

TRP Practice, p. 70

TRP Reteaching, p. 44

TRP Transparency 35

OBJECTIVE

- Solve problems that involve buying on an installment plan.

TEACHING THE LESSON

WARM-UP Write the following exercises on the chalkboard or an overhead transparency.

Find the answer.

1. $\frac{1}{12} \times \$960$ (\$80)
2. 12 × \$88.70 (\$1,064.40)
3. \$695.30 − \$576.70 (\$118.60)
4. $\frac{1}{24} \times \$2,304$ (\$96)
5. 24 × \$114.72 (\$2,753.28)

INTRODUCTION Ask students to tell what they know about buying on an installment plan. Ask volunteers to share information about installment purchases that their families may have made.

INSTRUCTION As you discuss the introductory example, you may wish to mention that the finance charge is, in effect, interest on the money that the store has advanced or "lent" to the customer.

Guide students to realize that each of the 12 monthly installments is not $\frac{1}{12}$ of the cash price; it is greater than $\frac{1}{12}$ of the cash price.

It may be worthwhile to note that there are different types of installment plans. In some installment plans, there is a down payment, and then equal monthly installments are paid. A revolving charge account involves a form of installment buying; however, the customer pays installments that are not necessarily equal, and the store charges interest on the balance.

You may wish to have students use calculators to help them solve the problems in this lesson.

PROBLEM Solving APPLICATION

6.13 CONSUMER: BUYING ON AN INSTALLMENT PLAN

Many stores offer customers different ways to pay for purchases. If you pay for an item all at once, you pay the **cash price.** If you buy the item using a number of payments over a period of time, you are buying the item in **installments.** In this case, the store is providing the credit, and there is a **finance charge** for this service.

You buy a couch and a chair at Helpful Harry's Furniture Company. The cash price of the furniture is \$840. You decide to pay in 12 monthly installments. The installments are \$76.30 per month. What is the total amount that you pay for the couch and the chair if you use the installment plan? What is the finance charge?

	NUMBER OF INSTALLMENTS	×	AMOUNT OF EACH INSTALLMENT	=	TOTAL AMOUNT
1. Multiply to find the total amount.	12	×	\$76.30	=	\$915.60
	TOTAL	−	**CASH PRICE**	=	**FINANCE CHARGE**
2. Subtract the cash price from the total to find the finance charge.	\$915.60	−	\$840.00	=	\$75.60

So, the total amount you pay is \$915.60. The finance charge is \$75.60.

Find the total amount and the finance charge.
Remember to estimate whenever you use your calculator.

	Cash Price	No. of Installments	Amount of Each Installment	Total Amount	Finance Charge
• **1.**	$ 480	12	$ 43.60	$523.20 ■	$43.20 ■
2.	$ 720	12	$ 65.40	$784.80 ■	$64.80 ■
3.	$ 540	12	$ 49.05	$588.60 ■	$48.60 ■
4.	$1,092	12	$ 99.19	$1,190.28 ■	$98.28 ■
5.	$2,400	24	$120.00	$2,880.00 ■	$480.00 ■
6.	$3,600	24	$175.00	$4,200.00 ■	$600.00 ■
7.	$5,000	36	$145.02	$5,220.72 ■	$220.72 ■
8.	$3,895	48	$109.87	$5,273.76 ■	$1,378.76 ■

Solve.

• **9.** The cash price of dining room furniture is $960. Mrs. Pannier buys it on an installment plan. She pays in 12 monthly installments, $87.20 per month. What is the total amount she pays? What is the finance charge? $1,046.40; $86.40

10. Doreen Washington buys a stereo system on an installment plan. The total amount of money she pays is $1,194. If each installment is $\frac{1}{6}$ of the total amount, how much is each installment? $199

11. Pat Chang buys a dining room set on an installment plan. He pays in 12 monthly installments, $71.35 per month. When he has finished paying the installments, the total amount he has paid will be $69.40 more than the cash price. What is the cash price? $786.80

ASSIGNMENTS

Level 1	1–6, 9
Level 2	1–11
Level 3	5–11

FOLLOW-UP ACTIVITY

APPLICATION Explain that in each of the following exercises, the installment plan involves a down payment and then monthly installments. Ask students to find the total amount paid and the finance charge.

1. Cash price: $500
Down payment: $75
Number of monthly payments: 12
Amount of each payment: $41.70
Total amount paid: ■ ($575.40)
Finance charge: ■ ($75.40)

2. Cash price: $800
Down payment: $120
Number of monthly payments: 12
Amount of each payment: $65.17
Total amount paid: ■ ($902.04)
Finance charge: ■ ($102.04)

SUPPLEMENTARY MATERIALS

TRP Practice, p. 71

TRP Applications, p. 11

TRP Transparency 36

OBJECTIVE

- Decide, given relevant information, whether to pursue a do-it-yourself plan.

TEACHING THE LESSON

WARM-UP Write the following exercises on the chalkboard or an overhead transparency.

Complete the table.

	Daily rate	Number of days	Total
1.	$ 40	6	
2.		9	$333
3.	$19.50	31	
		Total cost	

(**1.** $240 **2.** $37 **3.** $604.50 **4.** $1,177.50)

INTRODUCTION Present the problem situation to the class. Discuss do-it-yourself projects with them, and ask students what factors they think are important when deciding if a project is feasible. List the factors on the chalkboard.

INSTRUCTION Have students study the three estimates and note the differences they feel may be important. Also have them read and discuss the additional information regarding the family's experience and schedule.

Direct attention to the decision-making factors. Ask the students if there are any they would add to the list from those on the chalkboard.

Have students complete Exercises 1–8 in the table in order to compare the three estimates. They can also analyze any additional factors they have listed in the same way. Ask them to write conclusions that can be drawn from the data and discuss the appropriateness of these conclusions and any possible misinterpretations.

DECISION MAKING

6.14 DO IT YOURSELF

(Open-ended problem solving)

Some household projects may appear too difficult or time consuming to do them yourself. However, estimated costs for a professional job often cause people to reevaluate their options.

PROBLEM

The exterior of the Acevedo family's two-story frame house needs painting and should be done this spring. The estimates from the painting contractors seem high. Therefore, they have decided to develop a plan for painting the house themselves. First they collect all the available data.

Estimates:

Busy Bee Painters
Est. 1986

Paint: $150
Labor: $3,000
Sanding and scraping as necessary: $20 per hour
Time: 5 days

Ace Painters
Est. 1939

Paint: $200 for siding
$ 75 for trim
Labor: $3,500 (includes scraping and sanding)
Time: 3 days

Do It Ourselves

Paint: $225
Brushes: $60
Other supplies: $130
Ladder rental: $50/day (12 days)
Sander rental: $50/day (4 days)
Time: 6 weekends if Victor helps

Mr. and Mrs. Acevedo both have experience with interior painting. Their oldest son, Victor, who is seventeen, painted his bedroom last year. Fifteen-year-old Melissa and ten-year-old Danny have never done any painting. Both parents work full-time, and Victor and Melissa work after school and on Saturdays.

DECISION-MAKING FACTORS

Cost Time Experience

DECISION-MAKING COMPARISONS

Some answers may vary. Accept reasonable answers based on students' rationales.

Make a chart to compare the three estimates.

	Busy Bee	Ace	Ourselves
Cost	• 1. ■	• 2. ■	3. ■
Time	4. ■	5. ■	6. ■
Experience	Limited	7. ■	8. ■

About $3,150; $3,775; $1,215
5 days; 3 days; 12 days; over six weeks; Extensive; Very limited

MAKING THE DECISIONS

• **9.** Which of the three estimates is the highest? Lowest? Ace; Ourselves

• **10.** Is the difference in time between the two professional painters significant? no

11. Based on experience only, which group would probably do the best job? Ace

12. Based on the given estimates, about how much money would the family save by doing the project themselves? $1,935–$2,560

13. Victor makes about $75 a week, and two-thirds of this is earned on Saturday. About how much pay would he lose if he gave up his job for 6 weekends to help paint? $300

14. Should the pay Victor would lose be included in the cost of the project? yes

15. By working on weekends only, what factor must be considered when renting the ladder and sander? Picking them up and returning them each weekend

16. Mr. Acevedo could take 1 week of vacation to do the painting, but Mrs. Acevedo would have to take a week off without pay ($750). Since the children would be in school and not available to help, is this a good idea? Why? No; they would not finish the job in a week.

17. What differences between interior painting and exterior painting must the family consider? Size of the job, weather conditions, working on a high ladder

18. The house is two stories high, and the Acevedos would have to work on long ladders for the first time. What other factor should be considered? Safety

19. What factors could affect the time in any of the plans? Rain, wind, temperature, illness, accidents

20. Given all the factors, do you think the Acevedos should paint the house themselves? Why? Answers will vary.

Assign Exercises 9–20 and have students work in small groups. Encourage them to explain and defend their answers and decisions as necessary. Reinforce the fact that individual capabilities and situations strongly influence decisions of this type. Observe the groups as they work and interact as necessary to draw nonparticipating students into the discussion.

ASSIGNMENTS

Level 1	1–20
Level 2	1–20
Level 3	1–20

FOLLOW-UP ACTIVITY

APPLICATION Have students work in small groups and assign them the task of developing a plan for painting or somehow refurbishing a large room in your school. (Choose something like the auditorium, gymnasium, or cafeteria.) You may want to assign a different room to each group.

Their plan for the room should assume students will be doing it themselves and should include a description of the work to be done based on the group's determination of what is needed. The plan should include estimates for the factors covered in the lesson as well as others the group needs to address, such as the number of people needed to do the work.

Accept reasonable estimates for materials and equipment, and be sure to have the class discuss the feasibility or desirability of the plan. Some students may overestimate their own capabilities and plan to have two people paint a huge room with very high ceilings in a day. Others may plan to paint over decorative ceilings or natural woodwork which may or may not be desirable or necessary.

SUPPLEMENTARY MATERIALS

TRP Practice, p. 72
TRP Applications, p. 12
TRP Transparency 36

OBJECTIVES

- Review vocabulary.
- Practice key chapter concepts and skills.

USING THE REVIEW

The Chapter Review is designed to help students prepare for taking the Chapter Test. The first section focuses on vocabulary. It requires that students select words to complete statements. The second section presents practice exercises of key mathematical skills. Under each directive there is a sample exercise with the answer.

Each item on the review is referenced to the page on which the topic is taught in the Pupil's Edition. You may wish to have students refer to these pages to help review any concepts or skills they have not yet mastered.

It is suggested that students work in small-sized heterogeneous cooperative learning groups. Some cooperative learning methods that may be used are as follows:

1. After each student has independently completed the entire Chapter Review, a discussion should follow within each group about the solutions to the practice exercises.
2. The group can complete the entire Chapter Review by working together to discuss the sample exercises and then to answer the practice exercises.

End the lesson with an entire class discussion in which any questions brought up in group discussions are presented and answered.

CHAPTER REVIEW

Vocabulary Choose the letter of the word(s) that completes the statement.

1. If the product of two numbers equals 1, the numbers are ■. [150] b

   a. improper fractions **b.** reciprocals **c.** mixed numbers **d.** numerators

2. To multiply or divide mixed numbers, first change the mixed numbers to ■. [148, 152] c

   a. common factors **b.** addends **c.** improper fractions **d.** GCF

Skills Multiply. [146, 148]

$\frac{4}{9} \times \frac{3}{8}$ $\frac{1}{6}$

3. $\frac{2}{3} \times \frac{3}{4}$ $\frac{1}{2}$ **4.** $\frac{1}{6} \times \frac{3}{4}$ $\frac{1}{8}$ **5.** $\frac{7}{8} \times \frac{4}{5}$ $\frac{7}{10}$

6. $2\frac{2}{3} \times 1\frac{1}{6}$ $3\frac{1}{9}$ **7.** $4\frac{1}{2} \times 3\frac{2}{3}$ $16\frac{1}{2}$ **8.** $3\frac{3}{8} \times 2\frac{3}{4}$ $9\frac{9}{32}$ **9.** $\frac{8}{11} \times \frac{2}{5}$ $\frac{16}{55}$

Divide. [150, 152]

$\frac{3}{4} \div \frac{3}{8}$ 2

10. $\frac{3}{4} \div \frac{1}{2}$ $1\frac{1}{2}$ **11.** $\frac{4}{15} \div \frac{3}{5}$ $\frac{4}{9}$ **12.** $3\frac{3}{4} \div 5$ $\frac{3}{4}$

13. $2\frac{2}{5} \div 1\frac{3}{4}$ $1\frac{13}{35}$ **14.** $\frac{7}{10} \div \frac{15}{20}$ $\frac{14}{15}$ **15.** $1\frac{9}{15} \div 1\frac{4}{5}$ $\frac{8}{9}$ **16.** $12\frac{1}{2} \div \frac{5}{8}$ 20

Add. [158, 160]

$\frac{2}{3} + \frac{3}{5} = 1\frac{4}{15}$

17. $\frac{5}{9} + \frac{5}{6}$ $1\frac{7}{18}$ **18.** $\frac{9}{15} + \frac{4}{6}$ $1\frac{4}{15}$ **19.** $\frac{7}{16} + \frac{7}{8}$ $1\frac{5}{16}$

20. $8\frac{3}{4} + 7\frac{3}{8}$ $16\frac{1}{8}$ **21.** $12\frac{5}{8} + 24\frac{3}{5}$ $37\frac{9}{40}$ **22.** $4\frac{3}{5} + 2\frac{4}{9}$ $7\frac{2}{45}$ **23.** $6 + 5\frac{2}{3}$ $11\frac{2}{3}$

24. $6\frac{1}{4} + 8$ $14\frac{1}{4}$ **25.** $1\frac{2}{3} + 4\frac{4}{5}$ $6\frac{7}{15}$ **26.** $\frac{1}{2} + \frac{1}{6} + \frac{5}{8}$ $1\frac{7}{24}$ **27.** $\frac{3}{4} + \frac{1}{5} + \frac{7}{10}$ $1\frac{13}{20}$

Subtract. [162, 164]

$\frac{5}{6} - \frac{1}{4} = \frac{7}{12}$

28. $\frac{5}{6} - \frac{3}{8}$ $\frac{11}{24}$ **29.** $\frac{2}{3} - \frac{2}{5}$ $\frac{4}{15}$ **30.** $\frac{15}{16} - \frac{1}{8}$ $\frac{13}{16}$

31. $6\frac{1}{4} - 4\frac{6}{7}$ $1\frac{11}{28}$ **32.** $5 - 4\frac{2}{5}$ $\frac{3}{5}$ **33.** $15\frac{4}{7} - 11\frac{3}{8}$ $4\frac{11}{56}$ **34.** $4\frac{5}{8} - 1\frac{9}{10}$ 2[illegible]

CHAPTER TEST

Multiply. Write the product in simplest form.

1. $\frac{1}{3} \times \frac{3}{4}$ $\frac{1}{4}$ **2.** $\frac{4}{5} \times \frac{1}{8}$ $\frac{1}{10}$ **3.** $\frac{1}{2} \times \frac{1}{2}$ $\frac{1}{4}$ **4.** $\frac{7}{11} \times \frac{1}{7}$ $\frac{1}{11}$ **5.** $\frac{6}{7} \times \frac{2}{3}$ $\frac{4}{7}$ **6.** $\frac{5}{6} \times \frac{9}{15}$ $\frac{1}{2}$

7. $9 \times \frac{2}{3}$ 6 **8.** $\frac{1}{2} \times 8$ 4 **9.** $1\frac{1}{2} \times 6$ 9 **10.** $3\frac{3}{4} \times 2\frac{1}{2}$ $9\frac{3}{8}$ **11.** $2\frac{1}{3} \times 5\frac{3}{6}$ $12\frac{5}{6}$ **12.** $\frac{1}{3} \times 6\frac{1}{2}$ $2\frac{1}{6}$

Divide. Write the quotient in simplest form.

13. $\frac{2}{5} \div \frac{3}{4}$ $\frac{8}{15}$ **14.** $\frac{1}{4} \div \frac{1}{2}$ $\frac{1}{2}$ **15.** $\frac{3}{5} \div 5$ $\frac{3}{25}$ **16.** $\frac{3}{4} \div \frac{7}{8}$ $\frac{6}{7}$ **17.** $9 \div \frac{2}{3}$ $13\frac{1}{2}$ **18.** $\frac{4}{9} \div \frac{2}{3}$ $\frac{2}{3}$

19. $4 \div 2\frac{1}{2}$ $1\frac{3}{5}$ **20.** $9 \div 5\frac{2}{3}$ $1\frac{10}{17}$ **21.** $3\frac{3}{4} \div 2$ $1\frac{7}{8}$ **22.** $\frac{5}{9} \div 1\frac{2}{3}$ $\frac{1}{3}$ **23.** $2\frac{1}{8} \div \frac{3}{4}$ $2\frac{5}{6}$ **24.** $6\frac{4}{5} \div 2$ $3\frac{2}{5}$

Add. Write the sum in simplest form.

25. $\frac{4}{7} + \frac{2}{7}$ $\frac{6}{7}$ **26.** $\frac{5}{8} + \frac{5}{8}$ $1\frac{1}{4}$ **27.** $\frac{4}{9} + \frac{5}{9}$ 1 **28.** $\frac{2}{5} + \frac{2}{3}$ $1\frac{1}{15}$ **29.** $\frac{3}{4} + \frac{7}{12}$ $1\frac{1}{3}$ **30.** $\frac{2}{3} + \frac{4}{9}$ $1\frac{1}{9}$

31. $4\frac{1}{5} + 1\frac{3}{5}$ $5\frac{4}{5}$ **32.** $3\frac{3}{8} + 3\frac{7}{8}$ $7\frac{1}{4}$ **33.** $\frac{1}{3} + 5\frac{4}{6}$ 6 **34.** $9\frac{4}{9} + 3\frac{7}{9}$ $13\frac{2}{9}$ **35.** $12 + 5\frac{1}{6}$ $17\frac{1}{6}$ **36.** $3\frac{1}{4} + 4\frac{3}{4}$ 8

Subtract. Write the difference in simplest form.

37. $\frac{3}{4} - \frac{1}{4}$ $\frac{1}{2}$ **38.** $\frac{8}{9} - \frac{5}{9}$ $\frac{1}{3}$ **39.** $\frac{5}{6} - \frac{1}{3}$ $\frac{1}{2}$ **40.** $\frac{11}{12} - \frac{1}{4}$ $\frac{2}{3}$ **41.** $\frac{2}{3} - \frac{1}{8}$ $\frac{13}{24}$ **42.** $\frac{11}{12} - \frac{5}{6}$ $\frac{1}{12}$

43. $6\frac{4}{5} - 2\frac{1}{5}$ $4\frac{3}{5}$ **44.** $8\frac{7}{8} - 3\frac{3}{8}$ $5\frac{1}{2}$ **45.** $9\frac{2}{3} - 7\frac{1}{2}$ $2\frac{1}{6}$ **46.** $10 - 4\frac{5}{6}$ $5\frac{1}{6}$ **47.** $15\frac{7}{9} - 8\frac{2}{3}$ $7\frac{1}{9}$ **48.** $11 - 6\frac{7}{9}$ $4\frac{2}{9}$

Solve.

49. Karen has $2,500 in her checking account. She deposited $1,000 yesterday. Last week she deposited $500. How much money was in the account before she deposited the $500? $1,000

50. Terry Foster bought a couch on an installment plan. The total amount of money she paid is $775. If each installment is $\frac{1}{5}$ of the total amount, how much is each installment? $155

Use the rate table at the right to answer Exercises 51 and 52.

51. What is the cost of a taxi trip of 1 mi? $2.35

52. What is the cost of a taxi trip of 2 mi? $4.10

BRADLEY RATES
95¢ first $\frac{1}{5}$ mi
35¢ each additional $\frac{1}{5}$ mi

OBJECTIVE

- Evaluate achievement of the chapter objectives.

USING THE TEST

The Chapter Test may be used as a posttest to evaluate student achievement. However, you may wish to use the Chapter Posttest offered in the Teacher's Resource Package or to design your own chapter test. If this page is not used as a test, you may wish to assign it as additional review or practice.

The test items are correlated to the chapter objectives in the table below.

Chapter objectives	Test items
A. Multiply fractions.	1–12
B. Divide fractions.	13–24
C. Add fractions.	25–36
D. Subtract fractions.	37–48
E. Work backward to solve problems.	49
F. Apply computational skills to real-life situations.	50–52

SUPPLEMENTARY MATERIALS

TRP Ch. 6 Posttest Form A, pp. 1–2

TRP Ch. 6 Posttest Form B, pp. 1–2

OBJECTIVES

- Review and maintain previously taught concepts and skills.
- Practice taking tests in a multiple-choice format.

USING THE REVIEW

Assign the Cumulative Review to all students. Provide students with an answer sheet to record their answers.

Each Cumulative Review gives students an opportunity to practice taking tests that are written in multiple-choice format.

Provide appropriate remedial help for students having difficulty with any of the skills and concepts on these pages.

CUMULATIVE REVIEW

Choose the letter of the correct answer.

1. A(n) ■ shows how many times a number or base is used as a factor. a

a. exponent **b.** average **c.** product **d.** none of these

2. The ■ of a set of data is the same as the average. c

a. median **b.** mode **c.** mean **d.** none of these

3. Which of the following is greater than $\frac{5}{6}$? d

a. $\frac{3}{4}$ **b.** $\frac{5}{7}$ **c.** $\frac{19}{24}$ **d.** none of these

4. A ■ is a bar graph that shows the frequencies of intervals of data. c

a. frequency table **b.** pictograph **c.** histogram **d.** none of these

5. A(n) ■ fraction has a numerator equal to or greater than its denominator. c

a. equivalent **b.** mixed number **c.** improper **d.** none of these

Select the best estimated answer.

6. 892 × 114	d	**a.** 800,000	**b.** 9,000	**c.** 18,000	**d.** 90,000
7. 7,198 × 342	b	**a.** 210,000	**b.** 2,100,000	**c.** 3,800,000	**d.** 280,000
8. 16,263 ÷ 829	a	**a.** 20	**b.** 200	**c.** 2,000	**d.** 2
9. $4\frac{5}{8} \div \frac{3}{4}$	c	**a.** 9	**b.** 1	**c.** 5	**d.** 2
10. 432 + 1,983 + 96	a	**a.** 2,500	**b.** 1,000	**c.** 600	**d.** 1,500

Compute.

11. $125\overline{)200}$	c	**a.** 0.16	**b.** 16	**c.** 1.6	**d.** none of these
12. $\frac{5}{8} \times \frac{1}{3}$	b	**a.** $1\frac{7}{8}$	**b.** $\frac{5}{24}$	**c.** $1\frac{5}{72}$	**d.** none of these
13. $\frac{360,000}{9,000}$	a	**a.** 40	**b.** 400	**c.** 4,000	**d.** none of these
14. $8\overline{)3,245}$	a	**a.** 405 R5	**b.** 45 R5	**c.** 4,005 R5	**d.** none of these
15. 75,000 × 4,000	c	**a.** 300,000	**b.** 3,000,000	**c.** 300,000,000	**d.** none of these

Solve.

16. Sal's Taxi Service charges 75¢ for the first mile and 35¢ for each additional mile. What is the cost of a 3-mi trip? b

a. \$1.50 **b.** \$1.45 **c.** \$1.10 **d.** none of these

17. Jane wants to purchase a new car called the Gazelle which has a base price of \$16,826. She would like the optional air conditioning which costs \$552 and the automatic transmission which costs \$823. What is the total price of the car? a

a. \$18,201 **b.** \$17,378 **c.** \$17,649 **d.** none of these

18. A board of lumber is $7\frac{7}{8}$ ft long. Into how many $\frac{7}{8}$-ft sections can it be cut? c

a. $6\frac{11}{32}$ **b.** 8 **c.** 9 **d.** none of these

19. Dennis buys clothes through a mail order catalog. He orders 3 plaid shirts which cost \$16.75 each, 2 flannel shirts which are \$22.50 each, and 4 white dress shirts at \$24.95 each. Shipping costs are \$2.75. What is the total amount of his order? d

a. \$193.05 **b.** \$193.95 **c.** \$183.05 **d.** none of these

20. The Carini family made a budget. They allotted 0.24 for food, 0.28 for rent, 0.36 for daily expenses, and 0.12 for vacation expenses. Which expense is $\frac{3}{25}$ of their total budget? b

a. food **b.** vacation **c.** daily expenses **d.** none of these

THINKING ABOUT MATH See TE side column.

1. Mark, José, and Joel each belong to one of three sports teams. Mark and Joel are not on the basketball team. Joel does not play on the tennis team. Who plays on the baseball team? Describe the best method for solving the problem.

2. Tom ran the 100-yd dash several times. To find his average speed, would you calculate his median speed or his mean speed? Explain.

3. You need to make a graph showing the approximate number of cars manufactured in Detroit, Michigan, last year. Which graph would you use: a broken-line graph, a bar graph, or a pictograph? Why?

4. Explain how you can quickly tell that 10^8 is not the exponent form of 10,000,000.

THINKING ABOUT MATH

The questions in this section can be used in discussions with individual students or with small groups. Guide students in conveying their ideas clearly and precisely.

Listed below are expected student answers. However, accept any reasonable answer.

1. Students should make a table that will enable them to obtain a clearer picture of which student belongs to which club.
2. To find the average speed, calculate the mean speed. The median is only the middle value of a set of numbers.
3. Since the graph is showing approximate information, a pictograph would show the information most clearly.
4. Since 10,000,000 has 7 zeros, 10^7 is the correct answer.

SUPPLEMENTARY MATERIALS

TRP Cum. Test, Ch. 1–6, pp. 1–4

OBJECTIVE

- Test the prerequisite skills needed to learn the concepts developed in Chapter 7.

USING THE TEST

The Pre-Skills Test is designed to diagnose students' strengths and weaknesses on prerequisite skills necessary to study the mathematics in Chapter 7.

Have students take the Pre-Skills Test and provide reteaching and practice in those areas where individuals demonstrate a misunderstanding of a concept or weakness in a skill.

The items in the test are referenced to the pages on which the topics are taught in the Pupil's Edition. You may wish to have students refer to these pages for review.

The following table correlates the items on the Pre-Skills Test with the prerequisite skill and the lesson(s) in the chapter for which it is needed.

Item(s)	Prerequisite skill	Lesson(s)
1–12	Multiply whole numbers and decimals by powers of 10.	7.10
13–24	Divide whole numbers and decimals by powers of 10.	7.10
25–33	Multiply fractions, whole numbers, and mixed numbers.	7.1, 7.2, 7.4
34–42	Divide whole numbers, writing the remainder as a fraction.	7.1, 7.2, 7.4
43–50	Add and subtract whole numbers.	7.3, 7.4

Pre-Skills Test

Multiply. [68]

1. 100 × 35 3,500
2. 1,000 × 22 22,000
3. 100 × 2.4 240
4. 23.5 × 100 2,350
5. 200 × 10 2,000
6. 1,000 × 1,000 1,000,000
7. 2.73 × 1,000 2,730
8. 0.32 × 100 32
9. 0.006 × 100 0.6
10. 6.02 × 100 602
11. 10 × 0.075 0.75
12. 10,000 × 46.0041 460,041

Divide. [68]

13. 450 ÷ 10 45
14. 3,400 ÷ 100 34
15. 65 ÷ 10 6.5
16. 2,000 ÷ 1,000 2
17. 25 ÷ 1,000 0.025
18. 64.5 ÷ 100 0.645
19. 0.35 ÷ 10 0.035
20. 1 ÷ 1,000 0.001
21. 0.03 ÷ 10 0.003
22. 80 ÷ 100 0.8
23. 953 ÷ 10 95.3
24. 645.3 ÷ 10,000 0.06453

Multiply. [40, 148]

25. 4 × 16 64
26. 5 × 12 60
27. 3 × 1,760 5,280
28. 5,280 × 2 10,560
29. $8 \times \frac{3}{4}$ 6
30. $\frac{5}{8} \times 16$ 10
31. $24 \times 2\frac{1}{3}$ 56
32. $60 \times 1\frac{1}{4}$ 75
33. $\frac{1}{2} \times 52$ 26

Divide. Write the remainder as a fraction in simplest form. [46, 48]

34. 48 ÷ 8 6
35. 128 ÷ 16 8
36. 24 ÷ 16 $1\frac{1}{2}$
37. 30 ÷ 12 $2\frac{1}{2}$
38. 330 ÷ 60 $5\frac{1}{2}$
39. 70 ÷ 36 $1\frac{17}{18}$
40. 105 ÷ 7 15
41. 108 ÷ 3 36
42. 160 ÷ 32 5

Add or subtract. [30, 32]

43. 1,760 + 567 = 2,327
44. 1,567 + 1,365 = 2,932
45. 6,630 − 2,000 = 4,630
46. 2,060 − 500 = 1,560
47. 28,159 + 306 + 5,856 = +34,321
48. 9,050 − 8,962 = 88
49. 16,849 + 82 + 106 = 17,037
50. 20,147 − 16,586 = 3,561

Measurement units have not always been standard. What problems may develop if people use different systems of measurement?

OBJECTIVE

- Formulate problems using text and data.

USING THE CHAPTER OPENER

Begin by discussing the question in terms of the students' own experiences. They may focus on sports and realize that if their school kept track and field records in "zerks" and a competing school kept them in "zonks," they wouldn't have any idea if they were about to meet a better, worse, or comparable team.

Move on to trade problems (How much grain are we getting for our money?), communications (What time is the meeting?), and so on. Students should have little trouble seeing the value of standard systems.

COOPERATIVE LEARNING

Present the following challenge.

> Create a universal measurement system to replace both metric and customary systems. It must be easy to learn!

After the groups have developed their system, they should invent tools for measuring. Tell them that they should not use existing systems to describe their new system. (They should not say, for example, that 1 "zonk" = 1 in.) They should instead select a base unit they feel is useful and then multiply or divide it to get larger and smaller units. Remind the groups that the names of the various units must be easy to remember.

When the systems and tools are complete, have the groups measure the following to test their systems' usefulness and convenience.

- The thickness of a dime
- The length of their desk
- The length of the classroom
- The distance from home to school

SUPPLEMENTARY MATERIALS

TRP Group Projects, p. 7

OBJECTIVES

- Find equivalent customary units of length.
- Measure objects to the nearest $\frac{1}{16}$ inch.

TEACHING THE LESSON

WARM-UP Present the following exercises on the chalkboard or an overhead transparency.

Multiply or divide.

1. $5{,}280 \times 3$ **2.** $3\frac{1}{2} \times 12$

3. 35.3×36 **4.** $24.6 \div 3$

5. $1{,}760 \div 100$ **6.** $54 \div 36$

(**1.** 15,840 **2.** 42 **3.** 1,270.8 **4.** 8.2 **5.** 17.6 **6.** 1.5)

INTRODUCTION Write:

■ in. = 1 ft; ■ ft = 1 yd; ■ in. = 1 yd;

Use a ruler and a yardstick to review these basic units of customary length. Extend the activity by having students use yardsticks to find:

18 in. = ■ ft   2 ft = ■ in.

INSTRUCTION In Example 1, encourage the following method of thinking:

Greater unit ⟶ smaller unit
There will be more. Multiply.

Example 2 shows the opposite case. Students should think:
Smaller unit ⟶ greater unit
There will be fewer. Divide.
Emphasize the importance of first determining the direction of change when converting units of measure.

In Example 3, you may wish to provide students with rulers marked to the nearest 16th of an inch. Elicit from the students that the marks divide each inch into 16 equal parts. Thus, the distance between each mark represents a 16th of an inch.

7.1 CUSTOMARY UNITS OF LENGTH

One type of measurement used in the United States are the Customary Units. The table shows the basic units of length.

LENGTH
12 inches (in.) = 1 foot (ft)
36 in. = 1 yard (yd)
3 ft = 1 yd
5,280 ft = 1 mile (mi)
1,760 yd = 1 mi

To change from a larger unit to a smaller unit, multiply.

EXAMPLE 1 Complete:

a. 2 mi = ■ yd

THINK: 1 mi = 1,760 yd

2 mi = (2 × 1,760) yd

2 mi = 3,520 yd

b. $6\frac{1}{2}$ ft = ■ in.

THINK: 1 ft = 12 in.

$6\frac{1}{2}$ ft = $\left(6\frac{1}{2} \times 12\right)$ in.

$6\frac{1}{2}$ ft = 78 in.

To change from a smaller unit to a larger unit, divide.

EXAMPLE 2 Complete:

a. 12 ft = ■ yd

THINK: 1 yd = 3 ft

12 ft = (12 ÷ 3) yd

12 ft = 4 yd

b. 54 in. = ■ yd

THINK: 1 yd = 36 in.

54 in. = (54 ÷ 36) yd

54 in. = $1\frac{1}{2}$ yd

When measuring, smaller units give more precise measurements.

EXAMPLE 3 What is the length of the pencil measured to the nearest $\frac{1}{16}$ in.?

1. Put the eraser at the 0 mark of the scale.

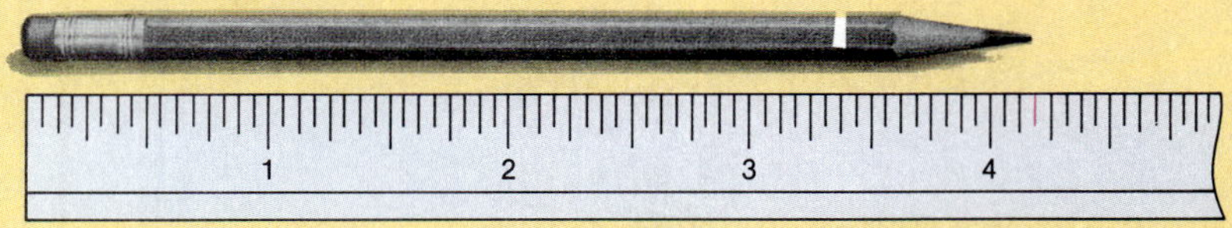

2. Read the mark on the ruler closest to the tip of the pencil.

The pencil is $4\frac{3}{16}$ in. long.

CHECKPOINT

Write the letter of the correct answer.

Complete.

1. $3\frac{1}{2}$ yd = ■ ft c **a.** $1\frac{1}{6}$ **b.** 7 **c.** $10\frac{1}{2}$

2. 48 in. = ■ yd a **a.** $1\frac{1}{3}$ **b.** 4 **c.** 1,728

PRACTICE EXERCISES

See Extra Practice, page 441.

Complete.

• **1.** 2 ft = ■ in. 24	**2.** 3 yd = ■ ft 9	**3.** 2 mi = ■ yd 3,520
4. 20 ft = ■ in. 240	**5.** 27 yd = ■ ft 81	**6.** 3 mi = ■ ft 15,840
7. 12 ft = ■ in. 144	**8.** 5 mi = ■ ft 26,400	**9.** 43 yd = ■ in. 1,548
• **10.** 36 in. = ■ ft 3	**11.** 27 ft = ■ yd 9	**12.** 5,280 yd = ■ mi 3
13. 144 in. = ■ ft 12	**14.** 2,640 ft = ■ mi $\frac{1}{2}$	**15.** 432 in. = ■ yd 12
16. 24 ft = ■ yd 8	**17.** 2,112 yd = ■ mi $1\frac{1}{5}$	**18.** 440 yd = ■ mi $\frac{1}{4}$
19. $2\frac{1}{2}$ ft = ■ in. 30	**20.** $35\frac{2}{3}$ yd = ■ in. 1,284	**21.** 66 in. = ■ yd $1\frac{5}{6}$
22. 32 in. = ■ ft $2\frac{2}{3}$	**23.** $\frac{1}{2}$ mi = ■ ft 2,640	**24.** 1,760 ft = ■ mi $\frac{1}{3}$
25. 880 yd = ■ mi $\frac{1}{2}$	**26.** 27 in. = ■ ft $2\frac{1}{4}$	**27.** $3\frac{2}{3}$ yd = ■ ft 11

Measure the line segment to the nearest $\frac{1}{16}$ in.

• **28.** __________

$\frac{3}{4}$ in.

• **29.** ______________________________

$2\frac{3}{16}$ in.

30. ____________________

$1\frac{1}{2}$ in.

31. _______________________

$1\frac{11}{16}$ in.

32. ____

$4\frac{9}{16}$ in.

Solve.

• **33.** The last event of the meet between Randolph and Fordham is the 440-yd relay. After 30 s, Randolph is at the 220-yd mark. Fordham has 760 ft left to go. Who is ahead? Randolph

34. In the 880-yd race, Frankie the Fordham Flash beats Joe "Turtle" James by 57 s. Slick Olivette beats James by 34 s. By how many seconds does Frankie beat Slick? 23 s

35. The Randolph High track team water bucket is 52 in. high. The Randolph mascot is 4 ft high. Is the mascot too tall to fit inside the bucket? no

36. For his 6-mi workout, Speed Vogel of Randolph runs 24 laps around the school track. How many yards are in each lap? 440 yd

CHECKPOINT Distribute to each student 3 index cards on which the answer choices a–c appear. After allowing students a few minutes, ask them to raise the letter that indicates their answer choice. Scan the cards and identify those students who need additional attention before proceeding to the Practice Exercises. The incorrect answer choices include common errors that students make.

COMMON ERROR

Students will sometimes use the wrong operation when converting customary units. For example, in item 2 of Checkpoint, they would select c. You may want to work through the exercises with the class, having them state whether they should multiply or divide to convert each measurement.

ASSIGNMENTS

Level 1	Even 1–18, 28–34
Level 2	Odd 5–27, 28–35
Level 3	Even 2–26, 28–36

FOLLOW-UP ACTIVITY

ESTIMATION Have students estimate, and then measure.

1. Themselves: height, nearest $\frac{1}{4}$ in.
2. Textbook: length, nearest $\frac{1}{8}$ in.
3. Desk: width, nearest $\frac{1}{2}$ in.
4. Quarter: thickness, nearest $\frac{1}{16}$ in.
5. Length: classroom, nearest foot

SUPPLEMENTARY MATERIALS

TRP Practice, p. 73

TRP Reteaching, p. 45

TRP Lesson Aids, p. 5

TRP Transparency 37

Objective

- Find equivalent customary units of capacity and weight.

Teaching the Lesson

WARM-UP Present the following exercises on the chalkboard or an overhead transparency.

Multiply or divide.

1. $3.4 \times 2{,}000$
2. $3\frac{3}{4} \times 16$
3. $2\frac{2}{3} \times 2$
4. $128 \div 32$
5. $16\frac{1}{2} \div 5$
6. $12.5 \div 2$

(**1.** 6,800 **2.** 60 **3.** $5\frac{1}{3}$ **4.** 4 **5.** 3.3 **6.** 6.25)

INTRODUCTION Display containers of the following sizes: gallon, quart, pint, cup. Write:

■ cups = 1 pt

■ pt = 1 qt

■ qt = 1 gal

Ask students if they recall how many cups equal 1 pt. Check by having a volunteer fill the cup container as many times as needed to fill the pint container. Repeat for pints and quarts and for quarts and gallons.

INSTRUCTION The computation in Example 1 is straightforward. You may wish to extend the instruction by guiding students through an example that involves computation with mixed numbers, such as:

6 qt = ■ gal $(1\frac{1}{2})$

$3\frac{1}{2}$ qt = ■ pt (7)

As you guide students through the computation in Example 2, pay particular attention to writing the remainder as a fraction.

Complete a similar example with the students:

44 oz = ■ lb $(2\frac{3}{4})$

7.2 CUSTOMARY UNITS OF CAPACITY AND WEIGHT

You can use these relationships to change from one unit to another.

CAPACITY	WEIGHT
8 fluid ounces (fl oz) = 1 cup (c)	16 ounces (oz) = 1 pound (lb)
2 c = 1 pint (pt)	2,000 lb = 1 ton (T)
2 pt = 1 quart (qt)	
4 qt = 1 gallon (gal)	

Dr. Thomas is driving a rebuilt 1964 convertible. She notices that the windshield washer is empty of fluid, so at the next gas station, she pulls over to refill it. She uses 6 c of water to refill the windshield washer. How many fluid ounces does she use?

To change from a larger unit to a smaller unit, multiply.

EXAMPLE 1 Complete: 6 c = ■ fl oz

THINK: 1 c = 8 fl oz

6 c = (6×8) fl oz

6 c = 48 fl oz

So, Dr. Thomas uses 48 fl oz of water.

To change from a smaller unit to a larger unit, divide.

EXAMPLE 2 At the gas station, Dr. Thomas remembers to buy a new oil filter for the car. The filter that Dr. Thomas buys weighs 24 oz. How many pounds does the filter weigh?

Complete: 24 oz = ■ lb

THINK: 1 lb = 16 oz

24 oz = $(24 \div 16)$ lb

24 oz = $1\frac{1}{2}$ lb

So, the filter weighs $1\frac{1}{2}$ lb.

CHECKPOINT

Write the letter of the correct answer.

Complete.

1. 3 gal = ■ qt — a — **a.** 12 **b.** 6 **c.** $\frac{3}{4}$
2. $2\frac{3}{4}$ T = ■ lb — c — **a.** 44 **b.** 2,750 **c.** 5,500
3. 28 oz = ■ lb — a — **a.** $1\frac{3}{4}$ **b.** $3\frac{1}{2}$ **c.** 448

PRACTICE EXERCISES

See Extra Practice, page 441.

Complete.

• **1.** 4 c = ■ fl oz 32
2. 5 gal = ■ qt 20
3. 2 pt = ■ c 4
4. 8 lb = ■ oz 128
5. 3 c = ■ fl oz 24
6. 2 T = ■ lb 4,000
7. 4 gal = ■ qt 16
8. 5 pt = ■ c 10
9. 4 qt = ■ pt 8
• **10.** $2\frac{1}{2}$ pt = ■ c 5
11. $4\frac{3}{8}$ lb = ■ oz 70
12. $\frac{3}{4}$ gal = ■ qt 3
13. 32 fl oz = ■ c 4
14. 4 pt = ■ qt 2
15. 8 c = ■ pt 4
16. 36 oz = ■ lb $2\frac{1}{4}$
17. 4,000 lb = ■ T 2
18. 12 qt = ■ gal 3
19. 20 fl oz = ■ c $2\frac{1}{2}$
20. 5 c = ■ pt $2\frac{1}{2}$
21. 2,500 lb = ■ T $1\frac{1}{4}$
22. 3,500 lb = ■ T $1\frac{3}{4}$
23. 16 qt = ■ gal 4
24. 12 pt = ■ qt 6
25. $2\frac{1}{4}$ lb = ■ oz 36
26. 3 T = ■ lb 6,000
27. 128 fl oz = ■ c 16
28. 1 oz = ■ lb $\frac{1}{16}$
29. $3\frac{3}{4}$ c = ■ fl oz 30
30. 15 qt = ■ gal $3\frac{3}{4}$
31. $6\frac{1}{2}$ gal = ■ qt 26
32. 12,000 lb = ■ T 6
33. 20 qt = ■ pt 40
34. $3\frac{3}{8}$ lb = ■ oz 54
35. $2\frac{1}{5}$ T = ■ lb 4,400
36. 4 fl oz = ■ c $\frac{1}{2}$
37. 1 pt = ■ fl oz 16
38. 1 qt = ■ fl oz 32
39. 1 gal = ■ fl oz 128
40. 6 c = ■ qt $1\frac{1}{2}$
41. 200 fl oz = ■ gal $1\frac{9}{16}$
42. 35 c = ■ gal $2\frac{3}{16}$

Solve.

• **43.** A 1964 convertible weighs 3,000 lb. How much does it weigh in tons? $1\frac{1}{2}$ T

44. The convertible's engine holds 6 qt of oil. How many pints is that? 12 pt

45. In 1964, gasoline sold for 30.4¢ per gallon. The convertible's tank holds 17 gal. How much did it cost to fill up the car in 1964? $5.17

46. In highway driving, the convertible uses up an average of 8 fl oz of gas each mile. How many miles per gallon does the convertible get? 16 mi/gal

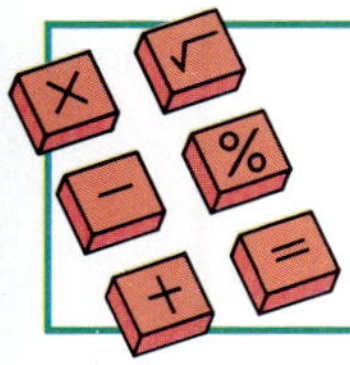

CALCULATOR

Mrs. Hoffman pulls into a gas station for some unleaded gasoline. The gasoline sells for $1.10 per gallon. Mrs. Hoffman buys $15 worth. How many gallons did she get? Round to the nearest tenth. 13.6 gal

CHECKPOINT The incorrect answer choices include common errors that students make.

CALCULATOR This is a situation from everyday life in which a calculator is very helpful. Ask how to find the answer without a calculator. Discuss ways of estimating.

COMMON ERROR

Some students will use the wrong operation when converting units. For example, in item 3 of Checkpoint, they would select c. Work individually with these students to isolate and remediate the source of difficulty. For instance, some students may not understand that the unit of measure to which they are converting is either greater or smaller than the given measurement.

ASSIGNMENTS

Level 1	Odd 1–27, 43–44, CA
Level 2	Odd 1–41, 43–45, CA
Level 3	Even 10–42, 43–46, CA

FOLLOW-UP ACTIVITY

COOPERATIVE LEARNING Have students work in small groups. Using newspapers, catalogs, flyers, and their own experience, have students list products that are sold by the pint, quart, gallon, ounce, pound, and ton. Have the groups share their lists in a follow-up discussion.

Extend the activity by having students find equivalent measures. For example, ask how many quarts would equal 3 gal of paint.

SUPPLEMENTARY MATERIALS

TRP Practice, p. 74

TRP Reteaching, p. 46

TRP Transparency 37

OBJECTIVE

- Add and subtract customary measures of length, weight, and capacity.

TEACHING THE LESSON

WARM-UP Present the following exercises on the chalkboard or an overhead transparency.

Complete.

1. 3 yd 5 ft = 4 yd ■ ft (2)
2. 6 gal 5 qt = 7 gal ■ qt (1)
3. 7 lb 17 oz = 8 lb ■ oz (1)
4. 6 T 450 lb = 5 T ■ lb (2,450)
5. 4 mi 300 yd = 3 mi ■ yd (2,060)

INTRODUCTION Write the following examples on the chalkboard.

```
  67        6 ft 7 in.
 +38       +3 ft 8 in.
```

Ask students whether they would add the same way in both examples.

INSTRUCTION In Example 1, you may wish to provide students with an alternate explanation of regrouping.

6 ft 16 in. = 6 ft + 16 in.
= 6 ft + (12 in. + 4 in.)
= (6 ft + 1 ft) + 4 in.
= 7 ft + 4 in.
= 7 ft 4 in.

In discussing Example 2, point out that to regroup 9 ft to smaller units, the following procedure can be used:

9 ft = 8 ft + 1 ft
= 8 ft + 12 in.
= 8 ft 12 in.

CHECKPOINT The incorrect answer choices include common errors that students make.

7.3 COMPUTING WITH CUSTOMARY MEASURES

The stage manager for the rock band *Deffening* has to make 2 shelves for guitar cases that are 3 ft 8 in. long. How much shelving does the stage manager need? If wood shelving is sold only in 9-ft lengths, how much wood will be left over?

You can add to find how much shelving the stage manager needs.

EXAMPLE 1 Add: 3 ft 8 in. + 3 ft 8 in.

1. Add the inches. Regroup.

THINK: 8 in. + 8 in. = 16 in.
16 in. = 1 ft 4 in.

```
   1 ft
  3 ft 8 in.
 +3 ft 8 in.
       4 in.
```

2. Add the feet.

```
   1 ft
  3 ft 8 in.
 +3 ft 8 in.
  7 ft 4 in.
```

So, the stage manager needs 7 ft 4 in. of shelving.

You can subtract to find how much wood will be left over.

EXAMPLE 2 Subtract: 9 ft − 7 ft 4 in.

1. Regroup.

THINK: 9 ft = 8 ft 12 in.

```
  8 ft  12 in.
  9 ft  0 in.
 −7 ft  4 in.
```

2. Subtract.

```
  8 ft  12 in.
  9 ft  0 in.
 −7 ft  4 in.
  1 ft  8 in.
```

There will be 1 ft 8 in. of wood left over.

CHECKPOINT

Write the letter of the correct answer.

Add or subtract.

	Exercise	Answer	a.	b.	c.
1.	5 ft 9 in. + 4 ft 7 in.	b	1 ft 2 in.	10 ft 4 in.	10 ft 6 in.
2.	4 yd 1 ft − 3 yd 2 ft	c	1 yd 2 ft	1 yd 1 ft	2 ft
3.	1 lb 12 oz + 12 oz	c	1 lb	2 lb 4 oz	2 lb 8 oz

PRACTICE EXERCISES

See Extra Practice, page 441.

Add or subtract. See page 478 for answers to Exercises 1–28.

1. 3 yd 1 ft +4 yd 1 ft	**2.** 4 lb 3 oz +7 lb 5 oz	**3.** 1 ft 3 in. +1 ft 8 in.	**4.** 6 gal 1 qt +18 gal 2 qt
5. 2 yd 2 ft +1 yd 2 ft	**6.** 6 gal 3 qt +7 gal 2 qt	**7.** 1 pt 1 c +1 pt 1 c	**8.** 3 mi 2,875 ft +6 mi 2,475 ft
9. 6 ft 2 in. −4 ft 1 in.	**10.** 3 yd 2 ft −2 yd 1 ft	**11.** 5 c 7 fl oz −3 c 3 fl oz	**12.** 4 mi 1,300 yd −3 mi 500 yd
13. 7 qt −5 qt 1 pt	**14.** 5 lb 6 oz −3 lb 7 oz	**15.** 2 gal 1 qt −1 gal 2 qt	**16.** 6 T 450 lb −4 T 780 lb
17. 5 T 750 lb −3 T 460 lb	**18.** 3 pt −1 pt 4 fl oz	**19.** 1 pt 5 fl oz +1 pt 3 fl oz	**20.** 8 mi 567 yd +7 mi 56 yd
21. 8 lb 5 oz −3 lb 8 oz	**22.** 7 lb 4 oz +9 lb 8 oz	**23.** 5 T 1,500 lb +4 T 375 lb	**24.** 5 mi 1,350 ft −2 mi 2,000 ft
25. 1 c 3 fl oz + 6 fl oz	**26.** 12 T 1,100 lb − 5 T 1,500 lb	**27.** 7 yd 1 ft −3 yd 2 ft	**28.** 1 ft 9 in. +1 ft 9 in.

Solve.

29. *Deffening* is giving a concert so loud that you can hear it 6 mi away. The Morosco sisters are late. They have been able to hear the band for the last 3 mi 650 yd. How far are they from the arena? 2 mi 1,110 yd

30. Lead guitarist Ogre Krebs starts the show wearing 12 lb 6 oz of gold. During a solo, he takes off a gold chain weighing 5 lb 12 oz. How much gold is he wearing now? 6 lb 10 oz

31. When the Morosco sisters arrive, they order refreshments. Rosie buys a 2-qt Big Deal, while Rhonda orders a 72-fl oz Thirst Thrasher. Which sister ordered more? Rhonda

32. The *Deffening* stage speakers consist of 3 ft 7 in. tweeters stacked up on top of woofers that are twice as high. How tall is each speaker? 10 ft 9 in.

TIME OUT

This is *Deffening*'s "My Sink Says Drip" Tour. The stage show features a giant 20-ft tall kitchen sink that drips during the 210-min concert from a 35-gal water tank hidden behind the faucet. The faucet drips at a constant rate and is timed to empty the tank just as the show ends. How much water drips from the faucet each minute? How many gallons are left in the tank after the band has played for 120 min? $\frac{1}{6}$ gal/min; 15 gal left

TIME OUT Guide students in writing the fraction $\frac{35}{210}$ to find the number of gallons per minute that drip from the faucet. Then students can apply their knowledge of multiplying fractions to find the number of gallons left after 120 min.

COMMON ERROR

When computing with customary units of measurement, some students will regroup as in base 10. For example, in item 1 of Checkpoint, they would select c.

For these students, provide more practice in renaming customary units of measure before assigning the exercises. For instance, have students match a list of equivalent measurements such as 2 ft 6 in. and 1 ft 18 in. After completing such exercises successfully, have them practice regrouping measurements with exercises similar to those in the Warm-Up.

ASSIGNMENTS

Level 1	Odd 1–31
Level 2	Even 2–28, 29–31, TO
Level 3	Odd 5–27, 29–32, TO

FOLLOW UP ACTIVITY

COOPERATIVE LEARNING Have students work in small groups and plan the menu for a birthday or graduation party. Instruct them to estimate the amount of food or beverage needed to serve 24 people. Have the groups share their menus with the class. Discuss the methods students used to estimate the amounts of each item on their menus.

SUPPLEMENTARY MATERIALS

TRP Practice, p. 75

TRP Reteaching, p. 47

TRP Transparency 38

OBJECTIVES

- Find equivalent units of time.
- Compute elapsed time.

TEACHING THE LESSON

WARM-UP Present the following exercises on the chalkboard or an overhead transparency.

Multiply or divide.

1. 2 × 365 **2.** 4 × 52

3. 15 × 60 **4.** 104 ÷ 52

5. 120 ÷ 60 **6.** 48 ÷ 24

(**1.** 730 **2.** 208 **3.** 900 **4.** 2 **5.** 2 **6.** 2)

INTRODUCTION Ask volunteers to describe their methods for solving this problem.

You are driving to visit your best friend who moved away 6 months ago. You leave your house at 8:30 A.M. and arrive at your friend's house at 1:15 P.M. How long was the trip? (4 h 45 min)

INSTRUCTION In Example 1, work through the following steps for finding elapsed time:

1. Change the times to hours and minutes.
2. If one time is 1:00 P.M. or later, add 12 hours to the time.
3. Subtract the like units, regrouping if necessary.

Some students may become confused when changing P.M. times to hours and minutes. Point out that times such as 13.00 and 18.30 exist in the 24-hour clock used by the military and in many foreign countries. Explain how the 24-hour clock works.

Use the computation in Example 2 as a starting point for more practice in regrouping units of time. Have students complete similar examples:

3 wk 4 d = 2 wk ■ d (11 d)

75 s = ■ min ■ s (1 min 15 s)

2 d 6 h = 1 d ■ h (30 h)

7.4 COMPUTING WITH TIME

The standard unit of measurement for time is the day. A **day** is the time it takes the Earth to rotate one full turn on its axis. A **year** is the time it takes the Earth to revolve around the sun.

The 24 h of the day are separated into two 12-h time periods: before noon (A.M.) and after noon (P.M.).

To find elapsed time, subtract. Sometimes it helps to use a 24-h clock.

TIME	
60 seconds (s) =	1 minute (min)
60 min =	1 hour (h)
24 h =	1 day (d)
7 d =	1 week (wk)
365 d =	1 year (y)
52 wk =	1 y
12 months (mo) =	1 y

EXAMPLE 1 At the Memphis railway station, Trevor Rivera boards the City of New Orleans at 8:52 A.M. He arrives in Jackson at 1:19 P.M. How long did the trip take?

1. Change both times to hours and minutes.

THINK: 1:19 P.M. = 1 h 19 min + 12 h = 13 h 19 min

```
   13 h 19 min
 −  8 h 52 min
```

2. Regroup. Subtract.

THINK: 13 h 19 min = 12 h 79 min

```
   12 h  79 min
   1̸3 h 1̸9 min
 −  8 h 52 min
    4 h 27 min
```

The trip took 4 h 27 min.

To find the end time, add.

EXAMPLE 2 The City of New Orleans pulls into Kankakee at 9:32 A.M. to pick up some freight. The loading takes 4 h 40 min. At what time is the last of the freight loaded aboard?

1. Change the starting time to hours and minutes.

THINK: 9:32 A.M. = 9 h 32 min

```
   9 h 32 min
 + 4 h 40 min
```

2. Add the minutes and regroup.

72 min = 1 h 12 min

```
   1 h
   9 h 32 min
 + 4 h 40 min
       12 min
```

3. Add the hours. Change to clock time.

14 h 12 min − 12 h = 2 h 12 min = 2:12 P.M.

```
   1 h
   9 h 32 min
 + 4 h 40 min
  14 h 12 min
```

The last of the freight is loaded aboard at 2:12 P.M.

FOR DISCUSSION See TE side column.

European train schedules all use the 24-h clock. So does the military. What are some of its advantages?

PRACTICE EXERCISES See Extra Practice, page 442.

You get on the train at 8:35 A.M. How long does your trip take if you arrive at:

• **1.** 8:52 A.M.? 17 min **2.** 9:17 A.M.? 42 min **3.** 9:26 A.M.? 51 min **4.** 10:47 A.M.? 2 h 12 min.

You get on the train at 12 noon. How long does it take to get home if you arrive at:

5. 4:29 P.M.? 4 hr 29 min **6.** 7:17 P.M.? 7 h 17min **7.** 12 midnight? 12 h **8.** 8:39 A.M.? 20 h 39 min

The time is now 9:45 P.M. What will be the correct time in:

• **9.** $\frac{3}{4}$ hour? 10:30 P.M. **10.** 3 hours 15 minutes? 1:00 A.M. **11.** 5 hours 25 minutes? 3:10 A.M.

The time is now 5:20 A.M. What was the correct time:

12. 2 h 47 min ago? 2:33 A.M. **13.** $8\frac{1}{2}$ h ago? 8:50 P.M. **14.** 33 min ago? 4:47 A.M.

Complete the schedule of the Northeast Corridor Rail Service.

	Route	Departure Time	Arrival Time	Length of Trip
•15.	Boston–New York	7:32 A.M.	11:05 A.M.	3 h 33 min ■
16.	Boston–Washington	8:29 A.M.	3:47 P.M.	7 h 18 min ■
•17.	Philadelphia–Miami	10:27 A.M.	■ 5:54 A.M.	19 h 27 min
18.	New York–Atlanta	1:40 P.M.	4:36 A.M.	14 h 56 min ■
19.	Savannah–Richmond	■ 3:12 P.M.	12 midnight	8 h 48 min
20.	Buffalo–Cleveland	9:52 P.M.	1:06 P.M.	15 h 14 min ■

Solve.

• **21.** The City of New Orleans stops at 11:37 A.M. to allow another train to pass. The City of New Orleans starts up again 1 h 6 min later. At what time does the train start up again? 12:43 P.M.

22. There is a jukebox in the dining car. Rivera puts in a quarter and picks 2 songs that are each 2 min 50 s long. How long does the jukebox play Rivera's songs? 5 min 40 s

23. The City of New Orleans runs on a special summer schedule from May 24 until September 6. How many weeks does the train run on its special schedule? 15 wk

24. It took Rivera 18 min to get from the train to a taxi and another 24 min to get to his hotel. If the train arrived at 1:19 P.M., when did Rivera get to his hotel? 2:01 P.M.

FOR DISCUSSION Students should explain that the chief advantage of the 24-hour clock is to eliminate confusion between A.M. and PM. times.

COMMON ERROR

Many students will make errors when subtracting to find the elapsed time. Subtraction errors occur when one time is in the A.M. period and the other in the P.M. period.

Counting forward is an alternate method that may reduce this error. For example, to find how much time has elapsed from 11:50 A.M. to 2:06 P.M.:

- Count forward 10 minutes to get to noon.
- Count forward 2 hours to get to 2:00 P.M.
- Add the remaining 6 minutes to get to 2:06 P.M.

The elapsed time is 2 h 16 min.

ASSIGNMENTS

Level 1	1–14, Odd 15–19, 21–22
Level 2	Even 2–20, 21–24
Level 3	9–24

FOLLOW-UP ACTIVITY

MENTAL MATH Have students write the corresponding P.M. times by subtracting 12 mentally from the following times.

14.00 hours (2 P.M.)
17.30 hours (5:30 P.M.)
20.40 hours (8:40 P.M.)
18.35 hours (6:35 P.M.)
13.01 hours (1:01 P.M.)
23.55 hours (11:55 P.M.)

SUPPLEMENTARY MATERIALS

TRP Practice, p. 76
TRP Reteaching, p. 48
TRP Transparency 38

OBJECTIVES

- Solve problems that involve time zones.
- Solve problems that involve reading and interpreting travel schedules.

TEACHING THE LESSON

WARM-UP Read the following exercises aloud.

How many hours and minutes pass between:

1. 7:15 A.M. and 8:45 A.M.? (1h 30 min)
2. 8:45 A.M. and 11:00 A.M.? (2 h 15 min)
3. 11:00 A.M. and 1:35 P.M.? (2 h 35 min)
4. 1:35 P.M. and 3:12 P.M.? (1 h 37 min)

INTRODUCTION Ask a student what time it is. Then, using the name of a large American city in a different time zone, ask whether the time is the same in that city. Ask a volunteer to explain why the time is different in that city.

Then ask students if they have ever traveled by plane. Encourage them to discuss their experiences using airline schedules.

INSTRUCTION As students read and discuss the text material, emphasize that a move west from one time zone to the next yields a time that is 1 hour earlier. A move east yields a time that is 1 hour later.

Before assigning the exercises, discuss the paragraph on page 189 that relates to Problems 11–15. Also discuss the flight schedule and the abbreviations shown below it.

PROBLEM Solving APPLICATION

7.5 CONSUMER: TIME ZONES AND TRAVEL SCHEDULES

When you call a person who lives in another part of the United States or in another country, be sure you know what time it is there. Remember, there are 24 time zones in the world.

You are working for a company in Washington, D.C. You want to call Mr. Jones as soon as he arrives at his office in Los Angeles, California. He arrives at his office at 8 A.M. Pacific Time. At what time should you make the call?

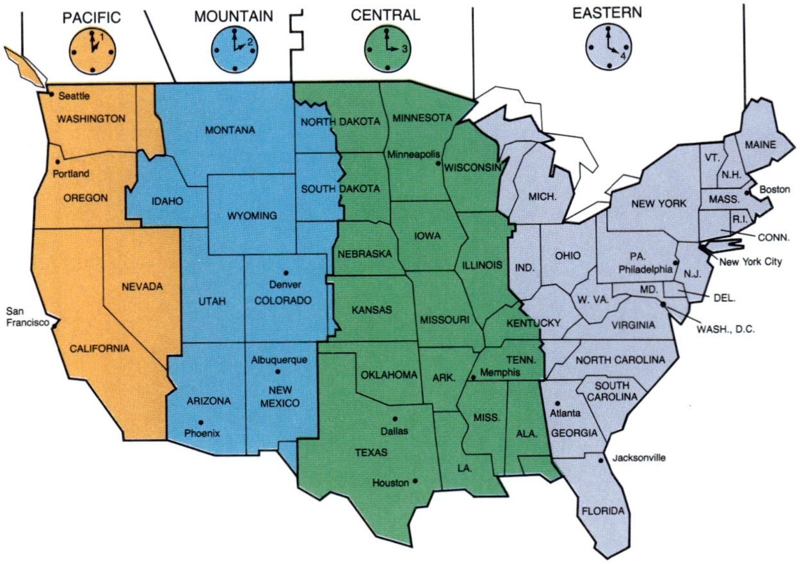

As the map shows, there is a 3-hour time difference between the Eastern zone and the Pacific zone. The time in Washington is 3 hours later than the time in Los Angeles.

So, to call Mr. Jones when it is 8 A.M. in Los Angeles, you should make the call at 11 A.M. Eastern Time.

Use the map above to answer Exercises 1–10.

If the time is 3 P.M. in New York, NY, what time is it in:

• **1.** Minneapolis, MN? 2 P.M. **2.** Albuquerque, NM? 1 P.M. **3.** Seattle, WA? 12 P.M.

If the time is 1 P.M. in Reno, NV, what time is it in:

4. Denver, CO? 2 P.M. **5.** Dallas, TX? 3 P.M. **6.** Atlanta, GA? 4 P.M.

Suppose the time is:	What time is it in:
• **7.** 9:30 A.M. in Boston, MA	Houston, TX? 8:30 A.M.
8. 7:05 P.M. in Phoenix, AZ	Jacksonville, FL? 9:05 P.M.
9. 1:15 P.M. in Philadelphia, PA	San Francisco, CA? 10:15 A.M.
10. 11:08 P.M. in Portland, OR	Memphis, TN? 1:08 A.M.

Often when you plan a trip by plane, you need to be aware of the different time zones. A flight schedule shows departures and arrivals in terms of the local time at each city.

Use the flight schedule at the right to answer Exercises 11–15.

PHILADELPHIA, PA				
Leave	Arrive	Flt.	Stops/Meal	
TO: Chicago, IL				
6:30a	7:56a	181	NS	M
8:00a	9:30a	197	NS	M
10:00a	11:16a	207	NS	S
12:00p	1:15p	219	NS	M
TO: Denver, CO				
6:30a	10:15a	181/317	C	M
10:00a	4:05p	207/365	C	M

a-A.M. p-P.M.
NS-Nonstop C-Connection
M-Meal S-Snack

• **11.** Mrs. Carelli plans to fly from Philadelphia to Denver. Does the schedule show any nonstop flights? no

12. Mrs. Carelli takes the 6:30 A.M. flight to Denver. How many hours will it take? 5 h 45 min

13. Is Flight 181 to Chicago less than 2 hours? If not, how long is it? no—2 h 26 min

14. How long is Flight 219? 2 h 15 min

15. How much longer is the 10 A.M. flight to Denver than the 6:30 A.M. flight? 2 h 20 min

ASSIGNMENTS

Level 1	1–8, 11–12
Level 2	1–15
Level 3	Even 2–10, 11–15

FOLLOW-UP ACTIVITY

ENRICHMENT Have the students use encyclopedias or other reference resources to find additional information about time zones and differences in time around the world. Challenge them to answer these questions by using the reference resources.

When it is 1:00 P.M. in New York, what time is it in:

1. London? (6:00 P.M.)

2. Moscow? (8:00 P.M.)

3. Tokyo? (3:00 A.M.)

4. Hawaii? (8:00 A.M.)

When it is 1:00 P.M. in Berlin, what time is it in:

5. Chicago? (6:00 A.M.)

6. Anchorage? (2:00 A.M.)

7. Manila? (8:00 P.M.)

8. Calcutta? (6:00 P.M.)

SUPPLEMENTARY MATERIALS

TRP Practice, p. 77

TRP Transparency 39

OBJECTIVE

- Evaluate student progress.

USING THE REVIEW

This page provides a means for informally evaluating students' understanding of the skills and concepts covered so far in this chapter.

Have the students look at the page to familiarize themselves with the various question formats that are presented. Discuss any questions that they may have. Then ask them to complete the page independently.

In addition to grading them individually, you may wish to review the answers to the questions collectively with the students.

Page references appear in brackets. They refer to pages on which a particular skill was introduced.

Before continuing on to the topics found in the remainder of the chapter, you may wish to have students review any skills or concepts in which they have demonstrated weakness.

MIDCHAPTER REVIEW

Complete. [180, 182, 186]

1. 5 ft = ■ in. 60 **2.** 3 T = ■ lb 6,000 **3.** 8 fl oz = ■ pt $\frac{1}{2}$

4. 60 oz = ■ lb $3\frac{3}{4}$ **5.** $2\frac{1}{2}$ yd = ■ ft $7\frac{1}{2}$ **6.** 2 gal = ■ fl oz 256

7. 7 qt = ■ c 28 **8.** 5,280 yd = ■ mi 3 **9.** 180 s = ■ min 3

10. 12 h = ■ d $\frac{1}{2}$ **11.** 5,000 lb = ■ T $2\frac{1}{2}$ **12.** 1 wk = ■ h 168

13. 25 ft = ■ yd $8\frac{1}{3}$ **14.** 64 c = ■ gal 4 **15.** 3 h = ■ min 180

16. $6\frac{3}{4}$ lb = ■ oz 108 **17.** $4\frac{1}{2}$ qt = ■ fl oz 144 **18.** 66 in. = ■ ft $5\frac{1}{2}$

Measure the line segment to the nearest $\frac{1}{16}$ in. [180]

19. ________________________ $2\frac{7}{16}$ in. **20.** ____ $\frac{3}{8}$ in.

21. __________ 1 in. **22.** ________________ $1\frac{9}{16}$ in.

Add or subtract. [184]

23. 7 T 500 lb − 6 T 350 lb = 1T 150 lb

24. 8 pt − 6 pt 1 c = 1 pt 1c

25. 3 gal 3 qt + 1 gal 2 qt = 5 gal 1 qt

26. 3 mi 855 yd + 3 mi 855 yd = 6 mi 1,710 yd

Complete the Cinema 6 schedule. [186] **27.** 1 h 35 min; **28.** 1 h 52 min; **29.** 1:02 A.M.; **30.** 2 h 56 m **31.** 10:12 P.M.; **32.** 12:22 A.M.

	Movie	Begins	Ends	Length of Film
27.	"Brown Cars, Silver Bridges"	1:30 P.M.	3:05 P.M.	■
28.	"Don't Worry, Baby"	8:20 P.M.	10:12 P.M.	■
29.	"In Concert: Deffening"	10:45 P.M.	■	2 hours 17 minutes
30.	"Lonely Is as Lonely Does"	12:40 P.M.	3:36 P.M.	■
31.	"Bach to Bach"	■	12 midnight	1 hour 48 minutes

Use the flight schedule at the right to answer Exercises 32–33. [188]

32. John Allen takes Flight 118. How long is his flight? 2 hours

33. Susan Raimes wants to arrive in Chicago as close to 11:00 A.M. as possible. Which flight should she take? 121

Leave	Arrive	Flt.	Stops/Meal	
TO: Chicago, IL				
8:00a	10:00a	118	NS	M
10:00a	11:40a	121	NS	S
11:00a	12:45p	201	NS	S

PROBLEM Solving APPLICATION

7.6 CONSUMER: HOURLY WAGES AND OVERTIME PAY

In some jobs, employees are paid an hourly rate for the first 40 hours that they work in a week. For every hour over 40 hours, they are paid an **overtime rate** calculated at "time and a half."

Suppose your rate of pay is $5.00 per hour and you work 42 hours one week. What is your total pay including overtime?

1. Multiply to find the regular pay. $40 \times \$5.00 = \$200.00$
2. Find the overtime pay.
   - Subtract to find the overtime hours. $42 - 40 = 2$
   - Multiply to find the overtime rate.

     THINK: $1\frac{1}{2} = 1.5$ $1.5 \times \$5.00 = \$7.50$
   - Multiply to find the total overtime pay. $2 \times \$7.50 = \$15.00$
3. Add to find the total pay. $\$200.00 + \$15.00 = \$215.00$

So, your total pay for the week is $215.00.

Find the total pay for the week including overtime pay.
Remember to estimate whenever you use your calculator.

	Number of Hours Worked		Rate per Hour		Number of Hours Worked		Rate per Hour
• **1.**	42	$210.70	$4.90	• **2.**	44	$239.20	$5.20
3.	48	$291.20	$5.60	**4.**	40	$192.00	$4.80
5.	41	$186.75	$4.50	**6.**	45	$413.25	$8.70
7.	47	$297.95	$5.90	**8.**	43.5	$298.65	$6.60

Solve.

• **9.** Your friend earns $4.80 per hour. One week he works 46 hours. What is his total pay including overtime? $235.20

10. Your cousin earns $485.00 per week for 40 weeks of the year. For the other 12 weeks, she works overtime and earns $533.50 per week. What is her total pay per year? What is her average total pay per week?
$25,802 per year; $496.19 per week

OBJECTIVE

- Solve problems that involve finding gross pay, including overtime.

TEACHING THE LESSON

WARM-UP Write the following exercises on the chalkboard or an overhead transparency.

Find the answer.

1. $9 \times \$5.40$ **2.** $40 \times \$4.60$
3. $\$192.00 + \$14.40$ **4.** $\$11{,}492 \div 52$

(**1.** $48.60 **2.** $184 **3.** $206.40 **4.** $221)

INTRODUCTION Ask a volunteer to explain the meaning of gross earnings. (Gross earnings, deductions, and net pay were introduced in Lesson 3.3.)

INSTRUCTION Give careful attention to the concept of "time and a half" paid for the hours over 40 hours that are worked in a week.

Make sure that students understand that there is no overtime pay in Problem 4, since the number of hours worked does not exceed 40. Also note that Problem 10 is a multistep problem that requires multiplication, addition, and division.

You may wish to have students use calculators to solve the problems in this lesson.

ASSIGNMENTS

Level 1	1–6, 9
Level 2	1–10
Level 3	5–10

SUPPLEMENTARY MATERIALS

TRP Practice, p. 78

TRP Transparency 39

OBJECTIVES

- Estimate lengths of objects in metric measures.
- Measure lengths of objects to the nearest centimeter and millimeter.

TEACHING THE LESSON

WARM-UP Draw the line segments of the following lengths on the chalkboard or overhead transparency.

Measure to the nearest sixteenth of an inch.

1. $5\frac{1}{2}$ in. **2.** $8\frac{1}{4}$ in.

3. $6\frac{5}{8}$ in. **4.** $7\frac{13}{16}$ in.

(Check students' drawings.)

INTRODUCTION You may wish to explain that the metric system was developed by a commission of French scientists in the eighteenth century. France adopted the metric system in 1799 and made its use compulsory in 1837. Since then, it has been revised several times. The present form was adopted in 1960 and is called the International System of Units.

INSTRUCTION Discuss the examples of kilometer, meter, centimeter, and millimeter. List other examples of these units.

For Example 1, point out that it is sometimes necessary to refer to a familiar measurement to determine the appropriate unit for measuring other objects.

You may wish to provide your students with metric rulers when discussing Example 2. Have them count the number of marks between 0 and 1, 1 and 2, etc.

FOR DISCUSSION Guide students to realize that a tape measure marked only in meters is used to measure to the nearest meter. Thus, everyone in the class would probably be 2 m tall. The tape measure marked in centimeters would give a more precise measure of a person's height.

7.7 METRIC UNITS OF LENGTH

The table below shows the most commonly used metric units of length.

kilometer (km) 1,000 m	meter (m) 1 m	centimeter (cm) 0.01 m	millimeter (mm) 0.001 m
	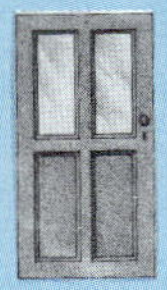 1 m	 1 cm	 1 mm
A kilometer is about the length of 12 city blocks.	A meter is about the width of a door.	A centimeter is about the width of a bandaid.	A millimeter is about the thickness of a dime.

To choose a reasonable estimate, determine the most likely unit of measure.

EXAMPLE 1 Choose the most reasonable estimate for the width of a shoelace.

a. 4 mm b. 4 cm c. 4 m d. 4 km

THINK: The width of a shoelace is several times thicker than a dime.

Since a dime is 1 mm thick, the most reasonable estimate for the width of a shoelace is 4 mm.

The smaller the unit, the more precise the measurement.

EXAMPLE 2 What is the length of the stamp measured to the nearest millimeter?

1. Put the 0 mark of the scale at the left edge of the stamp.

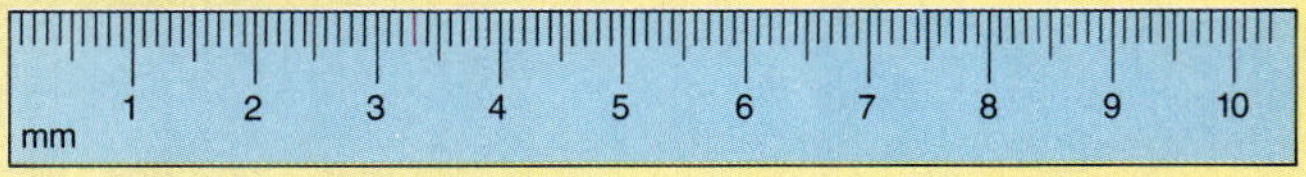

2. Read the mark closest to the right edge of the stamp.

The length of the stamp is 33 mm.

FOR DISCUSSION See TE side column.

To measure the height of several people, would you use a tape measure marked only in meters or one marked only in centimeters? Explain.

PRACTICE EXERCISES

See Extra Practice, page 442.

Choose the most reasonable estimate.

		Answer	a.	b.	c.
• 1.	length of a golf club	b	a. 1 km	b. 1 m	c. 1 cm
• 2.	height of an office building	b	a. 100 km	b. 100 m	c. 100 mm
3.	length of your textbook	b	a. 24 m	b. 24 cm	c. 24 mm
4.	thickness of your textbook	c	a. 30 m	b. 30 cm	c. 30 mm
5.	distance from Detroit to Miami	a	a. 2,000 km	b. 2,000 m	c. 2,000 cm
6.	height of a one-story house	b	a. 2 km	b. 4 m	c. 200 cm
7.	height of a ceiling	b	a. 0.5 km	b. 3 m	c. 122 cm
8.	length of a marathon course	a	a. 44 km	b. 1,800 m	c. 3,500 cm
9.	thickness of a wallet	c	a. 2 m	b. 5 cm	c. 12 mm
10.	length of a tennis racket	b	a. 2 m	b. 70 cm	c. 200 mm

Use a ruler to measure the line segment or object to the nearest centimeter and the nearest millimeter.

•11. ____________ 3 cm; 27 mm

•12. ____________________________ 7 cm; 66 mm

13. ___ 1 cm; 7 mm

14. ____________________ 5 cm; 45 mm

15. ____ 12 cm; 121 mm

Answers will vary for Exercises 16–19.

16. length of a pen

17. width of your desk

18. length of a new piece of chalk

19. length of a paper clip

Solve.

•20. Ralph Plotnick spills some coffee on the blueprint for his shed. Some labels are smudged. The blueprint lists the window size as 175 by 75. To which metric unit of length does the blueprint likely refer? centimeters

21. Ralph needs 6 sheets of paper towels to clean up the 8 fl oz of coffee he spilled. On average, how much coffee does each sheet pick up? $1\frac{1}{3}$ fl oz

22. The waitress gives Ralph a new cup of coffee and they begin to talk. She asks Ralph which metric unit he would be likely to use to measure the height of the letters on the menu. What should Ralph answer? millimeters

23. The cook prides himself on his skill in estimating the heights of customers while they are still seated. Ralph is about medium height. The cook estimates Ralph's height to be 245 cm. Is the estimate reasonable? no

COMMON ERROR

Some students will confuse the prefixes *kilo*, *centi*, and *milli*. Point out that *centi* can be thought of as *cent*, meaning one-hundredth (0.01) of a dollar. *Milli* is similar to *mill*, meaning one-thousandth (0.001) of a dollar.

ASSIGNMENTS

Level 1	1–21
Level 2	1–22
Level 3	1–23

FOLLOW UP ACTIVITY

APPLICATION Have students complete the following worksheet:

PERSONAL DATA

Age: ■ y

Height: ■ m

Weight: ■ kg

Length of reach: ■ cm

Length of stride: ■ cm

SUPPLEMENTARY MATERIALS

TRP Practice, p. 79

TRP Reteaching, p. 49

TRP Lesson Aids, p. 5

TRP Transparency 40

OBJECTIVES

- Choose an appropriate metric unit to measure the mass or capacity of an object.
- Estimate the mass or capacity of objects in metric units.

TEACHING THE LESSON

WARM-UP Write the following statement on the chalkboard or an overhead transparency.

The larger an object, the heavier it is.

Ask how many students think that this statement is true. Then write:

1. pillow or bowling ball
2. brick or inflatable raft
3. beach ball or can of paint
4. paper clip or feather

In each case, have students name the object that is heavier. Discuss why a smaller object can sometimes be heavier than a larger one. (It is density rather than size that determines an object's mass.)

INTRODUCTION Display an aquarium (or any other container that holds more than 1 L). Hold up a liter container and ask how many liters of water are needed to fill the aquarium. List the estimates and ask students to tell how they arrived at their estimates. Find the most reasonable estimate by filling the aquarium with water. Repeat, with containers of various sizes. Discuss the difficulty of estimating the capacity of the various containers and ways of making the estimation more accurate.

7.8 METRIC UNITS OF MASS AND CAPACITY

The table below shows the most commonly used metric units of mass. The gram (g) is the basic unit.

kilogram (kg)
1,000 g

A kilogram is about the mass of a textbook.

gram (g)
1 g

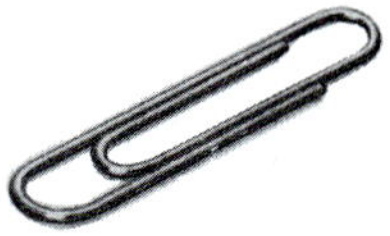

A gram is about the mass of a paper clip.

milligram (mg)
0.001 g

A milligram is about the mass of a seed.

To choose a reasonable estimate for the mass of an object, determine the most likely unit of measure.

EXAMPLE 1 Choose the most reasonable estimate for the mass of a quarter.

a. 10 kg b. 10 g c. 10 mg

THINK: The mass of a quarter is several times greater than the mass of a paper clip.

Since the mass of a paper clip is 1 g, then the most reasonable estimate for the mass of a quarter is 10 g.

The table shows the most commonly used metric units of capacity. The liter (L) is the basic unit.

kiloliter (kL)
1,000 L

A kiloliter is about the amount of water in a small pond.

liter (L)
1 L

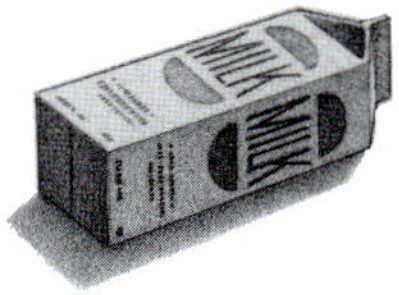

A liter is the amount of liquid in a liter container (or carton).

milliliter (mL)
0.001 L

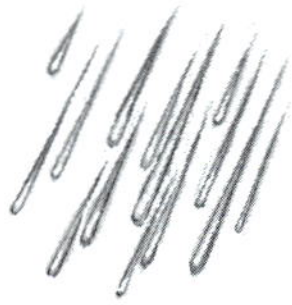

A milliliter is about the amount of water in a raindrop.

To choose a reasonable estimate for the capacity of an object, determine the most likely unit of measure.

EXAMPLE 2 Choose the most reasonable estimate for the amount of ink in a ballpoint pen. a. 125 kL b. 125 L c. 125 mL

THINK: The amount of ink in a ballpoint pen is greater than that of a raindrop, but less than that of a liter bottle.

Since the capacity of a raindrop is 1 mL, then the most reasonable estimate for the amount of ink in a ballpoint pen is 125 mL.

PRACTICE EXERCISES See Extra Practice, page 442.

Choose the most reasonable estimate.

•	**1.** mass of a light bulb	b	**a.** 40 kg	**b.** 40 g	**c.** 40 mg
	2. mass of a compact car	a	**a.** 1,000 kg	**b.** 1,000 g	**c.** 1,000 mg
	3. mass of a chicken leg	b	**a.** 125 kg	**b.** 125 g	**c.** 125 mg
	4. mass of a tennis ball	b	**a.** 80 kg	**b.** 80 g	**c.** 80 mg
	5. mass of a television	a	**a.** 25 kg	**b.** 25 g	**c.** 25 mg
	6. mass of a toothpick	c	**a.** 15 kg	**b.** 15 g	**c.** 15 mg
•	**7.** capacity of a vase	c	**a.** 800 kL	**b.** 800 L	**c.** 800 mL
	8. capacity of a swimming pool	a	**a.** 75 kL	**b.** 75 L	**c.** 75 mL
	9. capacity of a canteen	b	**a.** 0.6 kL	**b.** 0.6 L	**c.** 0.6 mL
	10. capacity of a sink	b	**a.** 8 kL	**b.** 8 L	**c.** 8 mL
	11. capacity of a hot-air balloon	a	**a.** 15 kL	**b.** 15 L	**c.** 15 mL
	12. amount of water in a lake	a	**a.** 5,000 kL	**b.** 5,000 L	**c.** 5,000 mL
	13. mass of a sandwich	b	**a.** 2 kg	**b.** 225 g	**c.** 200 mg
	14. mass of a tissue	c	**a.** 6 kg	**b.** 13 g	**c.** 875 mg
	15. mass of a bowling ball	a	**a.** 6 kg	**b.** 45 g	**c.** 900 mg
	16. mass of a fork	b	**a.** 45 kg	**b.** 90 g	**c.** 75 mg
	17. mass of a register receipt	c	**a.** 25 kg	**b.** 1 g	**c.** 5 mg
	18. amount of water in an ice cube	c	**a.** 0.3 kL	**b.** 0.8 L	**c.** 35 mL
	19. capacity of a bucket	b	**a.** 12 kL	**b.** 4 L	**c.** 12 mL
	20. amount of juice in a lemon	c	**a.** 25 kL	**b.** 9 L	**c.** 50 mL
	21. capacity of a water tower	a	**a.** 10 kL	**b.** 50 L	**c.** 3 mL
	22. capacity of a washing machine	b	**a.** 60 kL	**b.** 50 L	**c.** 40 mL

INSTRUCTION Discuss examples of estimating mass or capacity in everyday life. Emphasize the importance of relying on models of various units to estimate a measurement more accurately. Note that in the examples, the number of units is the same. This enables the students to focus on the type of unit to choose the most reasonable estimate.

Stress that estimating measurements often involves two steps:

1. Choosing the correct unit of measure.
2. Estimating the number of units. Use the following situation to emphasize this point. Write:

Iron	1 kg
Melon	2.5 kg
Bowling ball	7 kg
Television	25 kg
Compact car	1,200 kg

Have students use these references to estimate the mass of a school desk. Repeat with familiar objects in the classroom or in the home.

Use the same approach to help students estimate the capacity of familiar objects. For example, have students use the following references to estimate the capacity of a bathtub.

Sink	8 L
Car fuel tank	60 L
Swimming pool	75,000 L

TIME OUT Guide students to realize that there is no quick solution to this problem. As with many problems in everyday life, they will need to solve this one by guessing and checking.

ASSIGNMENTS

Level 1	Even 2–40, TO
Level 2	1–41, TO
Level 3	Odd 1–37, 39–42, TO

FOLLOW-UP ACTIVITY

COOPERATIVE LEARNING Have students copy the following table.

1 L	Water	Juice	Rice	Sand	Beans
Mass	▒ g	▒ g	▒ g	▒ g	▒ g

Tell them that they will fill a liter container to find the mass of each item. However, before beginning the experiment, have students list the items in order, starting with the one that they think has the greatest mass and ending with the one they think has the least mass. After the experiment is completed, have them check their estimates against the results.

SUPPLEMENTARY MATERIALS

TRP Practice, p. 80

TRP Reteaching, p. 50

TRP Transparency 40

MIXED REVIEW

Multiply or divide.

•**23.** $\frac{1}{4} \times \frac{2}{3}$ $\frac{1}{6}$

24. $\frac{4}{7} \times \frac{5}{8}$ $\frac{5}{14}$

25. $\frac{7}{8} \div \frac{7}{9}$ $1\frac{1}{8}$

26. $\frac{6}{7} \div \frac{3}{4}$ $1\frac{1}{7}$

27. $\frac{3}{7} \times \frac{7}{8}$ $\frac{3}{8}$

28. $\frac{3}{8} \times \frac{5}{6}$ $\frac{5}{16}$

29. $\frac{8}{9} \div \frac{2}{3}$ $1\frac{1}{3}$

30. $\frac{6}{7} \div \frac{7}{8}$ $\frac{48}{49}$

•**31.** $3\frac{1}{2} \times 2\frac{1}{4}$ $7\frac{7}{8}$

32. $4\frac{3}{8} \times 1\frac{6}{7}$ $8\frac{1}{8}$

33. $6\frac{7}{8} \div 1\frac{4}{7}$ $4\frac{3}{8}$

34. $5\frac{2}{3} \div 2\frac{5}{6}$ 2

35. $1\frac{5}{7} \times 4\frac{2}{3}$ 8

36. $3\frac{3}{5} \times 2\frac{2}{9}$ 8

37. $1\frac{1}{7} \div 1\frac{1}{7}$ 1

38. $4\frac{2}{9} \div 2\frac{5}{7}$ $1\frac{5}{9}$

Solve.

•**39.** Renaldo Thomas is an apprentice cook at a downtown cafeteria. The head chef tells Renaldo that the stew recipe starts with 45 kL of beef broth. Renaldo doesn't believe him. Why?
45 kL is clearly too much.

40. The head chef tells Renaldo to peel 22 lb of onions for the stew. The onions come in 32-oz bags. How many bags of onions does Renaldo have to peel? 11 bags

41. Renaldo empties a full bottle of hot sauce into the stew. He estimates that the bottle held 150 mL. Is his estimate reasonable? yes

42. Today's stew is a lamb stew. Each of the 200 servings includes about 250 g of lamb. How many kilograms of lamb are used in the stew altogether? 50 kg

TIME OUT

You have three buckets. The largest one is filled with 8 L of water. The other two buckets hold 5 L and 2 L. They are empty.

See page 478 for answers.

8 L   5 L   2 L

How can you pour water back and forth, without spilling, so that you end up with exactly 7 L in the largest bucket?

PROBLEM Solving APPLICATION

7.9 CAREER: DRAFTSPERSON

Ellen Jessup is a draftsperson. As part of her job, she makes mechanical drawings that show how to construct parts of machines. The drawings must be very precise. Otherwise, the parts of the machines will not be correct.

Find the measurement to the nearest tenth of a centimeter.

• **1.** 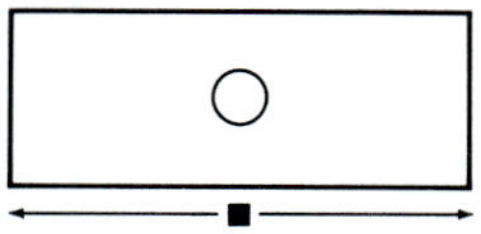

The actual machine part is 2.5 times as long as the drawing. What is the length of the actual part? 9.5 cm

2. 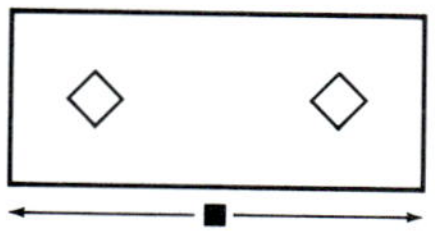

A metal strip is 4.5 times as long as the drawing. What is the length of the actual metal strip? 15.3 cm

3. 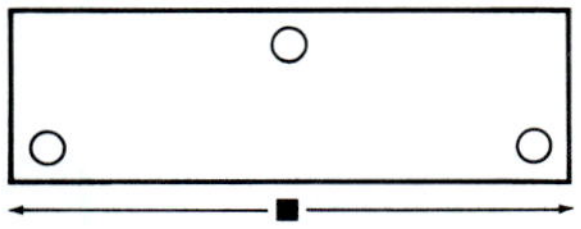

How long is the drawing? 4.6 cm

4. 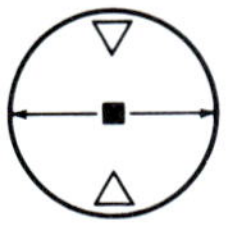

What is the diameter of the washer? 1.5 cm

Sometimes, large drawings are made for small machine parts. In Exercises 5–8, the actual part is 0.25 as long as the dimension of the given drawing.

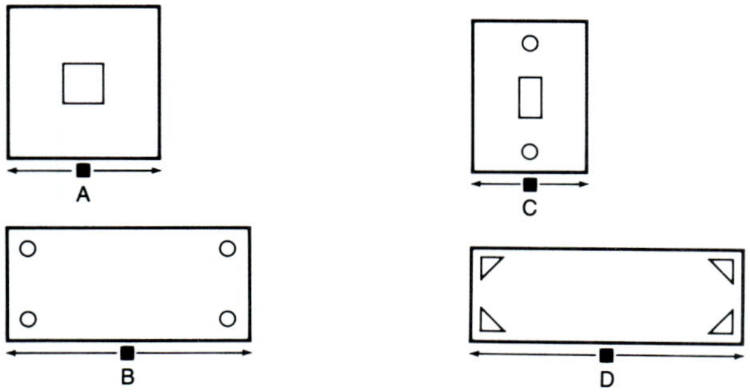

Find the missing numbers to the nearest tenth of a centimeter. Remember to estimate whenever you use your calculator.

	Drawing	Length of Drawing	Length of Part
• **5.**	A	■ 2 cm	■ 0.5 cm
6.	B	■ 3.2 cm	■ 0.8 cm
7.	C	■ 1.5 cm	■ 0.4 cm
8.	D	■ 3.6 cm	■ 0.9 cm

OBJECTIVE

- Solve problems that involve measurement skills and computation with fractions and decimals.

TEACHING THE LESSON

WARM-UP Write the following exercises on the chalkboard or an overhead transparency.

1. 3.5 × 6.8 (23.8)

2. 5.5 × 7.4 (40.7)

3. 0.5 × 12 (6)

4. 0.25 × 6 (1.5)

5. $8\frac{1}{4}$ × \$14.40 (\$118.80)

INSTRUCTION You may wish to use Problem 1 as a group activity to make sure that students understand how to solve Problems 2–4.

You may wish to have students use calculators to solve Problems 5–8.

ASSIGNMENTS

Level 1	1–6
Level 2	1–8
Level 3	3–8

SUPPLEMENTARY MATERIALS

TRP Practice, p. 81

TRP Transparency 41

OBJECTIVE

- Convert metric measurements of length, mass, and capacity.

TEACHING THE LESSON

WARM-UP Present the following exercises on the chalkboard or an overhead transparency.

Multiply or divide.

1. 0.35 × 100	**2.** 0.04 × 1,000
3. 0.01 × 1,000	**4.** 1 ÷ 1,000
5. 45.6 ÷ 10	**6.** 1,200 ÷ 100

(**1.** 35 **2.** 40 **3.** 10 **4.** 0.001 **5.** 4.56 **6.** 12)

INTRODUCTION Review the shortcuts for multiplying and dividing by powers of 10. Provide additional practice as needed.

INSTRUCTION Use the solution to the introductory problem to determine the approximate number of meters in 1 block. (80 m) Then have students estimate the distance they walk from home to school, from home to a shopping center, from home to a friend's house. Ask whether each distance is greater than or less than a kilometer.

Extend the instruction in Example 2 by having students estimate how long a tube of suntan lotion would last if it contains 100 g of lotion. Assume that a person uses about 0.8 g in each application (about 11 or 12 applications).

CHECKPOINT The incorrect answer choices include common errors that students make.

COMMON ERROR

Students will sometimes use the wrong operation when converting metric units of measurement. For example, in item 1 of Checkpoint, they would select c. You may want to work through the exercises with the class, having students state whether they should multiply or divide to convert the measurement.

7.10 CONVERTING METRIC UNITS

Terry Partridge walked the five blocks from her house to the Surf City swimming pool. The distance is 0.4 km. How many meters did she walk from her house to the pool?

To solve, you need to change 0.4 km to meters.

You can use these relationships to change from one unit to another.

LENGTH	MASS	CAPACITY
1 km = 1,000 m	1 kg = 1,000 g	1 kL = 1,000 L
1 m = 100 cm	1 g = 1,000 mg	1 L = 1,000 mL
1 m = 1,000 mm		

To change from a larger unit to a smaller unit, multiply.

EXAMPLE 1 Complete: 0.4 km = ■ m

THINK: 1 km = 1,000 m

0.4 km = (0.4 × 1,000) m
0.4 km = 400 m

So, Terry walked 400 m from her house to the pool.

To change from a smaller unit to a larger unit, divide.

EXAMPLE 2 After a short swim, Terry comes out of the pool and applies 800 mg of suntan lotion to her arms and legs. How many grams of suntan lotion does Terry apply?

Complete: 800 mg = ■ g

THINK: 1 g = 1,000 mg

800 mg = (800 ÷ 1,000) g
800 mg = 0.8 g

So, Terry applies 0.8 g of suntan lotion.

CHECKPOINT

Write the letter of the correct answer.

Complete.

1. 0.35 m = ■ cm	b	**a.** 350	**b.** 35	**c.** 0.0035
2. 4,500 L = ■ kL	c	**a.** 450	**b.** 45	**c.** 4.5
3. 0.04 g = ■ mg	b	**a.** 400	**b.** 40	**c.** 4

PRACTICE EXERCISES

See Extra Practice, page 443.

12. 1,000,000
21. 1,000,000

Complete.

• **1.** 2 m = ■ mm 2,000	**2.** 4 km = ■ m 4,000	**3.** 1 cm = ■ mm 10
• **4.** 4,000 mm = ■ m 4	**5.** 600 cm = ■ m 6	**6.** 9,000 m = ■ km 9
7. 500 mm = ■ m 0.5	**8.** 40 m = ■ km 0.04	**9.** 2,000 cm = ■ m 20
10. 4 g = ■ mg 4,000	**11.** 7 kg = ■ g 7,000	**12.** 1 kg = ■ mg
13. 9,000 g = ■ kg 9	**14.** 12,000 mg = ■ g 12	**15.** 21,000 g = ■ kg 21
16. 80 mg = ■ g 0.08	**17.** 140 g = ■ kg 0.14	**18.** 3 g = ■ kg 0.003
19. 9 L = ■ mL 9,000	**20.** 3 kL = ■ L 3,000	**21.** 1 kL = ■ mL
22. 8,000 L = ■ kL 8	**23.** 89,000 L = ■ kL 89	**24.** 2,000 mL = ■ L 2
25. 80 L = ■ kL 0.08	**26.** 750 mL = ■ L 0.75	**27.** 15 mL = ■ L 0.015
28. 0.01 m = ■ mm 10	**29.** 3.75 g = ■ mg 3,750	**30.** 23.45 mL = ■ L 0.02345
31. 10.5 kL = ■ L 10,500	**32.** 0.125 kg = ■ mg 125,000	**33.** 45.6 mm = ■ cm 4.56
34. 43 mg = ■ g 0.043	**35.** 2.225 km = ■ m 2,225	**36.** 13,321 L = ■ kL 13.321

MIXED REVIEW

Complete.

Data	Range	Mean	Median	Mode
12, 17, 19, 26, 43, 12, 11	•**37.** ■ 32	•**38.** ■ 20	•**39.** ■ 17	•**40.** ■ 12
212, 568, 345, 420, 345	**41.** ■ 356	**42.** ■ 378	**43.** ■ 345	**44.** ■ 345
3.6, 4.2, 7.4, 1.7, 4.2, 5.3, 3.7	**45.** ■ 5.7	**46.** ■ 4.3	**47.** ■ 4.2	**48.** ■ 4.2
2.5, 0.4, 3.4, 5.2, 6.7, 5.2	**49.** ■ 6.3	**50.** ■ 3.9	**51.** ■ 4.3	**52.** ■ 5.2

Solve.

•**53.** The Surf City swimming pool holds 75 kL of water. About 600 L a day are lost to evaporation, splashing, and so on. How many kiloliters are left in the pool at the end of the day before the pool is refilled? 74.4 kL

54. Felicia, the lifeguard, puts 350 mg of sun block on her nose each day. How many days will it take her to use up a 140-g tube of sun block? 400 d

55. The Surf City pool is open daily from 8:30 A.M. to 7:15 P.M. How long is the pool open each day? 10 h 45 min

56. Barbie is sunning herself on a towel that is 120 cm long. Barbie is 1.63 m tall. How much shorter is the towel than Barbie? 43 cm

ASSIGNMENTS

Level 1	Odd 1–35, 37–44, 53, 54
Level 2	Even 2–36, 37–55
Level 3	22–36, 45–56

FOLLOW-UP ACTIVITY

ENRICHMENT Distribute copies of this table.

Unit	Symbol	Prefix	Value
kilometer	km	kilo	1,000 m
hectometer	hm	hecto	100 m
dekameter	dam	deka	10 m
meter	m		1 m
decimeter	dm	deci	0.1 m
centimeter	cm	centi	0.01 m
millimeter	mm	milli	0.001 m

Guide students in using the table to convert one metric unit to another.

1. Find the original unit on the table.
2. Count the number of "moves" to the new unit.
3. Move the decimal point of the original number as many places as the number of moves. A move up the table means move the decimal point to the left. A move down means a move to the right.
4. Write the new number and unit.

Have students use this method for:

1. 3,400 m = ■ km
2. 2.25 km = ■ dam
3. 450 km = ■ hm
4. 85,000 mm = ■ km

(**1.** 3.4 **2.** 225 **3.** 4,500 **4.** 0.085)

SUPPLEMENTARY MATERIALS

TRP Practice, p. 82

TRP Reteaching, p. 51

TRP Transparency 41

OBJECTIVE

- Estimate temperatures in degrees Celsius and degrees Fahrenheit.

TEACHING THE LESSON

WARM-UP Present the following exercises on the chalkboard or an overhead transparency.

Name the missing numbers.

1. 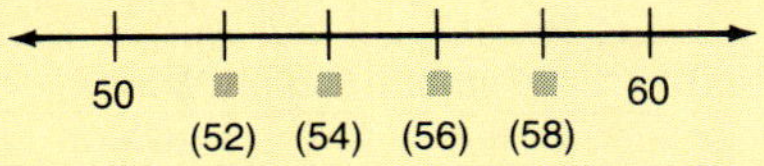

2. 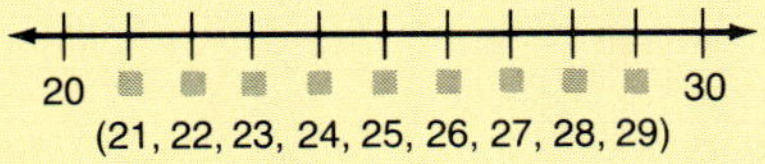

3. 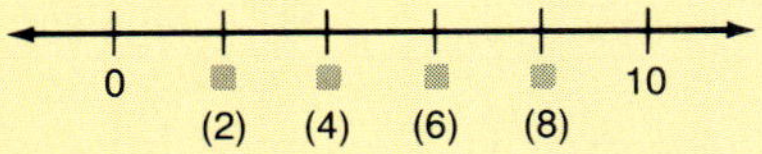

4. 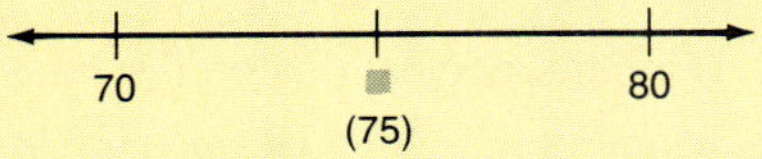

INTRODUCTION You may wish to point out that the Fahrenheit scale was devised in 1724 by the German physicist Gabriel Daniel Fahrenheit. He used a cold point, determined by a mixture of ice and salt, as zero degrees. On this scale, pure water started to freeze at 32 degrees and boiled at 212 degrees.

The Celsius scale was invented in 1742 by a Swedish astronomer, Anders Celsius. This scale uses zero degrees as the freezing point of pure water and 100 degrees as the boiling point of water.

INSTRUCTION In Example 1, call students' attention to the five marks from 50°C to 60°C. Explain that the value of each mark can be determined by dividing the size of the interval (10 degrees) by the number of marks within the interval (5). Therefore, each mark indicates a change in temperature of 2 degrees.

In Example 2, encourage students to memorize the key reference points of both scales.

7.11 MEASURING TEMPERATURE

Thermometers are used to measure temperature. Temperature can be measured in **degrees Celsius (°C)** or in **degrees Fahrenheit (°F).** To find the temperature, read the scale.

EXAMPLE 1 What is the temperature?

THINK: 50° + 8° = 58°

The temperature is 58°C.

The table below lists temperatures in both systems.

	°C	°F
Boiling point of water	100	212
Record high air temperature	58	136
Normal body temperature	37	98.6
Comfortable room temperature	20	68
Freezing point of water	0	32
Record low air temperature	−88	−127

Use the standard temperatures you already know to help you choose the most reasonable temperature.

EXAMPLE 2 Choose the most reasonable temperature for a very hot cup of tea.
a. 72°F b. 95°C c. 100°F

THINK: Water boils at 100°C or 212°F.

Compare. 95°C is close to 100°C.

So, the most reasonable temperature for a very hot cup of tea is 95°C.

FOR DISCUSSION

See TE side column.

Kim and Jane are talking on the phone. Kim lives in Southern California. She says that the temperature is 30° and that she's going swimming. Jane lives in Vermont and says that the temperature there is 30° but she's going ice skating. How can this be?

FOR DISCUSSION Students should explain that Kim is reading a Celsius thermometer while Jane is reading a Fahrenheit thermometer.

PRACTICE EXERCISES

See Extra Practice, page 443.

Write the temperature.

1. 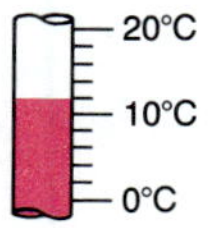

12°C

2. 

78°C

3. 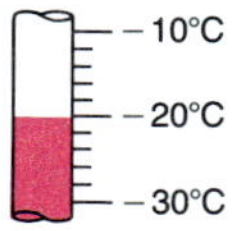

20°C

4. 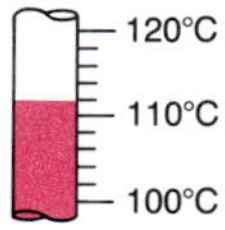

112°C

5. 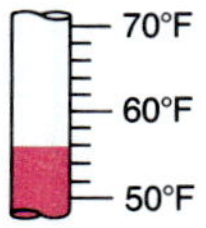

56°F

6. 

0°F

7. 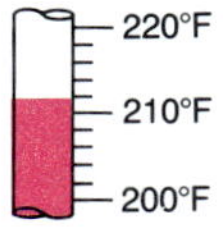

212°F

8. 

– 14°F

Choose the most reasonable temperature.

9.	hot oven	c	**a.** 80°C	**b.** 100°C	**c.** 200°C
10.	sunny summer day	b	**a.** 10°C	**b.** 30°C	**c.** 50°C
11.	ice cube	c	**a.** 22°C	**b.** 8°C	**c.** – 4°C
12.	refrigerator	c	**a.** – 12°C	**b.** 0°C	**c.** 4°C
13.	snowstorm	a	**a.** 28°F	**b.** 42°F	**c.** 112°F
14.	chicken roasting	c	**a.** 100°F	**b.** 212°F	**c.** 400°F
15.	swimming pool	a	**a.** 72°F	**b.** 32°F	**c.** 15°F
16.	heat wave	b	**a.** 185°F	**b.** 97°F	**c.** 65°F
17.	cool fall night	c	**a.** 20°F	**b.** 40°C	**c.** 55°F
18.	hot bathwater	b	**a.** 25°C	**b.** 100°F	**c.** 100°C
19.	aquarium	a	**a.** 75°F	**b.** 65°C	**c.** 10°C
20.	sweater weather	b	**a.** 75°C	**b.** 55°F	**c.** 35°C

Solve.

21. Horace Clark is going fishing. The weather report says that the temperature on the bay will be about 32°C. Does Horace need to wear anything heavier than a T-shirt?
no

22. The captain of the boat tells Horace that the ocean temperature is 62°. Does he mean degrees Celsius or degrees Fahrenheit?
degrees Fahrenheit

23. Horace catches a 4-lb flounder. After he cleans the fish, there are 54 oz of fish left. How many ounces of fish did Horace discard? 10 oz

24. Horace breads the flounder and fries it at about 250°. Is that degrees Celsius or degrees Fahrenheit?
degrees Fahrenheit

COMMON ERROR

Students will sometimes misread the scale on the thermometer and assume that each mark always represents a 1 degree change in temperature. Provide more practice with exercises similar to those in the Warm-Up to strengthen students' skill in determining the size of an interval.

ASSIGNMENTS

Level 1	1–22
Level 2	1–24
Level 3	1–8, Even 10–24

FOLLOW-UP ACTIVITY

ENRICHMENT Students should use a world map or globe to match each of the following cities with its average high temperature for February.

Cities	Temperatures °F
Cairo, Egypt	23°
Detroit, Michigan	78°
Singapore, Malaysia	69°
Sydney, Australia	32°
Moscow, Soviet Union	88°

In a follow-up discussion, guide students to realize that:

1. The closer the city is to the Equator, the warmer the temperature.
2. The seasons in the Northern Hemisphere are the opposite of those in the Southern Hemisphere.

SUPPLEMENTARY MATERIALS

TRP Practice, p. 83

TRP Reteaching, p. 52

TRP Transparency 42

OBJECTIVE

- Solve problems by drawing diagrams.

TEACHING THE LESSON

WARM-UP Write the following exercises on the chalkboard or an overhead transparency.

Solve.

1. 40 − 30 (10) **2.** 80 − 25 (55)

3. 2.4 + 2.3 + 2.8 (7.5)

4. 5.2 + 10.8 + 4.5 (20.5)

INTRODUCTION Review the directions north, south, east, and west. Draw a direction finder on the chalkboard.

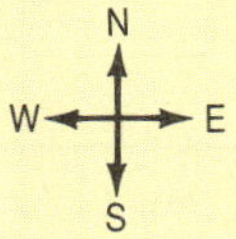

Ask a volunteer to stand at the front of the classroom. Refer to the student's position as point *A*. Tell the student to walk eight steps east. Then ask the student to turn around and walk five steps west. Ask students to find the number of steps that the student is from point *A*. (3 steps) Give similar directions for other students to follow.

INSTRUCTION Discuss the situation and the strategy with the class. As you work the subtraction in Step 2, you may wish to draw a diagram to show the 15 m.

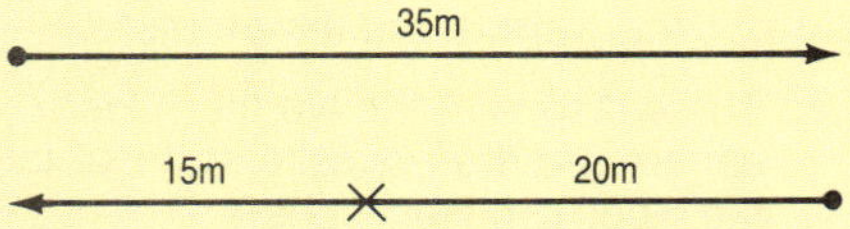

In Exercise 7, make sure students realize that three lengths of 1.5-m wire can be cut from each section.

PROBLEM Solving STRATEGY

7.12 DRAWING A DIAGRAM

Situation:

In filming a Western movie scene, the director asks an actor to ride his horse 35 m east from the barn and then stop. The next direction is to turn and trot 20 m west and stop again. How far must the actor ride the horse to get back to the barn?

Strategy:

Drawing a diagram can help you solve some problems.

Applying the Strategy:

1. Draw the horse and rider going east. Draw the horse and rider going west.
2. Subtract to find how far the actor must ride the horse to get back to the barn.

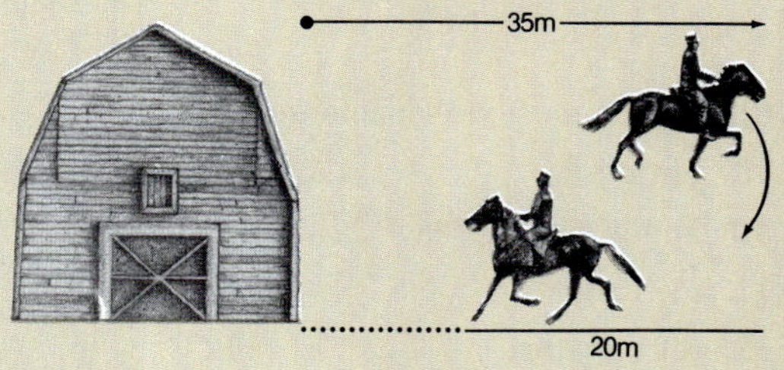

The actor must ride the horse 15 m to get back to the barn.

$$\begin{array}{r} 35 \\ -20 \\ \hline 15 \end{array}$$

FOR DISCUSSION

See TE side column.

A flag being used on the movie set has 9 stripes. The first stripe is blue. The second stripe is white. This pattern is repeated on the flag. What color is the seventh stripe?

1. What would you draw to help you solve the problem?

2. What is the solution?

PRACTICE EXERCISES

See Extra Practice, page 443.

Solve the problem by drawing a diagram. Check students' drawings.

• **1.** A set designer hangs a picture on the wall of a model room. The picture is 84 cm wide. The wall is 218 cm wide. If the picture is hung in the center of the wall, how much wall space is on each side of the picture? 67 cm

• **2.** A camera crew sets up a special light board. There are 25 light bulbs. There are five light bulbs in each row. The center and corner lights are yellow. How many yellow light bulbs are there in all? 5

Solve each problem by drawing a diagram.

3. A director tells an actor to walk 19 m from the stable to the barn. The next direction is to turn around and walk back 12 m to the well. How far from the stable is the actor? 7 m

4. The stage crew sets up 2 shelves on one wall of a movie set. The first shelf is 1.4 m off the ground. The second is 1.2 m above that and 1.8 m from the ceiling. What is the height of the set? 4.4 m

5. A set designer places a rug and a sofa on the living room floor. The rug has 12 stripes. The first stripe is white. The second is blue. The third is green. This pattern is repeated. How many times is the complete pattern repeated? 4

6. The camera crew moves around the stage for the filming of the Western. They move 6 m north, then 8.2 m east, then 10.5 m south, then 8.2 m west, and finally 4.5 m south. How far is the camera crew from its original location? 9 m

7. A sound technician needs wire cut into 1.5-m lengths. The technician has 6 sections of wire. Each section is 5 m long. How many 1.5-m lengths can the technician cut? 18 lengths

FOR DISCUSSION Students should explain that they would draw a series of nine stripes. They should label each stripe, using the blue-white pattern.

Elicit from students that the seventh stripe is blue.

ASSIGNMENTS

Level 1	1–5
Level 2	1–6
Level 3	3–7

FOLLOW-UP ACTIVITY

SITUATIONAL PROBLEM SOLVING Present students with the following problem. Your class is going to stage a musical comedy for the school. You are the director. Each class member must participate in the production. Here are some questions that you may need to consider to organize your production.

How will you choose the musical comedy?

How will roles be assigned?

Who will design the stage set and costumes?

How will the production be financed?

How will ticket sales be arranged?

Write a description of how you will stage your musical comedy.

SUPPLEMENTARY MATERIALS

TRP Practice, p. 84

TRP Applications, p. 13

TRP Transparency 42

OBJECTIVES

- Review vocabulary.
- Practive key chapter concepts and skills.

USING THE REVIEW

The Chapter Review is designed to help students prepare for taking the Chapter Test. The first section focuses on vocabulary. It requires that students select a word(s) to complete statements. The second section presents practice exercises of key mathematical skills. Under each directive there is a sample exercise with the answer.

Each item on the review is referenced to the page on which the topic is taught in the Pupil's Edition. You may wish to have students refer to these pages to help review any concepts or skills they have not yet mastered.

It is suggested that students work in small-sized heterogeneous cooperative learning groups. Some cooperative learning methods that may be used are as follows.

1. After each student has independently completed the entire Chapter Review, a discussion should follow within each group about the solutions to the practice exercises.
2. The group can complete the entire Chapter Review by working together to discuss the sample exercises, and then to answer the practice exercises.

End the lesson with an entire class discussion in which any questions brought up in group discussion are presented and answered.

CHAPTER REVIEW

Vocabulary Choose the letter of the word(s) that completes each statement.

1. The "A.M." in 7:00 A.M. indicates ■. [186] a

a. before noon **b.** noon **c.** after noon **d.** midnight

2. You would likely use centimeters to measure the width of a ■. [192] c

a. barn **b.** pin **c.** calculator **d.** canyon

Skills Complete. [180, 182, 186, 198]

7.45 L = *7,450 mL*

3. $2\frac{1}{2}$ lb = ■ oz 40

4. $2\frac{1}{3}$ ft = ■ in. 28

5. 7 yd = ■ ft 21

6. 500 g = ■ kg $\frac{1}{2}$

7. 10 qt = ■ gal $2\frac{1}{2}$

8. 6.5 m = ■ cm 650

9. 120 s = ■ h $\frac{1}{30}$

10. 3,800 mL = ■ L 3.8

Choose the most reasonable estimate. [192, 194, 198]

	the mass of a cotton ball	*a*	**a.** 750 mg	**b.** 5 g	**c.** 0.01 kg
11.	thickness of a scarf	a	**a.** 3 mm	**b.** 3 cm	**c.** 3 m
12.	capacity of a bathroom sink	b	**a.** 0.08 kL	**b.** 8 L	**c.** 800 mL
13.	mass of a credit card	b	**a.** 12 kg	**b.** 4 g	**c.** 75 mg
14.	temperature of boiling water	c	**a.** 100°F	**b.** 212°C	**c.** 212°F

Add or subtract. [184]

Sample: 3 qt 1 pt + 2 qt 1 pt = *6 qt*

15. 2 T 1,500 lb + 1 T 800 lb = 4 T 300 lb

16. 5 ft 3 in. − 3 ft 9 in. = 1 ft 6 in.

17. 3 c 4 fl oz − 6 fl oz = 2 c 6 fl oz

18. 4 yd 2 ft + 3 yd 2 ft = 8 yd 1 ft

19. 6 lb 3 oz − 4 lb 5 oz = 1 lb 14 oz

20. 9 gal − 2 gal 3 qt = 6 gal 1 qt

21. 5 mi 1,300 yd + 3 mi 800 yd = 9 mi 340 yd

Find the missing time. [186]

Start time: 12 noon
End time: 4:19 P.M.
Elapsed time: *4 h 19 min*

22. Start time: 6:28 P.M.
End time: ■
Elapsed time: 7 h 54 min
2:22 A.M.

23. Start time: 9:32 A.M.
End time: 3:17 P.M.
Elapsed time: ■
5 h 45 min

204 CHAPTER 7

CHAPTER TEST

Complete.

1. 45 s = ■ min $\frac{3}{4}$

2. 2.3 mm = ■ cm 0.23

3. 125 L = ■ kL 0.125

4. $9\frac{3}{4}$ gal = ■ qt 39

5. 21,120 ft = ■ mi 4

6. 1.05 g = ■ mg 1,050

7. 36 in. = ■ ft 3

8. $3\frac{1}{2}$ T = ■ lb 7,000

9. 7 wk = ■ d 49

Choose the most reasonable estimate.

10. The thickness of a notepad a — **a.** 10 mm **b.** 10 cm **c.** 10 m

11. The mass of a football player b — **a.** 250 g **b.** 100 kg **c.** 300 kg

12. The capacity of a bathtub b — **a.** 1 kL **b.** 180 L **c.** 250 mL

13. The temperature of ice water a — **a.** 0°C **b.** 0°F **c.** 4°C

Add or subtract.

14.
```
  2 ft 9 in.
+4 ft 7 in.
  7 ft 4 in.
```

15.
```
  6 c 3 fl oz
−4 c 5 fl oz
  1 c 6 fl oz
```

16.
```
  3 T 1,500 lb
+2 T   750 lb
  6 T 250 lb
```

Find the missing time.

17. Start time: 10:18 A.M.
End time: 6:32 P.M.
Elapsed time: ■ 8 h 14 min

18. Start time: 3:57 P.M.
End time: ■ 2:10 A.M.
Elapsed time: 10 hours 13 minutes

Write the temperature.

19. 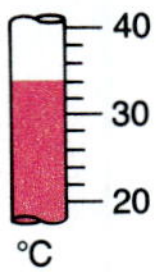

34°C

20. 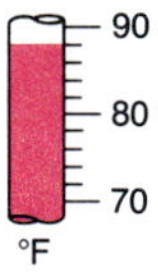

88°F

Find the length to the nearest $\frac{1}{16}$ in.

21. ———————— $1\frac{1}{4}$ in.

Find the length to the nearest millimeter.

22. ———————— 25 mm

Solve. Draw a diagram if necessary.

23. Isabel designed a scarf to knit. The scarf has 15 blocks. The first block is green, the second is blue, and the third is yellow. How many times is the complete pattern repeated? 5 times

24. Jack Cook earns $8.00 per hour. One week he works 48 hours. What is his total pay, including overtime? $416

OBJECTIVE

- Evaluate achievement of the chapter objectives.

USING THE TEST

The Chapter Test may be used as a posttest to evaluate student achievement. However, you may wish to use the Chapter Posttest offered in the Teacher's Resource Package or to design your own chapter test. If this page is not used as a test, you may wish to assign it as additional review or practice.

The test items are correlated to the chapter objectives in the table below.

Chapter objectives	Test items
A. Change measurement units within the customary and metric systems.	1–9
B. Choose the most reasonable estimate of measure.	10–13
C. Add or subtract customary measures.	14–16
D. Compute with time.	17–18
E. Read temperatures.	19–20
F. Measure lengths to the nearest $\frac{1}{16}$ inch or millimeter.	21–22
G. Draw a diagram to solve problems.	23
H. Apply computational skills in real-life situations.	24

SUPPLEMENTARY MATERIALS

TRP Ch. 7 Posttest Form A, pp. 1–2

TRP Ch. 7 Posttest Form B, pp. 1–2

OBJECTIVES

- Identify the more precise of two measurements.
- Find the greatest possible error and range of a measurement.

USING THE PAGE

Before beginning the lesson you may wish to use the following exercises to review computation with fractions and decimals.

$8 + \frac{1}{2}$ $(8\frac{1}{2})$	$10 - \frac{1}{2}$ $(9\frac{1}{2})$
$7\frac{1}{2} - \frac{1}{4}$ $(7\frac{1}{4})$	$12\frac{1}{2} + \frac{1}{4}$ $(12\frac{3}{4})$
$37 - 0.5$ (36.5)	$28 + 0.5$ (28.5)
$45.8 - 0.05$ (45.75)	$59.3 + 0.05$ (59.35)

Discuss the length of the nail shown at the top of the page. Then draw the following line segment. (48mm)

Have a volunteer measure the line segment to the nearest centimeter and then to the nearest millimeter. Emphasize once again that the smaller the unit, the more precise the measurement.

As you discuss greatest possible error and range of length, ask why the greatest possible error of 30 in. is $\frac{1}{2}$ in. and the greatest possible error of 41.5 km is 0.05 km.

Provide a classroom demonstration to reinforce the steps for finding the greatest possible error and the range of length for a given measurement. Have three volunteers stand against a wall and mark their heights. Have volunteers measure the height of the first student to the nearest inch, the second to the nearest half-inch, and the third to the nearest centimeter. Record the results on the chalkboard, and then find the greatest possible error and range of length for each measure.

ENRICHMENT PRECISION AND GREATEST POSSIBLE ERROR

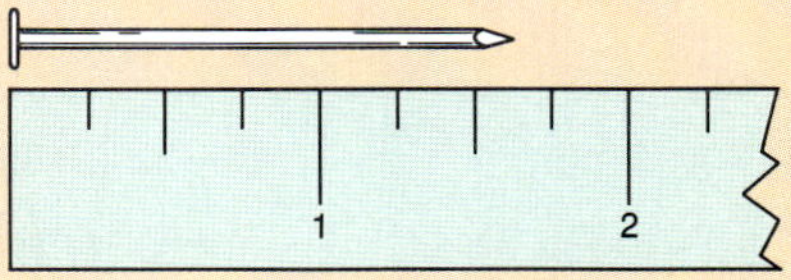

The length of this nail is closer in length to 2 in. than 1 in. So, the length of the nail is 2 in. to the nearest inch.

If you want to measure the nail more precisely, you can measure the nail to the nearest half-inch. The length of the nail is $1\frac{1}{2}$ in. to the nearest half-inch. The smaller the unit of measure, the more precise the measurement will be.

Examples Choose the more precise measurement.

8 cm or 8 mm

THINK: The smaller unit is millimeters.

The more precise measurement is 8 mm.

3 ft or 37 in.

THINK: The smaller unit is inches.

The more precise measurement is 37 in.

Since no measurement is exact, the greatest possible error is equal to one-half of the smallest unit of the measurement.

If your height is measured as $56\frac{1}{2}$ in., the greatest possible error is $\pm\frac{1}{4}$ in.
So, the range of your height is between $56\frac{1}{4}$ $\left(56\frac{1}{2} - \frac{1}{4}\right)$ and $56\frac{3}{4}$ $\left(56\frac{1}{2} + \frac{1}{4}\right)$ in.

Examples Find the greatest possible error of measurement. Then find the range of length.

30 in.

1. Find the greatest possible error.

   THINK: The smallest unit is 1 in.

   $\frac{1}{2} \times 1 = \frac{1}{2}$

   The greatest possible error is $\pm\frac{1}{2}$ in.

2. Find the range.

   $30 - \frac{1}{2} = 29\frac{1}{2}$ in.

   $30 + \frac{1}{2} = 30\frac{1}{2}$ in.

41.5 km

1. Find the greatest possible error.

   THINK: The smallest unit is 0.1 km.

   $\frac{1}{2} \times 0.1 = 0.05$

   The greatest possible error is ± 0.05 km.

2. Find the range.

   $41.5 - 0.05 = 41.45$ km

   $41.5 + 0.05 = 41.55$ km

Find the greatest possible error of measurement. Then find the range of length.

1. 42 in. $\frac{1}{2}$ in.; $41\frac{1}{2}$–$42\frac{1}{2}$ in.
2. 16 mm 0.5 mm; 15.5–16.5 mm
3. 84 km 0.5 km; 83.5–84.5 km
4. 93 mi $\frac{1}{2}$ mi; $92\frac{1}{2}$–$93\frac{1}{2}$
5. 32.4 m 0.05 m; 32.35–32.45 m
6. $36\frac{1}{2}$ in. $\frac{1}{4}$ in.; $36\frac{1}{4}$ in.–$36\frac{3}{4}$ in.
7. 56 ft $\frac{1}{2}$ ft; $55\frac{1}{2}$ ft–$56\frac{1}{2}$ ft
8. 89.2 cm 0.05 cm; 89.15–89.

OBJECTIVE

- Estimate the sum or difference of customary measures expressed in two units.

USING THE PAGE

The technique used to round measures is similar to that used to round numbers. For example:

Round 158 to the nearest 10.
Think: 8 is greater than 5. That is, 8 is greater than $\frac{1}{2}$ of 10.
158 rounds up to 160.

Round 15 ft 8 in. to the nearest foot.
Think: 8 in. is more than 6 in. That is, 8 in. is more than $\frac{1}{2}$ ft.
15 ft 8 in. rounds up to 16 ft.

In order to round the measures in the exercises correctly, students must recall equivalencies between pairs of customary units. You may wish to review these by having students complete the following statements orally.

1 hour = ■ minutes		(60)
1 minute = ■ seconds		(60)
1 foot = ■ inches		(12)
1 yard = ■ feet		(3)
1 yard = ■ inches		(36)
1 pound = ■ ounces		(16)
1 gallon = ■ quarts		(4)
1 pint = ■ cups		(2)
1 cup = ■ fluid ounces		(8)
1 pint = ■ fluid ounces		(16)

Students should understand that when we add or subtract with rounded numbers of units, the result may not be accurate to the nearest unit. For example, to estimate the sum of 3 ft 5 in. and 2 ft 4 in., we round 3 ft 5 in. to 3 ft and we round 2 ft 4 in. to 2 ft. Then the estimated sum is 5 ft. However, the actual sum is 5 ft 9 in., which is closer to 6 ft than to 5 ft.

ESTIMATION SKILLS — ESTIMATING WITH CUSTOMARY MEASURES

When you estimate the sum or difference of measures expressed in two units, round each measure to the greater unit.

Estimate the difference: 15 ft 2 in. − 8 ft 11 in.

1. Round each measure to the nearest number of feet.

   THINK: 2 in. < $\frac{1}{2}$ ft, so 15 ft 2 in. rounds to 15 ft.

   11 in. > $\frac{1}{2}$ ft, so 8 ft 11 in. rounds to 9 ft.

2. Subtract.

   15 − 9 = 6

The estimated difference is 6 ft.

Estimate the sum: 1 h 15 min + 5 h 37 min + 3 h 6 min

1. Round each measure to the nearest number of hours.

   THINK: 15 min < $\frac{1}{2}$ h, so 1 h 15 min rounds to 1 h.

   37 min > $\frac{1}{2}$ h, so 5 h 37 min rounds to 6 h.

   6 min < $\frac{1}{2}$ h, so 3 h 6 min rounds to 3 h.

2. Add.

   1 + 6 + 3 = 10

The estimated sum is 10 hours.

Estimate the sum or difference.

1. 18 ft 4 in. + 6 ft 9 in. 25 ft
2. 9 ft 5 in. − 3 ft 3 in. 6 ft
3. 4 h 8 min + 43 min 5 h
4. 16 h 40 min − 5 h 52 11 h
5. 9 lb 2 oz + 8 lb 3 oz 17 lb
6. 24 lb 3 oz − 15 lb 6 oz 9 lb
7. 2 gal 3 qt + 3 gal 1 qt 6 gal
8. 7 gal − 2 gal 1 qt 5 gal
9. 35 min 2 s + 20 min 8 s 55 min
10. 10 min 4 s − 5 min 53 s 4 min
11. 6 c 7 fl oz + 8 c 2 fl oz 15 c
12. 8 c 5 fl oz − 3 c 7 fl oz 5 c
13. 2 yd 32 in. + 3 yd 27 in. 7 yd
14. 10 yd 32 in. − 5 yd 5 in. 6 yd
15. 1 ft 5 in. + 3 ft 7 in. + 2 ft 9 in. 8 ft
16. 5 min 16 s + 4 min 42 s + 8 min 9 s 18 min
17. 2 lb 9 oz + 1 lb 3 oz + 15 oz 5 lb
18. 1 c 6 fl oz + 3 c 4 fl oz + 5 fl oz 6 c

OBJECTIVE

- Test the prerequisite skills needed to learn the concepts developed in Chapter 8.

USING THE TEST

The Pre-Skills Test is designed to diagnose students' strengths and weaknesses on prerequisite skills necessary to study the mathematics in Chapter 8.

Have students take the Pre-Skills Test. Allow the students to work together in pairs or small groups. Group members should help those who demonstrate a misunderstanding of a concept or a weakness in a skill.

The items in the test are referenced to the pages on which the topics are taught in the Pupil's Edition. You may wish to have students refer to these pages for review.

The following table correlates the items on the Pre-Skills Test with the prerequisite skill and the lesson(s) in the chapter for which it is needed.

Item(s)	Prerequisite skill	Lesson(s)
1–9	Multiply whole numbers.	8.2, 8.4, 8.5
10–19	Divide whole numbers.	8.4, 8.5
20–31	Write fractions in simplest form.	8.1, 8.2
32–39	Determine if fractions are equivalent.	8.1, 8.2
40–51	Write equivalent fractions.	8.1, 8.2

Chapter 8 Ratio & Proportion

Pre Skills TEST

Multiply. [40]

1. 5 × 27 135 **2.** 6 × 53 318 **3.** 8 × 96 768

4. 74 × 3 222 **5.** 87 × 4 348 **6.** 18 × 42 756

7. 39 × 15 585 **8.** 63 × 29 1,827 **9.** 84 × 57 4,788

Divide. [46, 48]

10. 4)72 18 **11.** 7)91 13 **12.** 5)85 17 **13.** 8)128 16 **14.** 9)306 34

15. 12)96 8 **16.** 15)225 15 **17.** 23)138 6 **18.** 47)360 7R31 **19.** 76)532 7

Write the fraction in simplest form. [126, 132]

20. $\frac{9}{12}$ $\frac{3}{4}$ **21.** $\frac{16}{20}$ $\frac{4}{5}$ **22.** $\frac{18}{24}$ $\frac{3}{4}$ **23.** $\frac{54}{63}$ $\frac{6}{7}$ **24.** $\frac{56}{72}$ $\frac{7}{9}$ **25.** $\frac{35}{42}$ $\frac{5}{6}$

26. $\frac{12}{3}$ 4 **27.** $\frac{26}{22}$ $1\frac{2}{11}$ **28.** $\frac{62}{48}$ $1\frac{7}{24}$ **29.** $\frac{112}{28}$ 4 **30.** $\frac{224}{178}$ $1\frac{23}{89}$ **31.** $\frac{500}{425}$ $1\frac{3}{17}$

Use cross products to determine if the fractions are equivalent. Write *yes* or *no*. [126]

32. $\frac{3}{7}$ ● $\frac{9}{24}$ no **33.** $\frac{5}{8}$ ● $\frac{35}{56}$ yes **34.** $\frac{24}{48}$ ● $\frac{4}{8}$ yes **35.** $\frac{56}{49}$ ● $\frac{7}{8}$ no

36. $\frac{12}{4}$ ● $\frac{48}{16}$ yes **37.** $\frac{56}{8}$ ● $\frac{118}{16}$ no **38.** $\frac{256}{16}$ ● $\frac{32}{2}$ yes **39.** $\frac{300}{250}$ ● $\frac{60}{50}$ yes

Complete. [126]

40. $\frac{7}{9}, \frac{14}{18}, \frac{21}{27}, \frac{■}{■}, \frac{■}{■}, \frac{■}{■}$ $\frac{28}{36}, \frac{35}{45}, \frac{42}{54}$ **41.** $\frac{5}{8}, \frac{10}{16}, \frac{15}{24}, \frac{■}{■}, \frac{■}{■}, \frac{■}{■}$ $\frac{20}{32}, \frac{25}{40}, \frac{30}{48}$

42. $\frac{3}{7}, \frac{6}{14}, \frac{9}{21}, \frac{■}{■}, \frac{■}{■}, \frac{■}{■}$ $\frac{12}{28}, \frac{15}{35}, \frac{18}{42}$ **43.** $\frac{4}{11}, \frac{8}{22}, \frac{12}{33}, \frac{■}{■}, \frac{■}{■}, \frac{■}{■}$ $\frac{16}{44}, \frac{20}{55}, \frac{24}{66}$

Find the missing term. [126]

44. $\frac{2}{3} = \frac{■}{12}$ 8 **45.** $\frac{7}{8} = \frac{■}{56}$ 49 **46.** $\frac{5}{9} = \frac{25}{■}$ 45 **47.** $\frac{25}{45} = \frac{■}{9}$ 5

48. $\frac{75}{15} = \frac{■}{3}$ 15 **49.** $\frac{■}{31} = \frac{124}{62}$ 62 **50.** $\frac{48}{45} = \frac{144}{■}$ 135 **51.** $\frac{153}{33} = \frac{■}{11}$ 51

Chapter 8 RATIO AND PROPORTION

Some patterns are easy to see; others are less obvious. How are patterns used? Why are patterns used?

OBJECTIVE

- Formulate problems using text and data.

USING THE CHAPTER OPENER

Discuss the question on the page. Most students are aware of patterns in fabrics, floor covering, brickwork, and so on; but they probably have not thought about why these patterns are used. Patterns that are purely decorative make an item attractive and therefore appealing to potential customers. The basic pattern in a brick wall, however, allows the wall to stand.

The use of mathematical patterns will be less obvious to students. Many charts show examples of patterns used in other commercial businesses—tax charts; yardage charts in fabric stores; charts for determining how much paint is needed for a room, carpeting for a floor, and so on. These charts aid the salespeople in these businesses and benefit the customers by reducing the margin for error.

COOPERATIVE LEARNING

Write the following on the chalkboard.

Tax chart of gasoline; 1 to 100 gal

Convert a recipe that serves 4 to 8, 16, . . ., 2,048

1 movie free for every 4 rented. How many free movies possible?

Assign groups to the task of developing a chart (using patterns) to aid in one of the situations listed. They are to investigate the situation, develop practical units for a scale, and complete the charts.

SUPPLEMENTARY MATERIALS

TRP Group Projects, p. 8

OBJECTIVES

- Write ratios.
- Write equivalent ratios.

TEACHING THE LESSON

WARM-UP Write the following exercises on the chalkboard or an overhead transparency.

Write two equivalent fractions for the fraction.

1. $\frac{2}{5}$ **2.** $\frac{3}{8}$ **3.** $\frac{6}{12}$ **4.** $\frac{5}{15}$

(Answers will vary.)

INTRODUCTION Place five pens and four pencils in a container. Demonstrate how to compare the number of pens to the number of pencils, using a fraction:

$$\frac{\text{Number of pens}}{\text{Number of pencils}} \longrightarrow \frac{5}{4}$$

Next, have the students compare the number of pencils to the number of pens:

$$\frac{\text{Number of pencils}}{\text{Number of pens}} \longrightarrow \frac{4}{5}$$

INSTRUCTION Use the introduction to help students realize that a ratio compares quantities of similar things. In Example 1, stress that the order of the quantities is important: 217 to 1,400 is not the same as 1,400 to 217. Point out that since a ratio can be written as a fraction, all the rules that apply to fractions apply to ratios.

Example 2 uses the rules for equivalent fractions to form equivalent ratios. Make sure that students understand that multiplying or dividing both terms by the same nonzero number is the same as multiplying by 1, the identity element for multiplication and division.

Example 3 uses the rules for reducing a fraction to lowest terms to write a ratio in simplest form. Stress that the simplest form of a ratio always has two terms. Elicit from students that a ratio cannot be written as a whole or mixed number.

8.1 RATIO

A **ratio** is a comparison of two numbers by division.
It can be written

In words	As a fraction	With a colon
5 to 6	$\frac{5}{6}$	5:6

You read each the same way: "5 to 6."

EXAMPLE 1 At Vic's Video, there are 1,400 videos in stock. The comedy section has 217 videos. What is the ratio of the number of comedy videos to the total number of videos?

THINK: 217 comedy videos out of 1,400 videos

WRITE:

In words	As a fraction	With a colon
217 to 1,400	$\frac{217}{1{,}400}$	217:1,400

So, the ratio is 217 comedy videos to 1,400 total videos.

The numbers in a ratio are called **terms.** When you multiply or divide both terms of a ratio by the same nonzero number, you get an equivalent ratio.

EXAMPLE 2 Complete to form an equivalent ratio.

a. $\frac{7}{8} = \frac{35}{■}$

THINK: $7 \times 5 = 35$
$8 \times 5 = ■$

$$\frac{7}{8} = \frac{7 \times 5}{8 \times 5} = \frac{35}{40}$$

So, $\frac{7}{8} = \frac{35}{40}$.

b. $\frac{15}{24} = \frac{■}{8}$

THINK: $24 \div 3 = 8$
$15 \div 3 = ■$

$$\frac{15}{24} = \frac{15 \div 3}{24 \div 3} = \frac{5}{8}$$

So, $\frac{15}{24} = \frac{5}{8}$.

A ratio is in simplest form when the greatest common factor of both terms is 1. To write a ratio in simplest form, divide both terms by the greatest common factor.

EXAMPLE 3 Write the ratio $\frac{24}{40}$ in simplest form.

THINK: The GCF of 24 and 40 is 8.

Divide each term by 8. $\frac{24}{40} = \frac{24 \div 8}{40 \div 8} = \frac{3}{5}$

So, the ratio $\frac{24}{40}$ is $\frac{3}{5}$ in simplest form.

PRACTICE EXERCISES

See Extra Practice, page 444.
1. $\frac{6}{1}$; 2. 6:1; 3. 5 to 8; 4. $\frac{5}{8}$ 5. 21 to 8; 6. 12:8; 7. $\frac{120}{250}$, 8. 120:250

Complete to show the ratio in three ways.

Words	Fraction	Colon	Words	Fraction	Colon
6 to 1	• 1. ■	• 2. ■	3. ■	4. ■	5:8
5. ■	$\frac{21}{8}$	6. ■	120 to 250	7. ■	8. ■

Complete to form an equivalent ratio.

• 9. $\frac{3}{5} = \frac{■}{25}$ 15 • 10. $\frac{4}{7} = \frac{■}{28}$ 16 11. $\frac{3}{10} = \frac{27}{■}$ 90 12. $\frac{2}{3} = \frac{16}{■}$ 24 13. $\frac{8}{13} = \frac{24}{■}$ 39

14. $\frac{18}{10} = \frac{■}{5}$ 9 15. $\frac{35}{42} = \frac{■}{6}$ 5 16. $\frac{27}{15} = \frac{9}{■}$ 5 17. $\frac{30}{140} = \frac{3}{■}$ 14 18. $\frac{32}{44} = \frac{8}{■}$ 11

19. $\frac{28}{48} = \frac{7}{■}$ 12 20. $\frac{5}{16} = \frac{■}{32}$ 10 21. $\frac{6}{13} = \frac{54}{■}$ 117 22. $\frac{14}{30} = \frac{■}{15}$ 7 23. $\frac{12}{51} = \frac{4}{■}$ 17

Use the denominator to write a ratio equivalent to $\frac{3}{5}$.

• 24. 15 9 • 25. 30 18 26. 20 12 27. 45 27 28. 60 36 29. 100 60 30. 500 300 31. 250 150

Use the numerator to write a ratio equivalent to $\frac{90}{120}$.

• 32. 9 12 • 33. 3 4 34. 6 8 35. 30 40 36. 180 240 37. 900 1,200 38. 450 600 39. 9,000 12,000

Write the ratio in simplest form.

• 40. $\frac{7}{21}$ $\frac{1}{3}$ • 41. $\frac{5}{30}$ $\frac{1}{6}$ 42. $\frac{6}{18}$ $\frac{1}{3}$ 43. $\frac{8}{38}$ $\frac{4}{19}$ 44. $\frac{9}{72}$ $\frac{1}{8}$ 45. $\frac{4}{36}$ $\frac{1}{9}$ 46. $\frac{2}{22}$ $\frac{1}{11}$ 47. $\frac{3}{36}$ $\frac{1}{12}$

48. $\frac{12}{28}$ $\frac{3}{7}$ 49. $\frac{15}{27}$ $\frac{5}{9}$ 50. $\frac{24}{36}$ $\frac{2}{3}$ 51. $\frac{18}{42}$ $\frac{3}{7}$ 52. $\frac{45}{63}$ $\frac{5}{7}$ 53. $\frac{24}{48}$ $\frac{1}{2}$ 54. $\frac{14}{35}$ $\frac{2}{5}$ 55. $\frac{18}{72}$ $\frac{1}{4}$

Solve.

• 56. Vic's Video has 420 adventure videos and 140 music videos. Write the ratio in simplest form of adventure videos to music videos. $\frac{3}{1}$

57. Mike rents 15 videos for every 6 that Sue rents. Write the ratio in simplest form of Sue's rentals to Mike's rentals. $\frac{2}{5}$

58. Joan rented 7 videos at $1.99 each. How much did she pay? $13.93

59. In 16 weeks, Sally has seen 96 videos. On an average, how many did she see a week? 6 videos

TIME OUT

The ratio of rock music videos to jazz music videos is 3 to 2. If the total number of videos is 85, how many rock music videos are there? How many jazz music videos are there?
51 rock music videos; 34 jazz music videos

TIME OUT Encourage students to use the guess-and-check strategy to solve the problem.

COMMON ERROR

Some students may interchange the terms in the ratio. Have these students write the ratio in words to avoid this error.

ASSIGNMENTS

Level 1	1–8, Odd 9–55, 56, 57
Level 2	1–8, Even 2–54, 56–59, TO
Level 3	Even 14–54, 56–59, TO

FOLLOW-UP ACTIVITY

APPLICATION Have the students work in heterogeneous groups of four or five students. Provide each group with pages from the sports section of a newspaper. Have each group find examples of ratios and report to the class how ratios are used.

SUPPLEMENTARY MATERIALS

TRP Practice, p. 85

TRP Reteaching, p. 53

TRP Transparency 43

OBJECTIVES

- Write rates.
- Find unit rates.

TEACHING THE LESSON

WARM-UP Write the following exercises on the chalkboard or an overhead transparency.

Write an equivalent fraction with a denominator of 1.

1. $\frac{10}{5}$ **2.** $\frac{21}{3}$ **3.** $\frac{32}{8}$ **4.** $\frac{50}{10}$

(**1.** $\frac{2}{1}$ **2.** $\frac{7}{1}$ **3.** $\frac{4}{1}$ **4.** $\frac{5}{1}$)

INTRODUCTION Ask students to name and discuss examples of rates. Examples may include: a rate of speed, such as 55 mph; an hourly wage, such as $9.00 an hour; fuel consumption, such as 28 mpg; prices, such as three cans for $0.99.

INSTRUCTION Point out that since a rate is a ratio that compares different units, all that is true for ratios is true for rates. Use Example 1 to write rates as fractions in simplest form.

Review the concept of comparing fractions to compare rates in Example 2. Point out that the order of the cross products for $\frac{a}{b} \times \frac{c}{d}$ must be *ad-bc* for the comparison to be correct.

Explain that when the second term of a rate is 1, the rate is called a unit rate.

Example 3 explains how to convert a rate into a unit rate. Point out that even though a unit rate is often written as a whole or decimal number, it still represents a quotient. For example, 88 km/hour is the same as $\frac{88 \text{ km}}{1 \text{ hour}}$.

In Lesson 8.3, students will learn how to compare unit prices, which are unit rates, to determine the better buy.

8.2 RATE

In 28 minutes, eight telephone calls were placed from Mount Wilson to Bayville. What was the rate of calls per minutes?

A **rate** is a ratio that compares different units. Rates, like ratios, can be written as fractions in simplest form.

EXAMPLE 1 Write the rate of calls per minutes as a fraction in simplest form.

THINK: The rate is $\frac{\textbf{8 calls}}{\textbf{28 minutes}}$

$$\frac{8 \text{ calls}}{28 \text{ minutes}} \begin{matrix}\rightarrow\\ \rightarrow\end{matrix} \frac{8}{28} = \frac{2}{7}$$

You can compare rates using cross products.

EXAMPLE 2 Phone rates in Bowville depend on the phone company. Dash charges $1.30 for 8 minutes. GME charges $0.50 for 3 minutes. Which company has the lower rate?

	Dash		**GME**
1. Write the rates.	$\frac{\$1.30}{8 \text{ minutes}}$		$\frac{\$0.50}{3 \text{ minutes}}$
2. Find the cross products.	$1.30 × 3		8 × $0.50
3. Compare.	$3.90	<	$4.00

Since $1.30 per 8 minutes is less than $0.50 for 3 minutes, Dash has the lower rate.

When the second term of a rate is 1, the rate is called a **unit rate.**

EXAMPLE 3 Write the rate of 272 calls for 8 phones as a unit rate.

THINK: To find a unit rate, divide both terms by the second term.

$$\frac{272 \text{ calls}}{8 \text{ phones}} = \frac{272 \div 8}{8 \div 8} = 34 \text{ calls per phone}$$

So, the unit rate is 34 calls per phone.

PRACTICE EXERCISES

See Extra Practice, page 444.

Write the rate as a fraction in simplest form.

• **1.** \$20 for 3 records $\frac{\$20}{3}$

• **2.** 7 laps in 9 minutes $\frac{7}{9}$

3. 3 points in 4 seconds $\frac{3}{4}$

4. 14 hits in 15 games $\frac{14}{15}$

5. 213 words in 6 minutes $\frac{71}{2}$

6. 420 km in 6 hours $\frac{70}{1}$

7. \$45 for 7 tickets $\frac{\$45}{7}$

8. 15 ft in 17 minutes $\frac{15}{17}$

9. 353 km on 2 L $\frac{353}{2}$

10. \$5.34 for 6 lb $\frac{\$0.89}{1}$

11. 15 m in 10 seconds $\frac{3}{2}$

12. 20 defects in 500 bulbs $\frac{1}{25}$

13. 44 books in 14 days $\frac{22}{7}$

14. 72 pages in 45 minutes $\frac{8}{5}$

15. \$46.20 for 22 balloons $\frac{\$2.10}{1}$

16. 35 cars in 14 minutes $\frac{5}{2}$

17. 125 people in 5 buses $\frac{25}{1}$

18. 266.6 mi in 8.6 gal $\frac{31}{1}$

Compare the rates. Write >, <, or =.

• **19.** $\frac{\text{4 cans}}{\$0.88}$ ● $\frac{\text{6 cans}}{\$1.32}$ =

• **20.** $\frac{\text{87 seats}}{\text{3 rows}}$ ● $\frac{\text{96 seats}}{\text{4 rows}}$ >

21. $\frac{\text{5 oz}}{\$0.39}$ ● $\frac{\text{8 oz}}{\$0.57}$ <

22. $\frac{\text{91 pens}}{\text{7 boxes}}$ ● $\frac{\text{84 pens}}{\text{6 boxes}}$ <

23. $\frac{\text{115 minutes}}{\text{5 tapes}}$ ● $\frac{\text{180 minutes}}{\text{8 tapes}}$ >

24. $\frac{\text{247 mi}}{\text{9 gal}}$ ● $\frac{\text{286 mi}}{\text{7 gal}}$ <

25. $\frac{\text{145 yd}}{\text{7 passes}}$ ● $\frac{\text{159 yd}}{\text{8 passes}}$ >

26. $\frac{\$10.38}{\text{11 gal}}$ ● $\frac{\$13.86}{\text{14 gal}}$ <

27. $\frac{\text{37 plants}}{\text{100 seeds}}$ ● $\frac{\text{68 plants}}{\text{150 seeds}}$ <

Write the rate as a unit rate. **28.** 8 pencils per pack; **29.** 28 people per bus; **30.** 12 mph **31.** \$0.65 per pound; **32.** 25 points per game

• **28.** $\frac{\text{32 pencils}}{\text{4 packs}}$

• **29.** $\frac{\text{84 people}}{\text{3 buses}}$

30. $\frac{\text{108 mi}}{\text{9 hours}}$

31. $\frac{\$3.25}{\text{5 lb}}$

32. $\frac{\text{750 points}}{\text{30 games}}$

33. $\frac{\$186}{\text{6 tickets}}$

34. $\frac{\text{112 players}}{\text{8 teams}}$

35. $\frac{\text{126 calls}}{\text{3 minutes}}$

36. $\frac{\text{2600 words}}{\text{13 pages}}$

37. $\frac{\$367.20}{\text{18 months}}$

33. \$31 per ticket **34.** 14 players per team **35.** 42 calls per minute **36.** 200 words per page **37.** \$20.40 per month

MIXED REVIEW

38. 3,280; **39.** 6.54; **40.** 3.7; **41.** 0.05; **42.** 9.86; **43.** 0.00548; **44.** 6.73; **45.** 0.0025

Multiply or divide.

• **38.** 100×32.8

39. 0.654×10

40. $1,000 \times 0.0037$

41. 100×0.0005

• **42.** $986 \div 100$

43. $5.48 \div 1,000$

44. $67.3 \div 10$

45. $0.25 \div 100$

Solve.

• **46.** A telephone operator quoted \$25.99 charges for a 23-minute call. How much did the call cost per minute? \$1.13

47. The telephone company installed 25,640,000,000 feet of cable. Write the word name for this number. twenty-five billion, six hundred forty million

48. BRC had a \$112.74 phone bill in October and a \$181.32 bill in November. How much more was the November bill? \$68.58

49. Dash charges \$2.38 for a 7-minute call. GME charges \$1.70 for a 5-minute call. Which company offers the lower rate? neither; same rate, \$0.34 per minute

ASSIGNMENTS

Level 1	1–9, 19–21, 28–32, 38–47
Level 2	10–18, 22–27, 33–48
Level 3	Odd 1–45, 46–49

FOLLOW-UP ACTIVITY

ENRICHMENT Have students work in pairs. Provide each pair of students with a page of classified ads from different sources. Have the students figure out the average number of characters per line, lines per ad, ads per column, and columns per page. Establish a rate for an ad, such as \$7.50 for the first 15 words, \$0.60 for each additional word, or \$2.50 per line. Have students write an ad and calculate its cost.

SUPPLEMENTARY MATERIALS

TRP Practice, p. 86

TRP Reteaching, p. 54

TRP Transparency 43

OBJECTIVES

- Solve problems that involve finding a unit price.
- Determine which of two items is a better buy.

TEACHING THE LESSON

WARM-UP Write the following exercises on the chalkboard or an overhead transparency.

Find the equivalent ratio.

1. $\frac{8}{2} = \frac{\blacksquare}{1}$ (4)

2. $\frac{15}{3} = \frac{\blacksquare}{1}$ (5)

3. $\frac{24}{2} = \frac{\blacksquare}{1}$ (12)

4. $\frac{88.8}{4} = \frac{\blacksquare}{1}$ (22.2)

INTRODUCTION Select an item, such as a box of raisins, from a large newspaper advertisement or flyer. Ask students how they can tell whether one item is a better buy than another similar item. Use the responses to lead into the concept of comparison shopping.

INSTRUCTION As you discuss the first example, be sure that students understand that the unit price is the price per standard amount.

Point out that sometimes items involve different units of measure; in that case, the items must be expressed in terms of the same unit of measure.

You may wish to have students use calculators to solve the problems. Note that Problem 15 involves different units of length measure (feet and yards); students should express both measurements in terms of the same unit before finding the unit price.

PROBLEM Solving APPLICATION

8.3 CONSUMER: COMPARISON SHOPPING

When you shop, you often need to decide which of two items is the better buy. In comparing prices, you need to use the unit price of each item. The **unit price** is the price per standard amount, such as the price per ounce or the price per dozen.

A 12-oz box of Superbran Flakes is selling for $1.92.
A 16-oz box of Superbran Flakes is selling for $2.44.
Which is the better buy?

THINK: You can use this rate: $\frac{\text{price of cereal}}{\text{amount of cereal}}$

The unit price is a unit rate, cost per ounce, or $\frac{\text{cost}}{1 \text{ oz}}$.

1. Find the unit price for the 12-oz box. $\frac{\$1.92}{12 \text{ oz}} = 12\overline{)\$1.92} = \$0.16$ per ounce
2. Find the unit price for the 16-oz box. $\frac{\$2.44}{16 \text{ oz}} = 16\overline{)\$2.44} = \$0.1525$ per ounce
3. Compare the unit prices. $\$0.1525 < \$0.16$

So, the 16-oz box is the better buy.

FOR DISCUSSION See TE side column.

When choosing between two different brands of a product, what other considerations are there besides the unit price of each box? Explain.

Find the unit price. Remember to estimate whenever you use your calculator.

- • **1.** Fruit: \$1.76 for a 16-oz can \$0.11 per ounce
- • **2.** Twine: \$3.25 for 50 ft \$0.065 per foot
- **3.** Cheese: 24 oz for \$2.88 \$0.12 per ounce
- **4.** Soap: 3 bars for \$1.26 \$0.42 per bar
- **5.** Apples: 6 for \$1.50 \$0.25 per apple
- **6.** Meat: 2.6 lb for \$8.06 \$3.10 per pound

Find the unit price to the nearest tenth of a cent.

- • **7.** An 8-oz box of Ultrawheat Flakes is selling for \$1.65. \$0.206 per ounce
- • **8.** 5 cans of fruit salad are selling for \$3.36. \$0.672 per can
- **9.** 3 green peppers are selling for \$1.07. \$0.357 per pepper
- **10.** 4 avocados are selling for \$3.50. \$0.875 per avocado
- **11.** A 12-oz bottle of fruit juice is selling for \$0.89. \$0.074 per ounce

Which is the better buy if price is the only consideration?

- • **12.** 3 lb of apples for \$1.89 or 5 lb of apples for \$3.17 3 lb for \$1.89
- • **13.** A 4-oz jar of olives for \$1.04 or a 6-oz jar of olives for \$1.53 6-oz jar for \$1.53
- **14.** 50 sheets of notebook paper for \$0.65 or 144 sheets for \$2.16 \$0.65 for 50 sheets
- **15.** A 15-ft extension cord for \$3.60 or a 4-yd extension cord for \$2.58 4-yd extension cord for \$2.58
- **16.** A 6-oz jar of tomato sauce for \$1.49 or a 16-oz jar for \$3.89 16-oz jar for \$3.89
- **17.** A 6-oz bag of grated mozzarella cheese for \$1.79 or a 16-oz piece for \$4.59 16-oz piece for \$4.59
- **18.** In many supermarkets and groceries, information is provided about the unit prices of various items. The weights of items and the total prices of the items are also indicated. Find such information about five different items. Then check to see that

$$\textit{weight} \times \textit{unit price} = \textit{total price of item.}$$

FOR DISCUSSION Students should mention factors such as quality of the item, practicality of size, and ease of packaging.

ASSIGNMENTS

Level 1	1–6, 12–15
Level 2	1–17
Level 3	Even 2–18

FOLLOW-UP ACTIVITY

APPLICATION Provide three or four different brands of the same item, such as three 16-oz jars of spaghetti sauce. Have students find the unit price for each, and then consider which item is the better buy. Encourage students to consider factors such as taste and packaging.

SUPPLEMENTARY MATERIALS

TRP Practice, p. 87

TRP Applications, p. 14

TRP Transparency 44

OBJECTIVE

- Write and solve proportions.

TEACHING THE LESSON

WARM-UP Write the following exercises on the chalkboard or an overhead transparency.

Write >, <, or =.

1. $4 \times 8 \bullet 3 \times 9$

2. $9 \times 7 \bullet 8 \times 8$

3. $2.25 \times 4 \bullet 3 \times 3$

4. $6.75 \times 4 \bullet 7 \times 3$

(**1.** > **2.** < **3.** = **4.** >)

INTRODUCTION Review equivalent ratios. Write:

a. $\frac{2}{3}$ and $\frac{12}{18}$ **b.** $\frac{3}{5}$ and $\frac{10}{6}$

Ask which ratios are equivalent. Stress that because $\frac{2}{3} \times \frac{6}{6} = \frac{12}{18}$, the fractions are equivalent. Point out that there is no similar relationship between $\frac{3}{5}$ and $\frac{10}{6}$.

INSTRUCTION After reviewing Example 1, you may wish to have students use cross products to determine if the following ratios form a proportion.

a. $\frac{3}{7} \bullet \frac{6}{14}$ (yes) **b.** $\frac{12}{20} \bullet \frac{4}{5}$ (no)

Example 2 introduces two new skills: writing a proportion and solving a proportion. Concentrate on building each skill separately. Guide students in solving the following proportions:

a. $\frac{3}{n} = \frac{15}{40}$ (8) **b.** $\frac{6}{10} = \frac{n}{30}$ (18)

CHECKPOINT Distribute to each student four index cards on which the answer choices a–d appear. After allowing students a few minutes, ask them to raise the letter that indicates their answer choice. Scan the cards and identify those students who need additional attention before proceeding to the Practice Exercises. The incorrect answer choices include common errors.

8.4 SOLVING A PROPORTION

A **proportion** is a statement that two ratios are equivalent.
To determine if two ratios form a proportion, check the cross products.
If the cross products of the ratios are equal, the ratios form a proportion.

EXAMPLE 1 Determine if the ratios form a proportion.

a. $\frac{8}{5} \stackrel{?}{=} \frac{2}{3}$

$\frac{8}{5} \times \frac{2}{3}$

$8 \times 3 \bullet 5 \times 2$

$24 \neq 10$

$\frac{8}{5}$ and $\frac{2}{3}$ do not form a proportion.

b. $\frac{6}{9} \stackrel{?}{=} \frac{5}{7.5}$

$\frac{6}{9} \times \frac{5}{7.5}$

$6 \times 7.5 \bullet 9 \times 5$

$45 = 45$

$\frac{6}{9}$ and $\frac{5}{7.5}$ form a proportion.

Many problems can be solved by writing a proportion and then finding the missing term.

EXAMPLE 2 Eric used a pancake mix that called for 3 eggs to make 16 pancakes. How many eggs would he need to make 48 pancakes?

1. Write a proportion.

THINK: **3 eggs is to 16 pancakes as *n* eggs is to 48 pancakes**

$\frac{3}{16} = \frac{n}{48}$ ← **eggs** / ← **pancakes**

2. Multiply to find cross products.

$3 \times 48 = 16 \times n$
$144 = 16 \times n$

3. Find n.

THINK: To find the missing factor, divide 144 by 16.

$144 \div 16 = n$
$9 = n$

So, 9 eggs are needed to make 48 pancakes.

CHECKPOINT

Write the letter of the correct answer.

Find n.

1. $\frac{n}{16} = \frac{4}{8}$	c	**a.** 2	**b.** 4	**c.** 8	**d.** 32
2. $\frac{6}{5} = \frac{n}{35}$	d	**a.** 7	**b.** 30	**c.** 36	**d.** 42
3. $\frac{9}{n} = \frac{3}{15}$	c	**a.** 5	**b.** 15	**c.** 45	**d.** 135
4. $\frac{12}{18} = \frac{2}{n}$	a	**a.** 3	**b.** 6	**c.** 24	**d.** 36

PRACTICE EXERCISES
See Extra Practice, page 444.

Determine if the ratios form a proportion. Write *yes* or *no*.

• **1.** $\frac{6}{7} \stackrel{?}{=} \frac{14}{12}$ no •**2.** $\frac{8}{3} \stackrel{?}{=} \frac{56}{21}$ yes **3.** $\frac{5}{9} \stackrel{?}{=} \frac{30}{54}$ yes **4.** $\frac{72}{35} \stackrel{?}{=} \frac{8}{5}$ no

5. $\frac{49}{14.7} \stackrel{?}{=} \frac{7}{2.1}$ yes **6.** $\frac{0.25}{0.3} \stackrel{?}{=} \frac{0.1}{0.12}$ yes **7.** $\frac{2.1}{30} \stackrel{?}{=} \frac{9}{1.5}$ no **8.** $\frac{4.8}{3.2} \stackrel{?}{=} \frac{9.4}{6.4}$ no

Find n.

• **9.** $\frac{3}{5} = \frac{n}{25}$ 15 •**10.** $\frac{7}{4} = \frac{n}{28}$ 49 **11.** $\frac{6}{15} = \frac{n}{45}$ 18 **12.** $\frac{9}{7} = \frac{n}{56}$ 72

13. $\frac{n}{48} = \frac{8}{12}$ 32 **14.** $\frac{n}{35} = \frac{8}{7}$ 40 **15.** $\frac{n}{26} = \frac{10}{13}$ 20 **16.** $\frac{n}{72} = \frac{3}{8}$ 27

17. $\frac{0.5}{0.8} = \frac{n}{3.2}$ 2 **18.** $\frac{n}{6.3} = \frac{0.4}{0.9}$ 2.8 **19.** $\frac{1.4}{0.3} = \frac{2.8}{n}$ 0.6 **20.** $\frac{4.5}{n} = \frac{6.3}{0.7}$ 0.5

21. $\frac{n}{15.9} = \frac{4.2}{12.6}$ 5.3 **22.** $\frac{n}{3.7} = \frac{18.9}{2.1}$ 33.3 **23.** $\frac{30.6}{n} = \frac{20.4}{4}$ 6 **24.** $\frac{5.2}{n} = \frac{3.5}{24.5}$ 36.4

MIXED REVIEW

Write the fraction or mixed number as a decimal.

•**25.** $\frac{25}{100}$ 0.25 •**26.** $\frac{45}{50}$ 0.9 **27.** $\frac{1}{8}$ 0.125 **28.** $4\frac{45}{60}$ 4.75 **29.** $30\frac{10}{16}$ 30.625 **30.** $7\frac{4}{5}$ 7.8

Use the table below for Exercises 31–34.

Item	Number	Mix	Milk	Eggs	Oil
Pancakes	25	$2\frac{1}{4}$ cups	2 cups	2	2 tbsp
Muffins	144	8 cups	$6\frac{1}{2}$ cups	4	$\frac{1}{4}$ cup

•**31.** How much milk is needed to make 175 pancakes? 14 cups

32. How much mix is needed to make 36 muffins? 2 cups

33. If you had 10 cups of milk, how much would be left after making 144 muffins? $3\frac{1}{2}$ cups

34. To triple the pancake recipe, how many cups of mix would you need? $6\frac{3}{4}$ cups

TIME OUT There are three plates. On one plate there are three pancakes stacked so that the largest one is on the bottom and the smallest one is on the top. You can only move one pancake at a time from plate to plate, and you can never place a larger pancake on a smaller one. What is the least number of moves that it will take to stack the pancakes in the same order on a different plate? 7 moves

TIME OUT You may wish to have students draw a picture to solve the problem. Students may find it helpful to make a separate drawing for each step so that they can clearly visualize the moves.

COMMON ERROR

Some students will forget to multiply by cross products when finding the missing term. For example, in Checkpoint 1, these students will select item d. Have these students draw arrows for the cross products to avoid this error.

ASSIGNMENTS

Level 1	1–4, 9–16, 25–32
Level 2	Odd 1–23, 25–34
Level 3	5–8, 17–34, TO

FOLLOW-UP ACTIVITY

ENRICHMENT Have students use cross products to check which of the following properties are true:

If $\frac{2}{3} = \frac{4}{6}$, then:

a. $\frac{2}{4} = \frac{3}{6}$. (yes)

b. $\frac{3}{2} = \frac{6}{4}$. (yes)

c. $\frac{3}{4} = \frac{2}{6}$. (no)

d. $\frac{2+3}{3} = \frac{4+6}{6}$. (yes)

SUPPLEMENTARY MATERIALS

TRP Practice, p. 88

TRP Reteaching, p. 55

TRP Transparency 44

OBJECTIVES

- Solve problems using estimation.
- Determine whether an overestimate or underestimate will best solve a problem.

TEACHING THE LESSON

WARM-UP Write the following exercises on the chalkboard or an overhead transparency.

Round each number to its greatest place.

1. 9.4 (9) **2.** 11.8 (12)

3. 226 (200) **4.** 381 (400)

5. 8,315 (8,000) **6.** 34,927 (30,000)

INTRODUCTION Ask students if they have ever been in situations when they needed to estimate prices. Encourage students to discuss the methods they used, and the results of their estimates. Then ask students to name other situations in which precise answers are not necessary, and estimated answers are adequate.

INSTRUCTION After you have discussed Example A, you may wish to present students with the following situation: Ties are priced at $7.29. You have $30. Can you buy 4 ties?

Ask students to overestimate the cost. (4 × $8 = $32) Note that it is greater than the amount they have to spend. Many students will incorrectly interpret this result to mean that they do not have enough money to buy the ties. Make sure students understand that this result only tells them that they now need to find the exact answer to be sure. In fact, they will discover they have enough money, since 4 × $7.29 = $29.16.

PROBLEM *Solving* STRATEGY

8.5 USING ESTIMATION

Situation:

At Byrne's Automotive Shop, front floor mats are priced at $9.29 each. Rear floor mats are priced at $6.45 each.

Strategy:

Sometimes an estimate is all that is needed to solve a problem.

Applying the Strategy:

A. Chris has $25. Does he have enough money to buy one front mat and two rear mats?

THINK: When you want to be sure that you have enough money to make a purchase, make an **overestimate.**

1. Estimate by rounding up. $9.29 → $10 $6.45 → $7

2 × $7 = $14 $10 + $14 = $24

Since $25 is greater than the overestimated total, Chris will have enough to buy the mats.

B. Last week Byrne's Shop sold 26 front floor mats. Were the total sales of front floor mats at least $200 during the week?

THINK: When you want to be sure that you have reached a minimum total amount, make an **underestimate.**

Estimate by rounding down. 26 → 20 $9.29 → $9 20 × $9 = $180

Since the underestimated total is less than $200, you cannot be sure that the sales reached $200.

Find the exact answer. 26 × $9.29 = $241.54

The total sales were at least $200 during the week.

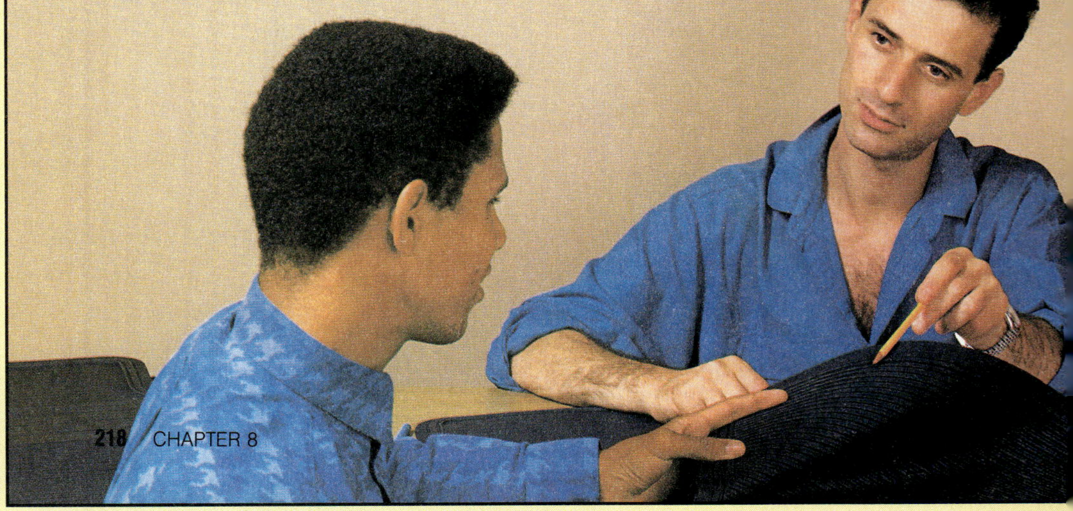

PRACTICE EXERCISES See Extra Practice, page 445.

Which is the better estimate?

1. Amy wants to buy 3 car polishing cloths and a flashlight. Each cloth costs $2.49. A flashlight costs $5.40. Will $12 be enough? b

   a. 3 × $2 = $6
   $6 + $5 = $11

   b. 3 × $3 = $9
   $9 + $6 = $15

Use the advertisement to answer Exercises 2–5.

Byrne's Sale

Paste Wax	$6.39
Safety Flares	$8.65
Dry Gas	$1.25
Vinyl Cleaner	$3.79

2. Will $8 be enough to buy 2 containers of vinyl cleaner? yes
3. Will $15 be enough to buy safety flares and paste wax? no
4. Will $25 be enough to buy 6 bottles of dry gas and safety flares? yes
5. Is $20 enough to buy one of each item? no

Use the table of projected profits to answer Exercises 6–9.

Byrne's Automotive Shop can hire another cashier if the projected profits are over $100,000 for this 4-month period.

BYRNE'S AUTOMOTIVE PROJECTED PROFIT RECORD

July	$28,750
August	$36,109
September	$24,930
October	$41,885

6. Should you overestimate or underestimate the projected profits to determine if the goal of $100,000 has been reached? underestimate
7. What is the projected estimated profit for this 4-month period? $110,000
8. What is the projected profit for this 4-month period? $131,674
9. Can Byrne's Automotive hire another cashier? yes

In Example B, students examine a situation where an underestimated total is less than the minimum amount needed and therefore they must find an exact answer to be sure if the minimum was reached. Make sure students understand that when an underestimated total is greater than the minimum amount needed, it assures that the minimum amount has been reached.

ASSIGNMENTS

Level 1	1–8
Level 2	1–8
Level 3	1–8

FOLLOW-UP ACTIVITY

SITUATIONAL PROBLEM SOLVING Have students work in small groups. Present the following problem:

Byrne's Automotive Shop is sponsoring a contest. Each contestant must design a "Car of the Future." The designer of the prize-winning car receives $5,000.

Have students work together to design a car. Have them include special features, a diagram of what the car would look like, and the name of the car.

SUPPLEMENTARY MATERIALS

TRP Practice, p. 89

TRP Applications, p. 15

TRP Transparency 45

OBJECTIVE

- Solve problems by using proportions.

TEACHING THE LESSON

WARM-UP Write the following exercises on the chalkboard or an overhead transparency.

Find the missing factor.

1. $4 \times n = 24$ (6)
2. $6 \times n = 90$ (15)
3. $45 = 5 \times n$ (9)
4. $154 = 7 \times n$ (22)
5. $2.5 \times n = 10$ (4)
6. $10.5 = 3 \times n$ (3.5)

INTRODUCTION Have students discuss trips that they have taken by car, or trips that they would like to take by car. How are car trips planned? How can you estimate how long a car trip will take? What are the advantages and disadvantages of traveling by car?

INSTRUCTION Point out that to solve these common problems about travel, students can apply their knowledge of solving proportions. Note that the greatest difficulty lies in writing the correct proportion. In Example A, emphasize that each ratio must have *distance* as the first term, and *time* as the second term. In Example B, each ratio must have *distance* as the first term and *gallons* as the second term.

Encourage students to make sure their answers are reasonable. For instance, in Example A, students should think: "6 hours is greater than 2 × 2.5 hours, so the number of miles must be greater than 2 × 125, or 250 miles. Therefore, 300 miles is a reasonable answer." When students find an answer that is not reasonable, encourage them to review the proportion they wrote, then check their computation.

PROBLEM Solving APPLICATION

8.6 USING PROPORTIONS TO SOLVE PROBLEMS

Speed, time, and distance can be calculated using proportion. When you travel, it takes a certain amount of time to go a certain number of miles. If you know your rate of speed, it is easy to calculate how far you can travel in a certain amount of time.

A. Your family is on a vacation trip.
In highway driving, you travel 125 miles in 2.5 hours.
If you continue to travel at this rate,
how many miles will you travel in 6 hours?

A proportion can be used to solve the problem.

THINK: Use the rate, $\frac{\text{distance}}{\text{time}}$.

You know that $\frac{125}{2.5}$ is the rate of speed.

1. Write a proportion.

THINK: 125 mi is to 2.5 hours as *n* mi is to 6 hours.

$$\frac{125}{2.5} = \frac{n}{6} \begin{matrix}\leftarrow \text{miles} \\ \leftarrow \text{hours}\end{matrix}$$

2. Multiply to find the cross products.

$$125 \times 6 = 2.5 \times n$$
$$750 = 2.5 \times n$$

3. Find n.

THINK: To find the missing factor, divide 750 by 2.5.

$$n = 750 \div 2.5$$
$$n = 300$$

You will travel 300 miles in 6 hours.

B. Suppose a car travels 112 miles on 3.5 gal of gasoline.
How many gallons are needed to travel 272 miles?

THINK: Use the rate $\frac{\text{distance}}{\text{gallons}}$ to write a proportion.

1. Write a proportion.

THINK: 112 mi is to 3.5 gal as 272 mi is to *n* gal.

$$\frac{112}{3.5} = \frac{272}{n} \begin{matrix}\leftarrow \text{miles} \\ \leftarrow \text{gallons}\end{matrix}$$

2. Multiply to find the cross products.

$$112 \times n = 3.5 \times 272$$
$$112 \times n = 952$$

3. Find n.

THINK: To find the missing factor, divide 952 by 112.

$$n = 952 \div 112$$
$$n = 8.5$$

8.5 gal are needed to travel 272 miles.

Solve.

How far can you travel in 5 hours if you travel at the rate of:

• **1.** 174 mi in 3 hours? 290 mi • **2.** 130 mi in 2.5 hours? 260 mi **3.** 320 mi in 6.4 hours? 250 mi

How far can you travel in 9.2 hours if you travel at the rate of:

4. 162 mi in 3 hours? 496.8 mi **5.** 216 mi in 4.5 hours? 441.6 mi **6.** 234 mi in 5.2 hours? 414 mi

If a car travels:	It can travel:
• **7.** 84 mi on 3 gal of gasoline	140 mi on ■ gal. 5
• **8.** 99 mi on 4.5 gal of gasoline	275 mi on ■ gal. 12.5
9. 207 mi on 6.9 gal of gasoline	441 mi on ■ gal. 14.7
10. 72 mi on 4 gal of gasoline	■ mi on 10 gal. 180
11. 156 mi on 6.5 gal of gasoline	■ mi on 8.5 gal. 204
12. 153 mi on 5.1 gal of gasoline	■ mi on 9.6 gal. 288

• **13.** One morning, you drive 216 mi in 4 hours. At that rate, how many miles will you drive in 7 hours? 378 mi

14. The Carruthers family drives 96 mi in 2.4 hours. At that rate, how long will it take them to drive 48 mi? 1.2 hours

15. The Carruthers family car travels 221 mi on 8.5 gal of gasoline. At that rate, how many miles will the car travel on 14.5 gal? 377 mi

16. The Carruthers family paid $13.44 for 12 gal of gas for their car. Their tank holds 16 gal. At that rate, how much would a full tank cost? $17.92

ASSIGNMENTS

Level 1	1–9, 13–14
Level 2	1–16
Level 3	Even 2–12, 13–16

FOLLOW-UP ACTIVITY

ENRICHMENT Introduce Hooke's Law as another example of the use of rates and proportions.

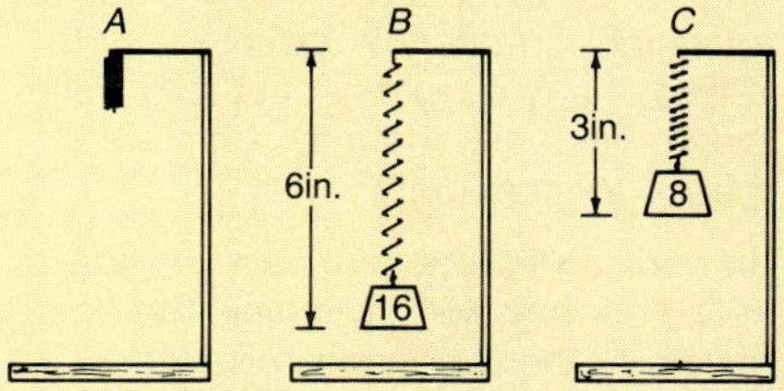

Each drawing shows the same spring, with different weights and the distances that it stretches. Hooke's Law states that for a given spring, the ratio of weight to distance stretched remains the same (so long as the spring remains elastic).

1. If a 12-lb weight is attached to the spring, how many inches will it stretch? (4.5 in.)

2. If a certain weight is attached to the spring, it stretches 7.5 in. How many pounds is the weight? (20 lb)

SUPPLEMENTARY MATERIALS

TRP Practice, p. 90

TRP Transparency 45

OBJECTIVE

- Evaluate student progress.

USING THE REVIEW

This page provides a means for informally evaluating students' understanding of the skills and concepts covered so far in this chapter.

Have the students look at the page to familiarize themselves with the various question formats that are presented. Discuss any questions that they may have. Then ask them to complete the page independently.

In addition to grading them individually, you may wish to review the answers to the questions collectively with the students.

Page references appear in brackets. They refer to pages on which a particular skill was introduced.

Before continuing on to the topics found in the remainder of the chapter, you may wish to have students review any skills or concepts in which they have demonstrated weakness.

MIDCHAPTER REVIEW

Write the ratio in two ways. [210]

	Colon	Fraction
75 oz to 29 oz	**1.** ■ 75:29	**2.** ■ $\frac{75}{29}$
32 days to 76 days	**3.** ■ 32:76	**4.** ■ $\frac{32}{76}$
\$14.12 for 3 tapes	**5.** ■ \$14.12:3	**6.** ■ $\frac{\$14.12}{3}$
329 mi on 13 gal	**7.** ■ 329:13	**8.** ■ $\frac{329}{13}$
4.5 cups to 8 oz	**9.** ■ 4.5:8	**10.** ■ $\frac{4.5}{8}$

Write the ratio in simplest form. [210]

11. $\frac{7}{28}$ $\frac{1}{4}$ **12.** $\frac{5}{45}$ $\frac{1}{9}$ **13.** $\frac{20}{32}$ $\frac{5}{8}$ **14.** $\frac{14}{63}$ $\frac{2}{9}$ **15.** $\frac{9}{21}$ $\frac{3}{7}$

Write as a unit rate. [212]

16. $\frac{24 \text{ pens}}{8 \text{ boxes}}$ 3 pens per box **17.** $\frac{35 \text{ pages}}{7 \text{ chapters}}$ 5 pages per chapter **18.** $\frac{28 \text{ people}}{14 \text{ seats}}$ 2 people per seat **19.** $\frac{30 \text{ miles}}{6 \text{ gallons}}$ 5 miles per gallon **20.** $\frac{120 \text{ calls}}{40 \text{ minutes}}$ 3 calls per minute

Determine if the ratios form a proportion. Write *yes* or *no*. [216]

21. $\frac{3}{5} \stackrel{?}{=} \frac{9}{15}$ yes **22.** $\frac{4}{9} \stackrel{?}{=} \frac{18}{36}$ no **23.** $\frac{5}{7} \stackrel{?}{=} \frac{15}{24}$ no **24.** $\frac{8}{11} \stackrel{?}{=} \frac{32}{44}$ yes **25.** $\frac{7}{3} \stackrel{?}{=} \frac{9}{21}$ no

26. $\frac{18}{21} \stackrel{?}{=} \frac{6}{7}$ yes **27.** $\frac{45}{15} \stackrel{?}{=} \frac{9}{5}$ no **28.** $\frac{30}{20} \stackrel{?}{=} \frac{60}{40}$ yes **29.** $\frac{7.5}{2.5} \stackrel{?}{=} \frac{1.8}{0.6}$ yes **30.** $\frac{4.8}{1.6} \stackrel{?}{=} \frac{22.6}{7.2}$ no

Find n. [216]

31. $\frac{n}{6} = \frac{25}{30}$ 5 **32.** $\frac{14}{n} = \frac{4}{8}$ 28 **33.** $\frac{18}{21} = \frac{n}{7}$ 6 **34.** $\frac{24}{15} = \frac{8}{n}$ 5 **35.** $\frac{54}{72} = \frac{n}{12}$ 9

Solve. [214, 218, 220]

36. Which is the better buy if price is the only consideration: an 8-oz box of spinach pasta for \$0.89 or a 12-oz box of pasta for \$1.40? 8-oz box for \$0.89

37. Beth wants to buy three cans of soup for \$0.89 a can, one loaf of wheat bread for \$1.09, and 4 lb of fish at \$4.59 per pound. Will \$20 be enough? no

38. How far can you go in 6 hours if you are traveling at a rate of 275 mi in 5 hours? 330 mi

39. A car goes 256 mi on 16 gal of gas. How far can it go on 10 gal? 160 mi

PROBLEM Solving APPLICATION

8.7 CAREER: PHOTOGRAPHER

Juan Mendoza is a photographer. He takes photographs for large corporations, magazines, and models. Sometimes Juan enlarges or reduces a photograph he has taken.

The relationships of the dimensions can be written in equivalent ratios.

$$\frac{\text{LENGTH OF ORIGINAL}}{\text{LENGTH OF ENLARGEMENT}} = \frac{\text{WIDTH OF ORIGINAL}}{\text{WIDTH OF ENLARGEMENT}}$$

A photograph is 7 in. long and 5 in. wide. Juan makes an enlargement that is 10.5 in. long. What is the width of the enlargement?

1. Write a proportion.

THINK: 7 in. is to 10.5 in. as 5 in. is to n in.

$$\frac{7}{10.5} = \frac{5}{n} \begin{array}{l}\leftarrow \textbf{original (in.)}\\ \leftarrow \textbf{enlargement (in.)}\end{array}$$

2. Multiply to find the cross products.

$7 \times n = 10.5 \times 5$

$7 \times n = 52.5$

3. Find n.

THINK: To find the missing factor, divide 52.5 by 7.

$n = 52.5 \div 7$

$n = 7.5$

The width of the enlargement is 7.5 in.

Find the missing dimension.

Original photograph		Enlargement		Original photograph		Reduction	
Length	Width	Length	Width	Length	Width	Length	Width
• **1.** 7 in.	5 in.	17.5 in.	■ in. 12.5	**2.** 9 in.	6 in.	4.5 in.	■ in. 3
3. 3 in.	5 in.	■ in. 6.6	11 in.	**4.** 10 in.	8 in.	■ in. 7.5	6 in.
5. 8 in.	5.6 in.	20 in.	■ in. 14	**6.** 11 in.	8.5 in.	4.4 in.	■ in. 3.4

Solve.

• **7.** A photograph has a length of 6 in. and a width of 4 in. Juan makes an enlargement that has a length of 13.5 in. What is the width of the enlargement? 9 in.

8. In Exercises 1, 3, and 5, is the ratio of the length to the width in the enlargement equivalent to the ratio of the length to the width in the original photograph? yes

OBJECTIVE

- Solve problems that involve enlargements or reductions of photographs.

TEACHING THE LESSON

WARM-UP Write the following exercises on the chalkboard or an overhead transparency.

Solve the proportion.

1. $\frac{8}{6} = \frac{12}{n}$ (9)
2. $\frac{6}{15} = \frac{n}{35}$ (14)
3. $\frac{9}{5} = \frac{13.5}{n}$ (7.5)
4. $\frac{3}{2} = \frac{n}{12.4}$ (18.6)

INTRODUCTION Point out that a photographer, an artist, or an architect often needs to reduce or enlarge a photograph, a drawing, or a blueprint.

INSTRUCTION The idea of figures that are the "same shape" should be familar to students. After you discuss the example, you may wish to point out that just as the length of the original photograph is 1.4 times as great as the width, so the length of the enlargement is 1.4 times as great as its width.

$7 \div 5 = 1.4$

$10.5 \div 7.5 = 1.4$

ASSIGNMENTS

Level 1	1–7
Level 2	1–7
Level 3	1–8

SUPPLEMENTARY MATERIALS

TRP Practice, p. 91

TRP Transparency 46

OBJECTIVE

- Use a proportion to find actual size, scale size, or scale.

TEACHING THE LESSON

WARM-UP Draw line segments of the following lengths on the chalkboard or an overhead transparency.

Have students measure each segment to the nearest centimeter and to the nearest millimeter.

1. 68mm **2.** 92mm **3.** 44mm

4. 85mm

(Check students' drawings.)

INTRODUCTION Pose the following problem:

You are an engineer who has designed a new airplane that you want to sell to the airlines. How can you show them what the plane will look like without building the plane?

Guide students to realize the need for a scale model. Emphasize that for a scale model to be accurate, it must be in proportion to the full-size plane. Review the meaning of *in proportion to.* Discuss other situations in which scale models or drawings are useful.

INSTRUCTION Explain that a proportion that compares the scale, or model size, to the actual size is a simple way to find measurements. Students should note that the ratio for one scale is usually written in the colon form. For instance, the scale in Example 1 is written as 3 in. : 40 ft.

Point out that the purpose of Example 1 is to find a scale measure. Conversely, the purpose of Example 2 is to find an actual measure. Emphasize that in both situations, the method of finding the missing measurement is the same.

Remind students to first measure, then solve a proportion to find the actual measurements in Exercises 1–16.

8.8 USING A SCALE

A **scale drawing** uses a ratio to compare an object's size on a drawing to its actual size. The ratio of the drawing size to the actual size is called the **scale.**

When you know the scale and the actual size, you can write and solve a proportion to find the scale size.

EXAMPLE 1 The Spruce Goose has the largest wingspan of any plane. It measures almost 320 ft tip to tip. For a scale model of the plane, a scale where 3 in. represents 40 ft was used. How long is the wingspan of the scale model of the plane?

1. Write a proportion. $\frac{3}{40} = \frac{n}{320}$ ← **model (in.)** ← **actual (ft)**
2. Multiply to find the cross products. $3 \times 320 = 40 \times n$; $960 = 40 \times n$
3. Find n. $960 \div 40 = n$; $24 = n$

So, the wingspan of the scale model is 24 in.

Scale drawings show something that is too large or too small to be shown easily in its actual size. You can also use a proportion to find the actual size.

EXAMPLE 2 Find the distance from the foul line to the basket.

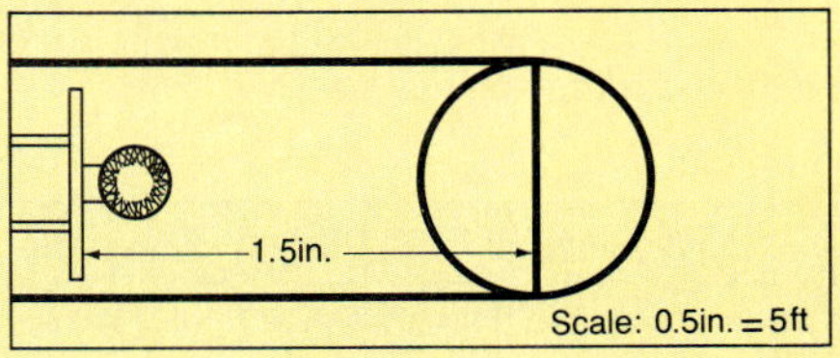

1. Write the proportion. $\frac{0.5}{5} = \frac{1.5}{n}$ ← **scale (in.)** ← **actual (ft)**
2. Multiply to find the cross products. $0.5 \times n = 5 \times 1.5$; $0.5 \times n = 7.5$
3. Find n. $n = 7.5 \div 0.5$; $n = 15$

So, the distance is 15 ft.

PRACTICE EXERCISES

See Extra Practice, page 445.

Use the floor plan to complete the table.

	Room	Actual measure	
• 1.	Living room (a)	■	8 m
• 2.	Living room (b)	■	6 m
3.	Family room (c)	■	8 m
4.	Family room (d)	■	8 m
5.	Kitchen (e)	■	6 m
6.	Laundry room (f)	■	3 m
7.	Laundry room (g)	■	4 m
8.	Bath (h)	■	4 m
9.	Bath (i)	■	3 m
10.	Closet (j)	■	2 m
11.	Dining room (k)	■	4 m
12.	Dining room (l)	■	7 m
13.	Hall (m)	■	2 m
14.	Living/family (n)	■	14 m
15.	Bath/laundry (o)	■	8 m
16.	Laundry/ kitchen/ family (p)	■	17 m

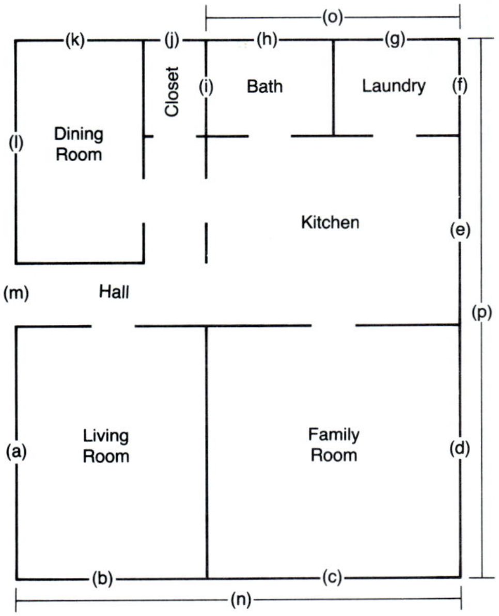

Scale: 0.5 cm = 1 m

A model train was made to the scale of 0.5 in.: 2 ft. Find the actual measure.

Car	Engine	Caboose	Flat car	Freight	Passenger
Scale measure	6 in.	4.5 in.	10 in.	5.5 in.	10.5 in.
Actual measure	• 17. ■	• 18. ■	19. ■	20. ■	21. ■

17. 24 ft
18. 18 ft
19. 40 ft
20. 22 ft
21. 42 ft

Solve.

• **22.** A scale drawing of an insect is 4.2 cm long. The scale is 6 cm : 1 cm. How long is the insect? 0.7 cm

23. A scale of 2 cm : 5 mm is used to draw a bolt. The bolt is 7.4 mm long. How long is the drawing? 2.96 cm

24. On a map, two towns are 5.8 in. apart. The actual distance is 40.6 mi. What is the scale? 1 in. : 7 mi

25. A scale drawing uses 13 cm to represent 390 km. What scale is being used? 1 cm : 30 km

ASSIGNMENTS

Level 1	1–16, 17–23
Level 2	1–16, 17–24
Level 3	1–16, 17–25

FOLLOW-UP ACTIVITY

APPLICATION Have students measure different rooms in the school and the furniture or equipment in them. Using a scale of 1 in. : 6 ft, have them make scale drawings of the rooms.

SUPPLEMENTARY MATERIALS

TRP Practice, p. 92

TRP Reteaching, p. 56

TRP Transparency 46

OBJECTIVE

- Solve problems by using a map.

TEACHING THE LESSON

WARM-UP Write the following exercises on the chalkboard or an overhead transparency.

Solve the proportion.

1. $\frac{4}{150} = \frac{6}{n}$ (225)

2. $\frac{6}{200} = \frac{15}{n}$ (500)

3. $\frac{2.5}{150} = \frac{7.5}{n}$ (450)

4. $\frac{1}{200} = \frac{4.5}{n}$ (900)

INTRODUCTION If possible, display a large map of the United States. Ask the students to recall situations in which they have used maps to estimate or calculate distances between cities. Let a few volunteers tell what procedure they used.

INSTRUCTION Be sure the students understand that a map is a scale drawing. On the map of Florida, the scale is 1.8 cm = 40 km.

In the completed example, note that the ratios use centimeters/kilometers rather than the same unit. However, that is the way such problems are generally solved. In effect, the idea of a *rate* is applied to maps. (Of course, the scale of the map can also be expressed in ratio form, using the same unit. In that case, the scale would be 1.8:4,000,000. However, if this ratio were used, extra steps would be needed at the end of the calculation.)

In the exercises, Problems 1–8 are similar to the completed example. Problems 9–18 involve finding or comparing several air distances. Problem 19 asks for total flight time. In Problem 20 the students must use reference resources that involve a different sort of map—that is, a road map.

PROBLEM *Solving* APPLICATION

8.9 CONSUMER: USING MAPS

A map is a scale drawing. You can use a map to determine distances between places. In the map of the Florida panhandle shown below the scale is 1.8 cm = 40 km. Each 1.8 cm on the map represents 40 km of air distance.

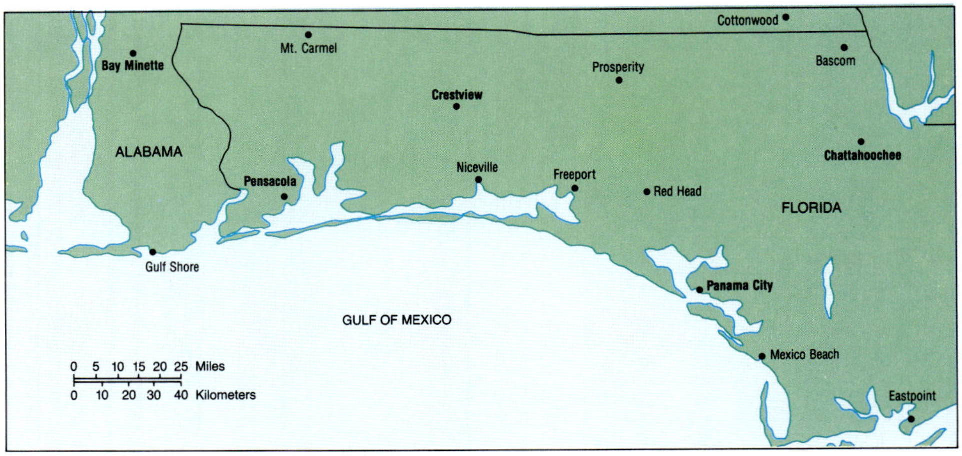

You are on a business trip for the IBQ Corporation. A company plane will fly you from Pensacola to Panama City. What is the air distance between these cities?

1. Measure the distance from Pensacola to Panama City on the map. — The distance is 7 cm to the nearest tenth of a centimeter.
2. Write a proportion.

   THINK: 1.8 cm is to 40 km as 7 cm is to *n* km

   $\frac{1.8}{40} = \frac{7}{n}$ ← **scale (cm)** ← **actual (km)**

3. Solve the proportion.

$$1.8 \times n = 40 \times 7$$
$$1.8 \times n = 280$$
$$n = 280 \div 1.8$$
$$n = 155.6$$

So, the air distance from Pensacola to Panama City is about 156 km.

FOR DISCUSSION

See TE side column.

Which of the following would be best for the scale of a state road map? Why?

1 in. represents: 100 ft, 100 yd, or 100 mi

Use the map on page 226. Measure the distance to the nearest tenth of a centimeter. Then find the air distance in kilometers between the Florida cities below.

	Distance on Map	Air Distance
Panama City to Crestview	• 1. ■ 5.0 cm	• 2. ■ 111 km
Crestview to Mexico Beach	3. ■ 6.5 cm	4. ■ 144 km
Mexico Beach to Bascom	5. ■ 5.2 cm	6. ■ 116 km
Bascom to Niceville	7. ■ 6.3 cm	8. ■ 140 km

Find the air distance in kilometers.

• **9.** Cottonwood, AL, to Freeport, FL 100 km

10. Pensacola, FL, to Prosperity, FL 131 km

11. Mt. Carmel, FL, to Bay Minette, AL 67 km

12. Eastpoint, FL, to Red Head, FL 127 km

Solve.

• **13.** What is the total air distance if you fly from Chattahoochee, FL, to Cottonwood, AL? 53 km

14. What is the total air distance if you fly from Pensacola, FL, to Crestview, FL, and on to Panama City, FL? 182 km

Suppose you fly from Bay Minette, AL, to Prosperity, FL, and on to Eastpoint, FL.

15. What is the total air distance? 342 km

16. About how many kilometers longer is the distance than if you fly directly from Bay Minette to Eastpoint? 29 km

Suppose you fly from Gulf Shores, AL, to Pensacola, FL, to Mexico Beach, FL.

• **17.** What is the total air distance? 238 km

18. About how many kilometers longer is the distance than if you fly directly from Gulf Shores to Mexico Beach? 11 km

19. If the plane flies at an average of 140 mph, about how long will it take to fly from Gulf Shores to Mexico Beach? about 1 hour

20. In a reference library, find a road map of the Florida-Alabama border. Compute the road distance of a trip from Gulf Shores, AL, to Pensacola, FL, to Mexico Beach, FL. Check students' answers.

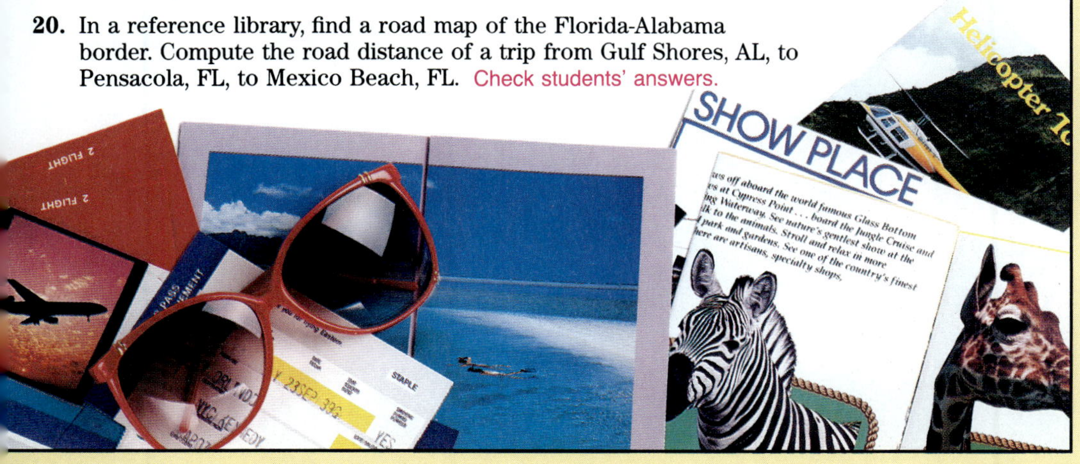

FOR DISCUSSION Students should explain that since a road map is used by automobile drivers, miles would be the appropriate unit for the scale.

ASSIGNMENTS

Level 1	1–14
Level 2	1–20
Level 3	Even 2–20

FOLLOW-UP ACTIVITY

MENTAL MATH Tell the students that a certain map has the scale of 1 cm = 150 km. Then have them answer these questions orally.

What is the air distance if the distance on the map is:

1. 2 cm? (300 km)

2. 5 cm? (750 km)

3. 4.1 cm? (615 km)

4. 5.5 cm? (825 km)

What is the map distance if the air distance is:

5. 600 km? (4 cm)

6. 900 km? (6 cm)

7. 375 km? (2.5 cm)

8. 75 km? (0.5 cm)

SUPPLEMENTARY MATERIALS

TRP Practice, p. 93

TRP Transparency 47

OBJECTIVES

- Review vocabulary.
- Practice key chapter concepts and skills.

USING THE REVIEW

The Chapter Review is designed to help students prepare for taking the Chapter Test. The first section focuses on vocabulary. It requires that students select a word(s) to complete statements. The second section presents practice exercises of key mathematical skills. Under each directive there is a sample exercise with the answer.

Each item on the review is referenced to the page on which the topic is taught in the Pupil's Edition. You may wish to have students refer to these pages to help review any concepts or skills they have not yet mastered.

It is suggested that students work in small-sized heterogeneous cooperative learning groups. Some cooperative learning methods that may be used are as follows:

1. After each student has independently completed the entire Chapter Review, a discussion should follow within each group about the solutions to the practice exercises.
2. The group can complete the entire Chapter Review by working together to discuss the sample exercises and then to answer the practice exercises.

End the lesson with an entire class discussion in which any questions brought up in group discussions are presented and answered.

CHAPTER REVIEW

Vocabulary Choose the letter of the word that completes each statement.

1. A comparison of two numbers by division is a ■. [210] c
a. quotient **b.** colon **c.** ratio **d.** proportion

2. You can compare rates by comparing ■. [212] d
a. means **b.** extremes **c.** units **d.** cross products

3. In a proportion there are two equivalent ■. [216] b
a. equations **b.** ratios **c.** extremes **d.** means

Skills Write the ratio in simplest form. [210]

$\frac{18}{42}$ *$\frac{3}{7}$*

4. $\frac{10}{25}$ $\frac{2}{5}$ **5.** $\frac{12}{36}$ $\frac{1}{3}$ **6.** $\frac{8}{28}$ $\frac{2}{7}$ **7.** $\frac{21}{30}$ $\frac{7}{10}$ **8.** $\frac{18}{63}$ $\frac{2}{7}$ **9.** $\frac{28}{56}$ $\frac{1}{2}$ **10.** $\frac{32}{48}$ $\frac{2}{3}$

Compare the rates. Write >, <, or =. [212]

$\frac{2\text{ in.}}{3\text{ ft}}$ ● $\frac{5\text{ in.}}{6\text{ ft}}$ *<*

11. $\frac{6\text{ lb}}{12\text{ oz}}$ ● $\frac{5\text{ lb}}{10\text{ oz}}$ = **12.** $\frac{3\text{ in.}}{8\text{ ft}}$ ● $\frac{7\text{ in.}}{9\text{ ft}}$ <

13. $\frac{30\text{ mi}}{10\text{ gal}}$ ● $\frac{240\text{ mi}}{6\text{ gal}}$ < **14.** $\frac{\$1.68}{5\text{ cans}}$ ● $\frac{\$2.12}{7\text{ cans}}$ > **15.** $\frac{49\text{ cars}}{20\text{ minutes}}$ ● $\frac{63\text{ cars}}{30\text{ minutes}}$ >

Write as a unit rate. [212]

$\frac{250\text{ mi}}{10\text{ gal}}$ *25 mi per gal*

16. $\frac{91\text{ chairs}}{7\text{ rows}}$ 13 chairs per row **17.** $\frac{132\text{ points}}{4\text{ games}}$ 33 points per game **18.** $\frac{\$0.90}{3\text{ cans}}$ \$0.30 per can

19. $\frac{150\text{ pencils}}{6\text{ boxes}}$ 25 pencils per box **20.** $\frac{235\text{ coins}}{5\text{ banks}}$ 47 coins per bank **21.** $\frac{400\text{ mi}}{8\text{ hours}}$ 50 mph **22.** $\frac{126\text{ yd}}{9\text{ rushes}}$ 14 yd per rush

Determine if the ratios form a proportion. Write *yes* or *no.* [216]

$\frac{5}{9} \stackrel{?}{=} \frac{18}{10}$ *no*

23. $\frac{4}{7} \stackrel{?}{=} \frac{18}{28}$ no **24.** $\frac{3}{5} \stackrel{?}{=} \frac{15}{9}$ no **25.** $\frac{2}{9} \stackrel{?}{=} \frac{12}{54}$ yes

26. $\frac{35}{14} \stackrel{?}{=} \frac{5}{2}$ yes **27.** $\frac{1.2}{2.1} \stackrel{?}{=} \frac{0.3}{0.5}$ no **28.** $\frac{15}{0.24} \stackrel{?}{=} \frac{0.5}{0.8}$ no **29.** $\frac{3.6}{4.8} \stackrel{?}{=} \frac{0.6}{0.8}$ yes

Find n. [216]

$\frac{n}{8} = \frac{21}{56}$ *3*

30. $\frac{n}{12} = \frac{5}{6}$ 10 **31.** $\frac{13}{n} = \frac{26}{8}$ 4 **32.** $\frac{5}{9} = \frac{n}{63}$ 35

33. $\frac{3}{8} = \frac{21}{n}$ 56 **34.** $\frac{10}{n} = \frac{100}{130}$ 13 **35.** $\frac{15}{24} = \frac{n}{8}$ 5 **36.** $\frac{28}{49} = \frac{4}{n}$ 7

CHAPTER TEST

Write the ratio in two ways.

		Colon	Fraction	
14 L water to 15 L ammonia	14:15	1. ■	2. ■	$\frac{14}{15}$
5 cm to 13 m	5:13	3. ■	4. ■	$\frac{5}{13}$
32 mi to 4 mi	32:4	5. ■	6. ■	$\frac{32}{4}$
29 hours to 25 hours	29:25	7. ■	8. ■	$\frac{29}{25}$

Write the ratio in simplest form.

9. $\frac{12}{48}$ $\frac{1}{4}$ **10.** $\frac{16}{24}$ $\frac{2}{3}$ **11.** $\frac{42}{28}$ $\frac{3}{2}$ **12.** $\frac{45}{27}$ $\frac{5}{3}$ **13.** $\frac{33}{72}$ $\frac{11}{24}$

Write as a unit rate.

14. 21 seeds per pack **15.** 43 candles per box **16.** 67 albums per box **17.** 167 students per class **18.** 94 hours per month

14. $\frac{\text{84 seeds}}{\text{4 packs}}$ **15.** $\frac{\text{129 candles}}{\text{3 boxes}}$ **16.** $\frac{\text{469 albums}}{\text{7 boxes}}$ **17.** $\frac{\text{835 students}}{\text{5 classes}}$ **18.** $\frac{\text{564 hours}}{\text{6 months}}$

Compare the rates. Write >, <, or =.

19. $\frac{\text{2 in.}}{\text{3 ft}}$ ● $\frac{\text{7 in.}}{\text{8 ft}}$ < **20.** $\frac{\$1.52}{\text{7 oz}}$ ● $\frac{\$0.98}{\text{4 oz}}$ < **21.** $\frac{\text{280 mi}}{\text{14 gal}}$ ● $\frac{\text{360 mi}}{\text{18 gal}}$ =

22. $\frac{\text{142 words}}{\text{2 minutes}}$ ● $\frac{\text{201 words}}{\text{3 minutes}}$ > **23.** $\frac{\text{220 mi}}{\text{4.5 in.}}$ ● $\frac{\text{305 mi}}{\text{3.2 in.}}$ < **24.** $\frac{\text{3 cm}}{\text{125 km}}$ ● $\frac{\text{7 cm}}{\text{342 km}}$ >

Determine if the ratios form a proportion. Write *yes* or *no.*

25. $\frac{2}{3} \stackrel{?}{=} \frac{8}{9}$ no **26.** $\frac{5}{7} \stackrel{?}{=} \frac{25}{35}$ yes **27.** $\frac{6}{11} \stackrel{?}{=} \frac{42}{48}$ no **28.** $\frac{4}{9} \stackrel{?}{=} \frac{32}{72}$ yes

Find n.

29. $\frac{n}{5} = \frac{8}{20}$ 2 **30.** $\frac{7}{n} = \frac{21}{33}$ 11 **31.** $\frac{15}{9} = \frac{n}{3}$ 5 **32.** $\frac{42}{18} = \frac{7}{n}$ 3

33. $\frac{n}{6} = \frac{7}{8}$ 5.25 **34.** $\frac{5}{n} = \frac{4}{9}$ 11.25 **35.** $\frac{3}{10} = \frac{n}{23}$ 6.9 **36.** $\frac{6}{7} = \frac{9}{n}$ 10.5

Solve.

37. On a map, the distance from one city to another measures 3.5 in. According to the scale of the map, 1 in. represents 15 miles. How many actual miles is the distance? 52.5 mi

38. Wendy wants to buy two ball-point pens, two tablets of paper, and one notebook. The pens cost $1.19 each, the tablets of paper cost $2.49 each, and the notebook costs $2.79. Will $10 be enough? no

OBJECTIVE

- Evaluate achievement of the chapter objectives.

USING THE TEST

The Chapter Test may be used as a posttest to evaluate student achievement. However, you may wish to use the Chapter Posttest offered in the Teacher's Resource Package or design your own chapter test. If this page is not used as a test, you may wish to assign it as additional review or practice.

The test items are correlated to the chapter objectives in the table below.

Chapter objectives	Test items
A. Write ratios.	1–8
B. Write ratios in simplest form	9–13
C. Find unit rates.	14–18
D. Compare rates.	19–24
E. Determine if two ratios form a proportion.	25–28
F. Write and solve proportions.	29–36
G. Apply computational skills in real-life situations.	37
H. Solve problems using estimation.	38

SUPPLEMENTARY MATERIALS

TRP Ch. 8 Posttest Form A, pp. 1–2

TRP Ch. 8 Posttest Form B, pp. 1–2

OBJECTIVES

- Solve inverse proportions.
- Solve word problems with inverse proportions.

USING THE PAGE

Direct students' attention to the diagram. Guide them to realize that the greater the mass, the smaller the distance from the fulcrum. Conversely, the smaller the mass, the greater the distance from the fulcrum. Some students may find this relationship easier to understand if it is written as:

$$\text{mass}_1 \times \text{distance}_1 = \text{mass}_2 \times \text{distance}_2$$

$$12 \times 2 = 8 \times n$$

Provide guidance in writing a correct porportion. Encourage students to label the given values of x and y as x^1, x^2, and y^1. Students should also realize that x^1 and y^1 form a cross product; so if x^1 is the numerator of one ratio, then y^1 must be the denominator of the other ratio. The same is true for x^2 and y^2.

Encourage students to check the reasonableness of their answers. Instruct them to reread each problem after they find the answer. Then, as in the case of Exercise 3, they should ask themselves questions such as, "If a trip takes 2.5 hours at 40 km/hour, will the trip be longer or shorter at a faster speed?"

ENRICHMENT INVERSE PROPORTIONS

Inverse means *opposite.* In an inverse proportion, as the value of x increases, the value of y decreases. Similarly, as the value of x decreases, the value of y increases.

Inverse proportions can be applied to many real-life situations.

Example A 12-kg stone is 2 meters from the fulcrum. How far will an 8-kg stone be from the fulcrum to balance it?

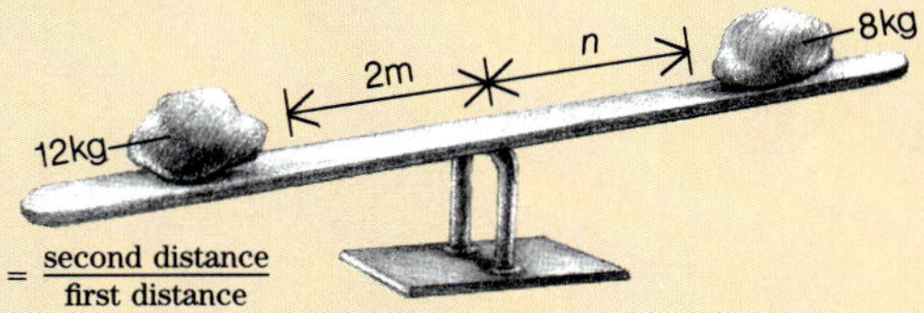

$$\frac{\text{first mass}}{\text{second mass}} = \frac{\text{second distance}}{\text{first distance}}$$

THINK: First mass = 12 kg, second mass = 8 kg
First distance = 2 m, second distance = n meters

$$\frac{12}{8} = \frac{n}{2}$$

$$12 \times 2 = 8 \times n$$

$$24 = 8 \times n$$

$$24 \div 8 = n$$

$$3 = n$$

So, the second stone will be 3 meters from the fulcrum.

Use an inverse proportion to solve.

1. A 32-kg mass is 10 cm from the fulcrum. How far will a 40-kg mass be from the fulcrum to balance it? 8 cm
2. A 40-kg stone is 9 cm from the fulcrum. To balance it, another stone is 12 cm away from the fulcrum. What is the mass of this stone? 30 kg
3. A trip takes 2.5 hours at 40 km/hour. How long will the same trip take at 50 km/hour? 2 hours
4. A trip took 6 hours at 50 mph. If the same trip took 4 hours, what was the rate of speed? 75 mph
5. Ten word processors can input a manuscript in 6 hours. How long would it take 5 word processors to do the same amount of work? 12 hours

230 CHAPTER 8

CALCULATOR SOLVING PROPORTIONS ON A CALCULATOR

You can use a calculator to solve proportions.

Solve for n: $\frac{51}{96} = \frac{n}{80}$

Think	Calculator Entry	Calculator Display
$96n = 51 \times 80$		
$n = \frac{51 \times 80}{96}$	5 1 × 8 0 ÷ 9 6 =	42.5
$n = 42.5$		

Solve for n: $\frac{3.5}{n} = \frac{6}{4.9}$ Round the answer to the nearest tenth.

Think	Calculator Entry	Calculator Display
$6n = 3.5 \times 4.9$		
$n = \frac{3.5 \times 4.9}{6}$	3 . 5 × 4 . 9 ÷ 6 =	2.8583333
$n = 2.9$ ← **Rounded to the nearest tenth**		

Solve for n. Use a calculator.
When necessary, round your answer to the nearest tenth.

1. $\frac{n}{95} = \frac{48}{57}$ 80 **2.** $\frac{63}{91} = \frac{45}{n}$ 65 **3.** $\frac{385}{n} = \frac{365}{292}$ 308

4. $\frac{23}{n} = \frac{115}{229}$ 45.8 **5.** $\frac{153}{594} = \frac{n}{1,485}$ 382.5 **6.** $\frac{n}{126} = \frac{273}{168}$ 204.8

7. $\frac{3.4}{10.2} = \frac{2.7}{n}$ 8.1 **8.** $\frac{n}{5.2} = \frac{6.8}{20.8}$ 1.7 **9.** $\frac{1.5}{0.9} = \frac{n}{19.5}$ 32.5

10. $\frac{1.9}{n} = \frac{0.2}{1.3}$ 12.4 **11.** $\frac{0.7}{0.5} = \frac{2.2}{n}$ 1.6 **12.** $\frac{2.5}{11.1} = \frac{n}{3.6}$ 0.8

Write a proportion to find the answer. Use a calculator to solve the proportion.

13. A carpenter earned $596.80 for a job that took 32 hours. At this rate, how much would he earn on a job that takes 25 hours? $466.25

14. A temporary office worker earns $12.65 per hour. Last week she earned $354.20. How many hours did she work last week? 28 hours

15. A sales representative drove his car 368 mi on 11.5 gal of gasoline. At this rate, how many gallons of gasoline would his car use on a 480-mi trip? 15 gal

16. A driver drove her school bus 285 mi on 30 gal of gasoline. At this rate, how many miles could she have driven on 12 gal of gasoline? 114 mi

OBJECTIVE

- Solve proportions using a calculator.

USING THE PAGE

When students use a calculator to solve a proportion, a common error is to divide by the wrong number. They may find it helpful to think of the pairs of factors that form the cross products. Example 1 may be used to illustrate this.

Two pairs of factors form the cross products. In one pair, *both* factors are numbers. Multiply them.

$\frac{\mathbf{51}}{96}$ ⇄ $\frac{n}{\mathbf{80}}$ Multiply: **51 × 80**

In the other pair, just *one* factor is a number. Divide by this number.

$\frac{51}{\mathbf{96}}$ ⇄ $\frac{n}{80}$ Divide: $\frac{51 \times 80}{\mathbf{96}}$

When students solve word problems involving proportions they may use their calculators correctly, but the proportions that they write may be set up incorrectly. Such errors can often be detected if answers are checked to see if they are reasonable. In Exercise 16, for instance, a student might think: 12 gal is somewhat less than $\frac{1}{2}$ of 30 gal. Therefore, the bus driver could travel somewhat less than $\frac{1}{2}$ of 285 mi. The incorrect proportion

$$\frac{285}{30} = \frac{12}{n}$$

yields the rounded answer of 1.3 mi, and it should be clear that this is not reasonable. Similarly, the incorrect proportion

$$\frac{n}{285} = \frac{30}{12}$$

yields the answer 712.5 mi, which does not make sense.

OBJECTIVE

- Test the prerequisite skills needed to learn the concepts developed in Chapter 9.

USING THE LESSON

The Pre-Skills Test is designed to diagnose students' strengths and weaknesses on prerequisite skills necessary to study the mathematics in Chapter 9.

Have students take the Pre-Skills Test. Allow the students to work together in pairs or small groups. Group members should help those who demonstrate a misunderstanding of a concept or a weakness in a skill.

The items in the test are referenced to the pages on which the topics are taught in the Pupil's Edition. You may wish to have students refer to these pages for review.

The following table correlates the items on the Pre-Skills Test with the prerequisite skill and the lesson(s) in the chapter for which it is needed.

Item(s)	Prerequisite skill	Lesson(s)
1–5	Rename fractions as hundredths.	9.2
6–10	Rename decimals as fractions in hundredths.	9.1
11–20	Write fractions in simplest form.	9.2, 9.13
21–30	Write fractions as decimals.	9.1, 9.3
31–34	Multiply decimals by whole numbers.	9.4, 9.5, 9.6
35–38	Multiply fractions and mixed numbers by whole numbers.	9.4
39–42	Divide whole numbers and decimals by decimals.	9.7, 9.8
43–46	Divide whole numbers by fractions or mixed numbers.	9.8
47–54	Solve proportions.	9.9

Pre-Skills Test

Write as a fraction with a denominator of 100. [126]

1. $\frac{1}{2}$ $\frac{50}{100}$ **2.** $\frac{3}{4}$ $\frac{75}{100}$ **3.** $\frac{7}{10}$ $\frac{70}{100}$ **4.** $\frac{4}{5}$ $\frac{80}{100}$ **5.** $\frac{13}{25}$ $\frac{52}{100}$

6. 0.25 $\frac{25}{100}$ **7.** 0.04 $\frac{4}{100}$ **8.** 0.3 $\frac{30}{100}$ **9.** 3.2 $\frac{320}{100}$ **10.** 5.42 $\frac{542}{100}$

Write as a fraction in simplest form. [126]

11. $\frac{25}{75}$ $\frac{1}{3}$ **12.** $\frac{3}{15}$ $\frac{1}{5}$ **13.** $\frac{5}{100}$ $\frac{1}{20}$ **14.** $\frac{24}{50}$ $\frac{12}{25}$ **15.** $\frac{60}{75}$ $\frac{4}{5}$

16. $\frac{24}{36}$ $\frac{2}{3}$ **17.** $\frac{32}{64}$ $\frac{1}{2}$ **18.** $\frac{35}{100}$ $\frac{7}{20}$ **19.** $\frac{72}{81}$ $\frac{8}{9}$ **20.** $\frac{40}{56}$ $\frac{5}{7}$

Write as a decimal. [136]

21. $\frac{3}{4}$ 0.75 **22.** $\frac{5}{8}$ 0.625 **23.** $\frac{9}{100}$ 0.09 **24.** $\frac{7}{10}$ 0.7 **25.** $\frac{17}{25}$ 0.68

26. $\frac{3}{5}$ 0.6 **27.** $\frac{7}{20}$ 0.35 **28.** $\frac{9}{40}$ 0.225 **29.** $\frac{3}{25}$ 0.12 **30.** $\frac{17}{50}$ 0.34

Multiply. [74, 146, 148]

31. $0.9 \times 5{,}000$ 4,500 **32.** 1.3×75 97.5 **33.** 0.875×24 21 **34.** 0.125×72 9

35. $\frac{3}{4} \times 24$ 18 **36.** $\frac{2}{9} \times 78$ $17\frac{1}{3}$ **37.** $1\frac{1}{2} \times 15$ $22\frac{1}{2}$ **38.** $2\frac{2}{3} \times 123$ 328

Divide. [80, 150, 152]

39. $4.8 \div 0.32$ 15 **40.** $21 \div 0.7$ 30 **41.** $287 \div 1.4$ 205 **42.** $11 \div 0.125$ 88

43. $300 \div \frac{2}{3}$ 450 **44.** $150 \div \frac{3}{8}$ 400 **45.** $525 \div 3\frac{1}{2}$ 150 **46.** $289 \div 2\frac{3}{7}$ 119

Solve for n. [216]

47. $\frac{3}{n} = \frac{21}{63}$ 9 **48.** $\frac{7}{9} = \frac{56}{n}$ 72 **49.** $\frac{5}{6} = \frac{n}{72}$ 60 **50.** $\frac{n}{12} = \frac{144}{24}$ 72

51. $\frac{2}{3} = \frac{n}{24}$ 16 **52.** $\frac{n}{100} = \frac{4}{5}$ 80 **53.** $\frac{3}{n} = \frac{48}{32}$ 2 **54.** $\frac{6}{10} = \frac{54}{n}$ 90

Many schools sponsor class trips for students. How could you plan a trip that all your classmates would enjoy?

OBJECTIVE

- Formulate problems using text and data.

USING THE CHAPTER OPENER

Begin the discussion of the question by asking various individuals where they think everyone in the class would like to go on a class trip. After each suggestion, determine by a show of hands if everyone agrees and keep a record of the votes for each idea. If there are no suggestions with class-wide appeal, ask for ideas for how to come to some compromise. This might involve selecting a place where there are a variety of things to do so everyone has something they like.

COOPERATIVE LEARNING

Write the following on the chalkboard.

CLASS TRIP
Chartered bus provided
3 days, 2 nights
50% educational, 50% recreational

Assign groups to the project explaining that they are to plan a trip that all their classmates will enjoy. They are limited only by the restrictions provided. Each group is to develop a plan for the entire trip including the place, travel schedules, tours, hotel accommodations, meals, and costs. This will require research on their part, and they should devise a plan within the group for sharing this task. If some data is not available, provide some reasonable estimates for them.

After the plans are presented to the class, have the students vote for the most popular, and try to determine if everyone in the class would enjoy such a trip.

SUPPLEMENTARY MATERIALS

TRP Group Projects, p. 9

OBJECTIVES

- Write a percent as a decimal.
- Write a decimal as a percent.

TEACHING THE LESSON

WARM-UP Present the following exercises on the chalkboard or an overhead transparency.

Write as a decimal.

1. $\frac{15}{100}$ **2.** $\frac{135}{100}$ **3.** $\frac{0.68}{100}$

4. $\frac{1}{4}$ **5.** $\frac{3}{8}$ **6.** $6\frac{1}{2}$

(**1.** 0.15 **2.** 1.35 **3.** 0.0068 **4.** 0.25 **5.** 0.375 **6.** 6.5)

INTRODUCTION Write the following statements on the chalkboard.

- Starting next month, your pay will be increased by 15%.
- Starting next month, your rent will be increased by 15%.

Ask which statement is good news and which is bad news. Discuss why. Ask what it means for someone's pay or rent to be increased by 15%.

INSTRUCTION In Examples 1 and 2, point out that moving the decimal point two places to the left is the same as dividing by 100.

Direct the students' attention to Example 1b. Point out that since 100% = 1, percents greater than 100% have decimal equivalents that are greater than 1. Example 1c shows the converse. Since 1% = 0.01, percents that are less than 1% have decimal equivalents that are less than 0.01.

In Example 2, stress that $\frac{1}{4}$% means $\frac{1}{4}$ of 1%, not $\frac{1}{4}$ of 100%.

In Example 3, emphasize that moving the decimal point two places to the right is the same as multiplying by 100. Call attention to Example 3b in which a zero has been written to place the decimal point correctly and to show that 1.7 = 170%, not 17%.

9.1 DECIMALS AND PERCENTS

Here are some examples of percents used in daily life:

Write the three percents as decimals.

THINK: **Percent (%)** means per hundred or hundredths.
15% means 15 per 100, or 15 hundredths. So, 15% = 0.15.
135% means 135 per 100, or 135 hundredths. So, 135% = 1.35.
0.74% means 0.74 per 100, or 0.74 hundredths. So, 0.74% = 0.0074.

You can use a shortcut to write a percent as a decimal. Move the decimal point two places to the left and drop the percent sign.

EXAMPLE 1 Write as a decimal.

a. 15%
15% = 15%
= 0.15

b. 135%
135% = 135%
= 1.35

c. 0.74%
0.74% = 000.74%
= 0.0074

Sometimes you need to write a fractional or mixed-number percent as a decimal. Rename the fraction or mixed number as a decimal.

EXAMPLE 2 Write as a decimal.

a. $\frac{1}{4}$%
$\frac{1}{4}$% = 0.25% = 000.25%
= 0.0025

b. $6\frac{1}{2}$%
$6\frac{1}{2}$% = 6.5% = 006.5%
= 0.065

To write a decimal as a percent, move the decimal point two places to the right and include the percent sign.

EXAMPLE 3 Write as a percent.

a. 0.54
0.54 = 0.54
= 54%

b. 1.7
1.7 = 1.70
= 170%

c. $0.11\frac{1}{9}$
$0.11\frac{1}{9}$ = $0.11\frac{1}{9}$
= $11\frac{1}{9}$%

CHECKPOINT

Write the letter of the correct answer.
See Extra Practice, page 445.

Complete.

1. 12% = ■ b **a.** 0.012 **b.** 0.12 **c.** 1.2 **d.** 1,200
2. $\frac{1}{2}$% = ■ a **a.** 0.005 **b.** 0.05 **c.** 5 **d.** 50
3. 0.29 = ■ c **a.** 0.29% **b.** 2.9% **c.** 29% **d.** 290%

PRACTICE EXERCISES

5. 0.31 **10.** 0.09 **15.** 1.57 **20.** 0.005 **24.** 0.1115 **25.** 2.006 **30.** 0.00375 **35.** 0.09375 **40.** 25% **45.** 5.5% **50.** 80% **55.** $22\frac{2}{9}$%

Write as a decimal.

• **1.** 28% 0.28	**2.** 65% 0.65	**3.** 42% 0.42	**4.** 10% 0.1	**5.** 31%
6. 8% 0.08	**7.** 4% 0.04	**8.** 6% 0.06	**9.** 3% 0.03	**10.** 9%
• **11.** 130% 1.3	**12.** 200% 2	**13.** 350% 3.5	**14.** 245% 2.45	**15.** 157%
• **16.** 0.35% 0.0035	**17.** 0.17% 0.0017	**18.** 0.51% 0.0051	**19.** 0.8% 0.008	**20.** 0.5%
21. 7.2% 0.072	**22.** 3.09% 0.0309	**23.** 90.4% 0.904	**24.** 11.15%	**25.** 200.6%
• **26.** $\frac{1}{4}$% 0.0025	**27.** $\frac{1}{5}$% 0.002	**28.** $\frac{3}{4}$% 0.0075	**29.** $\frac{4}{5}$% 0.008	**30.** $\frac{3}{8}$%
• **31.** $3\frac{1}{2}$% 0.035	**32.** $2\frac{3}{4}$% 0.0275	**33.** $8\frac{2}{5}$% 0.084	**34.** $10\frac{1}{4}$% 0.1025	**35.** $9\frac{3}{8}$%

Write as a percent.

• **36.** 0.23 23%	**37.** 0.45 45%	**38.** 0.15 15%	**39.** 0.37 37%	**40.** 0.25
41. 0.07 7%	**42.** 0.03 3%	**43.** 0.092 9.2%	**44.** 0.013 1.3%	**45.** 0.055
• **46.** 3.7 370%	**47.** 2.1 210%	**48.** 0.9 90%	**49.** 4.2 420%	**50.** 0.8
• **51.** $0.66\frac{2}{3}$ $66\frac{2}{3}$%	**52.** $0.16\frac{2}{3}$ $16\frac{2}{3}$%	**53.** $0.37\frac{1}{2}$ $37\frac{1}{2}$%	**54.** $0.83\frac{1}{3}$ $83\frac{1}{3}$%	**55.** $0.22\frac{2}{9}$

Solve.

• **56.** Carol Heilman plans to finance her new car. The dealer is offering an annual interest rate of 8.9%. The bank's offer is an annual interest rate of $8\frac{9}{11}$%. Which is the lower rate? $8\frac{9}{11}$%

57. Carol's new car gets 1.7 times more miles to the gallon than her old car. Write this decimal as a percent. 170%

58. Carol must pay a sales tax of $8\frac{1}{4}$% on her purchase. Write this percent as a decimal. 0.0825

CHECKPOINT Distribute to each student 4 index cards on which the answer choices a-d appear. After allowing students a few minutes, ask them to raise the letter that indicates their answer choice. Scan the cards and identify those students who need additional attention before proceeding to the Practice Exercises. The incorrect answer choices include common errors that students make.

COMMON ERROR

When changing percents to decimals and vice versa, some students may move the decimal point in the wrong direction. For example, in item 1 of Checkpoint, they would select item d.

Encourage these students to make a sign to serve as a reference.

Decimal ⟶ percent: Move right

0.56 = 56%

Percent ⟶ decimal: Move left

78% = 0.78

ASSIGNMENTS

Level 1	Odd 1–55, 56, 57
Level 2	Even 2–54, 56–58
Level 3	Even 2–54, 56–58

FOLLOW-UP ACTIVITY

CALCULATOR Have students use a calculator to match.

Percent	**Decimals**
1. 50% (a)	**a.** 0.5
2. $\frac{1}{2}$% (d)	**b.** 5.75
3. 0.2% (e)	**c.** 0.0575
4. $5\frac{3}{4}$% (c)	**d.** 0.005
5. 575% (b)	**e.** 0.002
6. 2% (f)	**f.** 0.02

SUPPLEMENTARY MATERIALS

TRP Practice, p. 94
TRP Reteaching, p. 57
TRP Transparency 47

OBJECTIVE

- Write a percent as a fraction in simplest form.

TEACHING THE LESSON

WARM-UP Present the following exercises on the chalkboard or an overhead transparency.

Write in simplest form.

1. $\frac{45}{100}$ **2.** $\frac{20}{100}$ **3.** $\frac{225}{100}$

4. $\frac{26}{100}$ **5.** $\frac{250}{100}$ **6.** $\frac{444}{100}$

(**1.** $\frac{9}{20}$ **2.** $\frac{1}{5}$ **3.** $2\frac{1}{4}$ **4.** $\frac{13}{50}$ **5.** $2\frac{1}{2}$ **6.** $4\frac{11}{25}$)

INTRODUCTION Write the following problem on the chalkboard.

The school bulletin board said that Kate Wilson was elected as class president with 60% of the vote. The school newspaper said that she won with $\frac{3}{5}$ of the vote. Both were right. How could this be?

Guide students to realize that $\frac{3}{5}$ = 60% by having them write $\frac{3}{5}$ as a decimal, then as a percent.

INSTRUCTION Remind students that *percent* means *hundredths.* In Example 1, point out that $\frac{67}{100}$ is in simplest form because 1 is the greatest number that divides both 67 and 100. $\frac{30}{100}$ is not in simplest form because 10 divides both 30 and 100.

In discussing Example 2b, you may wish to show an alternate method of renaming the percent.

$$\begin{aligned} 215\% &= 200\% + 15\% \\ &= \frac{200}{100} + \frac{15}{100} \\ &= 2 + \frac{3}{20} \\ &= 2\frac{3}{20} \end{aligned}$$

9.2 WRITING PERCENTS AS FRACTIONS

The Kennesaw Gazette conducted a poll to find out what its readers thought of their congressional representative. Of the 97% who had an opinion, 67% supported Representative Sherman and 30% opposed him. What fraction of those polled supported Sherman? What fraction opposed him?

To change a percent to a fraction, write the percent as a fraction with a denominator of 100. Then, simplify the fraction.

EXAMPLE 1 Write as a fraction.

a. 67%

$67\% = \frac{67}{100}$

Of those polled, $\frac{67}{100}$ supported Sherman.

b. 30%

$30\% = \frac{\overset{3}{\cancel{30}}}{\underset{10}{\cancel{100}}} = \frac{3}{10}$

Of those responding, $\frac{3}{10}$ opposed Sherman.

Percents greater than or equal to 100 can be written either as whole numbers or as mixed numbers.

EXAMPLE 2 Write as a whole number or a mixed number.

a. 300%

$300\% = \frac{\overset{3}{\cancel{300}}}{\underset{1}{\cancel{100}}} = 3$

b. 215%

$215\% = 2\frac{\overset{3}{\cancel{15}}}{\underset{20}{\cancel{100}}} = 2\frac{3}{20}$

CHECKPOINT

Write the letter of the correct answer.

Write as a fraction, whole number, or mixed number.

1. 23%	a	**a.** $\frac{23}{100}$	**b.** $\frac{23}{10}$	**c.** $\frac{100}{23}$	**d.** 2,300
2. 50%	b	**a.** $\frac{5}{100}$	**b.** $\frac{1}{2}$	**c.** $\frac{5}{10}$	**d.** 5,000
3. 6%	c	**a.** $\frac{60}{100}$	**b.** $\frac{6}{10}$	**c.** $\frac{3}{50}$	**d.** $\frac{0.06}{100}$
4. 350%	c	**a.** 35,000	**b.** $35\frac{1}{100}$	**c.** $3\frac{1}{2}$	**d.** $\frac{35}{100}$

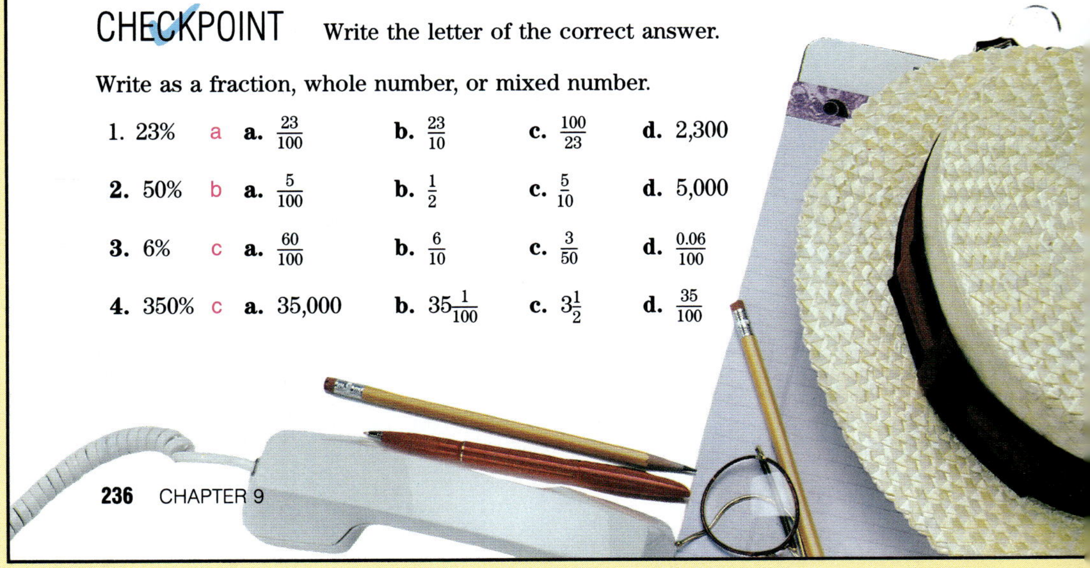

PRACTICE EXERCISES See Extra Practice, page 446.

Write as a fraction, whole number, or mixed number in simplest form.

• **1.** 3% $\frac{3}{100}$	**2.** 11% $\frac{11}{100}$	**3.** 17% $\frac{17}{100}$	**4.** 29% $\frac{29}{100}$	**5.** 61% $\frac{61}{100}$
6. 53% $\frac{53}{100}$	**7.** 33% $\frac{33}{100}$	**8.** 97% $\frac{97}{100}$	**9.** 59% $\frac{59}{100}$	**10.** 83% $\frac{83}{100}$
• **11.** 8% $\frac{2}{25}$	**12.** 10% $\frac{1}{10}$	**13.** 15% $\frac{3}{20}$	**14.** 25% $\frac{1}{4}$	**15.** 50% $\frac{1}{2}$
16. 85% $\frac{17}{20}$	**17.** 36% $\frac{9}{25}$	**18.** 75% $\frac{3}{4}$	**19.** 40% $\frac{2}{5}$	**20.** 95% $\frac{19}{20}$
• **21.** 200% 2	**22.** 500% 5	**23.** 300% 3	**24.** 100% 1	**25.** 700% 7
26. 400% 4	**27.** 600% 6	**28.** 900% 9	**29.** 800% 8	**30.** 1,000% 10
• **31.** 111% $1\frac{11}{100}$	**32.** 247% $2\frac{47}{100}$	**33.** 323% $3\frac{23}{100}$	**34.** 421% $4\frac{21}{100}$	**35.** 559% $5\frac{59}{100}$
36. 125% $1\frac{1}{4}$	**37.** 250% $2\frac{1}{2}$	**38.** 340% $3\frac{2}{5}$	**39.** 475% $4\frac{3}{4}$	**40.** 515% $5\frac{3}{20}$
41. 26% $\frac{13}{50}$	**42.** 240% $2\frac{2}{5}$	**43.** 9% $\frac{9}{100}$	**44.** 45% $\frac{9}{20}$	**45.** 98% $\frac{49}{50}$
46. 112% $1\frac{3}{25}$	**47.** 1% $\frac{1}{100}$	**48.** 49% $\frac{49}{100}$	**49.** 51% $\frac{51}{100}$	**50.** 2,000% 20
51. 30% $\frac{3}{10}$	**52.** 99% $\frac{99}{100}$	**53.** 280% $2\frac{4}{5}$	**54.** 444% $4\frac{11}{25}$	**55.** 1,500% 15

MIXED REVIEW

Solve the proportion.

• **56.** $\frac{5}{100} = \frac{n}{67}$ 3.35	• **57.** $\frac{6}{100} = \frac{n}{36}$ 2.16	**58.** $\frac{75}{100} = \frac{93}{n}$ 124	**59.** $\frac{n}{100} = \frac{16}{128}$ 12.5
60. $\frac{n}{100} = \frac{64}{24}$ $266\frac{2}{3}$	**61.** $\frac{250}{100} = \frac{n}{44}$ 110	**62.** $\frac{125}{100} = \frac{110}{n}$ 88	**63.** $\frac{25}{n} = \frac{78}{312}$ 100

Solve.

• **64.** The Kennesaw Gazette polled 327 voters by phone. Each call took an average of 4 minutes. How long did it take to conduct the poll? 21 hours 48 minutes

65. Representative Sherman has served in Congress for 9 years. Each term is 2 years long. How many times has she been re-elected? 4

66. Included in the results of the poll were 3% who had no opinion. What fractional part of those polled had no opinion? $\frac{3}{100}$

67. For its poll, the Gazette contacted 16% of Kennesaw's voters. What fractional part of Kennesaw's voters were contacted for the Gazette poll? $\frac{4}{25}$

TIME OUT In the Kennesaw mayoral election, Joan Bigelow beat Philip Oakley by 212 votes. A total of 21,628 ballots were cast. How many votes did each candidate receive? Bigelow: 10,920; Oakley: 10,708

CHECKPOINT The incorrect answer choices include common errors that students make.

COMMON ERROR

After naming a percent as a fraction in hundredths, students will make errors in writing the fraction in simplest form. For example, in item 2 of Checkpoint, they would select c. Use the following two-step approach to isolate the students' difficulty.

Give students a list of fractions, some of which are in simplest form and some of which are not. Have them identify those in simplest form.

Then, write fractions such as $\frac{6}{8}$, $\frac{9}{12}$ and $\frac{20}{80}$ and guide students in writing them in simplest form.

ASSIGNMENTS

Level 1	Odd 1–63, 64–65
Level 2	Even 1–66, TO
Level 3	41–67, TO

FOLLOW-UP ACTIVITY

ESTIMATION Have students tell whether the fraction is close to 1%, 50%, or 100%. Encourage them to explain their reasoning.

$\frac{5}{8}$ (50%)	$\frac{11}{12}$ (100%)	$\frac{1}{16}$ (1%)
$\frac{3}{50}$ (1%)	$\frac{4}{9}$ (50%)	$\frac{19}{20}$ (100%)
$\frac{11}{10}$ (100%)	$\frac{2}{25}$ (1%)	$\frac{29}{60}$ (50%)

Have students write two fractions that are close to 1%, two that are close to 50%, and two that are close to 100%.

SUPPLEMENTARY MATERIALS

TRP Practice, p. 95

TRP Reteaching, p. 58

TRP Transparency 48

OBJECTIVE

- Write a fraction as a percent.

TEACHING THE LESSON

WARM-UP Present the following exercises on the chalkboard or an overhead transparency.

Write as a decimal.

1. $\frac{1}{5}$ **2.** $\frac{3}{4}$ **3.** $\frac{7}{20}$

4. $\frac{2}{3}$ **5.** $\frac{1}{6}$ **6.** $4\frac{2}{5}$

(**1.** 0.2 **2.** 0.75 **3.** 0.35 **4.** $0.6\overline{6}$ **5.** $0.1\overline{6}$ **6.** 4.4)

INTRODUCTION Present the following problem: You made this sign to announce a sale at the sports shop.

PARKAS—$\frac{1}{2}$ OFF

Your boss tells you to replace the $\frac{1}{2}$ with the percent of discount. What do you write? Explain.

INSTRUCTION In Example 1, point out that when the denominator is a factor of 100, students can write the corresponding percent by first writing an equivalent fraction with a denominator of 100.

In Example 2, stress that when the denominator of the fraction is not a factor of 100, students must divide to find the equivalent decimal. Stress the importance of writing the remainder as a fraction.

In Example 3a, encourage students to think:

$$1 = 100\%, \text{ so } 5 = 500\%.$$

In Example 3b, emphasize that the first step is to rename the fraction as an equivalent decimal.

CHECKPOINT The incorrect answer choices include common errors that students make.

9.3 WRITING FRACTIONS AS PERCENTS

In the state finals of the high school basketball tournament, Visalia lost because of poor free-throw shooting. The team made only 11 of 20 attempts. What percent of its free throws did Visalia make?

THINK: The ratio of 11 out of 20 can be written as the fraction $\frac{11}{20}$.

EXAMPLE 1 Write $\frac{11}{20}$ as a percent.

1. Write the fraction as a decimal. $\frac{11}{20} \rightarrow 20\overline{)11.00}$ = 0.55

   THINK: Divide the numerator by the denominator.

2. Write the decimal as a percent. $0.55 = 55\%$

Visalia made 55% of its free throws.

Some fractions are written as mixed-number percents.

EXAMPLE 2 Write as a percent.

a. $\frac{1}{9}$

$\frac{1}{9} \rightarrow 9\overline{)1.00}$ = $0.11\frac{1}{9} = 11\frac{1}{9}\%$

b. $\frac{5}{6}$

$\frac{5}{6} \rightarrow 6\overline{)5.00}$ = $0.83\frac{1}{3} = 83\frac{1}{3}\%$

To write a whole number or mixed number as a percent, change the number to a decimal. Then, write the decimal as a percent.

EXAMPLE 3 Write as a percent.

a. 5

$5 = 5.00 = 500\%$

b. $2\frac{1}{2}$

$2\frac{1}{2} = 2.50 = 250\%$

CHECKPOINT

Write the letter of the correct answer.

Write as a percent.

		a.	b.	c.	d.
1. $\frac{3}{4}$	c	175%	$100\frac{3}{4}\%$	75%	0.75%
2. $\frac{1}{3}$	c	30%	33%	$33\frac{1}{3}\%$	$133\frac{1}{3}\%$
3. $2\frac{4}{5}$	b	2.8%	280%	24%	240%

PRACTICE EXERCISES

See Extra Practice, page 446. **6.** 62.5% **12.** 7.5% **18.** $88\frac{8}{9}\%$ **24.** $71\frac{3}{7}\%$ **30.** 1,000% **36.** 1,500% **42.** 560% **48.** 1,262.5% **54.** $116\frac{2}{3}\%$

Write as a percent.

• **1.** $\frac{1}{2}$ 50%	**2.** $\frac{2}{5}$ 40%	**3.** $\frac{3}{4}$ 75%	**4.** $\frac{3}{8}$ 37.5%	**5.** $\frac{4}{5}$ 80%	**6.** $\frac{5}{8}$
7. $\frac{9}{10}$ 90%	**8.** $\frac{3}{20}$ 15%	**9.** $\frac{6}{25}$ 24%	**10.** $\frac{7}{50}$ 14%	**11.** $\frac{2}{25}$ 8%	**12.** $\frac{3}{40}$
• **13.** $\frac{1}{3}$ $33\frac{1}{3}\%$	**14.** $\frac{4}{7}$ $57\frac{1}{7}$	**15.** $\frac{2}{9}$ 22^2	**16.** $\frac{5}{6}$ $83\frac{1}{3}\%$	**17.** $\frac{2}{3}$ $66\frac{2}{3}\%$	**18.** $\frac{8}{9}$
19. $\frac{4}{9}$ $44\frac{4}{9}\%$	**20.** $\frac{1}{7}$ $14\frac{2}{7}\%$	**21.** $\frac{5}{9}$ $55\frac{5}{9}\%$	**22.** $\frac{3}{7}$ $42\frac{6}{7}\%$	**23.** $\frac{2}{7}$ $28\frac{5}{7}\%$	**24.** $\frac{5}{7}$
• **25.** 3 300%	**26.** 1 100%	**27.** 9 900%	**28.** 8 800%	**29.** 6 600%	**30.** 10
31. 7 700%	**32.** 2 200%	**33.** 4 400%	**34.** 11 1,100%	**35.** 12 1,200%	**36.** 15
• **37.** $3\frac{3}{4}$ 375%	**38.** $5\frac{1}{2}$ 550%	**39.** $8\frac{3}{4}$ 875%	**40.** $1\frac{4}{5}$ 180%	**41.** $6\frac{3}{8}$ 637.5%	**42.** $5\frac{3}{5}$
43. $4\frac{2}{5}$ 440%	**44.** $3\frac{1}{6}$ $316\frac{2}{3}\%$	**45.** $6\frac{2}{3}$ $666\frac{2}{3}\%$	**46.** $2\frac{4}{9}$ $244\frac{4}{9}\%$	**47.** $9\frac{1}{3}$ $933\frac{1}{3}\%$	**48.** $12\frac{5}{8}$
49. $\frac{1}{8}$ 12.5%	**50.** $\frac{7}{9}$ $77\frac{7}{9}\%$	**51.** $1\frac{3}{5}$ 160%	**52.** 23 2,300%	**53.** $\frac{7}{8}$ 87.5%	**54.** $1\frac{1}{6}$

Solve.

• **55.** Attendance at the finals was 9,124. The semifinals drew 6,348. How many more people attended the finals? 2,776

56. Visalia made 4 of 10 three-point shots. What percent of its three-point shots did Visalia make? 40%

57. Steve Jerome was the high scorer for Athens, the team that beat Visalia. He made 8 of 15 field goals. What percent of his field goals did Jerome make? $53\frac{1}{3}\%$

58. Athens made 7 of 11 three-point shots and 23 two-point shots. How many points did Athens score? 67 points

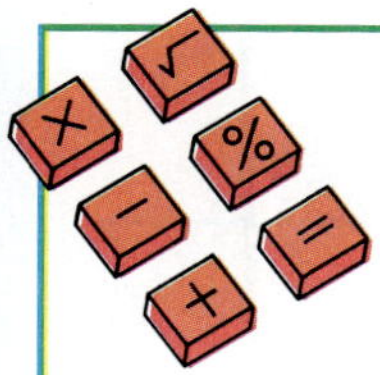

CALCULATOR

Copy and complete. Round to the nearest percent.

ATHENS TOURNAMENT FREE-THROW SHOOTING

	Free throws attempted	Free throws made	Percent made
Round of 32	25	21	1. ■ 84%
Round of 16	11	9	2. ■ 82%
Round of 8	17	15	3. ■ 88%
Semifinals	32	26	4. ■ 81%
Finals	12	7	5. ■ 58%

CALCULATOR Work through an example with the students to help them enter the numbers in correct order and to write the decimal as a percent.

COMMON ERROR

Some students may forget to move the decimal point when renaming a decimal as a percent. For example, item 1 of Checkpoint, they would select d. Review the procedure for writing a decimal as a percent. If needed, shade portions of a one-hundred square section of graph paper to highlight the difference between 75% ($\frac{75}{100}$) and 0.75% ($\frac{3}{4}$ of $\frac{1}{100}$).

ASSIGNMENTS

Level 1	Odd 1–47, 55–57
Level 2	Even 2–58, CA
Level 3	Odd 1–53, 55–58, CA

FOLLOW-UP ACTIVITY

MENTAL MATH Encourage students to memorize the percent equivalents of unit fractions. Show students how patterns can help them find the percent that corresponds to a fraction. For example,

$$\frac{1}{4} = 25\%, \text{ so } \frac{2}{4} = 2 \times 25 \text{ or } 50\%,$$
$$\frac{3}{4} = 3 \times 25 \text{ or } 75\%$$

Repeat for thirds, fifths, and tenths. Have students make flash cards and practice in pairs to improve their recall of corresponding percents and fractions.

SUPPLEMENTARY MATERIALS

TRP Practice, p. 96

TRP Reteaching, p. 59

TRP Transparency 48

OBJECTIVES

- Find a percent of a number.
- Estimate a percent of a number.

TEACHING THE LESSON

WARM-UP Present the following exercises on the chalkboard or an overhead transparency.

Multiply.

1. 0.46×10 **2.** 0.16×50

3. 0.075×60 **4.** $\frac{5}{6} \times 36$

5. $\frac{2}{7} \times 49$ **6.** $\frac{3}{8} \times 480$

(**1.** 4.6 **2.** 8 **3.** 4.5 **4.** 30 **5.** 14 **6.** 180)

INTRODUCTION Draw:

Ask a volunteer to cross out 60% of the circles and explain how he or she found the answer.

Repeat the activity to find:

- 40% of 5
- 75% of 8

INSTRUCTION In Example 1, emphasize that *of* means to *multiply* and *is* means *equals.* Point out that an alternate method is to change 80% to a fraction rather than a decimal.

$$n = \frac{4}{5} \times 30 = 24$$

In Example 2, you may wish to review the process for changing $37\frac{1}{2}\%$ to $\frac{3}{8}$. Point out that computing with mixed-number percents is easier if they are renamed as fractions.

To estimate the answer in Example 3, point out that the percent is rounded to the nearest 10%. Stress that this is not a rule. Emphasize the importance of using compatible numbers when estimating. Discuss how to estimate: 24.6% of 40, 34.2% of 66. (*Think:* 25% of 40, $33\frac{1}{3}\%$ of 66.)

9.4 FINDING A PERCENT OF A NUMBER

Audio Magazine's subscription price is 80% of its newsstand price. The newsstand price is \$30 per year. What is the subscription price?

To solve, you need to find what number is 80% of \$30.

EXAMPLE 1 What is 80% of 30?

1. Write a number sentence. What is 80% of 30?

 ↓ ↓ ↓ ↓ ↓

 THINK: 80% = 0.8 $n = 0.8 \times 30$

2. Find n. $n = 24$

The subscription rate is \$24 per year.

Sometimes it is easier to use a fraction to find the percent of a number.

TABLE OF FRACTIONAL EQUIVALENTS

$25\% = \frac{1}{4}$	$50\% = \frac{1}{2}$	$75\% = \frac{3}{4}$	
$20\% = \frac{1}{5}$	$40\% = \frac{2}{5}$	$60\% = \frac{3}{5}$	$80\% = \frac{4}{5}$
$16\frac{2}{3}\% = \frac{1}{6}$	$33\frac{1}{3}\% = \frac{1}{3}$	$66\frac{2}{3}\% = \frac{2}{3}$	$83\frac{1}{3}\% = \frac{5}{6}$
$12\frac{1}{2}\% = \frac{1}{8}$	$37\frac{1}{2}\% = \frac{3}{8}$	$62\frac{1}{2}\% = \frac{5}{8}$	$87\frac{1}{2}\% = \frac{7}{8}$

EXAMPLE 2 What is $37\frac{1}{2}\%$ of 240?

1. Write a number sentence. What is $37\frac{1}{2}\%$ of 240?

 ↓ ↓ ↓ ↓ ↓

 THINK: $\mathbf{37\frac{1}{2}\% = \frac{3}{8}}$ $n = \frac{3}{8} \times 240$

2. Find n. $n = 90$

So, $37\frac{1}{2}\%$ of 240 is 90.

You can estimate the percent of a number.

EXAMPLE 3 Estimate 47.5% of 48.

THINK: 47.5% is about 50% or $\frac{1}{2}$. 48 is about 50.

$$n = \frac{1}{2} \times 50$$

$$n = 25$$

So, 47.5% of 48 is about 25.

FOR DISCUSSION

See TE side column.

When will the percent of a number be greater than the number itself?

PRACTICE EXERCISES

See Extra Practice, page 446.

Find the percent of the number.

- **1.** 2% of 60 1.2 — **2.** 10% of 30 3 — **3.** 20% of 40 8 — **4.** 46% of 10 4.6
- **5.** 25% of 120 30 — **6.** 7.5% of 60 4.5 — **7.** 2.5% of 200 5 — **8.** 75% of 200 150
- **9.** $12\frac{1}{2}\%$ of 32 4 — **10.** $37\frac{1}{2}\%$ of 64 24 — **11.** $16\frac{2}{3}\%$ of 36 6 — **12.** $33\frac{1}{3}\%$ of 90 30
- **13.** $66\frac{2}{3}\%$ of 27 18 — **14.** $87\frac{1}{2}\%$ of 56 49 — **15.** $62\frac{1}{2}\%$ of 40 25 — **16.** $83\frac{1}{3}\%$ of 18 15

Compute.

- **17.** What is 2% of 200? 4 — **18.** What is $12\frac{1}{2}\%$ of 480? 60 — **19.** What is $16\frac{2}{3}\%$ of 360? 60
- **20.** What is 3% of 100? 3 — **21.** What is 5% of 50? 2.5 — **22.** What is 12.5% of 8? 1
- **23.** Find 75% of 648. 486 — **24.** Find $33\frac{1}{3}\%$ of 99. 33 — **25.** Find 40% of 200. 80
- **26.** Find $37\frac{1}{2}\%$ of 72. 27 — **27.** Find 250% of 12. 30 — **28.** Find 650% of 6. 39
- **29.** What number is $83\frac{1}{3}\%$ of 42? 35 — **30.** 9.5% of 20 is what number? 1.9

Estimate the percent of the number.

- **31.** 24.7% of 79 20 — **32.** 9.8% of 42 4 — **33.** 52.1% of 95 50 — **34.** 75.7% of 396 300
- **35.** 19.6% of 107 20 — **36.** 0.9% of 100 1 — **37.** 47.4% of 801 400 — **38.** 97% of 312 300

Solve.

39. *Audio Magazine*'s cover photo is reprinted at $83\frac{1}{3}\%$ of its original size. Write this percent as a fraction. $\frac{5}{6}$

40. A full-page ad in *Audio Magazine* costs $87\frac{1}{2}\%$ more than a half-page ad. A half-page ad costs $500. How much more does a full-page ad cost? $437.50

41. For every 3 copies sold on the newsstand, 2 copies are sold to subscribers. Circulation in June was 24,795 copies. How many went to subscribers? 9,918 copies

42. One of the tape decks advertised in *Audio Magazine* costs $275. Sales tax is 6%. What is the final cost of the tape deck? $291.50

FOR DISCUSSION Students should explain that when the percent of a number is greater than 100%, the answer is greater than the original number. Encourage them to give examples to verify this conclusion.

COMMON ERROR

Some students may forget to change the percent to a decimal or to a fraction prior to multiplying. The answer becomes a number that is greater than the given number. Use examples such as 20% of 45 to show that if the given percent is less than 100%, the answer should be less than the given number.

ASSIGNMENTS

Level 1	Odd 1–39
Level 2	Even 2–30, 31–40
Level 3	Odd 1–29, 31–42

FOLLOW-UP ACTIVITY

APPLICATION Distribute newspapers and mail-order catalogs. Each student uses a newspaper or catalog to write one problem that involves finding the percent of a number. Use the problems as additional practice. Have students take turns writing their problems on the chalkboard and demonstrating how to find the answers.

SUPPLEMENTARY MATERIALS

TRP Practice, p. 97

TRP Reteaching, p. 60

TRP Transparency 49

OBJECTIVE

- Solve problems that involve computing taxes.

TEACHING THE LESSON

WARM-UP Write the following exercises on the chalkboard or an overhead transparency.

In which set does each number belong?

Set A: 0–10,500

Set B: 10,501–21,000

Set C: 21,001–31,500

1. 4,789 **2.** 23,097 **3.** 19,999

4. 6,078

(**1.** A **2.** C **3.** B **4.** A)

INTRODUCTION Write the following on the chalkboard:

April 15

Ask the students if they know why this date is important to Americans. (It is the date when federal income tax returns must generally be filed.) Ask volunteers to tell the class what they know about federal income taxes.

INSTRUCTION In discussing the federal income tax, be sure the students understand that the taxable income for a person is not his or her gross income.

Among the allowable exemptions, adjustments, or deductions are a personal exemption, adjustments for certain retirement plans, and various possible deductions related to items such as mortgage interest.

Give careful attention to the introductory example. Point out that $2,677.50 is the tax on the first $17,850 of taxable income; $1,680 is the amount of tax on the extra $6,000 of taxable income.

You may wish to have students use calculators to help them solve the problems in this lesson.

PROBLEM Solving APPLICATION

9.5 CONSUMER: FEDERAL INCOME TAXES

Federal income tax is withheld from the paychecks of most people. However, when you prepare your tax return, you need to determine whether the amount that has been withheld from January 1 through December 31 is correct.

Marcie Graham is single and her taxable income is $23,850. During the year, she had $3,956.78 withheld from her paychecks. How much in additional taxes does Marcie need to pay?

Single Taxpayers

If taxable income is: over--	but not over--	The tax is:	of the amount over--
$ 0	$17,850	... 15%	$ 0
17,850	43,150	$2,677.50 + 28%	17,850
43,150	89,560	9,761.50 + 33%	43,150
89,560	----	... 28%	0

1. Read the table instructions.

   THINK: $23,850 is greater than $17,850 and less than $43,150.

2. Find the total tax.

   - Subtract to find out how much greater $23,850 is than $17,850. — $23,850 − $17,850 = $6,000
   - Multiply to find 28% of $6,000. **THINK: 28% = 0.28** — 0.28 × $6,000 = $1,680
   - Add to find the total tax. — $2,677.50 + $1,680.00 = $4,357.50

3. Subtract to find the amount owed. — $4,357.50 − $3,956.78 = $400.72

So, Marcie needs to pay $400.72 in additional taxes.

Use the table on page 242. Find the federal income tax for a single taxpayer. Remember to estimate whenever you use your calculator.

Taxable income	Federal income tax	Taxable income	Federal income tax
• **1.** $3,000	■ $450	• **2.** $7,000	■ $1,050
3. $15,000	■ $2,250	**4.** $24,850	■ $4,637.50
5. $98,500	■ $27,580	**6.** $49,150	■ $11,741.50
7. $19,350	■ $3,097.50	**8.** $42,950	■ $9,705.50
9. $46,050	■ $10,718.50	**10.** $88,950	■ $24,875.50
11. $17,890	■ $2,688.70	**12.** $43,153	■ $9,762.49

Find the tax for a single taxpayer. Then determine whether an amount is still owed or is to be refunded.

Taxable income	Federal income tax	Amount withheld	Amount owed	Amount to be refunded
•**13.** $6,575	■ $986.25	$856	$130.25 ■	■
•**14.** $43,100	■ $9,747.50	$5,850	$3,897.50 ■	■
15. $55,930	■ $13,978.90	$25,036	■	$11,057.10 ■
16. $108,426	■ $30,359.28	$42,865	■	$12,505.72 ■

Solve.

•**17.** Ann Caldor is a single taxpayer. Her taxable income is $32,450. What is her federal income tax? $6,765.50

18. Ben Tyson's taxable income is $20,000. Paul Wingate's is $40,000. Both are single. Is Paul's taxable income twice as great as Ben's? Is Paul's federal income tax twice as great as Ben's? Paul's income is twice as great but his taxes are more than twice as great.

19. Jessie Pelletier is single. She pays $8,250 in federal income tax. In which category is her taxable income? $17,850 to $43,150

ASSIGNMENTS

Level 1	1–8, 13–15, 17
Level 2	Even 2–16, 17–19
Level 3	7–19

FOLLOW-UP ACTIVITY

CALCULATOR Have the students compute the federal income tax for a single person with the given taxable income.

1. $43,149 ($9,761.22)

2. $43,150 ($9,761.50)

3. $43,151 ($9,761.78)

Challenge students to predict, without using the calculator, what the federal income tax is for a single person with a taxable income of $43,152. Ask them to explain how they decided what the tax is. ($9,762.06; for each extra dollar of income in Exercises 1–3, there is a $0.28 increase in the tax.)

FOR DISCUSSION
Indirect (hidden) taxes include taxes on sales of consumer goods or on goods in the process of production where the assumption (sometimes erroneous) is that the tax will be shifted to the consumer, who presumably has the ability to pay. Manufacturers and business owners actually pay the tax, but they add the cost to the prices they charge their customers. These sales taxes are a form of shifted taxes and include commodities such as gasoline, cigarettes, jewels, coffee, and airline tickets.

SUPPLEMENTARY MATERIALS

TRP Practice, p. 98

TRP Transparency 49

OBJECTIVE

- Evaluate student progress.

USING THE REVIEW

This page provides a means for informally evaluating students' understanding of the skills and concepts covered so far in this chapter.

Have the students look at the page to familiarize themselves with the various question formats that are presented. Discuss any questions that they may have. Then ask them to complete the page independently.

In addition to grading them individually, you may wish to review the answers to the questions collectively with the students.

Page references appear in brackets. They refer to pages on which a particular skill was introduced.

Before continuing on to the topics found in the remainder of the chapter, you may wish to have students review any skills or concepts in which they have demonstrated weakness.

MIDCHAPTER REVIEW

Write as a decimal. [234]

1. 6% 0.06 **2.** 15% 0.15 **3.** 56% 0.56 **4.** 150% 1.5 **5.** 14.7% 0.147

6. 0.1% 0.001 **7.** $16\frac{2}{3}\%$ $0.16\frac{2}{3}$ **8.** $37\frac{1}{2}\%$ 0.375 **9.** $166\frac{2}{3}\%$ $1.66\frac{2}{3}$ **10.** $\frac{1}{2}\%$ 0.005

Write as a fraction, whole number, or mixed number. [236]

11. 1% $\frac{1}{100}$ **12.** 4% $\frac{1}{25}$ **13.** 8% $\frac{2}{25}$ **14.** 24% $\frac{6}{25}$ **15.** 50% $\frac{1}{2}$

16. 80% $\frac{4}{5}$ **17.** 33% $\frac{33}{100}$ **18.** 480% $4\frac{4}{5}$ **19.** 300% 3 **20.** 2,000% 20

Write as a percent. [234, 238]

21. 0.34 34% **22.** 0.58 58% **23.** 0.04 4% **24.** 0.09 9% **25.** 0.3 30%

26. 0.2 20% **27.** 0.004 0.4% **28.** 0.001 0.1% **29.** 0.072 7.2% **30.** 1.4 140%

31. 5.82 582% **32.** $0.12\frac{1}{2}$ $12\frac{1}{2}\%$ **33.** $0.18\frac{2}{3}$ $18\frac{2}{3}\%$ **34.** $0.33\frac{1}{3}$ $33\frac{1}{3}\%$ **35.** $0.44\frac{4}{9}$ $44\frac{4}{9}\%$

36. $\frac{1}{5}$ 20% **37.** $\frac{1}{8}$ 12.5% **38.** $\frac{3}{4}$ 75% **39.** $\frac{5}{6}$ $83\frac{1}{3}\%$ **40.** $\frac{2}{3}$ $66\frac{2}{3}\%$

41. $\frac{3}{10}$ 30% **42.** $\frac{4}{25}$ 16% **43.** $\frac{17}{50}$ 34% **44.** $\frac{13}{20}$ 65% **45.** $\frac{8}{25}$ 32%

46. 8 800% **47.** 12 1,200% **48.** $6\frac{3}{8}$ 637.5% **49.** $1\frac{2}{3}$ $166\frac{2}{3}\%$ **50.** $7\frac{5}{6}$ $783\frac{1}{3}\%$

Find the percent of the number. [240]

51. 4% of 20 0.8 **52.** 25% of 16 4 **53.** 50% of 34 17 **54.** 200% of 125 250

55. $12\frac{1}{2}\%$ of 80 10 **56.** $16\frac{2}{3}\%$ of 12 2 **57.** $87\frac{1}{2}\%$ of 96 84 **58.** $66\frac{2}{3}\%$ of 93 62

Solve. Use the table on page 242 if necessary. [238, 240, 242]

59. Garth Tilson earned $24,000 last year. He saved 7% of his earnings. How much did Tilson save last year? $1,680

60. Garth Tilson spends $\frac{3}{8}$ of his monthly income on rent. What percent of his monthly income does Tilson spend on rent? 37.5%

61. Tilson's taxable income last year was $21,650. He filed as a single taxpayer. What was Tilson's federal income tax last year? $3,741.50

62. This year, Tilson earned $29,740 in taxable income. He had $6,303.96 withheld from his paychecks. Is Tilson entitled to a refund? If so, how much? If not, how much is owed? Tilson is entitled to a refund of $297.26.

244 CHAPTER 9

PROBLEM Solving APPLICATION

9.6 CAREER: STORE OWNER

Rosa Aquino owns a neighborhood clothing store. She buys merchandise at a certain cost and then sells it to people in small quantities.

Rosa must pay certain expenses such as rent, telephone, and electricity. To meet these expenses and make a profit, she must sell her merchandise for a greater amount than its cost.

To do this, Rosa marks up the merchandise at a rate of 50%. That is, the **markup** is 50% of the cost. The **selling price** is found by adding the markup to the cost. What would Rosa sell a pair of slacks for if they cost her $18?

1. Find the amount of the markup.

   THINK: 50% = 0.50 0.50 × $18 = $9

2. Add to the find the selling price. $18 + $9 = $27

So, Rosa would sell the pair of slacks for $27.

Find the markup and the selling price. Remember to estimate whenever you use your calculator.

Item	Cost of merchandise	Percent of markup	Amount of markup	Selling price
• **1.** A	$24	50%	▪ $12.00	▪ $36.00
• **2.** B	$15	30%	▪ $4.50	▪ $19.50
3. C	$25	40%	▪ $10.00	▪ $35.00
4. D	$60	45%	▪ $27.00	▪ $87.00
5. E	$8.50	43%	▪ $3.66	▪ $12.16
6. F	$19.75	36%	▪ $7.11	▪ $26.86
7. G	$25.20	$33\frac{1}{3}$%	▪ $8.40	▪ $33.60
8. H	$18.50	37.5%	▪ $6.94	▪ $25.44

Solve. Rosa buys one type of shirt at $192 per dozen.

9. What is the cost per shirt? $16 per shirt

10. What is the selling price per shirt if the markup is 50%? $24.00 per shirt

OBJECTIVE

- Solve problems that involve finding the cost, markup, and selling price of merchandise.

TEACHING THE LESSON

WARM-UP Write the following exercises on the chalkboard or an overhead transparency.

Find the answer.

1. 50% of $76 **2.** 60% of $80

3. 45% of $8.80 **4.** 56% of $12.75

(**1.** $38 **2.** $48 **3.** $3.96 **4.** $7.14)

INTRODUCTION Ask students if they think that customers pay the same price for merchandise that the owner pays. Discuss why owners charge more.

INSTRUCTION As you discuss the introductory problem, be sure the students understand that the cost is the cost to the owner of the store. The percent—50%—is the percent of markup. When we find 50% of $18, we are finding the amount of the markup.

Problems 1–8 require the students to find the amount of the markup and then the selling price. Problem 8 requires rounding to the nearest cent.

You may wish to have students use calculators to help them solve the problems in this lesson.

ASSIGNMENTS

Level 1	1–6
Level 2	1–10
Level 3	3–10

SUPPLEMENTARY MATERIALS

TRP Practice, p. 99

TRP Transparency 50

OBJECTIVES

- Find what percent one number is of another.
- Estimate what percent one number is of another.

TEACHING THE LESSON

WARM-UP Present the following exercises on the chalkboard or an overhead transparency.

Write as a percent.

1. 0.5 **2.** 0.875 **3.** $0.33\frac{1}{3}$
4. 2.5 **5.** $1.16\frac{2}{3}$ **6.** 9
(**1.** 50% **2.** 87.5% **3.** $33\frac{1}{3}$%
4. 250% **5.** $116\frac{2}{3}$% **6.** 900%)

INTRODUCTION Draw:

Ask what percent of the circles have been crossed out (60%) Have volunteers explain their answers. Repeat with similar examples.

INSTRUCTION In Example 1, elicit from the students that since 100% of 40 is 40, the missing percent must be less than 100%.

The opposite is true in Example 2. Since 100% of 16 is 16, the missing percent must be more than 100%. Encourage students to use this type of thinking to judge the reasonableness of their answers.

In Example 3, guide students in rounding to compatible number. Discuss the compatible numbers that would be used to estimate:

23.8 is what percent of 49.2?
What percent of 61 is 13.4?

25 is what percent of 50?
What percent of 60 is 10?

9.7 FINDING WHAT PERCENT ONE NUMBER IS OF ANOTHER

Of the first 40 Presidents, 22 came from one of the original 13 colonies. What percent of the first 40 Presidents came from one of the original 13 colonies?

To solve, you need to find what percent of 40 is 22.

EXAMPLE 1 What percent of 40 is 22?

1. Write a number sentence.

What percent of 40 is 22?
$n \times 40 = 22$

2. Find n.

THINK: To find the missing factor, divide 22 by 40.

$n \times 40 = 22$
$n = 22 \div 40$
$n = 0.55$

3. Rename the decimal as a percent. $0.55 = 55\%$

Of the first 40 Presidents, 55% came from one of the original 13 colonies.

Sometimes the missing percent is greater than 100%.

EXAMPLE 2 48 is what percent of 16?

1. Write a number sentence.

48 is what percent of 16?
$48 = n \times 16$

2. Find n.

THINK: To find the missing factor, divide 48 by 16.

$n \times 16 = 48$
$n = 48 \div 16$
$n = 3$

3. Rename the whole number as a percent. $3 = 300\%$

So, 48 is 300% of 16.

You can estimate the percent one number is of another.

EXAMPLE 3 Estimate the percent 3 is of 4.8.

THINK: 4.8 is about 5.

$n \times 5 = 3$
$n = 3 \div 5$
$n = 0.6$, or 60%

So, 3 is about 60% of 4.8.

FOR DISCUSSION See TE side column.

Which is the only reasonable answer to the following problem? Explain.

62 is what percent of 38? c **a.** 6% **b.** 63% **c.** 163%

PRACTICE EXERCISES

See Extra Practice, page 447.

Find the missing number.

- • **1.** What percent of 5 is 2? 40%
- **2.** What percent of 10 is 3? 30%
- **3.** What percent of 25 is 7? 28%
- **4.** What percent of 20 is 10? 50%
- • **5.** 3 is what percent of 4? 75%
- **6.** 8 is what percent of 32? 25%
- **7.** 16 is what percent of 32? 50%
- **8.** 36 is what percent of 48? 75%
- • **9.** What percent of 5 is 10? 200%
- **10.** What percent of 5 is 15? 300%
- **11.** What percent of 7 is 35? 500%
- **12.** What percent of 2 is 5? 250%
- **13.** 10 is what percent of 4? 250%
- **14.** 25 is what percent of 10? 250%
- **15.** 24 is what percent of 18? $133\frac{1}{3}\%$
- **16.** 18 is what percent of 12? 150%
- **17.** What percent of 500 is 2.5? 0.5%
- **18.** What percent of 200 is 3.5? 1.75%
- **19.** What percent of 25 is 120? 480%
- **20.** What percent of 125 is 25? 20%
- **21.** What percent of 140 is 70? 50%
- **22.** What percent of 48 is 43.2? 90%
- **23.** 67.2 is what percent of 24? 280%
- **24.** 19.8 is what percent of 99? 20%
- **25.** 37.8 is what percent of 14? 270%
- **26.** 147 is what percent of 245? 60%
- **27.** 148.8 is what percent of 124? 120%
- **28.** 218.4 is what percent of 260? 84%

Estimate.

- • **29.** What percent of 13 is 3.25? 20%
- **30.** What percent of 6 is 1.5? $33\frac{1}{3}\%$
- **31.** What percent of 45 is 4.5? 10%
- **32.** What percent of 12 is 3.8? $33\frac{1}{3}\%$
- • **33.** 26.8 is what percent of 50? 50%
- **34.** 14.5 is what percent of 50? 30%
- **35.** 21.6 is what percent of 10? 200%
- **36.** 45.2 is what percent of 30? 150%

Solve.

- • **37.** Of the first 28 Presidents, 8 were born in Virginia. What percent of the first 28 Presidents were born in Virginia? $28\frac{4}{7}\%$
- **38.** The youngest President ever, John F. Kennedy, was inaugurated on January 20, 1961. Kennedy was born on May 29, 1917. How old was Kennedy when he was inaugurated? 43 years old
- **39.** Franklin D. Roosevelt's 635 vetoes were the most by any President. About how many vetoes did Roosevelt average each year over his 12 years in office? 53 vetoes
- **40.** Of the first 40 Presidents, 25 were elected with 50% or more of the popular vote. What percent of the first 40 Presidents were elected with less than 50% of the popular vote? 37.5%

FOR DISCUSSION Students should explain that item c, 163%, is the only reasonable answer. They should realize that 100% of 38 is 38, so 62 must be more than 100% of 38.

COMMON ERROR

When finding the missing percent, some students will forget the final step of renaming the decimal answer as a percent. Encourage students to substitute the answer in the original question to see if the answer is reasonable.

ASSIGNMENTS

Level 1	Odd 1–39
Level 2	Even 2–36, 37–40
Level 3	17–40

FOLLOW-UP ACTIVITY

ENRICHMENT Have students refer to a map of the United States to find the percentage of the 50 states that:

- Border the Pacific Ocean. (10%)
- Border Canada. (22%)
- Border the Great Lakes. (16%)
- Border Mexico. (8%)
- Border the Atlantic Ocean. (28%)
- Are land-locked. (40%)

SUPPLEMENTARY MATERIALS

TRP Practice, p. 100

TRP Reteaching, p. 61

TRP Transparency 50

OBJECTIVE

- Find a number when a percent of it is known.

TEACHING THE LESSON

WARM-UP Present the following exercises on the chalkboard or an overhead transparency.

Divide.

1. $10 \div 0.1$ **2.** $18 \div 0.5$

3. $9.6 \div 0.08$ **4.** $10 \div \frac{1}{5}$

5. $16 \div \frac{2}{3}$ **6.** $10 \div 2\frac{1}{4}$

(**1.** 100; **2.** 36; **3.** 120; **4.** 50; **5.** 24; **6.** $4\frac{4}{9}$)

INTRODUCTION Write:

percent × whole = part

Have students refer to the statement to tell what is missing in the following situations:

1. Lakeview High School has 1,250 students. 85% of the student body attends a pep rally before a football game. (part)
2. At Lakeview's senior prom, 90% of the senior class shows up. There are 270 seniors at the prom. (whole)
3. Of the 445 juniors at Lakeview, 178 have their driver's licenses. (percent)

INSTRUCTION As an interesting side note, you may wish to discuss how information is stored in computers.

In Example 1, point out that since 120.9 represents a part of the missing number, the missing number must be greater than 120.9.

Similarly, in Example 2, the 8 represents only a part of the missing number. Thus, the missing number must be greater than 8. Encourage students to apply this type of thinking to check the reasonableness of their answers.

9.8 FINDING A NUMBER WHEN A PERCENT OF IT IS KNOWN

Barbara Samuels is a freelance journalist who uses a computer for her writing. Her Letters diskette is 65% full. It has 120.9 kilobytes (K) of information written onto it. How many kilobytes does the Letters diskette hold?

To solve, you need to find the number of which 65% is 120.9.

EXAMPLE 1 65% of what number is 120.9?

1. Write a number sentence.

65% of what number is 120.9?

$0.65 \times n = 120.9$

THINK: 65% = 0.65

2. Find n.

$0.65 \times n = 120.9$

$n = 120.9 \div 0.65$

$n = 186$

THINK: To find the missing factor, divide 120.9 by 0.65.

The Letters diskette holds 186K.

Sometimes it helps to change the percent to a fraction.

EXAMPLE 2 8 is $33\frac{1}{3}$% of what number?

1. Write a number sentence.

8 is $33\frac{1}{3}$% of what number?

$8 = \frac{1}{3} \times n$

THINK: $33\frac{1}{3}\% = \frac{1}{3}$

2. Find n.

$\frac{1}{3} \times n = 8$

$n = 8 \div \frac{1}{3}$

$n = 24$

THINK: To find the missing factor, divide 8 by $\frac{1}{3}$.

8 is $33\frac{1}{3}$% of 24.

CHECKPOINT

Write the letter of the correct answer.

Find the number.

1. 4% of what number is 8?	d	**a.** 0.32	**b.** 2	**c.** 20	**d.** 200
2. 3 is 20% of what number?	d	**a.** 0.6	**b.** 1.5	**c.** 6	**d.** 15
3. $87\frac{1}{2}$% of what number is 56?	d	**a.** 0.49	**b.** 6.4	**c.** 49	**d.** 64

PRACTICE EXERCISES See Extra Practice, page 447.

Find the number. Change the percent to a decimal.

- • **1.** 8% of what number is 4? 50
- • **2.** 1 is 5% of what number? 20
- **3.** 10% of what number is 6? 60
- **4.** 10 is 4% of what number? 250
- **5.** 1% of what number is 10? 1,000
- **6.** 18 is 50% of what number? 36
- **7.** 24% of what number is 48? 200
- **8.** 102 is 17% of what number? 600
- **9.** 27% of what number is 81? 300
- **10.** 9.6 is 8% of what number? 120
- **11.** 12% of what number is 38.4? 320
- **12.** 7.5 is 25% of what number? 30

Find the number. Change the percent to a fraction.

- • **13.** 10% of what number is 23? 230
- • **14.** 30 is 15% of what number? 200
- **15.** 20% of what number is 10? 50
- **16.** 18 is 3% of what number? 600
- **17.** 8% of what number is 28? 350
- **18.** 4 is 25% of what number? 16
- **19.** 12% of what number is 12? 100
- **20.** 12 is 40% of what number? 30
- **21.** $33\frac{1}{3}$% of what number is 7? 21
- **22.** 24 is $37\frac{1}{2}$% of what number? 64
- **23.** $62\frac{1}{2}$% of what number is 10? 16
- **24.** 21 is $87\frac{1}{2}$% of what number? 24

Find the number.

- • **25.** 100% of what number is 4.5? 4.5
- • **26.** 36 is 300% of what number? 12
- **27.** 200% of what number is 10? 5
- **28.** 140 is 350% of what number? 40
- **29.** 4.5% of what number is 4.5? 100
- **30.** 15 is 7.5% of what number? 200
- **31.** $12\frac{1}{2}$% of what number is 6? 48
- **32.** 16 is $66\frac{2}{3}$% of what number? 24
- **33.** 87.5% of what number is 56? 64
- **34.** 28.5 is 9.5% of what number? 300
- **35.** 6.25% of what number is 250? 4,000
- **36.** 3 is 0.5% of what number? 600
- **37.** 50% of what number is 36? 72
- **38.** 180 is 60% of what number? 300

Solve.

- • **39.** There are 8 bits in each byte and 1,000 bytes in each kilobyte. How many bits are there in a kilobyte? 8,000 bits
- **40.** Barbara paid $21.20 in sales tax for her printer. Sales tax is 8%. How much was the printer itself? $265.00
- **41.** The wholesale price for Barbara's computer is $1,050. Barbara paid 115% of the wholesale price. How much did she pay? $1,207.50
- **42.** Barbara is printing out a 40-page article. So far, 14 pages have been printed. What percent of the article is left to be printed? 65%

CHECKPOINT The incorrect answer choices include common errors that students make.

COMMON ERROR

Some students may write an incorrect number sentence to find the missing whole or total which, in turn, makes the subsequent computation incorrect. For example, in item 3 of Checkpoint, they would select c.

Encourage students who encounter difficulty to refer to the equation percent × whole = part when writing a number sentence to solve a percent problem.

ASSIGNMENTS

Level 1	Odd 1–37, 39–41
Level 2	Even 2–42
Level 3	9–16, 23–42

FOLLOW-UP ACTIVITY

CALCULATOR Point out that many statistics in sports involve percents. Have students use a calculator to find the missing statistics in the following table. Tell them to round to the nearest percent or whole number.

NFL LEADING LIFETIME PASSERS

Passer	Attempts	% Completed	Completions
K. Anderson	4,420	■ (59%)	2,627
D. Fouts	4,380	59%	■ (2,584)
S. Jurgensen	■ (4,268)	57%	2,433
L Dawson	3,741	57%	■ (2,132)
R. Staubach	2,958	■ (57%)	1,685
O. Graham	2,626	56%	■ (1,471)
J. Montana	■ (2,069)	64%	1,324
D. White	1,943	■ (59%)	1,155

SUPPLEMENTARY MATERIALS

TRP Practice, p. 101

TRP Reteaching, p. 62

TRP Transparency 51

OBJECTIVE

- Use proportions to solve the three cases of percent in real-life applications.

TEACHING THE LESSON

WARM-UP Present the following exercises in the chalkboard or an overhead transparency.

Solve for n.

1. $\frac{25}{100} = \frac{n}{16}$ **2.** $\frac{3}{100} = \frac{n}{48}$

3. $\frac{20}{100} = \frac{8}{n}$ **4.** $\frac{25}{100} = \frac{6}{n}$

5. $\frac{n}{100} = \frac{56}{64}$ **6.** $\frac{n}{100} = \frac{86}{43}$

(**1.** 4 **2.** 1.44 **3.** 40 **4.** 24 **5.** 87.5 **6.** 200)

INTRODUCTION Write the following problem on the chalkboard.

Carl Devlin is using his favorite picture frame for a new one that he is building. The original frame is 12 inches long. The new frame is 75% smaller than the original. How long will the new frame be?

Discuss ways of finding the answer. Test all reasonable suggestions. If needed, ask if a proportion could be used to find the answer. Ask a volunteer to come forward and write and solve the appropriate proportion. ($\frac{75}{100} = \frac{n}{12}$, $n = 9$)

INSTRUCTION Students will find that the most difficult aspect of this lesson is writing proportions correctly. Help them by labeling the ratios in Example 1.

Percent		Number (inches)
$\frac{72}{100}$	=	$\frac{n}{25}$ ←reduction ←actual

In Example 2, encourage students to use this model when writing proportions.

Percent		Number
$\frac{n}{100}$	=	$\frac{24}{60}$ ←Part ←Whole

9.9 USING PROPORTIONS TO SOLVE PERCENT PROBLEMS

Deborah Rabinowitz, the art director of an advertising agency, uses a copying machine to reduce the size of a piece of artwork. The machine is set to reduce the artwork to 72% of its original size. The artwork is 25 in. long. How long is the reduced copy of the artwork?

You can use a proportion to find the percent of a number.

EXAMPLE 1

What number is 72% of 25?

1. Write a proportion.

   THINK: 72% is to 100% as *n* inches are to 25 in. $\frac{72}{100} = \frac{n}{25}$

2. Multiply to find the cross products. $100 \times n = 72 \times 25$

   $100 \times n = 1{,}800$

3. Find n. $n = 1{,}800 \div 100$

   $n = 18$

The reduced copy of the artwork is 18 in. long.

You can use a proportion to find the percent one number is of another.

EXAMPLE 2

24 is what percent of 60?

1. Write a proportion.

   THINK: *n*% is to 100% as 24 is to 60. $\frac{n}{100} = \frac{24}{60}$

2. Multiply to find the cross products. $60 \times n = 24 \times 100$

   $60 \times n = 2{,}400$

3. Find n. $n = 2{,}400 \div 60$

   $n = 40\%$

You can use a proportion to find a number when a percent of it is known.

EXAMPLE 3

85% of what number is 17?

1. Write a proportion.

   THINK: 85% is to 100% as 17 is to *n*. $\frac{85}{100} = \frac{17}{n}$

2. Multiply to find the cross products. $85 \times n = 17 \times 100$

   $85 \times n = 1{,}700$

3. Find n. $n = 1{,}700 \div 85$

   $n = 20$

PRACTICE EXERCISES See Extra Practice, page 447.

Use a proportion to solve.

• **1.** What number is 18% of 40? 7.2

2. 25% of 16 is what number? 4

3. What number is 3% of 48? 1.44

4. 58% of 200 is what number? 116

5. What number is 120% of 50? 60

6. 250% of 18 is what number? 45

• **7.** 36 is what percent of 48? 75%

8. What percent of 60 is 15? 25%

9. 56 is what percent of 64? $87\frac{1}{2}\%$

10. What percent of 180 is 45? 25%

11. 21 is what percent of 12? 175%

12. What percent of 50 is 250? 500%

• **13.** 20% of what number is 8? 40

14. 18 is 36% of what number? 50

15. 25% of what number is 6? 24

16. 1 is 50% of what number? 2

17. 130% of what number is 39? 30

18. 90 is 225% of what number? 40

19. 200% of 68 is what number? 136

20. 150% of what number is 96? 64

21. 22.4 is what percent of 64? 35%

22. 86 is what percent of 43? 200%

23. 64% of what number is 384? 600

24. 100% of what number is 2? 2

25. 240 is what percent of 80? 300%

26. 70% of 490 is what number? 343

27. 12% of what number is 30? 250

28. 98 is what percent of 294? $33\frac{1}{3}\%$

29. 84 is what percent of 36? $233\frac{1}{3}\%$

30. 345% of what number is 138? 40

Solve.

• **31.** Bruce Kelvin, the assistant art director, just had his salary raised 11% to $19,425. What was Kelvin's salary before the raise? $17,500

32. Deborah and Bruce prepare 15 different logos for a dog food campaign. Of these, 3 will be used. What percent will be used? 20%

33. The agency has 135 employees, 60% of whom receive benefits. What fraction of the employees receive benefits? $\frac{3}{5}$

34. The agency gets a 12% discount on colored pencils. The pencils list for $1.29. How much does the agency save on an order of 2,000? $309.60

TIME OUT You are the owner of a store that sells artists' supplies. The wholesale price you pay for a drafting table is $150. Before you sell the table, you mark up that price by 60%. One week you put the table on sale for 25% off the marked-up price. In addition, you offer professional artists a 15% discount. If a professional artist buys the drafting table on sale, do you make or lose money? How much? You make $3.00.

Extend the instruction in Example 3 by helping students write proportions to find:

1. What number is 70% of 200? (140)
2. What percent of 50 is 10? (20%)
3. 16 is 25% of what number? (64)

TIME OUT This situation is typical of those that many merchants face. You may wish to have students use calculators, as merchants probably would, to find the answer.

ASSIGNMENTS

Level 1	Odd 1–29, 31–32
Level 2	Even 2–30, 31–34
Level 3	19–34, TO

FOLLOW-UP ACTIVITY

ESTIMATION Discuss tipping. Point out that in beauty shops, barber shops, restaurants, and taxicabs the standard tip is 15% of the total cost. Discuss ways of estimating 15% of the following costs.

Item	Cost	Tip
Lunch	$20.00	($3.00)
Haircut	$15.00	($2.25)
Taxi	$ 6.00	($1.00)
Dinner	$42.50	($6.50)
Perm	$37.25	($6.00)
Taxi	$17.75	($2.70)

SUPPLEMENTARY MATERIALS

TRP Practice, p. 102

TRP Reteaching, p. 63

TRP Transparency 51

OBJECTIVE

- Supply missing information to solve problems.

TEACHING THE LESSON

WARM-UP Write the following exercises on the chalkboard or on an overhead transparency.

1. 1 ft = ■ in. **2.** 1 yd = ■ ft

3. 1 day = ■ h **4.** 1 y = ■ wk

(**1.** 12; **2.** 3; **3.** 24; **4.** 52)

INTRODUCTION Read the following problems aloud. Have students explain what needed information is missing to answer each question.

1. Jane Cortez buys the newspaper every Sunday in May. How many newspapers does she buy? (the number of Sundays in May)
2. Fran DiFilippo waits $\frac{3}{4}$ of an hour for a bus. How many minutes does she wait? (the number of minutes in 1 hour)
3. Each of Bob Gardner's skis is 5 ft long. How many inches long is each ski? (the number of inches in a foot)

INSTRUCTION As you discuss Example A, stress that it is impossible to solve the problem without knowing the number of days in each of the months involved. Refer to a classroom calendar or the calendar given to find the number of days in each of the months December through March.

In Example B, guide students to realize that they can multiply to find the answer because they know the number of groups and the number in each group.

Before assigning independent work, remind students to refer to the table of measures on page 476 to obtain information as needed. You may wish to allow students to use calculators in solving these problems.

PROBLEM *Solving* STRATEGY

9.10 SUPPLYING MISSING INFORMATION

Situation:

Captain's Ridge Ski Lodge is open from December through March. Guests are shuttled to the lodge in minibuses that hold one dozen guests each.

Strategy:

Sometimes you have to use information you know or can find to help you solve the problem.

Applying the Strategy:

A. For how many days is Captain's Ridge Ski Lodge open?

Which months is the lodge open? December, January, February, March.

What information do you need to know? The number of days in each of the months.

Where could you find this information? In a calendar.

DECEMBER						
				1	2	3
4	5	6	7	8	9	10
11	12	13	14	15	16	17
18	19	20	21	22	23	24
25	26	27	28	29	30	31

JANUARY						
1	2	3	4	5	6	7
8	9	10	11	12	13	14
15	16	17	18	19	20	21
22	23	24	25	26	27	28
29	30	31				

FEBRUARY						
			1	2	3	4
5	6	7	8	9	10	11
12	13	14	15	16	17	18
19	20	21	22	23	24	25
26	27	28				

MARCH						
			1	2	3	4
5	6	7	8	9	10	11
12	13	14	15	16	17	18
19	20	21	22	23	24	25
26	27	28	29	30	31	

Add to find the total number of days. $31 + 31 + 28 + 31 = 121$

Captain's Ridge Ski Lodge is open for 121 days.

B. If 6 full minibuses arrive, how many guests are the minibuses carrying?

What information are you given? 6 minibuses arrive, a dozen guests in each.

What information do you need to know? There are 12 in a dozen.

Multiply to find the total number of guests. $6 \times 12 = 72$

The minibuses are carrying 72 passengers.

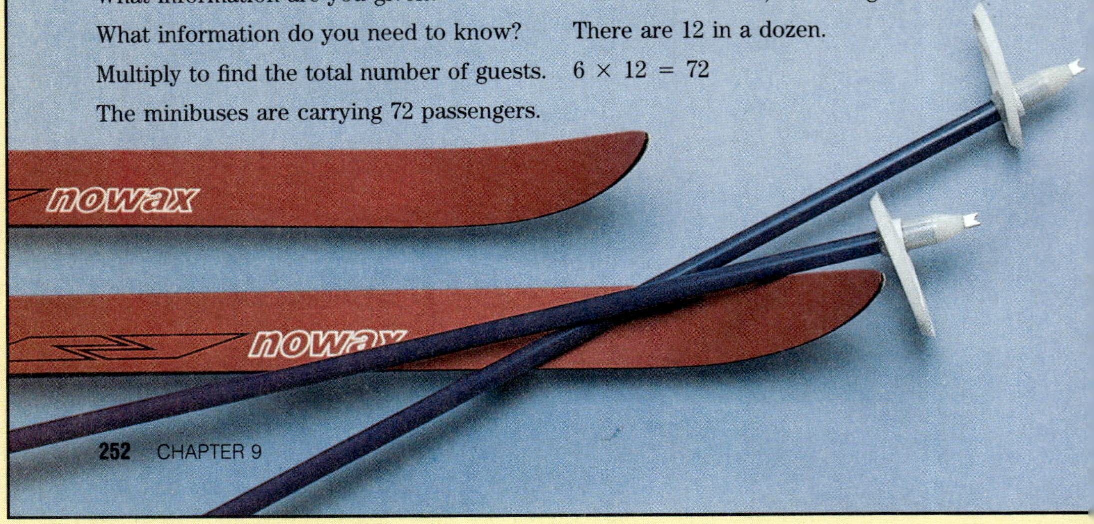

PRACTICE EXERCISES

See Extra Practice, page 448.

What information do you need to solve the problem?

1. The lodge makes sandwiches for the skiers to eat on the slopes. Each sandwich includes 3 oz of cheese. The cook made 96 sandwiches today. How many pounds of cheese did he use? b
   a. 8 fl oz = 1 cup
   b. 16 oz = 1 lb
   c. 2,000 lb = 1 ton

2. It takes 20 minutes to go up the mountain on the chair lift and then ski down again. How many times can a skier ski down the mountain in 2 hours? b
   a. 60 seconds = 1 minute
   b. 60 minutes = 1 hour
   c. 24 hours = 1 day

Solve by supplying the missing information.

3. The Captain ordered 9 dozen new ski poles for the rental shop. How many ski poles did he order? 108 ski poles

4. The waiter at the lodge is pouring hot apple cider from a 4-qt pitcher into mugs that hold 8 fl oz each. How many mugs can he fill? 16 mugs

5. The ski instructor skis the two expert trails, Beech and Echo, each morning before her lessons begin. Beech Trail is 875 yd long. Echo Trail is 2,500 ft long. How many feet longer is Beech Trail? 125 ft

6. Joan Galvin wanted to rent skis that were $5\frac{1}{2}$ ft long. The only skis available were 62 in. long. How many inches shorter were these skis than the ones Joan wanted to rent? 4 in.

7. Every Saturday, Ranger Frederickson is assigned to the ski patrol. He skis the length of the cross-country trail 8 times. The cross-country trail is 1,320 yd long. How many miles does the ranger ski each Saturday? 6 mi

ASSIGNMENTS

Level 1	1–5
Level 2	1–7
Level 3	Odd 3–7

FOLLOW-UP ACTIVITY

SITUATIONAL PROBLEM SOLVING Present students with the following problem. Your family is considering going on a 2-week ski vacation in Utah. You are in charge of researching the ski lodges available and choosing one. What questions would you want answered by your family and by the ski lodges before you made your decision? Develop your two sets of questions.

SUPPLEMENTARY MATERIALS

TRP Practice, p. 103

TRP Applications, p. 16

TRP Transparency 52

OBJECTIVE

- Solve problems that involve finding commissions and rates of commission.

TEACHING THE LESSON

WARM-UP Write the following exercises on the chalkboard or an overhead transparency.

Find the answer.

1. What number is 5% of \$3,000? (\$150)
2. What number is 4% of \$5,500? (\$220)
3. What percent of \$6,000 is \$600? (10%)
4. What percent of \$8,000 is \$400? (5%)

INTRODUCTION Ask students if they know people who sell real estate, computers, furniture, or some other type of item that has a high selling price. Encourage volunteers to tell whether such salespeople are paid only a weekly salary, or whether they are paid in different ways. Elicit that often these salespeople are paid a sum of money, or commission, for each item they sell.

INSTRUCTION Emphasize the difference between the commission and the rate of commission. A simple visual aid may help students remember the distinction:

Commission → \$
Rate of commission → %

As you discuss Examples A and B, give special attention to each first step, in which the question is translated into a number sentence.

You may wish to have students use calculators to help them solve the problems in this lesson.

PROBLEM *Solving* APPLICATION

9.11 CONSUMER: COMMISSION

You have a job as a real estate salesperson. You do not receive a weekly salary. Instead, you receive a **commission** on each house you sell. The **rate of commission** is a percent of the selling price of the house.

A. If your rate of commission is 3%, what is your commission on a selling price of \$90,000?

You need to find the number that is 3% of 90,000.

1. Write a number sentence.

THINK: 3% = 0.03

What number is 3% of 90,000?

$n = 0.03 \times 90{,}000$

2. Find n

$n = 2{,}700$

So, you receive a commission of \$2,700 on a selling price of \$90,000.

B. Your friend is a salesperson for another real estate company. She sells a home for \$150,000 and receives a \$6,000 commission. What is her rate (percent) of commission?

You need to find what percent of 150,000 is 6,000.

1. Write a number sentence.

Let n represent the rate of commission.

What percent of 150,000 is 6,000?

$n \times 150{,}000 = 6{,}000$

2. Find n.

$n = 6{,}000 \div 150{,}000$

$n = 0.04$, or 4%

So, your friend's rate of commission is 4%.

Find the commission. Remember to estimate whenever you use your calculator.

	Selling price	Rate of commission	Commission		Selling price	Rate of commission	Commission
1.	$80,000	3%	$2,400	2.	$110,000	4%	$4,400
3.	$176,000	4%	$7,040	4.	$158,000	3%	$4,740
5.	$190,000	4%	$7,600	6.	$232,500	3.5%	$8,137.50

Find the *rate* of commission.

	Selling price	Rate of commission	Amount of commission		Selling price	Rate of commission	Amount of commission
7.	$100,000	3%	$3,000	8.	$120,000	4%	$4,800
9.	$160,000	3.5%	$5,600	10.	$400,000	4.5%	$18,000

Solve.

11. Carl Marder sells boats. He receives a salary and a commission. His rate of commission is 2%. How much commission does he receive for a month in which he makes sales totaling $38,000? $760

12. Maria Gomez sells commercial real estate. Her rate of commission on sales is 5%. How much commission does she receive if she sells a building for $650,000? $32,500

13. Eddie Chang sells cars. On sales of $180,000, he receives commissions totaling $4,500. What is his rate of commission? 2.5%

14. One month Kelly Mandera receives $2,500 in commissions. Her rate of commission is 4%. What is the total amount of her sales that month? $62,500

ASSIGNMENTS

Level 1	Odd 1–9, 11–12
Level 2	1–14
Level 3	4–9, 11–14

FOLLOW-UP ACTIVITY

ENRICHMENT Have students solve these problems.

1. Amelia King sells boats. She receives a salary of $1,000 per month and a commission on the sales she makes. Her rate of commission is 1%. What are her total (gross) earnings for a month in which she makes sales totaling $42,000? ($1,420)

2. Juan Hernandez makes commissions based on this scale:

   4% of sales up to $5,000

   5% of sales between $5,000 and $10,000

   6% of sales greater than $10,000

How much commission does he earn on $7,000? ($300) How much commission does he earn on $11,000? ($510)

Enrichment (suggestion)
As a helpful hint to the students, tell them to figure the 1st $5,000 using 4%, the 2nd $5,000 using 5%, and any remaining using the 6%.

SUPPLEMENTARY MATERIALS

TRP Practice, p. 104

TRP Transparency 52

OBJECTIVE

- Solve problems that involve finding discounts, rates of discount, and sale prices.

TEACHING THE LESSON

WARM-UP Write the following exercises on the chalkboard or an overhead transparency.

Find the answer.

1. 25% of 160 = ▒ (40)
2. 30% of 280 = ▒ (84)
3. ▒% of 200 = 20 (10%)
4. ▒% of 480 = 168 (35%)

INTRODUCTION You may wish to display a large newspaper advertisement that shows a "10% OFF" or "25% OFF" sale at a department store. Ask volunteers to explain what the phrase "10% OFF" means.

INSTRUCTION Be sure that students understand the procedure in Example A: First find the discount in dollars; then subtract the discount from the original price.

In Example B, students find the percent of discount. As you discuss the example, use the phrase "rate of discount" to reinforce the phrase "percent of discount."

You may wish to have students use calculators to help them solve the problems in this lesson.

FOR DISCUSSION Students should realize that the size of the discount is not necessarily the most important element in buying an item. Other considerations are the quality of the item and the original price. If the original price is inflated or unreasonable, a large discount may not be significant.

PROBLEM Solving APPLICATION

9.12 CONSUMER: DISCOUNT

A **discount** is the amount that is subtracted from the list price of an item. The sign below shows the **rate of discount** or **percent of discount** for a sale at Eager Eddie's Electronics Mart.

A. You want to buy a VCR with a list price of $310. What is the amount of discount? What is the sale price?

THINK: List price = $310 **Percent of discount = 20% = 0.2**

LIST PRICE	×	PERCENT OF DISCOUNT	=	DISCOUNT
$310	×	0.20	=	$62

The discount is $62.

Subtract to find the sales price. $310 − $62 = $248

The sale price is $248.

B. Your friend received a discount of $60 on an item with a list price of $400. What was the percent of discount?

THINK: Discount = $60 **List price = $400**

DISCOUNT	÷	LIST PRICE	=	PERCENT OF DISCOUNT
$60	÷	$400	=	0.15, or 15%

The percent of discount was 15%.

FOR DISCUSSION See TE side column.

If two stores sell different brands of equipment, is it always a good idea to buy items in the store that has the greater percent of discount? Why?

Find the discount. Remember to estimate whenever you use your calculator.

	List price	Percent of discount	Amount of discount		List price	Percent of discount	Amount of discount
• **1.**	$80	50%	▪ $40	• **2.**	$90	10%	▪ $9
3.	$120	25%	▪ $30	**4.**	$160	30%	▪ $48
5.	$140	15%	▪ $21	**6.**	$200	35%	▪ $70
7.	$28.50	10%	▪ $2.85	**8.**	$39.90	20%	▪ $7.98
9.	$48.72	$12\frac{1}{2}\%$	▪ $6.09	**10.**	$59.97	$33\frac{1}{3}\%$	▪ $19.98

Find the discount and the sale price.

	List price	Percent of discount	Amount of discount	Sale price
• **11.**	$90	15%	▪ $13.50	▪ $76.50
• **12.**	$130	20%	▪ $26	▪ $104
13.	$180	35%	▪ $63	▪ $117
14.	$45.60	$37\frac{1}{2}\%$	▪ $17.10	▪ $28.50
15.	$90.00	$66\frac{2}{3}\%$	▪ $60.00	▪ $30.00

Solve.

• **16.** Your cousin buys a camera. The list price is $80. The percent of discount is 45%. What is the sale price? $44

17. Cary James says that he received a discount of $77 on an item listed at $220. What is the percent of discount? 35%

18. Your friend says that if the percent of discount is 20%, then the sale price must be 80% of the list price. Do you agree? Use two methods to find the sale price for this $760 refrigerator. $608

19. If the percent of discount is 35%, then the sale price is what percent of the list price? 65%

20. The sale price is 55% of the list price. What percent of the list price is the discount? 45%

ASSIGNMENTS

Level 1	Odd 1–15, 16–17
Level 2	Even 2–14, 16–20
Level 3	7–14, 16–20

FOLLOW-UP ACTIVITY

APPLICATION Explain to students that over a period of time a store may give more than one discount on the same item. For example, for an April sale, a store may offer a 20% rate of discount. Then, for a closeout sale in June, the store may offer a 10% rate of discount. (The 10% rate of discount is applied to the discounted price after the April sale.)

Have students calculate the final sale price:

	Original Price	First rate of discount	Second rate of discount
1.	$150	20%	10%
2.	$200	25%	10%
3.	$85	10%	10%

(**1.** $108 **2.** $135 **3.** $68.85)

SUPPLEMENTARY MATERIALS

TRP Practice, p. 105

TRP Transparency 53

OBJECTIVE

- Find the percent of increase or decrease given the original amount and the new amount.

TEACHING THE LESSON

WARM-UP Present the following exercises on the chalkboard or an overhead transparency.

Write as a percent.

1. $\frac{8}{24}$ **2.** $\frac{18}{27}$ **3.** $\frac{10}{50}$

4. $\frac{12}{48}$ **5.** $\frac{18}{30}$ **6.** $\frac{2}{9}$

(**1.** $33\frac{1}{3}\%$ **2.** $66\frac{2}{3}\%$ **3.** 20%;

4. 25%; **5.** 60%; **6.** $22\frac{2}{9}\%$)

INTRODUCTION Write the following problems on the chalkboard and discuss ways of solving them.

Last summer Gary Wilkins worked at Hi-Tech Tool and Dye. He earned $250 per week. This summer Gary will have the same job but will earn 10% more. How much will he earn each week this summer? ($275)

Gary decides to continue working part-time at the factory during the school year. His pay will drop from $275 to $137.50 a week. By what percent will his weekly pay decrease? (50%)

INSTRUCTION In Examples 1 and 2, the key to finding the percent of increase or decrease is in writing the correct ratio. Emphasize that the ratio involves comparing the amount of increase or decrease with the *original* amount, not the new amount. In writing the resulting fraction as a percent, remind students to first write the fraction in simplest form.

You may wish to extend the instruction by guiding students through the following examples.

1. The rent on Sandra Williams's apartment increased from $400 to $420 a month. What was the percent of increase? (5%)

9.13 PERCENTS OF INCREASE AND DECREASE

At the Griffin Book Store, Claire Martin makes $100 a week working part time. She gets a raise to $110 a week. What is the percent of increase in Claire Martin's salary?

To find the **percent of increase,** you first have to find the amount of increase.

EXAMPLE 1 Find the percent of increase in Claire Martin's salary.

1. Subtract to find the amount of increase. $\$110 - \$100 = \$10$
2. Write a ratio to show the amount of increase to the original amount. $\frac{10}{100}$ ← **amount of increase** / ← **original amount**
3. Rename the fraction as a percent. $\frac{10}{100} = 10\%$

The percent of increase in Claire Martin's salary is 10%.

To find the **percent of decrease,** you first have to find the amount of decrease.

EXAMPLE 2 In the fall, Claire Martin returns to school and cuts back on her hours from 40 to 28. What is the percent of decrease in the number of hours Claire Martin works?

1. Subtract to find the amount of decrease. $40 - 28 = 12$
2. Write a ratio to show the amount of decrease to the original amount. $\frac{12}{40}$ ← **amount of decrease** / ← **original amount**
3. Rename the fraction as a percent. $\frac{12}{40} = 30\%$

The number of hours Claire works is decreased by 30%.

FOR DISCUSSION See TE side column.

Yesterday, Amco stock rose 10%. Today, it fell 10%. Is today's closing price for Amco stock the same as yesterday's opening price? Explain.

See Extra Practice, page 448.

1. 50% inc. 2. $33\frac{1}{3}$% inc. 3. 400% inc. 4. 25% inc.
5. 150% inc. 6. 40% inc. 7. 40% dec. 8. 30% dec.
9. $33\frac{1}{3}$% dec. 10. $12\frac{1}{2}$% dec. 11. $66\frac{2}{3}$% dec. 12. 85% dec.
13. 14% dec. 14. 32% inc. 15. 35% inc. 16. 104% inc.
17. 4% dec. 18. 344% inc. 19. 98% dec. 20. 44% dec.

PRACTICE EXERCISES

Copy and complete the table.

	Original Amount	New Amount	Percent Change		Original Amount	New Amount	Percent Change
• 1.	200	300	■	• 2.	36	48	■
3.	12	60	■	4.	52	65	■
5.	60	150	■	6.	85	119	■
• 7.	1,000	600	■	• 8.	60	42	■
9.	24	16	■	10.	72	63	■
11.	27	9	■	12.	120	18	■
13.	85	73.1	■	14.	45	59.4	■
15.	112	151.2	■	16.	65	132.6	■
17.	2,500	2,400	■	18.	350	1,554	■
19.	72	1.44	■	20.	35	19.6	■
21.	50	■ 60	20% increase	22.	132	■ 165	25% increase
23.	200	■ 184	8% decrease	24.	250	■ 235	6% decrease
25.	600	■ 462	23% decrease	26.	275	■ 396	44% increase
27.	■ 1,220	305	75% decrease	28.	■ 800	804	0.5% increase
29.	■ 64	72	$12\frac{1}{2}$% increase	30.	■ 210	35	$83\frac{1}{3}$% decrease
31.	■ 120	252	110% increase	32.	79	■ 0	100% decrease

MIXED REVIEW

Complete.

• **33.** 5 d = ■ h 120 • **34.** 4 wk = ■ d 28 **35.** 90 s = ■ min $1\frac{1}{2}$

36. 30 min = ■ h $\frac{1}{2}$ **37.** 42 mo = ■ y $3\frac{1}{2}$ **38.** 6 min = ■ sec 360

Solve.

• **39.** Sales of art books at the Griffin totaled \$245.50 in March. In April, sales were \$279.87. What was the percent of increase in sales? 14%

40. On Labor Day, the Griffin had 3,260 textbooks in stock. By October, that stock had dwindled to 1,141. By how many copies did the stock fall? 2,119 copies

41. A customer buys a boxed set of reference books for \$23.95. Sales tax is 6.5%. What is the final total cost? \$25.51

42. March sales at the Griffin totaled 4,500 books. In April, sales were down 6%, but in May they were up 10%. What were sales in May? 4,653

2. The price of a rock cassette dropped from \$8.00 to \$6.40. What was the percent of decrease? (20%)

FOR DISCUSSION Students should use a numerical example to show that increasing and decreasing by the same percent is not the same as adding and subtracting by, or as multiplying and dividing by, the same number.

COMMON ERROR

Some students may have difficulty in differentiating between the original amount and the new amount. Work individually with students and have them identify the original amount in each exercise.

ASSIGNMENTS

Level 1	Odd 1–19, 33–41
Level 2	Even 2–40, 41, 42
Level 3	15–42

FOLLOW-UP ACTIVITY

APPLICATION Pose this question to your students: "Suppose your boss offered to increase your salary by 10% for one week as long as you were willing to have it decreased by 10% the following week. Would you accept the offer? Why or why not?"

Encourage students to experiment with a fictitious salary to find out what would happen. (The offer is not good because the amount of decrease is always greater than the amount of increase.)

SUPPLEMENTARY MATERIALS

TRP Practice, p. 106

TRP Reteaching, p. 64

TRP Transparency 53

OBJECTIVE

- Solve problems that involve finding simple interest and compound interest.

TEACHING THE LESSON

WARM-UP Write the following exercises on the chalkboard or an overhead transparency.

Find the answer.

1. $n = 96 \times \frac{1}{2}$ (48)
2. $n = 120 \times 1.5$ (180)
3. $n = 800 \times 0.14 \times \frac{1}{2}$ (56)
4. $n = 2{,}000 \times 0.16 \times 0.5$ (160)

INTRODUCTION Ask students if anyone in their family has ever borrowed money from a bank. Lead a discussion, asking questions such as the following: Did the bank charge extra money for making the loan? What is such a charge called?

INSTRUCTION In the first example, emphasize that time must be expressed in years, since rate is expressed as an annual rate. Thus, 6 months can be expressed as $\frac{6}{12}$ year, or $\frac{1}{2}$ year, or 0.5 year.

Give careful attention to the explanation of compound interest on page 261. Notice that in Problems 21–24, students continue the computations in the example.

You may wish to have students use calculators to help them solve the problems in this lesson.

FOR DISCUSSION Students should explain that banks seek additional accounts because they can invest the money in loans and other types of investments. These investments pay the banks more than the amount that they pay in interest.

PROBLEM Solving APPLICATION

9.14 CONSUMER: INTEREST

A charge for the use of money is called **interest.** If you borrow money from a bank, you will need to pay interest for the use of the money. On the other hand, when you have money in a savings account, the bank pays interest for the use of your money.

One type of interest is called **simple interest.** To compute simple interest, use this formula:

$$\text{INTEREST} = \text{PRINCIPAL} \times \text{RATE} \times \text{TIME}$$
$$I = p \times r \times t$$

I is the amount of interest

p is the principal, which is the amount of money that is borrowed or is in the savings account

r is the *yearly* rate of interest

t is the time in years

How much interest do you have to pay if you borrow $500.00 for 6 months at a rate of 12% per year?

THINK: Principal = $500
rate = 12%, or 0.12
time = 6 mo = $\frac{6}{12}$ **y, or** $\frac{1}{2}$ **y**

$I = \$500 \times 0.12 \times \frac{1}{2}$

$I = \$60.00 \times \frac{1}{2}$

$I = \$30.00$

So, you would have to pay $30.00 in interest.

FOR DISCUSSION See TE side column.

Some banks advertise their interest rates in newspapers. They offer free gifts to people who open accounts. Why should banks want additional money on which they will have to pay interest?

See page 478 for answers to Exercises 1–20.

Find the simple interest. Remember to estimate whenever you use your calculator.

	Interest	Principal	Rate	Time		Interest	Principal	Rate	Time
• 1.	■	$1,000	6%	1 y	• 2.	■	$800	7%	2 y
3.	■	$1,200	9%	6 mo	4.	■	$2,000	10%	1 y 3 mo
5.	■	$5,000	$7\frac{1}{2}$%	18 mo	6.	■	$3,000	5%	2 y 4 mo
7.	■	$4,500	11%	9 mo	8.	■	$250	8%	3 mo
9.	■	$8,000	$6\frac{1}{4}$%	4 y	10.	■	$1,750	5.5%	1 y
11.	■	$10,000	12%	60 mo	12.	■	$9,800	11%	4 y 6 mo
13.	■	$500	15%	9 mo	14.	■	$5,000	15%	9 mo
15.	■	$50,000	15%	9 mo	16.	■	$500,000	15%	9 mo
17.	■	$12,800	$9\frac{3}{4}$%	3 y	18.	■	$975	$4\frac{1}{4}$%	3 mo
19.	■	$15,225	8.75%	16 mo	20.	■	$3,788	23%	90 days

Most banks pay **compound interest** on savings accounts. This means the interest is computed two or more times a year. If you do not withdraw your interest, it is credited to your account.

How much is the principal in your account after 6 months if you deposit $500 in a savings account that pays 6% interest compounded quarterly?

THINK: Compounded quarterly means four times a year, or every 3 months.

First quarter earnings

$I = \$500 \times 0.06 \times \frac{1}{4}$

$I = \$30 \times \frac{1}{4}$

$I = \$7.50$

So, the new principal is $500 + $7.50, or $507.50.

Second quarter earnings

$I = \$507.50 \times 0.06 \times \frac{1}{4}$

$I = \$30.45 \times \frac{1}{4}$

$I = \$7.61$ (rounded to nearest cent)

So, the new principal is $507.50 + $7.61, or $515.11.

Use the compound-interest information given above.
Find the amount of money in your account:

•**21.** after 9 months. $522.84

22. after 12 months. $530.68

23. after 15 months. $538.64

24. after 18 months. $546.72

25. Find the principal after 1 year if $5,000 is deposited in a savings account that pays an annual interest rate of 8% compounded quarterly. $5,412.16

26. How much interest will you earn in 9 months if you deposit $10,000 into a savings account that pays an annual interest rate of 12% compounded quarterly? $927.27

ASSIGNMENTS

Level 1	Even 2–22
Level 2	Odd 1–19, 21–25
Level 3	11–26

FOLLOW-UP ACTIVITY

CALCULATOR Have students use their calculators to compute compound interest in the following situations. Instruct students to round the answer to the nearest cent.

1. $5,000 is deposited for 1 year at 8% compounded quarterly. ($5,412.16)
2. $2,000 is deposited for 2 years at 6% compounded quarterly. ($2,252.99)
3. $10,000 is deposited for 2 years at 10% compounded quarterly. ($12, 184.02)

SUPPLEMENTARY MATERIALS

TRP Practice, p. 107

TRP Transparency 54

OBJECTIVES

- Review vocabulary.
- Practice key chapter concepts and skills.

USING THE REVIEW

The Chapter Review is designed to help students prepare for taking the Chapter Test. The first section focuses on vocabulary. It requires that students select words to complete statements. The second section presents practice exercises of key mathematical skills. Under each directive there is a sample exercise with the answer.

Each item of the review is referenced to the page on which the topic is taught in the Pupil's Edition. You may wish to have students refer to these pages to help review any concepts or skills they have not yet mastered.

It is suggested that students work in small-sized heterogeneous cooperative learning groups. Some cooperative learning methods that may be used are as follows:

1. After each student has independently completed the entire Chapter Review, a discussion should follow within each group about the solutions to the practice exercises.
2. The group can complete the entire Chapter Review by working together to discuss the sample exercises and then to answer the practice exercises.

End the lesson with an entire class discussion in which any questions brought up in group discussion are presented and answered.

CHAPTER REVIEW

Vocabulary Choose the letter of the word(s) that completes the sentence.

1. Percent is another name for ■. [234] b
   a. hundreds **b.** hundredths **c.** proportion
2. To find the percent of a number, you would *not* change the percent to a ■. [240] c
   a. fraction **b.** decimal **c.** proportion
3. The ratio of the amount of increase to the original amount is the ■. [258] b
   a. proportion **b.** percent of increase **c.** percent of decrease

Skills Write as a percent. [234, 238]

0.37 = *37%*

4. 0.34 34% **5.** 0.7 70% **6.** 0.002 0.2% **7.** $0.87\frac{1}{2}$ $87\frac{1}{2}\%$

8. $\frac{3}{100}$ 3% **9.** $\frac{4}{5}$ 80% **10.** $\frac{5}{6}$ $83\frac{1}{3}\%$ **11.** 6 600% **12.** $3\frac{1}{3}$ $333\frac{1}{3}\%$

Find the number. [240, 246, 248, 250]

What number is 25% of 16? *4*

13. $83\frac{1}{3}\%$ of 36 is what number? 30

14. What percent of 40 is 30? 75%

15. 3 is what percent of 9? $33\frac{1}{3}\%$

16. 75% of what number is 18? 24

17. 12 is $37\frac{1}{2}\%$ of what number? 32

18. 160 is 25% of what number? 640

19. 338 is what percent of 845? 40%

Estimate. [240, 246]

What is 9.8% of 30? *3*

20. 11.6 is what percent of 48.4? 25%

21. What is 9.75% of 200? 20

22. 11.3 is what percent of 51? 20%

23. 49.7% of 59.9 is what number? 30

24. 19.8 is 79.3% of what number? 25

Find the percent of increase or decrease. [258]

Original amount: $4.50
New amount: $1.62
Percent of decrease: *64%*

Original amount: $225
New amount: $486
Percent of increase: *116%*

25. Original amount: $312.00
New amount: $265.20 15% decrease

26. Original amount: $65.00
New amount: $74.10 14% increase

CHAPTER TEST

Write as a decimal.

1. 12% 0.12 **2.** 24.5% 0.245 **3.** 33% 0.33 **4.** 0.6% 0.006

Write as a fraction, whole number, or mixed number.

5. 5% $\frac{1}{20}$ **6.** 1,200% 12 **7.** $12\frac{1}{2}$% $\frac{1}{8}$ **8.** 140% $1\frac{2}{5}$

Write as a percent.

9. 0.35 35% **10.** 2.5 250% **11.** 15 1,500% **12.** $\frac{3}{5}$ 60% **13.** $2\frac{2}{3}$ $266\frac{2}{3}$%

Find the number.

14. 30% of 45 is what number? 13.5

15. What number is 3.5% of 20? 0.7

16. 12 is what percent of 18? $66\frac{2}{3}$%

17. What percent of 8 is 4? 50%

18. 300% of what number is 16.2? 5.4

19. 20 is 4% of what number? 500

20. What percent of 140 is 7? 5%

21. What number is 150% of 30? 45

22. 60% of what number is 63? 105

23. What percent of 50 is 35? 70%

Estimate.

24. 21.65 is 25.3% of what number? 80

25. 39 is what percent of 49? 80%

26. 31.08 is 74% of what number? 40

27. What percent of 24.8 is 9.7? 40%

Find the percent of increase or decrease.

28. Original amount: $84.00
New amount: $31.50 62.5% decrease

29. Original amount: $125
New amount: $165 32% increase

Solve.

30. Gruen Shoes sells its own brand of black wingtips for $81. The store's markup is 35%. How much profit does Gruen make on each pair of black wing tips sold? $28.35

31. Each Gruen salesperson makes a base salary of $425 a week plus a commission bonus equal to 4% of his or her sales. If a salesperson sells $5,000 worth of shoes in a week, how much is that week's paycheck? $625

32. Gruen Shoes is expecting an order of 28 dozen pairs of sandals. How many pairs of sandals are to be shipped? 336

33. On Saturdays, a customer walks into Gruen Shoes an average of every 4 minutes. About how many customers enter the store during a 12-hour day? 180

OBJECTIVE

- Evaluate achievement of the chapter objectives.

USING THE TEST

The Chapter Test may be used as a post-test to evaluate the achievement of all students. However, you may wish to use the Chapter Posttest offered in the Teacher's Resource Package or design your own chapter test. If this page is not used as a test, you may wish to assign it as additional review or practice.

The test items are correlated to the chapter objectives in the table below.

Chapter objectives	Test items
A. Write a percent as a decimal.	1–4
B. Write a percent as a fraction, whole number, or mixed number.	5–8
C. Write a decimal, fraction, or mixed number as a percent.	9–13
D. Compute, using the three cases of percent.	14–23
E. Estimate with percents.	24–27
F. Find the percent of increase or decrease.	28–29
G. Apply computational skills in real-life situations.	30–31
H. Solve problems by supplying missing information.	32–33

SUPPLEMENTARY MATERIALS

TRP Ch. 9 Posttest Form A, pp. 1–2

TRP Ch. 9 Posttest Form B, pp. 1–2

OBJECTIVES

- Review and maintain previously taught concepts and skills.
- Practice taking tests in a multiple-choice format.

USING THE REVIEW

Assign the Cumulative Review to all students. Provide students with an answer sheet to record their answers.

Each Cumulative Review gives students an opportunity to practice taking tests that are written in multiple-choice format.

Provide appropriate remedial help for students having difficulty with any of the skills and concepts on these pages.

CUMULATIVE REVIEW

1. To determine if any two fractions are equivalent, you can compare the ■. c

a. numerators **b.** GCF **c.** cross products **d.** none of these

2. A ■ shows the amount of change over a period of time. b

a. bar graph **b.** broken-line graph **c.** pictograph **d.** none of these

3. A ■ means per hundred or hundredths. a

a. percent **b.** decimal **c.** ratio **d.** none of these

4. A(n) ■ is a ratio that compares different units. c

a. range **b.** estimate **c.** rate **d.** none of these

5. To change a mixed number to an improper fraction, you must first multiply the whole number by the ■. c

a. numerator **b.** fraction **c.** denominator **d.** none of these

Choose the most reasonable estimate for each of the following.

6. The temperature of dry ice	a	**a.** −80°C	**b.** 0°C	**c.** 140°C	**d.** −5°C
7. The capacity of a fish bowl	b	**a.** 15 kL	**b.** 5 L	**c.** 15 mL	**d.** 300 L
8. The weight of 4 apples	d	**a.** 2 oz	**b.** 10 lb	**c.** 18 lb	**d.** 32 oz
9. 19.76 − 4.13	a	**a.** 16	**b.** 10	**c.** 6	**d.** 1
10. 42.873 + 12.41	c	**a.** 20	**b.** 40	**c.** 50	**d.** 60

Find the answer.

11. $\frac{■}{18} = \frac{3}{6}$	c	**a.** 12	**b.** 6	**c.** 9	**d.** none of these
12. 0.45 = ■	a	**a.** $\frac{9}{20}$	**b.** $\frac{5}{9}$	**c.** $\frac{45}{1{,}000}$	**d.** none of these
13. $6\frac{2}{3} \times 5\frac{2}{5}$	b	**a.** 12	**b.** 36	**c.** $\frac{4}{9}$	**d.** none of these
14. $\frac{19}{40}$ = ■%	c	**a.** 4.75	**b.** 21	**c.** 47.5	**d.** none of these
15. 65% = ■	d	**a.** $\frac{3}{5}$	**b.** $\frac{11}{20}$	**c.** $\frac{2}{3}$	**d.** none of these

Solve.

16. Sally has a drawing that is 5 in. wide and 7 in. long. She wants the drawing enlarged. The photocopy she has made is 14 in. long. How wide is the photocopy? d

a. 19.6 in. **b.** 14 in. **c.** 12 in. **d.** none of these

17. A chicken dinner at the Bunkhouse Cafe costs $9.50. Cole slaw, chicken, corn, pie, and a beverage are included. Jim doesn't want dessert so he orders chicken for $7.25 plus side orders of corn and cole slaw, which cost $0.85 each. A glass of milk costs $0.90. How much would Jim have saved by ordering the complete dinner? a

a. $0.35 **b.** $1.25 **c.** $0.05 **d.** none of these

18. The Lucca family went to Sound-A-Rama. They bought a jazz album for $7.79. They gave the cashier a $10 bill. What is the least number of coins they could have received? b

a. 4 coins **b.** 3 coins **c.** 5 coins **d.** none of these

19. Last year Erik spent $4,200 on rent. This year his rent was increased to $367.50 per month. How much more will he spend this year on rent? c

a. $330.50 **b.** $200.00 **c.** $210.00 **d.** none of these

20. Carmen earns $5.50 per hour. She gets time and a half for every hour over 40 she works in a week. Last week her gross earnings were $277.75. How many hours overtime did she work? b

a. 6 hours **b.** 7 hours **c.** 10.5 hours **d.** none of these

THINKING ABOUT MATH See TE side column.

1. A store has a special on paper towels. One roll costs $0.89. If you buy a case of 20 rolls, each roll costs $0.65. Is buying a case of paper towels a better buy? Explain.

2. You need to measure the mass of an elephant. Which metric unit of measure would be the most appropriate? Why?

3. A photocopying machine can enlarge a picture by 100%. How much larger is the new picture compared to the original?

4. Explain why you would have a decimal quotient if you divided a 2-digit number by a 3-digit number.

5. Describe the method you would use to quickly determine if the ratios form a proportion.

THINKING ABOUT MATH

The questions in this section can be used in discussions with individual students or with small groups. Guide students in conveying their ideas clearly and precisely.

Listed below are expected student answers. However, accept any reasonable answer.

1. There is no correct answer. A case of paper towels is $4.80 cheaper per 20 rolls than buying one roll at a time. However, are 20 rolls needed at once? Also, it may be inconvenient to carry home a case of paper towels and to store it. Therefore, it may not be a better buy.
2. A metric ton would be the easiest unit of measurement to use because it would take fewer metric tons to describe the elephant's mass.
3. The new picture is twice as large as the original.
4. Whenever a smaller number is divided by a greater number, the answer is always less than 1, or a decimal.
5. Multiply cross products to tell whether two ratios form a proportion. If the cross products are equal, then the ratios form a proportion.

SUPPLEMENTARY MATERIALS

TRP Cum. Test, Ch. 1–9, pp. 1–4

OBJECTIVE

- Test the prerequisite skills needed to learn the concepts developed in Chapter 10.

USING THE TEST

The Pre-Skills Test is designed to diagnose students' strengths and weaknesses on prerequisite skills necessary to study the mathematics in Chapter 10.

Assign the Pre-Skills Test. Allow the students to work together in pairs or small groups. Group members should help those who demonstrate a misunderstanding of a concept or a weakness in a skill.

The items in the test are referenced to the pages on which the topics are taught in the Pupil's Edition. You may wish to have students refer to these pages for review.

The following table correlates the items on the Pre-Skills Test with the prerequisite skill and the lesson(s) in the chapter for which it is needed.

Item(s)	Prerequisite skill	Lesson(s)
1–6	Write ratios as fractions.	10.1, 10.3, 10.6
7–16	Write fractions in simplest form.	10.1, 10.3, 10.6
17–24	Add fractions.	10.6
25–36	Subtract fractions.	10.6
37–44	Multiply fractions.	10.6
45–50	Find percents of numbers.	10.3

Write the ratio as a fraction in simplest form. [210]

1. 2 blue socks to 18 socks $\frac{1}{9}$
2. 12 color TVs to 5 TVs $\frac{12}{5}$
3. 4 oriental rugs to 13 rugs $\frac{4}{13}$
4. 14 tulip bulbs to 35 bulbs $\frac{2}{5}$
5. 5 tee shirts to 8 shirts $\frac{5}{8}$
6. 15 jets to 27 planes $\frac{5}{9}$

Write the fraction in simplest form. [126]

7. $\frac{6}{9}$ $\frac{2}{3}$
8. $\frac{4}{12}$ $\frac{1}{3}$
9. $\frac{7}{14}$ $\frac{1}{2}$
10. $\frac{10}{18}$ $\frac{5}{9}$
11. $\frac{25}{30}$ $\frac{5}{6}$
12. $\frac{25}{50}$ $\frac{1}{2}$
13. $\frac{42}{49}$ $\frac{6}{7}$
14. $\frac{15}{30}$ $\frac{1}{2}$
15. $\frac{56}{63}$ $\frac{8}{9}$
16. $\frac{54}{81}$ $\frac{2}{3}$

Add. Write the answer in simplest form. [158]

17. $\frac{1}{6} + \frac{5}{6}$ 1
18. $\frac{3}{7} + \frac{2}{7}$ $\frac{5}{7}$
19. $\frac{5}{8} + \frac{3}{8}$ 1
20. $\frac{6}{11} + \frac{4}{11}$ $\frac{10}{11}$
21. $\frac{1}{2} + \frac{1}{3}$ $\frac{5}{6}$
22. $\frac{4}{5} + \frac{2}{7}$ $1\frac{3}{35}$
23. $\frac{7}{8} + \frac{5}{12}$ $1\frac{7}{24}$
24. $\frac{4}{9} + \frac{2}{3}$ $1\frac{1}{9}$

Subtract. Write the answer in simplest form. [162, 164]

25. $\frac{4}{5} - \frac{2}{5}$ $\frac{2}{5}$
26. $\frac{7}{8} - \frac{3}{8}$ $\frac{1}{2}$
27. $\frac{6}{7} - \frac{4}{7}$ $\frac{2}{7}$
28. $\frac{9}{11} - \frac{5}{11}$
29. $\frac{5}{8} - \frac{1}{2}$ $\frac{1}{8}$
30. $\frac{6}{7} - \frac{2}{3}$ $\frac{4}{21}$
31. $\frac{7}{8} - \frac{3}{4}$ $\frac{1}{8}$
32. $\frac{1}{2} - \frac{2}{5}$ $\frac{1}{10}$
33. $1 - \frac{3}{4}$ $\frac{1}{4}$
34. $1 - \frac{3}{8}$ $\frac{5}{8}$
35. $1 - \frac{5}{9}$ $\frac{4}{9}$
36. $1 - \frac{6}{13}$

Multiply. Write the answer in simplest form. [146]

37. $\frac{2}{5} \times \frac{4}{7}$ $\frac{8}{35}$
38. $\frac{6}{11} \times \frac{3}{7}$ $\frac{18}{77}$
39. $\frac{3}{8} \times \frac{2}{9}$ $\frac{1}{12}$
40. $\frac{3}{4} \times \frac{8}{9}$ $\frac{2}{3}$
41. $\frac{21}{25} \times \frac{40}{49}$ $\frac{24}{35}$
42. $\frac{18}{45} \times \frac{20}{36}$ $\frac{2}{9}$
43. $\frac{15}{63} \times \frac{28}{45}$ $\frac{4}{27}$
44. $\frac{32}{54} \times \frac{27}{56}$

Find the percent of the number. [240]

45. 35% of 1,020 357
46. 72% of 830 597.6
47. 69% of 2,500 1,725
48. 84% of 5,000 4,200
49. 60% of 3,180 1,908
50. 125% of 7,450 9,312.5

Chapter 10 PROBABILITY

Weather forecasters depend on interpreting data to make their predictions. Do you think they are usually right or wrong?

OBJECTIVE

- Formulate problems using text and data.

USING THE CHAPTER OPENER

Discuss the question on the page asking if students can remember specific examples of forecasts being right or wrong. Most people only remember when they are wrong. A favorite is usually 10 in. of snow when flurries were predicted. Extend the discussion to why accurate forecasts are important to some people. These people might include farmers, construction company supervisors, any people who work outdoors, or even people planning a picnic.

COOPERATIVE LEARNING

Assign groups the task of tracking the accuracy of weather forecasts. They are to use as many sources as feasible—newspapers, radio, television, even the Farmers' Almanac. (As an alternate, you may want to assign one of these sources to each group.) Each group is to choose a way to record and then present the data clearly.

As work progresses, the students should find that their sources may not agree on the forecast for a particular day, and this can also be included in their reports as an interesting bit of data. On the other hand, if they find identical forecasts, they should investigate further because this might indicate that these forecasts come from the same service. For example, if three radio stations use the same forecasting service, it will distort the students' data. Thus, the service itself should be tracked.

Have the groups follow the forecasts for at least a week, but a month will produce better data.

SUPPLEMENTARY MATERIALS

TRP Group Projects, p. 10

OBJECTIVES

- Find the probability of simple events.
- Find the odds in favor of or against events occurring.

TEACHING THE LESSON

WARM-UP Write the following exercises on the chalkboard or an overhead transparency.

Write the ratio as a fraction in simplest form.

1. 12 silver cars to 36 cars $(\frac{1}{3})$

2. 10 baseball caps to 25 hats $(\frac{2}{5})$

3. 18 goldfish to 24 fish $(\frac{3}{4})$

4. 8 box kites to 48 kites $(\frac{1}{6})$

5. 28 tennis balls to 40 balls $(\frac{7}{10})$

INTRODUCTION Write the following sets of letters on the chalkboard or an overhead transparency.

I. {A, E, I, O, U}

II. {B, F, M, P, T}

How many letters are:

1. in I? **2.** in II? **3.** in I or II?

How many letters in the word MATH are:

4. in I? **5.** in II? **6.** in I or II?

(**1.** 5 **2.** 5 **3.** 10 **4.** 1 **5.** 2 **6.** 3)

Lead students to see that the number in I or in II is always the sum of the number in I plus the number in II. (This is only true when the sets have no common members.)

INSTRUCTION Use Example 1 to apply the definition of probability to both single and multiple events. You may want to find $P(\text{H}) + P(\text{E})$ and compare it with the answer in Example 1b. This will reinforce the concept that an "or" probability is the sum of the separate probabilities when the events have nothing in common, or are *mutually exclusive.*

10.1 PROBABILITY

Suppose you turn all the cards shown at the right face down. If you pick a card at random, it is equally likely that you will pick any one of them. Each card represents a **possible outcome.**

For events that occur at random, **probability** measures the likelihood of a particular outcome. The probability of that event is a ratio:

$$P(\text{event}) = \frac{\text{number of favorable outcomes}}{\text{number of possible outcomes}}$$

EXAMPLE 1 Use the CHANCE cards to find the probability. Write the probability as a ratio in simplest form.

a. $P(\text{N}) = ■$

THINK: There is 1 letter N. There are 6 letters in all.

$\frac{1}{6}$ **← favorable outcomes / ← possible outcomes**

So, $P(\text{N}) = \frac{1}{6}$.

b. $P(\text{H or E}) = ■$

THINK: There is 1 letter H and 1 letter E. There are 6 letters in all.

$\frac{2}{6}$ **← favorable outcomes / ← possible outcomes**

So, $P(\text{H or E}) = \frac{2}{6}$ or $\frac{1}{3}$.

When no outcome is favorable, the event is **impossible.** The probability of an impossible event is 0.

When every outcome is favorable, the event is **certain.** The probability of a certain event is 1.

EXAMPLE 2 Use the CHANCE cards to find the probability. Write the probability as a ratio in simplest form.

a. $P(\text{a digit}) = ■$

THINK: None of the cards shows a digit. Picking a digit is an impossible event.

$P(\text{a digit}) = \frac{0}{6}$ or 0.

b. $P(\text{a letter}) = ■$

THINK: All the cards show a letter. Picking a letter is a certain event.

$P(\text{a letter}) = \frac{6}{6}$ or 1.

The **odds** of an event occurring is the ratio of the number of favorable outcomes to the number of unfavorable outcomes.

EXAMPLE 3 Use the CHANCE cards to find the odds of:

a. picking a C.

THINK: $\frac{\text{number of favorable outcomes}}{\text{number of unfavorable outcomes}} = \frac{2}{4}$ (two C's) / (not C's)

So, the odds of picking a C are 2 to 4 or 1:2.

b. not picking a C.

THINK: $\frac{\text{Number of favorable outcomes}}{\text{Number of unfavorable outcomes}} = \frac{4}{2}$ (not C's) / (two C's)

So, the odds of not picking a C are 4 to 2 or 2:1.

FOR DISCUSSION

See TE side column.

Why can P(event) never be greater than 1?

PRACTICE EXERCISES

See Extra Practice, page 449.

A box contains the 12 counters shown at the right. Each counter is identified by color and by number. A counter is picked at random.

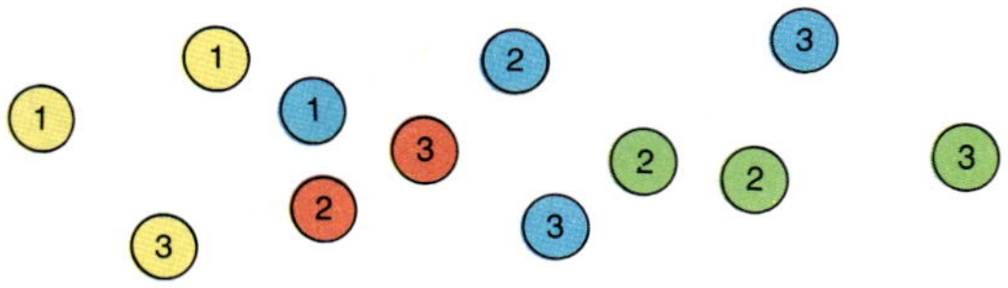

Find the probability.

- • **1.** P(green) $\frac{1}{4}$
- • **2.** P(red even) $\frac{1}{12}$
- **3.** P(yellow or blue) $\frac{7}{12}$
- **4.** P(odd) $\frac{2}{3}$
- **5.** P(less than 4) 1
- **6.** P(not orange) 1
- **7.** P(red or green) $\frac{5}{12}$
- **8.** P(yellow prime) $\frac{1}{4}$
- **9.** P(red, yellow, or green) $\frac{2}{3}$
- **10.** P(black even) 0
- **11.** P(white or blue) $\frac{1}{3}$
- **12.** P(blue less than 3) $\frac{1}{6}$

Look at the spinner shown at the right. It is spun once.

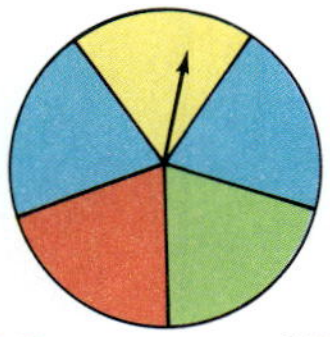

Find the odds of spinning:

- • **13.** green. 1:4
- • **14.** red or blue. 3:2
- **15.** red, yellow, or green. 3:2
- **16.** red. 1:4
- **17.** yellow. 1:4
- **18.** red, blue, or green. 4:1
- **19.** not red. 4:1
- **20.** not green. 4:1
- **21.** not blue and not green. 2:3
- **22.** blue. 2:3
- **23.** not yellow. 4:1
- **24.** green, blue, or yellow. 4:1

Ask students to give examples of certain and of impossible events. Elicit from students that in Example 2a the probability of an impossible event must be 0. In Example 2b, the probability written as a ratio must equal 1.

Be sure that students understand the distinction between a probability (a ratio of numbers of outcomes) and odds (a ratio of probabilities). In Example 3a, a simplified version of the formula for odds is:

$$\text{Odds} = \frac{P(C)}{P(\text{not } C)} = \frac{\frac{2}{6}}{\frac{4}{6}}$$

$$= \frac{2}{6} \div \frac{4}{6} = \frac{\cancel{2}^{1}}{\cancel{6}_{1}} \times \frac{\cancel{6}^{1}}{\cancel{4}_{2}} = \frac{1}{2}$$

Point out that in the odds ratio (before simplifying), the sum of the terms gives the number of possible outcomes. In Example 3a:

$$\text{For } \frac{2}{4},\ 2 + 4 = 6.$$

This addition provides a quick check on the correctness of the odds.

Example 3b shows that odds, unlike probability, can be greater than 1.

FOR DISCUSSION Students should explain that there can never be more favorable outcomes than possible outcomes.

TIME OUT Suggest students use three-dimensional drawings to determine where the letters will appear on the cube.

ASSIGNMENTS

Level 1	Odd 1–35, 37, 38
Level 2	Even 2–36, 37–40
Level 3	7–18, 31–40, TO

FOLLOW-UP ACTIVITY

ENRICHMENT Use the students' knowledge of probability and odds to find one when given the other.

1. Given odds of 3:7, find the probability.

   $\text{Odds} = \frac{\text{favorable}}{\text{unfavorable}} = \frac{3}{7}$

   $\text{Possible} = \text{favorable} + \text{unfavorable}$

   $\text{Probability} = \frac{\text{favorable}}{\text{possible}} = \frac{3}{3 + 7}$

   So, $P(E) = \frac{3}{10}$.

2. Given $P(E) = \frac{2}{5}$, find the odds.

   $\text{Probability} = \frac{\text{favorable}}{\text{possible}} = \frac{2}{5}$

   $\text{Unfavorable} = \text{possible} - \text{favorable}$

   $\text{Odds} = \frac{\text{favorable}}{\text{unfavorable}} = \frac{2}{5 - 2}$

   So, odds $= \frac{2}{3}$ or 2:3.

SUPPLEMENTARY MATERIALS

TRP Practice, p. 108

TRP Reteaching, p. 65

TRP Lesson Aids, p. 6

TRP Transparency 54

MIXED REVIEW

30. 1.0003

Complete.

•**25.** 18 L = ■ mL 18,000
•**26.** 38,000 g = ■ kg 38
27. 6,500 km = ■ m 6,500,000
28. 4,200 mg = ■ g 4.2
29. 7.9 cm = ■ mm 79
30. 1,000.3 mL = ■ L
31. 4,500 cm = ■ m 45
32. 3.1 kg = ■ g 3,100
33. 30 L = ■ kL 0.03
34. 4 km = ■ cm 400,000
35. 0.4 kL = ■ mL 400,000
36. 4,500 mg = ■ kg 0.0045

Solve.

•**37.** Cards with the vowels A, E, I, O, and U are removed from a set of alphabet cards. A consonant card is then picked at random. What is the probability of picking a letter found in the word CHANCE? $\frac{1}{7}$

38. How many sets of CHANCE cards can be made from a package of 50 index cards? How many index cards will be left over? 8 sets; 2 cards left over

39. A letter from the word CHANCE is picked at random. What is the probability of picking a vowel? $\frac{1}{3}$

40. Earl Llewellyn removes the letters in his name from a set of alphabet cards. How many cards are left? 19 cards

TIME OUT

The 6 letters in the word CHANCE are written on the alphabet block shown below. Figures A, B, C, and D are all different views of the same block. What is the missing letter in Figure D? C

A B C D

PROBLEM *Solving* APPLICATION

10.2 CONSUMER: LIFE INSURANCE

Life insurance companies use probability in determining the rates they charge. When people want to buy life insurance, a company needs to be able to estimate how many years the person is likely to live.

There are many types of tables that provide estimates about life expectancy. Below is a simplified table for a recent year. For example, the table indicates that a person at age 20 is likely to live 56.0 more years. In that case, the total life span would be 76.0 years.

EXPECTATION OF LIFE

At Age	Expectation of Life in Years
5	70.6
10	65.7
15	60.8
20	56.0
25	51.3
30	46.6
35	41.9
40	37.2
45	32.7
50	28.3
55	24.2
60	20.4
65	16.8
70	13.7
75	10.8
80	8.3

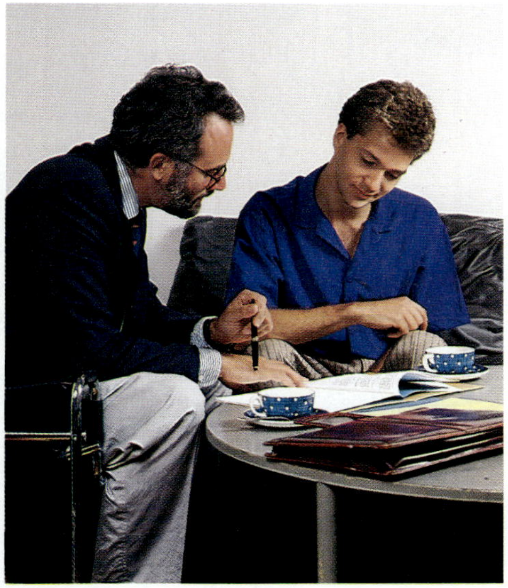

Use the table above. Find the life expectancy in years for a person at age:

1. 15. 60.8 **2.** 25. 51.3 **3.** 40. 37.2 **4.** 55. 24.2

Find the estimated *total* life span for a person of:

5. 30. 76.6 **6.** 45. 77.7 **7.** 60. 80.4 **8.** 80. 88.3

9. Jackie is likely to live 41.9 more years. How old is she? 35

10. Abe is 10. His father is 35. Abe's grandfather is 70. What are their life expectancies? Abe, 65.7; his father, 41.9; Abe's grandfather, 13.7

11. What are some of the factors that have caused an increase in life expectancy over the past 60 years? Accept any reasonable answer, such as advances in medical technology, discovery of new drugs.

OBJECTIVE

- Solve problems that involve using data from a life expectancy table.

TEACHING THE LESSON

WARM-UP Write the following exercises on the chalkboard or overhead transparency.

Add or subtract.

1. 15 + 56.8 (71.8)

2. 33.7 − 14.8 (18.9)

3. 28.3 + 45 (73.3)

4. 74.3 − 48.9 (25.4)

5. 60 − 14.8 (45.2)

6. 23.6 + 50 (73.6)

INSTRUCTION Discuss the table. Be sure the students understand that the entries in the second column indicate the expectation of additional years of life for persons of the given ages. Students should realize that the sum of the two numbers in any given row indicates the estimated total life span for a person of the given age.

Notice that Problems 1–4 only require that the students read the table correctly. Problems 5–10 require the students to add to find the total life span. Encourage a class discussion to answer Problem 11.

ASSIGNMENTS

Level 1	1–9
Level 2	1–11
Level 3	Odd 1–11

SUPPLEMENTARY MATERIALS

TRP Practice, p. 109

TRP Transparency 55

OBJECTIVE

- Predict the number of times an event will occur.

TEACHING THE LESSON

WARM-UP Write the following exercises on the chalkboard or an overhead transparency.

Write the fraction as a decimal.

1. $\frac{2}{5}$ **2.** $\frac{3}{4}$ **3.** $\frac{5}{6}$ **4.** $\frac{7}{8}$

(**1.** 0.4 **2.** 0.75 **3.** $0.8\overline{3}$ **4.** 0.875)

INTRODUCTION Ask students for the probability of getting a 5 if they were to roll a number cube with 1–6 on its faces. ($\frac{1}{6}$) Have student roll a number cube and record how often a 5 appears in 6 rolls, 12 rolls, 24 rolls, 48 rolls. The results should show that while a 5 does not necessarily appear once in every six rolls, it does appear about $\frac{1}{6}$ of the time.

INSTRUCTION Point out in Example 1 that probability can be written as a decimal: $p = \frac{3}{10} = 0.3$. Working with decimals will often make computation easier.

Use Example 2 to show how samples are used to predict events. Elicit from students why a random sample is necessary (to ensure against a biased sample) and how the size might affect the results (a larger sample usually gives more accurate results). Ask for examples of when samples are used and how they might be made to reflect bias (politicians could canvas just their supporters for a voter survey).

Explain that Example 2, which focuses on estimating or predicting outcome based on experience, is called empirical probability (as opposed to theoretical probability in Example 1).

10.3 PROBABILITY AND PREDICTIONS

A bag contains the 10 marbles shown at the right. A marble is picked at random and then replaced. This is repeated 500 times.

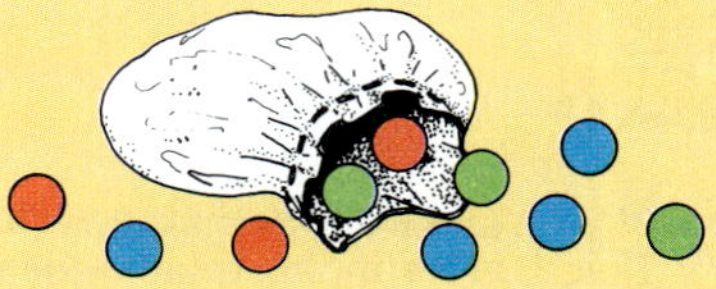

You can use probability to **predict** the number of times a particular event will occur.

EXAMPLE 1 Predict how many times you can expect a red marble to be picked.

1. Find the probability.

   THINK: $\frac{3}{10}$ ← **favorable outcomes** / ← **possible outcomes** $\quad P(\text{red}) = \frac{3}{10} = 0.3$

2. Multiply the probability by the number of picks. $\quad 0.3 \times 500 = 150$

So, you can expect a red marble to be picked 150 times.

The owners of the Amazing Adventure Amusement Park took a random survey of 100 visitors to the park. They summarized the results in the following table:

Type of pass		Favorite ride		Age	
One-day	20	Coast-to-Coaster	40	3–11	23
Two-day	27	Mud Slide	30	12–20	35
Three-day	15	Loop-A-Doo	18	21–59	26
Seasonal	38	Other	12	60 and over	16

Sometimes you can make predictions using data from past events.

EXAMPLE 2 A crowd of 5,000 visitors is expected at Amazing Adventure over the weekend. Predict how many of these visitors will be under 21.

1. Find the probability.

   THINK: **Age of visitors 3–11: 23**
   Age of visitors 12–20: 35
   58

   $P(\text{under } 21) = \frac{58}{100}$ or 0.58

2. Multiply the probability by the number of expected visitors. $\quad 0.58 \times 5{,}000 = 2{,}900$

So, Amazing Adventure can expect 2,900 of these visitors to be under 21.

PRACTICE EXERCISES

See Extra Practice, page 449.

Complete.

• **1.** A spinner is divided into four equal sections numbered 1 through 4. It is spun 100 times. Predict how many times a 3 will be spun. 25

2. A coin is flipped 80 times. Predict how many flips will be tails. 40

3. A spinner is divided into five equal sections numbered 1 through 5. Predict how many times out of 250 an odd number will be spun. 150

A drawer contains the 15 socks shown at the right. One sock is picked at random and then replaced. This is repeated 300 times. Predict how many times you can expect the color of the sock to be:

• **4.** brown. 160 •**5.** white. 60

6. not white. 240 **7.** blue or black. 60

Use the results of the survey on page 272 for Exercises 8–29.

For each attendance figure, predict how many of the visitors will choose each of the rides listed below as their favorite ride.

Attendance	Coast-to-Coaster	Mud Slide	Loop-A-Doo	Other
10,000	• **8.** ■ 4,000	• **9.** ■ 3,000	**10.** ■ 1,800	**11.** ■ 1,200
7,500	**12.** ■ 3,000	**13.** ■ 2,250	**14.** ■ 1,350	**15.** ■ 900
3,800	**16.** ■ 1,520	**17.** ■ 1,140	**18.** ■ 684	**19.** ■ 456

Over Labor Day weekend, 25,000 visitors are expected at Amazing Adventure. Predict how many will:

•**20.** be using 3-day passes? 3,750

•**21.** be 12–20 years old? 8,750

22. be older than 20? 10,500

23. be using seasonal passes? 9,500

24. be younger than 60? 21,000

25. be using a one- or two-day pass? 11,750

Solve.

•**26.** What is the probability that the favorite ride of the next visitor to Amazing Adventure will be either the Loop-A-Doo or the Mud Slide? $\frac{12}{25}$

27. A busload of 250 high school students arrives at Amazing Adventure. Predict how many of them are between 12 and 20 years old. 250

28. Of those whose favorite ride was not on the survey, half picked the Log Jam. What percent of those surveyed preferred the Log Jam? 6%

29. One Tuesday in May attendance at Amazing Adventure is 5,500. Predict how many of these people are using 1- or 2-day passes? 2,585

ASSIGNMENTS

Level 1	Odd 1–25
Level 2	1–27
Level 3	3–11, 20–29

FOLLOW-UP ACTIVITY

COOPERATIVE LEARNING

Have students work in small groups. Ask each student to make a prediction: If a coin is tossed 100 times, how many times will it land "heads up"? Have the students conduct 100 trials, and then 200 trials. Ask the groups to state their results. Students should find that as the number of trials increases, the results move closer to the prediction.

SUPPLEMENTARY MATERIALS

TRP Practice, p. 110

TRP Reteaching, p. 66

TRP Lesson Aids, p. 6

TRP Transparency 55

OBJECTIVE

- Solve problems that involve finding probabilities.

TEACHING THE LESSON

WARM-UP Write on the chalkboard or an overhead transparency:

two green marbles
one yellow marble
one purple marble

Then read these exercises aloud.

The four marbles are placed in a bag. You pick one without looking. Find the probability of picking:

1. a purple marble. $(\frac{1}{4})$
2. a green marble. $(\frac{1}{2})$
3. a marble that is *not* yellow. $(\frac{3}{4})$
4. a marble that is *not* green or purple. $(\frac{1}{4})$

INSTRUCTION Use this lesson to emphasize the important role that mathematics plays in science and other disciplines.

Problems 4–6 and 8 and 9 ask for the probability of an event *not* happening. Guide the students through the equations:

$$P(\text{event occurring}) + P(\text{event not occurring}) = 1$$
or
$$1 - P(\text{event occurring}) = P(\text{event not occurring})$$

ASSIGNMENTS

Level 1	1–6
Level 2	1–9
Level 3	Odd 1–9

SUPPLEMENTARY MATERIALS

TRP Practice, p. 111

TRP Transparency 56

PROBLEM Solving APPLICATION

10.4 CAREER: SCIENTIST

Beth McCaffrey is a scientist. She performs many scientific experiments. One of her experiments is described below.

- A number of plants with white flowers are crossbred with a number of plants with red flowers, and seeds are produced. When these seeds are planted, the resulting plants have pink flowers.
- The plants with pink flowers are bred with each other, and *second-generation* seeds are produced. When these second-generation seeds are planted, some of the resulting plants have flowers that are white; others have flowers that are red; and others have flowers that are pink.

The probability that a plant will have flowers of a given color is as follows:

Probability of white: $\frac{1}{4}$

Probability of red: $\frac{1}{4}$

Probability of pink: $\frac{1}{2}$

Solve.

Suppose 1,000 of the second-generation seeds are planted. About how many of the plants will have flowers that are:

- **1.** red? 250 **2.** pink? 500 **3.** white? 250

Suppose 2,500 of the second-generation seeds are planted. About how many of the plants will have flowers that are:

- **4.** not red? 1,875 **5.** not pink? 1,250 **6.** neither white nor pink? 625

Write the probability as a decimal. Use the probabilities for second-generation seeds.

- **7.** Probability of a plant with red flowers 0.25

8. Probability of a plant that does not have pink flowers 0.50

9. Probability of a plant that does not have pink flowers, red flowers, or white flowers 0

OBJECTIVE

- Evaluate student progress.

USING THE REVIEW

This page provides a means for informally evaluating students' understanding of the skills and concepts covered so far in this chapter.

Have the students look at the page to familiarize themselves with various question formats that are presented. Discuss any questions that they may have. Then ask them to complete the page independently.

In addition to grading them individually, you may wish to review the answers to the questions collectively with the students.

Page references appear in brackets. They refer to pages on which a particular skill was introduced.

Before continuing on to the topics found in the remainder of the chapter, you may wish to have students review any skills or concepts in which they have demonstrated weakness.

MIDCHAPTER REVIEW

A bag contains the 20 marbles shown. One marble is picked at random.

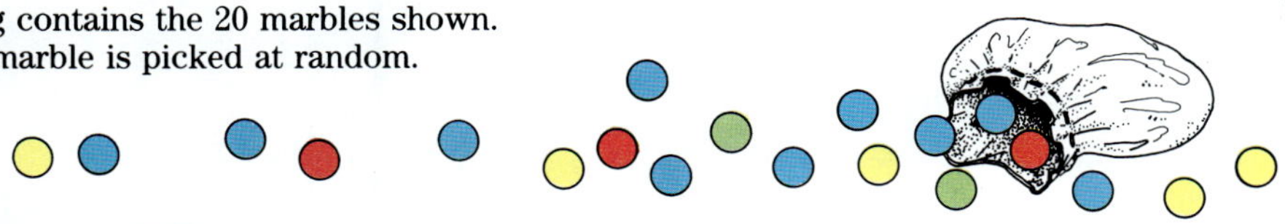

Find the probability. [268]

1. P(yellow) $\frac{1}{4}$ **2.** P(blue) $\frac{1}{2}$ **3.** P(green) $\frac{1}{10}$

4. P(red) $\frac{3}{20}$ **5.** P(orange) 0 **6.** P(green or blue) $\frac{3}{5}$

7. P(not yellow) $\frac{3}{4}$ **8.** P(red or yellow) $\frac{2}{5}$ **9.** P(not yellow and not blue) $\frac{1}{4}$

Find the odds of picking a marble that is: [268]

10. yellow. 1:3 **11.** blue. 1:1 **12.** red. 1:1

13. not green. 9:3 **14.** red or yellow. 2:3 **15.** not black and not blue. 1:1

16. black or blue. 1:1 **17.** not red. 17:3 **18.** green, red, or yellow. 1:1

The spinner shown at the right is spun 300 times. Predict how many times you can expect to spin: [272]

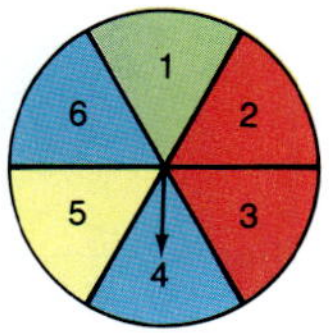

19. green. 50 **20.** red. 100 **21.** odd. 150

22. not even. 150 **23.** even or yellow. 200 **24.** blue or odd. 250

25. more than 6. 0 **26.** green or red. 150 **27.** 3 or 4. 100

Solve. Use the table on page 271 if necessary. [268, 271, 272]

28. Harold Evans is 35 years old. How many more years can he reasonably expect to live? 41.9 years

29. A recent study shows that out of 100 monkeys tested, 34 have the gene for a long tail. What is the probability of having this gene? $\frac{17}{50}$

30. Jane Kramer is 50 years old. What is her estimated total life span? 78.3 years

31. The probability that a monkey will have the gene for a short tail is $\frac{12}{25}$. Out of 500 monkeys, predict how many will have this gene. 240 monkeys

OBJECTIVE

- Use a tree diagram or multiplication to find the number of possible outcomes.

TEACHING THE LESSON

WARM-UP Write the following exercises on the chalkboard or an overhead transparency.

Multiply mentally.

1. $7 \times 5 \times 8$ **2.** $20 \times 9 \times 5$

3. $18 \times 10 \times 3$ **4.** $25 \times 17 \times 4$

(**1.** 280 **2.** 900 **3.** 540 **4.** 1,700)

Encourage the students to use the associative and commutative properties.

INTRODUCTION Ask students how many different ways they can arrange the following items: white shirt; blue shirt; plain tie; striped tie, plaid tie. Elicit suggestions for ways to find all arrangements, such as making a list.

INSTRUCTION Use the method of making a list to lead into Example 1. A tree diagram is a systematic and orderly way of showing all possible outcomes. Make sure students understand the pattern for a tree diagram. (List one category and then link each of those options to each option in the next category. Continue through all possible categories.)

Explain that when the categories have numerous options (and only the number of outcomes is needed), a tree diagram is not the most efficient method. Use Example 2 to show students how multiplication can be used. When successive choices are made, the product of each choice and the next choice in the series gives the total number of choices. (This is called the Fundamental Counting Principle.) Stress that this method gives only the number of outcomes, whereas a tree diagram gives the actual outcomes.

10.5 POSSIBLE OUTCOMES

Berkeley Airlines offers its flight attendants a variety of uniform styles from which to choose. The jackets come in both red and blue. The blouses come in red, white, and blue. How many uniforms are possible?

A **sample space** is a listing of all the possible outcomes in a given situation. You can use a **tree diagram** to show all the possible outcomes. Each "branch" represents a choice or possible outcome.

EXAMPLE 1 Make a tree diagram to show all the possible uniforms.

	Jacket	**Blouse**	**Outcome**
Uniform	Red	Red	Red jacket, red blouse
		White	Red jacket, white blouse
		Blue	Red jacket, blue blouse
	Blue	Red	Blue jacket, red blouse
		White	Blue jacket, white blouse
		Blue	Blue jacket, blue blouse

So, there are 6 possible uniforms.

Sometimes when you have a series of events, you need to find only the number of possible outcomes and not what each of them is.

To find the total number of possible outcomes of a series of events, multiply the number of ways each event can occur.

EXAMPLE 2 The Berkeley Airlines baggage tags use 3-letter city codes. The first and third letters of each code are always consonants and the second letter is always a vowel. How many different codes are possible?

THINK: 21 consonants, 5 vowels.

consonant		**vowel**		**consonant**	
↓		↓		↓	
21	×	5	×	21	= 2,205

So, 2,205 different codes are possible.

ASSIGNMENTS

Level 1	1–7, Even, 8–18, 20–21
Level 2	1–23
Level 3	Odd 1–19, 20–23

PRACTICE EXERCISES

See Extra Practice, page 449.

Complete the following tree diagram.

• **1.**

BERKELEY AIRLINES SEATING CHOICES IN COACH

Coach
- Smoking
  - Aisle — Smoking, aisle
  - Center — Smoking, center
  - Window — Smoking, window
- Nonsmoking
  - Aisle — Nonsmoking, aisle
  - Center — Nonsmoking, center
  - Window — Nonsmoking, window

Use the diagram above to answer Exercises 2–7.
How many different types of the seat are available?

• **2.** window 2 • **3.** smoking 3 **4.** nonsmoking and center 1

5. aisle and smoking 1 **6.** coach 6 **7.** nonsmoking 3

Berkeley Airlines tickets have 4-digit codes. How many codes are possible if the digits are:

• **8.** any 4 digits? 10,000 • **9.** only odd digits? 625 **10.** 2, 3, 5, or 7? 256

11. all multiples of 3? 81 **12.** less than 3? 81 **13.** 2 even then 2 odd? 625

Berkeley Airlines routing codes are made up of digits and letters.
How many codes are possible if each code is made up of:

• **14.** 3 digits? 1,000 • **15.** 3 letters? 17,576

16. 3 consonants? 9,261 **17.** 2 odd digits then 2 letters? 16,900

18. 3 even digits then 1 letter? 3,250 **19.** 3 digits then 2 vowels? 25,000

Solve.

• **20.** On Flight 1106, 60 passengers paid first-class fares, 277 paid coach fares, and 63 paid excursion fares. What is the probability that a passenger picked at random did not pay a first-class fare? $\frac{17}{20}$

21. Captain Max Gomez is filing a route plan for a Miami-to-Memphis flight stopping in Atlanta. There are 3 routes from Miami to Atlanta and 4 routes from Atlanta to Memphis. How many route plans are possible? 12

22. Jon Galvin's ticket originally cost $373. It carried with it a 25% penalty for change or cancellation. Jon changed his return flight. What was the ticket's final cost? $466.25

23. The airlines' fashionable baggage handlers wear pants, shirts, and jackets that each come in the same 4 colors. Pants and jacket must match. How many uniforms are possible? 16

FOLLOW-UP ACTIVITY

ENRICHMENT Introduce Pascal's triangle to students. Explain that the frame of the triangle is all ones and that each number in the triangle is found by adding the two numbers above it.

Row	
	1
1	1 1
2	1 2 1
3	1 3 3 1
4	1 4 6 4 1

One property of Pascal's triangle is that the chance of getting heads or tails or any combination of them on a given number of coin tosses can be read. For example, to find the number of heads you can get by tossing three coins and the probability for each, use Row 3 as follows:

Number of heads:	3	2	1	0
Number of ways: (from Row 3)	1	3	3	1 = 8
Probability:	$\frac{1}{8}$	$\frac{3}{8}$	$\frac{3}{8}$	$\frac{1}{8}$

Have students extend Pascal's triangle to Row 6. Use the triangle to find probabilities.

SUPPLEMENTARY MATERIALS

TRP Practice, p. 112

TRP Reteaching, p. 67

TRP Transparency 56

OBJECTIVE

- Find the probability of independent and dependent events.

TEACHING THE LESSON

WARM-UP Write the digits 0 through 9 on the chalkboard. Have students use them to write the answers for the following exercises.

Find the probability.

1. $P(\text{odd})$ $(\frac{1}{2})$ **2.** $P(\text{not } 4)$ $(\frac{9}{10})$

3. $P(<11)$ (1) **4.** $P(>2)$ $(\frac{7}{10})$

INTRODUCTION Elicit from students the familiar meanings of the words *independent* and *dependent.* Then show how this relates to probability (events are independent or dependent depending upon whether one outcome affects another). Give examples of events and have students determine if they are independent or dependent. Some examples might be: tossing a coin; picking a card from a deck with and without replacing it; rolling a numbered cube.

INSTRUCTION You may want to show students how Example 1 is an application of the method used in Lesson 10.5 (The Fundamental Counting Principle). For the series of events, find the total number of favorable outcomes and the total number of possible outcomes.

	ones		fours	
Favorable =	1	×	1	= 1
	digits		digits	
Possible =	8	×	8	= 64

So, $P(1, 4) = \frac{1}{64}$.

Multiplying successive probabilities in Example 1 gives the same result. Point out that the order of the events does not affect the outcome.

$P(4, 1) = \frac{1}{64}$.

10.6 INDEPENDENT AND DEPENDENT EVENTS

At the annual Royce High School Fair, Max Lasher is in charge of the Spin-to-Win booth. The wheel is shown at the right. What is the probability of spinning a 1 and then a 4?

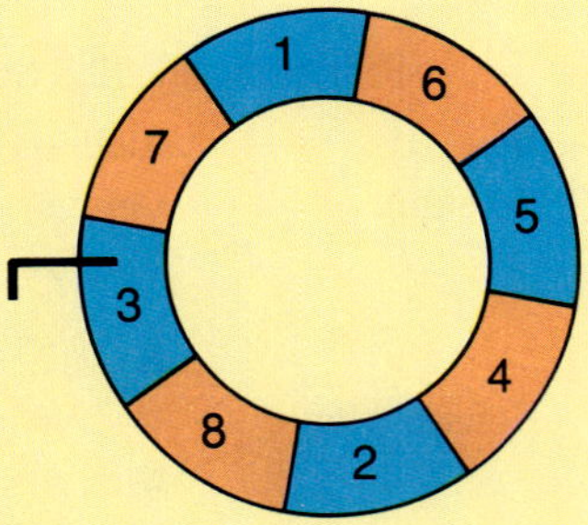

The outcome of the first spin does not affect the outcome of the second spin. These are **independent events.**

To find the probability of a series of independent events, multiply the probabilities of the individual events.

EXAMPLE 1 Find $P(1, \text{then } 4)$

THINK: $P(1, \text{then } 4) = P(1) \times P(\text{then } 4)$

$$P(1, \text{then } 4) = \frac{1}{8} \times \frac{1}{8} = \frac{1}{64}$$

So, the possibility of spinning a 1 and then a 4 is $\frac{1}{64}$.

When the outcome of the first event affects the outcome of the second event, the events are **dependent events.**

EXAMPLE 2 Sally Teicher runs the Match Game. Twelve cards are mixed and placed face down. The object is to pick the two red cards.

One card is picked, left face up, and then another card is picked. What is the probability of winning?

1. Find the probability of the first event. $P(\text{red}) = \frac{2}{12}$, or $\frac{1}{6}$.

2. Find the probability of the second event.

   THINK: The first card is face up. There is one less possible outcome. $P(\text{then red}) = \frac{1}{11}$

3. Multiply the probabilities. $\frac{1}{6} \times \frac{1}{11} = \frac{1}{66}$

So, the probability of winning is $\frac{1}{66}$.

PRACTICE EXERCISES See Extra Practice, page 450.

Write *independent* or *dependent* to describe the series of events.

- • **1.** Spinning a spinner 3 times independent
- **2.** Picking 3 cards without replacing any of them dependent
- **3.** Spinning a spinner, covering that number, and spinning again dependent

Wheel A is spun.

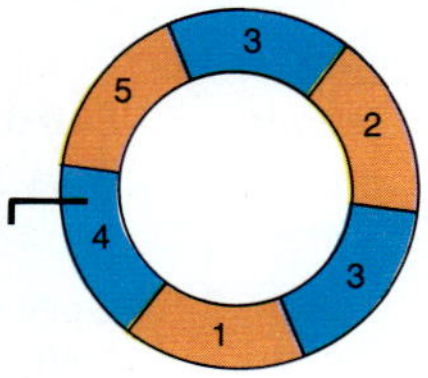

Then Wheel B is spun.

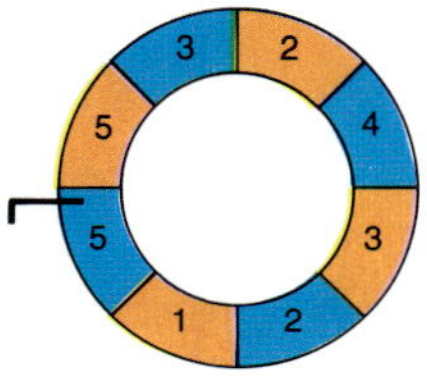

Find the probability.

- • **4.** $P(1, \text{ then } 4)$ $\frac{1}{48}$
- • **5.** $P(3, \text{ then } 3)$ $\frac{1}{12}$
- **6.** $P(\text{less than } 4, \text{ then } 4)$ $\frac{1}{12}$
- **7.** $P(\text{even, then } 6)$ 0
- **8.** $P(\text{even, then even})$ $\frac{1}{8}$
- **9.** $P(\text{a factor of } 6, \text{ then odd})$ $\frac{5}{12}$
- **10.** $P(5, \text{ then } 1)$ $\frac{1}{48}$
- **11.** $P(2, \text{ then even})$ $\frac{1}{16}$
- **12.** $P(\text{less than } 1, \text{ then } 3)$ 0

The cards shown below are placed face down. One at a time, cards are picked at random and then not replaced.

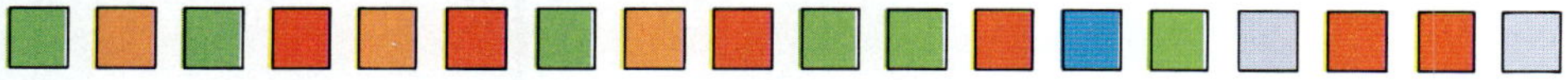

Find the probability.

- • **13.** $P(\text{green, then red})$ $\frac{2}{17}$
- • **14.** $P(\text{orange, then blue})$ $\frac{1}{102}$
- **15.** $P(\text{orange, then orange})$ $\frac{1}{51}$
- **16.** $P(\text{blue, then red or green})$ $\frac{2}{51}$
- **17.** $P(\text{purple, then blue or green})$ $\frac{7}{153}$
- **18.** $P(\text{red, then red or blue})$ $\frac{2}{17}$
- **19.** $P(\text{red, then green, then green})$ $\frac{5}{136}$
- **20.** $P(\text{orange, then white, then green})$ 0
- **21.** $P(\text{blue, then green, then red or blue})$ $\frac{1}{136}$
- **22.** $P(\text{red, then red, then red, then red})$ $\frac{1}{204}$

MIXED REVIEW

For each set of data, find the mean, median, mode, and range.

	Mean	Median	Mode	Range
24, 36, 27, 24, 26, 28, 24, 27	• **23.** ▇ 27	• **24.** ▇ 26.5	**25.** ▇ 24	**26.** ▇ 12
1.4, 1.8, 2.1, 1.7, 1.8, 1.5, 1.6	**27.** ▇ 1.7	**28.** ▇ 1.7	**29.** ▇ 1.8	**30.** ▇ 0.7
110, 104, 120, 100, 100, 100, 110, 120	**31.** ▇ 108	**32.** ▇ 107	**33.** ▇ 100	**34.** ▇ 20

Example 2 focuses on finding the probability of dependent events happening in succession. Stress that the denominator decreases with each successive event because the cards that are drawn are not replaced. Rework Example 2 as independent events. Elicit from students which type of event, independent or dependent, will have the greater probability (dependent) and why (the denominator will be smaller).

ASSIGNMENTS

Level 1	1–3, Even 4–34
Level 2	Odd 1–33
Level 3	7–18, 31–34

FOLLOW-UP ACTIVITY

COOPERATIVE LEARNING Divide the class into groups of three or four students. Provide each group with a well-shuffled standard deck of 52 cards.

Have each group do the following:

1. Pick four cards, one at a time, replacing the card each time.
2. Record the cards that were picked.
3. Repeat steps 1 and 2 nine more times.

Have the students find probabilities such as:

1. $P(\text{J, Q, K, A})$ **2.** $P(\text{B, B, B, B})$
3. $P(\text{R, B, R, B})$ **4.** $P(\text{all spades})$

You may wish to have students use calculators. Ask students to compare the results of their group work.

SUPPLEMENTARY MATERIALS

TRP Practice, p. 113

TRP Reteaching, p. 68

TRP Transparency 57

OBJECTIVE

- Solve problems by making an organized list.

TEACHING THE LESSON

WARM-UP Write the following numbers on the chalkboard or an overhead transparency.

Order from greatest to least.

8,340; 8,430; 3,480; 3,840

(8,430; 8,340; 3,840; 3,480)

INTRODUCTION Discuss situations where order is important (arrangements). An example may be naming the order of winners in a race. Then discuss situations where order is not important. An example may be selecting three students from eight volunteers to be on a school committee.

INSTRUCTION As students read the opening situation, elicit from students that order is important in naming the different ways that the players can be organized. Point out that they are being asked to find arrangements.

As you read each question in the example, use the chalkboard to show students how to make a list to organize the answers.

Note that in Exercises 1 and 2 the number of arrangements to be made is given. In Exercises 3 and 4, students must first list the possible arrangements and then tell how many there are. In Exercises 5 and 6, the number of combinations to be made is given. In Exercise 7, students must first list the possible combinations and then tell how many there are. In Exercise 8, students list the possible combinations.

Throughout the lesson emphasize the distinction between *arrangements* in which order is important and *combinations* in which order is not important.

PROBLEM Solving STRATEGY

10.7 MAKING AN ORGANIZED LIST

Situation:

Dale, Keva, and Cory are players on a softball team. The coach plans to have them bat in the first 3 positions in the batting order. In how many different ways can the coach arrange these 3 players?

Strategy:

Making an organized list can help solve some problems.

Applying the Strategy:

	Batting Order		
	First	**Second**	**Third**
THINK: How many different batting orders are possible if Dale bats first?	• Dale	Keva	Cory
	• Dale	Cory	Keva
How many different batting orders are possible if Keva bats first?	• Keva	Dale	Cory
	• Keva	Cory	Dale
How many different batting orders are possible if Cory bats first?	• Cory	Dale	Keva
	• Cory	Keva	Dale
How many different batting orders are possible in all?	6		

The coach can arrange these 3 players in 6 different ways.

FOR DISCUSSION
See TE side column.

Beth, Helen, and Tom all want to run in a race. A team consists of two runners. Which list shows all the possible combinations of runners? a

a. Beth and Helen
Beth and Tom
Helen and Tom

b. Beth and Helen — Helen and Beth
Beth and Tom — Tom and Beth
Helen and Tom — Tom and Helen

PRACTICE EXERCISES

Make an organized list to solve each problem. See page 478 for lists.

• **1.** Jed is picking a 2-digit number for his football jersey. He must choose from the digits 3, 5, and 7. List all 6 possible arrangements.

2. The coach plans to use Joe, Al, and Sam in the outfield. List all 6 ways the coach can arrange them in left field, right field, and center field.

3. Phil is picking a 3-digit number for a license plate. He must choose from the digits 2, 4, 6, and 8. List all the possible arrangements. How many are there? 64

4. Ben, Ari, Joan, and Sarah go bowling. List all the possible orders in which they can bowl. How many are there? 24

• **5.** In a round-robin singles tennis tournament, Cathy, Ann, Wes, and Glen all play each other. List all 6 of the matches they play against one another.

6. A basketball team is practicing defense. Each of the 5 players practices with each of the other players. List all 10 possible practice pairings. (*Hint:* You can label the players 1, 2, 3, 4, and 5.)

7. There are 6 sprinters on the track team. List all the possible 4-runner relay teams the coach can form. How many are there? 15

8. Rudi won 4 medals at the track championships. At least 2 of his medals were silver. What are the possible arrangements of gold, silver, and bronze medals Rudi could have won? (2G, 2S) (2B, 2S) (4S) (1G, 3S) (1B, 3S) (1G, 2S, 1B)

FOR DISCUSSION After reading the situation, ask students whether or not order is important in naming the teams. (no) Emphasize that they are being asked to find combinations. Students should explain that List A is correct because List B shows each combination twice. For example, the team of Beth and Helen is the same team as Helen and Beth.

ASSIGNMENTS

Level 1	1, 2, 5, 6
Level 2	1–7
Level 3	Even 2–8

FOLLOW-UP ACTIVITY

SITUATIONAL PROBLEM SOLVING Present students with the following problem:

Your baseball team wants new uniforms. You decide to hold a contest to come up with a new design. Your team must earn the money to have the new uniforms made. Develop a plan that explains how you would organize the contest and earn the money.

SUPPLEMENTARY MATERIALS

TRP Practice, p. 114

TRP Applications, p. 17

TRP Transparency 57

OBJECTIVE

- Solve problems by using proportions.

TEACHING THE LESSON

WARM-UP Write the following exercises on the chalkboard or overhead transparency.

Solve for *n*.

1. $\frac{4}{100} = \frac{n}{2,000}$ (80)
2. $\frac{8}{400} = \frac{n}{5,000}$ (100)
3. $\frac{15}{500} = \frac{n}{3,000}$ (90)
4. $\frac{13}{300} = \frac{n}{7,200}$ (312)

INTRODUCTION Tell students that the lesson talks about the quality control of manufactured goods, which is one instance of random sampling. Discuss political polls and TV ratings as other forms of random sampling. Ask the students for other examples.

INSTRUCTION As you discuss the text material, emphasize that the sample is considered to be representative of the total number of items produced. So, the ratios $\frac{3}{100}$ and $\frac{n}{5,000}$ are assumed to be equivalent and we solve $\frac{3}{100} = \frac{n}{5,000}$.

Note that Problems 6–10 involve percents. You may wish to remind students that any percent, $n\%$, can be written as $\frac{n}{100}$.

PROBLEM Solving APPLICATION

10.8 SAMPLING AND QUALITY CONTROL

Manufacturers frequently monitor the quality of their products by checking a random **sample** of the items produced. They determine the ratio of defective items to the total number of sample items. They then predict that in the larger group, there is an equivalent ratio of defective items to the total number of items.

The Alpha Corporation produces electrical appliances. In one month it manufactures 5,000 toasters. A random sample of 200 toasters is checked, and 6 are found to be defective. Predict the total number of defective toasters likely to be manufactured during the month.

THINK: In the sample, 6 out of 200 items are defective.
So, the ratio is $\frac{6}{200}$, or $\frac{3}{100}$.

Use a proportion to solve the problem.

1. Write a proportion.
   Let *n* represent the total number of defective items. $\quad \frac{3}{100} = \frac{n}{5,000}$
2. Find *n*.

$$\begin{aligned} 3 \times 5,000 &= 100 \times n \\ 15,000 &= 100 \times n \\ 15,000 \div 100 &= n \\ 150 &= n \end{aligned}$$

So, it is likely that 150 defective toasters will be manufactured during the month.

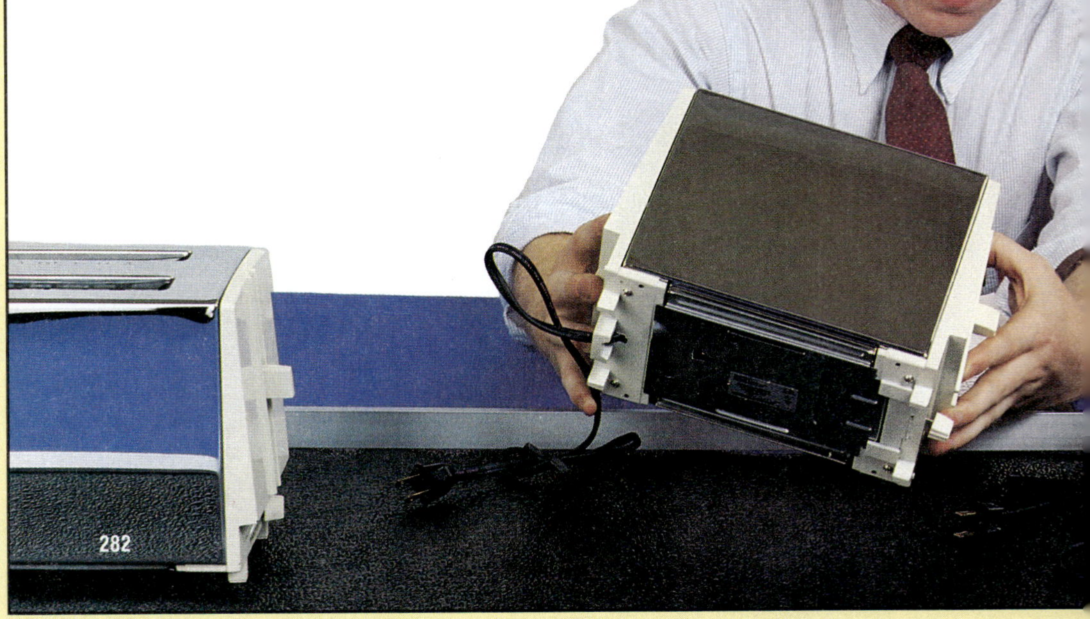

Predict the total number of defective items manufactured each month.

Item	No. in Sample	No. Defective in Sample	Total No. Produced	No. Defective
• **1.** Blender	100	2	3,000	■ 60
2. Electric knife	500	25	12,500	■ 625
3. Radio	400	16	8,400	■ 336
4. Cassette player	300	11	6,300	■ 231

Solve. Remember to estimate whenever you use your calculator.

• **5.** The Omega Corporation manufactures 16,000 Model A calculators in one month. A random sample of 900 calculators is checked. 20 are defective. Predict the total number of defective Model A calculators likely to be manufactured during the month. 356 calculators

• **6.** In one month the Omega Corporation manufactures 4,500 Model B calculators. A random sample of these calculators is checked. 2% are found to be defective. Predict the total number of Model B calculators that are not likely to be defective. 4,410 calculators

7. The Gamma Company manufactures 4,000 Super-robots in one month. A random sample of these toys is checked, and 3% are found to be defective. Predict the total number of Super-robots that are likely to be defective. 120 Super-robots

8. In one month the Gamma Company manufactures 5,680 Mini-robots. A random sample of these robots is checked, and 2.5% are found to be defective. Predict the total number of Mini-robots that are likely to be defective. 142 Mini-robots

9. A quality control technician examines 900 electronic toys. Of these, 18 are defective. About what percent are defective? 2%

10. A quality control technician examines 550 learning devices. Of these, 15 are defective. About what percent are defective? 3%

Problem 6 asks for the number of calculators that are *not* defective. Some students will solve this problem by finding the number of items that *are* defective, and then subtracting that number from 4,500. Other students will decide that 98% of the calculators are *not* defective and use the proportion $\frac{98}{100} = \frac{n}{4,500}$.

You may wish to allow students to use calculators to help them solve Problems 5–10.

ASSIGNMENTS

Level 1	1–6
Level 2	1–10
Level 3	Even 2–10

FOLLOW-UP ACTIVITY

ENRICHMENT Have the students conduct their own random sampling of favorite television shows. Check their results for accuracy. Discuss the possibility of a built-in bias if the samples are far off.

SUPPLEMENTARY MATERIALS

TRP Practice, p. 115

TRP Transparency 58

OBJECTIVE

- Decide, given relevant information, which mode of transportation to use.

TEACHING THE LESSON

WARM-UP Write the following exercises on the chalkboard or an overhead transparency.

	One-way fare	Round-trip fare	Number of people	Total cost
1.	$179	$358	6	
2.		$295	3	
3.	$98.50			$1,379

(**1.** $2,148 **2.** $147.50, $885 **3.** $197, 7)

INTRODUCTION Present the problem situation to the class. Discuss various modes of transportation and ask what factors are important when deciding how to travel.

INSTRUCTION Have students read the data provided for the three modes of transportation and discuss any other information students may feel is important based on their own experiences with travel.

Direct attention to the decision-making factors. Ask the students if there are any factors they would add to the list from those written on the chalkboard.

Have students complete Exercises 1–18 in the table in order to compare the three plans. Ask them to write conclusions that can be drawn from the data and discuss the appropriateness of these conclusions and any possible misinterpretations.

Assign Exercises 19–28 and have students work in small groups. Encourage them to explain and defend their answers and decisions as necessary. Continually reinforce the fact that individual preferences and experiences strongly influence decisions of this type. Observe the groups as they work and interact as necessary to draw nonparticipating students into the discussion.

DECISION MAKING

10.9 CHOOSING TRANSPORTATION

(Open-ended problem solving)

Choosing a vacation spot can be a fun decision. Deciding how to get there requires research and some careful considerations.

Karen, her parents, and her grandmother will spend their vacation at her cousin's house in Orlando, Florida. The adults have decided to allow 10 days for the vacation, and have asked Karen to help them choose the mode of transportation for the trip from New York. Their options are to fly, drive, or take a bus.

Karen made the following notes.

Plane: $2\frac{1}{2}$-hour flight each way
$298 round-trip air fare per person
Taxi to and from airport: $23 each way

Car: 3 days each way
One-way expenses:
Gasoline and tolls: about $100
Motel: 2 nights, $48 per room, per night
(2 rooms required)
Meals: 7, averaging $30 a meal for all 4 people
(Car is a compact model; 4 people can drive)

Bus: 2 days each way
$ 184 one way, 10% discount for senior citizens
(motel for one night included)
Meals: 5, averaging $ 30 a meal for all 4 people

DECISION-MAKING FACTORS

Cost Time Comfort (fatigue)

DECISION-MAKING COMPARISONS Some answers may vary. Accept reasonable answers based on students' rationales.

Compare the three modes of transportation.

	Plane	Car	Bus
Cost (round trip for all 4 people)			
Transportation	•1. ■ $1,238	2. ■ $ 200	3. ■ $1,043
Motel	•4. ■ 0	5. ■ $ 384	Included
Food	•6. ■ 0	7. ■ $ 420	8. ■ $ 300
Total cost	•9. ■ $1,238	10. ■ $1,004	11. ■ $1,435.20
Time			
Traveling	12. ■ 5 hours	13. ■ 6 days	14. ■ 4 days
In Orlando	10 days	15. ■ 4 days	16. ■ 6 days
Comfort	17. ■	18. ■	Could be tiring

17. Not tiring 18. Very tiring

MAKING THE DECISIONS

Which mode of transportation should Karen's family use:

• **19.** if cost were the only factor? Car

• **20.** if time were the only factor? Plane

21. if comfort were the only factor? Plane

22. How much could the family save by driving instead of flying? $234

23. If the family decided to drive but bring food for the trip, about how much would they save on meals?
Any amount less than $420 (the difference being the cost of purchasing food)

24. If the aim of the trip was to see part of the East coast, which mode of transportation would be best? Car or bus

25. If the family had never been to Florida, which would be the best way to travel? Why? Plane; they could spend more time there

26. What if Karen's cousin did not have a car the family could use. How could that affect their decision? They might drive so they could have a car, or they could rent one and thereby increase expenses.

27. In terms of meals, how could driving be better than taking the bus? When driving they could select the restaurants and meal times.

28. If you were making the decision, which mode of transportation would you choose? Why? Answers will vary.

ASSIGNMENTS

Level 1	1–28
Level 2	1–28
Level 3	1–28

FOLLOW-UP ACTIVITY

APPLICATION Ask students to collect travel brochures, accommodations information, transportation schedules, and so on from local travel agencies. Also collect newspaper ads with travel information and any other material that may be available. Gather the material and make it available as a reference for all students.

Have students work in groups. Each group is to choose a vacation spot and plan a 2-week trip there. They should develop their plan based on a list of factors they agree upon. Encourage them to look for hidden costs. (Will they need new clothing, shots, a visa, passport, and so on?)

In evaluating the plans, consider that some groups will choose exotic or exclusive locales for which actual costs may not be available. Accept reasonable estimates and look for thoroughness as a measure of their work.

SUPPLEMENTARY MATERIALS

TRP Practice, p. 116

TRP Transparency 58

OBJECTIVES

- Review vocabulary.
- Practice key chapter concepts and skills.

USING THE REVIEW

The Chapter Review is designed to help students prepare for taking the Chapter Test. The first section focuses on vocabulary. It requires that students select words to complete statements. The second section presents practice exercises of key mathematical skills. Under each directive there is a sample exercise with answer.

Each item on the review is referenced to the page on which the topic is taught in the Pupil's Edition. You may wish to have students refer to these pages to help review any concepts or skills they have not yet mastered.

It is suggested that students work in small-sized heterogeneous cooperative groups. Some cooperative learning methods that may be used are as follows:

1. After each student has independently completed the entire Chapter Review, a discussion should follow within each group about the solutions to the practice exercises.
2. The group can complete the entire Chapter Review by working together to discuss the sample exercises and then to answer the practice exercises.

End the lesson with an entire class discussion in which any questions brought up in group discussions were presented and answered.

CHAPTER REVIEW

Vocabulary Choose the letter of the word(s) that completes the statement.

1. The probability of an event is a ■. [268] c
a. possible outcome **b.** favorable outcome **c.** ratio

2. A tree diagram lists all the ■. [276] a
a. possible outcomes **b.** favorable outcomes **c.** odds

3. When one event has no effect on a second, the events are ■. [278] b
a. dependent **b.** independent **c.** impossible

Skills A letter is picked at random from the word MATHEMATICS. Find the probability. [268]

P(M) $\frac{2}{11}$

4. *P*(A) $\frac{2}{11}$ **5.** *P*(a vowel) $\frac{4}{11}$ **6.** *P*(a consonant) $\frac{7}{11}$

7. *P*(M or T) $\frac{4}{11}$ **8.** *P*(A, B, or C) $\frac{3}{11}$ **9.** *P*(K) 0

A box contains 12 counters. Of these, 2 are white, 3 are green, 3 are blue, and 4 are red. Find the odds of picking at random one that is: [268]

red. *1:2*

10. white. 1:5 **11.** blue. 1:3 **12.** green. 1:3

13. red or blue. 7:5 **14.** not green. 3:1 **15.** red or green. 7:5 **16.** not white. 5:1

A recent survey showed that $\frac{13}{100}$ of those questioned were born on a Saturday. For each sample, predict the number born on a Saturday. [272]

5,000 *650*

17. 12,000 1,560 **18.** 20,000 2,600 **19.** 3,500 455

A bag contains 18 marbles. Of these, 2 are orange, 3 are green, 5 are blue, and 8 are red. A marble is picked at random and then replaced. Find the probability. [278]

P(green, then red) $\frac{2}{27}$

20. *P*(orange, then blue) $\frac{5}{162}$ **21.** *P*(purple, then green) 0

22. *P*(red, then red) $\frac{16}{81}$ **23.** *P*(blue, then green) $\frac{5}{108}$

Find the probability if the marble is not replaced. [278]

P(green, then red) $\frac{4}{51}$

24. *P*(blue, then red) $\frac{20}{153}$ **25.** *P*(orange, then green) $\frac{1}{51}$

26. *P*(blue, then blue) $\frac{10}{153}$ **27.** *P*(red, then orange) $\frac{8}{153}$

Chapter Test

Jurors in Hibbing are selected at random from the voting rolls. The distribution by political party is shown at the right. Find the probability that the next juror will be:

SURVEY RESULTS
Party Membership of 200 Voters

Republican: 65 Democratic: 75
Conservative: 24 Liberal: 20
Independent: 16

1. Conservative. $\frac{3}{25}$ **2.** Liberal. $\frac{1}{10}$ **3.** not a Democrat. $\frac{5}{8}$

Find the odds that the next juror will be:

4. Republican. 13:27 **5.** not a Liberal. 9:1 **6.** Conservative or Democrat. 99:101

The results of a recent survey of 75 visitors to Graceland, Elvis Presley's home in Memphis, Tennessee, are shown in the table below.

Number of Elvis albums owned	0	1–3	4–7	8 or more
Number of visitors	9	39	12	15

If 1,200 people visited Graceland yesterday, predict how many own:

7. no albums. 144 **8.** less than 8 albums. 960 **9.** from 1 to 3 albums. 624

10. A coin is tossed 3 times. How many possible outcomes are there for this series of tosses? 8

A bag contains 12 marbles. Of these, 1 is white, 2 are blue, 3 are red, and 6 are green. A marble is picked at random and then replaced. A second marble is picked. Find the probability.

11. *P*(green, then blue) $\frac{1}{12}$ **12.** *P*(red, then red) $\frac{1}{16}$ **13.** *P*(white, then not white) $\frac{11}{144}$

Find the probability if the first marble is not replaced.

14. *P*(red, then green) $\frac{3}{22}$ **15.** *P*(blue, then blue) $\frac{1}{66}$ **16.** *P*(white, then white) 0

Solve.

17. There are 4 books on a shelf. Make a list to show the different ways they can be ordered.
Check students' lists.

18. Delta pricing codes are made up of 3 even digits. Make a list to show all the possible codes.
Check students' lists.

19. Of a random sample of 900 fans, 15 do not work. AMX Inc. orders 30,000. Predict how many do not work. 500

20. A random sample of 750 mugs shows 25 to be cracked. Out of 18,000 mugs in stock, predict how many are cracked. 600

OBJECTIVE

- Evaluate achievement of the chapter objectives.

USING THE TEST

The Chapter test may be used as a posttest to evaluate student achievement. However, you may wish to use the Chapter Posttest offered in the Teacher's Resource Package or to design your own chapter test. If this page is not used as a test, you may wish to assign it as additional review or practice.

The test items are correlated to the chapter objectives in the table below.

Chapter objectives	Test items
A. Find the probabilities of simple events.	1–3
B. Find the odds in favor of or against events occurring.	4–6
C. Predict the number of times an event will occur.	7–9
D. Find the number of possible outcomes.	10
E. Find the probability of independent and dependent events.	11–16
F. Make an organized list to solve problems.	17–18
G. Apply computational skills in real-life situations.	19–20

SUPPLEMENTARY MATERIALS

TRP Ch. 10 Posttest Form A, pp. 1–2

TRP Ch. 10 Posttest Form B, pp. 1–2

OBJECTIVE

- Compute the number of permutations.

USING THE PAGE

As you discuss the example, you may wish to draw a diagram to show that the number of choices for each leg of the race decreases by 1 each time. Stress that order is very important when finding the number of permutations. For example, if *A, B, C,* and *D* stand for the four swimmers, the sequence *ABCD* is not the same as the sequence *BCDA.*

The number of permutations is 4! (four **factorial**). A **factorial** of a given number is the product of all natural numbers up to and including a given number.

The formal notation for finding the number of permutations is $_nP_r$, in which n indicates which factor to begin with and r stands for the number of factors to be used.

Only when $n = r$ is the permutation equivalent to a factorial of n. A factorial is defined as follows:

> For $n>0$, $n!$ means to multiply all the consecutive numbers from n to 1.

Factorials can be used to find the number of permutations. Note that permutations sometimes use factorials that are incomplete or that do not always use all the numbers from n to 1. (see examples)

Extend the instruction by guiding students in writing and finding the number of permutations for the following situations.

- If no digit can be used more than once, how many 3-digit numbers can be formed from the digits 0 through 9: ($_{10}P_3 = 720$)
- How many ways can a coach position 12 players on a 5-position basketball team? ($_{12}P_5 = 95{,}040$)

ENRICHMENT PERMUTATIONS

The swimming team at Eisenhower High School is practicing for the league championship. Coach Barker has chosen his four best swimmers for the 400-m freestyle relay. Each swimmer will swim a 100-m leg of the relay. In how many different ways can the coach arrange the order of the swimmers?

An arrangement of objects in which order is important is a **permutation.** You can use the following notation to find the number of permutations in the problem above.

$$_4P_4$$

Number of choices for first position (left 4) — **Number of positions** (right 4)

Multiply to find the answer. $_4P_4 = 4 \times 3 \times 2 \times 1 = 24$

THINK: **Number of choices for the first leg** → 4
Number of choices for the second leg → 3
Number of choices for the third leg. → 2
Number of choices for the fourth leg → 1

So, Coach Barker can arrange the order of the swimmers in 24 different ways.

Examples

Symbol	**Meaning**	**Solution**
$_5P_3$	The number of permutations of 5 objects taken 3 at a time.	$_5P_3 = 5 \times 4 \times 3 = 60$
$_8P_5$	The number of permutations of 8 objects taken 5 at a time.	$_8P_5 = 8 \times 7 \times 6 \times 5 \times 4 = 6{,}720$

Complete.

1. $_5P_4 = 5 \times 4 \times ■ \times ■ = ■$ 3 2 120

2. $_6P_3 = 6 \times ■ \times ■ = ■$ 5 4 120

3. $_3P_3 = ■ \times ■ \times ■ = ■$ 3 2 1 6

4. $_7P_4 = ■ \times ■ \times ■ \times ■ = ■$ 7 6 5 4 840

Find the number of permutations.

5. $_4P_2$ 12 **6.** $_5P_3$ 60 **7.** $_6P_4$ 360 **8.** $_7P_3$ 210

9. $_7P_5$ 2,520 **10.** $_8P_2$ 56 **11.** $_9P_3$ 504 **12.** $_{10}P_2$ 90

Solve.

13. In how many ways can 4 flags be ordered on 2 flagpoles? 12

14. In how many ways can 9 players be ordered in 6 positions? 60,480

MENTAL MATH PROBABILITIES AS FRACTIONS

When a probability is expressed as a simple fraction, you may be able to do calculations mentally.

One marble is chosen at random from a bag containing 200 marbles. The probability that the marble is red is $\frac{1}{4}$. How many marbles are red?

Find $\frac{1}{4}$ of 200. **THINK:** $\frac{1}{4}$ of 200 is $\frac{1}{4} \times 200$.

To multiply by $\frac{1}{4}$, divide by 4. $\frac{1}{4} \times 200 = 200 \div 4 = 50$

There are 50 red marbles.

The probability that the marble is blue is $\frac{2}{5}$. How many marbles are blue?

Find $\frac{2}{5}$ of 200. **THINK:** $\frac{2}{5}$ of 200 is $\frac{2}{5} \times 200$.

To multiply by $\frac{2}{5}$:

1. Divide by 5. $200 \div 5 = 40$
2. Multiply by 2. $2 \times 40 = 80$

There are 80 blue marbles.

Compute mentally.

1. $\frac{1}{4}$ of 40 10 **2.** $\frac{1}{10}$ of 120 12 **3.** $\frac{1}{8}$ of 240 30 **4.** $\frac{1}{3}$ of 165 55

5. $\frac{3}{5}$ of 50 30 **6.** $\frac{3}{4}$ of 80 60 **7.** $\frac{2}{3}$ of 210 140 **8.** $\frac{5}{8}$ of 320 200

Solve the problems mentally.

One card is chosen at random from a group of 400 marked cards.

The probability that the card shows a book title is $\frac{1}{5}$.

The probability that it shows a song title is $\frac{3}{4}$.

9. How many cards show a book title? 80 **10.** How many show a song title? 300

One of 240 envelopes is chosen at random.

The probability that the envelope is empty is $\frac{1}{8}$.
The probability that it contains a $1 bill is $\frac{2}{3}$.

11. How many envelopes are empty? 30 **12.** How many contain a $1 bill? 160

OBJECTIVE

- Multiply whole numbers by fractions mentally.

USING THE PAGE

Point out that the techniques illustrated in the lesson are useful not only in probability problems, but also in other situations in which a whole number is multiplied by a fraction. It might also be noted that it is often easier to multiply a whole number mentally by a fraction than by a decimal. For example, it is easier to multiply by $\frac{1}{4}$ (that is, divide by 4) than to multiply by 0.25.

In the examples and the exercises, each whole number is a multiple of the denominator of the fraction by which it is multiplied. You might wish to point out that when the whole number is *not* a multiple of the denominator of the fraction, it is usually easier to multiply first and then divide. The following example illustrates this.

Find $\frac{3}{4}$ of 11.

THINK: $\frac{3}{4}$ of 11 is $\frac{3}{4} \times 11$.

To multiply by $\frac{3}{4}$:

1. Multiple mentally by 3. $\frac{3}{4} \times 11 = \frac{3 \times 11}{4}$

2. Divide by 4. $= \frac{33}{4}$

$= 8\frac{1}{4}$

Students can practice this technique by finding the following numbers mentally.

1. $\frac{2}{3}$ of 8 $(5\frac{1}{3})$ **2.** $\frac{2}{5}$ of 12 $(4\frac{4}{5})$

3. $\frac{5}{6}$ of 7 $(5\frac{5}{6})$ **4.** $\frac{3}{8}$ of 23 $(8\frac{5}{8})$

OBJECTIVE

- Test the prerequisite skills needed to learn the concepts developed in Chapter 11.

USING THE TEST

The Pre-Skills Test is designed to diagnose students' strengths and weaknesses on prerequisite skills necessary to study the mathematics in Chapter 11.

Have students take the Pre-Skills Test. Allow the students to work together in pairs or small groups. Group members should help those who demonstrate a misunderstanding of a concept of a weakness in a skill.

The items in the test are referenced to the pages on which the topics are taught in the Pupil's Edition. You may wish to have students refer to these pages for review.

The following table correlates the items on the Pre-Skills Test with the prerequisite skill and the lesson(s) in the chapter for which it is needed.

Item(s)	Prerequisite skill	Lesson
1–16	Add or subtract 2- and 3-digit numbers.	11.3, 11.7, 11.10
17–20	Measure line segments.	11.4, 11.5
21–28	Multiply with fractions, mixed numbers, and decimals.	11.6
29–36	Find the square of a number.	11.9, 11.10
37–44	Solve a proportion.	11.14, 11.15

Pre-Skills Test

Add. [30]

1. 35 + 55 90
2. 47 + 86 133
3. 98 + 82 180
4. 98 + 39 137
5. 124 + 35 159
6. 119 + 61 180
7. 151 + 13 164
8. 45 + 123 168

Subtract. [32]

9. 90 − 45 45
10. 180 − 46 134
11. 180 − 87 93
12. 90 − 67 23
13. 90 − 76 14
14. 180 − 32 148
15. 90 − 69 21
16. 180 − 117 63

Measure to the nearest half-inch. [180]

17. $1\frac{1}{2}$ in.
18. 2 in.

Measure to the nearest centimeter. [192]

19. 2 cm
20. 1 cm

Multiply. [40, 74, 146, 148]

21. 2×9 18
22. 2×5.8 11.6
23. $2 \times 1\frac{3}{4}$ $3\frac{1}{2}$
24. $2 \times 3\frac{1}{2}$ 7
25. $\frac{1}{2} \times 12$ 6
26. $\frac{1}{2} \times 3$ $1\frac{1}{2}$
27. $\frac{1}{2} \times 4.6$ 2.3
28. $\frac{1}{2} \times 2\frac{3}{4}$ $1\frac{3}{8}$

Find the product. [18]

29. 3^2 9
30. 8^2 64
31. 12^2 144
32. 15^2 225
33. 13^2 169
34. 20^2 400
35. 25^2 625
36. 28^2 784

Find n. [216]

37. $\frac{3}{5} = \frac{n}{15}$ 9
38. $\frac{2}{7} = \frac{8}{n}$ 28
39. $\frac{n}{6} = \frac{20}{24}$ 5
40. $\frac{3}{n} = \frac{9}{21}$ 7
41. $\frac{5}{8} = \frac{n}{24}$ 15
42. $\frac{4}{9} = \frac{16}{n}$ 36
43. $\frac{n}{4} = \frac{9}{12}$ 3
44. $\frac{4}{n} = \frac{24}{30}$ 5

All builders use geometric shapes in their constructions. What factors determine the size and shape of a building?

OBJECTIVE

- Formulate problems using text data.

USING THE CHAPTER OPENER

In discussing the question, students should come up with some basic ideas such as available space, customer needs, community needs, and so on. They may or may not be aware of building codes that restrict size and placement of buildings, or even design, in some communities. Most cities have zoning laws that designate land for industrial, commercial, or residential use. Discuss the pros and cons of such codes and laws. The pros are usually obvious. Most people do not want someone to build a factory next to their home. The cons might include a housing shortage because the community prohibits or limits apartment houses.

COOPERATIVE LEARNING

For this activity, groups are to observe the buildings in their community and select one or more of the most interesting for further study. They could choose these buildings based on their interesting shapes, beauty, historic significance, or some other criteria. Then each group is to build a model of the building they select. They should choose a medium that is practical. (Foam-core board is a good choice and is available at most art supply stores.) Remind students that their goal in producing the models is to provide examples of geometric shapes in architecture.

The completed models should be presented along with a brief written description of the building and why the group chose it for their project. You may want to arrange to have the models displayed at the local library, city hall, or other community building.

SUPPLEMENTARY MATERIALS

TRP Group Projects, p. 11

OBJECTIVE

- Identify and name points, lines, line segments, rays, angles, and planes.

TEACHING THE LESSON

INTRODUCTION Discuss the various ways that people express their impressions of real objects. For example, an automobile can be described as a four-wheeled vehicle that provides transportation and is usually powered by a gasoline engine. A simple drawing of an automobile or the word *automobile* conveys the same idea.

Have students select familiar objects and describe them using the object's properties, a drawing, a word, or a symbol. Tell students that in this lesson they will learn how to express ideas about geometric figures by using properties, drawings, and symbols.

INSTRUCTION Encourage students to identify physical objects that are models of the geometric figures shown in the table. For example, a beam of light from a flashlight is a model of a ray; the surface of a lake on a calm day takes on the properties of a plane.

In Example 1, discuss the difference between a ray and a line. Stress that a line extends in opposite directions without end while a ray has one endpoint and thus only extends in one direction. Emphasize that $\overrightarrow{CD}$ is only a part of $\overleftrightarrow{CD}$.

In Example 2, make sure students understand that $\overline{EF}$ is a part of $\overrightarrow{EF}$. Point out that while a line segment has two endpoints and thus a measurable length, a ray extends without end in one direction and cannot be measured.

FOR DISCUSSION

1. Students should explain that one and only one line can be drawn through both points. Thus, we say that two points determine a line.

11.1 POINTS, LINES, AND PLANES

Geometric figures are made up of sets of points. You can use a drawing or a symbol to represent these figures.

Definition	Drawing	Symbol
A **point** is a location in space with no size.		A
A **plane** is a flat surface extending in all directions without end.	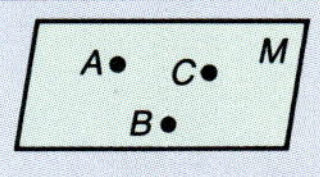	plane ABC or plane M
A **line** is a set of points that forms a straight path extending in opposite directions without end.	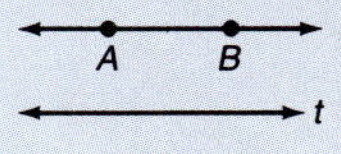	$\overleftrightarrow{AB}$ or $\overleftrightarrow{BA}$ line t
A **line segment** is a part of a line with two endpoints.		$\overline{AB}$ or $\overline{BA}$
A **ray** is part of a line with one endpoint.	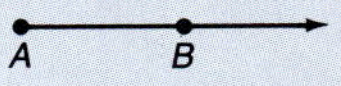	$\overrightarrow{AB}$
An **angle** is made up of two rays with the same endpoint. The endpoint is called the **vertex.**	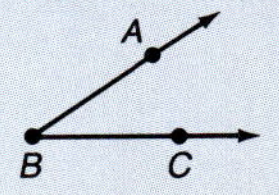	$\angle ABC$

EXAMPLE 1 Draw and label the figure.

a. $\overleftrightarrow{CD}$

THINK: line

b. $\overrightarrow{CD}$

THINK: ray

Some geometric figures can be named in different ways. Look at the following angle.

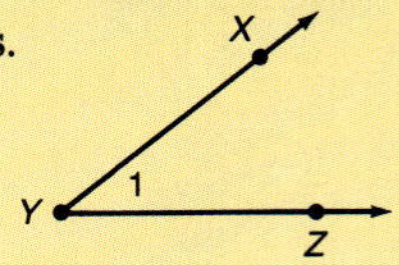

It can be named as $\angle XYZ$, $\angle ZYX$, $\angle Y$, or $\angle 1$.

EXAMPLE 2 Name the figure.

a. 

line segment *EF*, or $\overline{EF}$

b. 

ray *EF*, or $\overrightarrow{EF}$

FOR DISCUSSION See TE side column.

1. How many lines can pass through points A and B?

$\bullet$ A $\qquad\qquad$ $\bullet$ B

2. Explain why some angles must be named by three points and not just the vertex point.

PRACTICE EXERCISES See Extra Practice, page 450.

Draw and label the figure. Check students' drawings.

• 1. Plane RST • 2. $\overleftrightarrow{RS}$ 3. $\angle RST$ 4. $\overline{RS}$ 5. $\overrightarrow{RS}$ 6. $\angle SRT$

Identify the geometric figure.

• 7. $\angle T$ angle T • 8. $\overline{BC}$ line segment BC 9. $\overleftrightarrow{UV}$ line UV 10. $\overrightarrow{MP}$ ray MP 11. Z point Z 12. $\angle RST$ angle RST

Name the figure.

• 13. 

line P

• 14. 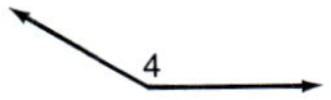

angle 4

15. 

point E

16. 

line segment MN

17. 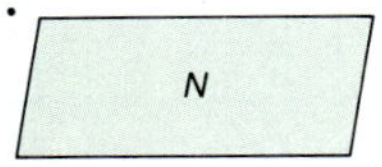

plane N

18. 

ray IH

19. 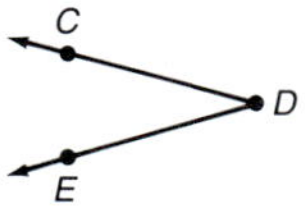

angle CDE

20. 

ray FG

Name the ray in three ways.

• 21. 

$\overrightarrow{AD}$, $\overrightarrow{AC}$, $\overrightarrow{AB}$

22. 

$\overrightarrow{UR}$, $\overrightarrow{US}$, $\overrightarrow{UT}$

23. Name six line segments.

$\overline{AB}$, $\overline{BC}$, $\overline{CD}$, $\overline{AC}$, $\overline{AD}$, $\overline{BD}$

24. Name the angle in four ways.

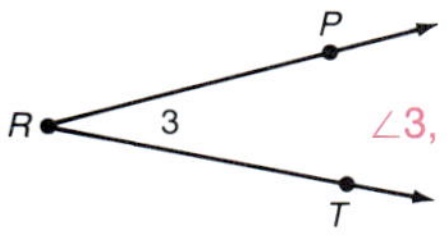

$\angle 3$, $\angle R$, $\angle PRT$, $\angle TRP$

Write *true* or *false*.

• 25. A line can be measured. false

26. A line segment can be measured. true

27. A ray extends in two directions without end. false

28. Three points will always lie in the same plane. true

2. Draw the following diagram on the chalkboard.

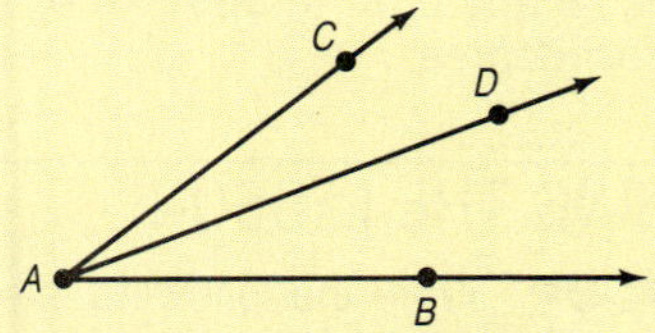

Have students note that $\angle A$ could mean any of three different angles. Thus, in cases like this, three letters should be used to name a specific angle: $\angle CAD$, $\angle DAB$, $\angle CAB$.

COMMON ERROR

In writing the symbol for a ray, some students will point the arrow in the direction the ray is pointing. For example, in Exercise 20, ray FG could be incorrectly named as $\overleftarrow{GF}$ or $\overleftarrow{FG}$ rather than $\overrightarrow{FG}$. Point out that when writing the symbol for a ray, the endpoint is written first.

ASSIGNMENTS

Level 1	1–16, Odd 21–27
Level 2	1–16, 21–28
Level 3	1–6, 17–28

FOLLOW-UP ACTIVITY

APPLICATION Name a geometric figure, for example, ray DC. Ask two students to go to the chalkboard. Have one student draw and label the geometric figure and the other student write the symbol. Continue with similar examples.

SUPPLEMENTARY MATERIALS

TRP Practice, p. 117

TRP Reteaching, p. 69

TRP Transparency 59

OBJECTIVE

- Measure and draw angles to the nearest degree.

TEACHING THE LESSON

WARM-UP Write the following exercises on the chalkboard or an overhead transparency.

Draw and label the figure.

1. $\angle QRS$ **2.** $\overline{CD}$ **3.** $\overrightarrow{GH}$ **4.** $\overleftrightarrow{YZ}$

(Check students' drawings.)

INTRODUCTION Distribute protractors and discuss their use. Have students locate the point on each protractor that would be placed at the vertex when measuring an angle. Have them locate the following marks:

0°, 45°, 90°, 135°, 180°

INSTRUCTION You may wish to use a transparent protractor with an overhead projector to show students how to measure and to draw angles. Demonstrate these techniques in a discussion of Examples 1 and 2 and with other angles.

Point out that sometimes the sides of an angle may have to be extended so that it can be measured more easily. Explain that extending the sides does not change the measure of the angle.

Extend the instruction in Example 2 by having students draw angles that measure 60°, 90°, and 145°.

COMMON ERROR

Many protractors have two scales. When measuring angles, students may read the wrong scale. For example, in Exercise 13, some students may find the measure to be 115° instead of 65°. Help students to place one side of the angle along the 0° mark on the protractor. If the angle opens to the right, use the scale that begins with 0° on the right. If the angle opens to the left, use the scale that begins with 0° on the left.

11.2 MEASURING AND DRAWING ANGLES

You can use a protractor to measure angles. The **degree** (°) is the standard unit of measure.

EXAMPLE 1 Find the measure of $\angle ABC$.

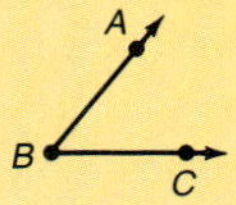

1. Place the center of the protractor on B. Then move the protractor so that the 0° mark is along $\overrightarrow{BC}$.

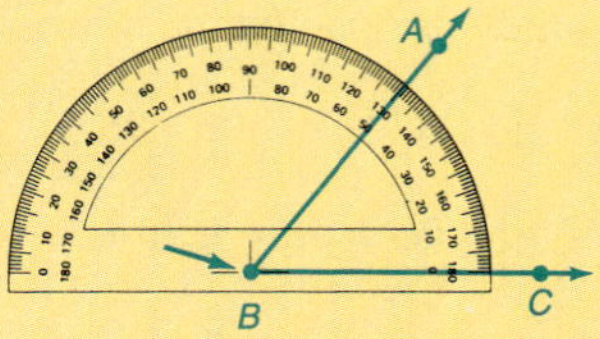

2. Read the measure where $\overrightarrow{BA}$ crosses the same scale as $\overrightarrow{BC}$.

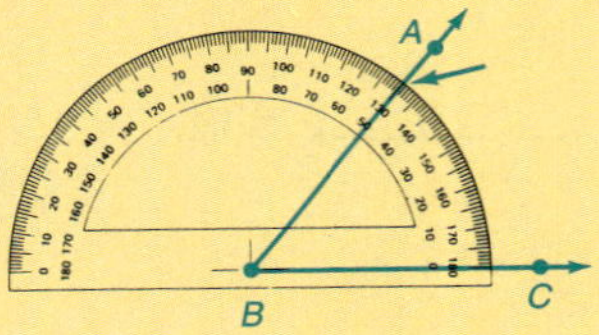

So, the measure of $\angle ABC$ is 50°. This is written $m\angle ABC = 50°$.

A protractor may also be used to draw an angle.

EXAMPLE 2 Draw an angle that measures 120°.

1. Draw $\overrightarrow{AB}$. Place the center of the protractor on A so that the 0° mark is along $\overrightarrow{AB}$.

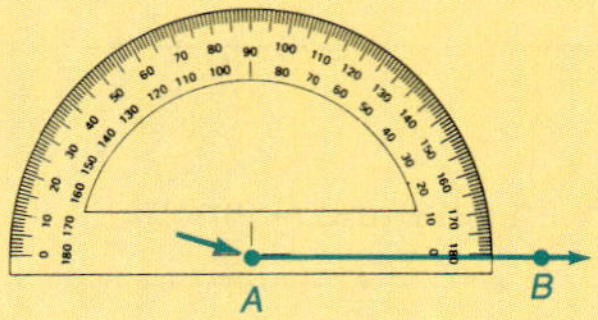

2. Find the 120° mark on the same scale as $\overrightarrow{AB}$. Draw point C. Then draw $\overrightarrow{AC}$.

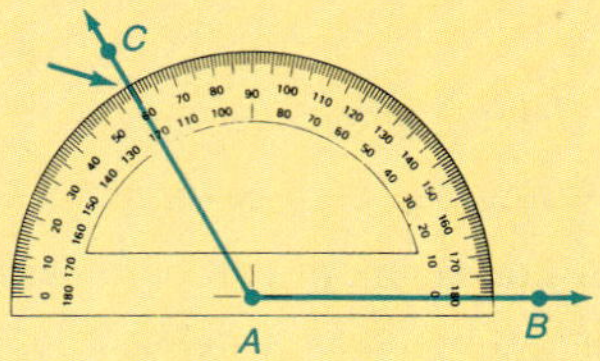

PRACTICE EXERCISES

1. 30° 2. 45° 3. 90° 4. 135° 5. 150° 6. 180° 7. 30° 8. 45° 9. 135°

Use the figure at the right for Exercises 1–9.
Measure the angle.

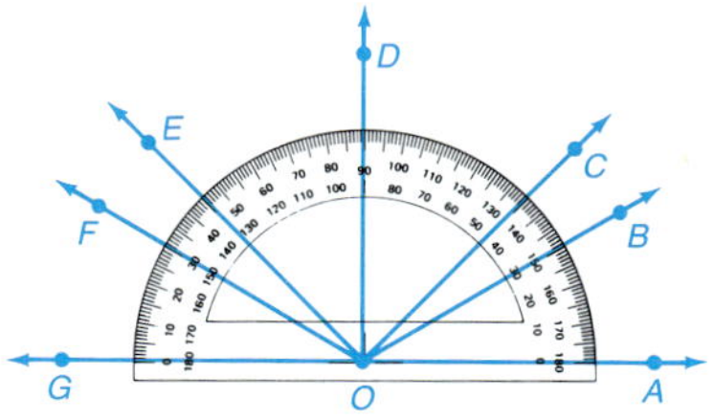

• 1. ∠AOB • 2. ∠AOC 3. ∠AOD
4. ∠AOE 5. ∠AOF 6. ∠AOG
7. ∠GOF 8. ∠GOE 9. ∠GOC

Measure the angle.

• 10.

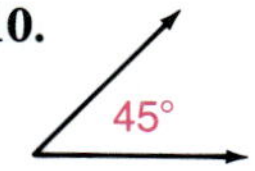

• 11.

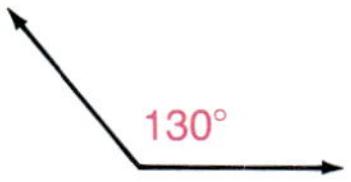

12.

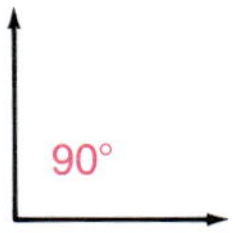

13.

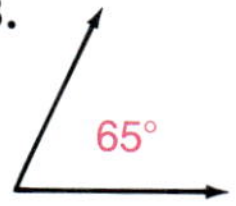

14.

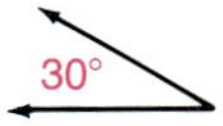

15.

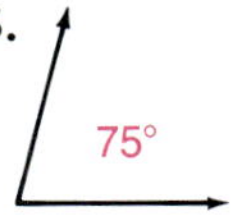

16.

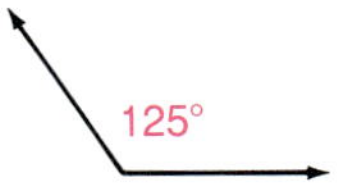

17.

180°

Draw an angle with the following measure. Check students' drawings.

• 18. 20° • 19. 130° 20. 90° 21. 55° 22. 145° 23. 70°

Measure the angle.

• 24. ∠BAC 110°

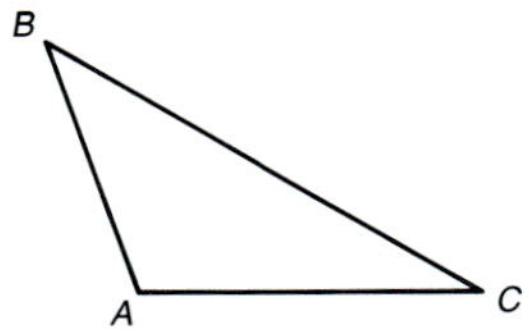

25. ∠ADC 70°

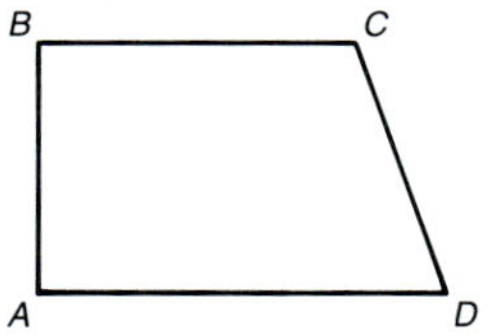

26. ∠BCD 110°

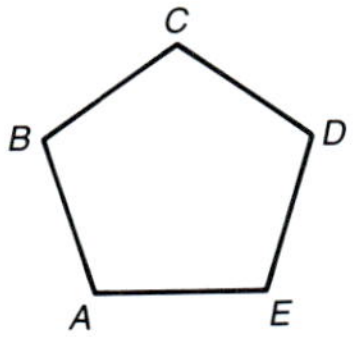

ASSIGNMENTS

Level 1	1–9, Odd 11–25
Level 2	Even 4–26
Level 3	Even 10–22, 24–26

FOLLOW-UP ACTIVITY

COOPERATIVE LEARNING Give each student a copy of the following:

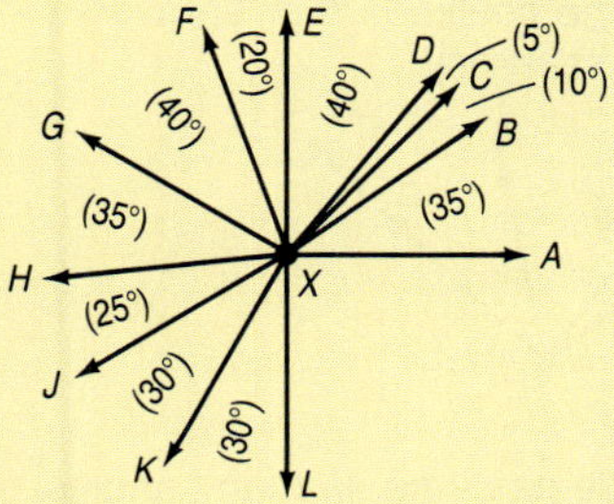

Have students work in pairs to complete the magic square below by measuring the corresponding angles. Each row, column, and diagonal should add up to 180°.

∠EXF (20°)	∠BXC (10°)	∠AXE (90°)	∠LXJ (60°)
∠HXK (55°)	∠EXL (90°)	∠DXC (5°)	∠JXK (30°)
∠DXA (50°)	∠GXJ (60°)	∠HXJ (25°)	∠CXA (45°)
∠BXE (55°)	∠FXE (20°)	∠DXF (60°)	∠CXE (45°)

SUPPLEMENTARY MATERIALS

TRP Practice, p. 118
TRP Reteaching, p. 70
TRP Lesson Aids, p. 7
TRP Transparency 59

OBJECTIVES

- Classify angles as *acute, right, obtuse,* or *straight.*
- Classify pairs of angles as *complementary* or *supplementary.*
- Compute the complement or the supplement of an angle.

TEACHING THE LESSON

WARM-UP Draw the following diagram on the chalkboard or an overhead transparency.

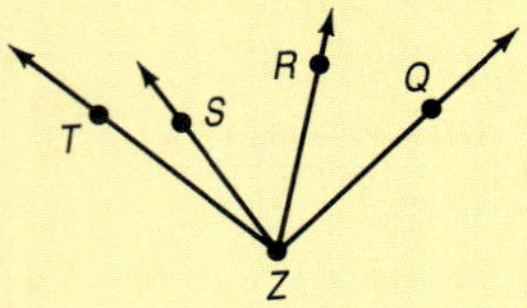

Have students name as many angles as possible.

(*∠QZR, ∠QZS, ∠QZT, ∠RZS, ∠RZT, ∠SZT*)

INSTRUCTION Discuss the classification of angles by their measures. In Example 1, point out that the position of an angle does not affect its classification. Stress that classification depends upon an angle's measure.

In discussing complementary and supplementary angles in Example 2, be careful to point out that the angles do not have to be adjacent.

Extend the instruction in Example 3 by having students find the complement of an angle that measures:

20° (70°) 53° (37°) 84° (6°)

Have them find the supplement of an angle that measures:

110° (70°) 40° (140°) 166° (14°)

11.3 CLASSIFYING ANGLES

You can classify angles by their measures.

Angle or angle pairs	Name	Definition
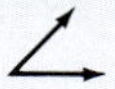	Acute angle	An **acute angle** measures greater than 0° and less than 90°.
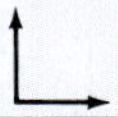	Right angle	A **right angle** measures 90°.
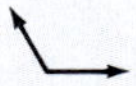	Obtuse angle	An **obtuse angle** measures greater than 90° and less than 180°.
	Straight angle	A **straight angle** measures 180°.
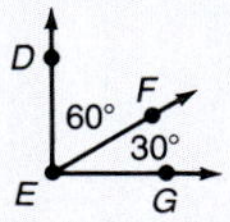 m∠*DEF* + m∠*FEG* = 90°	Complementary angles	Angles are **complementary** when the sum of their measures is 90°.
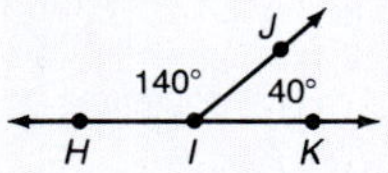 m∠*HIJ* + m∠*JIK* = 180°	Supplementary angles	Angles are **supplementary** when the sum of their measures is 180°.

EXAMPLE 1 Classify ∠*ABC*.

THINK: m∠*ABC* = 135°
135° is greater than 90°
and less than 180°.

∠*ABC* is an obtuse angle.

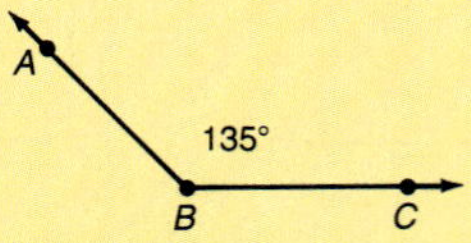

You can classify pairs of angles by the sum of their measures.
The measures (m) of complementary angles add up to 90°.
The measures (m) of supplementary angles add up to 180°.

EXAMPLE 2
a. Which pair of angles is complementary?
b. Which pair of angles is supplementary?

m∠1 = 40° m∠2 = 130° m∠3 = 50°

a. **THINK: m∠1 + m∠3 = 40° + 50°**
= 90°

So, ∠1 and ∠3 are complementary angles.

b. **THINK: m∠2 + m∠3 = 130° + 50°**
= 180°

So, ∠2 and ∠3 are supplementary angles.

You can find the complement or supplement of an angle by subtracting.

EXAMPLE 3 $\angle A$ is 60°.

a. Find the complement of $\angle A$.

THINK: ■ + 60° = 90°
■ = 90° − 60°
■ = 30°

So, the complement is 30°.

b. Find the supplement of $\angle A$.

THINK: ■ + 60° = 180°
■ = 180° − 60°
■ = 120°

So, the supplement is 120°.

PRACTICE EXERCISES See Extra Practice, page 451.

Classify the angle as *acute, right, obtuse,* or *straight.*

• **1.**  obtuse

•**2.** 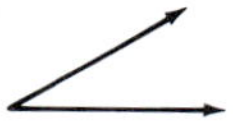 acute

3.  straight

4.  acute

• **5.** 28° acute
•**6.** 45° acute
7. 95° obtuse
8. 180° straight
9. 5° acute
10. 90° right

Tell whether the pair of angles are *complementary, supplementary,* or *neither.*

•**11.** 40°, 30° neither
•**12.** 10°, 80° complementary
13. 35°, 55° complementary
14. 110°, 80° neither
15. 145°, 35° supplementary

Use the figure to the right for Exercises 16–19. Tell whether the pair of angles are *complementary, supplementary,* or *neither.*

•**16.** $\angle AOB$ and $\angle BOD$ complementary

17. $\angle AOE$ and $\angle EOG$ supplementary

18. $\angle AOC$ and $\angle COG$ supplementary

19. $\angle AOF$ and $\angle EOG$ neither

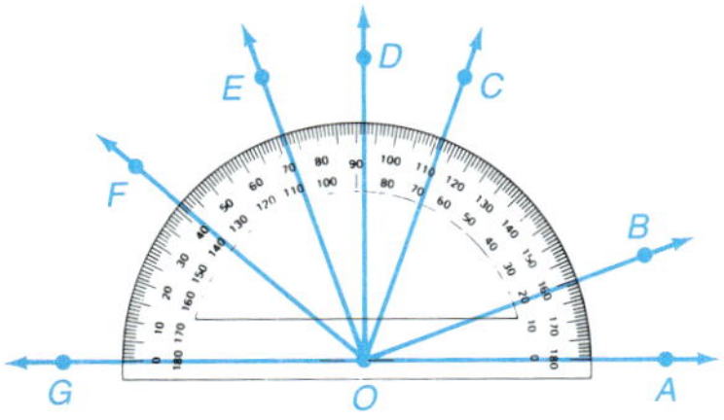

Find the complement of the angle.

•**20.** 20° 70°
•**21.** 55° 35°
22. 18° 72°
23. 67° 23°
24. 5° 85°
25. 38° 52°

Find the supplement of the angle.

•**26.** 60° 120°
•**27.** 140° 40°
28. 95° 85°
29. 15° 165°
30. 134° 46°
31. 87° 93°

Solve.

32. Two angles are complementary and have the same measure. What is the measure of each angle? 45°

33. $\angle AOD$ and $\angle DOE$ are supplementary angles. $\angle DOE$ and $\angle COD$ are complementary angles. If m$\angle AOD$ = 145°, what are the measures of $\angle DOE$ and $\angle COD$?
m$\angle DOE$ = 35°; m$\angle COD$ = 55°

ASSIGNMENTS

Level 1	1–15, Even 16–32
Level 2	1–32
Level 3	Odd 1–31, 32, 33

FOLLOW-UP ACTIVITY

MENTAL MATH On each of 30 index cards, write an angle measurement and either of the words *complement* or *supplement.*

Divide the class into two teams. Have the first player on one team draw a card and display it for everyone to see. The corresponding player on the other team then mentally computes the complement or the supplement of the given measurement. Award a point for each correct answer. Have the teams reverse roles and repeat the procedure. The team with the greatest number of points after a specified period of time wins.

SUPPLEMENTARY MATERIALS

TRP Practice, p. 119

TRP Reteaching, p. 71

TRP Transparency 60

OBJECTIVES

- Construct a line segment congruent to a given line segment.
- Construct an angle congruent to a given angle.
- Construct an angle bisector.

TEACHING THE LESSON

WARM-UP Provide time for students to practice making arcs with their compasses. Explain that an arc is a curve that is a set of points all the same distance from a given point.

INTRODUCTION Discuss the instruments used in constructing geometric figures. A *straightedge* is like a ruler without a scale. It cannot be used to draw a line segment of a specific length. Demonstrate how to open a compass to measure and transfer specific lengths.

Have students draw a ray and mark off distances such as 1 in., 2 in., and 3 in. Encourage students to use several sheets of paper as a backing to avoid damaging their desks.

INSTRUCTION Begin by reviewing the difference between the equals sign (=) and the congruence symbol (≅).

Demonstrate the constructions on the chalkboard or an overhead projector. Students should use a compass and a straightedge and follow the steps at their desks. You may wish to prepare worksheets with various angles and segments for students to copy.

Students may use a ruler or protractor to check their constructions. Remind them to allow for a small difference in the measurement between the original figure and the constructed figure.

11.4 CONSTRUCTING CONGRUENT LINE SEGMENTS AND ANGLES

Geometric figures that have the same shape and size are **congruent.**

$\overline{ST}$ is congruent to $\overline{UV}$

$\overline{ST} \cong \overline{UV}$

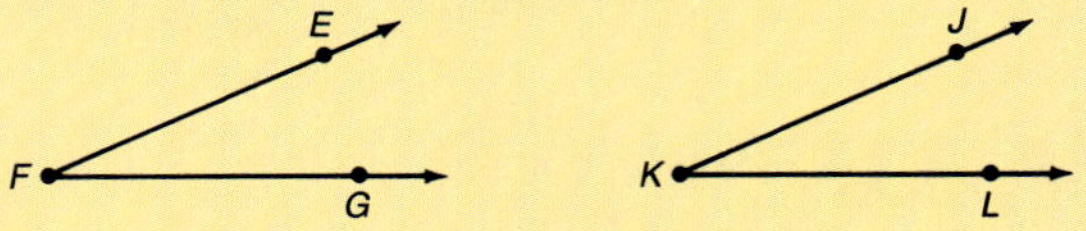

$\angle EFG$ is congruent to $\angle JKL$

$\angle EFG \cong \angle JKL$

You can construct congruent line segments or congruent angles using a straightedge and a compass.

EXAMPLE 1

Construct a line segment congruent to $\overline{MN}$.

1. Draw a ray. Label the endpoint X.
2. Open the compass to the length of $\overline{MN}$. Using the same compass opening, place the compass point on X and draw a small arc that crosses the ray. Label the point of intersection Y.

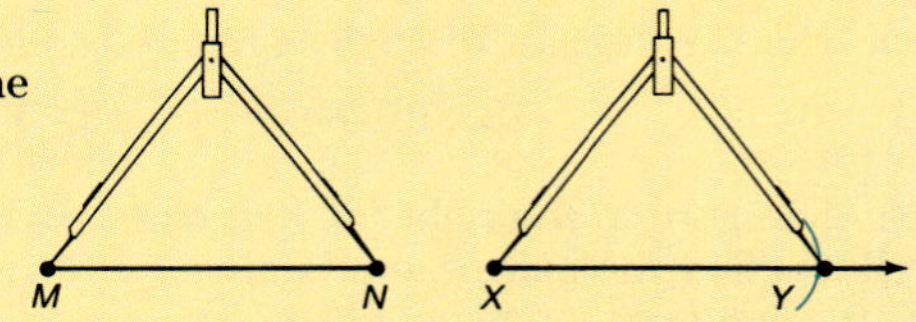

$\overline{XY}$ has the same length as $\overline{MN}$. $\overline{XY} \cong \overline{MN}$

EXAMPLE 2

Construct an angle congruent to $\angle BAC$.

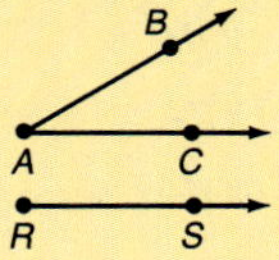

1. Draw $\overrightarrow{RS}$.
2. Draw an arc that intersects both sides of $\angle BAC$. Label these points J and K. Using the same compass opening, place the compass point on R and draw an arc that intersects $\overrightarrow{RS}$. Label the point U.

3. Open the compass to measure the distance between J and K. Using the same compass opening, place the compass point on U and draw an arc that intersects the arc through U. Label this point T.

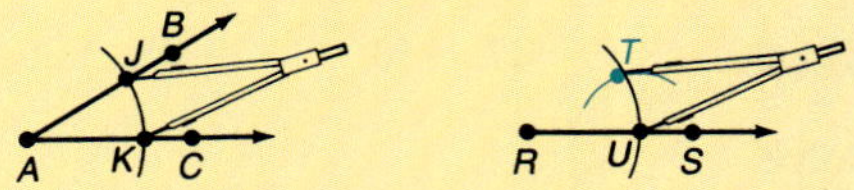

4. Draw $\overrightarrow{RT}$.

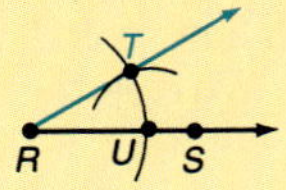

$\angle BAC$ has the same measure as $\angle TRS$. $\angle BAC \cong \angle TRS$

A geometric figure is **bisected** if it is separated into two congruent parts.

$m\angle XYW = 30°$ $m\angle WYZ = 30°$

$\angle XYW \cong \angle WYZ$

$\overrightarrow{YW}$ bisects $\angle XYZ$.

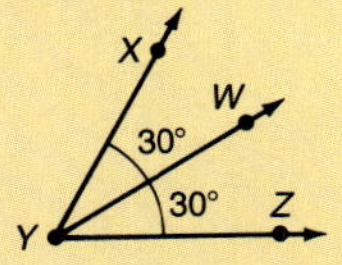

EXAMPLE 3 Construct the **angle bisector** of $\angle ABC$.

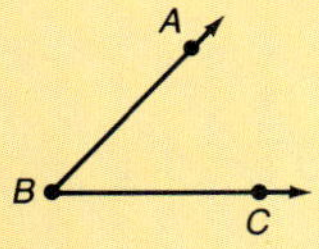

1. Draw an arc that intersects both sides of $\angle ABC$. Label these points D and E.

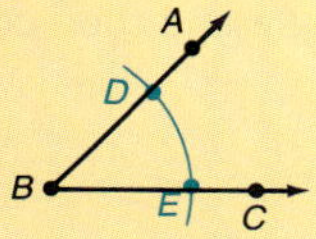

2. Place the compass point on D. Draw an arc in the interior of the angle. Use the same compass opening. Place the compass point on E and draw an arc that intersects the first arc. Label the intersection F.

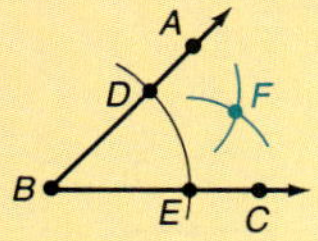

3. Draw $\overrightarrow{BF}$.

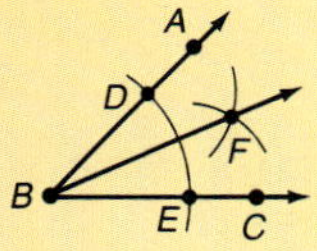

$\angle ABF \cong \angle FBC$. So, $\overrightarrow{BF}$ is the angle bisector of $\angle ABC$.

FOR DISCUSSION

See TE side column.

1. Can two angles be congruent even though the drawn sides of one angle are longer than the drawn sides of the other angle? Explain.
2. Is it possible for an angle to have two different bisectors? Explain.

PRACTICE EXERCISES

See Extra Practice, page 451.

Check students' constructions

Trace the line segment and then construct a congruent line segment.

1. 

2. 

3. 

4. 

In discussing angle bisectors in Example 3, elicit from the students the meaning of *bisect* (to separate into two congruent parts). Note that an angle bisector is a ray and that there is only one angle bisector for any given angle.

FOR DISCUSSION

1. Students should explain that the sides of an angle are rays that extend endlessly regardless of the length of the sides in the drawing. Consequently, two angles can be congruent even though the sides are drawn to different lengths.
2. Students should explain that there is only one angle bisector for a given angle.

TIME OUT Guide students to realize that they should begin constructing the triangle by drawing the given side first. After students are finished, have them compare their constructions.

Point out that their triangles are congruent. Ask them to repeat the constructions but, this time, begin by drawing the given angles and ignoring the length of the given side. Have them compare this set of triangles. Students should see that the triangles are not always congruent.

COMMON ERROR

The compass openings may change when students move the compass from the given figure to the construction. Emphasize that once they have used the compass to measure a distance or draw an arc on the given figure, they must be careful to keep the same compass opening when moving the compass to the construction. It may be helpful for some students to use a compass where the opening can be locked in place.

ASSIGNMENTS

Level 1	Odd 1–15, 17–32, 33
Level 2	Even 2–32, 33–34
Level 3	Odd 1–31, 33–34, TO

FOLLOW-UP ACTIVITY

COOPERATIVE LEARNING Separate students into groups of four. Have each group make a set of 20 cards with 10 pairs of congruent line segments or angles. (Pairs of congruent figures need not be in the same position on matching cards.)

Have students play *Memory* in groups of three or four. Provide each group with a set of cards that one of the other groups has made, as well as a ruler and a protractor.

Have a player shuffle the cards and place them face down in rows of four. The first player turns over two cards. If both figures are line segments or angles, the player measures the figures to determine whether they are congruent. If so, the player keeps the cards and tries again. If not, the cards are placed face down and the next student plays. The player with the greatest number of cards at the end of the game wins.

SUPPLEMENTARY MATERIALS

TRP Practice, p. 120

TRP Reteaching, p. 72

TRP Transparency 60

Trace the angle and then construct a congruent angle. Check students' constructions.

• **5.** 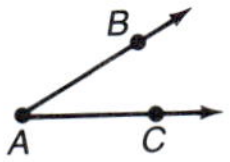

6. 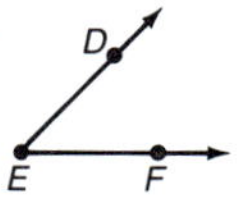

7. 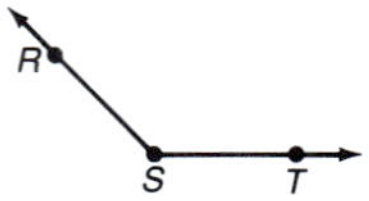

8. 

9. 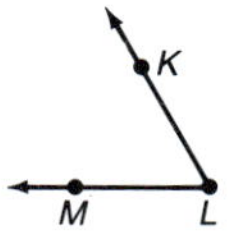

10. 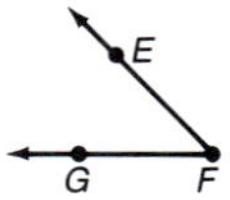

Trace the angle and then construct the angle bisector. Check students' constructions.

• **11.** 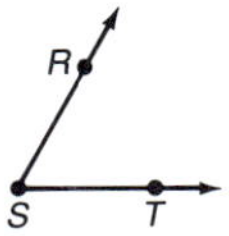

12. 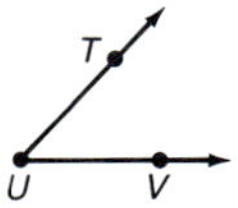

13. 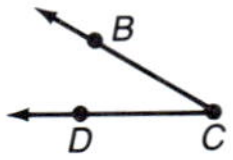

14. 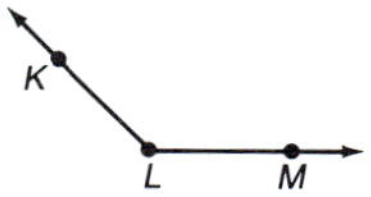

15. 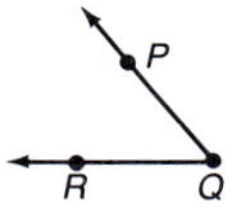

16. 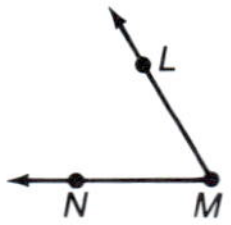

MIXED REVIEW

Complete.

• **17.** 3 ft = ■ in. 36 • **18.** 7 qt = ■ pt 14 **19.** 32 oz = ■ lb 2 **20.** 28 qt = ■ gal 7

21. $8\frac{1}{3}$ yd = ■ ft 25 **22.** 144 in. = ■ yd 4 **23.** $3\frac{1}{2}$ h = ■ min 210 **24.** 75 in. = ■ ft $6\frac{1}{4}$

25. 4.2 m = ■ cm 420 **26.** 0.8 L = ■ mL 800 **27.** 243 g = ■ kg 0.243 **28.** 72 mm = ■ cm 7.2

29. 9.1 mm = ■ m 0.0091 **30.** 83.4 kg = ■ g 83,400 **31.** 800 m = ■ km 0.8 **32.** 750 mL = ■ L 0.75

Solve. Check students' constructions.

• **33.** Use a ruler to draw a 4-cm line segment. Label it $\overline{AB}$. Construct $\overline{CD}$ so that $\overline{CD}$ is twice as long as $\overline{AB}$.

34. Use a protractor to draw a 40° angle. Label it $\angle MNO$. Construct $\angle XYZ$ so that m$\angle XYZ$ is twice the m$\angle MNO$.

TIME OUT

Draw a 3-in. line segment and label it $\overline{AB}$. Draw a 30° angle and label it $\angle A$. Draw a 45° angle and label it $\angle B$. Draw a triangle using the two angles and the line segment. Check students' drawings.

11.5 CONSTRUCTING PERPENDICULAR AND PARALLEL LINES

Two lines in a plane may or may not intersect. If the lines do not intersect, they are **parallel.** If the lines intersect to form right angles, they are **perpendicular.**

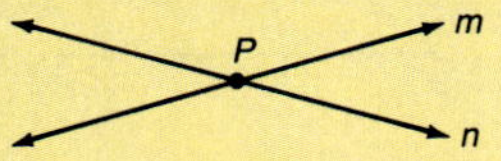

Intersecting lines
m and n intersect at P

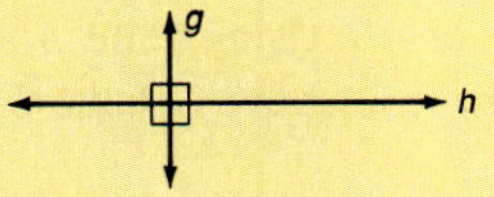

Perpendicular lines
$g \perp h$ and $h \perp g$

Parallel lines
$r \parallel s$ and $s \parallel r$

EXAMPLE 1

Construct a line perpendicular to $\overleftrightarrow{AB}$ through point P on the line.

1. Using the point P as a center, draw an arc that intersects $\overleftrightarrow{AB}$ at points R and S.
2. Open the compass to a distance greater than half the length of $\overline{RS}$. Using points R and S as centers, draw two arcs. Label the point of intersection C.
3. Draw $\overleftrightarrow{CP}$. $\overleftrightarrow{CP}$ is perpendicular to $\overleftrightarrow{AB}$ at P. This is written as $\overleftrightarrow{CP} \perp \overleftrightarrow{AB}$ at P.

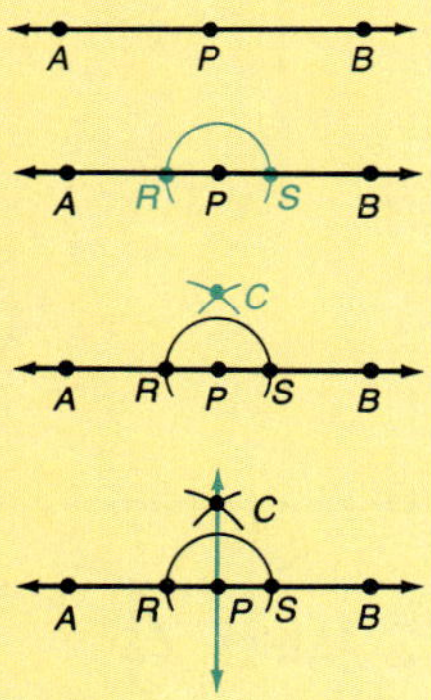

EXAMPLE 2

Construct a line perpendicular to $\overleftrightarrow{AB}$ through point P not on the line.

1. Using point P as a center, draw an arc that intersects $\overleftrightarrow{AB}$ at points R and S.
2. Open the compass to a distance greater than half the length of $\overline{RS}$. Using points R and S as centers, draw two arcs. Label the point of intersection C.
3. Draw $\overleftrightarrow{PC}$. Label the point of intersection D. $\overleftrightarrow{PC}$ is perpendicular to $\overleftrightarrow{AB}$ at D. This is written as $\overleftrightarrow{PC} \perp \overleftrightarrow{AB}$ at D.

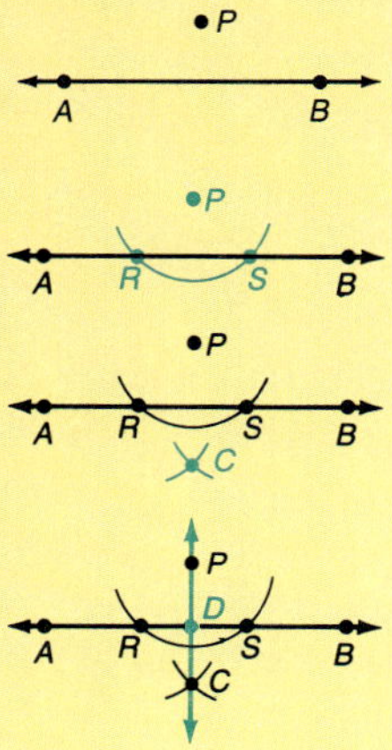

OBJECTIVES

- Construct a line perpendicular to a given line.
- Construct the perpendicular bisector of a line segment.
- Construct a line parallel to a given line.

TEACHING THE LESSON

WARM-UP Write the following exercises on the chalkboard or an overhead transparency.

Draw the angle. Then classify it as *acute, right, obtuse,* or *straight.*

1. 120° **2.** 90° **3.** 65° **4.** 180°

(Check students' drawings.
1. obtuse **2.** right **3.** acute
4. straight)

INTRODUCTION Have students draw a model of two lines that:

- cross
- cross to form right angles
- never cross

Have volunteers display their drawings. Discuss examples from everyday life of these types of lines.

INSTRUCTION Demonstrate the four constructions on the chalkboard or an overhead projector. Have students follow the steps with their straightedges and compasses. Encourage them to make their drawings large. The tendency is for students to make their drawings too small, which results in inaccurate constructions.

In Examples 1 and 2, discuss the similarities in the two constructions. Students should realize that the only difference between the constructions is the location of point *P*.

In Example 3, discuss the concept of a bisector of a segment. Point out that if a line contains the midpoint of a line segment, then it bisects the segment. Draw a line segment and its midpoint on the chalkboard. Have students draw lines that contain the midpoint and thus bisect the line segment. Emphasize that there are an infinite number of lines that bisect a segment, but that there is only one *perpendicular* bisector.

In Example 4, you may wish to remind students that since $\overleftrightarrow{AB}$ and $\overleftrightarrow{EF}$ are both perpendicular to $\overleftrightarrow{CD}$, their intersection with *CD* forms right angles. Point out that the right angles are congruent and that lines *EF* and *AB* are parallel.

FOR DISCUSSION

1. Students should explain that the perpendicular bisector of a line would have to contain the line's midpoint. Since a line has no endpoints, it has no midpoint. Thus, it cannot have a perpendicular bisector.
2. Students should explain that they would first bisect the line segment. Then they would bisect the resulting two line segments. Finally they would bisect each of the four line segments to create eight congruent line segments.

COMMON ERROR

Some students may not open the compass far enough when constructing a perpendicular bisector. Demonstrate that the arcs will not intersect unless the compass opening is greater than half the length of the line segment. Guide students in determining a proper compass opening.

The midpoint of a line segment divides it into two congruent parts. A **perpendicular bisector** is a line perpendicular to a line segment that intersects it at the midpoint.

EXAMPLE 3

Construct the perpendicular bisector of $\overline{AB}$.

1. Open the compass to a distance greater than half the length of $\overline{AB}$. Using point A as a center, draw one arc above $\overline{AB}$ and one arc below $\overline{AB}$.

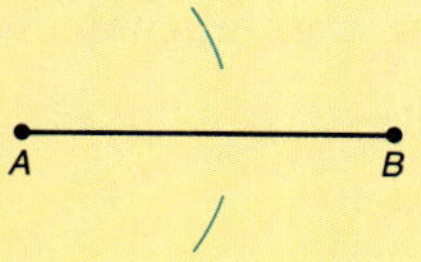

2. Use the same compass opening. Using point B as a center, draw one arc above $\overline{AB}$ and one arc below $\overline{AB}$. Label the points of intersection C and D.

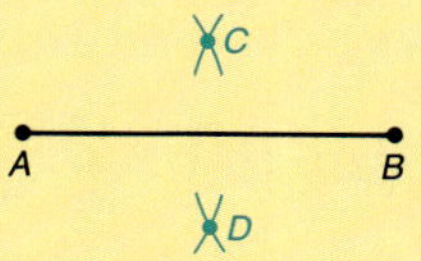

3. Draw $\overleftrightarrow{CD}$. Label the point where $\overleftrightarrow{CD}$ intersects $\overline{AB}$ as M, the midpoint. This is written as $\overleftrightarrow{CD} \perp \overline{AB}$ at M.

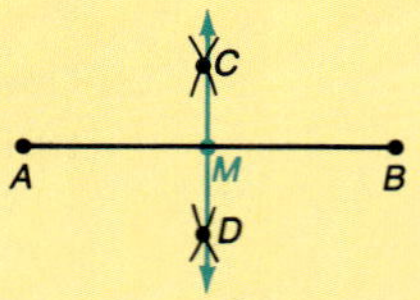

EXAMPLE 4

Given $\overleftrightarrow{AB}$, construct a line parallel to $\overleftrightarrow{AB}$.

1. Mark a point C on $\overleftrightarrow{AB}$.

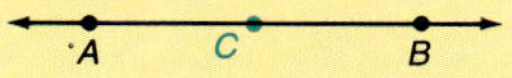

2. Construct $\overleftrightarrow{DC} \perp \overleftrightarrow{AB}$.

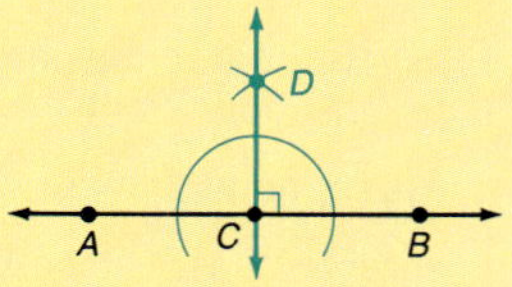

3. Mark a point E on $\overleftrightarrow{DC}$. Construct $\overleftrightarrow{EF} \perp \overleftrightarrow{DC}$. $\overleftrightarrow{EF}$ is parallel to $\overleftrightarrow{AB}$. This is written as $\overleftrightarrow{EF} \parallel \overleftrightarrow{AB}$.

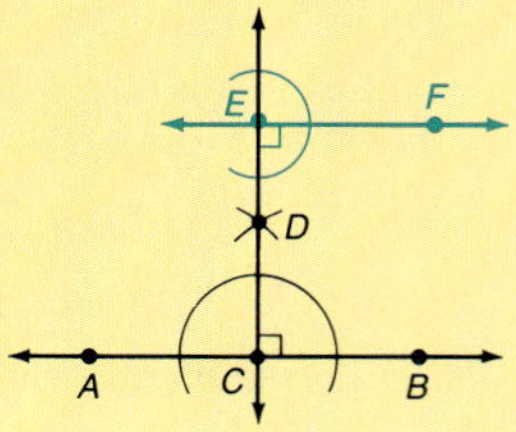

FOR DISCUSSION See TE side column.

1. Can a perpendicular bisector of a line be constructed? Explain.

2. How can you divide a line segment into eight congruent line segments?

PRACTICE EXERCISES See Extra Practice, page 451.

Trace the line and then construct a line perpendicular to it through the point on the line. Check students' constructions.

• 1. 

2. 

3. 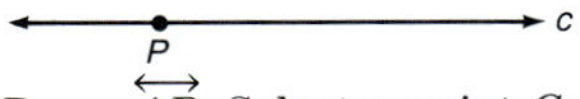

4. 

5. Draw $\overleftrightarrow{AB}$. Select a point C not on the line. Construct a line perpendicular to $\overleftrightarrow{AB}$ through point C. Check students' constructions.

Trace the line segment and then construct the perpendicular bisector. Check students' constructions.

• 6. 

7. 

8. 

9. 

Trace the line and construct a line parallel to it. Check students' constructions.

•10. 

11. 

12. 

Solve.

13. Draw a 4-in. line segment. Label it $\overline{AB}$. Use a compass and straightedge to bisect $\overline{AB}$ at point C. Bisect $\overline{AC}$ and $\overline{CB}$ at points E and F. The length of $\overline{AE}$ is what fractional part of the length of $\overline{AB}$? $\frac{1}{4}$

14. Draw a 6-in. line segment. Label it $\overline{XY}$. Use a compass and straightedge to bisect $\overline{XY}$ at point Z. Bisect $\overline{ZY}$ at point P. What is the length in inches of $\overline{XP}$? $4\frac{1}{2}$ in.

ASSIGNMENTS

Level 1	Even 2–12
Level 2	Odd 1–13
Level 3	Odd 5–11, 13–14

FOLLOW-UP ACTIVITY

COOPERATIVE LEARNING
Write the following on the chalkboard.

$\overleftrightarrow{CD}$ intersects $\overleftrightarrow{AB}$ at E, the midpoint of $\overline{AB}$. $\overleftrightarrow{XY}$ is parallel to $\overleftrightarrow{AB}$ and intersects $\overline{CD}$ at Z, the midpoint of $\overline{XY}$. $\overline{XZ}$ is congruent to $\overline{YZ}$ and to $\overline{AE}$. $\overrightarrow{BY}$ is parallel to $\overleftrightarrow{CD}$.

Have students work in pairs to construct the figure described above.

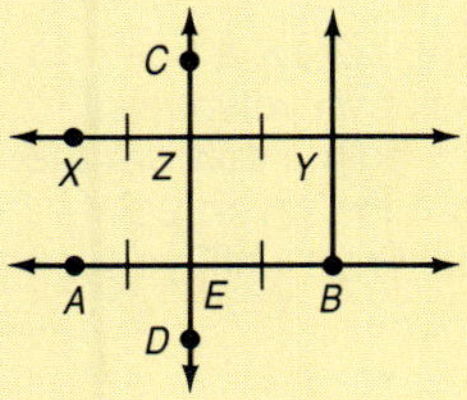

SUPPLEMENTARY MATERIALS

TRP Practice, p. 121

TRP Reteaching, p. 73

TRP Transparency 61

OBJECTIVES

- Identify and define polygons and regular polygons.
- Classify polygons by the number of sides.
- Identify and define a circle and its parts.

TEACHING THE LESSON

WARM-UP Write the following exercises on the chalkboard or on an overhead transparency.

Find the complement of the angle.

1. 50° (40°) **2.** 74° (16°)
3. 19° (71°)

Find the supplement of the angle.

4. 145° (35°) **5.** 22° (158°)
6. 93° (87°)

INTRODUCTION Draw a point on the chalkboard. Ask two volunteers to use a ruler to draw points that are the same distance from the given point. Instruct them to continue until you tell them to stop.

INSTRUCTION Draw several different polygons on the chalkboard. Point out that a diagonal is a line segment that connects two nonadjacent vertices. Have volunteers draw as many diagonals as they can for each polygon.

Discuss how polygons are classified. You may wish to point out that the prefix for each word tells how many sides are in the polygon. Point out that not all polygons have names. Mathematicians refer to many-sided polygons as *n-gons* where *n* represents the number of sides in the polygon. Thus, a 15-sided polygon would be called a *15-gon.*

Call attention to the students who are drawing points on the chalkboard. Guide the class in discovering the definition of a circle. Review the terms *radius, diameter, chord,* and *central angle.*

11.6 POLYGONS AND CIRCLES

A **polygon** is a simple closed curve consisting entirely of line segments. This is polygon *ABCD*. It is named by its four vertices. $\overline{AC}$ is a diagonal.

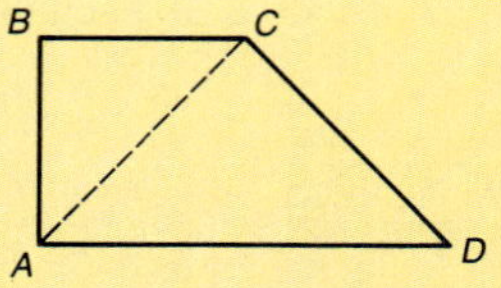

A **regular polygon** has congruent sides and congruent angles. Polygon *EFGHIJ* is a regular polygon.

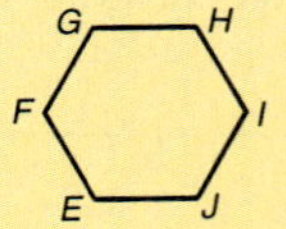

Here are some polygons.

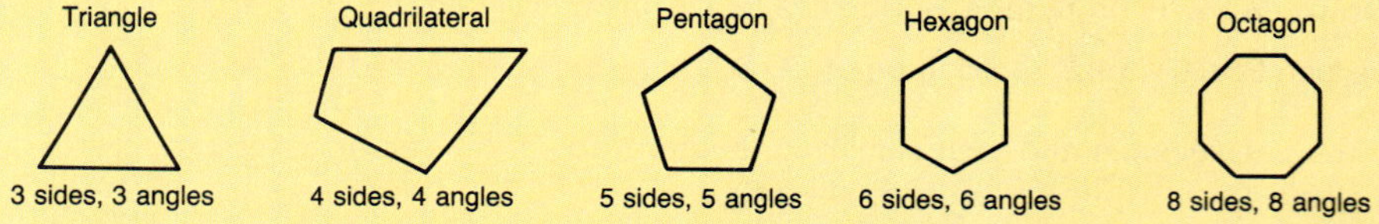

EXAMPLE 1 Name the polygon.

a.

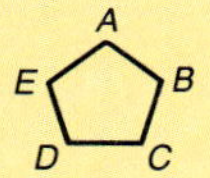

THINK: 5 sides
Vertices: *A, B, C, D, E*

Pentagon *ABCDE*

b.

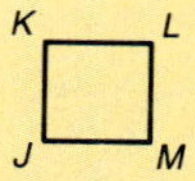

THINK: 4 sides
Vertices: *J, K, L, M*

Quadrilateral *JKLM*

A **circle** is the set of all points in a plane that are the same distance from the center. In this circle:

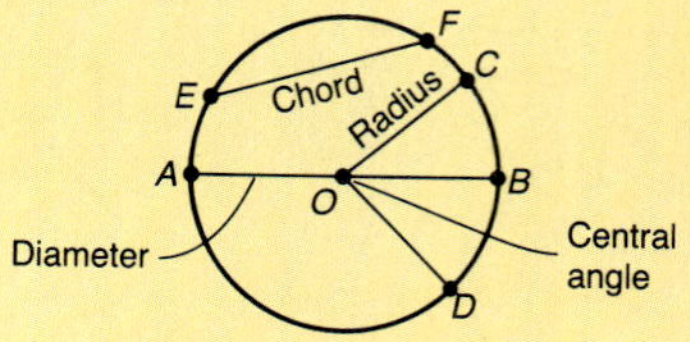

O is the **center.**

$\overline{EF}$ is a chord. A **chord** is a line segment that has endpoints on the circle.

$\overline{AB}$ is a diameter. A **diameter** is a chord that passes through the center of a circle.

$\overline{OC}$ is a radius. A **radius** is a line segment that goes from the center of the circle to a point on it.

$\angle BOD$ is a central angle. A **central angle** is an angle whose vertex is at the center of the circle.

You can find the length of either the diameter or radius of a circle if you know the length of the other.

EXAMPLE 2 Find the missing length.

a. $r = 5.2$ cm, $d = ■$

THINK: $d = 2 \times r$
$= 2 \times 5.2$
$= 10.4$

The diameter is 10.4 cm.

b. $d = 3$ in., $r = ■$

THINK: $r = \frac{1}{2} \times d$
$= \frac{1}{2} \times 3$
$= 1\frac{1}{2}$

The radius is $1\frac{1}{2}$ in.

PRACTICE EXERCISES

See Extra Practice, page 452.

Name the polygon.

• **1.**

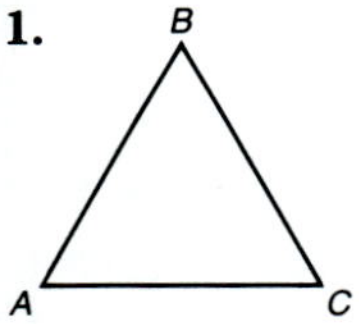

triangle ABC

2.

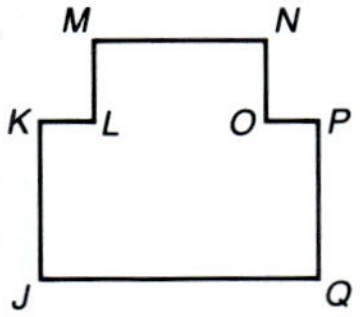

octagon $JKLMNOPQ$

3.

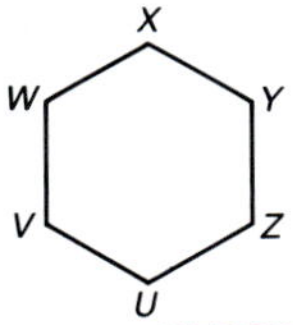

hexagon $UVWXYZ$

4.

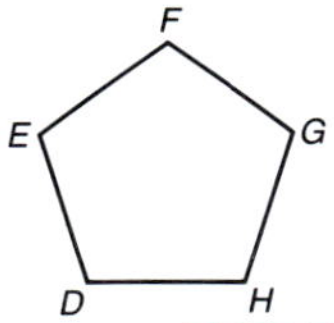

pentagon $DEFGH$

Use the figure to name the part of the circle.

• **5.** $\overline{BD}$ diameter **6.** $\overline{AB}$ chord **7.** $\overline{OC}$ radius **8.** $\angle COD$ central angle

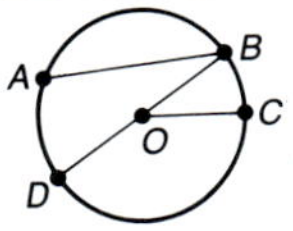

Find the missing length.

• **9.** $r = 5$ cm, $d = ■$ 10 cm **10.** $r = 15$ in., $d = ■$ 30 in. **11.** $r = 3.5$ cm, $d = ■$ 7 cm

12. $r = 12.4$ cm, $d = ■$ 24.8 cm **13.** $r = 8\frac{1}{2}$ ft, $d = ■$ 17 ft **14.** $r = 4\frac{1}{4}$ in., $d = ■$ $8\frac{1}{2}$ in.

• **15.** $d = 8$ ft, $r = ■$ 4 ft **16.** $d = 58$ mm, $r = ■$ 29 mm **17.** $d = 4.8$ cm, $r = ■$ 2.4 cm

18. $d = 5$ in., $r = ■$ $2\frac{1}{2}$ in. **19.** $d = 1.3$ m, $r = ■$ 0.65 m **20.** $d = 12\frac{1}{2}$ ft, $r = ■$ $6\frac{1}{4}$ ft

Solve.

• **21.** The schoolyard is in the shape of a quadrilateral. How many sides does the school yard have? 4

22. The diameter of a circle is 18 in. What is the length of the radius? 9 in.

23. One side of a regular octagon measures 2 cm. What is the measure of each of the other sides? 2 cm

24. The radius of circle A is 3 ft. The radius of circle B is twice as long. Find the length of the diameter of circle B. 12 ft

ASSIGNMENTS

Level 1	1–8, Odd 9–19, 21, 22
Level 2	1–23
Level 3	1–8, 15–24

FOLLOW-UP ACTIVITY

COOPERATIVE LEARNING

Have students work in pairs. Each pair of students needs a compass and straightedge. Write the following instructions on the chalkboard.

- Draw a large circle.
- Without changing the compass setting, start at any point on the circle and divide the circle into six congruent parts.
- Connect every other point to form an equilateral triangle.
- Connect the other three points to form a second equilateral triangle.

Have students refer to their diagrams to answer the following questions.

1. How many triangles are there in all? (8)
2. How many hexagons are there? (1)
3. How many pentagons are there? (6)
4. How many squares are there? (0)

SUPPLEMENTARY MATERIALS

TRP Practice, p. 122

TRP Reteaching, p. 74

TRP Transparency 61

OBJECTIVES

- Classify triangles by the number of congruent sides.
- Classify triangles by the measures of their angles.
- Find the measure of the third angle of a triangle.

TEACHING THE LESSON

WARM-UP Write the following exercises on the chalkboard or an overhead transparency.

Compute.

1. 180 − (63 + 17)
2. 180 − (58 + 75)
3. 180 − (135 + 19)
4. 180 − (120 + 35)
5. 180 − (112 + 64)
6. 180 − (19 + 34)

(**1.** 100 **2.** 47 **3.** 26 **4.** 25 **5.** 4 **6.** 127)

INTRODUCTION Draw the following illustration on the chalkboard. Discuss the symbols used to mark the congruent sides and angles of a triangle.

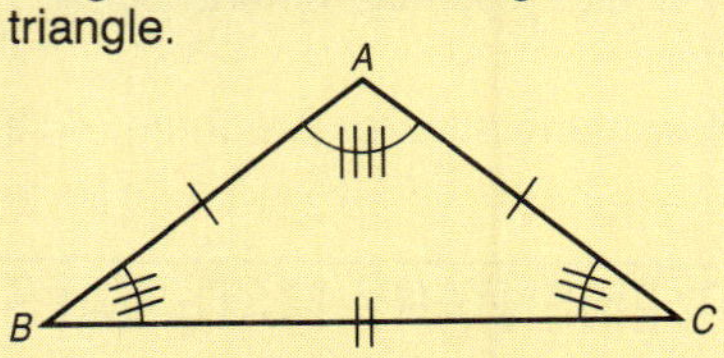

In the figure, $\angle B = \angle C$, and $\overline{AB} = \overline{AC}$. Point out that if the marks are different, then the sides or angles have different measures. Conversely, if the marks are the same, then the sides or angles have the same measure.

INSTRUCTION While presenting Examples 1 and 2, you may wish to discuss the various types of triangles by including some of their properties.

11.7 TRIANGLES

Triangles may be classified by the number of congruent sides.

Equilateral triangle	Isosceles triangle	Scalene triangle
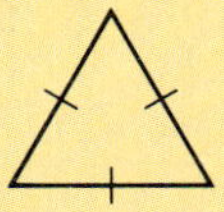	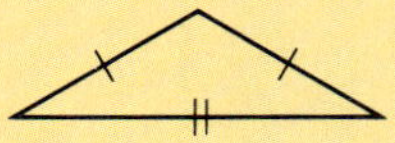	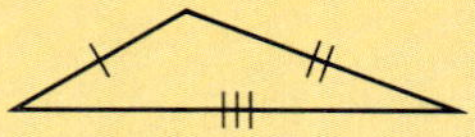
3 congruent sides, 3 congruent angles	2 congruent sides, 2 congruent angles	no congruent sides, no congruent angles

EXAMPLE 1 Identify the triangle as *equilateral*, *isosceles*, or *scalene*.

a.

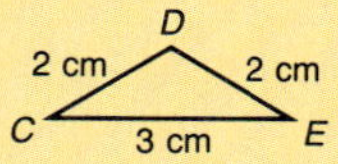

THINK: 2 sides congruent

So, $\triangle CDE$ is an isosceles triangle.

b.

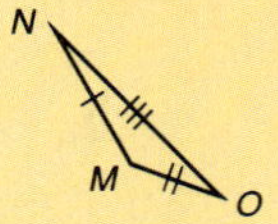

THINK: no sides congruent

So, $\triangle MNO$ is a scalene triangle.

Triangles may also be classified by their angles.

Equiangular triangle	Acute triangle	Obtuse triangle	Right triangle
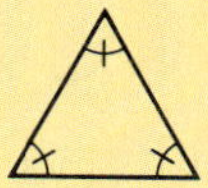	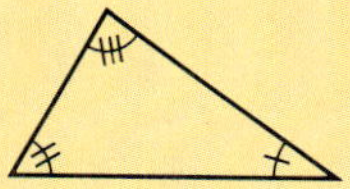	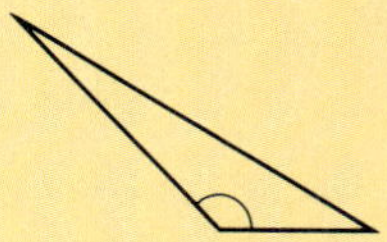	
all congruent angles	all acute angles	1 obtuse angle	1 right angle

EXAMPLE 2 Identify the triangle as *equiangular*, *acute*, *obtuse*, or *right*.

a.

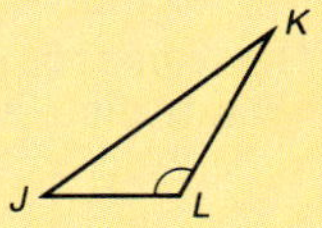

THINK: 1 obtuse angle

So, $\triangle JKL$ is an obtuse triangle.

b.

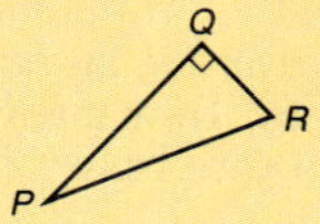

THINK: 1 right angle

So, $\triangle PQR$ is a right triangle.

In all triangles, the sum of the measures of the three angles is 180°. If you know the measure of two angles of a triangle, you can find the measure of the third angle.

EXAMPLE 3 In $\triangle ABC$, $m\angle A = 65°$ and $m\angle B = 35°$. Find $m\angle C$.

1. Add the measures of the two known angles. $65° + 35° = 100°$
2. Subtract the sum from 180°. $180° - 100° = 80°$

So, $m\angle C = 80°$.

PRACTICE EXERCISES

Identify the triangle as *equilateral*, *isosceles*, or *scalene*.

• **1.**
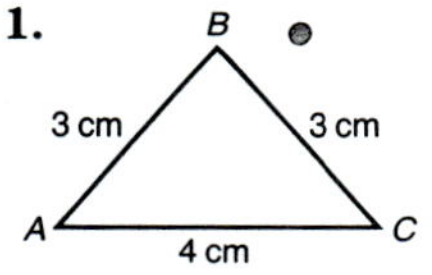

isosceles

• **2.**
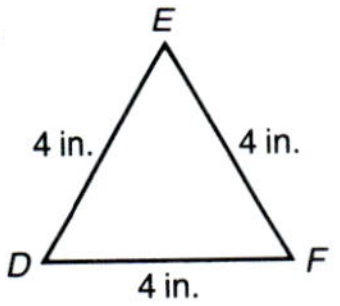

equilateral

3.
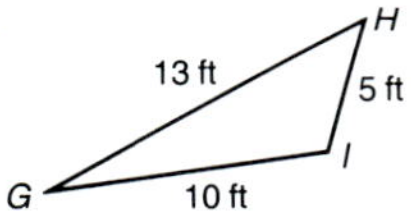

scalene

4.
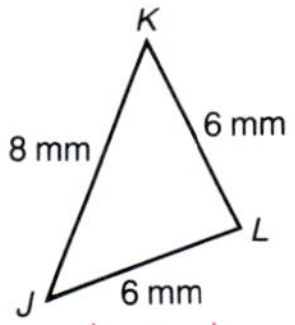

isosceles

5.
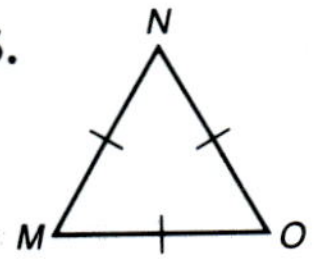

equilateral

6.
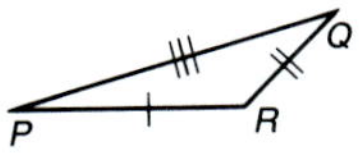

scalene

7.
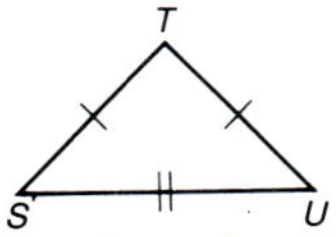

isosceles

8.
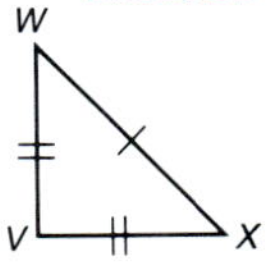

isosceles

Identify the triangle as equiangular, acute, obtuse, or right.

• **9.**
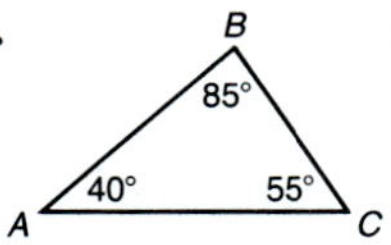

acute

• **10.**
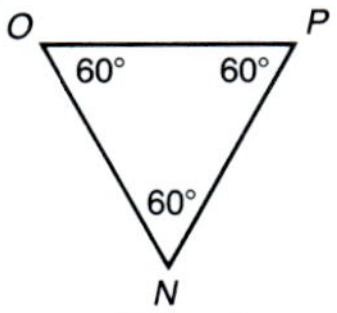

equiangular

11.
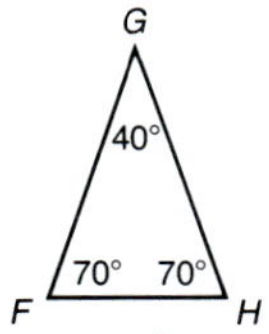

acute

12.
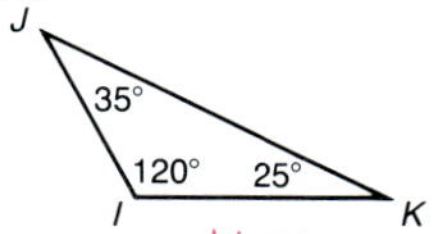

obtuse

13.
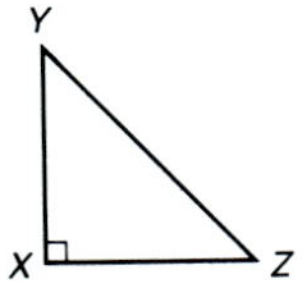

right

14.
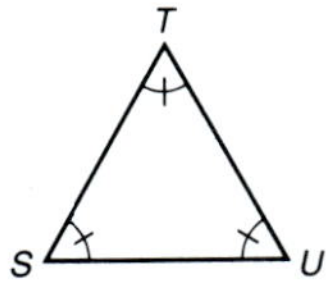

equiangular

15.
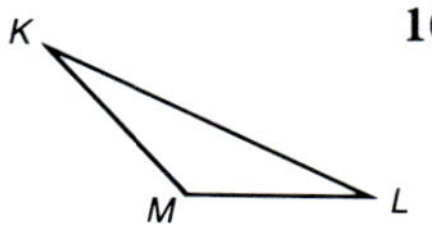

obtuse

16.
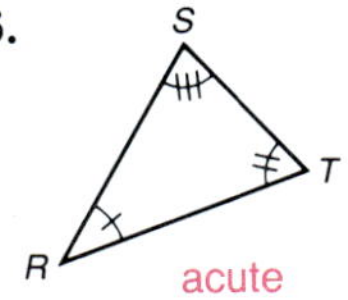

acute

Draw the triangle. Check students' drawings.

• **17.** Equilateral triangle
• **18.** Acute isosceles triangle
19. Obtuse isosceles triangle
20. Right isosceles triangle
21. Obtuse scalene triangle
22. Acute scalene triangle

For example:

- All equilateral triangles are equiangular.
- In an isosceles triangle, the angles opposite the congruent sides are congruent.
- In a scalene triangle, none of the angles are congruent.

Extend the instruction by asking:

- Are all equilateral triangles isosceles? (yes)
- Are all isosceles triangles equilateral? (no)
- Are all equiangular triangles equilateral? (yes)

In Example 3, stress that in every triangle, the sum of the measures of the three angles is always 180°. Thus, if the measures of two angles of a triangle are known, the measure of the third can always be found.

TIME OUT Guide students through the activity on page 308. Encourage students to experiment with a right triangle first. Have them extend one of the perpendicular sides, tear off the two acute angles, and place them along the extended side. Point out that together all three angles form a straight angle. Thus, the sum of the angle measures must be 180°. Then have the students repeat the activity with an acute and an obtuse triangle.

COMMON ERROR

Students may have difficulty recognizing right triangles in nonstandard positions. Have students draw right triangles in the more familiar positions and then rotate their papers to observe the changing appearance of the triangles.

ASSIGNMENTS

Level 1	Odd 1–43
Level 2	Even 2–46, TO
Level 3	5–8, 13–16, Odd 17–45, 46, TO

FOLLOW-UP ACTIVITY

CALCULATOR Write:

62° + 47° + ▩° = 180°

Encourage students to experiment with the memory keys on the calculator to find the measure of the third angle above. If needed, provide the following hint:

180 − (62 + 47) = ▩ (71°)

Have students demonstrate their methods of finding the answer. Continue with these exercises:

1. 117° + 48° + ▩° = 180° (15°)
2. ▩° + 93° + 25° = 180° (62°)
3. 79° + ▩° + 42° = 180° (59°)

SUPPLEMENTARY MATERIALS

TRP Practice, p. 123

TRP Reteaching, p. 75

TRP Lesson Aids, p. 8

TRP Transparency 62

The measures of two angles of a triangle are given. Find the measure of the third angle.

•**23.**

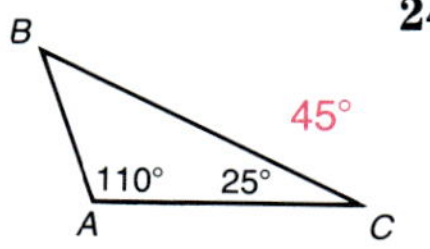

45°

24.

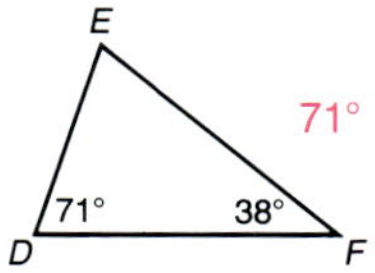

71°

25.

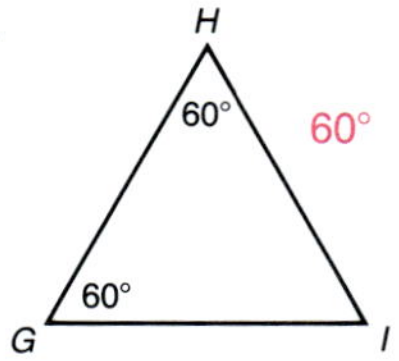

60°

26.

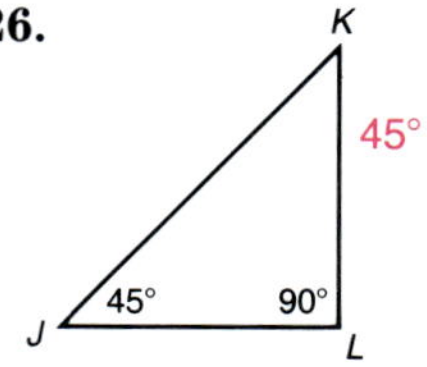

45°

•**27.** 40°, 75° 65° •**28.** 120, 45° 15° **29.** 45°, 45° 90° **30.** 110°, 23° 47°

31. 81°, 92° 7° **32.** 100°, 13° 67° **33.** 51°, 113° 16° **34.** 90°, 10° 80°

MIXED REVIEW

Find the number.

•**35.** 28% of 15 4.2 •**36.** 15% of 50 7.5 **37.** 32% of 90 28.8 **38.** 68% of 64 43.52

39. 8 is what percent of 32? 25%

40. 44 is what percent of 80? 55%

41. 25 is 12.5% of what number? 200

42. 9.9 is 10% of what number? 99

Solve.

•**43.** One of the congruent angles of an isosceles triangle is 35°. Find the measures of the other two angles. 35°; 110°

44. One angle of an isosceles triangle is 120°. Find the measures of the other two congruent angles. 30°; 30°

45. The radius of a circle is 3.2 cm. What is the length of the diameter? 6.4 cm

46. How many angles and sides does a regular pentagon have? 5 angles; 5 sides

TIME OUT

The following activity can be performed to find the sum of the measures of the triangle.

Draw any triangle. Cut out the triangle, and tear off the three angles.

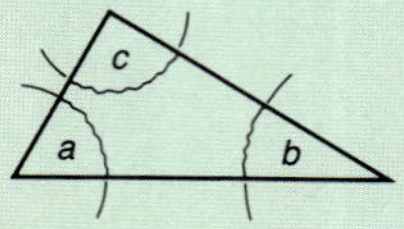

Put the angles together to form a straight angle as in the figure. Remember, a straight angle measures 180°.

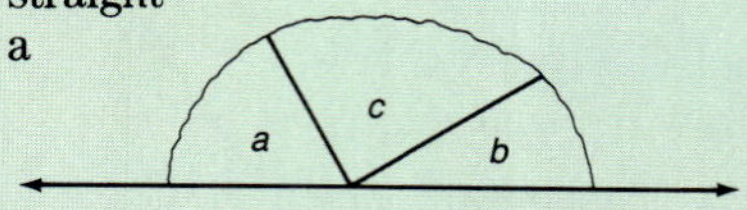

So, the sum of the measures of the angles of a triangle is 180°.

MIDCHAPTER REVIEW

Name the geometric figure. [292]

1. $\angle T$ angle T **2.** $\overline{BC}$ line segment BC **3.** $\overrightarrow{AB}$ ray AB **4.** $.P$ point P **5.** $\overleftrightarrow{EF}$ line EF

Use the figure at the right for Exercises 6–19.

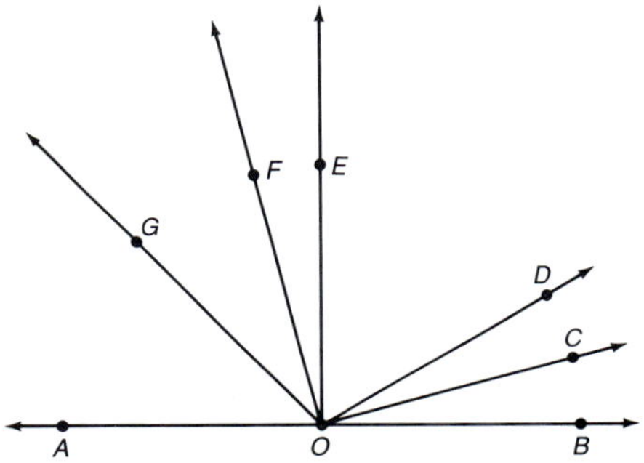

Measure the angle. [294]

6. $\angle AOC$ 165° **7.** $\angle AOE$ 90° **8.** $\angle AOG$ 45°

9. $\angle GOF$ 30° **10.** $\angle GOC$ 120° **11.** $\angle GOB$ 135°

Identify the angle as *acute, right, obtuse,* or *straight.* [296]

12. $\angle AOC$ obtuse **13.** $\angle GOE$ acute **14.** $\angle AOD$ obtuse

15. $\angle AOE$ right **16.** $\angle BOA$ straight **17.** $\angle GOC$ obtuse

Tell whether the pair of angles are *complementary, supplementary,* or *neither.* [296]

18. $\angle AOB$ and $\angle BOD$ neither

19. $\angle AOC$ and $\angle COB$ supplementary

Trace the line segment.
Check students' constructions.

20. Construct a line segment congruent to $\overline{AB}$. [298]

21. Construct a line parallel to $\overline{AB}$. [301]

Identify the triangle as *equilateral, isosceles,* or *scalene.* [306]

22.

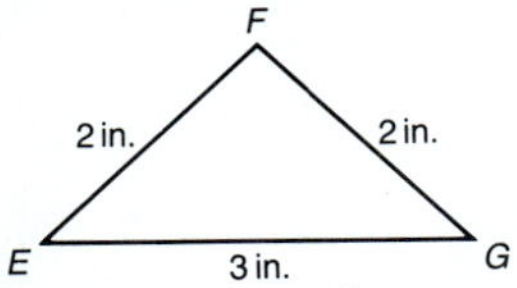

isosceles

23.

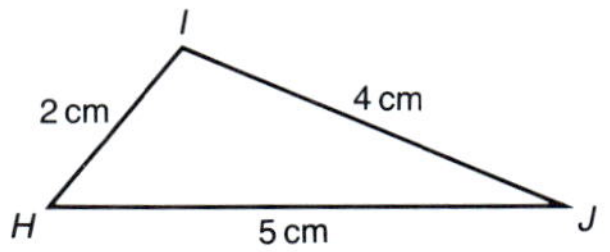

scalene

24.

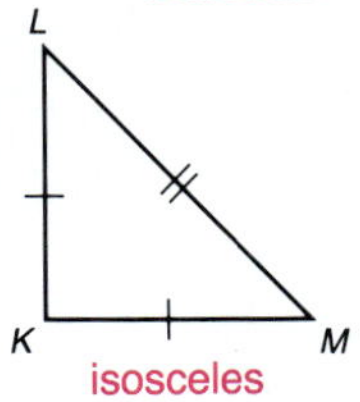

isosceles

25.

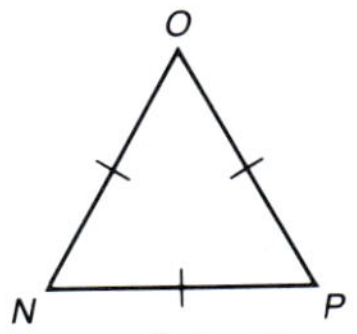

equilateral

OBJECTIVE

- Evaluate student progress.

USING THE REVIEW

This page provides a means for informally evaluating students' understanding of the skills and concepts covered so far in this chapter.

Have the students look at the page to familiarize themselves with the various question formats that are presented. Discuss any questions that they may have. Then ask them to complete the page independently.

In addition to grading them individually, you may wish to review the answers to the questions collectively with the students.

Page references appear in brackets. They refer to pages on which a particular skill was introduced.

Before continuing on to the topics found in the remainder of the chapter, you may wish to have students review any skills or concepts in which they have demonstrated weakness.

OBJECTIVE

- Use patterns to solve word problems.

TEACHING THE LESSON

WARM-UP Write the following exercises on the chalkboard or an overhead transparency.

Complete the pattern.

1. 1, 4, 7, 10, ■, ■ (13, 16)

2. 50, 40, 30, ■, ■ (20, 10)

3. 5, 15, 45, ■, ■ (135, 405)

4. 1, 3, 7, ■, ■ (15, 31)

5. 4, 13, 40, ■, ■ (121, 364)

INTRODUCTION Point out that patterns exist among some objects that are so familiar to us that we are unaware of the patterns. In a group discussion, decide whether patterns exist in the placement of the following objects in your town.

- Bus stops
- Traffic lights
- Speed-limit signs
- Mailboxes
- Entrances to expressways
- Trash containers

Discuss the advantages of such patterns.

INSTRUCTION In discussing the problem-solving situation, point out that to find a pattern, the first step is to discover the relationship between each number in the pattern. Emphasize that drawing a diagram may help to discover the pattern. Make sure that students check the rule for all numbers in the pattern.

PROBLEM Solving STRATEGY

11.8 LOOKING FOR A PATTERN

Situation:

Fred and Lynne Shaw own a landscape company. They have planned a border of shrubs. The first shrub is planted 4 ft from the road. The second shrub is 9 ft from the road. The third shrub is 16 ft from the road. The next plant is 20 ft from the road, followed by another plant 25 ft from the road. How many feet from the road are the next two shrubs?

Strategy:

Finding a pattern can often help you to solve a problem.

Applying the Strategy:

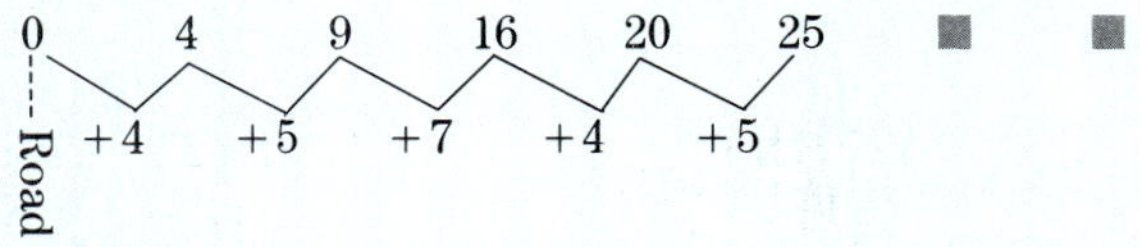

By adding, you can see a pattern: +4, +5, +7, +4, +5.
The pattern repeats. The next shrub is +7 ft away from the last plant.

So, the next shrub is 25 + 7 = 32 ft from the road.
The following shrub is 32 + 4 = 36 ft from the road.

PRACTICE EXERCISES

Alex Cook has planned a display around a fountain. The first ring has 16 flowers. The second ring has 32 flowers. The third ring has 64 flowers. How many flowers should be planted in the next two rings?

• **1.** What is the pattern? multiply by 2

• **2.** Complete the pattern: 16, 32, 64, ■, ■ 128, 256

Solve the problem by finding the pattern.

3. Jane doodles as she plans her display. One day this was the sketch on her drawing pad.

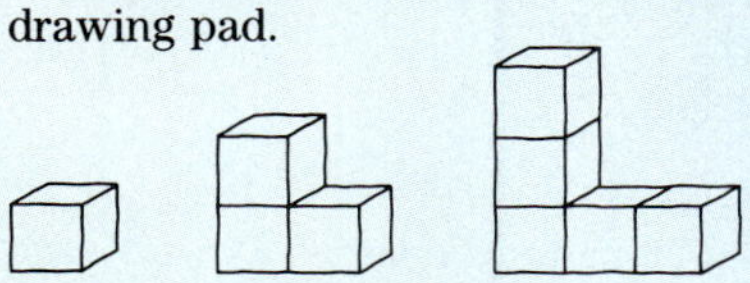

How many blocks should there be in the next picture to continue the pattern? Draw a possible configuration. 7; Check students' drawings.

4. One day Fred found this drawing on Jane's drawing pad.

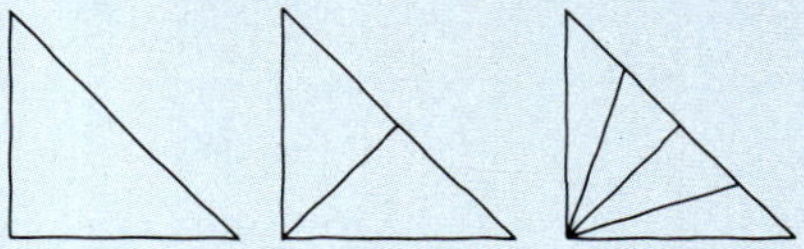

How many small triangles should be drawn in the next triangle to continue the pattern? 8

5. Marie was planting bushes. The first bush was 12 ft from the street. The second bush was twice the distance from the street. The third bush was 48 ft away from the street. How far from the street was the fourth bush? 96 ft

96 ft

6. James was placing rocks in a rock garden. The distances between the rocks were: 3 ft, 6 ft, 16 ft, 19 ft, 29 ft. What is the distance between the fifth and sixth rock? The sixth and seventh rock? 32 ft; 42 ft

ASSIGNMENTS

Level 1	1–4
Level 2	1–6
Level 3	3–6

FOLLOW-UP ACTIVITY

SITUATIONAL PROBLEM SOLVING Have students design a small park for the center of a town. Have the students discuss and agree on the overall size of the park. Then have them work individually or in small groups to plan the park. They should use at least three patterns in their plans.

SUPPLEMENTARY MATERIALS

TRP Practice, p. 124

TRP Applications, p. 18

TRP Transparency 62

OBJECTIVES

- Find the square root of a number that is a perfect square.
- Use a table of squares and square roots to find the square root of a number.

TEACHING THE LESSON

WARM-UP Write the following exercises on the chalkboard or an overhead transparency.

Multiply mentally.

1. 8×8 **2.** 10×10

3. 30×30 **4.** 50×50

5. 70×70 **6.** 100×100

(**1.** 64 **2.** 100 **3.** 900 **4.** 2,500 **5.** 4,900 **6.** 10,000)

INTRODUCTION Extend the activity in the Warm-Up by having students name the number that will make each of the following sentences true.

$\blacksquare \times \blacksquare = 25$ (5)

$\blacksquare \times \blacksquare = 81$ (9)

$\blacksquare \times \blacksquare = 144$ (12)

$\blacksquare \times \blacksquare = 400$ (20)

$\blacksquare \times \blacksquare = 1{,}600$ (40)

INSTRUCTION In discussing Example 1, stress the importance of using basic multiplication and division facts in finding whole-number square roots.

In Example 2, remind students that the table reads from left to right and is grouped in sets of three columns: *N*, N^2, and $\sqrt{N}$. Ask students to use the table to find the square root of 51. (7.141)

In discussing Example 3, ask students to name the greatest value listed in the column labeled *N*. (100) Elicit from them that this limit is the reason for using the N^2 column to find the square roots of some greater numbers. Extend the instruction by having students find the square roots of 289 (17) and 8,649 (93).

11.9 SQUARE ROOT

For the graduation ceremony, 400 chairs are arranged so that the number of chairs in each row is the same as the number of rows. How many chairs are in each row and how many rows are there?

To solve this problem you need to find the number that when multiplied by itself equals 400. That number is the **square root** of 400. The symbol $\sqrt{\ }$ means square root.

EXAMPLE 1 Find the square root of 400.

THINK: $2 \times 2 = 4$, so $20 \times 20 = 400$.

$\sqrt{400} = 20$

So, there are 20 chairs per row and 20 rows of chairs.

Sometimes the square root is not a whole number. You may need to use a table of squares and square roots like the one on page 313.

EXAMPLE 2 Find $\sqrt{15}$.

Read the table on page 313.

$\sqrt{15}$ is approximately equal to 3.873. $3.873 \approx 3.9$

You can use the column marked n^2 to find some square roots.

EXAMPLE 3 Find $\sqrt{529}$.

Read the table on page 313.

Look for 529 in the n^2 column and then find n.

THINK: $n^2 = 529$

$n = \sqrt{529}$

$n = 23$

So, $\sqrt{529} = 23$.

TABLE OF SQUARES AND SQUARE ROOTS

n	n^2	$\sqrt{n}$	n	n^2	$\sqrt{n}$	n	n^2	$\sqrt{n}$	n	n^2	$\sqrt{n}$
1	1	1	26	676	5.099	51	2,601	7.141	76	5,776	8.718
2	4	1.414	27	729	5.196	52	2,704	7.211	77	5,929	8.775
3	9	1.732	28	784	5.292	53	2,809	7.280	78	6,084	8.832
4	16	2	29	841	5.385	54	2,916	7.348	79	6,241	8.888
5	25	2.236	30	900	5.477	55	3,025	7.416	80	6,400	8.944
6	36	2.449	31	961	5.568	56	3,136	7.483	81	6,561	9
7	49	2.646	32	1,024	5.657	57	3,249	7.550	82	6,724	9.055
8	64	2.828	33	1,089	5.745	58	3,364	7.616	83	6,889	9.110
9	81	3	34	1,156	5.831	59	3,481	7.681	84	7,056	9.165
10	100	3.162	35	1,225	5.916	60	3,600	7.746	85	7,225	9.220
11	121	3.317	36	1,296	6	61	3,721	7.810	86	7,396	9.274
12	144	3.464	37	1,369	6.083	62	3,844	7.874	87	7,569	9.327
13	169	3.606	38	1,444	6.164	63	3,969	7.937	88	7,744	9.381
14	196	3.742	39	1,521	6.245	64	4,096	8	89	7,921	9.434
15	225	3.873	40	1,600	6.325	65	4,225	8.062	90	8,100	9.487
16	256	4	41	1,681	6.403	66	4,356	8.124	91	8,281	9.539
17	289	4.123	42	1,764	6.481	67	4,489	8.185	92	8,464	9.592
18	324	4.243	43	1,849	6.557	68	4,624	8.246	93	8,649	9.644
19	361	4.359	44	1,936	6.633	69	4,761	8.307	94	8,836	9.695
20	400	4.472	45	2,025	6.708	70	4,900	8.367	95	9,025	9.747
21	441	4.583	46	2,116	6.782	71	5,041	8.426	96	9,216	9.798
22	484	4.690	47	2,209	6.856	72	5,184	8.485	97	9,409	9.849
23	529	4.796	48	2,304	6.928	73	5,329	8.544	98	9,604	9.899
24	576	4.899	49	2,401	7	74	5,476	8.602	99	9,801	9.950
25	625	5	50	2,500	7.071	75	5,625	8.660	100	10,000	10

PRACTICE EXERCISES See Extra Practice, page 452.

Find the square root.

• **1.** $\sqrt{4}$ 2 • **2.** $\sqrt{9}$ 3 **3.** $\sqrt{25}$ 5 **4.** $\sqrt{64}$ 8 **5.** $\sqrt{81}$ 9 **6.** $\sqrt{100}$ 10

7. $\sqrt{10}$ 3.162 **8.** $\sqrt{37}$ 6.083 **9.** $\sqrt{58}$ 7.616 **10.** $\sqrt{72}$ 8.485 **11.** $\sqrt{87}$ 9.327 **12.** $\sqrt{95}$ 9.747

Find the square root. Round to the nearest tenth.

• **13.** $\sqrt{15}$ 3.9 • **14.** $\sqrt{18}$ 4.2 **15.** $\sqrt{29}$ 5.4 **16.** $\sqrt{35}$ 5.9 **17.** $\sqrt{37}$ 6.1 **18.** $\sqrt{45}$ 6.7

19. $\sqrt{50}$ 7.1 **20.** $\sqrt{55}$ 7.4 **21.** $\sqrt{63}$ 7.9 **22.** $\sqrt{78}$ 8.8 **23.** $\sqrt{89}$ 9.4 **24.** $\sqrt{94}$ 9.7

Find the square root. Use the table.

• **25.** $\sqrt{196}$ 14 • **26.** $\sqrt{576}$ 24 **27.** $\sqrt{1,225}$ 35 **28.** $\sqrt{1,521}$ 39 **29.** $\sqrt{1,681}$ 41

30. $\sqrt{2,304}$ 48 **31.** $\sqrt{3,249}$ 57 **32.** $\sqrt{4,225}$ 65 **33.** $\sqrt{6,241}$ 79 **34.** $\sqrt{9,216}$ 96

ASSIGNMENTS

Level 1	1–6, Odd 7–33
Level 2	1–34
Level 3	7–34

FOLLOW-UP ACTIVITY

CALCULATOR Have students use calculators to find the square roots of the following numbers without using the square-root key.

2,500 (50)

4,096 (64)

10,000 (100)

14,641 (121)

1,440,000 (1,200)

SUPPLEMENTARY MATERIALS

TRP Practice, p. 125

TRP Reteaching, p. 76

TRP Transparency 63

OBJECTIVES

- Determine if a triangle is a right triangle, given the lengths of its sides.
- Find the measure of the length of the hypotenuse or the length of a leg of a right triangle.

TEACHING THE LESSON

WARM-UP Write the following exercises on the chalkboard or an overhead transparency.

Find the value of n.

1. $n^2 = 25$ (5) **4.** $n^2 = 100$ (10)
2. $n^2 = 49$ (7) **5.** $n^2 = 225$ (15)
3. $n^2 = 64$ (8) **6.** $n^2 = 400$ (20)

INTRODUCTION Write:

Find z.

$$x^2 + y^2 = z^2$$

$$x = 6, \quad y = 8 \quad (z = 10)$$

Work together to discover the steps needed to find z. Encourage students to apply their knowledge of finding the square and the square root of a number. Repeat for $x = 5$, $y = 12$.

INSTRUCTION In discussing Example 1, emphasize that since the sides of the triangle did not satisfy the equation $c^2 = a^2 + b^2$, it was not a right triangle. Provide students with another example such as:

$$a = 9, b = 12, \text{ and } c = 15$$

Ask them to determine if this is a right triangle. They should note that because $225 = 81 + 144$, the triangle is a right triangle.

Note the alternate forms of the Pythagorean theorem used in Examples 2a and 2b. You may wish to provide students with other alternate forms of the Pythagorean theorem such as:

$$a^2 = c^2 - b^2 \text{ and } b^2 = c^2 - a^2$$

Encourage them to use whichever form is the easiest for a particular problem.

11.10 THE PYTHAGOREAN THEOREM

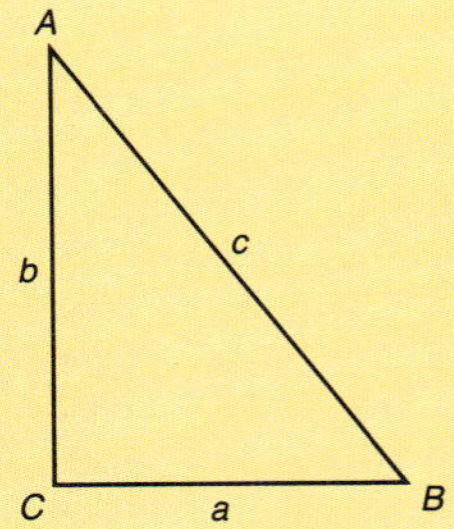

Triangle ABC is a right triangle.

$\angle C$ is the right angle.

The side opposite the right angle is called the **hypotenuse.**

In this triangle, side c is the hypotenuse. Sides a and b are **legs.**

There is a relationship between the hypotenuse and the legs of a right triangle.

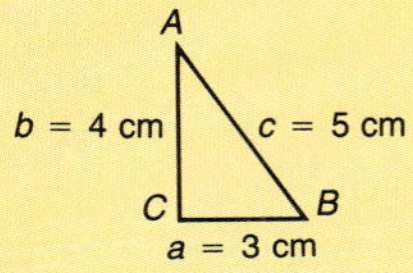

Square the length of each side.

$$a^2 = 3 \times 3 = 9$$
$$b^2 = 4 \times 4 = 16$$
$$c^2 = 5 \times 5 = 25$$

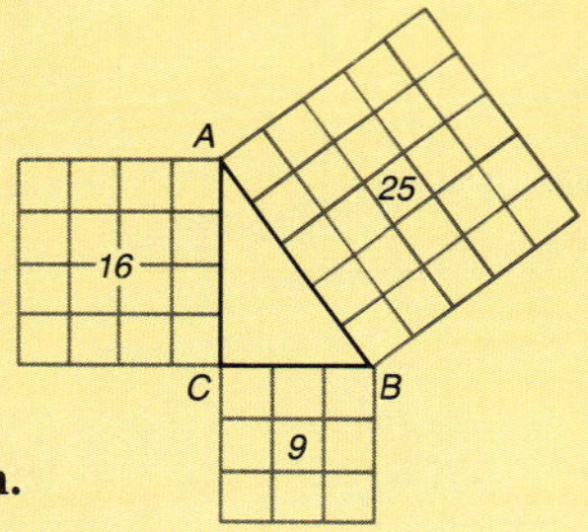

Notice that the square of the length of the hypotenuse is equal to the sum of the squares of the lengths of the legs.

$$\begin{array}{ccccc} 25 & = & 9 & + & 16 \\ \downarrow & & \downarrow & & \downarrow \\ c^2 & = & a^2 & + & b^2 \end{array}$$

This relationship is called the **Pythagorean theorem.** It is true for every right triangle. For right triangle ABC,

$$c^2 = a^2 + b^2 \text{ or } a^2 + b^2 = c^2.$$

Use the Pythagorean theorem to determine if a triangle is a right triangle.

EXAMPLE 1 In $\triangle ABC$, $a = 10$, $b = 12$, and $c = 15$. Is $\triangle ABC$ a right triangle?

THINK: $c^2 = a^2 + b^2$

$$15^2 \stackrel{?}{=} 10^2 + 12^2$$
$$225 \stackrel{?}{=} 100 + 144$$
$$225 \neq 244$$

The triangle is not a right triangle.

You can also use the Pythagorean theorem to find the length of the hypotenuse or a leg of a right triangle.

EXAMPLE 2 $\triangle ABC$ is a right triangle. Find the missing length.

a. $a = 6, b = 8, c = ■$

THINK: $c^2 = a^2 + b^2$

$c^2 = 6^2 + 8^2$

$c^2 = 36 + 64$

$c^2 = 100$

$c = \sqrt{100}$, or 10

So, the hypotenuse c is equal to 10.

b. $c = 13, b = 12, a = ■$

THINK: $a^2 + b^2 = c^2$

$a^2 + 12^2 = 13^2$

$a^2 + 144 = 169$

$a^2 = 169 - 144$

$a^2 = 25$

$a = \sqrt{25}$, or 5

So, the leg a is equal to 5.

PRACTICE EXERCISES

See Extra Practice, page 452.

Do sides of the following lengths form a right triangle? Write *yes* or *no*.

• **1.** $a = 6$, $b = 8$, $c = 10$ yes

• **2.** $a = 0$, $b = 12$, $c = 15$ no

3. $a = 24$, $b = 15$, $c = 17$ no

4. $a = 5$, $b = 12$, $c = 13$ yes

5. $a = 7$, $b = 10$, $c = 12$ no

Find the missing length.

• **6.**

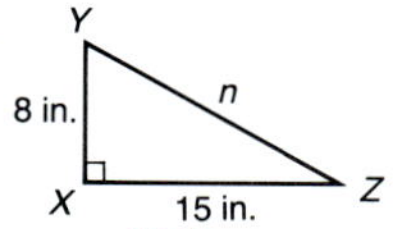

17 in.

7.

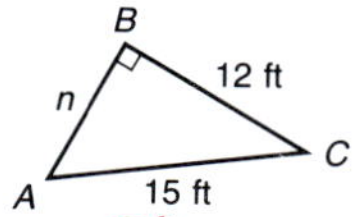

9 ft

8.

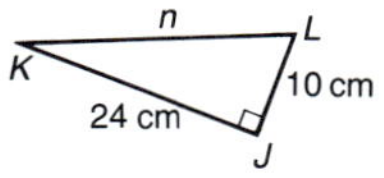

26 cm

9.

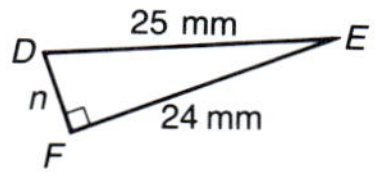

7 mm

• **10.** $a = 15$, $b = 20$, $c = ■$ 25

11. $a = 9$, $b = ■$ 40, $c = 41$

12. $a = ■$ 20, $b = 48$, $c = 52$

13. $a = 28$, $b = 45$, $c = ■$ 53

14. $a = 16$, $b = ■$ 30, $c = 34$

MIXED REVIEW

Multiply.

• **15.** $\frac{1}{2} \times 4$ 2

• **16.** $\frac{1}{4} \times 16$ 4

17. $4\frac{1}{2} \times \frac{1}{3}$ $1\frac{1}{2}$

18. $3\frac{5}{6} \times \frac{2}{5}$ $1\frac{8}{15}$

19. $\frac{1}{2} \times 4\frac{3}{8}$ $2\frac{3}{16}$

20. $\frac{1}{15} \times \frac{30}{48}$ $\frac{1}{24}$

21. $\frac{1}{8} \times 6\frac{2}{3}$ $\frac{5}{6}$

22. $\frac{1}{7} \times 5\frac{4}{9}$ $\frac{7}{9}$

Solve.

23. The foot of a ladder is 10 ft from the base of a building. The top of the ladder is resting against the building, 24 ft above the ground. How long is the ladder? 26 ft

COMMON ERROR

Some students will assume that the Pythagorean theorem works with all triangles. You may need to give these students some additional examples of triangles that are not right triangles and have them verify that the lengths of the sides do not satisfy the conditions of the theorem.

ASSIGNMENTS

Level 1	Odd 1–21
Level 2	Even 2–22
Level 3	Odd 1–23

FOLLOW-UP ACTIVITY

ENRICHMENT A *Pythagorean triple* is a set of numbers that satisfies the equation $c^2 = a^2 + b^2$.

Show how to generate Pythagorean triples. Pick any pair of nonzero whole numbers m and n such that $m > n$, and substitute them into the following equations.

For example, for $m = 2$ and $n = 1$:

$a = m^2 - n^2$ $\quad b = 2mn$ $\quad c = m^2 + n^2$

$a = 2^2 - 1^2$ $\quad b = 2 \times 2 \times 1$ $\quad c = 2^2 + 1^2$

$a = 4 - 1$ $\quad c = 4 + 1$

$a = 3$ $\quad b = 4$ $\quad c = 5$

Have students generate Pythagorean triples for the following values of m and n.

1. $m = 3, n = 1$ (8, 6, 10)

2. $m = 3, n = 2$ (5, 12, 13)

3. $m = 4, n = 1$ (15, 8, 17)

4. $m = 4, n = 2$ (12, 16, 20)

5. $m = 5, n = 1$ (24, 10, 26)

6. $m = 5, n = 3$ (16, 30, 34)

SUPPLEMENTARY MATERIALS

TRP Practice, p. 126

TRP Reteaching, p. 77

TRP Transparency 63

OBJECTIVE

- Solve problems that involve right triangles and indirect measurement.

TEACHING THE LESSON

WARM-UP Write the following exercises on the chalkboard or an overhead transparency.

Find the product.

1. $\frac{1}{3} \times 420$ (140) 2. $\frac{1}{2} \times 310$ (155)
3. $\frac{3}{5} \times 450$ (270) 4. $\frac{2}{5} \times 80$ (32)

INSTRUCTION In discussing the fact about right triangles with a 30° angle, it will be helpful to make a large sketch of the triangle on the chalkboard. Then you can identify the side opposite the 90° angle and the side opposite the 30° angle. (The 60° angle and the side opposite that angle might also be identified.)

Similarly, a large sketch of a right triangle with a 37° angle would also be useful in discussing the information about that type of right triangle.

In the 37°–53°–90° triangle, the side opposite the 37° angle is 0.6018 of the length of the side opposite the 90° angle. This decimal is correct to the nearest ten thousandth. For computational convenience, we are using the fraction $\frac{3}{5}$, which is a very close approximation of 0.6018.

You may wish to have students use calculators to help them solve the problems.

ASSIGNMENTS

Level 1	1–4
Level 2	1–6
Level 3	4–8

SUPPLEMENTARY MATERIALS

TRP Practice, p. 127

TRP Transparency 64

PROBLEM Solving APPLICATION

11.11 CAREER: SURVEYOR

Janet Soldana is a surveyor. By measuring angles and certain distances that can be conveniently measured, she is able to calculate distances that cannot be directly measured.

Janet uses an instrument called a **transit** to measure angles. A transit is a telescope set up on a three-legged stand. The transit also includes two protractors—one for measuring angles in a horizontal plane and the other for measuring angles in a vertical plane.

There are two facts about right triangles that help Janet perform her job:

- In any right triangle with a 30° angle, the side opposite that angle is $\frac{1}{2}$ of the length of the side opposite the 90° angle.
- In any right triangle with a 37° angle, the side opposite that angle is about $\frac{3}{5}$ of the length of the side opposite the 90° angle.

Find n. Remember to estimate whenever you use your calculator.

• 1.
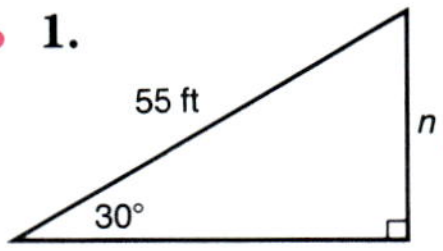

27.5 ft

• 2.
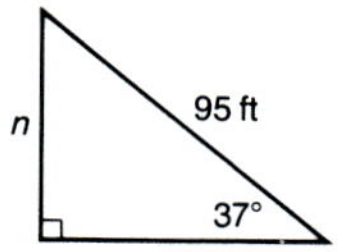

57 ft

3.
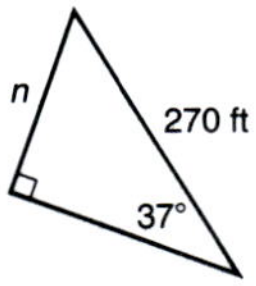

162 ft

4.
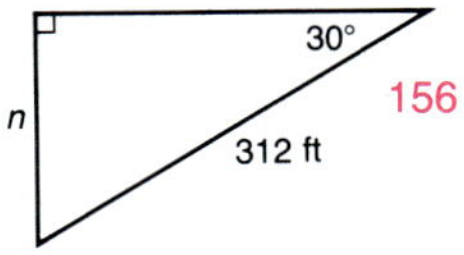

156 ft

5.
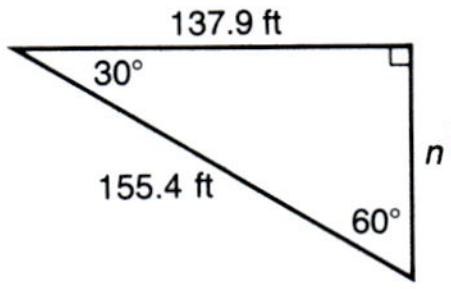

77.7 ft

6.
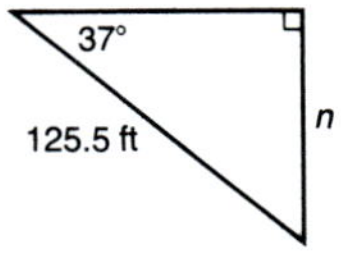

75.3 ft

7.
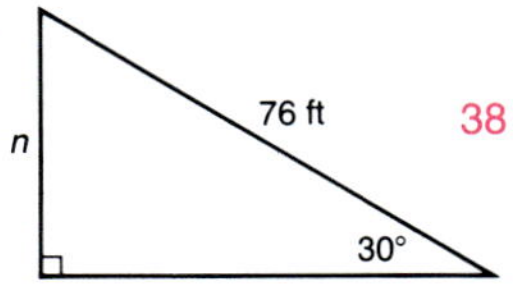

38 ft

8.
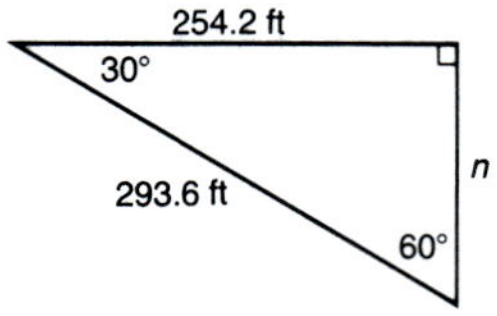

146.8 ft

PROBLEM Solving APPLICATION

11.12 CONSTRUCTING CIRCLE GRAPHS

You plan to discuss this year's budget at the next meeting of the Community Center Youth Group. You decide to display the data in a circle graph.

YOUTH GROUP BUDGET

Newsletter	15%	Special Activities	55%
Supplies	20%	Other Expenses	10%

To make a circle graph,

1. Find the measure of each central angle.

THINK: There are 360° in a circle.

Newsletter: 15% of 360° = 54°
Supplies: 20% of 360° = 72°
Special Activities: 55% of 360° = 198°
Other Expenses: 10% of 360° = 36°
Total: = 360°

2. Draw a circle. Then draw central angles with the measures you found in Step 1.

3. Label each part of the graph. Remember to title the graph.

YOUTH GROUP BUDGET

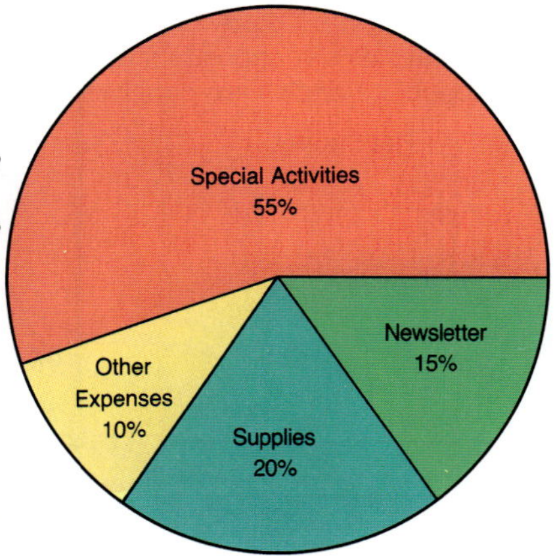

Use the circle graph to answer Exercises 1 and 2. If the total Youth Group budget is $575, how much will be spent on:

• **1.** the newsletter? $86.25

• **2.** special activities? $316.25

3. Your own personal budget includes the following categories and percents: school supplies, 25%; snacks, 5%; recreation, 60%; savings, 10%. Make a circle graph for this data. If necessary, round angle measures to the nearest degree.
Supplies—90°; snacks—18°; recreation—216°; savings—36°. Check students' graphs.

4. If your yearly budget is $800, how much less per month is spent on snacks than on school supplies? Use the information in Exercise 3.
$13.33

5. Lucy decides to make a yearly budget for $1,050. She decides to spend $\frac{1}{5}$ on clothes, $\frac{1}{4}$ on food, $\frac{3}{10}$ on sports events, $\frac{1}{20}$ on movies, and $\frac{1}{5}$ on savings. Determine the percents and make a circle graph for this data.
Clothes—20%, 72°; food—25%, 90°; sports events—30%, 108°; movies—5%, 18°; savings—20%, 72°. Check students' graphs.

OBJECTIVES

- Construct a circle graph.
- Solve problems that involve a circle graph.

TEACHING THE LESSON

WARM-UP Write the following exercises on the chalkboard or an overhead transparency.

Draw angles with the following measures.

1. 25° **2.** 65° **3.** 105°

4. 80° **5.** 200° **6.** 45°

(Check students' drawings.)

INTRODUCTION Remind the students that circle graphs are used to show parts of a whole. Discuss the advantages of using a circle graph to display data.

INSTRUCTION The lesson guides the students through the steps for making a circle graph.

Problems 1, 2, and 4 require students to interpret data from completed circle graphs. Problems 3 and 5 require students to construct circle graphs.

Students may use calculators when appropriate.

ASSIGNMENTS

Level 1	1–3
Level 2	1–5
Level 3	3–5

SUPPLEMENTARY MATERIALS

TRP Practice, p. 128

TRP Lesson Aids, p. 7

TRP Transparency 64

OBJECTIVE

- Classify a quadrilateral as a parallelogram or trapezoid.
- Classify a parallelogram as a rectangle, rhombus, or square.

TEACHING THE LESSON

WARM-UP Draw pairs of lines such as the following on the chalkboard or overhead transparency. Have students identify whether they are parallel, perpendicular, or intersecting.

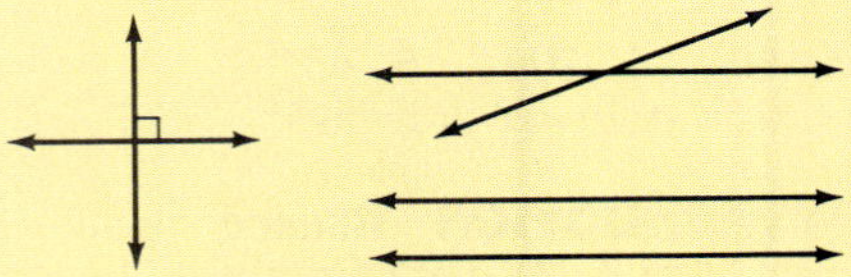

INTRODUCTION Draw:

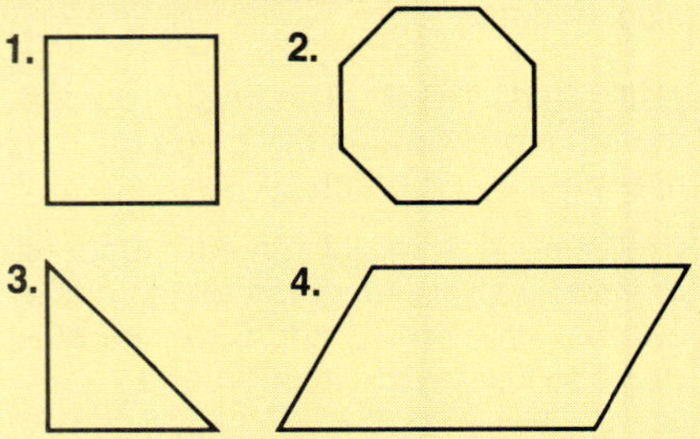

Ask students which of the figures have four sides (a, d). List other names students can think of for these quadrilaterals.

INSTRUCTION Review the symbols used to denote congruent line segments and right angles.

In Examples 1a and 1b, ask which pairs of sides and angles are congruent for each quadrilateral.

(Example 1a, all sides and angles congruent; Example 1b, $\angle J = \angle L$, $\angle K = \angle M$, all sides congruent.)

To learn more about the relationships among various types of parallelograms, have students complete these statements.

11.13 QUADRILATERALS

Quadrilaterals can be classified according to their sides and angles.

Trapezoid

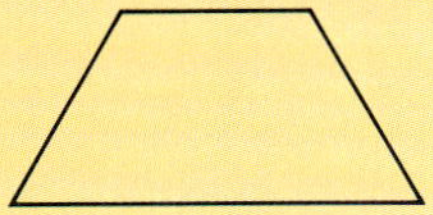

1 pair of parallel sides

Parallelogram

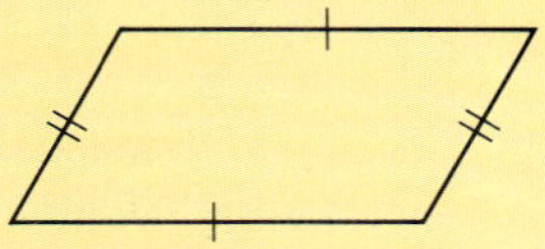

opposite sides parallel,
opposite sides congruent,
opposite angles congruent

There are several types of parallelograms. In each of these quadrilaterals the opposite sides are parallel and congruent. The opposite angles are congruent.

Rectangle

all right angles

Rhombus

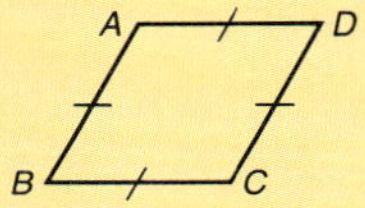

all sides congruent

Square

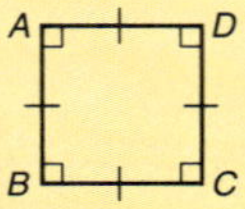

all right angles,
all sides congruent

EXAMPLE

Name the quadrilateral.

a.

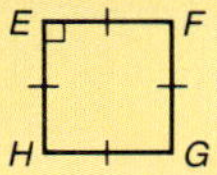

THINK: opposite sides parallel, all sides congruent, all right angles

So, quadrilateral *EFGH* is a square.

b.

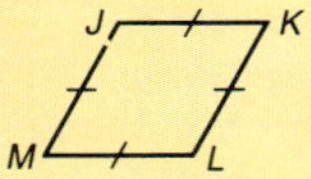

THINK: opposite sides parallel, all sides congruent, opposite angles congruent

So, quadrilateral *JKLM* is a rhombus.

FOR DISCUSSION See TE side column.

In parallelogram *ABCD*, tell how you can determine the measure of $\angle ABC$.

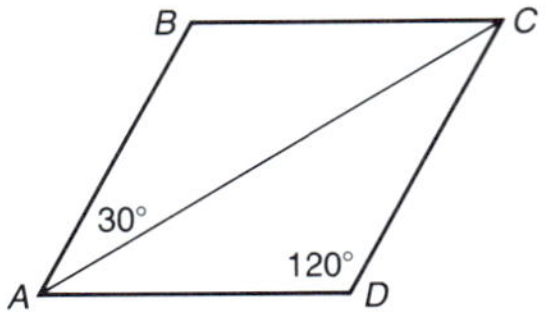

PRACTICE EXERCISES

Name the quadrilateral.

• **1.** 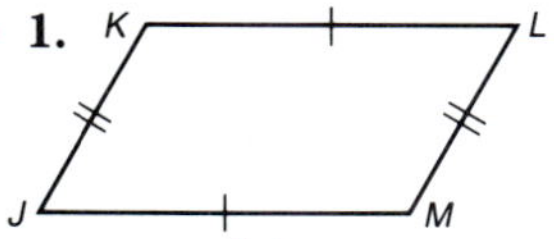

parallelogram

• **2.** 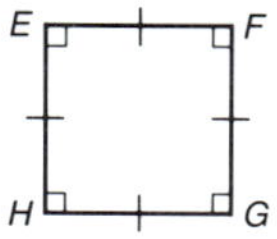

square

3. 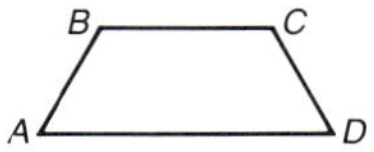

trapezoid

4. 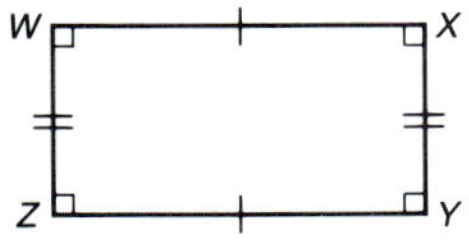

rectangle

5. 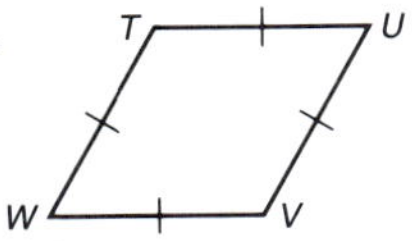

rhombus

6. 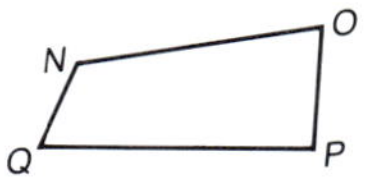

quadrilateral

Tell which quadrilateral(s) have the following properties.

• **7.** Four sides all

• **8.** Only one pair of parallel sides trapezoid

9. Opposite sides parallel

10. All right angles rectangle; square

11. All sides congruent square; rhombus

12. All right angles and all sides congruent square

9. parallelogram; rectangle; square; rhombus

Solve.

13. $\angle A$ and $\angle B$ are opposite angles of a parallelogram. If $m\angle A = 60°$, find $m\angle B$. 60°

14. Find the sum of all the angles of a square. 360°

15. The sum of the measures of adjacent angles in a parallelogram is 180°. In a rhombus *ABCD*, $m\angle A = 60°$. Find the measures of the other three angles.
$m\angle D = 120°$; $m\angle C = 60°$; $m\angle B = 120°$

TIME OUT

This diagram shows the relationship between the different types of quadrilaterals.

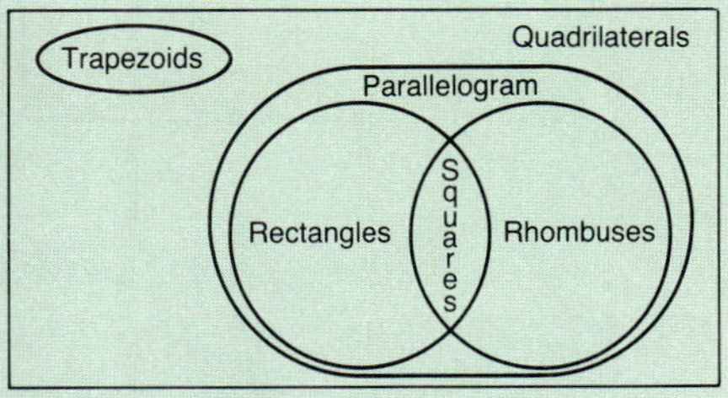

Tell whether the statement is true or false.

1. If a figure is a trapezoid, then it must be a quadrilateral. true

2. If a figure is a parallelogram, then it must be a rectangle. false

3. If a figure is a rhombus, then it must be a square. false

4. If a figure is a rectangle and a rhombus, then it must be a square. true

1. A square is a rhombus with four ▆ angles. (right)

2. A square is a rectangle with four ▆ sides. (congruent)

FOR DISCUSSION Students should explain that because the figure is a parallelogram the opposite angles are congruent. Therefore, $m\angle B = m\angle D = 120°$. Since the sum of the measures of the angles of a triangle equals 180°, $m\angle ACB + 120° + 30° = 180°$. Therefore, $m\angle ACB = 30°$.

TIME OUT Make sure that students know how to interpret a Venn diagram. Have students point to the section labeled Squares and note that all squares are also rectangles, rhombuses, parallelograms, and quadrilaterals. Point out that only some rectangles and some rhombuses are squares.

ASSIGNMENTS

Level 1	1–13
Level 2	1–14, TO
Level 3	4–15, TO

FOLLOW-UP ACTIVITY

COOPERATIVE LEARNING
Have students work in pairs to construct a square using only a compass and a straightedge.

(Draw $\overline{AB}$. Construct perpendiculars at *A* and *B*. Set the compass opening to $\overline{AB}$. Place the point at *A*, and where the arc intersects the perpendicular, mark *D*. Repeat to find *C* on the perpendicular at *B*. Draw $\overline{CD}$ to complete square *ABCD*.)

SUPPLEMENTARY MATERIALS

TRP Practice, p. 129
TRP Reteaching, p. 78
TRP Transparency 65

OBJECTIVES

- Determine if two figures are similar.
- Determine the missing lengths of sides in similar figures.

TEACHING THE LESSON

WARM-UP Write the following exercises on the chalkboard or an overhead transparency.

Find *n*.

1. $\frac{8}{6} = \frac{4}{n}$ (3) **2.** $\frac{4}{7} = \frac{8}{n}$ (14)

3. $\frac{n}{8} = \frac{9}{12}$ (6) **4.** $\frac{n}{14} = \frac{5}{10}$ (7)

5. $\frac{12}{n} = \frac{2}{5}$ (30) **6.** $\frac{10}{15} = \frac{n}{6}$ (4)

INTRODUCTION Make a list of examples from everyday life in which figures or objects have the same shape but different sizes. Possible answers include scale drawings, photographic enlargements and reductions, and a set of canisters.

INSTRUCTION You may wish to copy quadrilaterals *ABCD* and *EFGH* on the chalkboard using different colors of chalk to highlight the corresponding sides.

In discussing Example 1, emphasize that both conditions must be satisfied in order for the two figures to be similar. You may wish to draw pictures of a rhombus and a square on the chalkboard. Note that although their sides are proportional, they are not similar because their corresponding angles are not congruent.

In Example 2, encourage students to begin the proportion with the unknown value (side *RT*), and its corresponding side *(UW)*. Then, any other side with the unknown value can then be chosen with its corresponding side to complete the proportion. Stress that the order in which the ratios are written must be maintained.

11.14 SIMILAR FIGURES

Similar figures have the same shape but not the same size.
Triangles *ABC* and *DEF* are similar figures.
This is written as $\triangle ABC \sim \triangle DEF$.

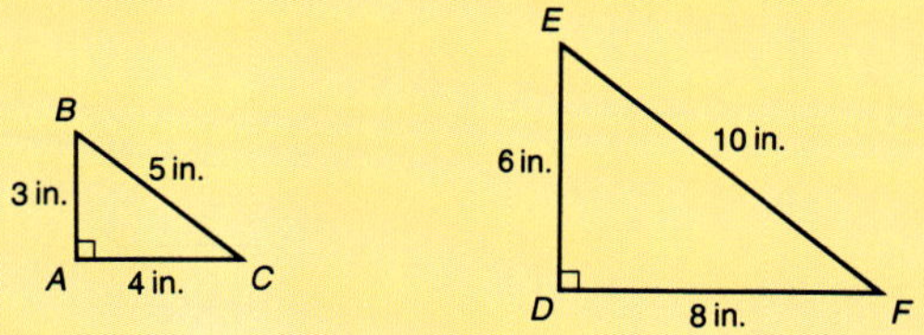

$\angle A \cong \angle D$ $\angle B \cong \angle E$ $\angle C \cong \angle F$

$\frac{\overline{AB}}{\overline{DE}} = \frac{3}{6} = \frac{1}{2}$ $\frac{\overline{BC}}{\overline{EF}} = \frac{5}{10} = \frac{1}{2}$ $\frac{\overline{AC}}{\overline{DF}} = \frac{4}{8} = \frac{1}{2}$

If two figures are similar:

1. The corresponding angles are congruent.
2. The ratios of the lengths of the corresponding sides are equal.

You can determine if two figures are similar by checking to see that they meet both conditions.

EXAMPLE 1 Are rectangles *ABCD* and *EFGH* similar?

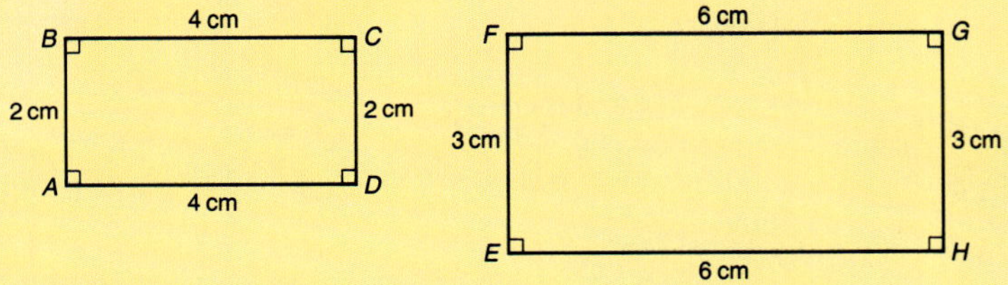

THINK: All angles are right angles.

$\angle A \cong \angle E$, $\angle B \cong \angle F$, $\angle C \cong \angle G$, $\angle D \cong \angle H$

All corresponding angles are congruent.

THINK: $\frac{\overline{AB}}{\overline{EF}} = \frac{2}{3}$ $\frac{\overline{BC}}{\overline{FG}} = \frac{4}{6} = \frac{2}{3}$ $\frac{\overline{CD}}{\overline{GH}} = \frac{2}{3}$ $\frac{\overline{DA}}{\overline{HE}} = \frac{4}{6} = \frac{2}{3}$

The ratios of the lengths of the corresponding sides are equal.

So, $ABCD \sim EFGH$.

You can find the length of the missing side of a similar figure using a proportion.

EXAMPLE 2 $\triangle RST \sim \triangle UVW$. Find the length of $\overline{RT}$.

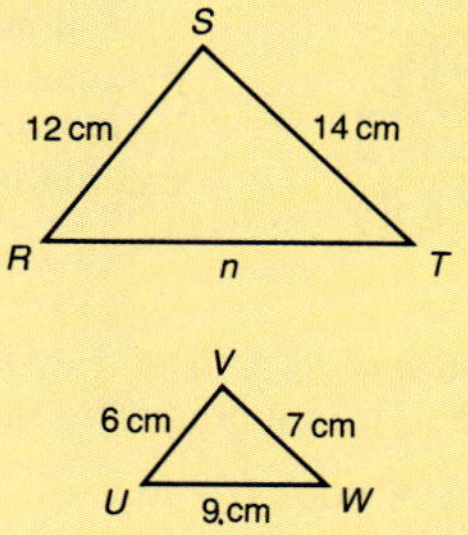

1. Write a proportion.

   length of $\overline{RT}$ $\rightarrow$ $\frac{n}{9} = \frac{12}{6}$ $\leftarrow$ **length of $\overline{RS}$**
   length of $\overline{UW}$ $\rightarrow$ $\leftarrow$ **length of $\overline{UV}$**

2. Solve.

$$n \times 6 = 12 \times 9$$
$$n \times 6 = 108$$
$$n = 108 \div 6$$
$$n = 18$$

So, the length of $\overline{RT}$ is 18 cm.

PRACTICE EXERCISES

Assume that $\triangle RST \sim \triangle XYZ$. Complete the statement.

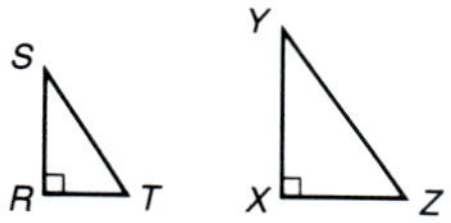

• **1.** $\angle R \cong$ ■ $\angle X$ • **2.** $\angle S \cong$ ■ $\angle Y$ **3.** $\angle T \cong$ ■ $\angle Z$

4. $\frac{\overline{RT}}{\overline{XZ}} = \frac{■}{\overline{YZ}}$ $\overline{ST}$ **5.** $\frac{\overline{RS}}{\overline{XY}} = \frac{\overline{ST}}{■}$ $\overline{YZ}$ **6.** $\frac{\overline{XY}}{■} = \frac{\overline{YZ}}{\overline{ST}}$ $\overline{RS}$

State why the figures are, or are not, similar.

• **7.**

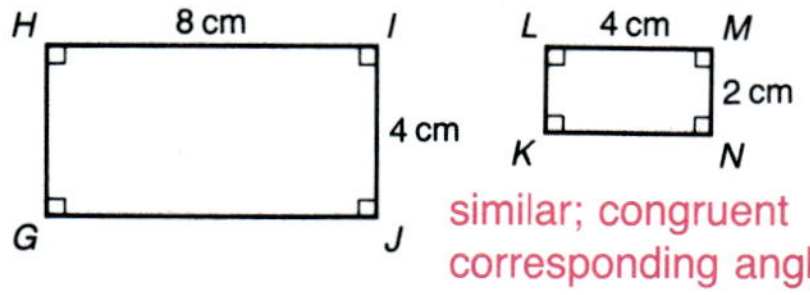

similar; congruent corresponding angles; ratio of sides equal

8.

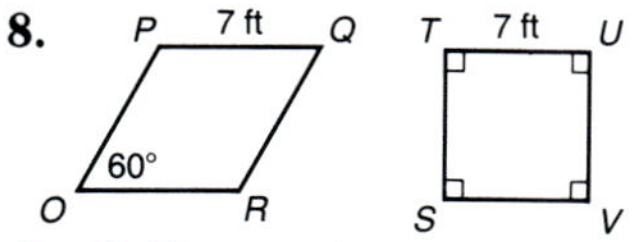

not similar; unequal corresponding angles

Find the length of the missing side for the pair of similar figures.

• **9.**

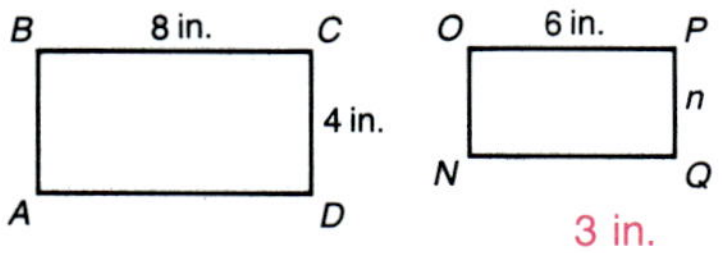

3 in.

10.

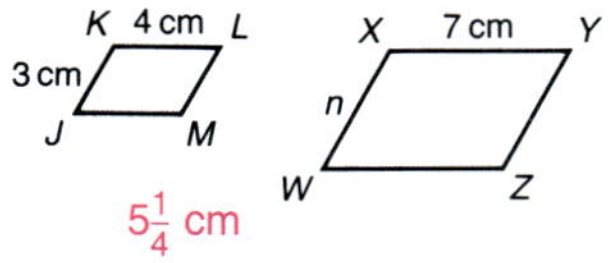

$5\frac{1}{4}$ cm

11.

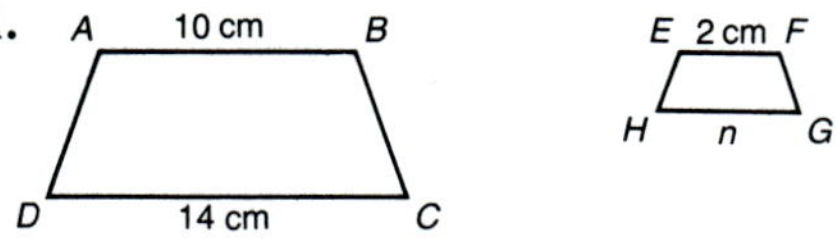

2.8 cm

12.

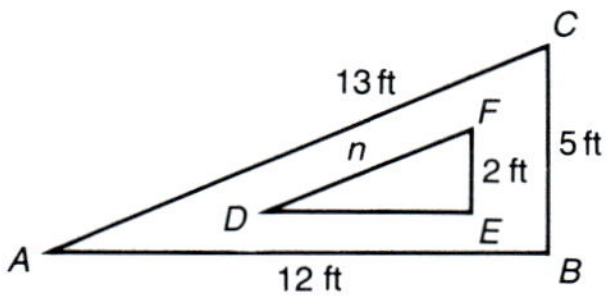

$5\frac{1}{5}$ ft

COMMON ERROR

Some students may reverse the ratios of corresponding sides when writing their proportions. For example, in Exercise 12 they may write

$\frac{n}{13} = \frac{5}{2}$ rather than $\frac{n}{13} = \frac{2}{5}$.

Remind students to check the reasonableness of their answers to make sure that the sides of the larger figure are longer than the corresponding sides of the smaller figure.

ASSIGNMENTS

Level 1	1–10
Level 2	1–11
Level 3	1–12

FOLLOW-UP ACTIVITY

ENRICHMENT Have students use the following tangram to construct three similar figures. The first should be $\frac{1}{2}$ the size of the original, the second should be $\frac{1}{4}$ the size, and the third $\frac{1}{8}$ the size.

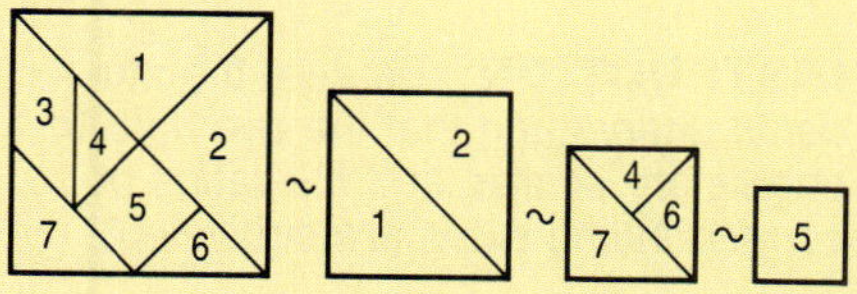

$\frac{1}{2}$ $\frac{1}{4}$ $\frac{1}{8}$

SUPPLEMENTARY MATERIALS

TRP Practice, p. 130

TRP Reteaching, p. 79

TRP Lesson Aids, p. 8

TRP Transparency 65

OBJECTIVE

- Solve problems that involve indirect measurement.

TEACHING THE LESSON

WARM-UP Write the following exercises on the chalkboard or an overhead transparency.

Solve the proportion.

1. $\frac{14}{20} = \frac{35}{n}$ (50)
2. $\frac{7.5}{10.0} = \frac{22.5}{n}$ (30)
3. $\frac{1.6}{2.4} = \frac{2.8}{n}$ (4.2)
4. $\frac{8.6}{25.8} = \frac{30.1}{n}$ (90.3)

INTRODUCTION Remind students that in Lesson 11.11 they learned about surveying and indirect measurement. Have a volunteer give an informal description of indirect measurement. (Indirect measurement is a procedure used when we cannot measure an object conveniently. Instead, we measure other objects and then calculate the measure in which we are interested.)

INSTRUCTION Be sure the students understand that the two right triangles are similar, and the ratios of corresponding sides are equivalent.

Note that although Problem 7 does not involve shadow reckoning or right triangles, it does involve similar triangles and indirect measurement.

Students may use calculators to help them solve the problems.

PROBLEM Solving APPLICATION

11.15 INDIRECT MEASUREMENT

The dimensions of an object can be found by direct measurement. Indirect measurement can also be used to find the height or length of an object by measuring other objects and then calculating the measurement.

Suppose you and your friend want to know the height of a flagpole. You can calculate its height by measuring:

- The height of your friend.
- The shadows of your friend and the flagpole.

This method, called **shadow reckoning,** uses what you have learned about similar triangles.

A flagpole casts a shadow 13.5 m long. Your friend, who is 1.6 m tall, casts a shadow that is 2.0 m long. What is the flagpole's height?

This drawing shows the two right triangles with the measurements that were made.

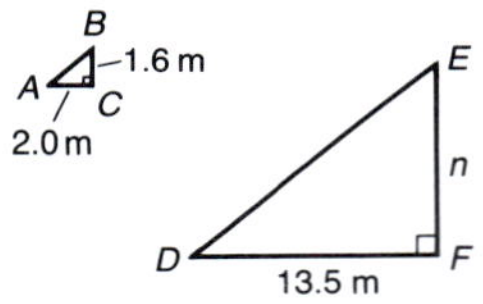

Use a proportion to solve the problem.

1. Write a proportion.
   Let n represent the height of the flagpole. $\frac{2.0}{13.5} = \frac{1.6}{n}$

2. Find n.
$$2.0 \times n = 13.5 \times 1.6$$
$$2.0 \times n = 21.6$$
$$n = 21.6 \div 2.0$$
$$n = 10.8$$

So, the height of the flagpole is 10.8 m.

On several days Jeff Corbin uses shadow reckoning to find the heights of trees in a park. He uses a stick 1.4 m long in his experiments. Find the height of each tree.

	Length of Shadow of Stick	Length of Shadow of Tree	Height of Stick	Height of Tree
• 1.	1.2 m	7.8 m	1.4 m	■ 9.1 m
2.	1.6 m	11.2 m	1.4 m	■ 9.8 m
3.	2.1 m	15.0 m	1.4 m	■ 10 m
4.	1.75 m	10.50 m	1.4 m	■ 8.4 m

Solve.

• **5.** A building casts a 24-m shadow at the same time that a 2-m street sign casts a 4-m shadow. What is the height of the building? 12 m

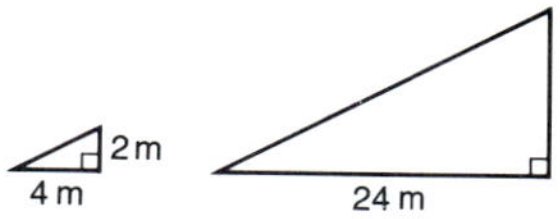

6. A flagpole casts a 16.8-m shadow at the same time that a 2.4-m shadow is cast by a person who is 1.8 m tall. What is the height of the flagpole? 12.6 m

7. Some campers wanted to find the distance across a pond. They laid out two similar triangles by pacing the distances from a tree at point *T*. (The measurements in the drawing are given in paces rather than meters.) In the two triangles, $\overline{TY}$ and $\overline{TQ}$ are corresponding sides. What is the distance across the pond? 108 paces

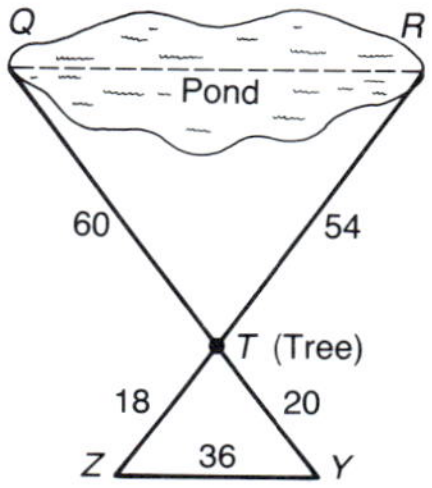

8. Work in teams of two students for this experiment. Use shadow reckoning to find the heights of three trees or other objects. Compare the results of your team with the results of another team. Check students' results.

ASSIGNMENTS

Level 1	1–3, 5–6
Level 2	1–8
Level 3	3–8

FOLLOW-UP ACTIVITY

ESTIMATION If Jim measures 2 m and casts a 3-m shadow, which of the following is the most reasonable estimate of the height of a tree that casts a:

1. 12.2 m shadow.
a. 20 m **b.** 12 m **c.** 8 m

2. 23.6 m shadow.
a. 16 m **b.** 13 m **c.** 10 m

3. 8.9 m shadow.
a. 18 m **b.** 12 m **c.** 6 m

4. 35.8 m shadow.
a. 30 m **b.** 24 m **c.** 18 m

5. 31.3 m shadow.
a. 20 m **b.** 24 m **c.** 28 m

(**1.** c **2.** a **3.** c **4.** b **5.** a)

SUPPLEMENTARY MATERIALS

TRP Practice, p. 131

TRP Lesson Aids, p. 7

TRP Transparency 66

OBJECTIVES

- Review vocabulary.
- Practice key chapter concepts and skills.

USING THE REVIEW

The Chapter Review is designed to help students prepare for taking the Chapter Test. The first section focuses on vocabulary. It requires that students select a word(s) to complete statements. The second section presents practice exercises of key mathematical skills. Under each directive there is a sample exercise with the answer.

Each item of the review is referenced to the page on which the topic is taught in the Pupil's Edition. You may wish to have students refer to these pages to help review any concepts or skills they have not yet mastered.

It is suggested that students work in small-sized heterogeneous cooperative learning groups. Some cooperative learning methods that may be used are as follows:

1. After each student has independently completed the entire Chapter Review, a discussion should follow within each group about the solutions to the practice exercises.
2. The group can complete the entire Chapter Review by working together to discuss the sample exercises and then to answer the practice exercises.

End the lesson with an entire class discussion in which any questions brought up in group discussions are presented and answered.

CHAPTER REVIEW

Vocabulary Choose the letter of the word(s) that correctly completes each statement.

1. A(n) ■ is part of a line with one endpoint. [292] a
a. ray **b.** line segment **c.** angle **d.** plane

2. A(n) ■ angle measures greater than 0° and less than 90°. [296] a
a. acute **b.** right **c.** obtuse **d.** straight

3. ■ angles are two angles the sum of whose measure is 180°. [296] c
a. Straight **b.** Complementary **c.** Supplementary **d.** Similar

4. A ■ is a line segment that goes from the center of a circle to a point on it. [304] b
a. diameter **b.** radius **c.** chord **d.** central angle

Skills Find the complement of the angle. [296]

40° *50°*

5. 30° 60° **6.** 55° 35° **7.** 60° 30° **8.** 70° 20°

Find the supplement of the angle. [296]

40° *140°*

9. 35° 145° **10.** 75° 105° **11.** 110° 70° **12.** 130° 50°

Use the figure to name the part of the circle. [304]

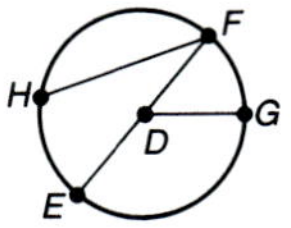

$\overline{EF}$ *diameter*

13. $\overline{HF}$ chord

14. $\angle EDG$ central angle

15. $\overline{DE}$ radius

Find the missing length. [304]

$r = 3.2$ cm, $d =$ ■
d = 6.4 cm

16. $r = 12.5$ in., $d =$ ■ 25 in.

17. $d = 7$ ft, $r =$ ■ $3\frac{1}{2}$ ft

Find the missing length in the right triangle. [314]

$a = 6, b = 8, c =$ ■
c = 10

18. $a = 12, b =$ ■, $c = 15$ 9

19. $a =$ ■, $b = 20, c = 25$ 15

20. Trace AB. Construct a line segment congruent to AB. [298]

21. Trace AB. Construct the perpendicular bisector. [301]

Check students' constructions.

CHAPTER TEST

Draw and label the figure. Check students' drawings.

1. $\overrightarrow{AB}$ **2.** $\overleftrightarrow{CD}$ **3.** $\angle EFG$ **4.** $\overline{HI}$

Use $\angle ABC$ for Exercises 5–10.

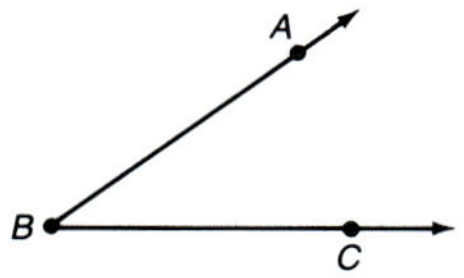

5. Find the measure of $\angle ABC$. 35°

6. Classify $\angle ABC$. acute

7. Find the complement of $\angle ABC$. 55°

8. Find the supplement of $\angle ABC$. 145°

9. Construct $\angle RST \cong \angle ABC$.
Check students' constructions.

10. Construct $\overrightarrow{BD}$, the bisector of $\angle ABC$.
Check students' constructions.

Find the missing length.

11. $r = 7$ in., $d =$ ■ 14 in.

12. $r = 5.2$ cm, $d =$ ■ 10.4 cm

13. $d = 15$ ft, $r =$ ■ $7\frac{1}{2}$ ft

Identify the triangle as *equilateral*, *isosceles*, or *scalene*.

14.

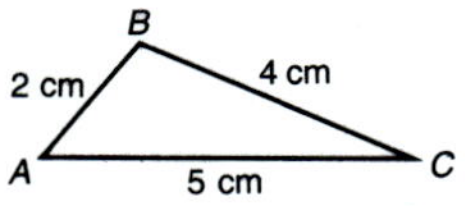

scalene

15.

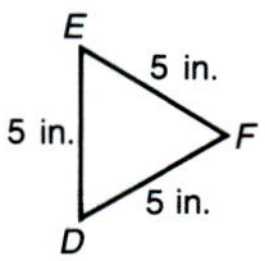

equilateral

16.

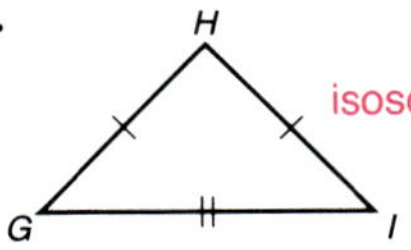

isosceles

Find the measure of the third angle.

17. 35°, 110° 35°

18. 70°, 90° 20°

Find the missing length in the right triangle.

19. $a = 12$
$b =$ ■ 16
$c = 20$

20. $a = 5$
$b = 12$
$c =$ ■ 13

21. $a = 7$
$b = 24$
$c =$ ■ 25

Find the length of the missing side for the pair of similar triangles.

22.

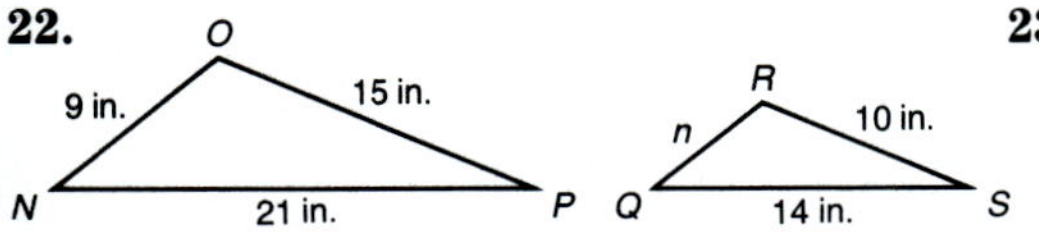

6 in.

23.

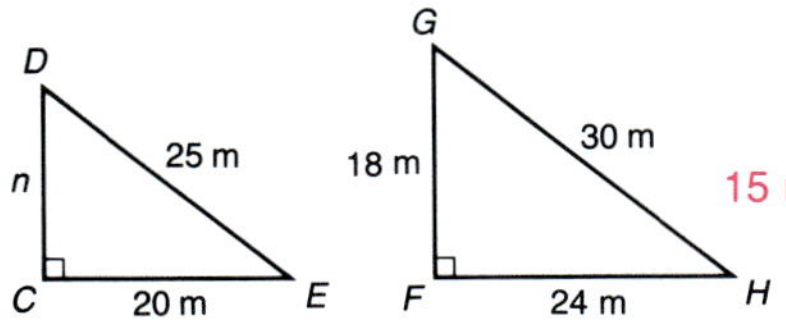

15 m

Solve.

24. A square table seats 4 people. Two tables placed end to end seat 6 people. How many people can be seated when 3 tables are placed end to end? 8 people

25. A building casts a 45-m shadow at the same time that a 3-m pole casts a 9-m shadow. What is the height of the building? 15 m

OBJECTIVE

- Evaluate achievement of the chapter objectives.

USING THE TEST

The Chapter Test may be used as a post-test to evaluate the achievement of all students. However, you may wish to use the Chapter Post-Test offered in the Teacher's Resource Package or design your own chapter test. If this page is not used as a test, you may wish to assign it as additional review or practice.

The test items are correlated to the chapter objectives in the chart below.

Chapter objectives	Test items
A. Draw geometric figures.	1–4
B. Classify and measure angles.	5–6
C. Find the complement or supplement of an angle.	7–8
D. Construct a congruent angle and an angle bisector.	9–10
E. Find the length of the diameter or radius of a circle.	11–13
F. Classify a triangle as equilateral, isosceles, or scalene.	14–16
G. Find the measure of the third angle of a triangle.	17–18
H. Find the missing side of a right triangle.	19–21
I. Find the missing sides of similar figures.	22–23
J. Find patterns to solve problems.	24
K. Apply computational skills in real life situations.	25

SUPPLEMENTARY MATERIALS

TRP Ch. 11 Posttest Form A, pp. 1–2

TRP Ch. 11 Posttest Form B, pp. 1–2

OBJECTIVES

- Identify corresponding and supplementary angles of parallel lines.
- Find the measure of corresponding angles of parallel lines.

USING THE PAGE

Begin by reviewing the definition of parallel lines.

Draw figure $\overleftrightarrow{AB} \parallel \overleftrightarrow{CD}$ on the chalkboard as on the lesson page. Have a volunteer point out each pair of corresponding angles and explain why they are corresponding. For example, ∠1 and ∠5 are corresponding because each angle is above the line and to the left of the transversal.

Before finding the measures of angles, you may want to review how to find the measure of a supplementary angle. Have a volunteer illustrate the following: If ∠*a* and ∠*b* are supplementary, and the m∠*A* = 45°, find the measure of ∠*b*. (135°)

Draw $\overleftrightarrow{GH} \parallel \overleftrightarrow{IJ}$ on the chalkboard. Have one volunteer list all the corresponding angles and another list all the supplementary angles. Then discuss how to find the measures of all the angles.

Before assigning the exercises, you may want the students to draw $\overleftrightarrow{WX} \parallel \overleftrightarrow{YZ}$ cut by transversal $\overleftrightarrow{AB}$ on their paper. Have them label the eight angles and list all the corresponding angles. Then, have each student use a protractor to measure each angle. List the pairs of corresponding angles and the pairs of supplementary angles on the chalkboard.

ENRICHMENT CORRESPONDING ANGLES

In the diagram, $\overleftrightarrow{EF}$ is a **transversal.** It intersects the parallel lines $\overleftrightarrow{AB}$ and $\overleftrightarrow{CD}$.

Eight angles are formed. ∠*1* and ∠*5*, ∠*2* and ∠*6*, ∠*3* and ∠*7*, and ∠*4* and ∠*8* are **corresponding angles.** Each pair of corresponding angles is congruent.

∠*1* and ∠*2* and ∠*1* and ∠*3* are supplementary angles. The sum of supplementary angles always equals 180°.

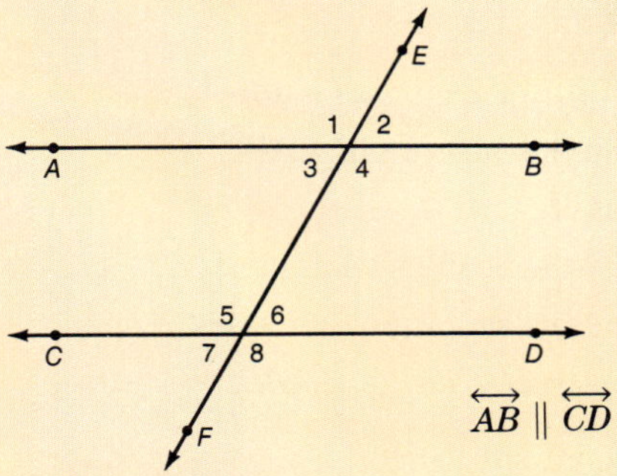

$\overleftrightarrow{AB} \parallel \overleftrightarrow{CD}$

Example In the diagram above, if m∠*1* = 120°, what are the measures of all the other angles?

m∠*2* = 60° because ∠*1* and ∠*2* are supplementary angles.
m∠*3* = 60° because ∠*1* and ∠*3* are supplementary angles.
m∠*4* = 120° because ∠*2* and ∠*4* are supplementary angles.
m∠*5* = 120° because ∠*1* and ∠*5* are corresponding angles.
m∠*6* = 60° because ∠*2* and ∠*6* are corresponding angles.
m∠*7* = 60° because ∠*3* and ∠*7* are corresponding angles.
m∠*8* = 60° because ∠*4* and ∠*8* are corresponding angles.

Using the diagram to the right, identify the pairs of angles as *corresponding* or *supplementary.*

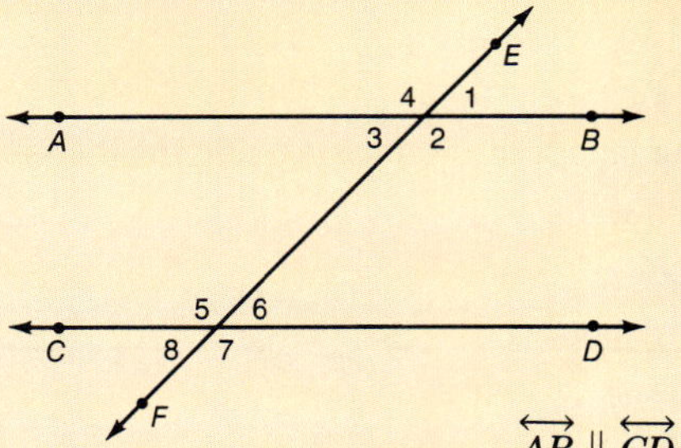

$\overleftrightarrow{AB} \parallel \overleftrightarrow{CD}$

1. ∠*1* and ∠*6* corresponding
2. ∠*2* and ∠*3* supplementary
3. ∠*3* and ∠*8* corresponding
4. ∠*1* and ∠*4* supplementary
5. ∠*5* and ∠*6* supplementary
6. ∠*2* and ∠*7* corresponding

In the diagram to the right, if m∠*1* = 45°, what is:

7. m∠*2*? 135°
8. m∠*6*? 45°
9. m∠*7*? 135°
10. m∠*3*? 45°
11. m∠*5*? 135°
12. m∠*8*? 45°

326 CHAPTER 11

ESTIMATION SKILLS ESTIMATING ANGLE MEASURES

To estimate angle measures, make mental comparisons with right angles.

Estimate the measure of $\angle ABC$.

THINK: **$\angle ABC$ is about $\frac{1}{3}$ the size of a right angle.**

$\frac{1}{3} \times 90 = 30$

$\angle ABC$ measures about 30°.

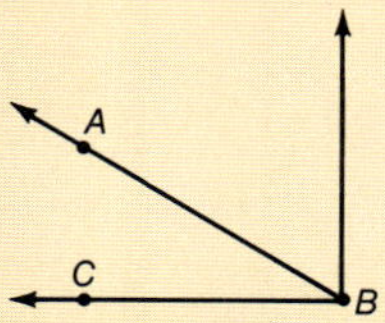

Estimate the measure of $\angle DEF$.

THINK: **$\angle DEF$ is the size of a right angle, plus about $\frac{1}{2}$ of a right angle.**

$\frac{1}{2} \times 90 = 45$

$90 + 45 = 135$

$\angle DEF$ measures about 135°.

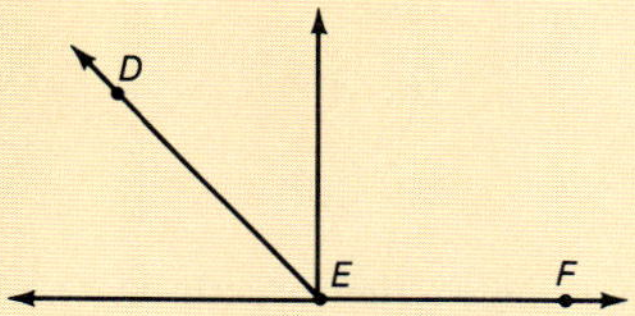

Which is the angle measure?

1.

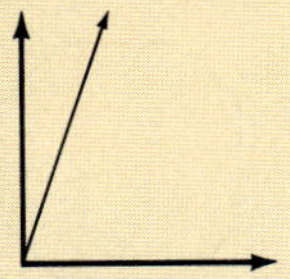

a. 30° **b.** 70° **c.** 90°

b

2.

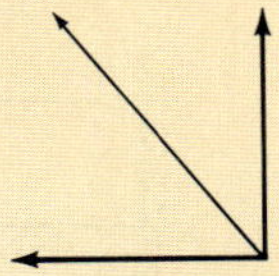

a. 20° **b.** 50° **c.** 80°

b

3.

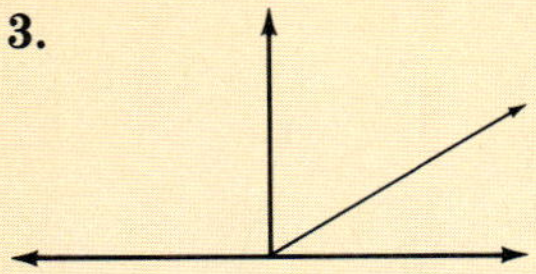

a. 105° **b.** 125° **c.** 150°

c

4.

a. 18° **b.** 41° **c.** 66°

a

5.

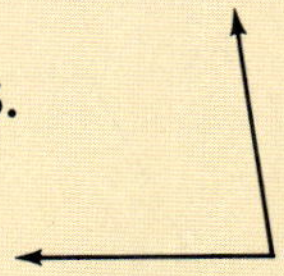

a. 44° **b.** 59° **c.** 82°

c

6.

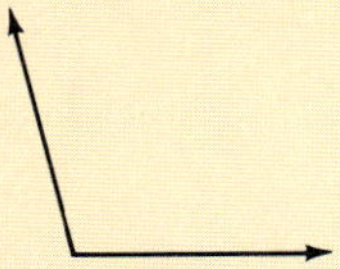

a. 103° **b.** 127° **c.** 162°

a

Estimate the measures of the angles. Use a protractor to check.

7.

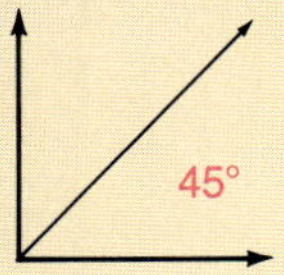

8.

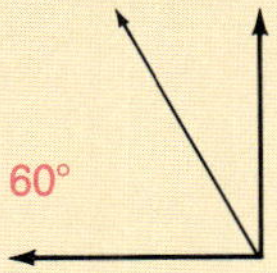

9.

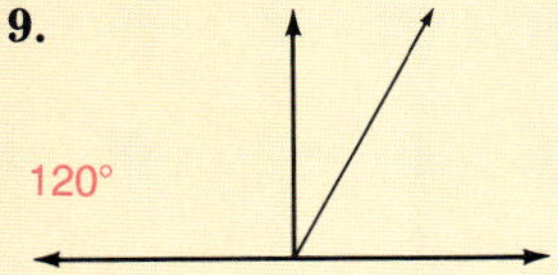

OBJECTIVE

- Estimate angle measures.

USING THE PAGE

Accept reasonable answers for Exercises 7–9. Students may think in different ways when they estimate angle measures. In Exercise 8, for example, one student may think: The angle is a little greater than $\frac{1}{2}$ of a right angle—say, 55°. Another student may think: The angle is about $\frac{1}{3}$ less than a right angle, so it's about 60°. Either way of estimating is perfectly valid.

It is generally the case that greater facility in visual estimation is acquired through practice. In order to provide additional practice, you might draw several polygons on the board. Have students estimate the measures of the angles. Then have volunteers check the estimates by measuring the angles with a protractor.

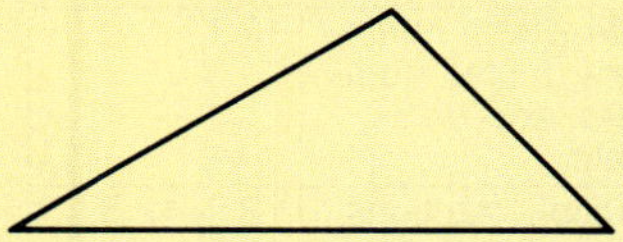

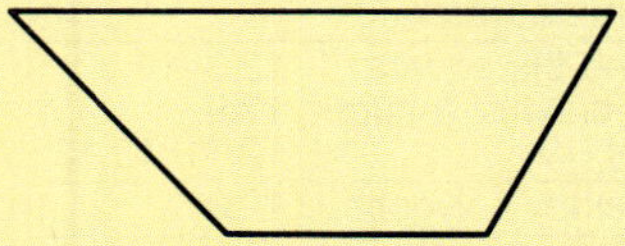

If time permits, have students sketch angles with given measures, using only a pencil and straightedge. If the required angle is less than 90°, it is helpful to sketch a right angle first. If the required angle is greater than 90°, students can start by sketching a pair of perpendicular lines. Suggest that they check the accuracy of their work with a protractor.

OBJECTIVE

- Test the prerequisite skills needed for the concepts developed in Chapter 12.

USING THE TEST

The Pre-Skills Test is designed to diagnose students' strengths and weaknesses on prerequisite skills necessary to study the mathematics in Chapter 12.

Have students take the Pre-Skills Test. Allow the students to work together in pairs or small groups. Group members should help those who demonstrate a misunderstanding of a concept or a weakness in a skill.

The following table correlates the items on the Pre-Skills Test with the prerequisite skill and the lesson(s) in the chapter for which it is needed.

Item(s)	Prerequisite skill	Lesson(s)
1–6	Add whole numbers, mixed numbers, and decimals.	12.1
7–4	Multiply fractions, whole numbers, mixed numbers and decimals.	12.2–12.5, 12.9, 12.11–12.13
15–22	Find the square or cube of a number.	12.3–12.5, 12.9, 12.11–12.13
23–30	Identify polygons.	12.8

Pre-Skills Test

Add. [30, 62, 158]

1. $15 + 22 + 15 + 22$ 74

2. $208 + 315 + 476 + 512 + 826$ 2,337

3. $16.2 + 9.3 + 16.2 + 9.3$ 51

4. $5.23 + 7.1 + 3.45 + 7.7 + 3.59$ 27.07

5. $1\frac{3}{4} + 2\frac{1}{2} + 1\frac{3}{4} + 2\frac{1}{2}$ $8\frac{1}{2}$

6. $5\frac{2}{3} + 3\frac{5}{9} + 3\frac{2}{3}$ $12\frac{8}{9}$

Multiply. [36, 74, 146, 148]

7. 4×80 320

8. 4×4.03 16.12

9. 3×17.5 52.5

10. $6 \times 3.14 \times 19.2$ 361.728

11. $4 \times \frac{1}{2}$ 2

12. $14 \times 3\frac{1}{7}$ 44

13. $6 \times \frac{22}{7}$ $18\frac{6}{7}$

14. $2 \times \frac{22}{7} \times \frac{1}{4}$ $1\frac{4}{7}$

Write as a product. [18]

15. 5^2 25

16. 8^2 64

17. 30^2 900

18. 600^2 360,000

19. 7^3 343

20. 10^3 1,000

21. 12^3 1,728

22. 20^3 8,000

Match each geometric figure with its name. [304, 318]

23. Square f

24. Rectangle c

25. Triangle b

26. Regular hexagon e

27. Trapezoid a

28. Parallelogram g

29. Pentagon h

30. Regular octagon d

a. 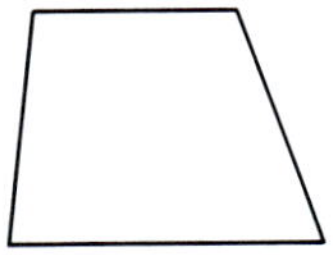

b. 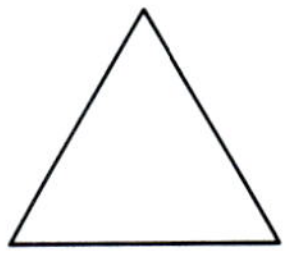

c. 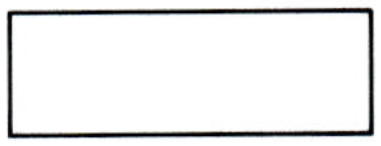

d. 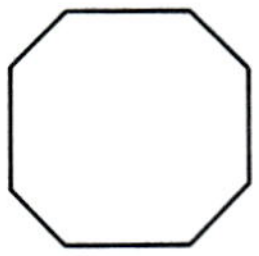

e. 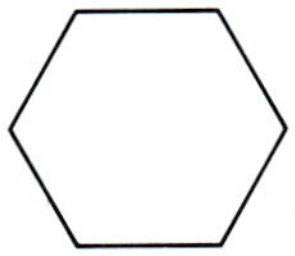

f. 

g. 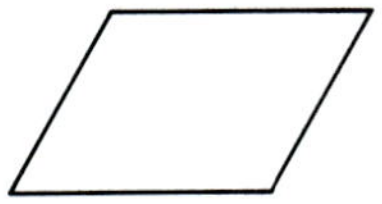

h. 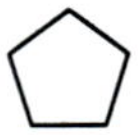

Road maps are very useful when traveling in unfamiliar areas. How are other kinds of maps used, and why are they important?

OBJECTIVE

- Formulate problems using text and data.

USING THE CHAPTER OPENER

Discuss the question on the page beginning with various uses of road maps. Then move on to other common maps that students may be aware of. This might include nautical maps which are used by weekend sailors as well as major transoceanic shipping firms; geological maps which are used by mining companies and oil drilling companies. Focus on why accurate maps are important to the people who use them. (They save time, money, and in some cases provide a measure of safety.) You may want to continue the discussion by turning to less obvious forms of maps. X-rays could be considered maps; so could blueprints, flowcharts, and so on.

COOPERATIVE LEARNING

Assign groups to the project and explain that they will be making a tourist map of their community. If your school is in a large city that normally attracts tourists, assign a specific area of the city that does not contain the usual tourist attractions. Discuss the idea that different people are interested in different kinds of sites. Groups could gear their map to any one of a number of groups. These might include people who are interested in architecture, art, history, shopping, eating out, visiting parks, and so on. Another approach might be to make a map for a specific age group. For example, a senior citizens' map might show every place that gives discounts to this age group; or a children's map could show playgrounds, parks, and toy stores.

SUPPLEMENTARY MATERIALS

TRP Group Projects, p. 12

OBJECTIVE

- Find the perimeter of a polygon.

TEACHING THE LESSON

WARM-UP Write the following exercises on the chalkboard or an overhead transparency.

How many congruent sides does each polygon have?

1. rhombus
2. regular pentagon
3. regular decagon
4. regular hexagon
5. regular octagon
6. square

(**1.** 4, **2.** 5, **3.** 10, **4.** 6, **5.** 8, **6.** 4)

INTRODUCTION Write the following problem on the chalkboard:

After practice, the Hawthorne football team runs 10 laps around the school as shown below. How far is each lap?

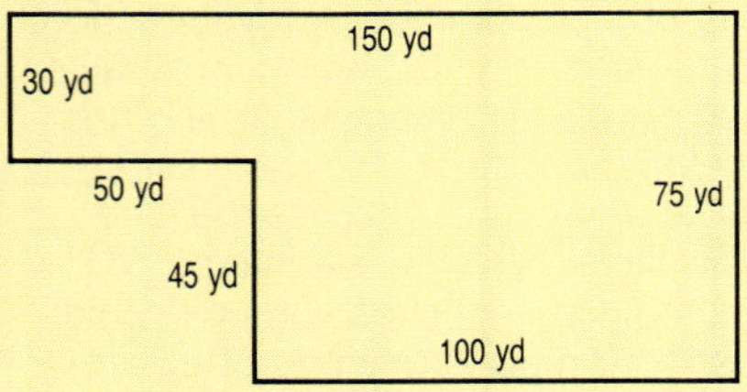

Discuss ways of solving the problem. Guide students to realize that they need to add together the lengths of each side of the school to find the total distance.

INSTRUCTION In discussing Example 1, remind students that all measurements must have the same units.

In discussing Example 2, point out that the opposite sides of a parallelogram are also congruent. Therefore, the formula for the perimeter of a rectangle can also be used to find the perimeter of a parallelogram.

12.1 PERIMETER

The Culver Construction Company has just finished paving a new downtown parking lot. A diagram of the lot is shown at the right. The manager will use steel fencing to enclose the lot. How much fencing should he order?

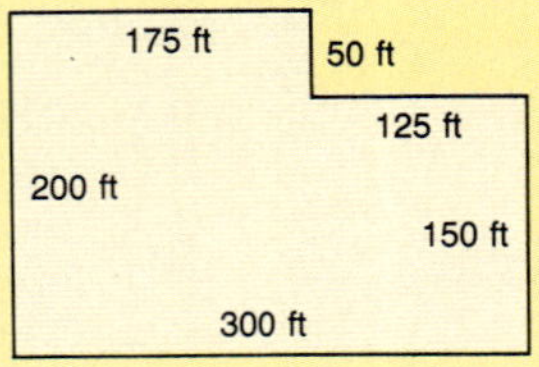

To find the correct amount of fencing, the manager needs to know the **perimeter,** or the distance around the parking lot.

EXAMPLE 1 Find the perimeter of the parking lot.

THINK: Add the lengths of the sides.

$P = 300 + 150 + 125 + 50 + 175 + 200$

$P = 1{,}000$

So, the manager should order 1,000 ft of fencing.

The opposite sides of a rectangle are congruent.

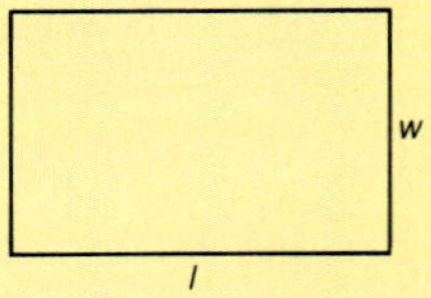

To find the perimeter of a rectangle:

1. Multiply the length by 2. $2 \times l$, or $2l$
2. Multiply the width by 2. $2 \times w$, or $2w$
3. Add the products. $P = 2l + 2w$

EXAMPLE 2 Find the perimeter.

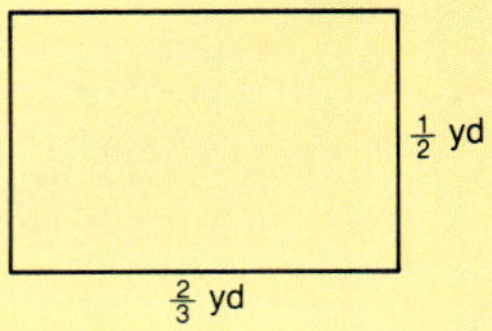

THINK: $P = 2l + 2w$

$P = 2\left(\frac{2}{3}\right) + 2\left(\frac{1}{2}\right)$

$P = 1\frac{1}{3} + 1 = 2\frac{1}{3}$

The perimeter of the rectangle is $2\frac{1}{3}$ yd.

In a regular polygon, all the sides are congruent.

To find the perimeter of a regular polygon, multiply the length of one side by the number of sides.

EXAMPLE 3 Find the perimeter.

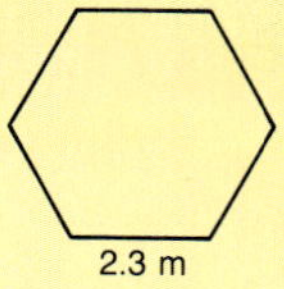

THINK: Each of the 6 sides is 2.3 m long.

$P = 6 \times 2.3 = 13.8$

So, the perimeter of the hexagon is 13.8 m.

PRACTICE EXERCISES See Extra Practice, page 453.

Find the perimeter.

• **1.** 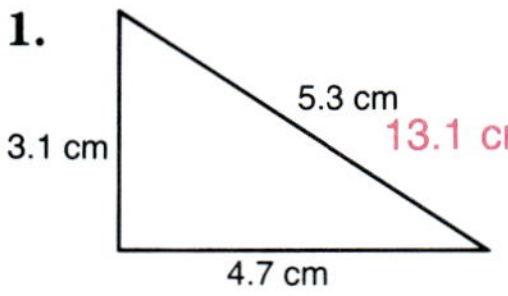

13.1 cm

2. 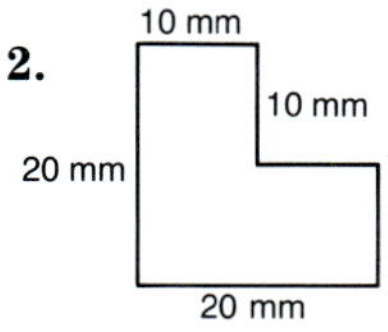

80 mm

3. 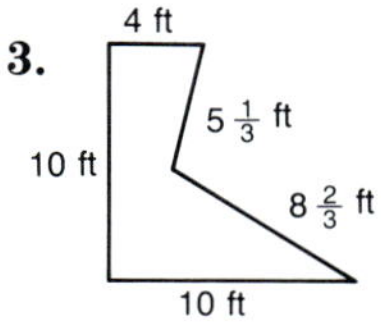

38 ft

• **4.** 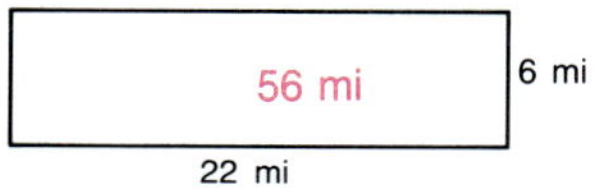

56 mi

5. 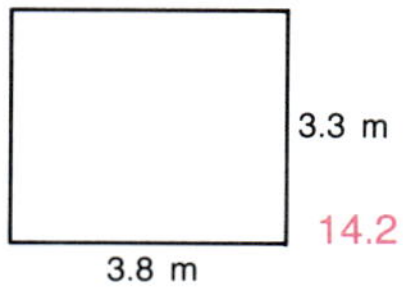

14.2 m

6. 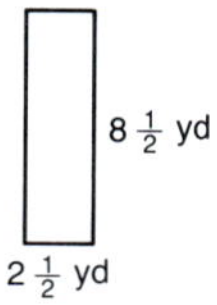

22 yd

• **7.** 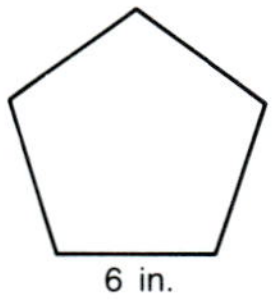

30 in.

8. 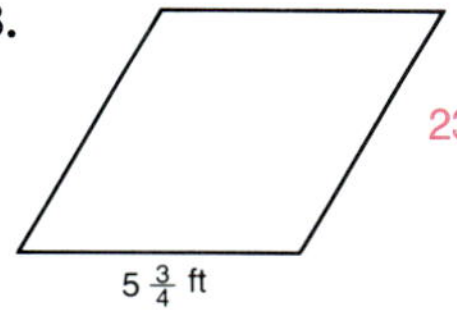

23 ft

9. 

37.38 mm

• **10.** Rectangle: $l = 3\frac{3}{4}$ ft, $w = 2\frac{1}{8}$ ft $11\frac{3}{4}$ ft

11. Rectangle: $l = 7$ in., $w = 1\frac{1}{4}$ in. $16\frac{1}{2}$ in.

12. Rectangle: $l = 3.7$ mm, $w = 12$ mm 31.4 mm

13. Rectangle: $l = 4.65$ m, $w = 1.34$ m 11.98 m

• **14.** Square: $s = 23$ ft 92 ft

15. Regular pentagon: $s = 123$ yd 615 yd

16. Regular octagon: $s = 4.5$ m 36 m

17. Regular hexagon: $s = 16.3$ mm 97.8 mm

Find the length of a side.

• **18.** Square: $P = 36$ m 9 m

19. Regular octagon: $P = 2\frac{1}{2}$ yd $\frac{5}{16}$ yd

Solve.

• **20.** How much fencing would it take to enclose a tennis court with a width of 16.7 m and a length of 35.3 m? 104 m

21. Culver Construction is building a patio in the shape of a regular octagon. Each side is $4\frac{3}{4}$ yd long. What is the perimeter of the patio? 38 yd

22. Culver charges $1.25 per foot to install molding. How much would it cost to have Culver install molding in a room that is 72 ft around? $90

23. Last year, Culver had gross earnings of $3.2 million. Culver's net profit was $800,000. What percent of gross earnings was Culver's net profit? 25%

In discussing Example 3, show how the formula for the perimeter of a regular polygon is derived. First, find the perimeter of the hexagon by adding together the lengths of its 6 congruent sides. Then use the definition of multiplication to arrive at the given equation.

You may want to allow the use of calculators throughout this chapter. Calculators will free students from cumbersome computation and allow them to concentrate on understanding the concepts of perimeter, area, and volume.

ASSIGNMENTS

Level 1	1–11, 14–15, 20–21
Level 2	1–9, Even 10–18, 20–22
Level 3	Odd 1–17, 18–23

FOLLOW-UP ACTIVITY

ENRICHMENT Display the following figure. Ask whether enough data is provided to find its perimeter.

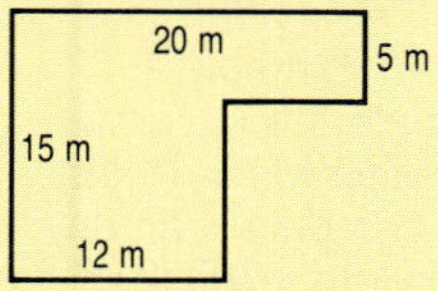

After the discussion, point out that the displayed figure has the same perimeter as the rectangle shown below.

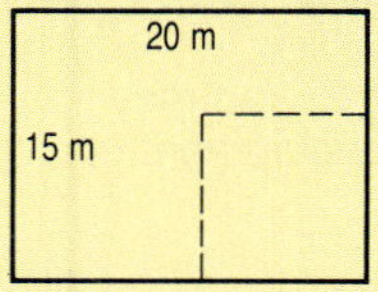

Finally, have volunteers fill in the unmarked lengths of the original figure and find its perimeter. (70 m)

SUPPLEMENTARY MATERIALS

TRP Practice, p. 132

TRP Reteaching, p. 80

TRP Transparency 66

OBJECTIVE

- Find the circumference of a circle.

TEACHING THE LESSON

WARM-UP Write the following exercises on the chalkboard or an overhead transparency.

Multiply.

1. 3.14×14 **2.** 3.14×100

3. $\frac{22}{7} \times 21$ **4.** $\frac{22}{7} \times \frac{7}{8}$

(**1.** 43.96 **2.** 314 **3.** 66 **4.** $2\frac{3}{4}$)

INTRODUCTION Display a circular object such as a paper plate. Measure its diameter with a ruler. Then use a piece of string to measure the distance around its edge. Write the two measurements on the chalkboard. Have students compare them. Emphasize that (1) the distance around the circle can be expressed as a straight line, and (2) the ratio of the measurements is 1 to a little more than 3. Repeat, with volunteers measuring other circular objects.

INSTRUCTION In Example 1, have students estimate the circumference using 3 as a value for π. Point out that although a computer can determine the value of π to more than 100,000 decimal places, π can never be expressed exactly. Therefore, calculations involving π are always approximations.

In discussing Example 2, you may wish to provide this development of the formula:

$$C = \pi \times d$$
$$C = \pi \times \overbrace{2 \times r}$$
$$C = 2 \times \pi \times r$$

Also, review the canceling involved.

$$2 \times \frac{\overset{11}{\cancel{22}}}{7} \times \frac{1}{\underset{2}{\cancel{4}}} = \frac{22}{14} = \frac{11}{7}$$

12.2 CIRCUMFERENCE

The diameter of a bicycle tire is 65 cm. The tire makes one complete revolution. How far has the bicycle traveled?

To find the answer, you need to find the distance around the tire.

The distance around a circle is called the **circumference.** For all circles, the ratio of the circumference to the diameter is the same. The Greek letter π (pi) shows this ratio.

$$\pi = \frac{\text{circumference of a circle}}{\text{length of the diameter}} \qquad \pi \text{ is approximately } 3.14 \text{ or } \frac{22}{7}.$$

To find the circumference of a circle, multiply the diameter by π.

EXAMPLE 1 Find the circumference of the tire. Use 3.14 for π.

THINK: $C = \pi \times d$

$C \approx 3.14 \times 65$

$C \approx 204.1$

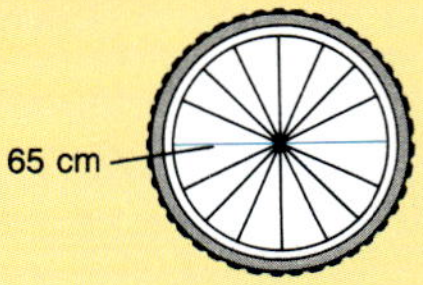

The bicycle has traveled approximately 204.1 cm.

A radius is half the length of a diameter. To find the circumference of a circle when you know only its radius, multiply the radius by π and then double the answer.

With fractions and mixed numbers, use $\frac{22}{7}$ for π.

EXAMPLE 2 Find the circumference of the circle shown below.

THINK: $C = 2 \times \pi \times r$

$C \approx 2 \times \frac{22}{7} \times \frac{1}{4}$

$C \approx \frac{11}{7} = 1\frac{4}{7}$

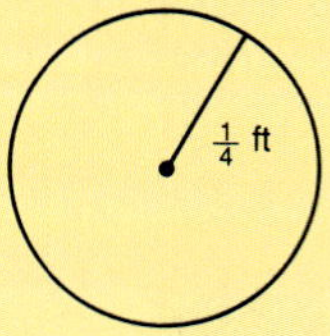

The circumference is approximately $1\frac{4}{7}$ ft.

CHECKPOINT Write the letter of the correct answer.

Find the circumference.

1. $d = 8$ mm b **a.** 11.14 mm **b.** 25.12 mm **c.** 50.24 mm **d.** 64 mm

2. $r = 1\frac{3}{4}$ ft c **a.** $3\frac{1}{2}$ ft **b.** $5\frac{1}{2}$ ft **c.** 11 ft **d.** 22 ft

PRACTICE EXERCISES See Extra Practice, page 453.

Find the circumference. Use 3.14 for π with whole numbers and decimals. Use $\frac{22}{7}$ for π with fractions and mixed numbers.

• **1.** 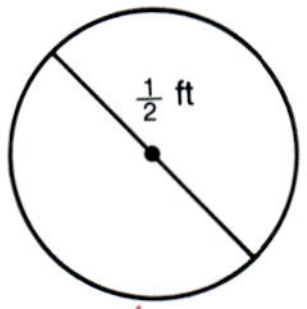

$1\frac{4}{7}$ ft

2. 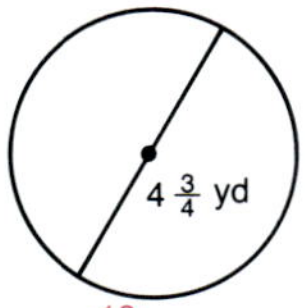

$14\frac{13}{14}$ yd

3. 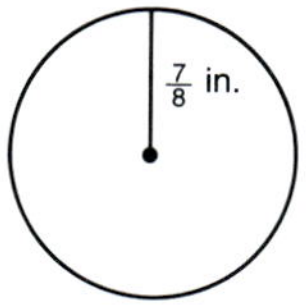

$5\frac{1}{2}$ in.

4. 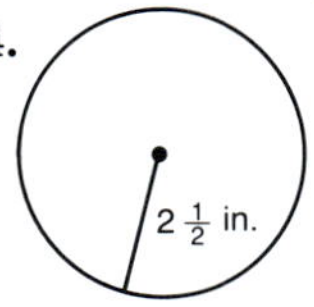

$15\frac{5}{7}$ in.

5. 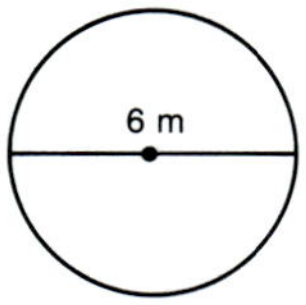

18.84 m

6. 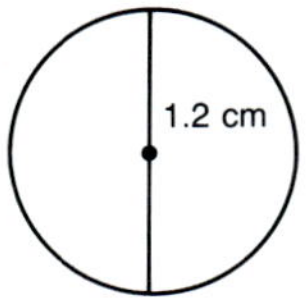

3.768 cm

7. 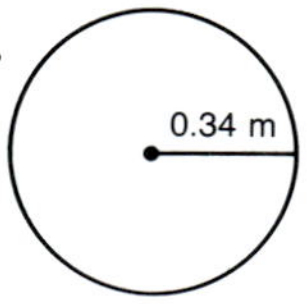

2.1352 m

8. 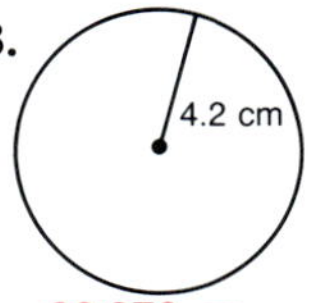

26.376 cm

• **9.** $d = \frac{7}{8}$ mi $2\frac{3}{4}$ mi

10. $d = \frac{2}{3}$ yd $2\frac{2}{21}$ yd

11. $d = 10\frac{1}{2}$ ft 33 ft

12. $d = 6$ km 18.84 km

13. $d = 14.7$ m 46.158 m

14. $d = 15.65$ cm 49.141 cm

• **15.** $r = 2\frac{1}{3}$ yd $14\frac{2}{3}$ yd

16. $r = \frac{5}{8}$ mi $3\frac{13}{14}$ mi

17. $r = 3\frac{1}{2}$ ft 22 ft

18. $r = 12.5$ cm 78.5 cm

19. $r = 7$ m 43.96 m

20. $r = 0.5$ km 3.14 km

Solve.

• **21.** Campbell Green is a rectangular park 2,300 yd long and 3,250 yd wide. How far is it to ride around Campbell Green? 11,100 yd

22. Bicycle races take place in arenas called velodromes. The Campbell Green Velodrome has a circular track. Its diameter is 100 m. How many meters are there in one lap? 314 m

23. The Campbell Rally is a 36-mi road race for bicycles. This year's winner averaged 30 mph. What was the winning time? 1 h 12 min

24. Inside Campbell Green is a circular bicycle path 2,200 m around. What is its radius to the nearest meter? 350 m

CHECKPOINT Distribute to each student 4 index cards on which the answer choices a–d appear. After allowing students a few minutes, ask them to raise the letter that indicates their answer choice. Scan the cards and identify those students who need additional attention before proceeding to the Practice Exercises. The incorrect answer choices include common errors that students make.

COMMON ERROR

Students sometimes confuse the circumference formulas and use $C = 2\pi d$ and $C = \pi r$. For example, in item 1 of Checkpoint, they would choose *c*. Encourage these students to write the correct formulas at the top of their papers and to refer to them.

ASSIGNMENTS

Level 1	1–8, 21–22
Level 2	1–8, Even 10–20, 21–23
Level 3	Odd 1–19, 21–24

FOLLOW-UP ACTIVITY

ENRICHMENT To show that diameter can be deduced from circumference, guide students through the following:

$C = 15.7$ m $\qquad C = \pi \times d$

$$15.7 = 3.14 \times d$$
$$15.7 \div 3.14 = d$$
$$5 = d$$

So, the diameter is 5 m.

Next, have the students find the diameter of the following circles:

1. $C = 18.84$ cm (6 cm)

2. $C = 3.454$ km (1.1 km)

SUPPLEMENTARY MATERIALS

TRP Practice, p. 133

TRP Reteaching, p. 81

TRP Transparency 67

OBJECTIVES

- Find the area of a rectangle.
- Find the area of a square.
- Find the area of a parallelogram.

TEACHING THE LESSON

WARM-UP Write the following exercises on the chalkboard or an overhead transparency.

Multiply.

1. 4.3×2 **2.** 0.8×4.1

3. $8\frac{3}{4} \times 4$ **4.** $5\frac{2}{3} \times 6$

(**1.** 8.6 **2.** 3.28 **3.** 35 **4.** 34)

INTRODUCTION Write the following problem on the chalkboard:

The school superintendent plans to repaint the walls of this classroom. How can he determine the amount of paint to be ordered?

Draw a diagram. Discuss ways of finding the amount of wall space to be painted. Guide students to realize that they need to know the amount of flat space, or area, of each wall.

INSTRUCTION When discussing Example 1, have students use graph paper to draw rectangles that measure 5 units by 3 units. Then have them count the number of squares within the rectangle to verify the area.

Also, point out that units of measure are squared just as numbers are:

$$5\text{ ft} \times 3\text{ ft} = (5 \times 3)(\text{ft} \times \text{ft})$$
$$= 15\text{ ft}^2$$

In discussing Example 2, emphasize that a square is a special type of rectangle. Then show the derivation of the formula for the area of a square.

$$A = l \times w$$
$$A = s \times s$$
$$A = s^2$$

Also, review the computation shown below.

$$(4\tfrac{1}{2})^2 = (\tfrac{9}{2})^2 = \tfrac{81}{4} = 20\tfrac{1}{4}$$

12.3 AREAS OF RECTANGLES, SQUARES, AND PARALLELOGRAMS

Peter Rios works for a company that remodels kitchens. In one kitchen, he builds the rectangular island shown at the right. How much butcher block does he need to cover its countertop?

To find the amount of butcher block, you need to find the area of the rectangle. **Area** is the number of square units needed to cover the surface of a plane figure.

To find the area of a rectangle, multiply the length by the width.

EXAMPLE 1 Find the area of the countertop.

THINK: $A = l \times w$

$A = 5 \times 3 = 15$

So, Peter needs 15 ft^2 to cover its countertop.

A square is a rectangle whose length and width both have the same measure. To find its area, square the length of one side.

EXAMPLE 2 Find the area of the square.

THINK: $A = s^2$

$A = (4\frac{1}{2})^2 = 4\frac{1}{2} \times 4\frac{1}{2} = 20\frac{1}{4}$

The area of the square is $20\frac{1}{4}$ ft^2.

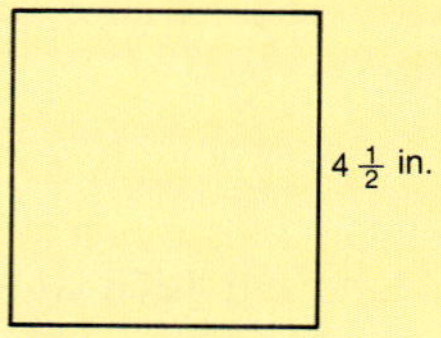

If a parallelogram and a rectangle have equal bases and equal heights, they are equal in area.

EXAMPLE 3 Find the area of the parallelogram.

THINK: Area (A) = base (b) × height (h)

$A = 8.3 \times 5.2 = 43.16$

The area of the parallelogram is 43.16 cm^2.

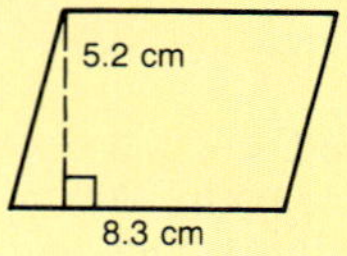

PRACTICE EXERCISES See Extra Practice, page 453.

Find the area of the rectangle.

1.
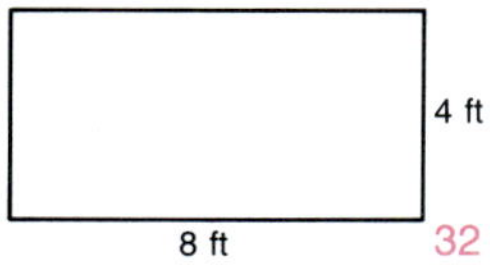

32 ft²

2.
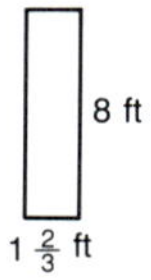

$13\frac{1}{3}$ ft²

3.
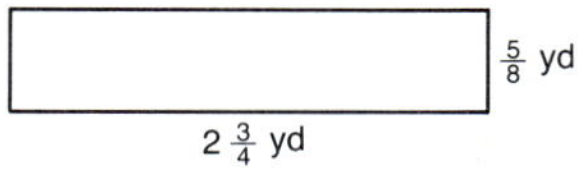

$1\frac{23}{32}$ yd²

4. $l = 4.3$ mm, $w = 2.7$ mm 11.61 mm²

5. $l = 3.4$ m, $w = 2.1$ m 7.14 m²

6. $l = 0.4$ m, $w = 2.2$ m 0.88 m²

Find the area of the square.

7.
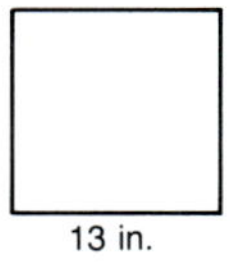

169 in.²

8.

$\frac{1}{16}$ in.²

9.
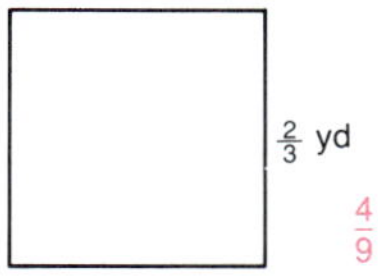

$\frac{4}{9}$ yd²

10. $s = 12$ m 144 m²

11. $s = 0.9$ cm 0.81 cm²

12. $s = 2.3$ km 5.29 km²

Find the area of the parallelogram.

13.
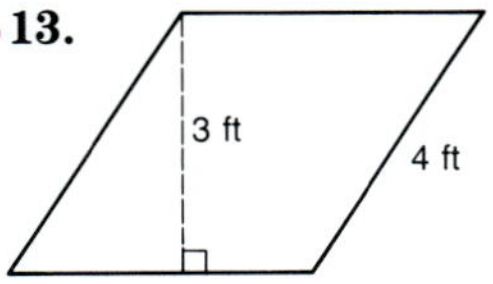

12 ft²

14.
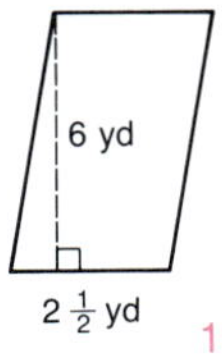

15 yd²

15.
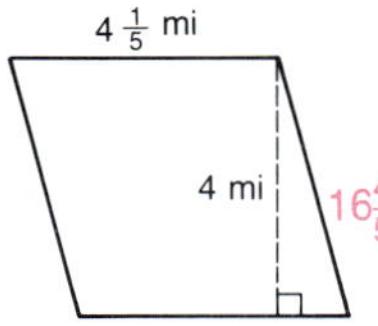

$16\frac{4}{5}$ mi²

16. $b = 49$ m, $h = 3.2$ m 156.8 m²

17. $b = 0.8$ cm, $h = 4.1$ cm 3.28 cm²

18. $b = 4.3$ m, $h = 5.1$ m 21.93 m²

Solve.

19. Peter remodels a rectangular kitchen 12 ft by 8 ft. How many square yards of tile are needed to cover its floor? (*Hint:* Change feet to yards.) $10\frac{2}{3}$ yd²

20. The blueprint for remodeling this kitchen uses a scale of 1:8. The actual height of the kitchen ceiling is 7 ft. What is this distance on the blueprint? $\frac{7}{8}$ ft

21. One quart of paint covers 80 ft². How many gallons of paint are needed to cover 1,200 ft²? $3\frac{3}{4}$ gal

22. Peter is installing an unusual square sink. Its area is 576 in.² What is the sink's perimeter? 96 in.

TIME OUT The first floor of a town house has 929 ft² of floor space. Using graph paper, design a floor plan that includes the following rooms: living room (360 ft²), den (225 ft²), dining room (114 ft²), kitchen (120 ft²), bathroom (80 ft²). Check students' plans.

In discussing Example 3, it may be helpful to have students verify the formula by cutting apart a parallelogram drawn on paper or cardboard and reassembling it as a rectangle as shown in the figure.

TIME OUT Students should take into account the proper location and use of each room. For instance, the dining room should be next to the kitchen, and the bathroom should not be 2 ft by 40 ft.

ASSIGNMENTS

Level 1	1–3, 7–9, 13–15, 19–20
Level 2	Odd 1–17, 19–21, TO
Level 3	Even 2–18, 19–22, TO

FOLLOW-UP ACTIVITY

ESTIMATION Distribute maps of your state along with rulers and tracing paper. Explain that every polygon has an area, not only rectangles, squares, and parallelograms.

Discuss methods of estimating the area of the state. Guide students to realize that the most effective way is to use the mileage scale as the basis for a grid drawn on the tracing paper.

Students should place the grid over the map. To estimate the area, they simply count the number of squares and partial squares that lie within the state, the area of each square being defined by the mileage scale.

SUPPLEMENTARY MATERIALS

TRP Practice, p. 134

TRP Reteaching, p. 82

TRP Lesson Aids, p. 4

TRP Transparency 67

OBJECTIVES

- Find the area of a triangle.
- Find the area of a trapezoid.

TEACHING THE LESSON

WARM-UP Write the following exercises on the chalkboard or an overhead transparency.

Multiply.

1. $\frac{1}{2} \times 14 \times 12$

2. $\frac{1}{2} \times (5\frac{2}{5} + 12\frac{3}{5})$

3. $\frac{1}{2} \times 5.2 \times 4$

4. $\frac{1}{2} \times (6.2 + 5.2) \times 18$

(**1.** 84 **2.** 9 **3.** 10.4 **4.** 102.6)

INTRODUCTION Distribute copies of various parallelograms. Have students cut each parallelogram along the diagonal to form two congruent triangles. Elicit the idea that the area of each triangle is $\frac{1}{2}$ the area of a parallelogram having the same base and height.

INSTRUCTION In discussing Example 1, remind the students that the height of a triangle is the perpendicular distance from a base to the opposite vertex. Therefore it depends upon which side is considered the base. Point out that the height of a right triangle whose base is not the hypotenuse is one of its legs.

When discussing Example 2, show how a distributive property is used to combine the two base terms.

You may also wish to demonstrate that the formula for the area of a trapezoid can also be developed from the formula for the area of a parallelogram.

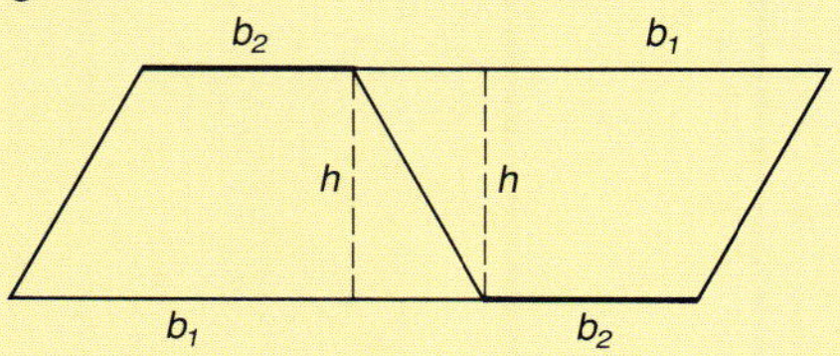

12.4 AREAS OF TRIANGLES AND TRAPEZOIDS

Laura Bergman is buying grass seed to repair the triangular patch of lawn shown at the right. How many square yards of lawn does she have to repair?

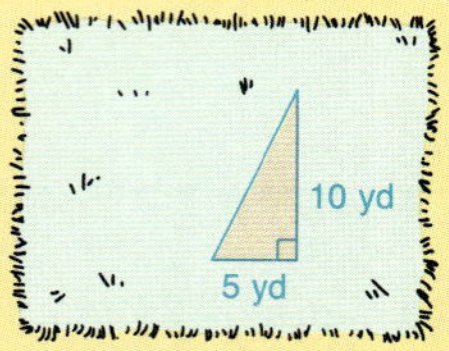

To find the amount of lawn to be repaired, you need to find the area of the triangle.

The figure at the right shows a parallelogram divided into two congruent triangles. Each triangle has an area equal to one-half the area of the parallelogram, or one-half the base times the height.

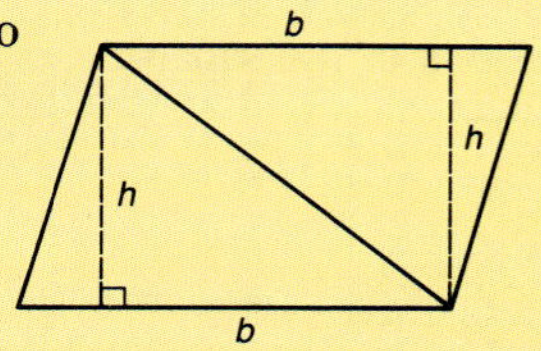

So, the area of a triangle is $\frac{1}{2} \times b \times h$.

EXAMPLE 1 Find the area of the triangular patch.

THINK: $A = \frac{1}{2} \times b \times h$

$A = \frac{1}{2} \times 5 \times 10$

$A = 25$

Laura Bergman has to repair 25 yd^2 of lawn.

In the figure below, a trapezoid is divided into two triangles. The area of the trapezoid is equal to the sum of the areas in the triangles.

Area ($\triangle_1$) $= \frac{1}{2} \times b_1 \times h$

Area ($\triangle_2$) $= \frac{1}{2} \times b_2 \times h$

Area (trapezoid) $= \frac{1}{2} \times (b_1 + b_2) \times h$

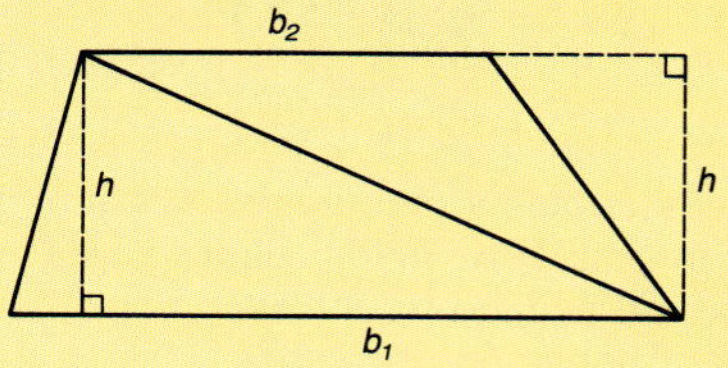

EXAMPLE 2 Find the area of the trapezoid.

THINK: $A = \frac{1}{2} \times (b_1 + b_2) \times h$

$A = \frac{1}{2} \times (7 + 3) \times 4$

$A = \frac{1}{2} \times 10 \times 4 = 20$

The area of the trapezoid is 20 m^2.

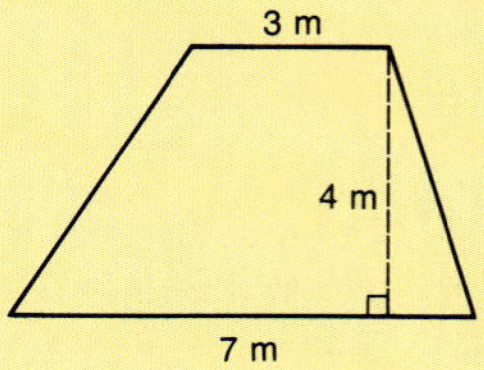

The area of the parallelogram is $(b_1 + b_2) \times h$. Therefore, the area of each of the two congruent trapezoids is $\frac{1}{2} \times (b_1 + b_2) \times h$.

COMMON ERROR

Some students may assume that a triangle has only one base, namely, the side upon which it appears to rest. Others may mistake an adjacent side for the height. Display the following figure and have students name each base and the corresponding altitude.

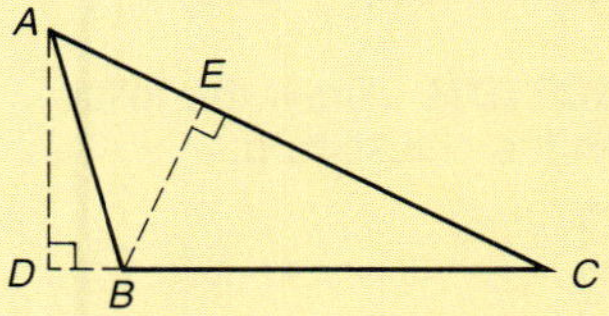

ASSIGNMENTS

Level 1	1–3, 7–9, 14–21
Level 2	1–4, 7–11, 14–22
Level 3	3–6, 9–13, 20–23

FOLLOW-UP ACTIVITY

MENTAL MATH Have students mentally compute the area of each of the following figures:

1. Square: $s = \frac{1}{4}$ in.
2. Rectangle: $l = 6$ m, $w = 1.1$ m
3. Parallelogram: $b = 8$ ft, $h = \frac{1}{2}$ ft
4. Triangle: $b = 3$ yd, $h = 12$ yd

(**1.** $\frac{1}{16}$ in.^{2} **2.** 6.6 m^2 **3.** 4 ft^2 **4.** 18 yd^2)

SUPPLEMENTARY MATERIALS

TRP Practice, p. 135

TRP Reteaching, p. 83

TRP Lesson Aids, p. 4

TRP Transparency 68

PRACTICE EXERCISES

See Extra Practice, page 454.

Find the area of the triangle.

1.

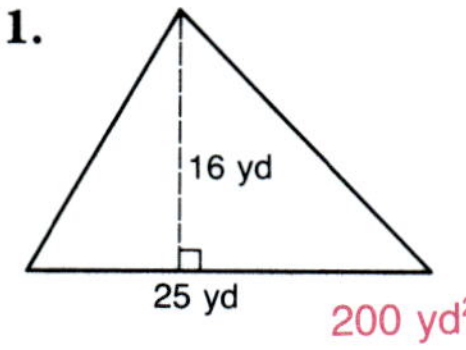

200 yd^2

2.

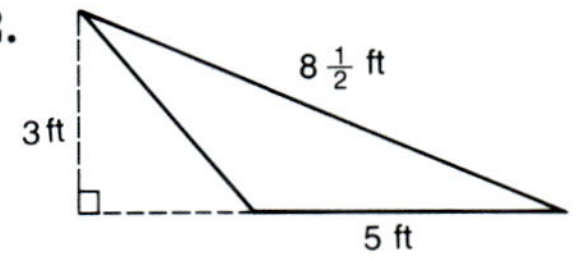

$7\frac{1}{2}$ ft^2

3.

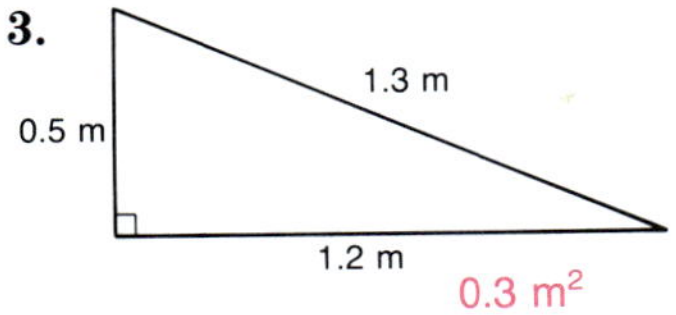

0.3 m^2

4. $b = 14$ in., $h = 12$ in. 84 in.^{2}

5. $b = 15.2$ cm, $h = 4$ cm 30.4 cm^2

6. $b = 2\frac{3}{8}$ yd, $h = 4$ yd $4\frac{3}{4}$ yd^2

Find the area of the trapezoid.

7.

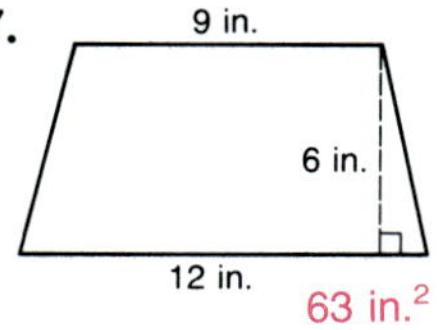

63 in.^{2}

8.

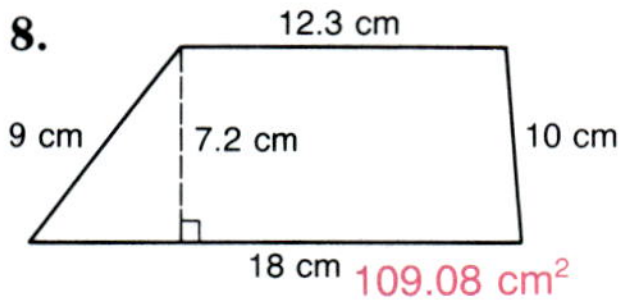

109.08 cm^2

9.

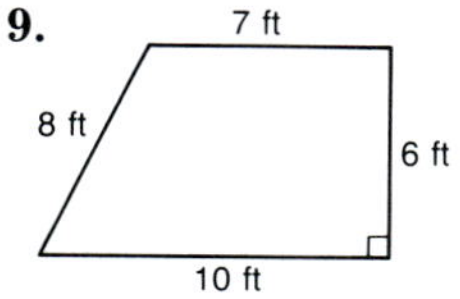

51 ft^2

10. $b_1 = 3$ ft, $b_2 = 4$ ft, $h = 2$ ft 7 ft^2

11. $b_1 = 5$ yd, $b_2 = 13$ yd, $h = 12$ yd 108 yd^2

12. $b_1 = 1.8$ m, $b_2 = 6.2$ m, $h = 4.2$ m 16.8 m^2

13. $b_1 = 3\frac{1}{4}$ ft, $b_2 = 6\frac{3}{4}$ ft, $h = 8\frac{2}{3}$ ft $43\frac{1}{3}$ ft^2

MIXED REVIEW

Find the number.

14. What number is 30% of 68? 20.4

15. 10% of what number is 8? 80

16. What percent of 12 is 8? $66\frac{2}{3}$%

17. 24 is 75% of what number? 32

18. 12 is 4% of what number? 300

19. 25 is what percent of 100? 25%

20. What number is 25% of 36? 9

21. What percent of 35 is 28? 80%

Solve.

22. Laura keeps a $3\frac{1}{2}$ ft-high hedge clipped in the shape of a trapezoid 8 ft along the bottom and 6 ft on the top. When she sprays one side, what area does she have to cover? $24\frac{1}{2}$ ft^2

23. When Laura fertilizes her zinnias, she mixes 1 oz of concentrate with enough water to make 1 gal of spray. In this spray, what is the ratio of concentrate to water? 1:127

24. Laura's flower bed is a regular octagon 4 ft on a side. She borders it with solid pieces cut from railroad ties 6 ft long. How many feet of scrap are left over? 16 ft

25. Laura walks off a patch of garden for tomatoes. She walks 12 ft north, 5 ft west, and then 13 ft straight back to where she started. What is the area of Laura's tomato patch? 30 ft^2

OBJECTIVE

- Find the area of a circle.

TEACHING THE LESSON

WARM-UP Write the following exercises on the chalkboard or an overhead transparency.

Multiply.

1. 3.14×3^2 **2.** 3.14×10^2

3. $\frac{22}{7} \times (\frac{1}{11})^2$ **4.** $\frac{22}{7} \times (1\frac{3}{4})^2$

(**1.** 28.26 **2.** 314 **3.** $\frac{2}{77}$ **4.** $9\frac{5}{8}$)

INTRODUCTION Draw the following figure on the chalkboard:

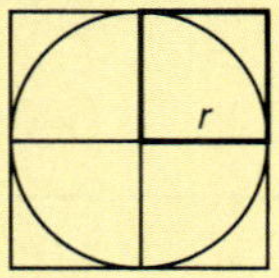

Point out that each of the fourths of the square has an area of r^2, so that the area of the square is $4 \times r^2$. Then draw an inscribed square as shown below.

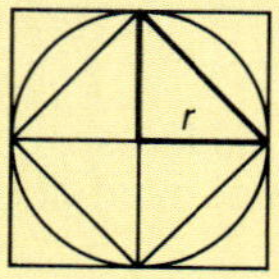

Point out that the area of each of the four right triangles thus created is $\frac{1}{2} \times r^2$, so that the area of the inscribed square is $2 \times r^2$. Elicit the idea that the area of the circle is less than the area of the outer square and greater than that of the inner square.

INSTRUCTION When discussing Examples 1 and 2, have students first estimate the area of the circle using 3 as an approximate value for π.

12.5 AREAS OF CIRCLES

Kevin Woollam is a landscaper. He is designing an underground sprinkler system for one of his clients. Each rotating sprinkler can water 40 ft of lawn in any direction. How many square feet of lawn can each sprinkler cover?

To find the number of square feet each sprinkler can cover, you need to find the area of the circle, which is $\pi \times r^2$.

EXAMPLE 1 Find the area of lawn each sprinkler can cover. Use 3.14 for π.

THINK: $A = \pi \times r^2$

$A \approx 3.14 \times 40^2$

$A \approx 5{,}024$

Each sprinkler can cover approximately 5,024 ft^2 of lawn.

With fractions and mixed numbers, use $\frac{22}{7}$ for π.

EXAMPLE 2 Find the area of a circle whose radius is $5\frac{1}{4}$ ft.

THINK: $A = \pi \times r^2$

$A \approx \frac{22}{7} \times \left(5\frac{1}{4}\right)^2$

$A \approx \frac{\overset{11}{\cancel{22}}}{\underset{1}{\cancel{7}}} \times \frac{\overset{3}{\cancel{21}}}{\underset{2}{\cancel{4}}} \times \frac{21}{4}$

$A \approx 86\frac{5}{8}$

The area of the circle is approximately $86\frac{5}{8}$ ft^2.

CHECKPOINT

Write the letter of the correct answer.

Find the area.

1. $r = 4$ km a **a.** 50.24 km^2 **b.** 25.12 km^2 **c.** 12.56 km^2 **d.** 6.28 km^2

2. $r = 3\frac{1}{2}$ ft a **a.** $38\frac{1}{2}$ ft^2 **b.** 25 ft^2 **c.** 11 ft^2 **d.** $5\frac{1}{2}$ ft^2

PRACTICE EXERCISES See Extra Practice, page 454.

Find the area. Use 3.14 for π with whole numbers and decimals. Use $\frac{22}{7}$ for π with fractions and mixed numbers.

• **1.**

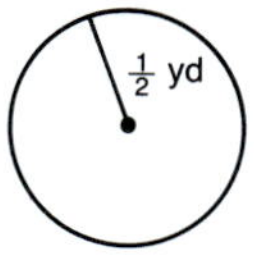

$\frac{11}{14}$ yd^2

•**2.**

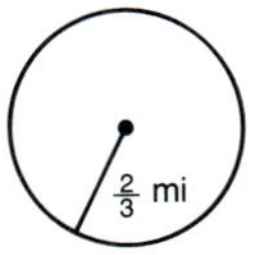

$1\frac{25}{63}$ mi^2

3.

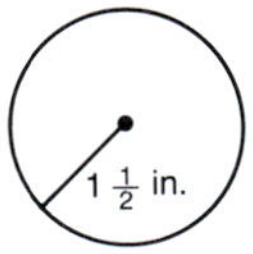

$7\frac{1}{14}$ in.^{2}

4.

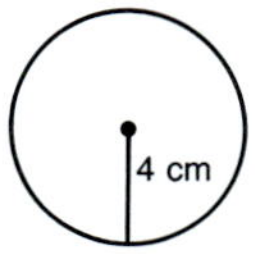

50.24 cm^2

5.

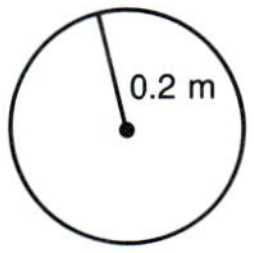

0.1256 m^2

6.

55.3896 mm^2

• **7.** $r = 1$ in. 3.14 in.^{2} •**8.** $r = 4$ mi 50.24 mi^2 **9.** $r = 10$ ft 314 ft^2 **10.** $r = 7$ yd 153.86 yd^2

11. $r = \frac{2}{3}$ ft $1\frac{25}{63}$ ft^2 **12.** $r = 2\frac{3}{4}$ yd $23\frac{43}{56}$ yd^2 **13.** $r = \frac{1}{8}$ in. $\frac{11}{224}$ in.^{2} **14.** $r = 3\frac{1}{4}$ ft $33\frac{11}{56}$ ft^2

15. $r = 3$ cm 28.26 cm^2 **16.** $r = 0.1$ m 0.0314 m^2 **17.** $r = 0.5$ km 0.785 km^2 **18.** $r = 13$ mm 530.66 mm^2

•**19.** $d = 10$ yd 78.5 yd^2 **20.** $d = 0.1$ m 0.00785 m^2 **21.** $d = 12$ in. 113.04 in.^{2} **22.** $d = 5$ m 19.625 m^2

MIXED REVIEW

Complete.

•**23.** 2 ft = ■ in. 24 •**24.** 12 qt = ■ gal 3 **25.** 6 T = ■ lb 12,000

26. 2,500 m = ■ km 2.5 **27.** 8 g = ■ mg 8,000 **28.** 3.5 cm = ■ mm 35

Kevin Woollam's next project will be a fish pond for the Bergers. Use the diagram for Exercises 29–31.

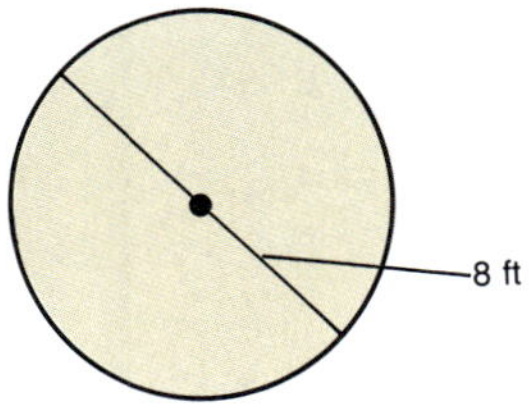

29. The Bergers have bought 60 ft^2 of tile for the pond. Do they have enough tile to cover the entire bottom of the pond? yes

30. Kevin plans to plant flowers along the border of the pond. What is the distance around the pond? $25\frac{1}{7}$ ft

31. The Bergers want to stock the pond with goldfish so that there are 2 goldfish for every 5 ft^2 of pond. How many fish do they need? 20 fish

In Example 2, you can also have them round the radius to the nearest whole number. Encourage students to use this type of estimation to check their answers.

Also, review canceling of fractions during multiplication, if necessary.

CHECKPOINT The incorrect choices include common errors.

COMMON ERROR

When using the area formula, some students may multiply by 2 instead of squaring the radius. For example, in item 1 of Checkpoint, they would choose *b*. Have these students use the alternate form, $A = \pi \times r \times r$.

ASSIGNMENTS

Level 1	1–10, 19–20, 23–29
Level 2	Odd 1–9, 11–20, 23–30
Level 3	Even 2–18, 19–31

FOLLOW-UP ACTIVITY

APPLICATION Write the following problem on the chalkboard:

The radar on the *USS Forrest* monitors the surrounding area up to a radius of 12 mi. Its rotating hand takes 1 minute to complete 1 revolution. In 5 seconds, how much area does the radar sweep? Round the answer to the nearest square mile. (38 mi^2)

Encourage students to draw a diagram to understand the problem. They should realize that because 5 seconds is $\frac{1}{12}$ of 1 minute, the area surveyed in 5 seconds is $\frac{1}{12}$ of the area surveyed in 1 minute.

SUPPLEMENTARY MATERIALS

TRP Practice, p. 136

TRP Reteaching, p. 84

TRP Lesson Aids, p. 4

TRP Transparency 68

OBJECTIVE

- Solve problems by writing simpler problems.

TEACHING THE LESSON

WARM-UP Write the following exercises on the chalkboard or an overhead transparency.

1. 3×19 **2.** $162.50 − $87.96

3. $624 \div 24$ **4.** $5\frac{1}{4} \times 2\frac{1}{2}$

(**1.** 57 **2.** $74.54 **3.** 26 **4.** ($13\frac{1}{8}$))

INTRODUCTION Read the following problems aloud. Have students explain which operation they would use to solve each problem.

1. There are 400 packages of seeds. Each rack holds 25 packages. How many racks are needed? (division)
2. Jeff sold 15 pine trees to Mr. Moran. Later that day, Mr. Moran came back to buy 8 more pine trees. How many pine trees did Mr. Moran buy? (addition)
3. A box of garden tools contained 4 different types of tools. There were 12 of each type of tool. How many tools were in the box? (multiplication)
4. Anne was making table arrangements. She bought 45 red carnations and 24 pink carnations. How many more red carnations did she buy? (subtraction)

INSTRUCTION After reading the introductory problem, discuss the three-step process that the students will follow: first, writing subproblems; second, substituting rounded numbers; third, solving the problem using the actual numbers given. Point out to students that it will be helpful to compare results using rounded numbers and actual numbers to check the reasonableness of their answers. Since 1,710 is close to 1,600, the answer is reasonable.

PROBLEM Solving STRATEGY

12.6 SOLVING A SIMPLER PROBLEM

Situation:

Delray Condominiums hires Green Valley Contractors to plant 3,990 ivy cuttings along a stone boundary wall. In one day, 4 Green Valley workers plant 570 ivy cuttings each. How many cuttings remain to be planted?

Strategy:

Sometimes it helps to solve a simpler problem. Write subproblems using rounded numbers.

Applying the Strategy:

Subproblem 1: How many ivy cuttings were planted in all?

THINK: NUMBER OF WORKERS × NUMBER OF CUTTINGS

4 × 600 = 2,400

Subproblem 2: How many ivy cuttings remain to be planted?

THINK: TOTAL NUMBER OF CUTTINGS − CUTTINGS ALREADY PLANTED

4,000 − 2,400 = 1,600

Now use the actual numbers.

Subproblem 1: 4 × 570 = 2,280 *Subproblem 2:* 3,990 − 2,280 = 1,710

There are 1,710 ivy cuttings that remain to be planted.

FOR DISCUSSION

See TE side column.

Green Valley needs 115 yd of molding for the Delray lobby. The molding costs $3.85 per yard. Green Valley's installation fee is $1.25 per yard. How much will Delray pay Green Valley for the lobby molding?

1. What rounded numbers would you use to solve the problem?
2. What subproblems would you use?
3. What is the actual solution?

PRACTICE EXERCISES See Extra Practice, page 454.

Write the letter of the better plan for simplifying the problem.

• **1.** A parking lot at Delray needs a new fence. The lot is 52.5 m long and 36.75 m wide. Will the lot need more than 250 m of fencing? b

a. *Step 1:* $50 \times 40 = 2{,}000\ m^2$
Step 2: $2{,}000\ m^2 > 250\ m$

b. *Step 1:* $2 \times 50 + 2 \times 40 = 180\ m$
Step 2: $180\ m < 250\ m$

• **2.** It took 5 gardeners $2\frac{3}{4}$ hours to prepare a lawn area of 1,650 m^2 for seeding. How long would it take one gardener working at the same speed at prepare 330 m^2 of lawn for seeding? a

a. *Step 1:* $1{,}500 \div 3 = 500$
Step 2: $500 \div 5 = 100$
Step 3: $300 \div 100 = 3$ hours

b. *Step 1:* $300 \times 3 = 900$
Step 2: $900 \div 5 = 180$
Step 3: $180 \div 3 = 60$ hours

Solve by writing a simpler problem.

3. Green Valley bought 50 pine trees for \$89.75 each. They sold the trees to Delray Condominiums for \$118.95 each. How much profit did Green Valley make on the pine trees? \$1,460

4. Green Valley is planting 192 juniper bushes along a slope. One morning, 3 workers planted 18 bushes each. How many juniper bushes remain to be planted? 138 juniper bushes

5. It takes $2\frac{1}{4}$ cans of preservative to treat 3 dozen railroad ties. Green Valley plans to treat 512 ties. How much preservative will be needed? 32 cans

6. Green Valley has workers paint the fence around the pool. The painters need $8\frac{1}{4}$ gal of paint for each 210 ft of fence. How much paint is needed to cover 840 ft of fence? 33 gal

7. Delray plans to add a new lawn of 775 m^2. The price to seed is \$1.95 per square meter. The price of sod is \$4.75 per square meter. How much money will be saved if seed is used instead of sod? \$2,170

8. Green Valley uses a special soil mix for its planters. For each 120 lb of topsoil, Green Valley uses $1\frac{3}{4}$ bales of peat moss. A bale weighs 40 lb. How many pounds of peat moss are needed for 420 lb of topsoil? 245 lb

You may wish to point out to students that Exercise 2 involves more than two steps. Note that Exercise 3 can be solved two ways. Students can multiply by 50 and then subtract the total amounts of money, or subtract first to find the difference in price and then multiply by 50. Exercise 8 involves more than two steps.

FOR DISCUSSION

1. Students should explain that 115 rounds to 100; \$3.85 rounds to \$4.00; and \$1.25 rounds to \$1.00.
2. Students should explain that the subproblems are (a) finding the charge to Delray for each yard of molding installed; and (b) finding the total installation charge for all 115 yd of molding.
3. The solution is \$586.50.

ASSIGNMENTS

Level 1	Odd 1–7
Level 2	1–7
Level 3	3–8

FOLLOW-UP ACTIVITY

SITUATIONAL PROBLEM SOLVING Present students with the following problem:

You have been asked to design a condominium complex for 7,000 people. There are 20 acres of land available. What would your housing units look like? How would you use the 20 acres of land? Develop your plan and draw a picture of the condominium complex.

SUPPLEMENTARY MATERIALS

TRP Practice, p. 137

TRP Applications, p. 19

TRP Transparency 69

OBJECTIVE

- Evaluate student progress.

USING THE REVIEW

This page provides a means for informally evaluating students' understanding of the skills and concepts covered so far in this chapter.

Have the students look at the page to familiarize themselves with the various question formats that are presented. Discuss any questions that they may have. Then ask them to complete the page independently.

In addition to grading them individually, you may wish to review the answers to the questions collectively with the students.

Page references appear in the brackets. They refer to pages on which a particular skill was introduced.

Before continuing on to the topics found in the remainder of the chapter, you may wish to have students review any skills or concepts in which they have demonstrated weakness.

MIDCHAPTER REVIEW

Find the perimeter and the area. [330, 334, 336]

1. $P = ■$ 14.4 m **2.** $A = ■$ 12.15 m^2

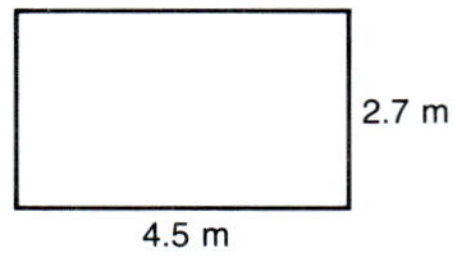

3. $P = ■$ 3 in. **4.** $A = ■$ $\frac{9}{16}$ $in.^2$

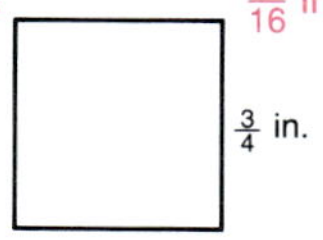

5. $P = ■$ 18 yd **6.** $A = ■$ 12 yd^2

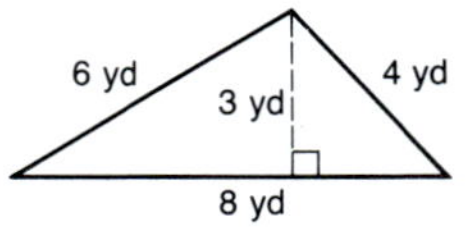

7. $P = ■$ $20\frac{1}{2}$ ft **8.** $A = ■$ 22 ft^2

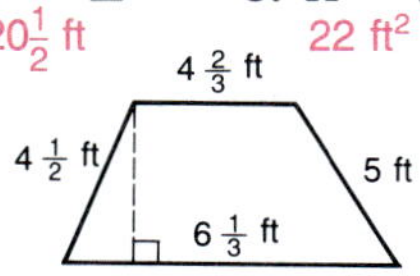

9. $P = ■$ 30 cm **10.** $A = ■$ 41.5 cm^2

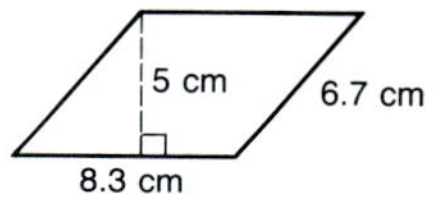

11. $P = ■$ 66 mm **12.** $A = ■$ 110 mm^2

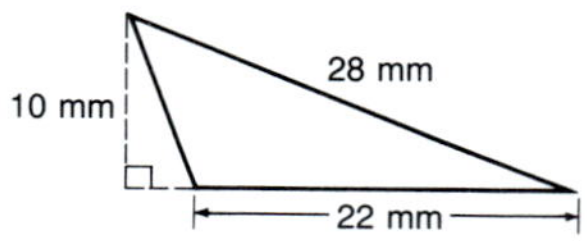

Find the circumference and the area. Use 3.14 for π with whole numbers and decimals. Use $\frac{22}{7}$ for π with fractions and mixed numbers. [332, 338]

13. $C = ■$ 12.56 m **14.** $A = ■$ 12.56 m^2

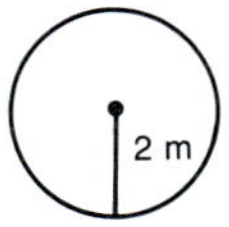

15. $C = ■$ $2\frac{2}{21}$ ft **16.** $A = ■$ $\frac{22}{63}$ ft^2

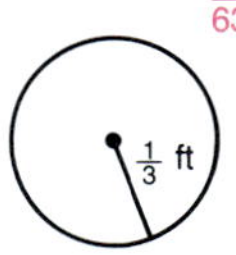

17. $C = ■$ 13.188 cm **18.** $A = ■$ 13.8474 cm^2

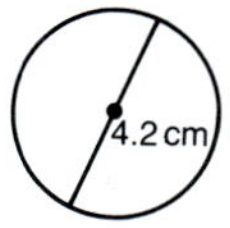

Solve. [340]

19. The Golden Hind is a deep-sea fishing boat out of Provincetown. It takes $2\frac{3}{4}$ hours for the Golden Hind to travel the 33 mi to the fishing grounds. How far can the Golden Hind travel in 2 hours? 24 mi

20. The mate who runs the snack bar uses $13\frac{3}{4}$ oz of cheese to make 5 ham and cheese sandwiches. Assuming he has enough ham and bread, how many sandwiches can he make from $92\frac{1}{2}$ oz of cheese? 33 sandwiches

21. It takes the hook 42 seconds to reach the ocean floor 315 ft below. How long would it take the hook to reach the ocean floor at a depth of $247\frac{1}{2}$ ft? 33 seconds

22. A 6-lb cod produces two $2\frac{1}{4}$-lb fillets. The Mull family leaves the Golden Hind with $31\frac{1}{2}$ lb of cod. This is the equivalent of how many 6-lb fish? seven 6-lb fish

PROBLEM Solving APPLICATION

12.7 CAREER: CARPET INSTALLER

Charles Monet is a carpet installer. The ability to measure accurately is important in wall-to-wall carpeting. Often, a room may consist of more than one rectangular section. Sometimes several rooms and a hallway are carpeted with the same type of carpet.

Find the number of square yards of carpeting needed.

1.

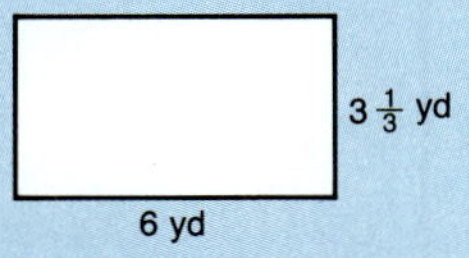

20 yd^2

2.

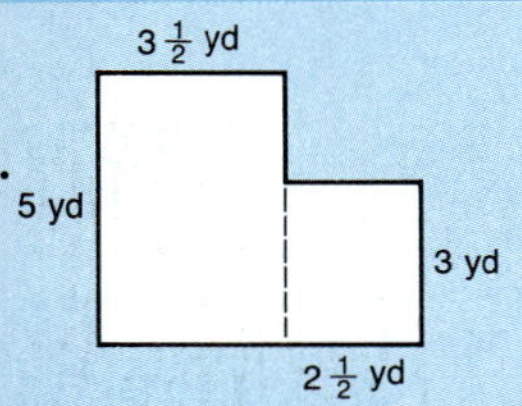

25 yd^2

3.

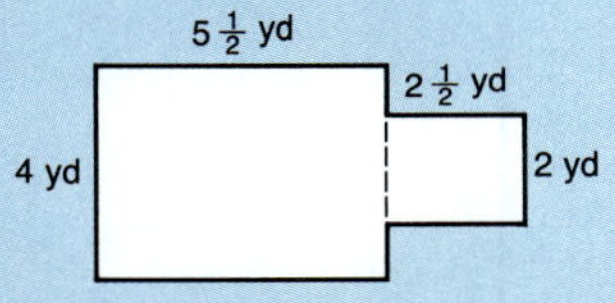

27 yd^2

4.

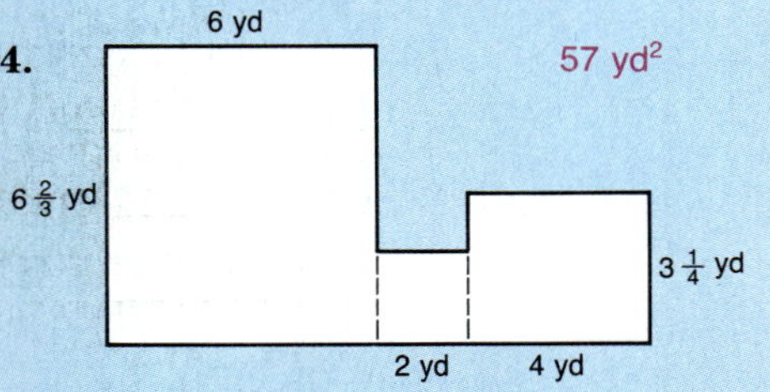

57 yd^2

Solve.

5. Charles installs wall-to-wall carpeting in a rectangular room that is 6.5 yd long and 4.0 yd wide. What is the area covered by the carpet? 26 yd^2

6. In the same house, Charles carpets a rectangular room 5.8 yd long and 4.2 yd wide. How much less is the area covered by this carpet than the area covered in Problem 5? 1.64 yd^2 less

7. During a period of one week, Charles worked a total of 35 hours. Of this time, 24.5 hours were spent installing carpets. 10.5 hours were spent traveling to and from the clients' homes. What percent of Charles' time was spent traveling? 30%

Charles' cousin, Ann, decides to buy wall-to-wall carpeting for a rectangular room 21 ft long and 18 ft wide.

8. What is the area of the floor in square yards? 42 yd^2

9. When Ann orders the carpeting, should she order an amount that is less than, equal to, or greater than the area of the floor? Why?
Greater than, to ensure allowance for waste and actual layout of rooms.

OBJECTIVE

- Solve problems that involve finding the area of floors.

TEACHING THE LESSON

WARM-UP Write the following exercises on the chalkboard or an overhead transparency.

Find the product.

1. $6 \times 2\frac{1}{2}$ (15) **2.** $8 \times 5\frac{1}{4}$ (42)

3. $7\frac{1}{3} \times 5$ $(36\frac{2}{3})$ **4.** 5.5×4.0 (22.0)

INSTRUCTION Point out that the diagrams in Problems 1–4 show rectangular rooms and halls.

Note that Problem 7 includes extraneous data (the 24.5 hours is not needed). In Problem 8, the students need to express 21 ft and 18 ft as yards before finding the area.

Problem 9 can be used as a springboard for discussion of the fact that there is some waste involved in carpeting rooms. One estimates in ordering carpeting, and orders *more than enough.* Emphasize the importance of estimation especially when the measurements include mixed numbers.

You may wish to have students use calculators to help them solve the problems.

ASSIGNMENTS

Level 1	1–6
Level 2	1–9
Level 3	3–9

SUPPLEMENTARY MATERIALS

TRP Practice, p. 138

TRP Transparency 69

OBJECTIVES

- Name solid figures
- Count the numbers of faces, edges, and vertices of solid figures.

TEACHING THE LESSON

WARM-UP Write the following exercises on the chalkboard or an overhead transparency.

Name each figure.

1. 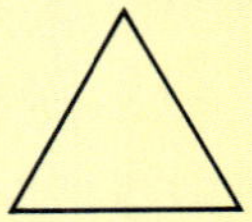

2. 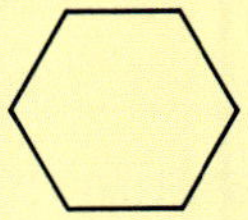

3. 

4. 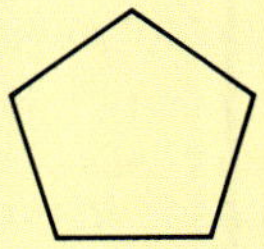

(**1.** triangle **2.** hexagon **3.** square **4.** pentagon)

INTRODUCTION Explain that so far the chapter has dealt only with flat figures that have only two dimensions, length and width. Display a model of a square cut from paper and a toy block or photo cube.

Ask how the cube differs from the square. Point out that the cube is a three-dimensional figure. Continue by asking what a three-dimensional rectangle and what a three-dimensional circle would look like. Draw or display models of a rectangular prism and a sphere.

INSTRUCTION When discussing Example 1, point out that prisms are polygons "pulled" through space. Demonstrate by cutting out two congruent paper triangles and attaching congruent lengths of string between corresponding vertices. Lay the triangles on top of each other to represent the polygonal base. Then pull them apart until the string is taut.

After discussing Example 2, you may wish to have students experiment with three-dimensional drawing by asking them to draw a rectangular pyramid.

12.8 SOLID FIGURES

Objects in the real world are called three-dimensional if they have length, width, and height. In mathematics, three-dimensional objects are known as **solid figures.**

A **polyhedron** is a solid figure whose sides are all polygons. Each side is called a **face.** Two faces intersect in a line segment called an **edge.** Two edges intersect in a point called a **vertex.**

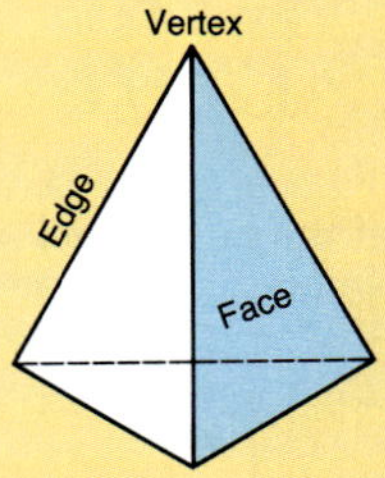

A **prism** is a polyhedron whose bases are congruent polygons and whose other faces are parallelograms. A prism is named according to the shape of its bases. A cube is a prism whose faces are all squares.

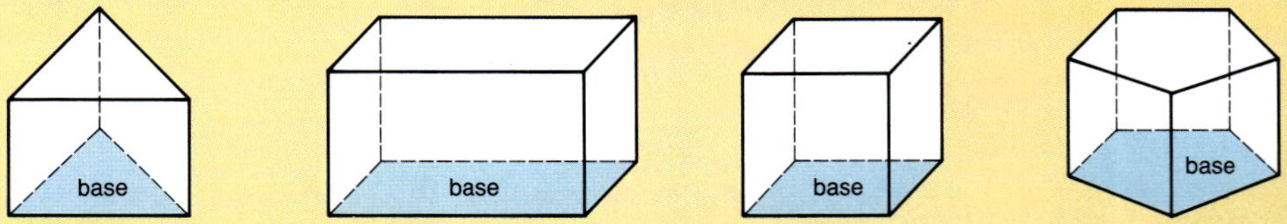

Triangular prism
Rectangular prism
Cube
Pentagonal prism

EXAMPLE 1

Identify the figure. Count the numbers of faces, edges, and vertices.

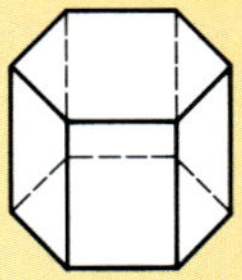

THINK: The sides are all polygons.
The figure is a prism.
Its bases are hexagons.

The figure is a hexagonal prism.
It has 8 faces, 18 edges, and 12 vertices.

A **pyramid** is a polyhedron whose base is any polygon and whose other faces are triangles that all meet at a common vertex. A pyramid is also named according to the shape of its base. The figure shown at the right is a square pyramid.

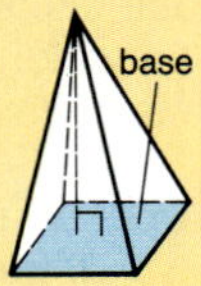

EXAMPLE 2

Identify the figure. Count the numbers of faces, edges, and vertices.

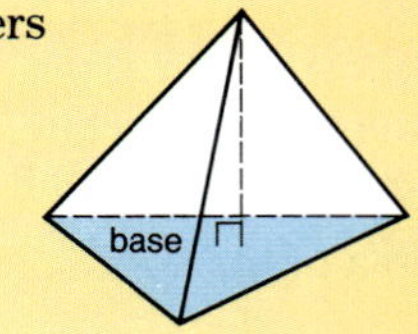

THINK: The sides are all triangles..
The figure is a pyramid.
Its base is a triangle.

The figure is a triangular pyramid.
It has 4 faces, 6 edges, and 4 vertices.

Other solid figures have curved surfaces.

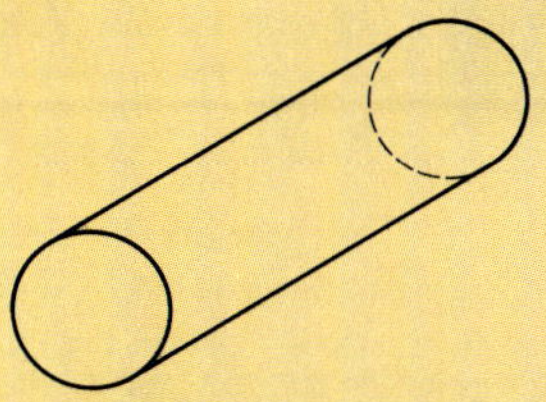

A **cylinder** has two circular bases that are congruent and parallel.

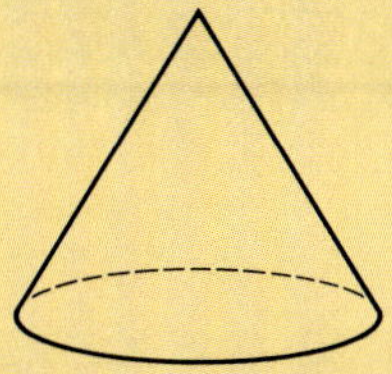

A **cone** has one circular base.

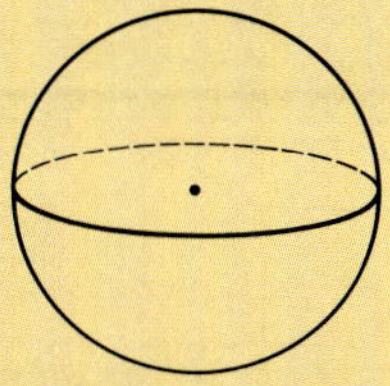

A **sphere** is a curved surface of points that are all the same distance from the center.

EXAMPLE 3 Write whether the following statement is *true* or *false*.

The circular bases of a cylinder can have different sizes.

THINK: According to the definition of a cylinder, the circular bases must be congruent.

The statement is false.

FOR DISCUSSION See TE side column.

Name objects from everyday life that have the following shapes: a rectangular prism, a cube, a square pyramid, a cylinder, a cone, a sphere.

PRACTICE EXERCISES See Extra Practice, page 455.

Identify the figure.

• **1.**

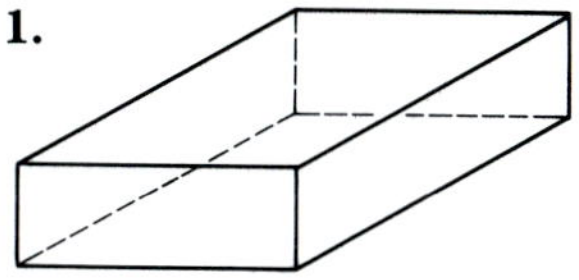

rectangular prism

• **2.**

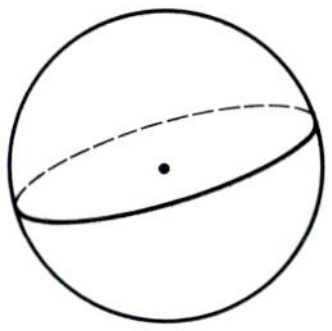

sphere

3.

cylinder

4.

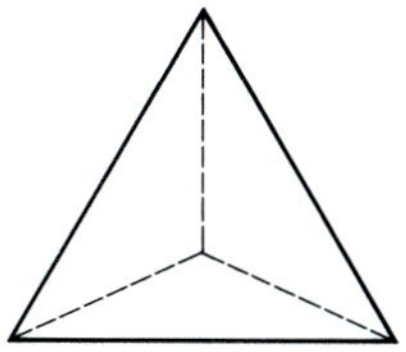

triangular pyramid

5.

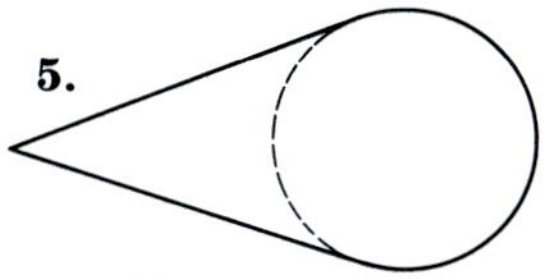

cone

6.

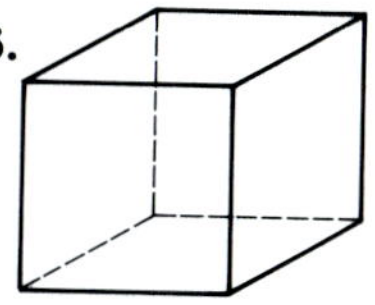

cube

Pay particular attention to the difficulty of counting unseen faces, edges, and vertices. Extend the instruction by guiding students in counting the numbers of faces, edges, and vertices in Exercises 1 and 4.

When discussing Example 3, point out that all these solids have some faces that are not polygons. Therefore, none are polyhedra.

Call attention to the difference between a circle and a sphere. Emphasize that a circle is the set of all points in a plane equidistant from a given point, whereas a sphere is the set of all points in space.

FOR DISCUSSION Students should explain that possible examples are a shoe box (rectangular prism), an ice cube (cube), the top of the Washington Monument (pyramid), a can of soup (cylinder), an ice-cream cone (cone), and a basketball (sphere).

TIME OUT The key here is to consider the third dimension. If students encounter difficulty, tell them that the title of this lesson is an important hint.

ASSIGNMENTS

Level 1	1–15, 19–27, TO
Level 2	1–27, TO
Level 3	Odd 7–27, TO

FOLLOW-UP ACTIVITY

APPLICATION Some students may be interested in classifying polyhedra by the number of faces:

4 faces = tetrahedron

6 faces = hexahedron

8 faces = octahedron

12 faces = dodecahedron

Encourage students to build models of the above polyhedra by gluing straws or toothpicks together.

SUPPLEMENTARY MATERIALS

TRP Practice, p. 139

TRP Reteaching, p. 85

TRP Lesson Aids, pp. 9–11

TRP Transparency 70

Copy and complete the table.

Polyhedron	Number of vertices	Number of faces	Number of edges
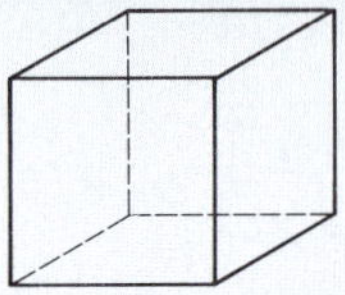	• 7. ■ 8	• 8. ■ 6	• 9. ■ 12
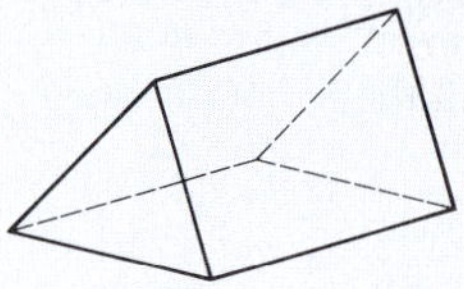	10. ■ 6	11. ■ 5	12. ■ 9
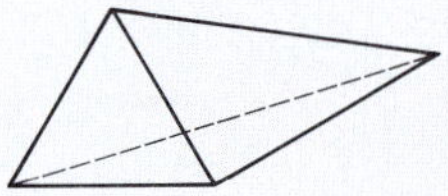	13. ■ 4	14. ■ 4	15. ■ 6
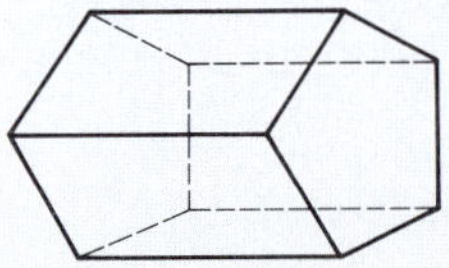	16. ■ 10	17. ■ 7	18. ■ 15

Write whether the statement is *true* or *false*.

• **19.** A sphere is a type of prism. false

• **20.** A cone is a circular pyramid. false

21. All prisms have 6 faces and 8 vertices. false

22. An octagonal prism has 8 edges. false

23. A prism is a type of polyhedron. true

24. A cylinder is a type of polyhedron. false

25. A pyramid is named by the shape of its base. true

26. A cube is a prism with 6 congruent square faces. true

27. The faces of a polyhedron must all be rectangles. false

TIME OUT Using 12 unbroken matchsticks touching end to end, it is possible to form 6 squares. How? by forming a cube

346 CHAPTER 12

12.9 SURFACE AREA

Deborah Kline owns an antique shop. She is refinishing the trunk shown at the right. The can of varnish she is using has enough liquid left in it to cover 30 ft^2. Is there enough varnish left in the can to refinish the trunk?

To find whether there is enough varnish left, you need to know the surface area of the trunk.

The **surface area** of a polyhedron is equal to the sum of the areas of its faces.

EXAMPLE 1 Find the surface area of the trunk.

1. Find the area of each face.

THINK: The faces are rectangles. Use $A = l \times w$.

Top/bottom: $3 \times 2 = 6$ ft^2 each
Front/back: $3 \times 1 = 3$ ft^2 each
Sides: $2 \times 1 = 2$ ft^2 each

2. Add the areas of the faces.

$(2 \times 6) + (2 \times 3) + (2 \times 2) = 22$ ft^2
top/bottom **front/back** **sides**

The surface area of the trunk is 22 ft^2.
There is enough varnish left to refinish the trunk.

Because the 6 faces of a cube are all congruent squares, the surface area of a cube is 6 times the area of one face, or $6 \times s^2$.

EXAMPLE 2 Find the surface area of the cube.

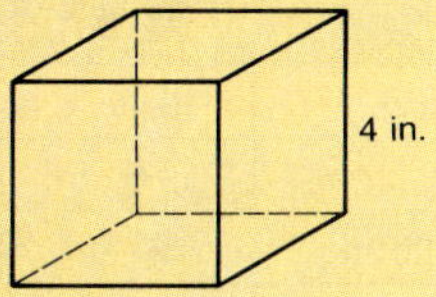

THINK: $A = 6 \times s^2$

$A = 6 \times 4^2 = 96$

The surface area of the cube is 96 $in.^2$.

You can use rounding to estimate the surface area of a prism.

EXAMPLE 3 Estimate the surface area of the cube.

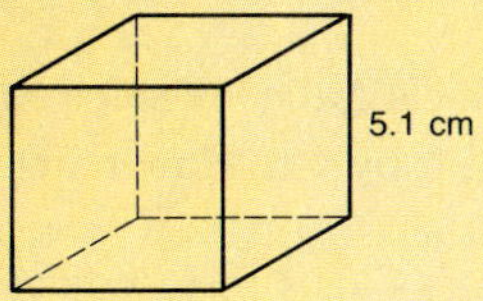

THINK: 5.1 rounds to 5.

Estimate: $A = 6 \times s^2$

$A = 6 \times 5^2 = 150$

The surface area of the cube is about 150 cm^2.

OBJECTIVES

- Find the surface area of a rectangular prism or cube.
- Estimate the surface area of a rectangular prism or cube.

TEACHING THE LESSON

WARM-UP Write the following exercises on the chalkboard or an overhead transparency.

Multiply.

1. 6×3^2 **2.** $4 \times 7\frac{1}{2}$ **3.** 6.2×4.1

4. $(2 \times 3.2) + (2 \times 1.3) + (2 \times 4)$

(**1.** 54 **2.** 30 **3.** 25.42 **4.** 17)

INTRODUCTION Display a shoebox. Ask what solid figure it resembles. (rectangular prism) Ask which faces are congruent. Guide students to realize that opposite faces are congruent and that their areas are congruent. Repeat with rectangular prisms of different sizes.

INSTRUCTION In discussing Example 1, make sure students understand that the factor of 2 in each term of Step 2 represents the pair of congruent faces. In other words, 6 ft^2 is the area of either the top or the bottom; 2×6 ft^2 is the total area of both.

In addition, you may wish to point out that surface area can also be found by adding together the areas of three different faces and then multiplying by 2.

$A = 2 \times [(3 \times 1) + (3 \times 2) + (2 \times 1)]$

$A = 2 \times (3 + 6 + 2) = 2 \times 11 = 22$ ft^2

When discussing Example 2, remind students that s^2 means $s \times s$.

In Example 3, note that the numbers are rounded to the nearest whole number. Point out that the result is only an approximation but that it can be useful as a check when using a calculator to compute the answer.

COMMON ERROR

When the dimensions of a rectangular prism are given in different units, students may sometimes ignore the units and merely multiply the numbers. To remedy this, remind students that the length, width, and height must all be converted to the same units before they can find the surface area.

ASSIGNMENTS

Level 1	1–6, 10–11, 14
Level 2	Odd 1–13, 14–15
Level 3	Even 2–12, 14–15

FOLLOW-UP ACTIVITY

COOPERATIVE LEARNING Draw the following diagram of the Great Pyramid of Cheops on the chalkboard:

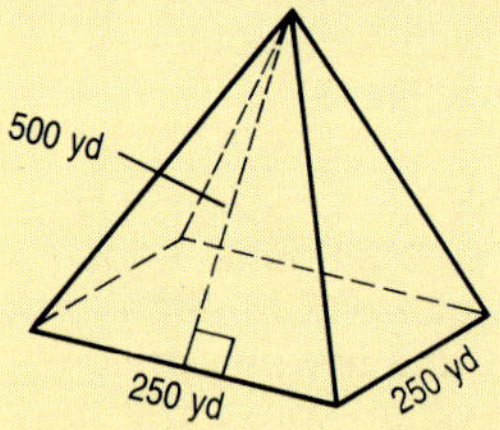

Have students work in small groups to find the surface area of the Great Pyramid. (312,500 yd^2) Provide the following hints if needed:

- The triangular faces are congruent. A shortcut is to find the area of one face, then multiply by 4.
- The base is in the shape of a square.

SUPPLEMENTARY MATERIALS

TRP Practice, p. 140

TRP Reteaching, p. 86

TRP Lesson Aids, pp. 9–10

TRP Transparency 70

PRACTICE EXERCISES

See Extra Practice, page 455.

Find the surface area.

• **1.**

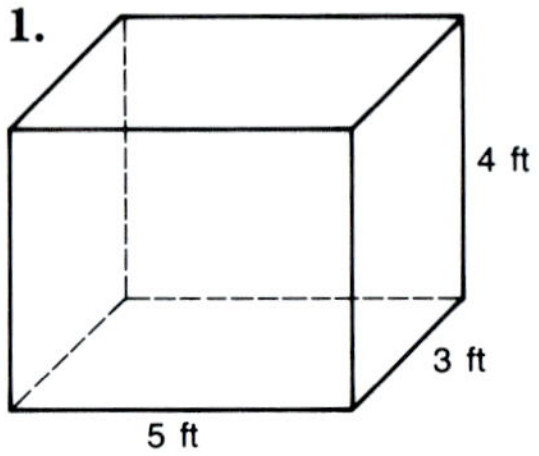

94 ft^2

2.

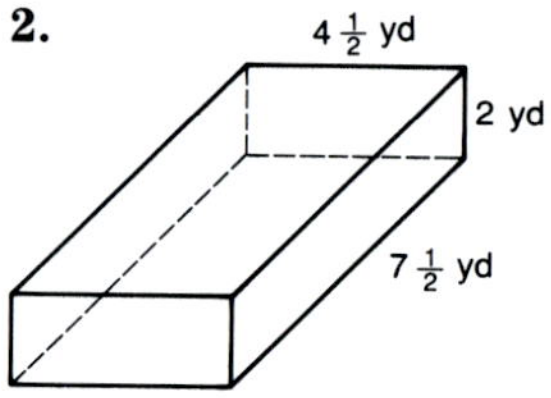

$115\frac{1}{2}$ yd^2

3.

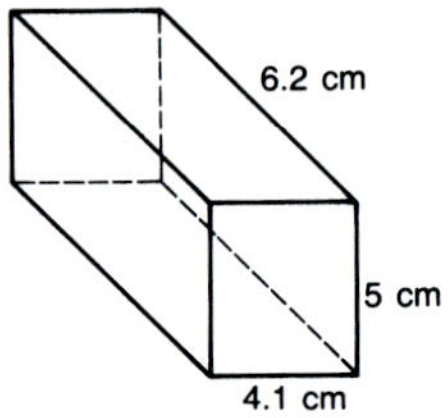

153.84 cm^2

• **4.**

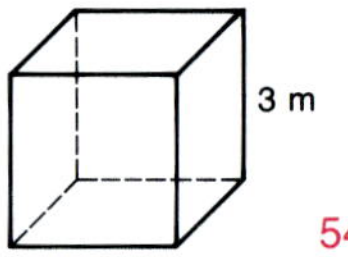

54 m^2

5.

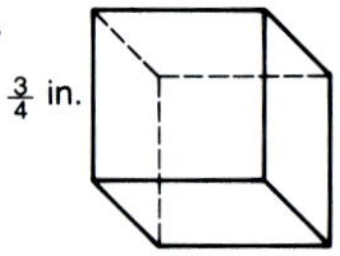

$3\frac{3}{8}$ $in.^2$

6.

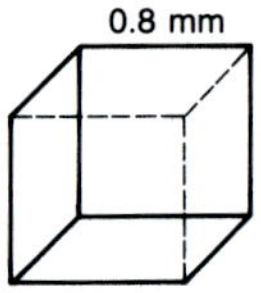

3.84 mm^2

7.

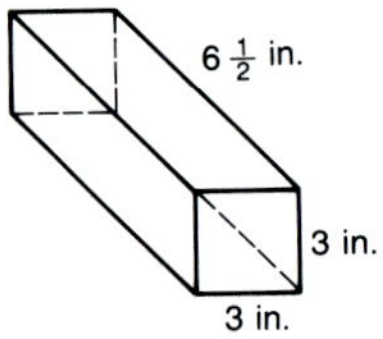

96 $in.^2$

8.

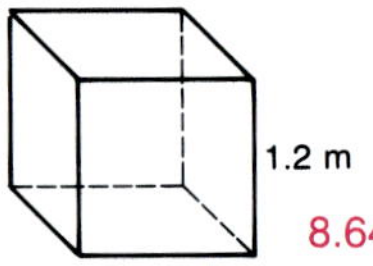

8.64 m^2

9.

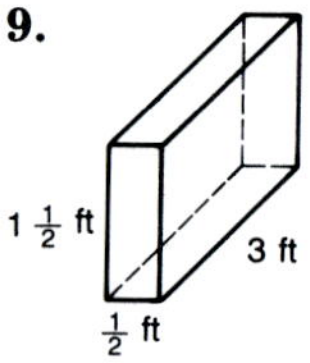

$13\frac{1}{2}$ ft^2

Complete the table.
Each set of measurements describes a rectangular prism.

Length	Width	Height	Estimated surface area	
• **10.** 62 ft	19 ft	101 ft	■	18,400 ft^2
• **11.** 10.2 cm	19.6 cm	72 cm	■	4,600 cm^2
12. 3.06 m	8.97 m	6.18 m	■	200 m^2
13. $\frac{7}{16}$ in.	$\frac{7}{16}$ in.	$\frac{7}{16}$ in.	■	$1\frac{1}{2}$ $in.^2$

Solve.

14. Kline's Antiques ships sofas in plywood crates measuring 3 ft by $10\frac{1}{2}$ ft by 4 ft. How much plywood does it take to make each crate? 171 ft^2

15. Kline's sells restored spice mills. Each mill is a cube 8 in. on a side. A pint of stain will cover 64 ft^2. How many mills will that same pint cover? (*Hint:* 1 ft^2 = 144 $in.^2$) 24 mills

PROBLEM Solving APPLICATION

12.10 CONSUMER: GIFT WRAPPING A PACKAGE

A package has the shape of a rectangular prism, with the dimensions shown. You want to gift wrap the package. When wrapping a package, you need more than the minimum surface area. You need to use an additional 20% to allow paper for folding and overlapping. How many square inches of wrapping paper do you need to wrap the package?

1. Find the total surface area.

Top and bottom: $2 \times (5 \times 3) =$	30
Front and back: $2 \times (6 \times 5) =$	60
Sides: $2 \times (6 \times 3) =$	36
Total:	126

2. Find 20% of the total surface area to allow for extra paper needed.

THINK: $20\% = 0.20$ $\qquad$ $0.20 \times 126 = 25.2$

3. Add to find the total amount of paper needed. $\qquad$ $126 + 25.2 = 151.2 \approx 151$

So, you need about 151 in.^{2} of gift wrapping paper.

Find the total amount, in square inches, of gift wrapping paper needed. Remember to allow for extra paper. Round the answer to the nearest in.^{2}

• **1.**

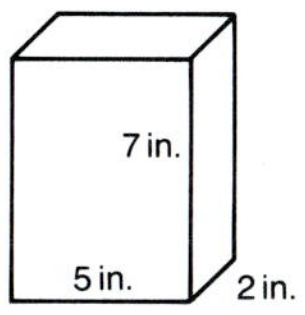

142 in.^{2}

2.

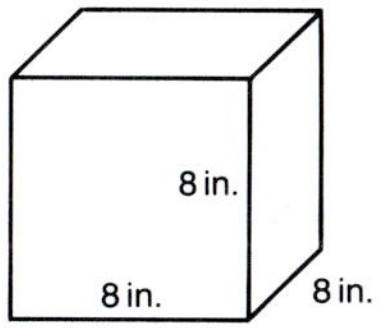

461 in.^{2}

3.

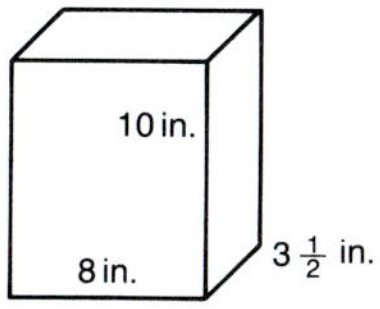

343 in.^{2}

• **4.** A box with a length of 2 ft, a width of 1 ft, and a height of 1 ft. 1,728 in.^{2}

5. A box with a length of 1 ft 6 in., a width of 1 ft 6 in., and a height of 8 in. 1,469 in.^{2}

You want to wrap a package with $l = 10$ in., $w = 6$ in., and $h = 4$ in.

6. Will 300 in.^{2} of wrapping paper cover the package? Will you have any paper left over? If so, how much? yes; yes; 2 in.^{2}

7. Will a sheet of wrapping paper measuring 12 in. by 15 in. cover the package? How many inches short or over will the sheet be? no; 118 in.^{2} short

OBJECTIVE

- Solve problems that involve finding the surface area of a rectangular prism.

TEACHING THE LESSON

WARM-UP Write the following exercises on the chalkboard or an overhead transparency.

Find the area of the rectangle.

1. $l = 8$ in., $w = 6$ in. (48 in.^{2})

2. $l = 12$ in., $w = 6$ in. (72 in.^{2})

3. $l = 5$ in., $w = 5$ in. (25 in.^{2})

INTRODUCTION Discuss the students' experiences in gift wrapping packages. Ask what difficulties they encountered and how they could avoid them in the future.

INSTRUCTION If necessary, use the introductory example to review the procedure for finding the surface area of a rectangular prism.

In Problem 5, the students should express the dimensions of the prism in a common unit before computing the areas of the faces. Some students may express the dimensions as $1\frac{1}{2}$ ft, $1\frac{1}{2}$ ft, and $\frac{2}{3}$ ft, and compute the areas of the faces in square feet. Others may express the dimensions as 18 in., 18 in., and 8 in., and compute the areas of the faces in square inches.

You may wish to encourage students to use a calculator, including the memory function, in solving the problems.

ASSIGNMENTS

Level 1	1–4, 6
Level 2	1–7
Level 3	3–7

SUPPLEMENTARY MATERIALS

TRP Practice, p. 141

TRP Transparency 71

OBJECTIVES

- Find the volume of a rectangular prism.
- Find the volume of a pyramid.

TEACHING THE LESSON

WARM-UP Write the following exercises on the chalkboard or an overhead transparency.

Multiply.

1. $10 \times 10 \times 10$

2. $8 \times 4\frac{1}{2} \times 4$

3. $\frac{1}{3} \times (2.8 \times 3 \times 3.2)$

4. $\frac{2}{3} \times \frac{2}{3} \times \frac{2}{3}$

(**1.** 1,000 **2.** 144 **3.** 8.96 **4.** $\frac{8}{27}$)

INTRODUCTION Use cubic blocks to construct a rectangular prism 4 blocks long, 3 blocks wide, and 2 blocks high. Ask students how many blocks there are in this solid figure. (24) Then ask them how they got the answer. Point out that $4 \times 3 \times 2$ also equals 24, that 4×3 represents the area of the base and 2 represents the number of bases used, or layers.

INSTRUCTION When discussing Example 1, point out that the formula $V = B \times h$ works for any prism. Draw a pentagonal prism on the chalkboard. Tell the students that the pentagonal base has an area of 12 m^2 and that the prism is 6 m high. Have them find its volume. (72 m^3)

12.11 VOLUMES OF PRISMS AND PYRAMIDS

Denise Grant keeps tropical fish. A model of her aquarium is shown at the right. Each fish needs at least 0.5 ft^3 of water. What is the maximum number of fish she can keep in the aquarium?

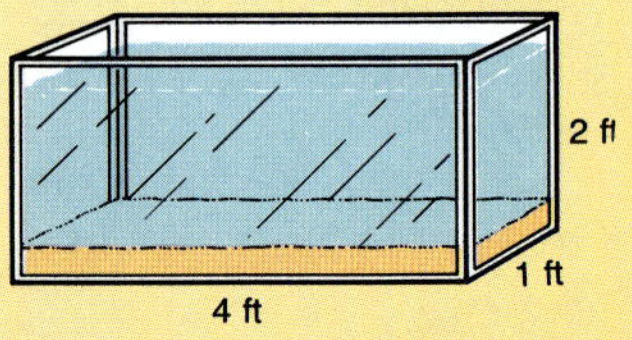

To find the answer, you need to find the volume of the aquarium.

Volume is the amount of space inside a solid figure. It is measured in cubic units such as cubic meters (m^3), cubic centimeters (cm^3), cubic yards (yd^3), cubic feet (ft^3), and cubic inches ($in.^3$).

To find the volume of a prism, multiply the area of its base by its height.

EXAMPLE 1 Find the volume of the aquarium.

THINK: The aquarium is a rectangular prism.

$V =$ area of base (B) $\times$ height (h)

$V = (l \times w) \times h$

$V = (4 \times 1) \times 2$

$V = 8$

The aquarium holds 8 ft^3 of water. Since each fish needs at least 0.5 ft^3 of water, the aquarium can hold a maximum of 16 fish ($8 \div 0.5 = 16$).

The volume of a pyramid is $\frac{1}{3}$ the volume of the prism that has the same base and height.

EXAMPLE 2 Find the volume of the pyramid.

THINK: $V = \frac{1}{3} \times B \times h$

$V = \frac{1}{3} \times (l \times w) \times h$

$V = \frac{1}{3} \times (10 \times 6) \times 4$

$V = 80$

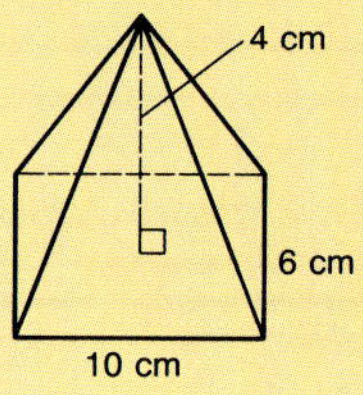

So, the volume of the pyramid is 80 cm^3.

FOR DISCUSSION See TE side column.

A number raised to the third power is called the cube of that number. Why is it called the cube?

PRACTICE EXERCISES See Extra Practice, page 455.

Find the volume of the rectangular prism.

• **1.** 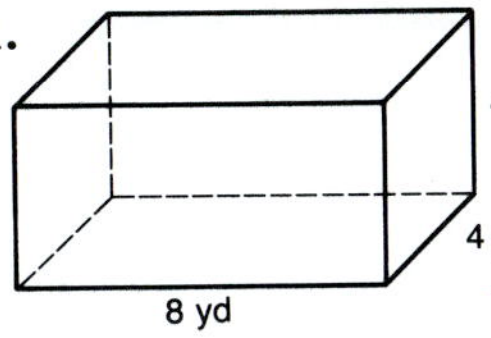

128 yd^3

2. 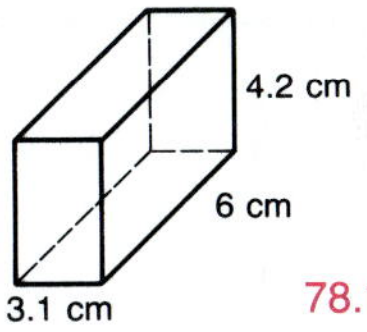

78.12 cm^3

3. 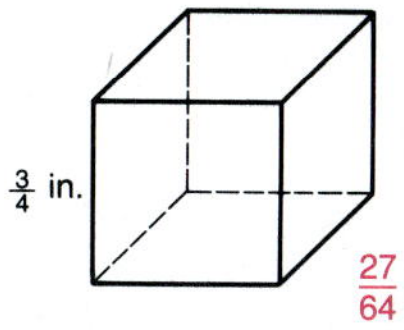

$\frac{27}{64}$ $in.^3$

	Length	Width	Height		Length	Width	Height
• **4.**	5 in.	4 in.	6 in. 120 $in.^3$	• **5.**	9 ft	9 ft	9 ft 729 ft^3
6.	$1\frac{2}{3}$ yd	6 yd	$2\frac{1}{2}$ yd 25 yd^3	**7.**	$\frac{1}{3}$ ft	$\frac{1}{3}$ ft	$\frac{1}{3}$ ft $\frac{1}{27}$ ft^3
8.	6 cm	6 cm	6 cm 216 cm^3	**9.**	9 m	2 m	5 m 90 m^3
10.	3.4 cm	2 cm	5.2 cm 35.36 cm^3	**11.**	12 m	4.5 m	0.5 m 27 m^3

Find the volume of the rectangular pyramid.

• **12.** 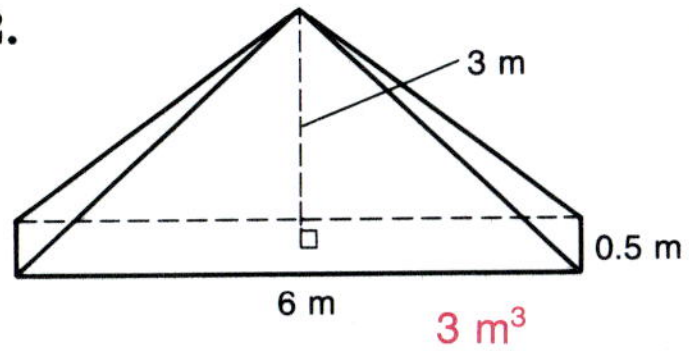

3 m^3

13. 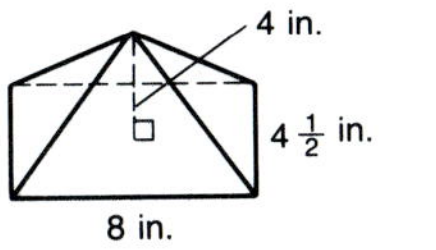

48 $in.^3$

14. 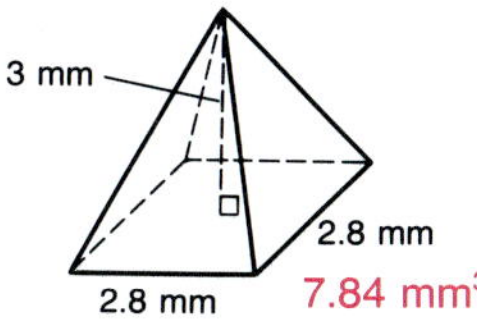

7.84 mm^3

	Length	Width	Height		Length	Width	Height
• **15.**	9 ft	5 ft	13 ft 195 ft^3	• **16.**	12 yd	3 yd	7 yd 84 yd^3
17.	$2\frac{1}{2}$ in.	4 in.	3 in. 10 $in.^3$	**18.**	$\frac{3}{4}$ ft	$\frac{3}{4}$ ft	16 ft 3 ft^3
19.	8 m	3 m	4 m 32 m^3	**20.**	6 cm	6 cm	6 cm 72 cm^3
21.	5.7 cm	3 cm	3.2 cm 18.24 cm^3	**22.**	0.5 m	1.6 m	6 m 1.6 m^3

Solve.

• **23.** The filter Denise Grant uses with her aquarium cleans 15 gal of water an hour. What part of a gallon does it clean in a minute? $\frac{1}{4}$ gal

24. The largest aquarium on sale at Fish U.S.A. is the 132-ft^3 Seascape model. The Seascape is 6 ft long and 4 ft high. How wide is the Seascape? $5\frac{1}{2}$ ft

25. A display aquarium at Fish U.S.A. measures 36 in. by 24 in. by 18 in. To the nearest gallon, how many gallons does the tank hold? (*Hint:* 231 $in.^3$ hold 1 gal.) 67 gal

26. Denise buys a saltwater tank 4 ft long, 2 ft wide, and 30 in. high. What is the area of the lamp hood that fits on top? 8 ft^2

When discussing Example 2, emphasize that, as in the formula for the volume of a prism, the *B* stands for the area of the figure's base. Were the figure a triangular pyramid, for instance, the formula would be

$$\frac{1}{3} \times (\frac{1}{2} \times b \times h) \times h$$

FOR DISCUSSION Students should explain that a number raised to the third power is called a cube because it is numerically equivalent to the volume of the cube whose side has a length equal to that number.

COMMON ERROR

Some students may confuse the *B* in the volume formulas with the *b* from the area formulas covered in previous lessons. This leads them to substitute lengths when areas are called for. Stress that the capital *B* stands for an area, while the lowercase *b* always stands for linear measures.

ASSIGNMENTS

Level 1	1–7, 12–18, 23–24
Level 2	Odd 1–21, 23–25
Level 3	Even 2–22, 23–26

FOLLOW-UP ACTIVITY

APPLICATION Present the students with the following problem:

The Kearns House in Ancram has an unusual octagonal shower. To cover the floor, $1\frac{1}{4}$ boxes of tile were used. Each box contained 20 ft^2 of tile. The stall is 7 ft high. What is its volume? (175 ft^3)

SUPPLEMENTARY MATERIALS

TRP Practice, p. 142

TRP Reteaching, p. 87

TRP Lesson Aids, pp. 9–10

TRP Transparency 71

OBJECTIVES

- Find the volume of a cylinder.
- Find the volume of a cone.

TEACHING THE LESSON

WARM-UP Write the following exercises on the chalkboard or an overhead transparency.

Multiply.

1. $3.14 \times 2^2 \times 10$ (125.6)
2. $\frac{22}{7} \times (\frac{7}{4})^2 \times \frac{1}{7}$ $(1\frac{3}{8})$
3. $\frac{1}{3} \times 3.14 \times 0.3^2 \times 10$ (0.942)
4. $\frac{1}{3} \times \frac{22}{7} \times (\frac{1}{2})^2 \times 7$ $(1\frac{5}{6})$

INTRODUCTION Place a stack of 10 identical coins on a flat surface. Discuss how to find the volume of this cylinder. Suggest that, as in a prism, each coin is a layer and the number of coins is therefore the number of layers.

INSTRUCTION When discussing Example 1, explain that $\pi \times r^2$ is inserted because it represents the area of the base B of the cylinder, which is a circle.

You may also wish to expand the renaming of cubic feet as gallons by including the converting of units as shown below.

$$7\frac{1}{2} \text{ gal/ft}^2 \times 616 \text{ ft}^2 = 4{,}620 \text{ gal}$$

In discussing Example 2, point out that the relationship between a cone and a cylinder is the same as that between a pyramid and a prism—a cone is $\frac{1}{3}$ the volume of the cylinder.

CHECKPOINT The incorrect choices include common errors students make.

12.12 VOLUME OF CYLINDERS AND CONES

Erik Henson spends his summers working for the Homer Pool Company. The pool shown at the right is the Odyssey model. What is its volume? What is its capacity in gallons? (*Hint:* 1 ft³ holds $7\frac{1}{2}$ gal.)

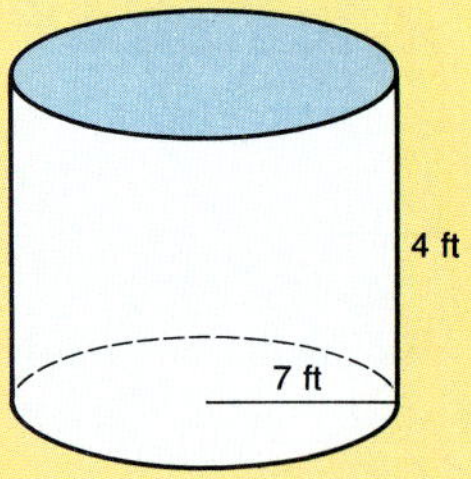

The pool is a cylinder. To find the capacity of the pool, you need to find the volume of the cylinder.

To find the volume of a cylinder, multiply the area of its base by its height.

EXAMPLE 1 Find the volume of the pool.

THINK: V = area of base (B) × height (h)

$V = (\pi \times r^2) \times h$

$V \approx \left(\frac{22}{7} \times 7^2\right) \times 4$

$V \approx 616$

The volume of the swimming pool is approximately 616 ft³. Its capacity is approximately $7\frac{1}{2} \times 616$, or 4,620, gal.

The volume of a cone is $\frac{1}{3}$ the volume of the cylinder that has the same base and height.

EXAMPLE 2 Find the volume of the cone.

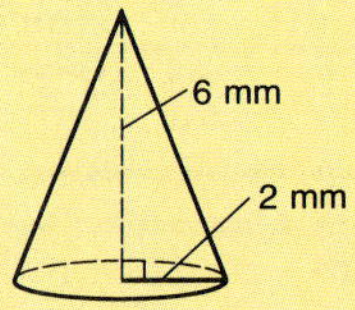

THINK: $V = \frac{1}{3} \times B \times h$

$V = \frac{1}{3} \times (\pi \times r^2) \times h$

$V \approx \frac{1}{3} \times (3.14 \times 2^2) \times 6$

$V \approx 25.12$

The volume of the cone is approximately 25.12 mm³.

CHECKPOINT

Write the letter of the correct answer.

Find the volume of the figure.

1. Cylinder: r = 3 mi, h = 7 mi b

a. 462 mi³ **b.** 198 mi³
c. 154 mi³ **d.** 66 mi³

2. Cone: r = 10 m, h = 3 m c

a. 94.2 m³ **b.** 282.6 m³
c. 314 m³ **d.** 942 m³

PRACTICE EXERCISES See Extra Practice, page 456.

Find the volume of the cylinder. Use 3.14 for π with whole numbers and decimals. Use $\frac{22}{7}$ for π with fractions and mixed numbers.

• **1.**

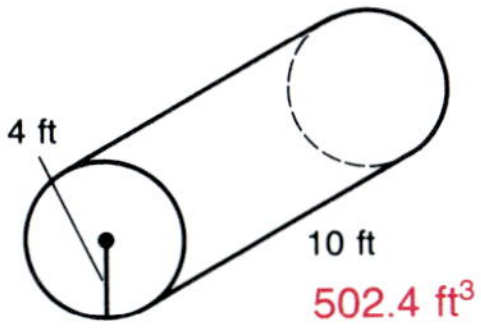

502.4 ft^3

2.

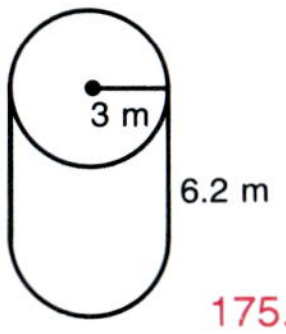

175.212 m^3

3.

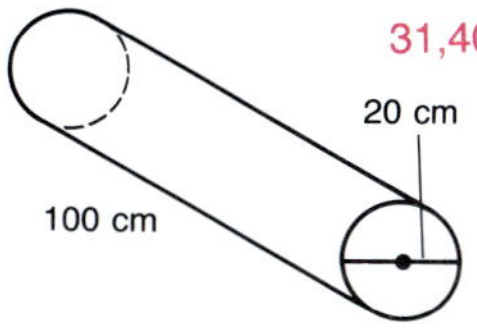

31,400 cm^3

	Radius	Height		Radius	Height		Radius	Height
• **4.**	2 ft	3 ft 37.68 ft^3	• **5.**	$\frac{3}{4}$ yd	2 yd $3\frac{15}{28}$ yd^3	**6.**	1 yd	$3\frac{1}{2}$ yd 11 yd^3
7.	10 km	4.3 km 1,350.2 km^3	**8.**	9 cm	3.2 cm 813.888 cm^3	**9.**	0.5 m	5 m 3.925 m^3

Find the volume of the cone. Use 3.14 for π with whole numbers and decimals. Use $\frac{22}{7}$ for π with fractions and mixed numbers.

• **10.**

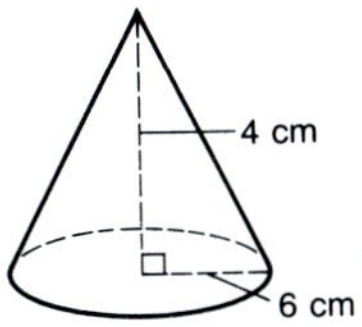

150.72 cm^3

11.

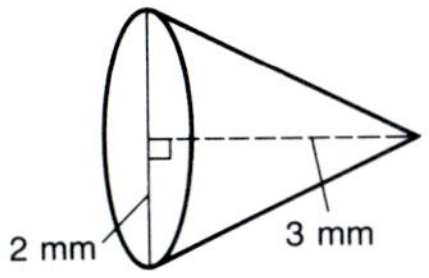

12.56 mm^3

12.

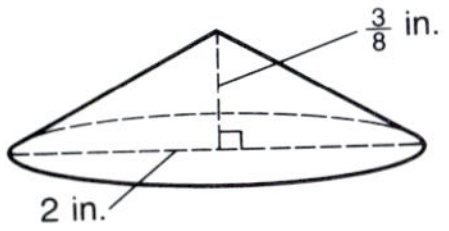

$\frac{11}{28}$ $in.^3$

	Radius	Height		Radius	Height		Radius	Height
• **13.**	3 ft	3 ft 28.26 ft^3	• **14.**	$\frac{3}{4}$ yd	3 yd $1\frac{43}{56}$ yd^3	**15.**	24 ft	$1\frac{1}{2}$ ft $905\frac{1}{7}$ ft^3
16.	3 m	6.3 m 59.346 m^3	**17.**	6 m	3.3 m 124.344 m^3	**18.**	0.3 cm	6 cm 0.5652 cm^3

Solve.

• **19.** The Iliad is also a cylindrical swimming pool. Its capacity is 168 m^3. If the area of the bottom is 112 m^2, what is the depth of the pool? 1.5 m

20. The in-ground Hector is a rectangular prism 60 yd long, 30 yd wide, and 11 yd deep. How many square yards of tile are needed to cover the bottom? 1,800 yd^2

21. The Homer Pool Company fills pools at the rate of \$0.015 per gallon. What would the charge be to fill a pool that holds 314 ft^3 of water? (*Hint:* 1 ft^3 holds $7\frac{1}{2}$ gal.) \$35.33

22. The Penelope is a rectangular in-ground pool. It is 50 ft long and 17 ft wide. What is the perimeter of the Penelope? 134 ft

COMMON ERROR

Some students may confuse the cylinder formula with the cone formula. For example, in item 1 of Checkpoint, they would choose *d.* Draw models of a cone and cylinder that have the same base. Emphasize that because the cone comes to a point, its volume is one-third of the volume of the cylinder, and not vice-versa.

ASSIGNMENTS

Level 1	1–3, 10–12, 19–20
Level 2	Odd 1–17, 19–21
Level 3	Even 2–18, 19–22

FOLLOW-UP ACTIVITY

ENRICHMENT A cylinder is inserted into a prism as shown below:

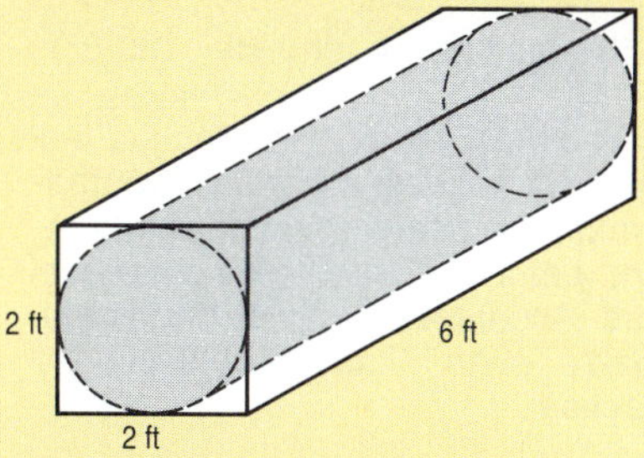

Have students find the ratio of the volume of the cylinder to that of the prism. (3:4)

SUPPLEMENTARY MATERIALS

TRP Practice, p. 143

TRP Reteaching, p. 88

TRP Lesson Aids, p. 11

TRP Transparency 72

OBJECTIVE

- Solve problems that involve the use of surface area in home improvement situations.

TEACHING THE LESSON

WARM-UP Write the following exercises on the chalkboard or an overhead transparency.

Find the surface areas.

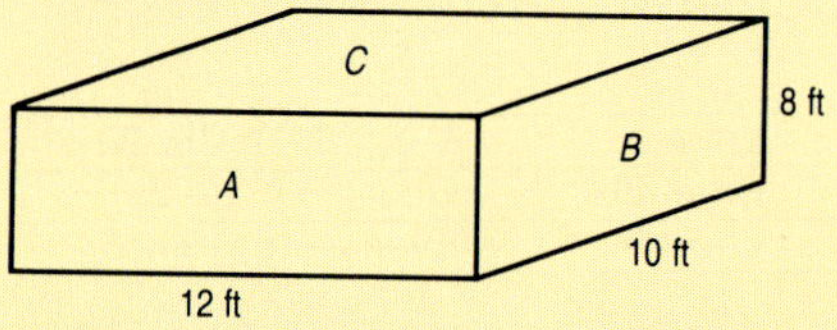

1. Face A (the front) ($96\ ft^2$)
2. Face B (the right side) ($80\ ft^2$)
3. Face C (the top) ($120\ ft^2$)
4. The back face ($96\ ft^2$)
5. The face at the left side ($80\ ft^2$)

INTRODUCTION Ask the students to identify situations in which geometry might be useful in home improvement. Let a few volunteers share their ideas. Accept any answer that is reasonable, even if it is not actually "practical."

INSTRUCTION Discuss the introductory example carefully. Be sure the students understand the basic plan for solving the problem: there are five surfaces—not six—to be painted. We find the combined area (in square feet) of the five surfaces, and then we divide by 450 to find the number of gallons required from a mathematical standpoint. Then the quotient is rounded upward so that the answer is sensible and realistic.

You may wish to allow students to use calculators to help them solve the problems.

PROBLEM Solving APPLICATION

12.13 CONSUMER: DECORATING AN APARTMENT

You want to paint your room, which is pictured below. It has the shape of a rectangular prism. Usually, a gallon of paint will cover about $450\ ft^2$. About how many gallons of paint will you need for your room?

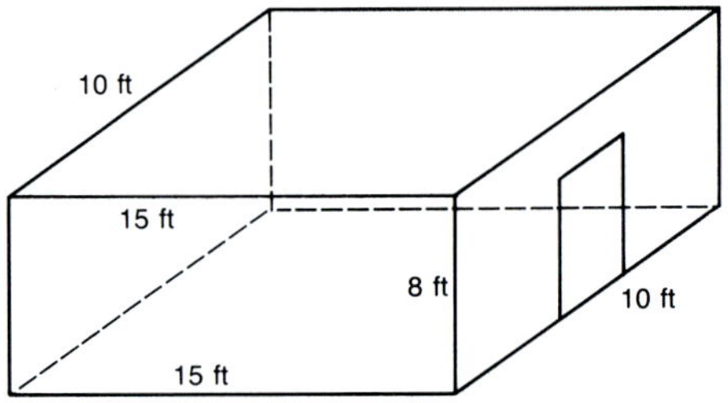

1. Find the total surface area.

Ceiling:	15×10	$= 150$
Longer walls:	$2 \times (8 \times 15)$	$= 240$
Shorter walls:	$2 \times (8 \times 10)$	$= 160$
Total:		550

2. Divide to find the approximate number of gallons needed. $550 \div 450 = 1.2$
3. Round up to next higher number. $1.2 \rightarrow 2$

So, you should buy 2 gal of paint for your room.

FOR DISCUSSION See TE side column.

A book about "do-it-yourself" home improvement says that preparation *before* painting should take 50% of your time. What do you think this means? What kinds of preparation are needed?

The area to be painted is given. Approximately how many gallons of paint are needed for one coat of paint? Give your answer to the nearest tenth.

• **1.** 1,280 ft^2 2.8 gal

2. 3,200 ft^2 7.1 gal

3. 860 ft^2 1.9 gal

4. 5,450 ft^2 12.1 gal

Approximately how many gallons of paint are needed? Give your answer to the nearest tenth.

• **5.** 1 coat of paint 1 gal

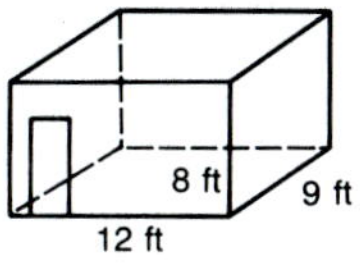

6. 2 coats of paint 4.3 gal

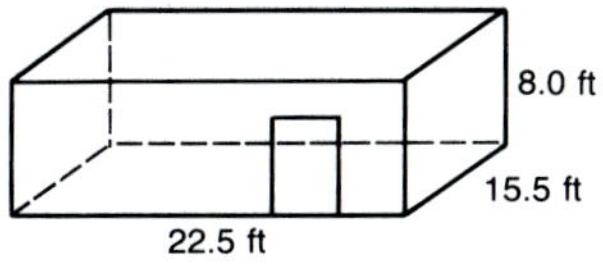

7. 3 coats of paint 2.8 gal

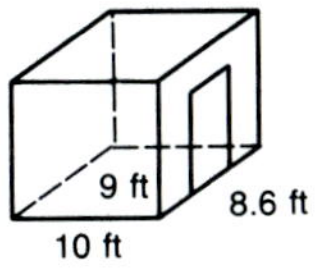

8. 2 coats of paint 4.5 gal

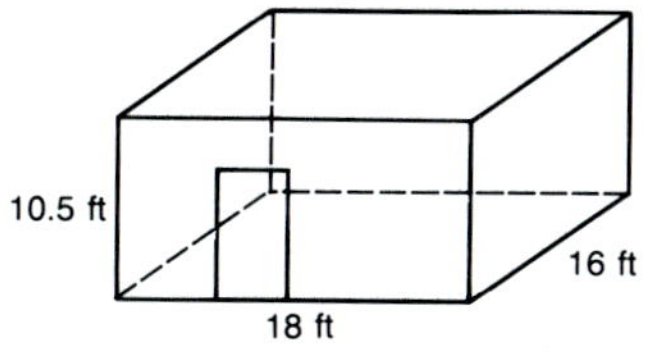

Tell how many 1-gal cans of paint you would buy for the room in:

• **9.** Exercise 1. 3

• **10.** Exercise 2. 8

11. Exercise 3. 2

12. Exercise 4. 13

13. Exercise 5. 1

14. Exercise 6. 5

15. Exercise 7. 3

16. Exercise 8. 5

Suppose you want to wallpaper the walls of your room. Wallpaper comes in different lengths and widths. However, a roll of American wallpaper usually contains about 20 ft^2 of usable wallpaper.

How many rolls of wallpaper are needed to wallpaper the room in:

17. Exercise 5? 17

18. Exercise 7? 17

Find the total cost.

• **19.** Paint the room in Exercise 6. Paint is \$16.87 a gallon. \$84.35

20. Wallpaper the walls in Exercise 6. Wallpaper is \$8.25 a roll. \$255.75

21. Paint the room in Exercise 1. Paint is \$9.40 a gallon. \$28.20

22. Wallpaper the walls in Exercise 8. Wallpaper is \$14.75 a roll. \$531.00

FOR DISCUSSION Students should explain that before painting, cracks in walls or ceilings may have to be repaired. If there was old wallpaper on the walls, it would have to be removed first, and then the walls would have to be repaired.

Preparation may also include tasks such as organizing materials.

ASSIGNMENTS

Level 1	1–6, 9–18
Level 2	Even 2–18, 19–22
Level 3	Odd 1–17, 19–22

FOLLOW-UP ACTIVITY

COOPERATIVE LEARNING Have the students work in groups of two or three to learn more about painting and wallpapering. Challenge them to find out how many rolls of wallpaper would be needed for the room in Problem 5 if the room had 8 windows and 2 doors—and no part of the windows or doors were being wallpapered. Do-it-yourself home improvement books or wallpaper stores can provide information about how to make allowances for windows and doors.

SUPPLEMENTARY MATERIALS

TRP Practice, p. 144

TRP Transparency 72

OBJECTIVE

- Decide, given relevant information, what to include in a budget.

TEACHING THE LESSON

WARM-UP Write the following exercises on the chalkboard or an overhead transparency.

BUDGET: MONTHLY EARNINGS OF $1,600

		Amount	Percent of budget
1.	Food	$320	
2.	Housing	$752	
3.	Telephone		3%
4.	Transportation	$128	
5.	Clothing		8%
6.	Miscellaneous	$144	
7.	Savings	$ 80	
8.	Total		

(**1.** 20% **2.** 47% **3.** $48 **4.** 8% **5.** $128 **6.** 9% **7.** 5% **8.** $1,600, 100%)

INTRODUCTION Present the problem situation to the class. Ask students what factors they think must be considered when developing such a budget. Remind them that the budget has to be approved by a committee of teachers and students who are responsible for distributing funds to all clubs. List the factors on the chalkboard.

INSTRUCTION Have students read the suggested activities and discuss the merits of each. Point out, if necessary, that there are 37 members in the club out of a total school population of 1,250. Students may or may not see the relative size of the group as a factor.

Direct attention to the decision-making factors. Ask students if there are any factors they would add to the list from those written on the chalkboard.

DECISION MAKING

12.14 DEVELOPING A BUDGET

(Open-ended problem solving)

Developing a budget for a club or organization requires group decisions, an understanding of the situation, and often compromise.

PROBLEM

The newly elected student officers of the Science Club are meeting to develop a budget for next year's activities. The budget must then be presented to the school's Funding Committee for approval. Since money is limited, the budget must be carefully done and include complete explanations of how the funds will be used. The students think that they can probably get more than the $500 the club got last year. Several of the 1,250 students in the school have submitted suggestions for what they would like to see done next year. The officers are studying each and estimating costs.

- Bus trip to the Science Museum for the Science Club members.

  Charter bus: $175 Admission tickets: $2 each for 37 members

- Help students pay for supplies for their Science Fair exhibits.

  About 40 students: $10 per student

- Sponsor a Health Fair at the school using volunteers from various health professions.

  Publicity: $50 Decorations, displays: $50 Refreshments: $150

- Build a simple greenhouse for students interested in horticulture.

  Materials: $200

- Plant trees in front of the school.

  Cost: 10 trees at about $25 each.

DECISION-MAKING FACTORS

Expense | Educational value

Benefits to student body | Benefits to the school

DECISION-MAKING COMPARISONS Some answers may vary. Accept reasonable answers based on students' rationales.

Complete the table.

	Expense	Educational value	Benefits to student body	Benefits to school
Bus trip	• 1. ■	Excellent	• 2. ■	3. ■
Science Fair Aid	4. ■	5. ■	Limited	6. ■
Health Fair	7. ■	8. ■	9. ■	Yes
Greenhouse	$200	10. ■	11. ■	12. ■
Trees	13. ■	14. ■	15. ■	16. ■

1. $249
2. Limited
3. No
4. $400
5. None
6. No
7. $250
8. Excellent
9. Extensive
10. Good
11. Limited
12. Yes
13. $250
14. None
15. None
16. Yes

MAKING THE DECISIONS

Which items should be in the budget:

• **17.** if the plan was to benefit the school? Health Fair, Greenhouse, Trees

• **18.** if the plan was to benefit the most students? Health Fair

19. if the plan was to provide the most educational value? Bus trip, Health Fair

20. The officers assumed that the Science Fair participants will compete even if they don't get money for supplies. What if half of them can't compete without financial aid. How does this affect the factor of "benefit to the school"? Aid could be a benefit since more students would be able to represent the school.

21. What benefits could planting trees provide? Answers will vary. Examples: beauty, shade

22. What benefits could building a greenhouse provide? Answers will vary. Examples: study plants, grow plants to sell

Select items and make a circle graph.
Answers will vary. Check students' graphs.

23. A budget that will most benefit all the students in the school.

24. A budget that will most benefit the school.

25. A budget that will most benefit the Science Club members.

Have students complete Exercises 1–16 in the table in order to compare the five options. Ask them to write conclusions that can be drawn from the data and discuss the appropriateness of these conclusions. Note that Science Fair aid is not considered a benefit to the school because it is assumed that students will participate anyway. (See Exercise 20.)

Assign Exercises 17–25 and have students work in small groups. Encourage them to explain and defend their answers and decisions. Remind them that their own preferences and experiences will influence their answers. Ask students to support their budgets for Exercises 23–25.

ASSIGNMENTS

Level 1	1–25
Level 2	1–25
Level 3	1–25

FOLLOW-UP ACTIVITY

APPLICATION Have groups of students develop a budget for a Mathematics Club. They are to come up with ideas for activities, determine what factors are important in their situation, and develop the budget. You may want to limit them to $500 (or $1,000). If they have difficulty in thinking of activities tell them about the Math Olympics.

Each group should present a detailed report that includes a circle graph to show how the budget is to be distributed.

SUPPLEMENTARY MATERIALS

TRP Practice, p. 145

TRP Applications, p. 20

TRP Transparency 73

OBJECTIVES

- Review vocabulary.
- Practice key chapter concepts and skills.

USING THE REVIEW

The Chapter Review is designed to help students prepare for taking the Chapter Test. The first section focuses on vocabulary. It requires that students select a word(s) to complete statements. The second section presents practice exercises of key mathematical skills. Under each directive there is a sample exercise with the answer.

Each item of the review is referenced to the page on which the topic is taught in the Pupil's Edition. You may wish to have students refer to these pages to help review any concepts or skills they have not yet mastered.

It is suggested that students work in small-sized heterogeneous cooperative learning groups. Some cooperative learning methods that may be used are as follows:

1. After each student has independently completed the entire Chapter Review, a discussion should follow within each group about the solutions to the practice exercises.
2. The group can complete the entire Chapter Review by working together to discuss the sample exercises and then to answer the practice exercises.

End the lesson with an entire class discussion in which any questions brought up in group discussions are presented and answered.

CHAPTER REVIEW

Vocabulary Choose the letter of the word(s) that completes each statement.

1. Polygon is to perimeter as circle is to ■. [330, 332] b

a. area **b.** circumference **c.** volume **d.** diameter

2. A solid figure whose faces are all polygons is a ■. [344] c

a. cylinder **b.** sphere **c.** polyhedron **d.** polygon

3. The amount of space inside a solid figure is its ■. [350] d

a. perimeter **b.** surface area **c.** height **d.** volume

Skills Find the area. [334, 336, 338]

Square: $s = 2.5$ km *6.25 km²*

4. Rectangle: $l = 5$ in., $w = 7$ in. 35 in.^{2}

5. Parallelogram: $b = 13$ ft, $h = 4$ ft 52 ft^2

6. Triangle: $b = 63$ cm, $h = 18$ cm 567 cm^2

7. Circle: $r = 7$ ft 154 ft^2

8. Circle: $r = 3$ m 28.26 m^2

Find the surface area. [347]

108 ft²

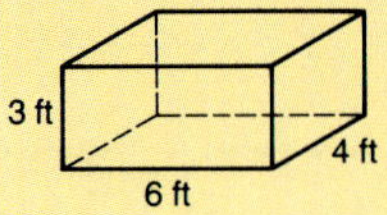

9.

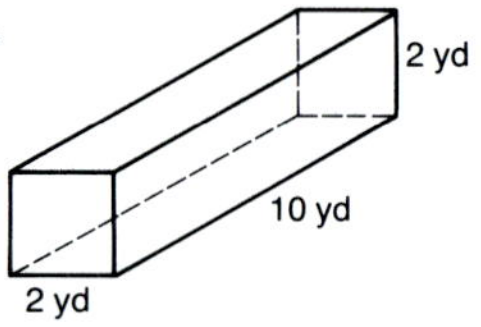

88 yd^2

10.

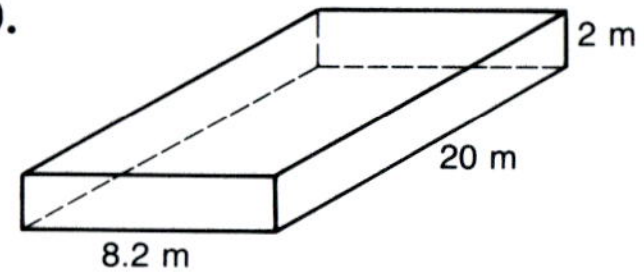

440.8 m^2

Find the volume. [350, 352]

37.68 in.³

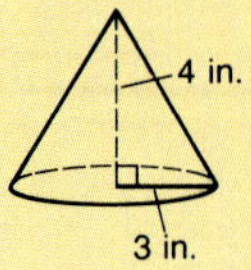

11.

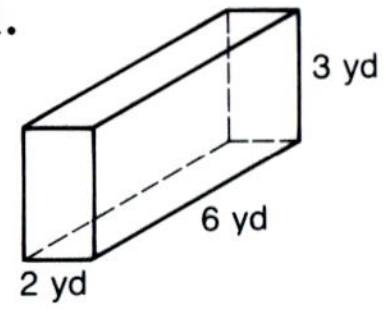

36 yd^3

12.

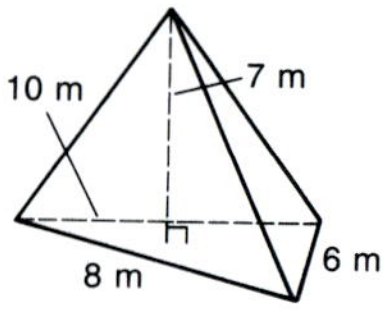

56 m^3

13. Cone: $r = 6$ ft, $h = 4$ ft $150\frac{6}{7}$ ft^3

14. Square pyramid: $s = 6$ ft, $h = 3$ ft 36 ft^3

15. Cylinder: $r = 1$ m, $h = 4.2$ m 13.188 m^3

16. Cube: $s = 0.1$ m 0.001 m^3

CHAPTER TEST

2. 6 faces, 12 edges, 8 vertices

Use Figure A above for Exercises 1–6.

1. Name the figure. rectangular prism

2. Count the faces, edges, and vertices.

3. Find the perimeter of face ABCD. 28.6 m

4. Find the area of base DCGH. 30.6 m^2

5. Find the surface area. 169.44 m^2

6. Find the volume. 125.46 m^3

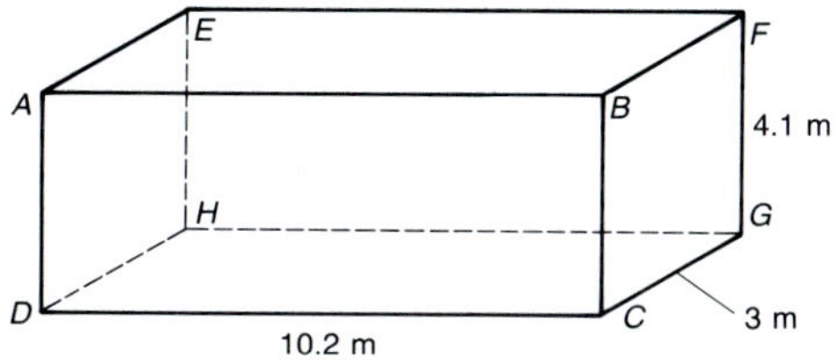

Figure A

Use Figure B above for Exercises 7–10.

7. Name the figure. square pyramid

8. Find the perimeter of the base. 24 ft

9. Find the area of face ABC. 33 ft^2

10. Find the volume. 96 ft^3

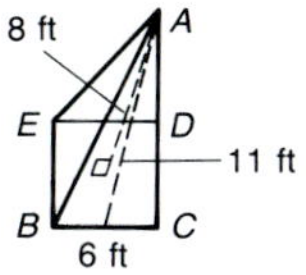

Figure B

Use Figure C above for Exercises 11–14.

11. Name the figure. cylinder

12. Find the circumference of the base. $3\frac{1}{7}$ ft

13. Find the area of the base. $\frac{11}{14}$ ft^2

14. Find the volume. $5\frac{1}{2}$ ft^2

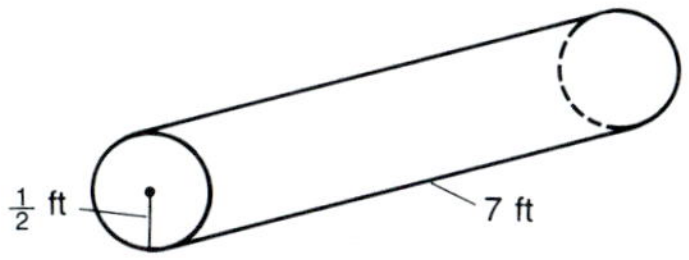

Figure C

Use Figure D above for Exercises 15–16.

15. Name the figure. cone

16. Find the volume. 150.72 m^3

Figure D

Solve.

17. John Hutchins' library is 12 ft by 16 ft with an 8-ft ceiling. How many gallons of paint would he need to repaint it? (*Remember:* 1 gal covers about 450 ft^2.) 1.4 gal

18. John uses 7 nails to put up each 4-ft length of ceiling molding. How many nails would he need to put up molding all around the library? 98 nails

OBJECTIVE

- Evaluate achievement of the chapter objectives.

USING THE TEST

The Chapter Test may be used as a posttest to evaluate the achievement of all students. However, you may wish to use the Chapter Posttest offered in the Teacher's Resource Package or design your own chapter test. If this page is not used as a test, you may wish to assign it as additional review or practice.

The test items are correlated to the chapter objectives in the table below.

Chapter objectives	Test items
A. Name solid figures.	1, 7, 11, 15
B. Count the numbers of faces, edges, and vertices of a prism.	2
C. Find the perimeters of polygons.	3, 8
D. Find the areas of polygons.	4, 9
E. Find the circumferences and areas of circles.	12–13
F. Find the surface areas of rectangular prisms.	5
G. Find the volumes of rectangular prisms, pyramids, cylinders, or cones.	6, 10, 14, 16
H. Apply computational skills in real-life situations.	17
I. Solve a problem by writing a simpler problem.	18

SUPPLEMENTARY MATERIALS

TRP Ch. 12 Posttest Form A, pp. 1–2

TRP Ch. 12 Posttest Form B, pp. 1–2

OBJECTIVES

- Review and maintain previously taught concepts and skills.
- Practice taking tests in a multiple-choice format.

USING THE REVIEW

Assign the Cumulative Review to all students. Provide students with an answer sheet to record their answers.

Each Cumulative Review gives students an opportunity to practice taking tests that are written in multiple-choice format.

Provide appropriate remedial help for students having difficulty with any of the skills and concepts on these pages.

CUMULATIVE REVIEW

Choose the letter of the correct answer.

1. The sum of the measures of ■ angles is equal to 90°. b

a. acute **b.** complementary **c.** supplementary **d.** none of these

2. A pair of ■ lines do not intersect. c

a. perpendicular **b.** bisecting **c.** parallel **d.** none of these

3. A(n) ■ triangle has three angles each of which is less than 90°. c

a. obtuse **b.** right **c.** acute **d.** none of these

4. The relationship of a scale drawing to an object's actual size can be expressed as a ■. a

a. proportion **b.** range **c.** mean **d.** none of these

5. The ■ is the distance around a circle. d

a. radius **b.** chord **c.** diameter **d.** none of these

Select the best estimated answer.

6. $\frac{7}{8} + \frac{5}{7}$	a	**a.** 2	**b.** $\frac{7}{8}$	**c.** 4	**d.** $\frac{1}{4}$
7. $18\frac{2}{3} - 7\frac{4}{5}$	d	**a.** 25	**b.** 27	**c.** 1	**d.** 12
8. $2\frac{7}{8} \times 2\frac{1}{8}$	a	**a.** 6	**b.** 1	**c.** $\frac{3}{16}$	**d.** $\frac{1}{2}$
9. 4,875 × 53	b	**a.** 25,000	**b.** 250,000	**c.** 20,000	**d.** 2,500,000
10. 18% of 200	c	**a.** 4,000	**b.** 400	**c.** 40	**d.** 0.40

Find the answer.

11. $\frac{11}{8} \div 2\frac{1}{8}$	a	**a.** $\frac{11}{17}$	**b.** $\frac{8}{17}$	**c.** $\frac{22}{17}$	**d.** none of these
12. 5 m = ■ mm	c	**a.** 0.005	**b.** 500	**c.** 5,000	**d.** none of these
13. $7\frac{2}{3} - \frac{1}{2}$	c	**a.** $\frac{1}{6}$	**b.** $7\frac{1}{3}$	**c.** $7\frac{1}{6}$	**d.** none of these
14. $12\frac{1}{2}$% of 200	b	**a.** 2.5	**b.** 25	**c.** 0.125	**d.** none of these
15. $12\frac{1}{8} + 15\frac{1}{5}$	d	**a.** $27\frac{1}{40}$	**b.** $27\frac{3}{8}$	**c.** $27\frac{11}{40}$	**d.** none of these

Solve.

16. Rhonda buys a dishwasher at Eddie's Appliance Center for $410. She decides to pay in 12 monthly installments. The payments are $37.50 per month. What is the finance charge? d

a. $450 **b.** $410 **c.** $37.50 **d.** none of these

17. Dan wants to carpet a hall that is $4\frac{1}{4}$ ft wide and 12 ft long. He also wants to carpet a room 17 ft wide and 19 ft long. How many square feet of carpet does he need? a

a. 374 ft^2 **b.** $107\frac{2}{3}$ ft^2 **c.** $124\frac{2}{3}$ ft^2 **d.** none of these

18. Bella is 1.4 m tall. She is standing beside a telephone pole and casts a shadow 4.2 m long. If the telephone pole casts a shadow that is 31.5 m long, how tall is the pole? a

a. 10.5 m **b.** 18.6 m **c.** 9.8 m **d.** none of these

19. A quality control technician for Call-Mate checks a random sample of 1,700 telephone answering machines. Of these, 34 are defective. What percent of the sample is defective? b

a. 5% **b.** 2% **c.** 1.25% **d.** none of these

20. Vito buys a CD player that is listed in audio magazines at $265. He receives a 25% discount on the player at Claude's Audio Barn. How much money did Vito save? c

a. $56.25 **b.** $198.75 **c.** $66.25 **d.** none of these

THINKING ABOUT MATH See TE side column.

1. Explain how you would quickly calculate the two remaining angle measures of a parallelogram if you knew the measure of two adjacent angles.
2. You need to create a cataloging system for your compact discs. You have decided on a 3-digit code. Describe a method for determining the number of different codes that are possible with your system.
3. Would you prefer to receive simple or compound interest on your savings account? Why?
4. Can two obtuse angles ever be supplementary? Explain.
5. Explain why store owners might offer discounts on their marked-up prices.

THINKING ABOUT MATH

The questions in this section can be used in discussions with individual students or with small groups. Guide students in conveying their ideas clearly and precisely.

Listed below are expected student answers. However, accept any reasonable answer.

1. The four angles in a parallelogram are made up of two congruent pairs of angles. Opposite angles are congruent. Therefore, the two angles that are opposite the adjacent angles have the same measures as the adjacent angles.
2. There are three digits and each digit has 10 possibilities, 0 to 9. Multiply 10 × 10 × 10 to find that there are 1,000 possible combinations that range from 0 to 999.
3. Compound interest pays interest on the amount of savings as well as the interest earned on it. Simple interest, however, only pays interest on the amount of savings. Therefore, more money is earned with compound interest.
4. No. For two angles to be supplementary, the sum of their measures must equal 180°. Since the measure of an obtuse angle is greater than 90°, the sum of the measures of any two obtuse angles is greater than 180°.
5. Often, discounted prices encourage people to buy products that they normally would not buy. Although owners lower their profits per item, they may sell more items and, in the end, make a profit.

SUPPLEMENTARY MATERIALS

TRP Cum. Test, Ch. 1–12, pp. 1–4

Chapter 13 Rational Numbers

OBJECTIVE

- Test the prerequisite skills needed to learn the concepts developed in Chapter 13.

USING THE TEST

The Pre-Skills Test is designed to diagnose students' strengths and weaknesses on prerequisite skills necessary to study the mathematics in Chapter 13.

Have students take the Pre-Skills Test. Allow the students to work together in pairs or small groups. Group members should help those who demonstrate a misunderstanding of a concept or a weakness in a skill.

The items in the test are referenced to the pages on which the topics are taught in the Pupil's Edition. You may wish to have students refer to these pages for review.

The following table correlates the items on the Pre-Skills Test with the prerequisite skill and the lesson(s) in the chapter for which it is needed.

Item(s)	Prerequisite skill	Lesson(s)
1–3	Add whole numbers.	13.2
4–12	Add fractions and decimals.	13.11
13–16	Subtract whole numbers.	13.3
17–24	Subtract fractions and decimals.	13.11
25–28	Multiply whole numbers.	13.4
29–36	Multiply fractions and decimals.	13.11
37–40	Divide whole numbers.	13.5
41–48	Divide fractions and decimals.	13.11
49–52	Name points on a number line.	13.1, 13.8, 13.10

Pre-Skills TEST

Add. [30, 62, 158]

1. 24 + 76 100 **2.** 20 + 45 65 **3.** 365 + 294 659

4. $\frac{2}{3} + \frac{1}{5}$ $\frac{13}{15}$ **5.** $\frac{3}{4} + \frac{2}{3}$ $1\frac{5}{12}$ **6.** $\frac{5}{8} + \frac{1}{4}$ $\frac{7}{8}$

7. 0.34 + 1.2 1.54 **8.** 4.98 + 7.5 12.48 **9.** 0.2 + 7 7.2

10. 823 + 560 1,383 **11.** $\frac{3}{5} + \frac{7}{9}$ $1\frac{17}{45}$ **12.** 6.3 + 0.306 6.606

Subtract. [32, 64, 162]

13. 34 − 19 15 **14.** 86 − 7 79 **15.** 812 − 391 421 **16.** 639 − 472 167

17. $\frac{3}{4} - \frac{1}{8}$ $\frac{5}{8}$ **18.** $\frac{7}{8} - \frac{2}{5}$ $\frac{19}{40}$ **19.** $\frac{2}{3} - \frac{5}{9}$ $\frac{1}{9}$ **20.** $\frac{4}{5} - \frac{1}{3}$ $\frac{7}{15}$

21. 1.3 − 0.98 0.32 **22.** 34.7 − 5.67 29.03 **23.** 41.9 − 12.83 29.07 **24.** 53 − 18.21 34.79

Multiply. [36, 40, 42, 74, 146]

25. 23 × 93 2,139 **26.** 15 × 100 1,500 **27.** 59 × 26 1,534 **28.** 965 × 30 28,950

29. $\frac{7}{8} \times \frac{4}{5}$ $\frac{7}{10}$ **30.** $\frac{4}{7} \times \frac{3}{4}$ $\frac{3}{7}$ **31.** $\frac{1}{2} \times \frac{6}{8}$ $\frac{3}{8}$ **32.** $\frac{4}{5} \times \frac{2}{3}$ $\frac{8}{15}$

33. 12.3 × 0.8 9.84 **34.** 4.2 × 10.1 42.42 **35.** 5.23 × 7.7 40.271 **36.** 0.35 × 6.72 2.352

Divide. [36, 48, 80, 150]

37. 612 ÷ 17 36 **38.** 444 ÷ 37 12 **39.** 840 ÷ 24 35 **40.** 920 ÷ 10 92

41. $\frac{3}{7} \div \frac{4}{7}$ $\frac{3}{4}$ **42.** $\frac{7}{8} \div \frac{3}{4}$ $1\frac{1}{6}$ **43.** $\frac{4}{5} \div \frac{1}{3}$ $2\frac{2}{5}$ **44.** $\frac{5}{6} \div \frac{1}{4}$ $3\frac{1}{3}$

45. 27.9 ÷ 0.9 31 **46.** 0.12 ÷ 0.06 2 **47.** 18.5 ÷ 0.5 37 **48.** 6.3 ÷ 0.3 21

Use the number line below for Exercises 49–52. [122]

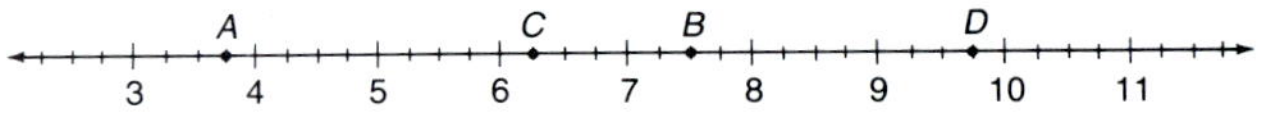

Write the number named by the point.

49. *A* $3\frac{3}{4}$ **50.** *B* $7\frac{1}{2}$ **51.** *C* $6\frac{1}{4}$ **52.** *D* $9\frac{3}{4}$

Chapter 13 RATIONAL NUMBERS

Some businesses are owned by many people who hold stock in the company. What are the advantages and disadvantages of this kind of shared ownership?

OBJECTIVE

- Formulate problems using text and data.

USING THE CHAPTER OPENER

In discussing the question on the page, begin by explaining the concept of private and public companies. Students should understand that private companies are owned by an individual or small group that controls all aspects of the firm. Public companies are owned by stockholders who invest their money in the company which is run by an elected board of directors.

An advantage of being a public company is the ability to raise capital through the sale of stock and thereby grow and expand operations and profits. A disadvantage could be the threat of takeover by another firm that buys up a controlling share of the stock.

COOPERATIVE LEARNING

Assign groups and explain that they are going to "invest" in the stock market. Each group will have $1,000 to spend. The only restrictions are that they must invest the entire $1,000 and they can buy only one share of stock in a particular company. Therefore, they will have to buy a variety of stocks. They are to choose their stocks from those listed on the New York Stock Exchange.

Explain that the project will last 10 working days. During that time they can sell stocks and buy others, but they must keep records of all transactions. They are to track each stock on a line graph showing increases and decreases on a daily basis. In their final report, the groups are to show each stock's activity during the period, and the overall profit or loss on their original investment.

SUPPLEMENTARY MATERIALS

TRP Group Projects, p. 13

OBJECTIVES

- Write and identify integers.
- Compare integers.
- Find the absolute value of an integer.

TEACHING THE LESSON

WARM-UP Write the following exercises on the chalkboard or an overhead transparency.

Name the opposite of each situation.

1. Right (left)
2. Over (under)
3. Top (bottom)
4. Up (down)
5. Loss of 13 yd (gain of 13 yd)
6. Win $52 (lose $52)
7. 3 lb more (3 lb less)
8. 3 minutes before (3 minutes after)

INTRODUCTION Of each pair of opposites discussed in the Warm-Up, ask students which they would associate with the plus (+) sign and which with the minus (−) sign.

INSTRUCTION You may wish to extend the instruction in Example 1 by having students write an integer to describe:

- a drop in temperature of 8 degrees (−8)
- a rise in altitude of 10,000 feet (+10,000)
- a loss of 25 points (−25)

When discussing Example 2, point out that zero is the only number that is the opposite of itself. Explain that the opposite of zero is written as 0, and not −0.

In discussing Example 3, make sure students understand that when comparing integers, the signs of the integers are crucial. In Example 3a, for instance, emphasize that though 6 would seem to be the greater digit, +2 is the greater integer.

13.1 INTEGERS

Whole numbers can be used to describe "how much" or "how many." When used with a positive sign (+) or a negative sign (−), whole numbers can describe opposite situations such as:

225 ft above sea level	+225	10 yd gain	+10
225 ft below sea level	−225	10 yd loss	−10

EXAMPLE 1 Write a positive or negative number to describe the situation.

a. Deposit $75 in your checking account.

THINK: You are increasing the amount of money in your account. Use the positive sign. +$75

b. Withdraw $100 from your checking account.

THINK: You are decreasing the amount of money in your account. Use the negative sign. −$100

Every positive whole number has an opposite, or negative whole number.

All positive and negative whole numbers make up a set of numbers called **integers.** Zero is also an integer, but it is neither positive nor negative.

Integers can be shown on a number line.

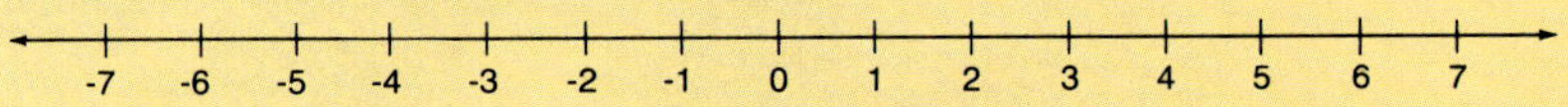

EXAMPLE 2 Name the opposite of the integer.

a. +4

THINK: +4 is 4 units to the right of zero. Find the integer that is 4 units to the left of zero.

The opposite of +4 is −4.

b. −6

THINK: −6 is 6 units to the left of zero. Find the integer that is 6 units to the right of zero.

The opposite of −6 is +6.

As you move right on the number line, integers increase in value. As you move left on the number line, integers decrease in value.

EXAMPLE 3 Compare the integers. Write > or <.

a. +2 ● −6

THINK: +2 is to the right of −6.

+2 > −6

b. −5 ● −1

THINK: −5 is to the left of −1.

−5 < −1

The **absolute value** of an integer is the distance the integer is from zero. The symbol for absolute value is | |.

EXAMPLE 4 Find the absolute value.

a. $|+7|$

THINK: +7 is 7 units from 0.

$|+7| = 7$

b. $|-3|$

THINK: −3 is 3 units from 0.

$|-3| = 3$

PRACTICE EXERCISES See Extra Practice, page 456.

Write the integer that best describes the situation.

• **1.** 40 degrees below zero −40	• **2.** a loss of \$15 −\$15
3. the opposite of −100 +100	**4.** 6 hours from now +6
5. 75 units to the left of zero −75	**6.** no gain on the play 0
7. 2 days ago −2	**8.** the opposite of +34 −34
9. 2 weeks late −2	**10.** a loss of 2 points −2

Write the integer named by the point.

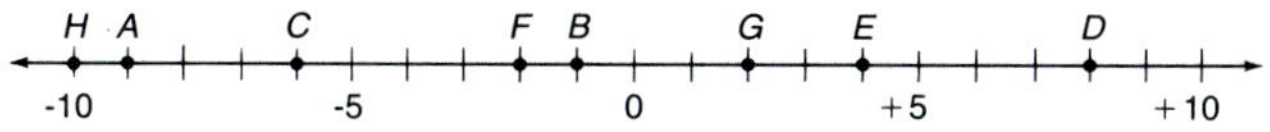

•**11.** *C* −6 •**12.** *A* −9 **13.** *E* +4 **14.** *B* −1 **15.** *D* +8 **16.** *G* +2 **17.** *H* −10 **18.** *F* −2

Write the opposite of the number.

•**19.** +16 −16 •**20.** −44 +44 **21.** −16 +16 **22.** +29 −29 **23.** +9 −9 **24.** −27 +27 **25.** +89 −89 **26.** −33 +33

Compare the integers. Write > or <.

•**27.** + 3 ● +5 < •**28.** −5 ● −6 > **29.** 0 ● −5 > **30.** −2 ● +3 < **31.** −7 ● 6 <

32. +23 ● −17 > **33.** −96 ● +100 < **34.** −33 ● −98 > **35.** 0 ● −45 > **36.** −19 ● −23 >

Write the absolute value.

•**37.** $|-1|$ 1 •**38.** $|-2|$ 2 **39.** $|0|$ 0 **40.** $|-5|$ 5 **41.** $|-7|$ 7

42. $|-9|$ 9 **43.** $|-6|$ 6 **44.** $|+89|$ 89 **45.** $|-24|$ 24 **46.** $|-17|$ 17

47. $|+27|$ 27 **48.** $|-8|$ 87 **49.** $|+76|$ 76 **50.** $|-91|$ 91 **51.** $|-101|$ 101

Use >, <, or = to make a true statement.

•**52.** $|-3|$ ● $|+3|$ = •**53.** $|+23|$ ● $|-32|$ < **54.** $|-5|$ ● $|0|$ > **55.** $|+12|$ ● $|-13|$ <

56. $|-17|$ ● $|+17|$ = **57.** −29 ● $|-29|$ < **58.** $|-71|$ ● $|+71|$ = **59.** $|0|$ ● $|-1|$ <

When discussing Example 4, stress that + and − signs have nothing to do with determining absolute value, which is the distance from zero in either direction. So, +7 and −7 have the same absolute values.

TIME OUT This is a question of semantics. If students encounter difficulty, point out that the question "How many months have only 28 days?" is not the one being asked.

COMMON ERROR

Students may sometimes have difficulty comparing negative and positive integers. For instance, they may think −999 > +1 because the digits 999 denote a great magnitude. Emphasize that any positive integer is greater than any negative integer. Using a number line can also help reinforce this point. Make sure students understand that numbers to the left are always smaller than numbers to the right.

ASSIGNMENTS

Level 1	Odd 1–45, 60–73, TO
Level 2	Odd 1–59, 60–74, TO
Level 3	Even 2–70, 72–75, TO

FOLLOW-UP ACTIVITY

APPLICATION Provide students with copies of the information listed below. Tell them that they are excerpts from the notebook of the chairman of the annual Volunteer Firemen's Labor Day Barbecue. Then have them write integers to describe each event.

- Paid out $132 for tables and chairs. (−132)
- Received $60 from Elks Lodge for tickets sold to members. (+60)
- Picked up donation of $102 from Mrs. Likins of Women's Auxiliary. (+102)
- Spent $398 on ground beef, hot dogs, buns, chicken. (−398)
- Charged $195 worth of paper plates, napkins, etc. (−195)
- Used check #243 to buy 32 cases of soda. Total: $287. (−287)
- Advance ticket sales of $1,045. (+1,045)
- Captain Glick reports same-day ticket sales of $420. (+420)

SUPPLEMENTARY MATERIALS

TRP Practice, p. 146

TRP Reteaching, p. 89

TRP Lesson Aids, p. 3

TRP Transparency 73

MIXED REVIEW

Identify the pair of angles as *complementary, supplementary,* or *neither.*

•**60.** 56°, 34° complementary
61. 112°, 78° neither
62. 81°, 19° neither
63. 120°, 60° supplementary

The measures of two angles of a triangle are given. Find the measure of the third angle.

•**64.** 40°

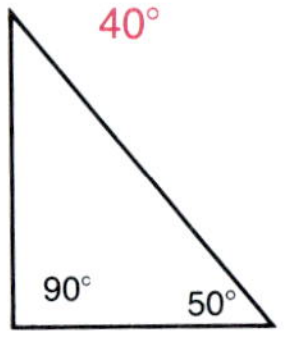

65.

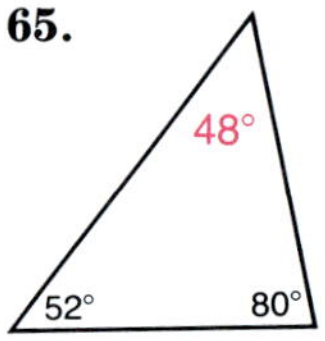

66.

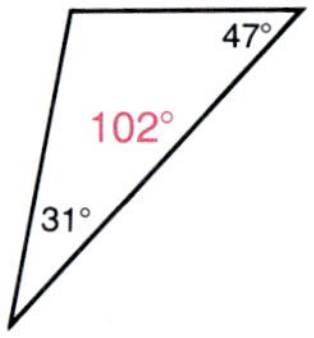

67. 33°

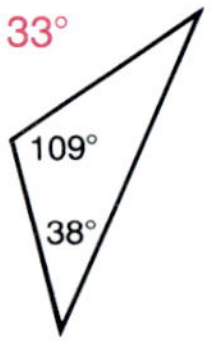

68. 46°, 69° 65°
69. 124°, 41° 15°
70. 90°, 80° 10°
71. 100°, 36° 44°

Solve.

•**72.** The highest point in the United States is Mount McKinley in Alaska. It is 20,320 ft above sea level. Write an integer to describe this altitude. +20,320

73. The lowest point in the United States is Death Valley, California. It is 282 ft below sea level. Write an integer to describe this altitude. −282

74. Borah Park in Idaho is 12,662 ft above sea level. Kings Peak in Utah is 13,528 ft above sea level. How much higher is Kings Peak than Borah Peak? 866 ft

75. The highest point in the State of Texas is Guadalupe Peak at 8,749 ft above sea level. The lowest point is at the Gulf of Mexico which is sea level. What is the range of altitude in the state? 8,749 ft

TIME OUT

Some months have 31 days.
Others have 30 days.
How many have 28 days? All 12 months have 28 days.

13.2 ADDING INTEGERS

During the first two plays of the game, the Miami Sharks gained 7 yd and lost 5 yd. What was the result of the two plays?

To find the answer, you can add +7 and −5.
Usually, the + sign is dropped when computing with positive integers.

EXAMPLE 1 Add 7 + (−5). Use the number line.

1. Locate 7.
2. Move 5 units to the left.

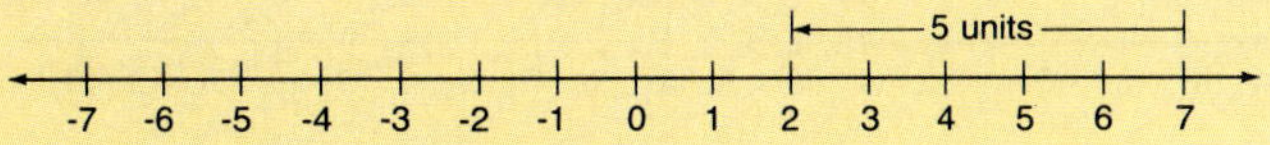

7 + (−5) = 2

So, the Sharks gained 2 yd.

You can add two integers without using a number line by following these rules.

- If the integers have the same signs, add their absolute values. The sum has the same sign as the addends.
- If the integers have different signs, subtract their absolute values. The sum has the sign of the addend with the greater absolute value.

EXAMPLE 2 Add the integers.

a. −6 + (−4)

1. Find the absolute values. $|-6| = 6$ $|-4| = 4$
2. Add. $6 + 4 = 10$
3. Write the correct sign in the sum.

THINK: The addends have the same signs. So, the sum has the same sign. $-6 + (-4) = -10$

b. 3 + (−5)

1. Find the absolute values. $|+3| = 3$ $|-5| = 5$
2. Subtract.

THINK: 5 > 3 $5 - 3 = 2$

3. Write the correct sign in the sum.

THINK: The addends have different signs. Since −5 has the greater absolute value, the sum is negative. $3 + (-5) = -2$

OBJECTIVE

- Add integers.

TEACHING THE LESSON

WARM-UP Write the following exercises on the chalkboard or an overhead transparency.

Find the absolute value.

1. $|-3|$ (3) **2.** $|+4|$ (4)
3. $|+11|$ (11) **4.** $|+17|$ (17)
5. $|-73|$ (73) **6.** $|-101|$ (101)

INTRODUCTION Write the following problem on the chalkboard and have students solve it:

On first down, the Lions gained 4 yd on a running play. On second down, the Lions lost 8 yd on a sack. How far are the Lions from the original line of scrimmage? (4 yd)

Encourage students to draw a diagram. Ask whether the answer (4 yd) represents a gain or a loss, a positive integer or a negative integer. (loss, negative)

INSTRUCTION In Example 1, point out that the Commutative Property of Addition also applies to integers. So, −5 + 7 also equals 2.

You may wish to extend the instruction by drawing a number line on the chalkboard and working through the following examples:

- −2 + (−5) (−7)
- −6 + 8 (2)
- −7 + 7 (0)
- 0 + (−3) (−3)

Emphasize that if the second integer is negative, move to the left. If the second integer is positive, move to the right. Use examples to stress that the sum of an integer and its opposite is zero.

In discussing Example 2, point out that adding integers on a number line is not always practical, especially when adding very large or very small numbers such as 243 and −584.

Also, explain that "subtract" in the second rule means to subtract the lesser absolute value from the greater.

Note that Exercises 29–38 have more than two addends. Before assigning these exercises, work through this example on the chalkboard:

$-4 + 8 + (-12) + 5$

$-16 + 13$

-3

CHECKPOINT Distribute to each student four index cards on which the answer choices a–d appear. After allowing students a few minutes, ask them to raise the letter that indicates their answer choice. Scan the cards and identify those students who need additional attention before proceeding to the Practice Exercises. The incorrect answer choices include common errors that students make.

CALCULATOR You may wish to have students estimate whether each sum is positive or negative before adding. This is a useful check when using a calculator to add integers.

COMMON ERROR

When adding integers with different signs, students will sometimes apply the wrong sign to the sum. For example, in item 2 of Checkpoint, they would choose *d*. Encourage students to estimate whether the sum is positive or negative by determining which integer has the greater absolute value.

CHECKPOINT

Write the letter of the correct answer.

Add.

1. $-8 + (-9)$ c **a.** −1 **b.** 17 **c.** −17 **d.** 1
2. $4 + (-7)$ a **a.** −3 **b.** −11 **c.** 11 **d.** 3
3. $-5 + 9$ b **a.** −4 **b.** 4 **c.** −14 **d.** 14

PRACTICE EXERCISES

See Extra Practice, page 456.

Add. Use the number line if necessary.

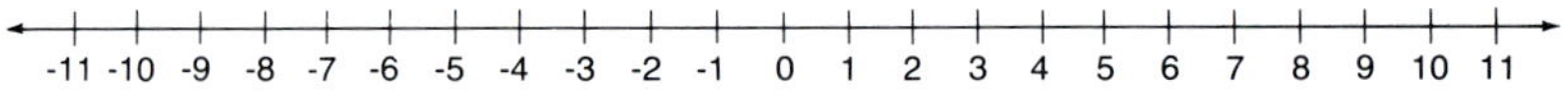

• **1.** $0 + 3$ 3 • **2.** $-3 + (-7)$ −10 **3.** $3 + (-3)$ 0 **4.** $(-9) + 1$ −8

5. $-3 + (-6)$ −9 **6.** $8 + 2$ 10 **7.** $-2 + (-4)$ −6 **8.** $-7 + 9$ 2

9. $-5 + 5$ 0 **10.** $9 + (-9)$ 0 **11.** $-8 + 11$ 3 **12.** $10 + (-9)$ 1

Add.

• **13.** $13 + 15$ 28 • **14.** $25 + 36$ 61 **15.** $-23 + (-24)$ −47 **16.** $-35 + (-76)$ −111

17. $6 + (-42)$ −36 **18.** $17 + (-6)$ 11 **19.** $-15 + 8$ −7 **20.** $-9 + 33$ 24

21. $-34 + 45$ 11 **22.** $-52 + 52$ 0 **23.** $72 + (-8)$ 64 **24.** $-60 + 46$ −14

25. $-88 + 76$ −12 **26.** $-100 + (-23)$ −123 **27.** $120 + (-75)$ 45 **28.** $99 + (-99)$ 0

Add. (*Hint:* Add the numbers with the same signs first.)

• **29.** $-3 + (-5) + 6$ −2 • **30.** $-5 + 9 + 5$ 9

31. $-6 + 7 + (-3)$ −2 **32.** $-2 + 8 + (-6) + 7$ 7

33. $-17 + 9 + (-16)$ −24 **34.** $19 + (-3) + (-1) + (-9)$ 6

35. $-45 + 13 + (-72)$ −104 **36.** $9 + (-8) + (-5) + 1$ −3

37. $-22 + (-34) + 22 + 34$ 0 **38.** $-14 + (-12) + (-26)$ −52

During the first six plays of the game, the Miami Sharks had the following results: 7-yd gain; 5-yd gain; 3-yd loss, 25-yd gain; 10-yd loss; 6-yd loss.

• **39.** Express the gains using positive integers. 37

40. Express the losses using negative integers. −19

41. Find the result of the first three plays. 9-yd gain

42. Find the result of the six plays. 18-yd gain

Solve.

43. When the football game began, the temperature was 52°F. During the game, the temperature rose 5°. What was the temperature at the end of the game? 57°F

44. Sue Keller spent $18 on her ticket for the football game. She also bought a ticket for a friend, and spent $8.50 on food. How much money did Sue spend altogether? $44.50

45. During the first three plays, the Detroit Dolphins gained 4 yd, lost 6 yd, and gained 2 yd. What was the result of the three plays? No gain

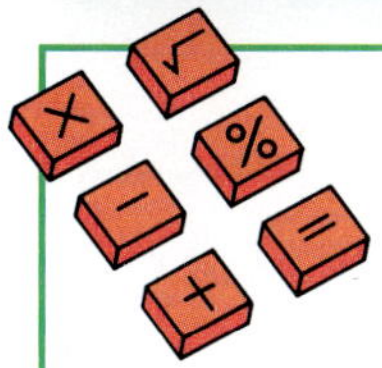

CALCULATOR

You can add integers on a calculator. Some calculators have a $+/-$ key that changes a positive integer to a negative integer. If your calculator has this key, follow these steps to add $-24 + 52$:

Enter: [2] [4] [+/−] [+] [5] [2] [=] [28.]

If your calculator does not have a $+/-$ key, follow these steps to add $-24 + 52$:

THINK: Press the − key before and after entering negative integers.

Enter: [−] [2] [4] [−] [+] [5] [2] [=] [28.]

Add. Use a calculator.

1. $7 + 12$ 19

2. $4 + (-6)$ −2

3. $5 + (-14)$ −9

4. $-4 + 8$ 4

5. $-5 + (-7) + (-12)$ −24

6. $8 + (-9) + (-3)$ −4

7. $17 + (-47) + (-98)$ −128

8. $-43 + 87 + (-19)$ 25

9. $125 + (-50) + (-12)$ 63

10. $30 + 158 + (-101)$ 87

ASSIGNMENTS

Level 1	1–20, 39–43, CA 1–5
Level 2	Odd 1–37, 39–44, CA
Level 3	21–45, CA

FOLLOW-UP ACTIVITY

COOPERATIVE LEARNING

Write the following expressions on the chalkboard or an overhead transparency:

1. $-30 + 15$ **2.** $7 + (-32)$

Have students make up word problems incorporating each expression. For example:

The floor of the main chamber of Jacob's Cavern is 30 ft underground. The entrance to the main chamber is 15 ft above the floor. Write an integer to describe the altitude of the entrance.

Then have pairs of students exchange papers, solve the problems, and return the papers to be checked. If an answer is wrong, have the author of the problem furnish and explain the correct answer.

SUPPLEMENTARY MATERIALS

TRP Practice, p. 147

TRP Reteaching, p. 90

TRP Transparency 74

OBJECTIVE

- Subtract integers.

TEACHING THE LESSON

WARM-UP Write the following exercises on the chalkboard or an overhead transparency.

Write the opposite of each integer.

1. 6 (-6) **2.** -3 (3)

3. 0 (0) **4.** -12 (12)

5. 32 (-32) **6.** -89 (89)

INTRODUCTION Provide the students with the following pattern:

$$2 - 0 = 2$$
$$2 - 1 = 1$$
$$2 - 2 = 0$$
$$2 - 3 = (-1)$$
$$2 - 4 = (-2)$$

Have them find the missing numbers. Repeat with similar number patterns. Guide them to realize that in a subtraction sentence, if the number being subtracted is greater, the difference will be negative.

INSTRUCTION Extend the instruction in Example 1 by having students tell in which direction and how many units on the number line they would move from the second integer to reach the first.

- $8 - 5$ (right 3 units)
- $-7 - (-5)$ (left, 2 units)
- $6 - 9$ (left, 3 units)
- $-2 - 7$ (left, 9 units)

Also, you may wish to give students an example such as $36 - (-78)$. Point out that the difference is obviously positive, but that using a number line to count the number of units in between is troublesome. Guide students to realize that there are 78 units between -78 and 0, and 36 units between 0 and 36. So, there are 114 units between -78 and 36.

Before discussing Examples 2 and 3, you may find it helpful to review how to find the opposite of an integer.

13.3 SUBTRACTING INTEGERS

At 3:00 A.M., the temperature was -2°C.
At 9:00 A.M., the temperature was 3°C.
What was the change in temperature from 3:00 A.M. to 9:00 A.M.?

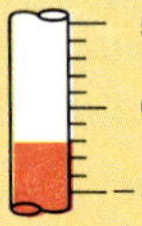

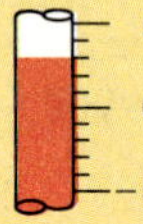

To find the temperature change, subtract -2° from 3°. You can subtract integers using a number line.

EXAMPLE 1 Subtract: $3 - (-2)$

1. Locate the integer being subtracted.
2. Count the number of units moved to reach the first integer.
3. A move to the right is positive; a move to the left is negative.

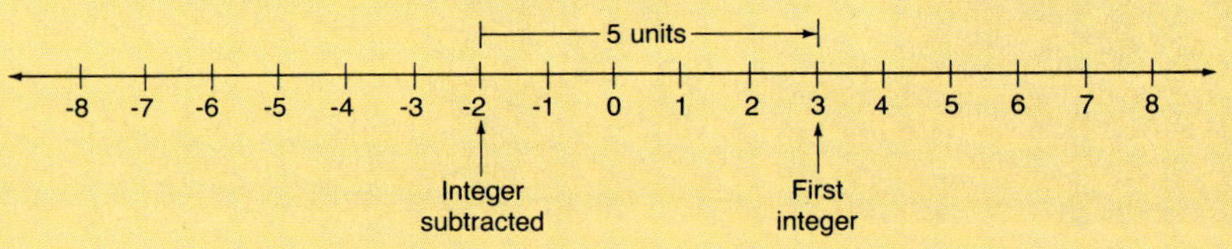

$$3 - (-2) = 5$$

So, the temperature rose 5°C.

Every subtraction example can be written as a related addition example. To write a related addition example, add the opposite of the number you are subtracting.

EXAMPLE 2 Write a related addition example.

a. $2 - 8$

THINK: The opposite of 8 is -8.

So, $2 - 8 = 2 + (-8)$.

b. $3 - (-4)$

THINK: The opposite of -4 is 4.

So, $3 - (-4) = 3 + 4$.

You can use opposites to subtract integers.

EXAMPLE 3 Subtract the integers.

a. $-10 - 5$

THINK: Add the opposite of the number being subtracted.

$-10 - 5$

$-10 + (-5) = -15$

So, $-10 - 5 = -15$.

b. $-6 - (-9)$

THINK: Add the opposite of the number being subtracted.

$-6 - (-9)$

$-6 + 9 = 3$

So, $-6 - (-9) = 3$.

CHECKPOINT

Choose the letter of the correct answer.

Subtract.

1. −3 − 7 b **a.** 10 **b.** −10 **c.** −4 **d.** 4
2. 6 − 9 a **a.** −3 **b.** 3 **c.** −15 **d.** 15
3. −1 − (−8) d **a.** −9 **b.** 9 **c.** −7 **d.** 7

PRACTICE EXERCISES

See Extra Practice, page 457.

Subtract. Use the number line if necessary.

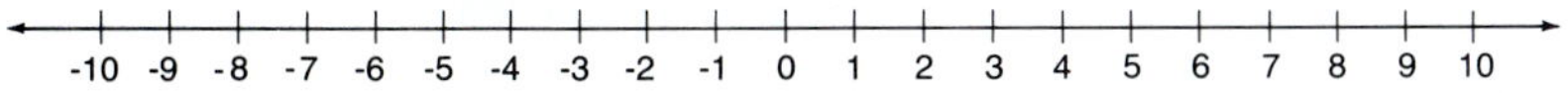

• **1.** 10 − 3 7 • **2.** 5 − 8 −3 **3.** 2 − 9 −7 **4.** 9 − 8 1

5. −3 − 5 −8 **6.** −6 − 3 −9 **7.** −2 − 9 −11 **8.** −9 − 9 −18

9. 4 − (−5) 9 **10.** 6 − (−7) 13 **11.** 9 − (−2) 11 **12.** 0 − (−6) 6

13. −3 − (−3) 0 **14.** −7 − (−9) 2 **15.** −1 − (−5) 4 **16.** −10 − (−4) −6

Complete.

• **17.** −6 − 2 = −6 + ■ −2 • **18.** 12 − 4 = 12 + ■ −4

19. 13 − (−28) = 13 + ■ 28 **20.** −27 − (−100) = −27 + ■ 100

21. 34 − 46 = 34 + ■ −46 **22.** −89 − 45 = −89 + ■ −45

Subtract.

• **23.** 13 − 15 −2 • **24.** 14 − 25 −11 **25.** 23 − 13 10 **26.** 19 − 29 −10

27. −14 − 17 −31 **28.** −16 − 13 −29 **29.** 0 − 17 −17 **30.** −19 − 18 −37

31. −17 − (−14) −3 **32.** −6 − (−19) 13 **33.** −22 − (−35) 13 **34.** −45 − (−100) 55

35. 17 − (−9) 26 **36.** −18 − 67 −85 **37.** 98 − 112 −14 **38.** −100 − (−134) 34

Use the table at the right to solve Exercises 39–41.

• **39.** What is the difference between the lowest temperature in Caribou and that of Kansas City? 20°

40. Which city has the lowest recorded temperature? Duluth

41. What is the difference between the lowest temperature in Chicago and that of Duluth? 8°

RECORDED LOW TEMPERATURES

City	Temperature, °F
Caribou, Maine	−27
Chicago, Illinois	−22
Denver, Colorado	−19
Duluth, Minnesota	−30
Kansas City, Missouri	−7

CHECKPOINT The incorrect choices include common errors students make.

COMMON ERROR

In writing a related addition sentence to subtract integers, some students may find the opposite of both integers. For example, in item 1 of Checkpoint, they would choose *c.* Guide these students carefully through the three-part process for subtracting integers. First have them circle the number to be subtracted and name its opposite. Then have them write the related addition problem. Finally, have them solve the related addition problem.

ASSIGNMENTS

Level 1	1–16, Odd 17–41
Level 2	9–41
Level 3	23–41

FOLLOW-UP ACTIVITY

ESTIMATION Have students estimate whether each of the following differences is negative or positive.

1. −4 − 6 (−)
2. 125 − (−56) (+)
3. −35 − (−37) (+)
4. 43 − 45 (−)
5. −121 − 120 (−)
6. 786 − 768 (+)
7. 43 − (−43) (+)
8. −22 − (−23) (+)

SUPPLEMENTARY MATERIALS

TRP Practice, p. 148

TRP Reteaching, p. 91

TRP Transparency 74

OBJECTIVE

- Multiply integers.

TEACHING THE LESSON

WARM-UP Write the following exercises on the chalkboard or an overhead transparency.

Multiply.

1. 32×20 (640)

2. 80×90 (7,200)

3. 310×25 (7,750)

4. $3 \times 2 \times 0 \times 5$ (0)

5. $10 \times 10 \times 10 \times 10$ (10,000)

INTRODUCTION Write the following on the chalkboard:

$-4 + (-4) + (-4) + (-4) + (-4)$

Have students find the sum. (-20) Then, using the definition of multiplication, guide students to realize that the expression above can also be written as $5 \times (-4)$. Point out that this product is also -20. Repeat with similar examples.

INSTRUCTION When discussing Example 1, point out that the product of two negative numbers is positive because both factors have the same sign (negative).

In addition, prior to assigning Exercises 1–20, you may wish to have students predict whether each product is positive or negative.

When discussing Example 2, you may wish to have students continue predicting the sign of the product using Exercises 21–29.

FOR DISCUSSION Students should explain that whenever zero is a factor, the product must also be zero.

13.4 MULTIPLYING INTEGERS

Karen DeSoto is in charge of all the statistics for the North Ridge High School football team. She needs to find the total loss of yards due to penalties in last week's game.

GAME 5		
Number of Penalties	Yards Lost Per Penalty	Total Loss
7	-10	

To find the total number of yards lost, multiply 7 and -10.

To multiply two integers:

- Multiply the absolute values.
- Determine whether the product is positive or negative.
  The product of two integers having the same sign is positive.
  The product of two integers having different signs is negative.

EXAMPLE 1 Multiply: $7 \times (-10)$

1. Multiply the absolute values. $|7| \times |-10| = 7 \times 10 = 70$
2. Determine whether the product is positive or negative.

THINK: The signs are different. The product is negative.

$7 \times (-10) = -70$

So, North Ridge High School lost 70 yd due to penalties in Game 5.

When there are more than two integers, count the number of negative integers.

- The product is positive if there is an even number of negative factors.
- The product is negative if there is an odd number of negative factors.

EXAMPLE 2 Multiply:

a. $-3 \times 5 \times (-2)$

$|-3| \times |5| \times |-2| = 30$

THINK: There is an even number of negative factors. The product is positive.

$-3 \times 5 \times (-2) = 30$

b. $3 \times (-2) \times (-4) \times (-5)$

$|3| \times |-2| \times |-4| \times |-5| = 120$

THINK: There is an odd number of negative factors. The product is negative.

$3 \times (-2) \times (-4) \times (-5) = -120$

FOR DISCUSSION

See TE side column.

Can you name this product at a glance? Explain.

$$-45 \times 36 \times (-78) \times (-86) \times 0 \times 14 \times (-9)$$

PRACTICE EXERCISES

See Extra Practice, page 457.

13. −2,550 15. 7,200 18. 12,222 20. −7,750 35. −1,000 39. −24 42. −50

Multiply.

• **1.** 5×8 40	• **2.** 8×11 88	**3.** 7×9 63	**4.** 23×42 966
5. $-7 \times (-2)$ 14	**6.** $-9 \times (-2)$ 18	**7.** $-6 \times (-14)$ 84	**8.** $-26 \times (-45)$ 1,170
9. $4 \times (-3)$ −12	**10.** $8 \times (-9)$ −72	**11.** -32×23 −736	**12.** -41×52 −2,132
13. $75 \times (-34)$	**14.** -83×61 −5,063	**15.** $-80 \times (-90)$	**16.** 76×89 6,764
17. 123×0 0	**18.** $-126 \times (-97)$	**19.** 250×10 2,500	**20.** $310 \times (-25)$

• **21.** $-1 \times (-2) \times 4$ 8	• **22.** $2 \times 3 \times (-3)$ −18	**23.** $-3 \times (-1) \times (-4)$ −12
24. $-2 \times (-4) \times 4$ 32	**25.** $3 \times 2 \times 5$ 30	**26.** $7 \times 0 \times (-9)$ 0
27. $3 \times (-1) \times 5$ −15	**28.** $-3 \times (-3) \times (-3)$ −27	**29.** $7 \times (-4) \times (-2)$ 56
30. $-2 \times (-4) \times (-7)$ −56	**31.** $3 \times 8 \times 7$ 168	**32.** $-1 \times 9 \times 78$ −702
33. $12 \times 0 \times (-19)$ 0	**34.** $-11 \times (-2) \times 1$ 22	**35.** $-10 \times (-10) \times (-10)$
• **36.** $-1 \times (-3) \times 4 \times (-2)$ −24	• **37.** $1 \times (-1) \times 1 \times (-1)$ 1	**38.** $3 \times (-2) \times 0 \times (-1)$ 0
39. $2 \times (-4) \times (-1) \times (-3)$	**40.** $4 \times (-2) \times 4 \times 2$ −64	**41.** $-2 \times (-3) \times 3 \times (-4)$ −72
42. $(-5) \times 5 \times (-1) \times (-2)$	**43.** $2 \times 4 \times 1 \times (-3)$ −24	**44.** $3 \times 3 \times 3 \times 3$ 81

• **45.** $1 \times (-1) \times 2 \times (-1) \times 2$ 4	• **46.** $(-2) \times (-2) \times (-2) \times (-2) \times (-2)$ −32
47. $3 \times (-2) \times 1 \times 0 \times (-7)$ 0	**48.** $(-4) \times (-1) \times 2 \times (-1) \times (-2)$ 16
49. $3 \times (-1) \times (-2) \times (-3) \times 1$ −18	**50.** $3 \times (-2) \times 0 \times (-5) \times (-5)$ 0

Solve.

• **51.** In a recent football game the North Ridge team gained an average of 6 yd every time that the ball was run instead of passed. The team ran with the ball 35 times during the game. Write an integer for the total number of yards gained. +210

52. Football coach Mueller came to North Ridge High School 12 years ago. On the average his teams have lost only 4 games out of 24 per season. Write a negative integer for the number of games lost since Coach Mueller was hired. −48

53. The place kicker on the football team averaged 36.5 yd per kick. If he kicked the ball a total of 292 yd, how many times did he kick? 8 times

54. The halfback gained an average of 7.8 yd per carry. If he carried the ball 15 times, how many yards did he gain for the team? 117 yd

ASSIGNMENTS

Level 1	1–12, 21–26, 36–41, 51–52
Level 2	Odd 1–49, 51–53
Level 3	Even 2–50, 51–54

FOLLOW-UP ACTIVITY

ENRICHMENT Have students find the missing factor.

1. $4 \times ■ = -72$ (−18)
2. $■ \times (-5) = 130$ (−26)
3. $■ \times 7 \times (-3) = -168$ (8)
4. $-6 \times (-2) \times ■ = -144$ (−12)
5. $9 \times ■ \times 10 = -990$ (−11)

SUPPLEMENTARY MATERIALS

TRP Practice, p. 149

TRP Reteaching, p. 92

TRP Transparency 75

OBJECTIVE

- Divide integers.

TEACHING THE LESSON

WARM-UP Write the following exercises on the chalkboard or an overhead transparency.

Divide.

1. 69 ÷ 23 (3) **2.** 88 ÷ 11 (8)
3. 91 ÷ 13 (7) **4.** 78 ÷ 13 (6)
5. 65 ÷ 13 (5) **6.** 0 ÷ 56 (0)

INTRODUCTION Write the following problem on the chalkboard:

In one week, the water level at the foot of Wabash Falls drops 14 ft. Write a negative integer to show the average drop per day.

Guide students through this solution:

14 ft drop ÷ 7 days = 2 ft drop/day
= −2 ft/day

Then explain that the following solution is also correct:

−14 ft ÷ 7 days = −2 ft/day

INSTRUCTION In discussing Examples 1 and 2, stress that the only difference between dividing integers and dividing whole numbers is that the sign of the quotient must be considered. Before assigning Exercises 1–56, you may wish to have students name the sign of each quotient.

CHECKPOINT The incorrect choices include common errors students make.

TIME OUT If students encounter difficulty, you might suggest that they use a table such as:

Visit	1	2	3	4
Member	$22	$24	$26	$28
Nonmember	$5	$10	$15	$20

13.5 DIVIDING INTEGERS

You already know that multiplication and division of whole numbers are related.

Multiplication
$6 \times 8 = 48$

Division
$48 \div 8 = 6$
$48 \div 6 = 8$

The same is true for multiplication and division of integers.

Multiplication
$(-7) \times (-5) = 35$
$6 \times (-9) = -54$

Division
$35 \div (-7) = -5$
$35 \div (-5) = -7$
$-54 \div (-9) = 6$
$-54 \div 6 = -9$

The sign rules for dividing integers are identical to those for multiplying integers.

EXAMPLE 1 Divide: $-4{,}500 \div (-150)$

$|-4{,}500| \div |-150| = 30$

THINK: The signs are the same. The quotient is positive.

$-4{,}500 \div (-150) = 30$

EXAMPLE 2 Divide: $-72 \div 9$

$|-72| \div |9| = 8$

THINK: The signs are different. The quotient is negative.

$-72 \div 9 = -8$

CHECKPOINT

Write the letter of the correct answer.

Divide.

1. $16 \div 4$	c	**a.** 20	**b.** −4	**c.** 4	**d.** 12
2. $-15 \div (-5)$	b	**a.** −3	**b.** 3	**c.** −10	**d.** −20
3. $-12 \div 3$	b	**a.** 4	**b.** −4	**c.** −9	**d.** −36
4. $0 \div (-9)$	d	**a.** 9	**b.** −9	**c.** −90	**d.** 0

PRACTICE EXERCISES See Extra Practice, page 457.

Divide.

• **1.** 6 ÷ 2 3	**2.** 4 ÷ 4 1	**3.** 6 ÷ 2 3	**4.** 10 ÷ 5 2
• **5.** −8 ÷ (−4) 2	**6.** 8 ÷ (−4) −2	**7.** 0 ÷ 9 0	**8.** −6 ÷ (−3) 2
• **9.** 12 ÷ (−3) −4	**10.** −15 ÷ 5 −3	**11.** 21 ÷ (−7) −3	**12.** −24 ÷ 6 −4
13. 36 ÷ (−6) −6	**14.** −28 ÷ 4 −7	**15.** 24 ÷ (−8) −3	**16.** −27 ÷ (−9) 3
17. −42 ÷ 7 −6	**18.** 42 ÷ (−6) −7	**19.** −39 ÷ 3 −13	**20.** 38 ÷ (−2) −19
21. 44 ÷ 4 11	**22.** −48 ÷ 8 −6	**23.** 48 ÷ 6 8	**24.** −49 ÷ (−7) 7
25. 56 ÷ (−8) −7	**26.** −64 ÷ (−8) 8	**27.** −72 ÷ 9 8	**28.** 85 ÷ 5 17
29. −24 ÷ 12 −2	**30.** 36 ÷ (−18) −2	**31.** −48 ÷ (−16) 3	**32.** 50 ÷ 10 5
33. −69 ÷ 23 −3	**34.** 60 ÷ (−15) −4	**35.** 34 ÷ 17 2	**36.** −91 ÷ (−13) 7
37. 98 ÷ (−14) −7	**38.** 99 ÷ 33 3	**39.** −96 ÷ 16 −6	**40.** 78 ÷ (−13) −6
41. −75 ÷ (−15) 5	**42.** −88 ÷ (−11) 8	**43.** 65 ÷ (−13) −5	**44.** 0 ÷ (−56) 0
45. 153 ÷ 9 17	**46.** −112 ÷ (−8) 14	**47.** 224 ÷ (−7) −32	**48.** −207 ÷ 9 −23
49. −156 ÷ (−13) 12	**50.** 210 ÷ 14 15	**51.** −408 ÷ 17 −24	**52.** 304 ÷ (−19) −16
53. −357 ÷ 21 −17	**54.** 448 ÷ 32 14	**55.** −656 ÷ (−41) 16	**56.** 627 ÷ (−3) −209

MIXED REVIEW

Write the product.

• **57.** 10^2 100	**58.** 10^4 10,000	**59.** 10^7 10,000,000	**60.** 10^3 1,000
61. 10^5 100,000	**62.** 10^8 100,000,000	**63.** 10^9 1,000,000,000	**64.** 10^6 1,000,000

Write in exponent form.

• **65.** 10,000 10^4 **66.** 100,000 10^5 **67.** 1,000,000 10^6

TIME OUT

You have an important decision to make. Should you join the new Flying Fins Swim Club? A membership costs $20. Each swim is $2 for members and $5 for nonmembers. How many times would you have to swim to make joining the club a better buy than not joining? 7 or more times

A more intuitive solution, however, involves recognizing that the member saves $3 on each visit. Therefore, 7 visits more than compensate for the $20 fee.

COMMON ERROR

When dividing integers, students may sometimes apply the wrong sign to the quotient. For example, in item 2 of Checkpoint, they would choose *a*. Emphasize the relationship between multiplication and division by having students write related multiplication examples such as:

−15 ÷ (−5) = ■ → −5 × ■ = −15

Encourage students to ask themselves questions such as: What number do I multiply −5 by to get −15?

ASSIGNMENTS

Level 1	1–36, 57–67
Level 2	Odd 1–55, 57–67, TO
Level 3	25–67, TO

FOLLOW-UP ACTIVITY

ENRICHMENT Have students apply their knowledge of computing with integers to the exercises below. Remind them to compute inside the brackets or parentheses first.

1. (−8 + 4) ÷ (−2) × (−6) − (−4) (−8)

2. [13 × (−3)] + (−32 ÷ 8) − 2 (−45)

3. [−9 − (−9)] × (−4 + 3) ÷ (−1) (0)

SUPPLEMENTARY MATERIALS

TRP Practice, p. 150

TRP Reteaching, p. 93

TRP Transparency 75

OBJECTIVE

- Solve problems that involve using data from a wind chill table.

TEACHING THE LESSON

WARM-UP Write the following exercises on the chalkboard or an overhead transparency.

Subtract.

1. 8 − (−7) **2.** 4 − (−9)

3. −12 − (−15) **4.** −5 − (−13)

(**1.** 15 **2.** 13 **3.** 3 **4.** 8)

INTRODUCTION Ask students to tell what they know about wind chill. Discuss questions such as:

- During which season do weather forecasters give information about wind chill?
- Why is information about wind chill valuable?

INSTRUCTION As you discuss the introductory example, be sure the students understand the procedure: first they find the appropriate entry in the wind chill table; then they find the difference between the actual temperature and the wind chill reading.

Note that Problems 1–6 require only that the students read the wind chill table correctly. Problem 10 involves reading the table, comparing integers, and computing as well.

ASSIGNMENTS

Level 1	1–8
Level 2	1–10
Level 3	5–10

SUPPLEMENTARY MATERIALS

TRP Practice, p. 151

TRP Transparency 76

PROBLEM *Solving* APPLICATION

13.6 CAREER: METEOROLOGIST

Ann Kurasawa is a meteorologist. As part of her work, she forecasts weather. Computers provide information about temperature, wind speeds, and barometric pressures.

Sometimes, the air outside feels colder than the actual temperature. As the wind speed increases, body heat is lost more rapidly and the air feels colder. This wind chill table shows how cold the outside air feels for various actual temperatures and wind speeds.

So, if the wind speed is 15 mph and the actual temperature is 10°F, then the wind chill is −18°F.

WIND CHILL TABLE

Wind speed (mph)	Temperatures (°F) 30	20	10	0	−10	−20	−30
5	27	16	7	−5	−15	−26	−36
15	9	−5	−18	−31	−45	−58	−72
25	1	−15	−29	−44	−59	−74	−88
35	−4	−20	−35	−52	−67	−82	−97
45	−6	−22	−38	−54	−70	−85	−102

Use the table. Find the wind chill.

Wind Speed	Temperature	Wind Chill	Wind Speed	Temperature	Wind Chill
25 mph	30°F	• **1.** ■ 1°F	35 mph	10°F	−35°F • **2.** ■
45 mph	20°F	**3.** ■ −22°F	5 mph	−10°F	−15°F **4.** ■
45 mph	0°F	**5.** ■ −54°F	15 mph	−30°F	−72°F **6.** ■

Solve.

• **7.** The thermometer outside Ann's window reads 20°F. She hears on the radio that the wind speed is 35 mph. What is the wind chill? −20°F

8. The actual temperature outside Ann's window is −20°F. A TV weather report says that the wind chill is −74°F. What is the wind speed?
25 mph

9. Wind speed is 25 mph. Actual temperature is 10°F. How many degrees less than actual temperature does the air feel? 39°F

10. Which of the following conditions has a lower wind chill reading:
Wind speed: 5 mph; temperature: −20°F
Wind speed: 15 mph; temperature: −10°F
How much lower is it?
15 mph; temperature: −10°F; 19° lower

MIDCHAPTER REVIEW

Write the integer named by the point. [364]

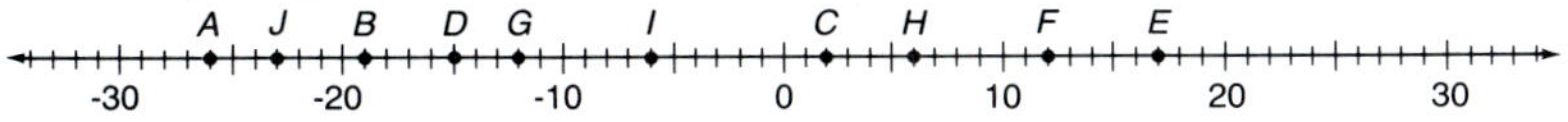

1. J −23 2. H 6 3. F 12 4. E 17 5. C 2

6. B −19 7. D −15 8. G −12 9. A −26 10. I −6

Compare. Write > or <. [364]

11. −3 ● 5 < 12. 6 ● −9 > 13. −3 ● −4 > 14. −12 ● −13 >

15. 35 ● 32 > 16. −54 ● −48 < 17. 23 ● −39 > 18. −42 ● −49 >

Add. [367]

19. −3 + 5 2 20. 6 + (−4) 2 21. 0 + 2 2 22. −12 + 4 −8

23. 25 + (−26) −1 24. 56 + (−62) −6 25. −32 + (−9) −41 26. −44 + (−12) −56

27. 3 + (−2) + (−8) −7 28. −4 + (−12) + 21 5 29. 2 + (−3) + 14 13

Subtract. [370]

30. 8 − 5 3 31. 7 − 9 −2 32. 3 − 5 −2 33. 5 − 12 −7

34. 5 − (−6) 11 35. 7 − (−3) 10 36. −12 − (−14) 2 37. −24 − (−39) 15

Multiply. [372]

38. 4 × (−3) −12 39. −6 × 12 −72 40. −4 × (−4) 16 41. 24 × (−3) −72

42. −4 × (−5) × 20 400 43. −5 × 9 × (−7) 315 44. −12 × (−4) × (−10) −480

Divide. [374]

45. 21 ÷ 7 3 46. −35 ÷ 5 −7 47. −45 ÷ (−9) 5 48. −84 ÷ (−7) 12

49. −250 ÷ 25 −10 50. 400 ÷ 80 5 51. −330 ÷ (−11) 30 52. −110 ÷ (−10) 11

Use the table for Exercises 53–54. [376]

53. The thermometer outside of Jerry's window reads 30°F. The weatherman on the television reports that the wind speed is 15 mph. What is the wind chill? 9°F

54. The temperature outside City Hall is 0°F. The wind speed is 25 mph. What is the wind chill? −44°F

WIND CHILL TABLE

Wind speed (mph)	Temperatures (°F) 30	20	10	0	−10	−20	−30
5	27	16	7	−5	−15	−26	−36
15	9	−5	−18	−31	−45	−58	−72
25	1	−15	−29	−44	−59	−74	−88

OBJECTIVE

- Evaluate student progress.

USING THE REVIEW

This page provides a means for informally evaluating students' understanding of the skills and concepts covered so far in this chapter.

Have students look at the page to familiarize themselves with the various question formats that are presented. Discuss any questions that they may have. Then ask them to complete the page independently.

In addition to grading them individually, you may wish to review the answers to the questions collectively with the students.

Page references appear in brackets. They refer to pages on which a particular skill was introduced.

Before continuing on to the topics found in the remainder of the chapter, you may wish to have students review any skills or concepts in which they have demonstrated weakness.

OBJECTIVES

- Write a standard numeral in scientific notation.
- Write scientific notation in standard form.

TEACHING THE LESSON

WARM-UP Write the following exercises on the chalkboard or an overhead transparency.

Write the missing factor.

1. $3.4 \times \blacksquare = 34$ (10)
2. $6.226 \times \blacksquare = 6{,}226$ (1,000)
3. $3 \times \blacksquare = 0.03$ (0.01)
4. $5 \times \blacksquare = 0.00005$ (0.00001)

INTRODUCTION Have students count the number of zeros in each:

- 30,000,000,000,000 (13)
- 0.00000000000000000000003 (23)

Point out how difficult it is to determine the size of these numbers at a glance. Explain that a method called scientific notation makes it much easier to compute with very large and very small numbers.

INSTRUCTION When discussing Examples 1 and 2, provide students with the following:

$1{,}000{,}000 = 10^6$	$0.000001 = 10^{-6}$
$100{,}000 = 10^5$	$0.00001 = 10^{-5}$
$10{,}000 = 10^4$	$0.0001 = 10^{-4}$
$1{,}000 = 10^3$	$0.001 = 10^{-3}$
$100 = 10^2$	$0.01 = 10^{-2}$
$10 = 10^1$	$0.1 = 10^{-1}$

Before assigning the exercises, guide students in writing the following numbers in scientific notation.

- 8,420,000,000 (8.42×10^9)
- 0.00063 (6.3×10^{-4})

FOR DISCUSSION Students should explain that Exercises 3 and 4 are not expressed in scientific notation because the first factors are not numbers between 1 and 10.

13.7 SCIENTIFIC NOTATION

Scientists frequently work with very large and very small numbers.

Distance from Earth to Alpha Centauri: About 30,000,000,000,000 mi

Mass of a molecule of water: About 0.00000000000000000000003 g

To simplify writing numerals for large and small numbers, scientists have developed a system called **scientific notation.** Numbers written in scientific notation have two factors. The first factor is a number between 1 and 10. The second factor is a power of 10.

EXAMPLE 1 Write in scientific notation.

a. 345,000 b. 0.00041

1. Write the first factor.

a. **THINK: Move the decimal point to form a number between 1 and 10.**

3.4 5, 0 0 0 → 3.45 ↑ **5 places**

b. **THINK: Move the decimal point to form a number between 1 and 10.**

0 0 0 0 4 1 → 4.1 ↑ **4 places**

2. Write the power of 10.

a. **THINK: The decimal point was moved left 5 places. The exponent is positive.**

$345{,}000 = 3.45 \times 10^5$

b. **THINK: The decimal point was moved right 4 places. The exponent is negative.**

$0.00041 = 4.1 \times 10^{-4}$

EXAMPLE 2 Write in standard form.

a. 2.16×10^7 b. 6.1×10^{-3}

a. **THINK: The exponent is positive. Move the decimal point 7 places to the right. Write additional zeros as needed.**

2 1 6 0 0 0 0 0. → 21,600,000

$2.16 \times 10^7 = 21{,}600{,}000$

b. **THINK: The exponent is negative. Move the decimal point 3 places to the left. Write additional zeros as needed.**

0.0 0 6 1

$6.1 \times 10^{-3} = 0.0061$

FOR DISCUSSION

See TE side column.

Is the number expressed in scientific notation? Explain.

1. 3×10^2
2. 8.2×10^{-5}
3. 14×10^2
4. 0.21×10^5

PRACTICE EXERCISES

See Extra Practice, page 458.

Write the missing exponent.

- • **1.** $34 = 3.4 \times 10^{\blacksquare}$ 1
- • **2.** $542 = 5.42 \times 10^{\blacksquare}$ 2
- **3.** $6{,}226 = 6.226 \times 10^{\blacksquare}$ 3
- **4.** $0.03 = 3 \times 10^{\blacksquare}$ −2
- **5.** $0.0091 = 9.1 \times 10^{\blacksquare}$ −3
- **6.** $0.00005 = 5 \times 10^{\blacksquare}$ −5
- **7.** $21{,}100 = 2.1 \times 10^{\blacksquare}$ 4
- **8.** $200{,}000 = 2 \times 10^{\blacksquare}$ 5
- **9.** $0.000043 = 4.3 \times 10^{\blacksquare}$ −5
- **10.** $0.00000000712 = 7.12 \times 10^{\blacksquare}$ −9

Write in scientific notation.

- • **11.** 25 2.5×10^1
- • **12.** 300 3×10^2
- **13.** 456 4.56×10^2
- **14.** 3,400 3.4×10^3
- **15.** 0.3 3×10^{-1}
- **16.** 0.41 4.1×10^{-1}
- **17.** 0.05
- **18.** 0.065 6.5×10^{-2}
- **19.** 30,000 3×10^4
- **20.** 100,000 1×10^5
- **21.** 125,000 1.25×10^5
- **22.** 0.000000002 2×10^{-9}

Write as a standard numeral. **26.** 94,300 **28.** 0.047 **29.** 0.00671 **30.** 0.000876

- • **23.** 2.4×10^1 24
- **24.** 3.1×10^2 310
- **25.** 1.43×10^3 1,430
- **26.** 9.43×10^4
- • **27.** 3.1×10^{-1} 0.31
- **28.** 4.7×10^{-2}
- **29.** 6.71×10^{-3}
- **30.** 8.76×10^{-4}
- **31.** 2.3×10^8 230,000,000
- **32.** 1.9×10^{-7} 0.00000019
- **33.** 6.54×10^{10} 65,400,000,000
- **34.** 2.43×10^{-9} 0.00000000243

Solve.

• **35.** Scientists estimate that the Sun is about 93,000,000 mi from Earth. Write this number in scientific notation. 9.3×10^7

36. By some estimates, the diameter of Mars is 4.2×10^3 mi. Write this number in standard form. 4,200 mi

37. At the closest point of its rotation, the Moon is 221,463 mi away from Earth. At its farthest point, the Moon is 252,710 mi away. What is the average distance that the Moon revolves around Earth? 237,086.5 mi

38. The surface gravity of Jupiter is 2.6 times that of Earth. If a man weighs 186 lb on Earth, what would he weigh on Jupiter? 483.6 lb

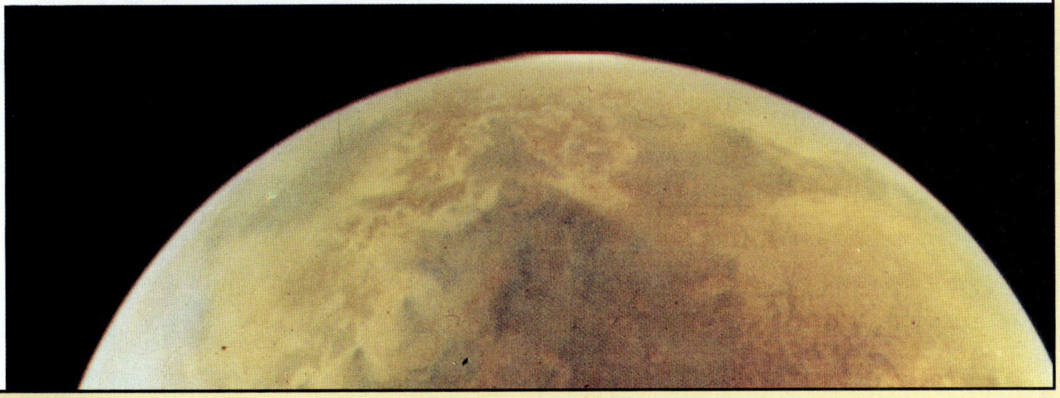

COMMON ERRORS

When writing a standard numeral in scientific notation, students may sometimes apply the incorrect sign to the exponent. Have students compare the first factor with the original number.

$687 = 6.87 \times 10^{\blacksquare}$ (10^2)

$0.045 = 4.5 \times 10^{\blacksquare}$ (10^{-2})

If the original number is greater, then the exponent must be positive. If the original number is lesser, then the exponent must be negative.

ASSIGNMENTS

Level 1	1–14, 23–26, 35–36
Level 2	Odd 1–33, 35–37
Level 3	15–22, 27–38

FOLLOW-UP ACTIVITY

APPLICATION Have students research the average distance each planet is from the sun. Have them make a table of the distances written in scientific notation.

Planet	Distance, mi
Mercury	(3.60×10^7)
Venus	(6.72×10^7)
Earth	(9.29×10^7)
Mars	(1.415×10^8)
Jupiter	(4.834×10^8)
Saturn	(8.860×10^8)
Uranus	(1.7820×10^9)
Neptune	(2.7920×10^9)
Pluto	(3.6750×10^9)

SUPPLEMENTARY MATERIALS

TRP Practice, p. 152

TRP Reteaching, p. 94

TRP Transparency 76

OBJECTIVES

- Identify points in a plane.
- Graph points in a plane.

TEACHING THE LESSON

WARM-UP Write the following exercises on the chalkboard or an overhead transparency.

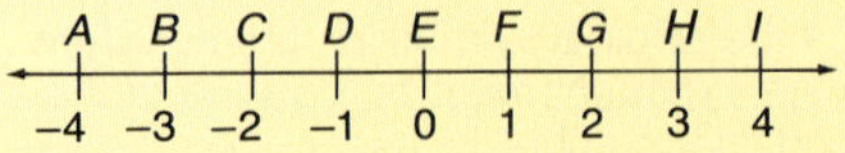

Write the integer named by the point.

1. *A* (−4) **2.** *H* (3) **3.** *C* (−2)

Name the point named by the integer.

4. −3 (*B*) **5.** 4 (*I*) **6.** 0 (*E*)

INTRODUCTION Draw Quadrant I of a coordinate plane on the chalkboard as though it were a city map. Mark specific landmarks such as a theater at (2, 3). Put City Hall at the origin. Ask how many blocks west of City Hall is the theater. (2 blocks) Ask how many blocks north. (3 blocks). Guide students to realize that this location can be written as (2 blocks west, 3 blocks north), or simply (2, 3). Have volunteers name ordered pairs for other landmarks.

INSTRUCTION In discussing Example 1, draw a coordinate plane on the chalkboard. Plot points and label them. Then have volunteers come to the chalkboard and use their index fingers to reinforce the correct sequence, move left or right, then up or down.

In discussing Example 2, use the same grid, but give students ordered pairs instead, and have them plot the points themselves, once again using fingers to count off units against each axis.

13.8 GRAPHING POINTS IN A PLANE

The President of the United States is visiting Lakeville. To monitor the progress of the President's motorcade from the airport to city hall, the police have divided the city into quadrants. There is a stakeout in each quadrant.

Every point on a coordinate plane can be named by using an **ordered pair** of numbers. In an ordered pair, the first number tells how far the point is to the right or left of the **origin.** The second number tells how far the point is up or down. The numbers in an ordered pair are called **coordinates.**

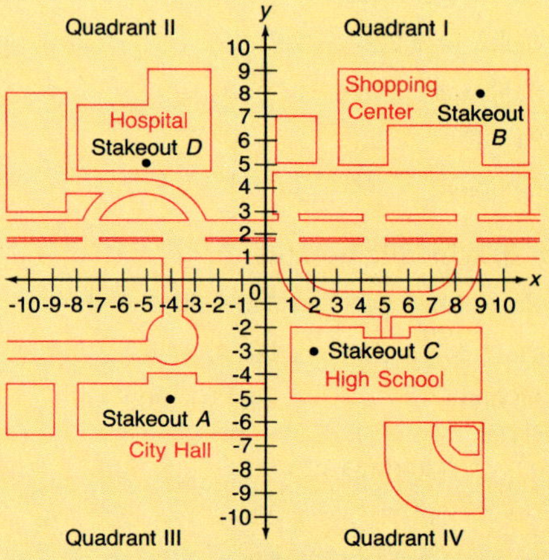

EXAMPLE 1 Use an ordered pair to name the stakeout in Quadrant III.

1. Find the x-coordinate.

   THINK: Move left 4 units from the origin. −4 is the x-coordinate.

2. Find the y-coordinate.

   THINK: Move down 5 units. −5 is the y-coordinate.

3. Write the ordered pair (x, y).

   THINK: The x-coordinate is −4. The y-coordinate is −5.

   The ordered pair is (−4, −5).

(−4, −5) names the location of the stakeout in Quadrant III.

You can use an ordered pair to locate a point on a coordinate plane.

EXAMPLE 2 What stakeout is located at (2, −3)?

1. Locate the x-coordinate.

   THINK: The x-coordinate is positive. Start at the origin. Move right two units.

2. Locate the y-coordinate.

   THINK: The y-coordinate is negative. Move down three units.

(2, −3) names the location of Stakeout C.

FOR DISCUSSION See TE side column.

What can you say about all the numbers in each ordered pair in:

1. Quadrant I?
2. Quadrant II?
3. Quadrant III?
4. Quadrant IV?

PRACTICE EXERCISES

Use the coordinate grid at the right for Exercises 1–24.
Write the ordered pair.

• 1. Point A (−6, 6)
• 2. Point B (1, 6)
3. Point C (−3, 5)
4. Point D (1, −6)
5. Point E (6, 5)
6. Point F (−3, −4)
7. Point G (0, −3)
8. Point H (−5, 1)

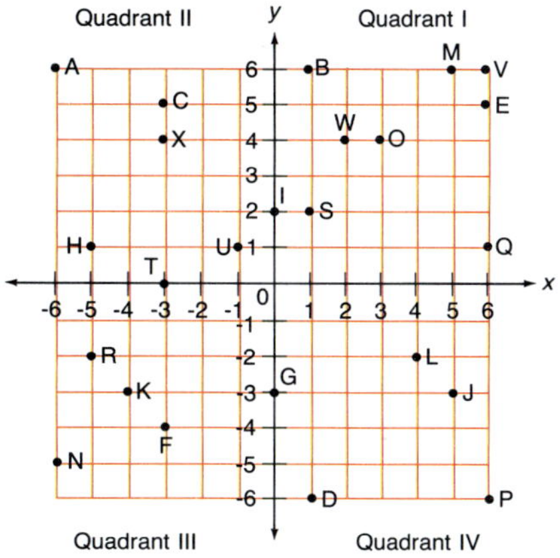

Write the point named by the ordered pair.

• 9. (1, 2) S
• 10. (3, 4) O
11. (5, 6) M
12. (6, 1) Q
13. (2, 4) W
14. (6, 6) V
15. (0, 2) I
16. (−3, 0) T
17. (−1, 1) U
18. (−3, 4) X
19. (6, −6) P
20. (5, −3) J
21. (4, −2) L
22. (−6, −5) N
23. (−4, −3) K
24. (−5, −2) R

Complete.

• 25. If a point is located on the y-axis, its first coordinate must be ■. 0
26. If a point is located on the x-axis, its second coordinate must be ■. 0
27. The coordinates of the origin are ■. (0, 0)
28. If the first coordinate of a point is negative and the second coordinate is positive, then the point lies in Quadrant ■. II
29. If the first coordinate of a point is positive and the second coordinate is negative, then the point lies in Quadrant ■. IV

Explain that because a line is one-dimensional, only one coordinate is necessary to name a point on a line. Because a plane has two dimensions, however, two coordinates are needed to name a point in a plane.

FOR DISCUSSION Students should explain that in Quadrant I, x and y are both positive. In Quadrant II, x is positive and y is negative. In Quadrant III, x and y are both negative. And in Quadrant IV, x is negative and y is positive.

ASSIGNMENTS

Level 1	1–24
Level 2	Odd 1–23, 25–29
Level 3	Even 2–24, 25–29

FOLLOW-UP ACTIVITY

APPLICATION Provide students with graph paper and these sets of points:

- (−5, −2), (−5, 2), (−3, 2), (−3, 0), (−5, 0)
- (1, 2), (−1, 2), (−1, −2), (1, −2), (1, 2)
- (3, 2), (3, −2), (5, −2)
- (9, −2), (7, 0), (7, −2), (7, 2), (7, 0), (9, 2)

Tell them that the President nicknamed "Young Hickory" is hidden in these coordinates. Have them find the name by plotting the points and joining each set in the proper order.

SUPPLEMENTARY MATERIALS

TRP Practice, p. 153
TRP Reteaching, p. 95
TRP Lesson Aids, p. 12
TRP Transparency 77

OBJECTIVE

- Solve problems that involve locating points by their latitude and longitude.

TEACHING THE LESSON

WARM-UP Draw this sketch on the chalkboard or an overhead transparency.

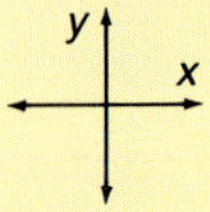

Then read these questions aloud.

Is a point above or below the *x*-axis if the *y*-coordinate is

1. +2 (above) **2.** −4 (below) **3.** −90 (below)

Is a point to the right or left of the *y*-axis if the *x*-coordinate is

4. −8 (left) **5.** +2 (right) **6.** −70 (left)

INTRODUCTION Display a large globe or map of the world. Ask the students whether any of them know what method is used to locate points on the globe or map. Let one or two volunteers give their responses. Some students are likely to recall the terms *latitude* and *longitude.*

INSTRUCTION As you discuss the way in which latitude and longitude are used to locate points, there is no need to emphasize starting at the prime meridian or the equator. Let students find the locations informally.

It would *not* be wise to overemphasize the "similarity" between the latitude/longitude system and the *x-y* coordinate plane. In the *x-y* plane, we list the *x*-coordinate first, whereas in the latitude/longitude situation, we list the north/south element first.

PROBLEM Solving APPLICATION

13.9 LATITUDE AND LONGITUDE

Suppose you find this scrap of paper in a drawer filled with old papers:

What does 10°N, 105°W mean? In what ocean is the sunken treasure ship?

The location of points on the earth can be described in terms of latitude and longitude.

Latitude is used to measure degrees north and south of the equator. If a point is on the equator, its latitude is 0°.

Longitude is used to measure degrees east and west of the prime meridian. The **prime meridian** is a line or circle that passes through Greenwich, England, and extends in a north–south direction. If a point is on the prime meridian, its longitude is 0°.

In using pairs of numbers to give latitude and longitude, the latitude is given first.

So, 10°N, 105°W means that the location is 10° north of the equator and 105° west of the prime meridian.

Look at the map on page 383. Find the line for 10° north latitude. Then move to the intersection of this line with the line for 105° west longitude.

So, 10°N, 105°W means that the sunken treasure is located in the Pacific Ocean.

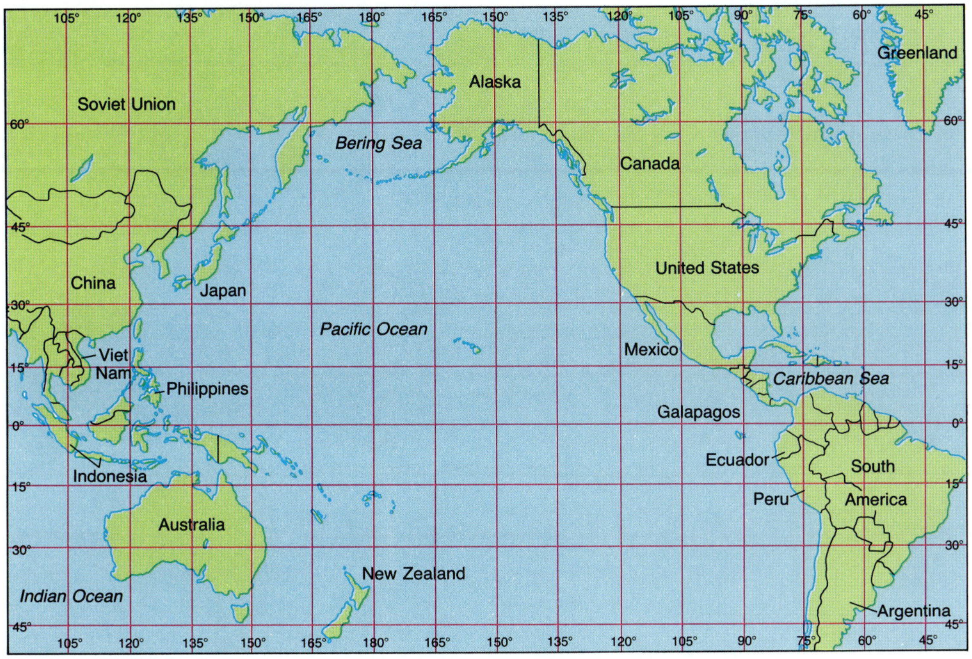

Use the map above to identify the country or body of water in which each point is located.

• **1.** 15°S, 120°W Pacific Ocean
• **2.** 45°S, 105°E Indian Ocean
3. 60°N, 105°W Canada
4. 30°N, 90°W United States
5. 30°N, 120°E China
6. 20°N, 105°W Mexico
7. 30°S, 120°E Australia
8. 60°N, 170°E Bering Sea
9. 60°N, 90°E Soviet Union

Use the map above to identify which point is closer to the equator.

• **10.** 10°N, 130°W or 30°N, 130°W 10°N, 130°W
11. 35°S, 120°E or 25°S, 120°E 25°S, 120°E
12. 20°N, 101°W or 1°N, 120°W 1°N, 120°W
13. 40°S, 120°E or 15°S, 130°E 15°S, 130°E

Use the map to find the country in which each point is located.

• **14.** 40°N, 100°W United States
• **15.** 45°S, 175°E New Zealand
16. 15°N, 120°E Philippines
17. 75°N, 45°W Greenland
18. 15°N, 105°E Thailand
19. 0°, 75°W Ecuador

Solve. Use library resources if necessary.

20. The north–south distance around the earth is about 24,860 mi. What is the distance from the North Pole to the South Pole? about 12,430 mi

21. West longitudes can be any number between 0° and 180°. Why is it not possible for the longitude of a place to be 200°W? East longitudes make up other half of circle.

ASSIGNMENTS

Level 1	1–6, 12–16
Level 2	Odd 1–19, 20
Level 3	Even 2–18, 20–21

FOLLOW-UP ACTIVITY

ENRICHMENT Have the students use reference resources to learn more about latitude and longitude. Ask students to determine the answers for questions such as the following.

1. Are two lines (circles) of latitude always the same distance apart? (yes)
2. Are two lines (circles) of longitude always the same distance apart? (no)
3. At 0° latitude, there are approximately 69 mi in 1° of longitude. Approximately how many miles are there in 1° of longitude at latitude 30°? (60 mi)
4. Approximately how many miles are there in 1° of longitude at latitude 60°? (35 mi)

SUPPLEMENTARY MATERIALS

TRP Practice, p. 154

TRP Transparency 77

OBJECTIVES

- Identify rational numbers on a number line.
- Compare rational numbers.
- Find the absolute value of rational numbers.

TEACHING THE LESSON

WARM-UP Write the following exercises on the chalkboard or an overhead transparency.

Compare. Write <, >, or =.

1. $\frac{3}{4}$ ● $\frac{4}{5}$ (<)

2. $\frac{1}{2}$ ● $\frac{4}{8}$ (=)

3. 0.75 ● $\frac{3}{4}$ (=)

4. 3.4 ● 3.99 (<)

INTRODUCTION Students may not have used rational numbers before, but they may have surmised that numbers such as −3.4 exist. Guide students to see rational numbers as a logical extension of integers by writing 0.7 on the chalkboard and asking a volunteer to name its opposite. (−0.7) Continue with examples such as:

- $\frac{1}{2}$ $(-\frac{1}{2})$
- $4\frac{3}{4}$ $(-4\frac{3}{4})$
- 6.78 (−6.78)
- 12.3 (−12.3)

INSTRUCTION Point out that in Examples 1, 2, and 3, students are applying skills they already know to a new situation.

You may find that some students do not understand why repeating decimals fit the definition. Remind them that most repeating decimals are simple fractions. For example:

$4.\overline{3} = 4\frac{1}{3}$ $\quad 0.\overline{21} = \frac{7}{33}$

13.10 RATIONAL NUMBERS

A **rational number** is a number that can be written as a ratio of two integers. The denominator cannot be zero. Rational numbers include:

Integers:	15, −7
Fractions and mixed numbers:	$\frac{5}{8}$, $-10\frac{3}{5}$
Terminating decimals:	0.75, −2.84
Repeating decimals:	$4.\overline{3}$, $-0.\overline{21}$

Every rational number can be shown on a number line.

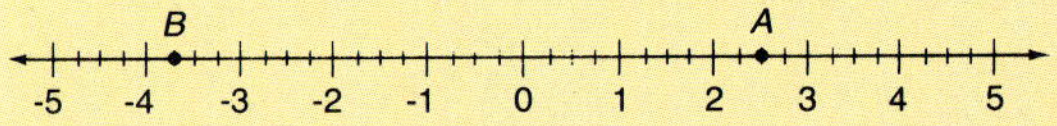

EXAMPLE 1 Write the rational number named by the point.

a. Point A

THINK: A is $\frac{1}{2}$ of the way between 2 and 3.

Point A names $2\frac{1}{2}$.

b. Point B

THINK: B is $\frac{3}{4}$ of the way between −3 and −4.

Point B names $-3\frac{3}{4}$.

You compare rational numbers the same way that you compare integers. Imagine a number line. As you move right, the numbers become greater. As you move left, the numbers become smaller.

EXAMPLE 2 Compare. Write <, >, or =.

a. 2 ● 0.75

THINK: 2 is to the right of 0.75.

$2 > 0.75$

b. $-1\frac{2}{3}$ ● −0.86

THINK: $-1\frac{2}{3}$ is to the left of −0.86.

$-1\frac{2}{3} < -0.86$

As with integers, the absolute value of a rational number is the distance the number is from zero.

EXAMPLE 3 Write the absolute value.

a. $|-6\frac{5}{8}|$

THINK: $-6\frac{5}{8}$ is $6\frac{5}{8}$ units from 0.

So, $|-6\frac{5}{8}|$ is $6\frac{5}{8}$.

b. $|12.7|$

THINK: 12.7 is 12.7 units from 0.

So, $|12.7|$ is 12.7.

PRACTICE EXERCISES

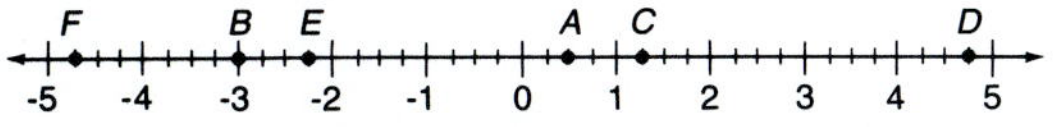

Use the number line above for Exercises 1–6.
Write the rational number named by the point.

• **1.** *A* $\frac{1}{2}$ • **2.** *B* -3 **3.** *C* $1\frac{1}{4}$ **4.** *D* $4\frac{3}{4}$ **5.** *E* $-2\frac{1}{4}$ **6.** *F* $-4\frac{3}{4}$

Compare. Write >, <, or =.

• **7.** $\frac{1}{3} \bullet \frac{2}{3}$ < **8.** $\frac{3}{4} \bullet \frac{1}{4}$ > **9.** $-\frac{6}{10} \bullet -\frac{3}{5}$ = **10.** $-\frac{1}{5} \bullet -\frac{4}{5}$ >

11. $\frac{3}{7} \bullet \frac{5}{7}$ < **12.** $-2\frac{3}{8} \bullet -2\frac{7}{8}$ > **13.** $-5\frac{4}{8} \bullet -5\frac{1}{2}$ = **14.** $-3\frac{6}{12} \bullet -3\frac{1}{2}$ =

• **15.** $-0.6 \bullet -0.600$ = **16.** $0.73 \bullet -0.87$ > **17.** $-0.15 \bullet 0.15$ < **18.** $-746 \bullet -0.699$ <

19. $2.34 \bullet -2.43$ > **20.** $-4.07 \bullet -4.09$ > **21.** $7.002 \bullet 6.092$ > **22.** $-100.3 \bullet -100.29$ <

• **23.** $-\frac{3}{4} \bullet -0.75$ = **24.** $2\frac{1}{4} \bullet -2.75$ > **25.** $-5.25 \bullet -5\frac{3}{8}$ > **26.** $4.75 \bullet -4\frac{7}{8}$ >

27. $-4 \bullet \frac{2}{3}$ < **28.** $5\frac{1}{2} \bullet -6.9$ > **29.** $-8 \bullet -\frac{3}{5}$ < **30.** $-5\frac{8}{9} \bullet -5.99$ >

31. $-\frac{3}{8} \bullet -\frac{6}{16}$ = **32.** $6\frac{5}{10} \bullet -6.50$ > **33.** $-34.098 \bullet -34.1$ > **34.** $7\frac{5}{8} \bullet 7.625$ =

Write the absolute value.

• **35.** $\left|\frac{3}{4}\right|$ $\frac{3}{4}$ • **36.** $\left|-\frac{2}{3}\right|$ $\frac{2}{3}$ **37.** $\left|-\frac{6}{7}\right|$ $\frac{6}{7}$ **38.** $\left|\frac{7}{9}\right|$ $\frac{7}{9}$ **39.** $\left|-\frac{9}{10}\right|$ $\frac{9}{10}$

40. $|0.9|$ 0.9 **41.** $|-0.3|$ 0.3 **42.** $\left|-13\frac{2}{3}\right|$ $13\frac{2}{3}$ **43.** $\left|4\frac{1}{2}\right|$ $4\frac{1}{2}$ **44.** $|0|$ 0

Draw a number line. Locate the following points. Check students' answers.

• **45.** Point *S* is an integer. It is neither positive nor negative.

46. Point *R* is located at -0.75.

47. Point *T* is located at the opposite of point *R*.

48. Point *V* is to the right of point *S*. The absolute value of point *V* is $2\frac{1}{2}$.

MIXED REVIEW

Find the number.

• **49.** 58% of 200 is what number? 116

• **50.** 14 is what percent of 40? 35

51. What is 40% of 200? 80

52. 102 is 17% of what number? 600

53. 147 is what percent of 245? 60%

54. What is 63% of 1,100? 693

55. 86 is what percent of 43? 200%

56. 187 is 85% of what number? 220

COMMON ERROR

In comparing negative rational numbers, students may sometimes compare their absolute values and neglect to consider their signs. For example, they may think that $-3.4 > -3.2$ because $3.4 > 3.2$. To preclude this error, emphasize that when comparing negative numbers, the larger the absolute value, the smaller the number. Students who continue to have difficulty should be encouraged to visualize the numbers on a number line and think which number is closer to zero. Of two negative numbers, the one closer to zero is the greater.

ASSIGNMENTS

Level 1	1–6, Odd 7–43, 49–56
Level 2	Odd 1–43, 45–56
Level 3	Even 8–44, 45–56

FOLLOW-UP ACTIVITY

MENTAL MATH Have students identify whether each number is any or all of the following: whole number, integer, rational number.

1. $\frac{5}{8}$ (rat.) **2.** 0.45 (rat.)

3. -4 (int., rat.) **4.** 132.6 (rat.)

5. $-\frac{4}{7}$ (rat.) **6.** -9.3 (rat.)

7. 67 (all three) **8.** -43 (int., rat.)

9. 0 (all three) **10.** $-9\frac{3}{4}$ (rat.)

SUPPLEMENTARY MATERIALS

TRP Practice, p. 155

TRP Reteaching, p. 96

TRP Transparency 78

OBJECTIVE

- Add, subtract, multiply, and divide rational numbers.

TEACHING THE LESSON

WARM-UP Write the following exercises on the chalkboard or an overhead transparency.

Compute.

1. $-3 + 18$	**2.** $4 + (-9)$
3. $5 - 9$	**4.** $-3 - (-8)$
5. $-6 \times (-4)$	**6.** -8×4
7. $24 \div (-4)$	**8.** $-36 \div (-9)$

(**1.** 15 **2.** −5 **3.** −4 **4.** 5 **5.** 24 **6.** −32 **7.** −6 **8.** 4)

INTRODUCTION Write the following on the chalkboard:

$-\frac{1}{2} + \frac{1}{4}$	$4.86 + (-2.2)$
$-\frac{1}{2} - \frac{1}{4}$	$4.86 - (-2.2)$
$-\frac{1}{2} \times \frac{1}{4}$	$4.86 \times (-2.2)$
$\frac{1}{2} \div \frac{1}{4}$	$4.86 \div (-2.2)$

For each example, ask whether the answer will be positive or negative. Have students explain their reasoning. Guide them to realize that the sign rules for computing with rational numbers are the same as those for computing with integers.

INSTRUCTION In discussing Examples 1 and 2, it may help students apply the sign rules more effectively if you have them first estimate the sign of the answer as in the Introduction.

CHECKPOINT The incorrect choices include common errors students make.

13.11 COMPUTING WITH RATIONAL NUMBERS

You can apply the rules for adding and subtracting integers to add and subtract rational numbers.

EXAMPLE 1

a. Add: $-34.56 + 125.8$

1. Subtract the absolute values.

THINK: $|-34.56| = 34.56$
$|125.8| = 125.8$ $\qquad 125.8 - 34.56 = 91.24$

2. Write the correct sign.

THINK: 125.8 has the greater absolute value. The sum is positive.

$-34.56 + 125.8 = 91.24$

b. Subtract: $-\frac{1}{5} - \left(-\frac{2}{3}\right)$

THINK: Add the opposite of the number being subtracted.

$-\frac{1}{5} - \left(-\frac{2}{3}\right) = -\frac{1}{5} + \frac{2}{3}$

$= -\frac{3}{15} + \frac{10}{15} = \frac{7}{15}$

$-\frac{1}{5} - \left(-\frac{2}{3}\right) = \frac{7}{15}$

You can also apply the rules for multiplying and dividing integers to multiply and divide rational numbers.

EXAMPLE 2

a. Multiply: $-5 \times \left(-\frac{4}{5}\right)$

THINK: The signs are the same. The product is positive.

$-\overset{1}{\cancel{5}} \times \left(-\frac{4}{\underset{1}{\cancel{5}}}\right) = 4$

b. Divide: $-0.34 \div 1.7$

THINK: The signs are different. The quotient is negative.

$-0.34 \div 1.7 = -0.2$

CHECKPOINT

Write the letter of the correct answer.

Compute.

			a.	b.	c.	d.
1.	$-\frac{3}{5} + \left(-2\frac{1}{10}\right)$	d	$-2\frac{4}{5}$	$2\frac{4}{5}$	$2\frac{7}{10}$	$-2\frac{7}{10}$
2.	$-7.65 - 5.36$	c	2.29	−2.29	−13.01	13.01
3.	$-31.2 \times (-0.07)$	c	21.84	−21.84	2.184	−2.184
4.	$-\frac{3}{4} \div \left(-\frac{2}{3}\right)$	d	$-\frac{1}{2}$	$\frac{1}{2}$	$-1\frac{1}{8}$	$1\frac{1}{8}$

PRACTICE EXERCISES

See Extra Practice, page 458.

Add or subtract.

- • 1. $-65 + 42.87$ — -22.13
- • 2. $\frac{1}{2} + \left(-\frac{7}{8}\right)$ — $-\frac{3}{8}$
- 3. $-4 + \frac{9}{16}$ — $-3\frac{7}{16}$
- 4. $-3\frac{1}{5} + \frac{5}{8}$ — $-2\frac{23}{40}$
- 5. $-12 + (-1.035)$ — -13.035
- • 6. $-75 - 23.08$ — -98.08
- 7. $2\frac{9}{12} - \frac{7}{8}$ — $1\frac{7}{8}$
- 8. $178.09 - (-45)$ — 223.09
- 9. $-1\frac{1}{9} - \left(-3\frac{1}{3}\right)$ — $2\frac{2}{9}$
- 10. $\frac{5}{6} - 2\frac{1}{3}$ — $-1\frac{1}{2}$
- • 11. $-\frac{2}{9} + \left(-3\frac{5}{6}\right)$ — $-4\frac{1}{18}$
- 12. $-15 + 75$ — 60
- 13. $-14 - 32$ — -46
- 14. $-0.75 + \frac{1}{2}$ — -0.25
- 15. $125 + (-178)$ — -53
- 16. $5 + \left(-\frac{2}{5}\right)$ — $4\frac{3}{5}$
- 17. $\frac{3}{7} - \frac{2}{14}$ — $\frac{2}{7}$
- 18. $9.9 + (-0.07)$ — 9.83
- 19. $\frac{8}{12} - \left(-\frac{3}{4}\right)$ — $1\frac{5}{12}$
- 20. $4.67 + (-8.02)$ — -3.35

Multiply or divide.

- • 21. 23×78 — $1{,}794$
- • 22. -45×-30 — $1{,}350$
- 23. $\frac{1}{2} \times \frac{3}{4}$ — $\frac{3}{8}$
- • 24. $-40 \times \frac{1}{8}$ — -5
- 25. $2\frac{3}{4} \times 4$ — 11
- 26. $100 \div 40$ — 2.5
- 27. $-0.24 \div \frac{1}{6}$ — -1.44
- 28. $-0.75 \div 0.25$ — -3
- 29. $\frac{2}{3} \div \frac{4}{5}$ — $\frac{5}{6}$
- • 30. $-5\frac{5}{9} \div \frac{1}{3}$ — $-16\frac{2}{3}$
- • 31. $-18 \div (-3)$ — 6
- 32. $-\frac{5}{6} \times \left(-\frac{6}{5}\right)$ — 1
- 33. $-3.75 \div (-1.5)$ — 2.5
- 34. $45 \times \left(-\frac{1}{3}\right)$ — -15
- 35. 25×7 — 175
- 36. $-2\frac{1}{2} \div \left(-\frac{1}{2}\right)$ — 5
- 37. $-14 \div \left(-\frac{1}{7}\right)$ — 98
- 38. $-0.99 \div 0.33$ — -3
- 39. $-1.25 \times \frac{1}{2}$ — -0.625
- 40. $-12 \div (-0.04)$ — 300

TIME OUT

Scientists and weather forecasters often have to change temperatures between the Celsius and Fahrenheit systems. The procedure is easy now that you can compute with rational numbers.

14°F = ■°C

1. Subtract 32 from the temperature. $14 - 32 = -18$
2. Multiply the difference by $\frac{5}{9}$. $-18 \times \frac{5}{9} = -10$

So, 14°F = −10°C.

−25°C = ■°F

1. Multiply the temperature by $\frac{9}{5}$. $-25 \times \frac{9}{5} = -45$
2. Add 32 to the product. $-45 + 32 = -13$

So, −25°C = −13°F.

Write the corresponding Celsius or Fahrenheit temperature.

1. 77°F — 25°C
2. 50°F — 10°C
3. −13°F — −25°C
4. 40°C — 104°F
5. −15°C — 5°F
6. −50°C — −58°F

TIME OUT You may wish to point out that the relationship between Fahrenheit and Celsius temperatures can also be represented as:

$°F = \left(\frac{9}{5} \times °C\right) + 32$ or

$°C = (°F - 32) \times \frac{5}{9}$

Have students use these formulas to verify that −40°C = −40°F.

COMMON ERROR

When computing with rational numbers, students may continue to make errors relating to fractions, such as neglecting to invert the divisor. For example, in item 4 of Checkpoint, they would choose *a*. Review the steps for multiplying and dividing fractions, using positive numbers only. Once these skills have been mastered, review the sign rules for integers before proceeding to examples with rational numbers.

ASSIGNMENTS

Level 1	1–10, 21–30
Level 2	1–15, 21–35, TO
Level 3	11–20, 31–40, TO

FOLLOW-UP ACTIVITY

ESTIMATION Have students estimate the answer.

1. $-\frac{1}{4} \times 407\frac{2}{3}$ (−100)
2. $0.3 \times (-1{,}042)$ (−300)
3. $-5\frac{7}{8} \times \left(-11\frac{7}{8}\right)$ (72)
4. $989 \div (-21)$ (−50)
5. $-49.2 \div (-6.88)$ (7)
6. $-3{,}013.5 \div 4.997$ (−600)

SUPPLEMENTARY MATERIALS

TRP Practice, p. 156

TRP Reteaching, p. 97

TRP Transparency 78

OBJECTIVE

- Choose one or more problem-solving strategies to solve word problems.

TEACHING THE LESSON

WARM-UP Write the following exercises on the chalkboard or an overhead transparency.

Solve.

1. 175 × $3 ($525)
2. 310 × $4.25 ($1,317.50)
3. $\frac{n}{16} = \frac{3}{12}$ (4)
4. $\frac{10}{25} = \frac{n}{20}$ (8)

INTRODUCTION Elicit from the students all the problem-solving strategies learned thus far this year. Have a volunteer make a list similar to the following on the chalkboard.

Problem-Solving Strategies

1. Identify Needed Information
2. Write Subproblems
3. Guess and Check
4. Make a Table
5. Interpret Remainders
6. Work Backward
7. Draw a Diagram
8. Use Estimation
9. Supply Missing Information
10. Make an Organized List
11. Look for a Pattern
12. Solve a Simpler Problem

INSTRUCTION Stress that there are many ways to solve word problems. In some cases, one or more problem-solving strategies may be used to solve word problems.

Have five volunteers come to the front of the classroom. Have them demonstrate the handshake problem on the lesson page. Make a list of all the possible handshakes on the chalkboard using the students' names.

PROBLEM Solving STRATEGY

13.12 CHOOSING A STRATEGY

Situation:

Suppose you walk into a party and see 4 of your friends. Each person shakes hands with the other 4 people. How many handshakes will there be among the 5 of you?

Strategy:

You can use more than one strategy to solve some problems.

Applying the Strategy:

A. You can solve the problem by drawing a diagram.

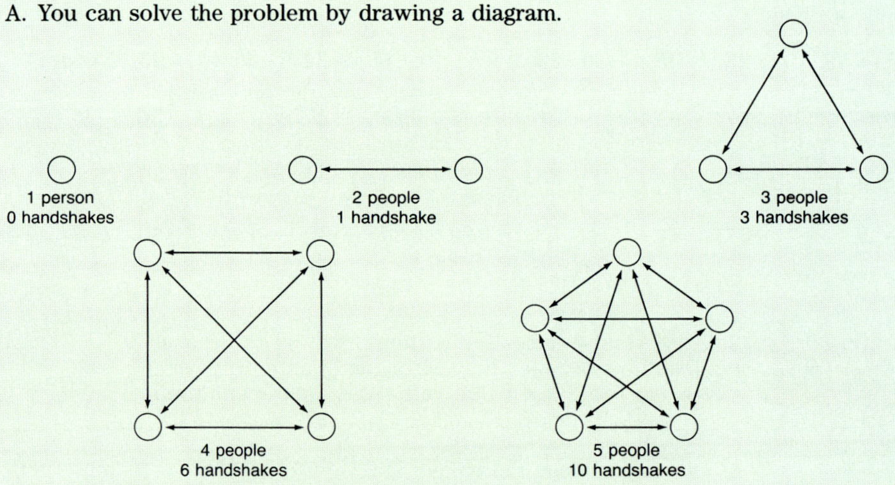

There will be 10 handshakes among the 5 of you.

B. You can also solve this problem by making a table and looking for a pattern.

Number of people	Number of handshakes	Pattern
1	0	0
2	1	1 + 0
3	3	2 + 1 + 0
4	6	3 + 2 + 1 + 0
5	10	4 + 3 + 2 + 1 + 0

There will be 10 handshakes among the 5 of you.

Before assigning Practice Exercises 5–8, you may want to discuss the problem-solving strategies that may be used to solve each problem.

Assignments

Level 1	1–4, 7–8
Level 2	1–8
Level 3	5–8

Follow-Up Activity

SITUATIONAL PROBLEM SOLVING Distribute copies of an ad from a local hardware store. Have students use the information in the ad to write problems for the class to solve. Write one of their problems on the chalkboard. Discuss and list the different strategies that could be used to solve it. Have students choose one of the strategies and solve the problem independently. For example:

Wallboards are on sale for $20 each. One wall of the classroom measures 20 ft × 8 ft. If each wallboard measures 4 ft × 8 ft, how many wallboards are needed to cover the wall? (5)

Supplementary Materials

TRP Practice, p. 157

TRP Transparency 79

PRACTICE EXERCISES See Extra Practice, page 459.

Choose the most reasonable strategy to solve the problem.

• **1.** Ms. Powell is planning a class trip to the museum for 56 students. The buses will cost $500. About how much money should Ms. Powell collect from each student to cover the cost of the buses? c
a. Draw a Diagram
b. Guess-and-Check
c. Use Estimation

2. The museum employs 50 people. 10% of them are cashiers; 60% of them are guards; 20% of them are maintenance workers; and 10% of them are managers. How many employees are there in each category? b
a. Interpret Remainders
b. Make a Table
c. Work Backward

3. In addition to 7 holidays, the museum is closed on Tuesdays. The average daily attendance is 1,543 people. How many days per year is the museum open? a
a. Identify Needed Information
b. Look for a Pattern
c. Write Subproblems

4. The museum offers guided tours of its exhibits for groups of 30 people or more. If there are 186 people who want a guided tour, how many guides are needed? b
a. Solve a Simpler Problem
b. Supply Missing Information
c. Make an Organized List

Solve the problem by using one or more problem-solving strategies.

• **5.** A laboratory activity uses 5 g of salt for every 10 mL of water. How many milliliters of water do you need for 10 g of salt? 20 mL

6. The area of the hamster cage is 600 cm^2. What are two possibilities for both the length and the width of the cage? Answers will vary. Possible answers include: 20 cm × 30 cm; 25 cm × 24 cm

7. The admission to the science museum was $2 per student and $2.75 for an adult. Dorothy collected $150 for admissions. Can 65 students and 10 adults be admitted to the museum? no

8. On Monday, the souvenir stand at the museum collected $3,500 and the cafeteria collected $5,775. Approximately the same amount of money was collected each day that week. About how much money was collected during the 5-day period? $50,000

OBJECTIVE

- Solve problems that involve operations with rational numbers.

TEACHING THE LESSON

WARM-UP Write the following exercises on the chalkboard or an overhead transparency.

Find the answer.

1. $16 - 2\frac{1}{2}$ $(13\frac{1}{2})$

2. $16 - (-2\frac{1}{2})$ $(18\frac{1}{2})$

3. $5 \times 1\frac{1}{4}$ $(6\frac{1}{4})$

4. $5 \times (-1\frac{1}{4})$ $(-6\frac{1}{4})$

5. $-1\frac{1}{8} + (-2\frac{1}{4}) + 1\frac{1}{4}$ $(-2\frac{1}{8})$

INTRODUCTION If possible, display a newspaper page showing the daily listings for the New York Stock Exchange or the American Stock Exchange. Have a volunteer tell what he or she knows about stocks.

You may also wish to point out that when a large number of shares of stock in a company are bought, the price of the stock usually rises. On the other hand, when a large number of shares of stock in a company are sold, the price usually falls.

INSTRUCTION Give careful attention to the meaning of each column in the table shown in the text. Be sure the students understand that the column marked "Stock" shows an abbreviation or acronym for the company name. The columns marked "High" and "Low" show the highest and lowest prices for the day; and the column marked "Last" gives the last or closing price for the day. So, the entry in the "Last" column *cannot* be greater than the high or less than the low for the day.

PROBLEM Solving APPLICATION

13.13 CONSUMER: TRADING STOCKS

If you buy shares of stock in a corporation, you own part of the corporation. The prices of many different stocks are listed each day in most newspapers.

The table below shows data about six corporations. Notice that the price (in dollars) of a share of stock is written as a mixed number or a whole number.

52-Week		Stock	High	Low	Last	Change
High	Low					
$9\frac{1}{8}$	$6\frac{7}{8}$	Nasco	$8\frac{1}{2}$	$7\frac{3}{4}$	$8\frac{1}{4}$	$+\frac{1}{4}$
$62\frac{3}{8}$	55	NES	60	$56\frac{7}{8}$	$57\frac{1}{8}$	$-1\frac{1}{4}$
$115\frac{1}{8}$	$100\frac{1}{2}$	Nostro	$110\frac{5}{8}$	$105\frac{3}{8}$	107	$+1\frac{5}{8}$
$92\frac{1}{4}$	$83\frac{1}{2}$	Noxco	$87\frac{1}{2}$	$85\frac{1}{2}$	$86\frac{1}{2}$	-1
$94\frac{1}{2}$	$80\frac{1}{4}$	NR	92	$86\frac{7}{8}$	87	$-2\frac{1}{2}$
$45\frac{1}{2}$	$34\frac{1}{4}$	NYA	$42\frac{1}{4}$	$40\frac{1}{4}$	42	$+1\frac{1}{2}$
The highest and lowest prices during the last 52 weeks		Names of corporations often abbreviated	The highest and lowest prices of today		Last closing price of day	The change from yesterday's price to today's price + increase/ − decrease

What was the closing price yesterday of a share of NYA stock?

THINK: To find yesterday's closing price, subtract the change from today's closing price.

Today's closing price of 1 share of NYA stock	−	Change from yesterday's price to today's price	=	Yesterday's closing price of 1 share of NYA stock
42	−	$1\frac{1}{2}$	=	$40\frac{1}{2}$

So, the closing price yesterday of a share of NYA stock was $40\frac{1}{2}$.

Find the closing price of the stock.

• **1.** Nasco $8\frac{1}{4}$ **2.** Noxco $86\frac{1}{2}$ **3.** NES $57\frac{1}{8}$ **4.** NR 87

Is today's price of a share of stock an increase or decrease from yesterday's?

• **5.** NYA increase **6.** Nostro increase **7.** Noxco decrease **8.** NES decrease

Solve.

• **9.** Which of the 6 stocks had the highest closing price today? Nostro

10. Which of the 6 stocks had the lowest closing price today? Nasco

11. How many of the 6 stocks decreased in price from yesterday to today? 3 stocks

12. Which stock increased in price by \$1.625 (1 dollar and $62\frac{1}{4}$ cents) from yesterday to today? Nostro

13. What were the 52-week high and low prices for Nasco? $9\frac{1}{8}$ high; $6\frac{7}{8}$ low

14. What was the difference between the 52-week high and low prices for Nasco? $2\frac{1}{4}$

15. What is the price of 10 shares of NYA stock at the end of the day? \$420

16. Was yesterday's closing price for Noxco higher than the closing price for NR? no

17. If you bought 100 shares of Nostro stock yesterday and sold it at the end of today, how much would your profit or loss be? \$162.50 profit

18. The price of NES changes by $-1\frac{3}{4}$ on each of the next 3 days. On each of the following 2 days, it changes by $+\frac{1}{8}$. What will the price of the stock be? $52\frac{1}{8}$

ASSIGNMENTS

Level 1	1–12
Level 2	Odd 1–17
Level 3	Even 2–18

FOLLOW-UP ACTIVITY

APPLICATION Tell the students that they have \$10,000 to invest in the stock market. Have them keep a log of their investments by checking the stock listings every day in the newspaper. At the end of 2 weeks, have students determine who has made the most profit.

SUPPLEMENTARY MATERIALS

TRP Practice, p. 158

TRP Transparency 79

OBJECTIVES

- Review vocabulary.
- Practice key chapter concepts and skills.

USING THE REVIEW

The Chapter Review is designed to help students prepare for taking the Chapter Test. The first section focuses on vocabulary. It requires that students select a word(s) to complete statements. The second section presents practice exercises of key mathematical skills. Under each directive there is a sample exercise with the answer.

Each item of the review is referenced to the page on which the topic is taught in the Pupil's Edition. You may wish to have students refer to these pages to help review any concepts or skills they have not yet mastered.

It is suggested that students work in small-sized heterogeneous cooperative learning groups. Some cooperative learning methods that may be used are as follows:

1. After each student has independently completed the entire Chapter Review, a discussion should follow within each group about the solutions to the practice exercises.
2. The group can complete the entire Chapter Review by working together to discuss the sample exercises and then to answer the practice exercises.

End the lesson with an entire class discussion in which any questions brought up in group discussions are presented and answered.

CHAPTER REVIEW

Vocabulary Choose the letter of the word(s) that completes each statement.

1. A number in scientific notation has a first factor between 1 and 10 and a second factor that is a(n) ■. [378] b

a. decimal **b.** power of 10 **c.** absolute **d.** square root

2. Every point on a coordinate plane can be named by using a(n) ■. [380] c

a. axis **b.** absolute value **c.** ordered pair **d.** quadrant pair

3. A number that can be written as a ratio of two integers and does not have a denominator of zero is a ■ number. [384] b

a. repeating **b.** rational **c.** rotating **d.** round

Skills Add. [367, 386]

Sample: $5 + 42$ *47*

4. $7 + (-13)$ −6 **5.** $-12 + (-28)$ −40 **6.** $-435 + 557$ 122 **7.** $89 + (-34)$ 55

8. $-8.7 + 4.58$ −4.12 **9.** $12.5 + (-5.4)$ 7.1 **10.** $\frac{3}{4} + \left(-\frac{4}{5}\right)$ $-\frac{1}{20}$ **11.** $-14 + (-0.2)$ −14.2 **12.** $\frac{1}{2} + (-1.25)$ −0.75

Subtract. [370, 386]

Sample: $5 - 6$ *−1*

13. $-12 - 16$ −28 **14.** $118 - (-101)$ 219 **15.** $332 - (-33)$ 365 **16.** $36 - (-34)$ 70

17. $2.3 - 1.2$ 1.1 **18.** $-43.5 - 22.8$ −66.3 **19.** $\frac{5}{6} - \frac{7}{9}$ $\frac{1}{18}$ **20.** $-\frac{3}{4} - (-2.75)$ 2 **21.** $-12.2 - 25.9$ −38.1

Multiply. [372, 386]

Sample: $25 \times (-3)$ *−75*

22. -75×2 −150 **23.** $0 \times (-25)$ 0 **24.** $-12 \times (-12)$ 144

25. $-20 \times (-2.5)$ 50 **26.** -10.25×3 −30.75 **27.** $-1 \times (-1.101)$ 1.101 **28.** 0×1.3 0

Divide. [374, 386]

Sample: $-20 \div (-5)$ *4*

29. $-40 \div (-40)$ 1 **30.** $63 \div (-9)$ −7 **31.** $250 \div (-50)$ −5

32. $-4\frac{3}{8} \div \frac{1}{2}$ −8.75 **33.** $100.25 \div (-5)$ −20.05 **34.** $\frac{5}{8} \div 3\frac{2}{5}$ $\frac{25}{136}$ **35.** $-20 \div \left(-1\frac{1}{5}\right)$ $16\frac{2}{3}$

Write in scientific notation. [378]

Sample: 350 *3.5×10^2*

36. 5,000 5×10^3 **37.** 250,000 2.5×10^5 **38.** 0.000005 5×10^{-6}

CHAPTER TEST

Complete. Use >, <, or = to make a true statement.

1. $|4| \bullet |-4|$ = **2.** $\left|-\frac{2}{3}\right| \bullet \left|\frac{3}{4}\right|$ < **3.** $|-45| \bullet |-44|$ > **4.** $\left|-2\frac{3}{8}\right| \bullet \left|\frac{1}{2}\right|$ > **5.** $|58.5| \bullet |-56.9|$ >

Add or subtract.

6. $12 + (-14)$ −2 **7.** $-35 + (-27)$ −62 **8.** $94 - (-63)$ 157 **9.** $-123 + 0 + (-47)$ −170

10. $-11 + (-15)$ −26 **11.** $-32 - (-54)$ 22 **12.** $-119 - 120$ −239 **13.** $-87 - (-78)$ −9

14. $\frac{6}{5} + \frac{8}{15}$ $1\frac{11}{15}$ **15.** $-1\frac{1}{4} + \frac{2}{3} + \frac{4}{5}$ $\frac{13}{60}$ **16.** $0.45 + (-2.4)$ −1.95 **17.** $4\frac{3}{5} - \left(-\frac{2}{5}\right)$ 5

18. $1.25 - \left(-\frac{7}{10}\right)$ 1.95 **19.** $-4.6 + 0.09$ −4.51 **20.** $12.45 - (-4.2)$ 16.65 **21.** $\frac{3}{7} - \left(-\frac{1}{2}\right)$ $\frac{13}{14}$

Multiply or divide.

22. 12×-40 −480 **23.** $6 \div (-30)$ −0.2 **24.** $-9 \times 12 \times (-5)$ 540 **25.** $25 \times 0 \times -16$ 0

26. $-125 \div (-25)$ 5 **27.** -10×50 −500 **28.** $200 \div (-25)$ −8 **29.** $-150 \div -5$ 30

30. $\frac{3}{10} + -\frac{5}{8}$ $-\frac{13}{40}$ **31.** $1\frac{1}{3} \times (-3)$ −4 **32.** 20×3.58 71.6 **33.** $15.5 \div -5$ −3.1

34. $\frac{7}{8} \div -1\frac{1}{3}$ $-\frac{21}{32}$ **35.** $0.03 \times -\frac{1}{2}$ −0.015 **36.** -45×-10.05 452.25 **37.** $-350 \div 0.07$ −5000

Write in scientific notation.

38. 10,000 1×10^4 **39.** 25,000 2.5×10^4 **40.** 0.0000007 7×10^{-7} **41.** 0.00026 2.6×10^{-4} **42.** 30,000,000 3×10^7

Write as a standard numeral.

43. 5×10^3 5,000 **44.** 4.2×10^{-5} 0.000042 **45.** 8.3×10^8 830,000,000 **46.** 9.5×10^{-4} 0.00095

Write the point named by the ordered pair.

47. (1, 3) *D* **48.** (2, −3) *F*

49. (−2, −4) *B* **50.** (4, −3) *E*

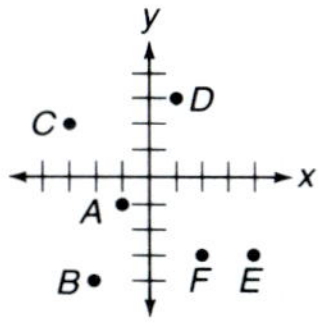

Solve.

51. ANEX stock sells for $60\frac{1}{2}$ per share today, up by $\frac{1}{2}$ from yesterday. If you bought 1,000 shares yesterday and sold it at the end of today, how much would your profit be? $500

52. You have won the Lucky Bucks Jackpot. You will receive \$1 on the first day, \$2 on the second, \$3 on the third day, and so on for 30-days. How much money will you have won in 30 days? \$465

OBJECTIVE

- Evaluate achievement of the chapter objectives.

USING THE TEST

The Chapter Test may be used as a posttest to evaluate the achievement of all students. However, you may wish to use the Chapter Posttest offered in the Teacher's Resource Package or design your own chapter test. If this page is not used as a test, you may wish to assign it as additional review or practice.

The test items are correlated to the chapter objectives in the table below.

Chapter objectives	Test items
A. Compare the absolute values of rational numbers.	1–5
B. Add and subtract integers.	6–13
C. Add and subtract rational numbers.	14–21
D. Multiply and divide integers.	22–29
E. Multiply and divide rational numbers.	30–37
F. Write a standard numeral in scientific notation and vice versa.	38–46
G. Name points in a coordinate plane.	47–50
H. Apply computational skills in real-life situations.	51
I. Solve a problem by selecting an appropriate strategy.	52

SUPPLEMENTARY MATERIALS

TRP Ch. 13 Posttest Form A, pp. 1–2

TRP Ch. 13 Posttest Form B, pp. 1–2

OBJECTIVES

- Multiply numbers in exponent form by adding exponents.
- Divide numbers in exponent form by subtracting exponents.

USING THE PAGE

As you guide students through the multiplication example, point out that adding exponents is much more efficient than writing all the factors and then multiplying. Stress, however, that this technique can only be used when the bases are the same. Students should realize that they cannot add exponents to multiply $4^3 \times 5^2$. The same is true for dividing with exponents.

Before assigning the exercises to be completed independently, guide students through the following examples. Emphasize that students can apply their knowledge of adding and subtracting integers to multiply and divide with negative exponents.

- $5^4 \times 5^2$ (5^6)
- $3^{-3} \times 3^8$ (3^5)
- $9^9 \div 9^2$ (9^7)
- $6^5 \div 6^{-3}$ (6^8)

You may wish to have students check some of their answers by multiplying and dividing in the long form as shown in the "Think" step of each example.

ENRICHMENT THE LAWS OF EXPONENTS

In the number 2^3, recall that 2 is the **base** and 3 is the **exponent.** The exponent tells you how many times the base occurs as a factor.

$$2^3 = 2 \times 2 \times 2$$

You can use the laws of exponents to help you multiply with exponents. When the bases are the same, add the exponents.

Example Multiply $8^3 \times 8^2$.

THINK: $8^3 \times 8^2 = (8 \times 8 \times 8) \times (8 \times 8) = 8^5$

So, $8^3 \times 8^2 = 8^{3+2}$ or 8^5.

To divide exponents with the same base, subtract the exponents.

Example Divide $7^5 \div 7^3$

THINK: $7^5 \div 7^3 = \frac{7 \times 7 \times 7 \times 7 \times 7}{7 \times 7 \times 7} = 7^2$

So, $7^5 \div 7^3 = 7^{5-3}$ or 7^2.

Multiply.

1. $10^3 \times 10^1$ 10^4 **2.** $10^6 \times 10^2$ 10^8 **3.** $10^5 \times 10^4$ 10^9 **4.** $10^{17} \times 10^3$ 10^{20}

5. $2^3 \times 2^3$ 2^6 **6.** $3^2 \times 3^3$ 3^5 **7.** $4^5 \times 4^3$ 4^8 **8.** $5^7 \times 5^2$ 5^9

9. $17^2 \times 17^{-3}$ 17^{-1} **10.** $21^{-3} \times 21^{-4}$ 21^{-7} **11.** $32^6 \times 32^{-3}$ 32^3 **12.** $100^{-5} \times 100^8$ 100^3

Divide.

13. $10^6 \div 10^3$ 10^3 **14.** $10^5 \div 10^4$ 10^1 **15.** $10^7 \div 10^5$ 10^2 **16.** $10^{19} \div 10^{11}$ 10^8

17. $4^3 \div 4^1$ 4^2 **18.** $5^7 \div 5^2$ 5^5 **19.** $6^9 \div 6^3$ 6^6 **20.** $9^8 \div 9^4$ 9^4

21. $15^4 \div 15^{-3}$ 15^7 **22.** $23^{-6} \div 23^4$ 23^{-2} **23.** $36^{-9} \div 36^5$ 36^{-14} **24.** $100^8 \div 100^{-4}$ 100^{12}

Multiply or divide.

25. $7^6 \div 7^3$ 7^3 **26.** $11^4 \times 11^9$ 11^{13} **27.** $5^6 \times 5^{-6}$ 5^0 **28.** $4^7 \div 4^3$ 4^4

29. $16^{-10} \div 16^8$ 16^{-18} **30.** $14^8 \times 14^{-2}$ 14^6 **31.** $9^{12} \div 9^{-5}$ 9^{17} **32.** $3^{-7} \times 3^6$ 3^{-1}

Write the missing exponent.

33. $2^2 \times 2^{■} = 2^6$ 4 **34.** $10^{■} \times 10^5 = 10^8$ 3

35. $8^6 \div 8^{■} = 8^3$ 3 **36.** $6^{■} \div 6^4 = 6^5$ 9

37. $7^{■} \times 7^{-3} = 7^8$ 11 **38.** $12^{10} \div 12^{■} = 12^{-2}$ 12

MENTAL MATH USING THE PROPERTIES OF RATIONAL NUMBERS

These properties apply to the addition and multiplication of rational numbers.

COMMUTATIVE PROPERTY	DISTRIBUTIVE PROPERTY
The order of addends does not change the sum. $-2 + 8 = 8 + (-2)$ The order of factors does not change the product. $3 \times (-6) = -6 \times 3$	To multiply a sum by a number, multiply each addend by the number, then add the products. $3 \times [(-2) + 6] = [3 \times (-2)] + (3 \times 6)$
ASSOCIATIVE PROPERTY	**PROPERTY OF OPPOSITES**
The grouping of addends does not change the sum. $-4 + [(-5) + 3] = [-4 + (-5)] + 3$ The grouping of factors does not change the product. $[8 \times (-6)] \times 7 = 8 \times [(-6) \times 7]$	The sum of any number and its opposite is zero. $2.5 + (-2.5) = 0$ **PROPERTY OF RECIPROCALS** The product of any number and its opposite is 1. $4 \times \frac{1}{4} = 1$
PROPERTY OF ZERO	**PROPERTY OF ONE**
The sum of a number and zero is the number. $-4.6 + 0 = -4.6$	The product of any number and 1 is the number. $1 \times (-9) = -9$

You can use these properties to help you compute mentally.

Compute: $-67 + (-89) + 89$

THINK: −89 and 89 are opposites; their sum is 0.

$$-67 + \underbrace{(-89) + 89}$$
$$-67 + 0 = -67$$

The sum of -67 and 0 is -67.

Compute: $-4 \times \left(\frac{1}{2} \times 3.5\right)$

THINK: Use the Associative Property.

$$-4 \times \left(\frac{1}{2} \times 3.5\right) = \left(-4 \times \frac{1}{2}\right) \times 3.5$$
$$= -2 \times 3.5$$
$$= -7$$

Compute mentally.

1. $47 + 12 + (-12)$ 47
2. $6.5 + 8.6 + (-6.5)$ 8.6
3. $-23 \times [5 + (-5)]$ 0
4. $\left(\frac{1}{5} \times 5\right) \times 24$ 24
5. $\frac{1}{4} \times [(-43) \times 4]$ −43
6. $\left[\left(-\frac{2}{3}\right) + \frac{2}{3}\right] \times 36$ 0
7. $8 \times \left(6 + \frac{1}{2}\right)$ 52
8. $-18 \times \left[\frac{1}{3} + (-1)\right]$ 12
9. $15 \times \left[\left(-\frac{1}{5}\right) + 2\right]$ 27
10. $24 \times \left(\frac{1}{2} + \frac{1}{3}\right)$ 20
11. $-12 \times \left(\frac{1}{6} + \frac{1}{4}\right)$ −5
12. $60 \times \left[\left(-\frac{1}{10}\right) + \frac{1}{3}\right]$ 14
13. $4 \times \left[\left(-\frac{3}{4}\right) + (-6)\right]$ −27
14. $-20 \times \left(\frac{3}{10} + 1\right)$ −26
15. $6 \times \left(\frac{2}{3} + \frac{1}{2}\right)$ 7

OBJECTIVE

- Compute mentally, using the properties of rational numbers.

USING THE PAGE

Students should recall the listed properties from earlier mathematics courses. It will be helpful, however, to review them at this time. Using the examples shown in the table, you can justify the commutative, associative, and distributive properties as follows.

Commutative Properties
$(-2) + 8 = 6$ and $8 + (-2) = 6$
So $(-2) + 8 = 8 + (-2)$.
$(-2) \times 8 = -16$ and $8 \times (-2) = -16$
So, $(-2) \times 8 = 8 \times (-2)$.

Associative Properties
$(-4) + [(-5) + 3] = (-4) + (-2) = -6$
and $[(-4) + (-5)] + 3 = (-9) + 3 = -6$
So, $(-4) + [(-5) + 3] = [(-4) + (-5)] + 3$.
$(-4) \times [(-5) \times 3] = (-4) \times (-15) = 60$
and $[(-4) \times (-5)] \times 3 = 20 \times 3 = 60$
So, $(-4) \times [(-5) \times 3] = [(-4) \times (-5)] \times 3$.

Distributive Property
$3 \times [(-2) + 6] = 3 \times 4 = 12$ and
$[3 \times (-2)] + (3 \times 6) = -6 + 18 = 12$
So, $3 \times [(-2) + 6] = [3 \times (-2)] + (3 \times 6)$.

In Exercises 1–6, students should keep in mind that addition and multiplication are commutative and associative. Therefore, it does not matter if the opposites are not consecutive addends or if the reciprocals are not consecutive factors.

The answers for Exercises 7–15 could be found by adding the numbers in parentheses and then multiplying. However, it is usually easier to use the distributive property.

Chapter 14 Equations

OBJECTIVE

- Test the prerequisite skills needed to learn the concepts developed in Chapter 14.

USING THE TEST

The Pre-Skills Test is designed to diagnose students' strengths and weaknesses on prerequisite skills necessary to study the mathematics in Chapter 14.

Assign the Pre-Skills Test. Allow the students to work together in pairs or small groups. Group members should help those who demonstrate a misunderstanding of a concept or a weakness in a skill.

Some items in the test are referenced to the pages on which the topics are taught in the Pupil's Edition. You may wish to have students refer to these pages for review.

The following table correlates the items on the Pre-Skills Test with the prerequisite skill and the lesson(s) in the chapter for which it is needed.

Item(s)	Prerequisite skill	Lesson
1–16	Add and subtract integers.	14.1, 14.3, 14.8, 14.10
17–32	Multiply and divide integers.	14.1, 14.3, 14.9, 14.10
33–40	Find the missing number in a number sentence.	14.6, 14.8, 14.9
41–46	Write an ordered pair to name a point.	14.11

Pre-Skills TEST

Add. [367]

1. $7 + (-12)$ -5 **2.** $-8 + 4$ -4 **3.** $-6 + (-5)$ -11 **4.** $11 + (-19)$ -8

5. $-4 + (-15)$ -19 **6.** $-13 + 8$ -5 **7.** $-4 + 10$ 6 **8.** $-2 + (-13)$ -15

Subtract. [370]

9. $3 - 9$ -6 **10.** $-6 - 7$ -13 **11.** $8 - (-5)$ 13 **12.** $-4 - (-2)$ -2

13. $-21 - 5$ -26 **14.** $0 - 15$ -15 **15.** $-3 - (-18)$ 15 **16.** $2 - (-34)$ 36

Multiply. [372]

17. $5 \times (-4)$ -20 **18.** -7×9 -63 **19.** $-3 \times (-6)$ 18 **20.** $8 \times (-6)$ -48

21. -12×0 0 **22.** $-2 \times (-13)$ 26 **23.** -9×4 -36 **24.** $-11 \times (-7)$ 77

Divide. [374]

25. $16 \div (-4)$ -4 **26.** $-35 \div 5$ -7 **27.** $-56 \div (-8)$ 7 **28.** $-63 \div 7$ -9

29. $-36 \div (-3)$ 12 **30.** $0 \div (-8)$ 0 **31.** $54 \div (-9)$ -6 **32.** $-48 \div (-8)$ 6

Find the missing number.

33. $n + 8 = 11$ 3 **34.** $n - 6 = 12$ 18 **35.** $n + 7 = 15$ 8 **36.** $n - 10 = 23$ 33

37. $5 \times n = 40$ 8 **38.** $-4 \times n = 28$ -7 **39.** $n \div 6 = 3$ 18 **40.** $-9 \div n = 45$ $-\frac{1}{5}$

Write the ordered pair that names the point. [380]

41. *A* (3, 1) **42.** *B* (−4, 3)

43. *C* (0, −1) **44.** *D* (4, −2)

45. *E* (−2, −4) **46.** *F* (−2, 0)

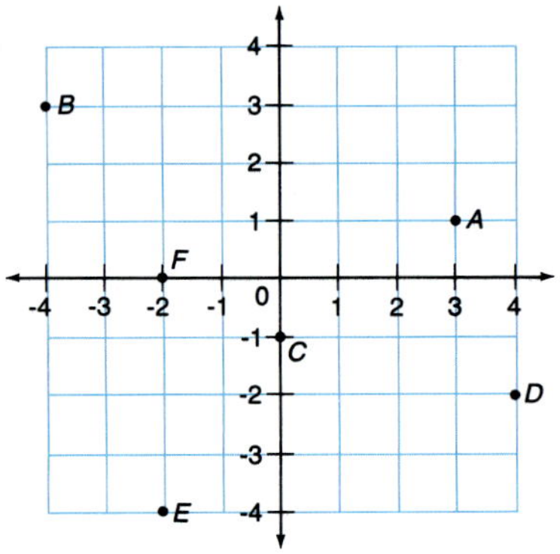

Chapter 14
EQUATIONS

People use a language, like English, to communicate. What other types of languages are in use today?

OBJECTIVE

- Formulate problems using text and data.

USING THE CHAPTER OPENER

Discuss the question on the page. Some types of languages might include computer languages (ask students to name specific languages such as BASIC, Logo, Pascal, etc.), mathematical expressions, scientific expressions, musical notation, and so on.

COOPERATIVE LEARNING

Write the following on the chalkboard.

supermarkets

banks

newspapers

manufacturing

music

Explain to the groups that they are to choose one of these topics (or another with your approval) to research. The focus of their research will be to find how computers are used in these areas. When they understand the systems for the particular area under study, they are to prepare a report in a graphic manner that is easy to understand. The groups should approach their report with the idea that it will be read by people with little or no understanding of the systems. They should include flowcharts, if they are appropriate, or other types of diagrams, in their reports.

Upon completion, each group should present its material to the class for discussion.

SUPPLEMENTARY MATERIALS

TRP Group Projects, p. 14

OBJECTIVE

- Simplify numerical expressions by using the order of operations.

TEACHING THE LESSON

WARM-UP Write the following exercises on the chalkboard or an overhead transparency.

Compute.

1. 72×5 (360) **2.** 4^3 (64)

3. $18 - (-6)$ (24) **4.** $216 \div 3$ (72)

INTRODUCTION Remind students that mathematics is a language with rules to follow. Write the following sentence on the chalkboard.

John Paul, Mary Lou, and I are in the same class.

Ask how many people are mentioned in the sentence. Show how commas can change the number of people:

4—John, Paul, Mary Lou, and I

5—John, Paul, Mary, Lou, and I

INSTRUCTION Use the opening problem to highlight the need for rules in any language. In mathematics, the rules for the order of operations ensure uniformity of answers in simplifying numerical expressions. Use Example 1 to show how the rules are applied. Remind students that ×, •, or ( ) can be used to show multiplication.

Example 2 extends the rules to include grouping symbols such as parentheses and brackets. A simple mnemonic device for remembering the steps in the order of operations is **Please Excuse My Dear Aunt Sally.**

The letters stand for:

P—parentheses
E—exponents
M—multiplication } must go L → R
D—division }
A—addition } must go L → R
S—subtraction }

14.1 ORDER OF OPERATIONS

During an inventory, Carlos Moreno found 20 single-sided disks and 25 double-sided disks. How many sides could be used for word processing?

To find how many sides, Carlos wrote:

$20 + 25 \cdot 2$ **The raised dot means multiply.**

To find the answer, apply the rules for the Order of Operations.

- Raise to a power.
- Multiply or divide in order from left to right.
- Add or subtract in order from left to right.

EXAMPLE 1

Simplify:	$20 + 25 \cdot 2$
1. Multiply.	$20 + 50$
2. Add.	70

There are 70 usable sides for word processing.

To simplify expressions with parentheses, work within the parentheses first.

EXAMPLE 2

Simplify:	$8 + 3(4^2 - 6)$
1. Work within parentheses.	
THINK: Raise to a power, then subtract.	$8 + 3(16 - 6)$ $8 + 3(10)$
2. Multiply.	
THINK: 3(10) means 3 · 10	$8 + 30$
3. Add.	38

So, $8 + 3(4^2 - 6) = 38$.

To simplify expressions with a division bar, first do the work above the bar, then below the bar.

EXAMPLE 3

Simplify:	$\frac{40 - 30}{16 \div 8 + 3}$
1. Work above the bar.	
THINK: Subtract.	$\frac{10}{16 \div 8 + 3}$
2. Work below the bar.	
THINK: Divide, then add.	$\frac{10}{2 + 3} = \frac{10}{5}$
3. Divide.	2

So, $\frac{40 - 30}{16 \div 8 + 3} = 2$.

PRACTICE EXERCISES See Extra Practice, page 459.

Write the operation and the number(s) you would compute first to simplify.

• **1.** $3 \cdot 5 + 6$ $3 \cdot 5$ •**2.** $3(5 + 6)$ $5 + 6$ **3.** $3 + 5 \cdot 6$ $5 \cdot 6$ **4.** $(3 + 5)6$ $3 + 5$

5. $12 - 4 - 6$ $12 - 4$ **6.** $(12 - 4)6$ $12 - 4$ **7.** $12 \cdot 4 \div 6$ $12 \cdot 4$ **8.** $(12 - 4) - 6$ $12 - 4$

9. $2^2 - 3 \cdot 5$ 2^2 **10.** $7 + 3^4 - 9$ 3^4 **11.** $\frac{6 - 4 + 3}{12 + 4^2}$ $6 - 4$ **12.** $\frac{8(5^2 + 1)}{9 - 2}$ 5^2

Simplify.

•**13.** $11 - 4 + 7$ 14 •**14.** $5 + 7 \cdot 2$ 19 **15.** $3^2 - 2^2 \div 4$ 8 **16.** $(-2)^3 \div 2^2$ -2

17. $2^3 + 3 \cdot 5$ 23 **18.** $4^2 + 16 \div 2$ 8 **19.** $8 + 3 \cdot 4^2$ 56 **20.** $5 - 4^3 \div 8$ -3

21. $5(8 - 3)$ 25 **22.** $(16 - 4) \div 3$ 4 **23.** $12 \div (8 - 5)$ 4 **24.** $(2 - 10) \div (3 + 5)$ -1

25. $3(5^2 - 20)$ 15 **26.** $12(2^3 - 4)$ 48 **27.** $(27 - 9) \div 3^2$ 2 **28.** $(6 - 2)^2 - (5 - 9)$ 20

29. $\frac{4 - 3}{6 + 2}$ $\frac{1}{8}$ **30.** $\frac{6(9 - 5)}{5 + 5^2}$ $\frac{4}{5}$ **31.** $\frac{7 - 2 \cdot 2}{4^2 - 1}$ $\frac{1}{5}$ **32.** $\frac{18 \div (3^2 - 18)}{3(6 - 5 + 2)}$ $-\frac{2}{9}$

33. $\frac{7^2 - 9}{30 + 2 \cdot 5}$ 1 **34.** $\frac{8 + 6 \cdot 2}{4^2 + 3^2}$ $\frac{4}{5}$ **35.** $\frac{16 - 12 \div 2 \cdot 3}{15 + 3(9 - 7)}$ $\frac{2}{3}$ **36.** $\frac{9 \div (6 - 15)}{6^2 - 5(4 + 2)}$ $-\frac{1}{6}$

Write parentheses to make the sentence true.

•**37.** $12 - 8 + 5 = -1$ $12 - (8 + 5) = -1$ **38.** $3 \cdot 2 - 6 \div 3 = 0$ $3 \cdot 2 - 6) \div 3 = 0$

39. $20 + 6 \cdot 2 \div 4 = 8$
$(20 + 6 \cdot 2) \div 4 = 8$

40. $8 - 2 + 8 \div 2 = 3$
$8 - (2 + 8) \div 2 = 3$

MIXED REVIEW

Measure the angle. Classify it as *acute*, *right*, *obtuse*, or *straight*.

•**41.**
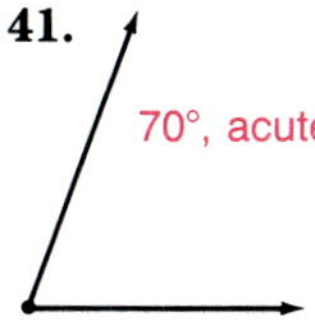
70°, acute

42.

180°, straight

43.

150°, obtuse

44.
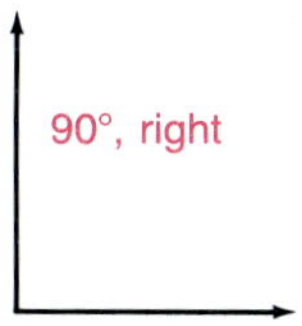
90°, right

TIME OUT

Everyone knows that 5 twos are 10, $5 \cdot 2 = 10$. But did you know that 5 twos can equal zero? Check students' answers. Sample answers below.

$$2 - 2 \div 2 - 2 \div 2$$

$$2 - \quad 1 \quad - \quad 1 \quad = 0$$

Use the symbols $+$, $-$, $\cdot$, and $\div$ and the rules for the Order of Operations to write expressions that show how 5 twos can equal 1, 2, then 3. $2 + 2 - 2 \div 2 = 1$ $2 + 2 + 2 - 2 \cdot 2 = 2$ $2 \cdot 2 \div 2 + 2 \div 2 = 3$

TIME OUT You may want to have students work in small groups to find the values of 1, 2, and 3. As a challenge, extend the activity to include finding the values of 4 to 9.

ASSIGNMENTS

Level 1	1–15, 21–23, 29–31, 41–44
Level 2	1–12, Odd 13–35, 41–44
Level 3	9–12, 16–20, 24–28, 32–44

FOLLOW-UP ACTIVITY

MENTAL MATH Here's a bit of math magic that students can use to impress their friends and families. Call forward a volunteer who has change in his pocket to follow these steps:

- Multiply his age by 2.
- Add 5.
- Multiply by 50.
- Add the change in his pocket less than 99¢.
- Subtract 250.

The first two digits of the answer name the student's age and the last two tell the amount of change.

Write the steps on the chalkboard for students to copy. Encourage them to practice with a classmate before performing in front of their families and friends.

SUPPLEMENTARY MATERIALS

TRP Practice, p. 159

TRP Reteaching, p. 98

TRP Transparency 80

OBJECTIVE

- Translate word phrases into algebraic expressions.

TEACHING THE LESSON

WARM-UP Write the following exercises on the chalkboard or an overhead transparency.

Find the number that is:

1. the sum of 7 and 5. (12)
2. the difference of 16 and 9. (7)
3. the product of 6 and 8. (48)
4. the quotient of 27 and 3. (9)

INTRODUCTION Ask what a translator does (takes words that are written or spoken in one language and expresses them in another language). Discuss how phrases and sentences can be translated into the language of mathematics by using letters and symbols.

INSTRUCTION Use the list of phrases to familiarize students with expressions associated with the basic operations. Ask students to think of similar answers such as "times for ×."

Use Example 1 to stress the thought process of first assigning a variable to the unknown quantity and then deciding on the operation. The phrase *less than* often presents difficulties, so make sure students understand why it is $n - 15$, not $15 - n$.

Example 2 extends the skill of writing an algebraic expression to include more than one operation.

Elicit other ways to express the answers in Example 3. For 3a, a student may prefer "the product of 3 and a number decreased by 5." Point out that there is more than one correct answer, but encourage students to write concise phrases.

14.2 ALGEBRAIC EXPRESSIONS

An **algebraic expression** uses a letter, or **variable,** to stand for one or more unknown numbers. Mathematical symbols are often used to translate word phrases into algebraic expressions.

Certain word phrases are associated with each operation.

+	−	×	÷
Sum of	**Difference of**	**Product of**	**Quotient of**
Increased by	**Decreased by**	**Multiplied by**	**Divided by**
More than	**Less than**	**Twice**	**Ratio**

EXAMPLE 1 In the state basketball tournament, Dondero High School scored 15 fewer points than Ferndale High School. Write an algebraic expression for the number of points scored by Dondero High School.

THINK: Let n represent Ferndale's points. Write: n
Subtract to get Dondero's points. $n - 15$

EXAMPLE 2 Write an algebraic expression for the word phrase. Let n represent the number.

a. 6 more than twice a number
$2 \cdot n + 6$ or
$2n + 6$

b. 18 less than $\frac{1}{2}$ a number
$\frac{1}{2} \cdot n - 18$ or
$\frac{1}{2}n - 18$

You can also translate an algebraic expression into a word phrase.

EXAMPLE 3 Write a word phrase for the algebraic expression.

a. $3(n - 5)$

3 times the difference of a number minus 5

b. $\frac{r + 8}{3}$

The sum of a number and 8 divided by 3

CHECKPOINT

Write the letter of the correct answer.

Write an algebraic expression for the word phrase.

1. 11 less than one-fourth of a number b **a.** $11 - 4n$ **b.** $\frac{1}{4}n - 11$ **c.** $11 - \frac{1}{4}n$
2. 19 times the difference of a number minus 2 a **a.** $19(n - 2)$ **b.** $19n - 2$ **c.** $19(2 - n)$
3. A number divided by the sum of 6 and 8 b **a.** $\frac{n}{6} + 8$ **b.** $\frac{n}{6 + 8}$ **c.** $\frac{6 + 8}{n}$

PRACTICE EXERCISES See Extra Practice, page 460.

Write an algebraic expression for the word phrase. Let n represent the number.

• **1.** 6 added to a number $n + 6$

• **2.** A number decreased by 14 $n - 14$

3. Twice a number that is cubed $2n^3$

4. 7 times a number, divided by 3 $7n \div 3$

5. 8 more than a number $n + 8$

6. 29 less than a number $n - 29$

7. The ratio of a number to 65 $\frac{n}{65}$

8. 34 divided by a number $34 \div n$

9. The sum of 10 and a number $10 + n$

10. The product of a number and 41 $41n$

11. 58 minus a number $58 - n$

12. A number divided by 17 $n \div 17$

13. 42 more than 8 times a number $8n + 42$

14. 79 decreased by twice a number $79 - 2n$

15. The quotient of a number divided by 20 $\frac{n}{20}$

16. A number doubled and decreased by 72 $2n - 72$

17. The quotient of a number plus 90, divided by 47 $\frac{n + 90}{47}$

18. The product of 56 and the sum of a number and 84 $56(n + 84)$

19. 3 times the sum of a number and 19, divided by 7 $\frac{3(n + 19)}{7}$

20. The fifth power of 2 increased by the product of a number and 4 $2^5 + n \cdot 4$

See page 478 for answers to Exercises 21–36.

Write a word phrase for the algebraic expression.

• **21.** $r + 3$

• **22.** $9 - a$

23. $64m$

24. $q - 14$

25. $17 + b$

26. $n - 6$

27. $5(3 + x)$

28. $4y - 6$

29. $11c - 2$

30. $13(9 - g)$

31. $28 - 7h$

32. $t^2 - 15$

33. $\frac{s - 7}{5}$

34. $\frac{3(x - 2)}{8}$

35. $\frac{60}{12 - z}$

36. $\frac{35}{2(r + 4)}$

Solve.

• **37.** The Hillside High School debate team scored 3 more than twice as many points as their opponents. Write an algebraic expression for the number of points they scored. $2n + 3$

38. The coach of the Tyler High School tennis team is 6 years older than twice the age of the youngest team member. Write the coach's age as an algebraic expression. $2n + 6$

39. At the Glenwood High School cheerleading tryouts, $\frac{1}{2}$ of the contestants were eliminated in the first round. In the second round, $\frac{1}{2}$ of the remaining contestants were eliminated. Another $\frac{1}{2}$ were eliminated in the third round. Six cheerleaders were finally chosen. How many students originally tried out for the squad? 48 students

CHECKPOINT Distribute to each student three index cards on which the answer choices a–c appear. After allowing students a few minutes, ask them to raise the letter that indicates their answer choice. Scan the cards and identify those students who need additional attention before proceeding to the Practice Exercises. The incorrect answer choices include common errors students make.

COMMON ERROR

Some students may confuse the "less than" translation. For example, in item 1 of Checkpoint, they would choose c. Encourage these students to read *less than* as *subtract from* to avoid this error.

ASSIGNMENTS

Level 1	Odd 1–31, 37, 38
Level 2	Even 2–36, 37–39
Level 3	Odd 11–31, 33–39

FOLLOW-UP ACTIVITY

ENRICHMENT To prepare students for word problems in algebra, distribute copies of the following worksheet.

Let *a* represent Martha's age now. Write an expression for:

1. her age in two years. $(a + 2)$
2. her age six years ago. $(a - 6)$
3. five times her age. $(5a)$
4. one-half her age. $(\frac{1}{2} a)$
5. the age of her sister, *s*, who is five years older. $(a + 5 = s)$
6. the age of her mother, *m*, who is four times her age five years ago. $(4(a - 5) = m)$

SUPPLEMENTARY MATERIALS

TRP Practice, p. 160

TRP Reteaching, p. 99

TRP Transparency 80

OBJECTIVE

- Evaluate algebraic expressions and formulas.

TEACHING THE LESSON

WARM-UP Write the following exercises on the chalkboard or an overhead transparency.

Simplify.

1. $3 \cdot 6 - 5$ (13)
2. $4^2 + 7 - 2$ (21)
3. $24 \div 8 - 5$ (-2)
4. $9(10 \div 5 - 3)$ (-9)

INTRODUCTION Write:

△ = 12 □ = 24 ○ = 4

☆ = 48 ⬡ = 6

Have students decode and simplify the expressions below.

1. △ + ○ (16)
2. □ − (☆ ÷ ⬡) (16)
3. □ / ○ (6)
4. (△ · ⬡) − ☆ (24)

Point out that a variety of symbols can be used to represent numbers. Stress, however, that no matter which symbols are used, the mathematical operations remain the same.

INSTRUCTION In Example 1, be sure students understand that 4*x* means $4 \cdot x$. Point out that when *x* is replaced with 5, 4*x* is not written as 45. Rather, the multiplication must be shown as 4(5) or $4 \cdot 5$.

Use Example 2 to show evaluating with multiple variables. Caution students to work carefully when replacing each variable with a number.

14.3 EVALUATING EXPRESSIONS AND FORMULAS

To **evaluate an expression,** first replace the variable by a given value. Then simplify the resulting numerical expression.

EXAMPLE 1 Evaluate: $4x + 7$ for $x = 5$

1. Replace *x* with 5. $4x + 7$
   $4(5) + 7$
2. Simplify. $20 + 7$
   27

Sometimes there is more than one variable in an expression.

EXAMPLE 2 Evaluate: $\frac{2x + y}{z}$ for $x = 2$, $y = 5$, and $z = 3$

1. Replace *x* with 2, *y* with 5, and *z* with 3. $\frac{2 \cdot 2 + 5}{3}$
2. Simplify. $\frac{4 + 5}{3} = 3$

A **formula** is an algebraic sentence. You can evaluate a formula by replacing the variables with a given value, then simplifying.

EXAMPLE 3 The pressure at a certain depth in the ocean can be approximated by the formula $P = 15 + \frac{1}{2}D$ in which *P* stands for the pressure in pounds per square inch and *D* is the depth in feet.

Find the amount of pressure at a depth of 40 ft.

1. Replace the variables with the given values. $P = 15 + \frac{1}{2}D$
   THINK: The depth *D* is 40 ft. $P = 15 + \frac{1}{2}(40)$
2. Simplify. $P = 15 + 20$
   $P = 35$

So, the water pressure at a depth of 40 ft is 35 lb per sq in.

CHECKPOINT Write the letter of the correct answer.

Evaluate.

1. $-3a + 4$ for $a = -2$ d **a.** -2 **b.** -1 **c.** 3 **d.** 10
2. $7 - (x - y)$ for $x = 5, y = -6$ b **a.** 4 **b.** -4 **c.** 6 **d.** 8

PRACTICE EXERCISES

See Extra Practice, page 460.

Evaluate for $x = 4$.

•**1.** $x + 7$ 11	•**2.** $x - 3$ 1	**3.** $5x$ 20	**4.** $-6x$ -24
5. $-x + 9$ 5	**6.** $x - 15$ -11	**7.** $-7x$ -28	**8.** $10x$ 40
9. $13 - x$ 9	**10.** $-22 + x$ -18	**11.** $-14 - x$ -18	**12.** $-3 + x - 4$ -3
13. $\frac{x}{2}$ 2	**14.** $\frac{16}{x}$ 4	**15.** $\frac{x}{-8}$ $-\frac{1}{2}$	**16.** $\frac{-x}{-2}$ 2

Evaluate for $a = -2$ and $b = 5$.

•**17.** $a + b$ 3	•**18.** $a - b$ -7	**19.** $b - a$ 7	**20.** ab -10
21. $6 - a + b$ 13	**22.** $3 + a - b$ -4	**23.** $2a - b$ -9	**24.** $3b + a$ 13
25. $\frac{ab}{10}$ -1	**26.** $\frac{a}{b}$ $-\frac{2}{5}$	**27.** $\frac{b}{a}$ $-\frac{5}{2}$	**28.** $\frac{a+b}{3}$ 1
29. $4a - b$ -13	**30.** $b - 2a$ 9	**31.** $5 + (b - a)$ 12	**32.** $3(a - b)$ -21

Evaluate for $x = 3$, $y = -1$, and $z = 2$.

•**33.** $x + y + z$ 4	•**34.** $x - y - z$ 2	**35.** $(x - y)z$ 8	**36.** $x - (y + z)$ 2
37. $\frac{x+z}{y}$ -5	**38.** $\frac{x+y}{z}$ 1	**39.** $\frac{y+z}{x}$ $\frac{1}{3}$	**40.** $\frac{xz}{y}$ -6
41. $5(x + y)$ 10	**42.** xyz -6	**43.** $x(y + z)$ 3	**44.** $z(x - y)$ 8
45. $4x + y + 2z$ 15	**46.** $-5xz$ -30	**47.** $8(z - y)$ 24	**48.** $7(2 + y)$ 7
49. $\frac{3(y-z)}{x}$ -3	**50.** $\frac{9(x-z)}{y+4}$ 3	**51.** $\frac{-4y+x}{1+3z}$ 1	**52.** $\frac{3x+y}{z}$ 4

The formula for the volume of a rectangular prism is $V = lwh$, where l is the length, w is the width, and h is the height. Find the volume.

•**53.** $l = 4$ m; $w = 2$ m; $h = 5$ m 40 m^3

54. $l = 0.5$ m; $w = 0.2$ m; $h = 1.6$ m 0.16 m^3

55. $l = 1.4$ cm; $w = 7$ cm; $h = 2.3$ cm 22.54 cm^3

56. $l = \frac{1}{2}$ ft; $w = \frac{3}{4}$ ft; $h = \frac{5}{6}$ ft $\frac{5}{16}$ ft^3

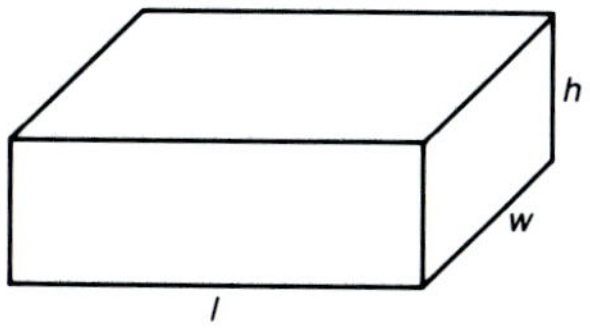

Example 3 extends the skill of evaluating expressions to formulas, as preparation for solving equations. Students have already worked informally with formulas throughout Chapter 12 for perimeter, area, and volume. Stress that a formula is an algebraic sentence because it includes the equals sign. Otherwise, there is no difference in the method of evaluating a formula from that of evaluating an expression.

CHECKPOINT The incorrect choices include common errors students make.

CALCULATOR This activity demonstrates another value of the calculator—to discover interesting number patterns with a degree of speed and ease that is not possible by hand.

COMMON ERROR

Some students will be so intent on replacing the variable that they may ignore the rules for computing with signed numbers. For example, in item 1 of Checkpoint, they would choose *a*. Have these students show every step and not compute mentally to save time.

ASSIGNMENTS

Level 1	Odd 1–63, CA
Level 2	Even 2–64, CA
Level 3	Odd 17–63, CA

FOLLOW-UP ACTIVITY

APPLICATION Discuss the factors that are involved in determining the driving range of a car. Guide students to discover:

Tank capacity		Miles per gallon		Driving range
C	$\times$	M	$=$	R

Have students use the formula to complete the table:

Tank capacity	mpg	Driving range
10.8 gal	20	■ (216 mi)
11.5 gal	24	■ (276 mi)
12.4 gal	26	■ (322.4 mi)
11.6 gal	25	■ (290 mi)
11.8 gal	31	■ (365.8 mi)

Have students find the driving range of their cars at home.

SUPPLEMENTARY MATERIALS

TRP Practice, p. 161

TRP Reteaching, p. 100

TRP Transparency 81

The formula for the area of a triangle is $A = \frac{1}{2}bh$ where b is the length of the base and h is the height. Find the area.

• **57.** $b = 14$ m; $h = 3$ m 21 m²

58. $b = 5.7$ m; $h = 6.2$ m 17.67 m²

59. $b = 4.9$ cm; $h = 8.2$ cm 20.09 cm²

60. $b = \frac{20}{3}$ in.; $h = \frac{9}{2}$ in. 15 in.²

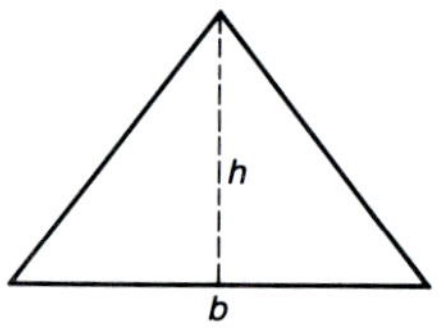

The formula for the area of a trapezoid is $A = \frac{1}{2}h(b_1 + b_2)$ where h is the height and b_1 and b_2 are the bases. Find the area.

• **61.** $h = 16$ m; $b_1 = 7$ m; $b_2 = 13$ m 160 m²

62. $h = 9.2$ mm; $b_1 = 3.8$ mm; $b_2 = 6.9$ mm 49.22 mm²

63. $h = 7.4$ cm; $b_1 = 4.9$ cm; $b_2 = 2.7$ cm 28.12 cm²

64. $h = \frac{26}{3}$ in.; $b_1 = \frac{43}{8}$ in.; $b_2 = \frac{58}{16}$ in. 39 in.²

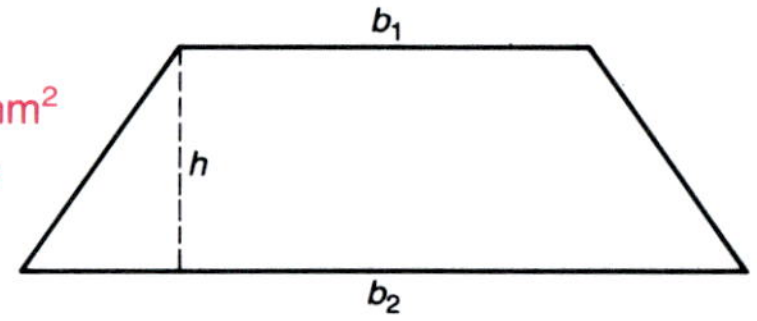

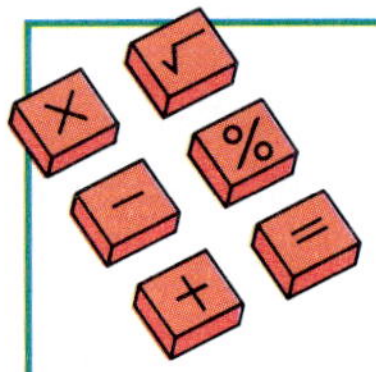

CALCULATOR

Calculators make it easy to discover number patterns. Try this experiment on your calculator.

1. Pick any number from 1 to 8. Divide it by 9, by 99, and by 999.

$8 \div 9 =$ ■ $0.\overline{8}$ $8 \div 99 =$ ■ $0.\overline{08}$ $8 \div 999 =$ ■ $0.\overline{008}$

Describe the pattern. One more zero occurs between the 8's each time.

2. Try two other numbers between 1 and 8. Does this pattern hold? yes

■ ÷ 9 = ■	■ ÷ 9 = ■
■ ÷ 99 = ■	■ ÷ 99 = ■
■ ÷ 999 = ■	■ ÷ 999 = ■

3. Does the pattern hold for 2-digit numbers? Only when the divisor is 99 or 999.

Try this	**Your turn**
$47 \div 9 =$ ■ $5.\overline{2}$	■ ÷ 9 = ■
$47 \div 99 =$ ■ $0.\overline{47}$	■ ÷ 99 = ■
$47 \div 999 =$ ■ $0.\overline{047}$	■ ÷ 999 = ■

4. Does the pattern hold if you divide a 2-digit number by 9,999? yes

PROBLEM Solving APPLICATION

14.4 USING THE DISTANCE FORMULA

A Metroliner train is traveling at the rate of 84 mph. What distance will the train travel in 2.5 hours?

You can use the distance formula to solve problems involving time, rate (average speed), and distance.

DISTANCE = RATE × TIME

$$d = rt$$

THINK: The rate and time are given. r = 84 mph, t = 2.5 h

1. Substitute the given numbers for each variable. $d = 84 \cdot 2.5$
2. Multiply. $d = 210$

So, the train travels 210 mi in 2.5 hours.

Find the missing amount.

• **1.** $r = 76$ mph, $t = 5$ hours, $d =$ ■ 380 mi

• **2.** $r = 52$ mph, $t = 6.5$ hours, $d =$ ■ 338 mi

3. $r =$ ■, $t = 3.2$ hours, $d = 384$ mi 120 mph

4. $r =$ ■, $t = 2\frac{1}{4}$ hours, $d = 126$ mi 56 mph

5. $r = 372$ mph, $t =$ ■, $d = 2{,}604$ mi 7 h

6. $r = 3\frac{1}{2}$ mph, $t =$ ■, $d = 6\frac{3}{10}$ mi 1.8 h

Solve. Remember to estimate whenever you use your calculator.

• **7.** A jet plane travels 2,210 mi in 3.4 hours. What is its rate in miles per hour? 650 mph

8. An Antelope ZR travels 459 mi at a rate of 54 mph. How long will the trip take? $8\frac{1}{2}$ h

9. The Concorde flew from Paris to New York in $3\frac{1}{2}$ h. Its average speed was 1,037.5 mph. What is the air distance between Paris and New York? 3,631.25 mi

10. A train travels 403.62 mi between Moscow and Leningrad in 4.65 hours. A train travels between Wilmington and Baltimore in 46 minutes at a rate of 89.2 mph. Which train is faster? By how much? Wilmington to Baltimore train is 2.4 mph faster.

11. A spacecraft orbiting the earth travels at a rate of 16,500 mph. How much more than 4 hours does it take to travel 80,850 mi? 0.9 hours

OBJECTIVE

- Solve problems that involve using the distance formula.

TEACHING THE LESSON

WARM-UP Write the following exercises on the chalkboard or an overhead transparency.

Evaluate the expression.

1. $a \times b$ if $a = 74$ and $b = 2.5$ (185)
2. $a \times b$ if $a = 58.8$ and $b = 3.5$ (205.8)
3. $d \div e$ if $d = 1{,}100$ and $e = 5.5$ (200)
4. $d \div e$ if $d = 640$ and $e = 200$ (3.2)

INSTRUCTION Emphasize the relationship between multiplication and division as you consider $d = r \times t$ and the two related division sentences: $r = d \div t$ and $t = d \div r$.

Note that Problems 1 and 2 involve finding d; Problems 3 and 4 involve finding r; and Problems 5 and 6 involve finding t.

You may wish to have students use their calculators to help them solve the problems.

ASSIGNMENTS

Level 1	1–7
Level 2	1–10
Level 3	5–11

SUPPLEMENTARY MATERIALS

TRP Practice, p. 162

TRP Transparency 81

OBJECTIVE

- Translate word sentences into equations and inequalities.

TEACHING THE LESSON

WARM-UP Write the following exercises on the chalkboard or an overhead transparency.

Write as an algebraic expression.

1. Seven more than a number $(7 + n)$
2. 45 less than a number $(n - 45)$
3. Three times a number, plus 6 $(3n + 6)$
4. The quotient of seventeen divided by a number $(\frac{17}{n})$

INTRODUCTION Write:

$n + 6 \qquad n + 6 = 14$

Ask which is an expression and which is a number sentence. Have students explain their reasoning. Write:

$8 < 12 \qquad 7 > 4 + 2$

Use these examples to show that number sentences do not always contain an equals sign.

INSTRUCTION In Example 1, help students see how to write an equation from the word problem. Students should note that the equation comes directly from the third sentence.

Point out that the two algebraic expressions, 120 and $50 + 2s$, are the left and right sides of the equation. Because the sides of an equation are interchangeable, some students may find it easier to solve if the variable is on the left. Encourage them to rewrite the equation as $50 + 2s = 120$ or, using the commutative property, $2s + 50 = 120$.

Example 2 shows that an inequality can also be translated directly from a word sentence to a number sentence. Emphasize that the sides are not interchangeable.

14.5 EQUATIONS AND INEQUALITIES

Jeff Chan is shopping for summer clothes. He wants to buy a jacket that costs $120. The cost of the jacket is fifty dollars more than the cost of two shirts he also wants to buy.

An **equation** is a mathematical sentence that states that two expressions are equal.

EXAMPLE 1 Write an equation to show the relationship between the price of the shirts and the cost of the jacket.

THINK: Cost of the jacket is $50 more than the cost of 2 shirts.

WRITE: $120 = 50 + 2s$

An **inequality** is a mathematical sentence that uses $<$, $>$, $\leq$, or $\geq$ to show the relationship between two expressions.

EXAMPLE 2 Write an inequality for the expression.

a. Five more than one-half of a number is greater than 13.

$5 + \frac{1}{2}n > 13$

b. 8 more than 3 times a number is less than or equal to 75.

$8 + 3n \leq 75$

FOR DISCUSSION See TE side column.

A rack of clothes had a sign that read, "No item more than $5." How could the sign be written as an inequality? Which of these price tags could appear on the clothes: $4, $7.50, $5, $3.95, $8.98?

PRACTICE EXERCISES

Write a mathematical sentence for the statement.

• **1.** 18 times a number is 36. $18n = 36$

• **2.** 7 more than a number is 15. $n + 7 = 15$

3. 6 less than a number is 3. $n - 6 = 3$

4. The quotient of a number divided by 9 is 5. $n \div 9 = 5$

5. 8 subtracted from 3 times a number is 16. $3n - 8 = 16$

6. The sum of two-thirds of a number and 6 is 18. $\frac{2}{3}n + 6 = 18$

7. 9 is 4 more than 3 times a number. $9 = 3n + 4$

8. 17 is 42 minus one-half a number. $17 = 42 - \frac{1}{2}n$

9. 8 times the sum of a number and 3 is 35. $8(n + 3) = 35$

10. 20 divided by the difference of 15 and a number is 3. $20 \div (15 - n) = 3$

11. 7 times a number is less than 21. $7n < 21$

12. 14 more than a number is greater than 56. $n + 14 > 56$

13. The quotient of a number divided by 14 is less than 93. $\frac{n}{14} < 93$

14. The difference between 87 and a number is greater than 34. $87 - n > 34$

15. 15 times the sum of a number and 96 is less than 28. $15(n + 96) < 28$

16. 13 and twice a number is less than or equal to 61. $13 + 2n \leq 61$

17. 4 more than a number is less than 38. $n + 4 < 38$

18. The quotient of 72 divided by a number is greater than or equal to 46. $\frac{72}{n} \geq 46$

19. 4 more than 5 times a number is less than or equal to 39. $4 + 5n \leq 39$

20. The product of 12 and a number, divided by 6, is greater than 58. $\frac{12n}{6} > 58$

MIXED REVIEW

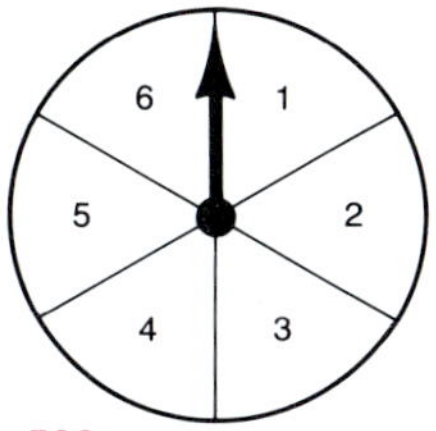

Find the probability of spinning:

• **21.** a 4. $\frac{1}{6}$

22. an even number. $\frac{1}{2}$

The spinner is spun 600 times.
Predict how many times the spinner will land on:

23. an odd number. 300

24. a number greater than 1. 500

TIME OUT In 1863 President Abraham Lincoln gave his famous Gettysburg Address. It starts "Fourscore and seven years ago. . . ." Fourscore and seven is equal to 87. How many years are in a score? What year was President Lincoln referring to? 20 years; 1776

Emphasize the distinction between the phrase, *3 more than a number* ($3 + n$), and the sentence, *3 is greater than a number* ($3 > n$). Similarly, *7 less than a number* ($n - 7$) is not the same as *7 is less than a number* ($7 < n$).

FOR DISCUSSION Students should explain that the phrase *no more than* is the same as *less than or equal to.* Therefore, the $5 price tag and the tags for any amount less than $5 could appear on the clothes.

TIME OUT Students should write the equation $4s + 7 = 87$ from the third sentence. Lead them to find the missing year by subtracting 87 from 1863 to get 1776. Ask what event President Lincoln was referring to.

ASSIGNMENTS

Level 1	Odd 1–19, 21–24
Level 2	Even 2–20, 21–24, TO
Level 3	10–24, TO

FOLLOW-UP ACTIVITY

COOPERATIVE LEARNING
Have students work in pairs and write equations based on data from their lives. For example, suppose student *A* is 14 years old and student *B* is 17. Student *A* would write $n + 3 = 17$ to show the relationship between her age and that of her partner. Conversely, student B would write $n - 3 = 14$.

Have partners write similar equations for the number of:

- hours they sleep each night.
- hours they work each week.
- blocks they live from school.
- dances they attend in a year.

SUPPLEMENTARY MATERIALS

TRP Practice, p. 163

TRP Reteaching, p. 101

TRP Transparency 82

OBJECTIVE

- Graph the solution to an equation or inequality.

TEACHING THE LESSON

WARM-UP Write the following exercises on the chalkboard or an overhead transparency.

Solve.

1. $x + 4 = 6$ **2.** $x - 7 = 11$

3. $x < 9$ **4.** $x \geq -1$

(**1.** 2 **2.** 18 **3.** any number < 9
4. any number ≥ -1)

INTRODUCTION Draw:

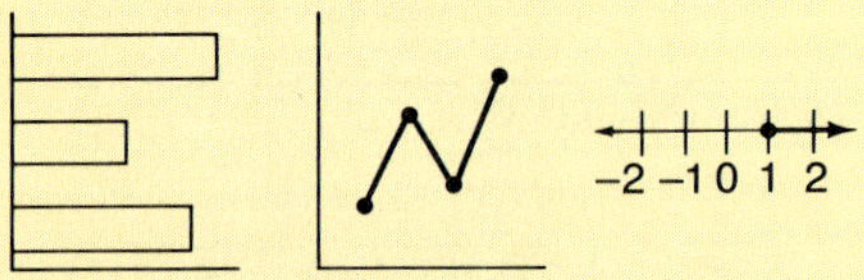

Review the purpose of graphs. Ask which of the drawings could represent a graph. Tell students that today they will learn how a number line can also serve as a graph.

INSTRUCTION In Example 1, stress that the solution to an equation is most often a single value and is graphed with a single dot.

Stress that the solution to an inequality can include all values on the number line to the left (<) or right (>) of a specific number. In Example 2, make sure students note the use of an open circle at 5 to indicate that 5 is not a member of the solution set. In Example 3, emphasize the use of a closed circle at −3 to indicate that −3 is a member of the solution set.

CHECKPOINT The incorrect choices include common errors students make.

CALCULATOR Here, the calculator eases the computation in a guess-and-check situation. Encourage students to begin by writing all the basic addition facts with a sum of 11.

14.6 GRAPHING POINTS ON A LINE

The solution to an equation or an inequality can be shown by a point or a set of points on a number line.

EXAMPLE 1 Graph the solution of $x + 6 = 8$.

1. Solve: $x + 6 = 8$

   THINK: $2 + 6 = 8$ $x = 2$

2. Graph:

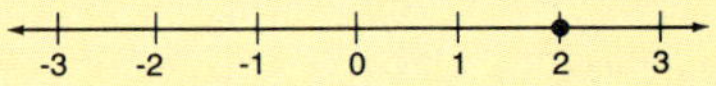

The solution of an inequality with the symbol > or < is shown by an open circle and a shaded line. The open circle indicates that the end point is not part of the solution.

EXAMPLE 2 Graph the solution of $x < 5$.

1. Solve: $x < 5$

   THINK: x can be any number less than 5.

2. Graph:

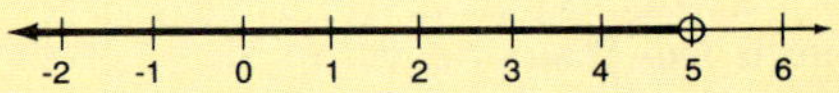

The solution to an inequality with the symbol ≥ or ≤ is shown by a closed circle and a shaded line.

EXAMPLE 3 Graph the solution of $x \geq -3$.

1. Solve: $x \geq -3$

   THINK: x is -3 or any number greater than -3.

2. Graph:

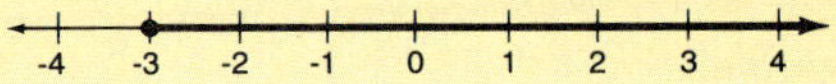

CHECKPOINT

Write the letter of the correct answer.

Choose the number sentence whose solution is graphed.

1.

c **a.** $x + 2 \geq 0$ **b.** $x < -2$ **c.** $x > -2$ **d.** $x \geq -2$

2.

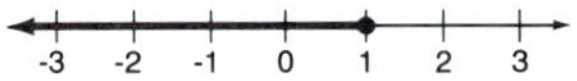

b **a.** $x + 8 \leq 9$ **b.** $x \leq 1$ **c.** $x > 1$ **d.** $x < 1$

PRACTICE EXERCISES

See Extra Practice, page 460.

Match the graph to the correct number sentence.

• 1. d 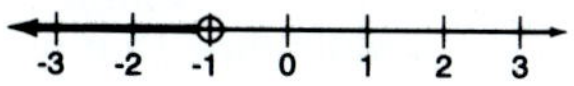

a. $x \leq -1$

• 2. e 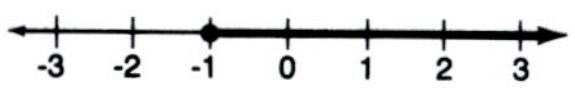

b. $x = -1$

3. a 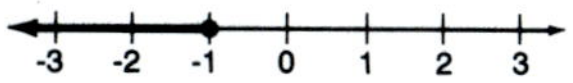

c. $x > -1$

4. b 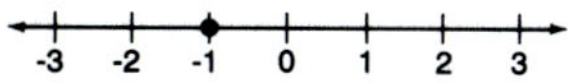

d. $x < -1$

5. c 

e. $x \geq -1$

Graph the solution. See page 478 for answers to Exercises 6–25.

• 6. $x + 5 = 7$ 7. $x + 1 = 9$ 8. $x + 4 = 10$ 9. $x + 8 = 12$

10. $x - 3 = 6$ 11. $x - 7 = 5$ 12. $x - 2 = 9$ 13. $x - 10 = 2$

• 14. $x < 6$ 15. $x > 4$ 16. $x > -1$ 17. $x < -5$

18. $x \geq -2$ 19. $x \geq 0$ 20. $x \leq 3$ 21. $x \leq -2$

22. $x \leq 3$ 23. $x \geq -3$ 24. $x \geq 2$ 25. $x \leq -4$

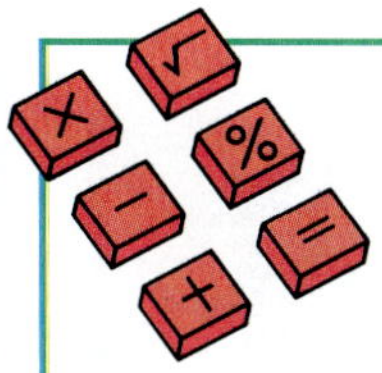

CALCULATOR

The sum of the digits of a 2-digit number is 11.
If the digits are reversed, the number formed is 27 less than the original number.
Find the original number.

Use a calculator and the guess-and-check method to find the number.

2-digit number	Reverse the digits. Subtract.	Difference of 27?
29, 2 + 9 = 11	92 − 29 = 63	No
38, 3 + 8 = 11	83 − 38 = 45	No

What is the number? 74

COMMON ERROR

Some students confuse the direction of the shaded line. For example, in item 1 of Checkpoint, they would choose *b.* Remind these students that the point of the inequality sign indicates the direction of the line. Note that this condition is true only when the variable is on the left side of the equation (all exercises in this lesson).

ASSIGNMENTS

Level 1	1–5, Odd 7–25
Level 2	1–5, Even 6–24, CA
Level 3	Odd 6–25, CA

FOLLOW-UP ACTIVITY

ENRICHMENT Draw:

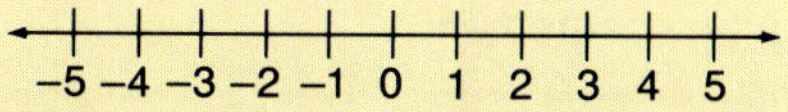

Have students copy the number line to use in finding four solutions to each of the following inequalities.

$x \geq 5$ or $x < 3$
(5, 6, 1, 2)

$x > -2$ and $x > 1$
(2, 3, 4, 5)

$x \leq 0$ or $x > 2$
(0, −1, 3, 4)

$x < 3$ and $x > -3$
(−2, −1, 0, 1)

Call attention to the words *and, or.* Emphasize that *and* means that any solution must satisfy the conditions of both inequalities. Conversely, *or* means that a solution has to satisfy only one of the inequalities.

SUPPLEMENTARY MATERIALS

TRP Practice, p. 164

TRP Reteaching, p. 102

TRP Lesson Aids, p. 3

TRP Transparency 82

OBJECTIVE

- Solve problems that involve solutions.

TEACHING THE LESSON

WARM-UP Write the following exercises on the chalkboard or an overhead transparency.

Find the answer.

1. What number is 10% of 140? (14)
2. What number is 3% of 200? (6)
3. What percent of 50 is 9? (18%)
4. What percent of 300 is 12? (4%)

INTRODUCTION Discuss with the students what a pharmacist does. Ask them why a pharmacist might need to convert a volume measurement into a weight measurement.

INSTRUCTION Work through the introductory problem carefully with the students. Problems 1–5 involve converting milliliters to grams. Problem 6 asks the students to find the percentage of a solution, and also requires the conversion of liters to milliliters. Problems 7 and 8 are multistep problems.

You may wish to allow students to use their calculators to help them solve the problems.

ASSIGNMENTS

Level 1	1–5
Level 2	1–6
Level 3	3–8

SUPPLEMENTARY MATERIALS

TRP Practice, p. 165

TRP Transparency 83

PROBLEM Solving APPLICATION

14.7 CAREER: PHARMACIST

Don Keefer is a pharmacist. He uses his knowledge of measurements and equations when preparing various solutions. For example, when working with grams and milliliters, Don uses the fact that 1 mL of water has a mass of 1 g.

How many grams of dextrose must Don use to prepare 3,000 mL of a solution that is 5% dextrose?

THINK: 5% of the 3,000-mL solution is dextrose. The number of grams of dextrose is equivalent to 5% of 3,000 g.

1. Write an equation. Let n represent the number of grams of dextrose. $n = 5\%$ of $3{,}000$
2. Multiply to find n. $n = 0.05 \cdot 3{,}000$; $n = 150$

So, Don must use 150 g of dextrose to make the 5% solution.

Find the number of grams of the chemical needed to make the solution.

	Chemical	Total volume of solution	% of chemical in solution	Number of grams needed
• **1.**	A	4,000 mL	3%	▩ 120 g
2.	B	2,000 mL	4%	▩ 80 g
3.	C	1,500 mL	2.5%	▩ 37.5 g
4.	D	3,500 mL	3.5%	▩ 122.5 g

Solve. Remember to estimate whenever you use your calculator.

• **5.** A 3,500-mL solution is 6% dextrose. How many grams of dextrose are used in the solution? 210 g

6. A 4,000-mL solution contains 2,800 L of alcohol. What percent of the solution is alcohol? 70%

7. Don has 120 mL of a solution that is 30% iodine. He adds 80 mL of water to the solution. What percent of the new solution is iodine? 18%

8. Don has 200 mL of a solution that is 40% iodine. He adds 20 g of iodine to the solution. What percent of the new solution is iodine? 50%

MIDCHAPTER REVIEW

Simplify. [398]

1. $(5 + 7) \div 2$ 6 **2.** $8 - (3 + 6)$ −1 **3.** $4 - (8 + 2) \div 2$ −1 **4.** $(5 - 4)3$ 3

5. $12 - 6 - 5$ 1 **6.** $9 + 4 - 2$ 11 **7.** $12 \div 2 - 1$ 5 **8.** $6 + 3 \cdot 5$ 21

9. $14 \div 7 + 3$ 5 **10.** $81 - 9 \div 3$ 78 **11.** $54 \div 3^2 \cdot 2$ 12 **12.** $72 - 6^2 \cdot 2$ 0

Write as an algebraic expression. [400]

13. 2 more than a number $n + 2$

14. 8 times a number $8n$

15. 19 less than a number $n - 19$

16. a number divided by 6 $n \div 6$

17. 5 added to twice a number $2n + 5$

18. half a number minus 11 $\frac{1}{2}n - 11$

Write as a word phrase. [400] See page 478 for answers to Exercises 19–26.

19. $3 + n$ **20.** $n - 12$ **21.** $2n - 4$ **22.** $73 - n$

23. $\frac{n-7}{5}$ **24.** $18 + n^3$ **25.** $16n + 9$ **26.** $12(n - 5)$

Evaluate for $x = 5$, $y = -3$. [402]

27. $x + y$ 2 **28.** $x - y$ 8 **29.** $y - x$ −8 **30.** xy −15

31. $2x - y$ 13 **32.** $2y - x$ −11 **33.** $3(x + y)$ 6 **34.** $(3 - y) - x$ 1

Write a number sentence. [406]

35. 6 more than 4 times a number is 28. $4n + 6 = 28$

36. 4 less than Kim's age is greater than 12. $n - 4 > 12$

37. The points in a game plus 10 are less than or equal to 64. $n + 10 \leq 64$

38. Fred's age 5 years ago was greater than or equal to 17. $n - 5 \geq 17$

Graph the solution on a number line. [408] See page 478 for answers to Exercises 39–42.

39. $x + 3 = 7$ **40.** $x - 2 = 8$ **41.** $x < 7$ **42.** $x \geq -5$

Solve. [405, 410]

43. Use the formula $d = rt$ to find the distance a jetliner would travel in 4.5 hours at an average speed of 560 mph. 2,520 mi

44. A 5,000-mL solution is 15% dextrose. How many grams of dextrose are used in the solution? Remember, 1 g = 1 mL. 750 g

OBJECTIVE

- Evaluate student progress.

USING THE REVIEW

This page provides a means for informally evaluating students' understanding of the skills and concepts covered so far in this chapter.

Have the students look at the page to familiarize themselves with the various question formats that are presented. Discuss any questions that they may have. Then ask them to complete the page independently.

In addition to grading them individually, you may wish to review the answers to the questions collectively with the students.

Page references appear in brackets. They refer to pages on which a particular skill was introduced.

Before continuing on to the topics found in the remainder of the chapter, you may wish to have students review any skills or concepts in which they have demonstrated weakness.

OBJECTIVE

- Solve equations using addition and subtraction.

TEACHING THE LESSON

WARM-UP Write the following exercises on the chalkboard or an overhead transparency.

Compute.

1. $6 + (-3)$ **2.** $-8 - 6$

3. $-5 + 9$ **4.** $3 - (-7)$

5. $-9 - (-2)$ **6.** $4 + (-8)$

(**1.** 3 **2.** −14 **3.** 4 **4.** 10 **5.** −7 **6.** −4)

INTRODUCTION Write:

■ + 9 = 17 (8) 6 + ■ = 11 (5)

■ − 5 = 2 (7) 10 − ■ = 4 (6)

Discuss how to find the missing number in each sentence. Repeat with more challenging examples.

−7 + ■ = −12 (−5)

■ + 3 = −6 (−9)

8 − ■ = 13 (−5)

■ − 4 = −11 (−7)

INSTRUCTION Equations are solved using properties of equality. Example 1 uses the subtraction property of equality while the addition property of equality is used in Example 2. You may want to use a balance scale to show that whatever is done to one side of the equation must also be done to the other side or the equality (balance) is lost. Stress the importance of checking the solution in the original equation.

Example 3 combines writing and solving an equation in a problem-solving situation. Stress the need to check the reasonableness of the answer. Point out that if an incorrect equation is solved correctly, the answer may not be reasonable.

14.8 SOLVING EQUATIONS BY ADDITION OR SUBTRACTION

Addition and subtraction are inverse operations. One undoes the other. To solve an equation, use an inverse operation to get the variable by itself.

When a number is attached to the variable by addition, you can unattach it by subtracting that number from both sides of the equation.

EXAMPLE 1 Solve: $x + 9 = 14$. Check the solution.

1. Solve. $x + 9 = 14$

   THINK: 9 is attached by addition. Subtract 9 from both sides.

   $x + 9 - 9 = 14 - 9$

   $x = 5$

2. Check. $x + 9 = 14$

   THINK: Substitute the value of x into the equation.

   $5 + 9 \stackrel{?}{=} 14$

   $14 = 14$ It checks.

When a number is attached to the variable by subtraction, you can unattach it by adding that number to both sides of the equation.

EXAMPLE 2 Solve: $x - 7 = 8$. Check the solution.

1. Solve. $x - 7 = 8$

   THINK: 7 is attached by subtraction. Add 7 to both sides.

   $x - 7 + 7 = 8 + 7$

   $x = 15$

2. Check. $x - 7 = 8$

   THINK: Substitute the value of x into the equation.

   $15 - 7 \stackrel{?}{=} 8$

   $8 = 8$ It checks.

You can write and solve an equation to solve a problem.

EXAMPLE 3 Carrie Collins and Tom Kowalski own Magic Carpet Cleaners. They had 12 customers this week, 8 fewer than last week. How many customers did they have last week?

1. Write an equation.

   THINK: Eight fewer than the number last week is 12.

   WRITE: $n - 8 = 12$

2. Solve.

   $n - 8 = 12$

   $n - 8 + 8 = 12 + 8$

   $n = 20$

3. Check.

   $n - 8 = 12$

   $20 - 8 \stackrel{?}{=} 12$

   $12 = 12$. It checks.

Last week Magic Carpet Cleaners had 20 customers.

CHECKPOINT Write the letter of the correct number.

Solve.

1. $x + 6 = 9$	b	**a.** -3	**b.** 3	**c.** -15	**d.** 15
2. $x - 4 = 12$	c	**a.** 8	**b.** -8	**c.** 16	**d.** -16
3. $-5 + x = 8$	a	**a.** 13	**b.** -13	**c.** -3	**d.** 3
4. $3 + x = 7$	c	**a.** 10	**b.** -10	**c.** 4	**d.** -4

PRACTICE EXERCISES See Extra Practice, page 461.

Solve and check.

• **1.** $x + 7 = 12$ 5 • **2.** $x + 9 = 16$ 7 **3.** $x + 4 = 23$ 19 **4.** $x + 5 = 13$ 8

5. $x + 2 = 50$ 48 **6.** $8 + x = 16$ 8 **7.** $3 + x = 27$ 24 **8.** $6 + x = 21$ 15

9. $2 + x = 19$ 17 **10.** $4 + x = 35$ 31 **11.** $x + 35 = 52$ 17 **12.** $x + 19 = 83$ 64

13. $x + 64 = 76$ 12 **14.** $x + 42 = 95$ 53 **15.** $x + 57 = 82$ 25 **16.** $56 + x = 84$ 28

17. $71 + x = 100$ 29 **18.** $27 + x = 65$ 38 **19.** $97 + x = 132$ 35 **20.** $87 + x = 113$ 26

21. $-5 + x = 7$ 12 **22.** $-9 + x = 8$ 17 **23.** $-6 + x = 13$ 19 **24.** $-7 + x = 15$ 22

• **25.** $x - 6 = 11$ 17 • **26.** $x - 3 = 8$ 11 **27.** $x - 4 = 12$ 16 **28.** $x - 8 = 11$ 19

29. $x - 2 = 18$ 20 **30.** $x - 6 = 16$ 22 **31.** $x - 32 = 18$ 50 **32.** $x - 71 = 35$ 106

33. $x - 58 = 29$ 87 **34.** $x - 64 = 57$ 121 **35.** $x - 23 = 46$ 69 **36.** $x - 83 = 105$ 188

37. $x + (-5) = 3$ 8 **38.** $x + \frac{1}{2} = \frac{3}{4}$ $\frac{1}{4}$ **39.** $x - (-6) = 10$ 4 **40.** $17.4 = x - 9.6$ 27

41. $x - \frac{3}{8} = \frac{1}{5}$ $\frac{23}{40}$ **42.** $-18 + x = -40$ -22 **43.** $x - (-26.8) = -18.2$ -45 **44.** $x + \frac{3}{4} = -\frac{9}{10}$ $-1\frac{13}{20}$

45. $3 + x + 4 = 11$ 4 **46.** $6 + x + 7 = 20$ 7 **47.** $2 + x + 9 = 18$ 7

48. $-4 + x - 18 = 37$ 59 **49.** $-4 + x - 8 = 34$ 46 **50.** $-3 + x - 2 = 41$ 46

Solve. Write an equation as needed.

• **51.** Magic Carpet Cleaners had $1,546.17 in its checking account. After Carrie made a deposit the balance was $1,823.39. How much money did Carrie deposit? $277.22

52. The odometer in one of the Magic Carpet Cleaners's vans read 794 on Monday. On Friday, it read 1,031. How many miles was the van driven that week? 237 mi

53. Kim was cleaning a carpet that was 85.7 ft^2. After she had cleaned 80% of the carpet, the machine broke. How many square feet of carpet were left to clean? 17.14 ft^2

54. A square rug and a rectangular rug are each 36 ft^2 in area. The length of the rectangular rug is twice that of the square rug. Find the dimensions of the rectangular rug.
l = 12 ft; w = 3 ft

CHECKPOINT The incorrect choices include common errors students make.

COMMON ERROR

In equations such as $-5 + x = 8$, some students may encounter difficulty in deciding whether to add or subtract the negative integer to isolate the variable. For example, in item 3 of Checkpoint, they would choose *d*. Have students use the Commutative Property to rewrite the equation:

$$-5 + x = 8$$
$$x + (-5) = 8$$

Encourage them to show all the work in solving the equation.

ASSIGNMENTS

Level 1	Odd 1–35, 51, 52
Level 2	Even 2–36, 51–54
Level 3	Odd 13–35, 37–54

FOLLOW-UP ACTIVITY

COOPERATIVE LEARNING Divide the class into small groups. Write the following statement on the chalkboard.

1 2 3 4 5 6 7 8 9 = 100

Explain that two addition signs and two subtraction signs have been omitted. Have students work in small groups to write the missing signs to form a true equation. Point out that the digits may be combined to form 2- and 3-digit numbers. Stress, however, that the order of the digits may not be changed. (123 + 45 − 67 + 8 − 9 = 100)

SUPPLEMENTARY MATERIALS

TRP Practice, p. 166

TRP Reteaching, p. 103

TRP Transparency 83

OBJECTIVE

- Solve equations using multiplication and division.

TEACHING THE LESSON

WARM-UP Write the following exercises on the chalkboard or an overhead transparency.

For each equation, write what is attached to the variable and how it is attached.

1. $7x = 21$ $(7, \times)$
2. $\frac{x}{3} = 5$ $(3, \div)$
3. $36 = -9x$ $(-9, \times)$
4. $2 = \frac{x}{-4}$ $(-4, \div)$

INTRODUCTION Write:

$\blacksquare \times 8 = -56$ (-7)

$-25 \times \blacksquare = 150$ (-6)

$\frac{63}{\blacksquare} = -7$ (-9)

$\frac{\blacksquare}{5} = -20$ (-100)

Discuss how to find the missing number in each sentence.

INSTRUCTION As in the previous lesson, students must use properties of equality to isolate the variable and get a simple equivalent equation of the form $x = \blacksquare$.

Example 1 uses the multiplication property of equality while the division property of equality is used in Example 2. Remind students that zero can never be a divisor.

The difficulty in Example 3 lies in writing the correct equation. Extend the instruction by guiding students in writing an equation for the following statements:

- Four times a number equals ninety-six. $(4n = 96)$
- The quotient of a number divided by six is three. $(\frac{n}{6} = 3)$

14.9 SOLVING EQUATIONS BY MULTIPLICATION OR DIVISION

Multiplication and division are also inverse operations.
When a number is attached to the variable by division, you can unattach it by multiplying both sides of the equation by that number.

EXAMPLE 1 Solve: $\frac{x}{6} = 7$. Check the solution.

1. Solve.

THINK: 6 is attached by division. Multiply both sides by 6.

$6 \cdot \frac{x}{6} = 6 \cdot 7$

$x = 42$

2. Check.

THINK: Substitute the value of x into the original equation.

$\frac{42}{6} \stackrel{?}{=} 7$

$7 = 7$ It checks.

When a number is attached to the variable by multiplication, you can unattach it by dividing both sides of the equation by that number.

EXAMPLE 2 Solve: $-4x = 20$. Check the solution.

1. Solve.

THINK: −4 is attached by multiplication. Divide both sides by −4.

$\frac{-4x}{-4} = \frac{20}{-4}$

$x = -5$

2. Check.

THINK: Substitute the value of x into the original equation.

$-4(-5) \stackrel{?}{=} 20$

$20 = 20$ It checks.

You can write and solve an equation to solve a problem.

EXAMPLE 3 At the end of his typing course, Steve typed 87 words per minute (wpm). This was 3 times faster than his rate at the beginning of the course. What was his rate at the beginning of the course?

1. Write an equation.

THINK: Three times a number equals 87 words per minute (wpm). $3n = 87$

2. Solve.

$3n = 87$

$\frac{3n}{3} = \frac{87}{3}$

$n = 29$

3. Check.

$3n = 87$

$3(29) \stackrel{?}{=} 87$

$87 = 87$ It checks.

Steve typed 29 words per minute at the beginning of the course.

CHECKPOINT Write the letter of the correct answer.

Solve and check.

1. $\frac{x}{3} = 9$ c **a.** 3 **b.** −3 **c.** 27 **d.** −27
2. $\frac{x}{0.34} = 7$ b **a.** −2.38 **b.** 2.38 **c.** 7 **d.** 20.58
3. $-2x = 8$ b **a.** 4 **b.** −4 **c.** 16 **d.** −16
4. $0.8x = 7.2$ c **a.** −9 **b.** 0.9 **c.** 9 **d.** 90

PRACTICE EXERCISES See Extra Practice, page 461.

Solve and check.

• **1.** $3x = 15$ 5 • **2.** $5x = 35$ 7 **3.** $2x = 12$ 6 **4.** $7x = 21$ 3 **5.** $6x = 30$ 5

6. $-8x = 56$ −7 **7.** $4x = -16$ −4 **8.** $-9x = 72$ −8 **9.** $5x = 85$ 17 **10.** $-3x = 96$ −32

• **11.** $\frac{x}{6} = 3$ 18 • **12.** $\frac{x}{4} = 8$ 32 **13.** $\frac{x}{7} = 11$ 77 **14.** $\frac{x}{2} = 9$ 18 **15.** $\frac{x}{9} = 5$ 45

16. $\frac{x}{5} = -12$ −60 **17.** $\frac{x}{-3} = 15$ −45 **18.** $\frac{x}{2} = -29$ −58 **19.** $\frac{x}{-7} = 32$ −224 **20.** $\frac{x}{8} = -18$ −144

21. $0.75x = 3$ 4 **22.** $0.5x = 7$ 14 **23.** $0.2x = 8$ 40 **24.** $0.8x = 20$ 25 **25.** $0.4x = 12$ 30

26. $0.6x = 18$ 30 **27.** $0.75x = 36$ 48 **28.** $0.3x = 27$ 90 **29.** $0.7x = 14$ 20 **30.** $0.04x = 3$ 75

31. $\frac{x}{0.14} = 5$ 0.7 **32.** $\frac{x}{0.08} = 12$ 0.96 **33.** $\frac{x}{0.73} = 4$ 2.92 **34.** $\frac{x}{0.29} = 6$ 1.74 **35.** $\frac{x}{0.92} = 8$ 7.36

36. $\frac{x}{0.31} = 0.4$ 0.124 **37.** $\frac{x}{0.65} = 0.2$ 0.13 **38.** $\frac{x}{0.97} = 1.6$ 1.552 **39.** $\frac{x}{0.42} = 5.3$ 2.226 **40.** $\frac{x}{0.86} = 3.9$ 3.354

Solve. Write an equation as needed.

• **41.** In typing class, the fastest student typed 60 wpm. That was three times as fast as the slowest student. How many words per minute did the slowest student type? 20 wpm

42. All the typing students were evenly divided among three classes. If each class had 23 students, how many students were taking typing? 69 students

43. In one week, Toni's typing rates were 63, 72, 59, 67, and 64 wpm. What was her average typing rate for the week? 65 wpm

CHECKPOINT The incorrect choices include common errors students make.

COMMON ERROR

Some students use the incorrect operation to solve an equation. For example, in item 1 of Checkpoint, they would choose *a*. Use the following procedure to remind students to use the inverse operation to isolate the variable.

Write: $8n = 64$ $\frac{n}{6} = 9$

For each equation ask:

- What number is attached to the variable?
- How is it attached?
- How can the number be unattached?

ASSIGNMENTS

Level 1	Odd 1–35, 41–42
Level 2	Even 2–40, 41–43
Level 3	21–43

FOLLOW-UP ACTIVITY

APPLICATION Discuss the meaning of batting average. Guide students in discovering the following formula:

Number of hits → $\frac{h}{t} = a$ ← Batting average
Number of times → at bat

Have students complete the table.

Player	Hits	Times at bat	Batting average
Rodriguez	(55)	200	0.275
Chapman	24	120	(0.200)
De Marco	70	(250)	0.280
Cohen	(36)	150	0.240

SUPPLEMENTARY MATERIALS

TRP Practice, p. 167

TRP Reteaching, p. 104

TRP Transparency 84

OBJECTIVE

- Solve two-step equations.

TEACHING THE LESSON

WARM-UP Write the following exercises on the chalkboard or an overhead transparency.

Solve.

1. $n + 6 = 11$ (5)
2. $3a = 24$ (8)
3. $x - 25 = 13$ (38)
4. $\frac{b}{8} = 7$ (56)
5. $-8 + y = -2$ (6)
6. $7t = 168$ (24)

INTRODUCTION For each equation, ask what is attached to the variable and how it is attached.

1. $3x - 2 = 7$
2. $\frac{x}{5} + 4 = 9$
3. $-5 + \frac{x}{-2} = 8$
4. $6 - 10x = 4$

(**1.** 3, × and 2, − **2.** 5, ÷ and 4, + **3.** −2, ÷ and −5, + **4.** −10, × and 6, +)

Lead students to realize that in each equation two different operations are needed to isolate the variable.

INSTRUCTION Help students use the third and fourth sentences of the introductory problem to write an equation. If needed, use the following diagram to show where 3*x* comes from.

Earnings

Day 1		Day 2		Day 3		Day 4		Total
x	+	x	+	x	+	32	=	200
				$3x$	+	32	=	200

Emphasize the importance of thinking of 3*x* as a single entity. For those who encounter difficulty with this concept, have them use a piece of paper to cover the variable. The equation would then appear as ■ + 32 = 200, which they should know how to solve. This technique also reinforces the rule that addition or subtraction must be unattached first.

14.10 SOLVING TWO-STEP EQUATIONS

Annette Coppola has a part-time job at the Frozen Fun Yogurt Stand. Last week she worked 4 days. She earned the same amount of money on each of the first 3 days and $32 on the fourth day. Her paycheck was $200. How much did Annette earn on each of the first 3 days?

You can write and solve an equation to find the answer.

EXAMPLE 1 Write and solve an equation to find how much money Annette earned on each of the first 3 days.

1. Write an equation.

THINK: Three times an amount plus $32 is $200. $3x + 32 = 200$

2. Solve: $3x + 32 = 200$

THINK: Unattach first by subtraction, then by division.

$$3x + 32 - 32 = 200 - 32$$
$$3x = 168$$
$$\frac{3x}{3} = \frac{168}{3}$$
$$x = 56$$

3. Check.

$$3x + 32 = 200$$
$$3(56) + 32 = 200$$
$$168 + 32 \stackrel{?}{=} 200$$
$$200 = 200 \quad \text{It checks.}$$

So, Annette earned $56 on each of the first three days.

CHECKPOINT Write the letter of the correct answer.

Solve.

1. $\frac{x}{4} - 7 = 1$ d **a.** −24 **b.** −2 **c.** 11 **d.** 32
2. $3x + 9 = 21$ b **a.** −2 **b.** 4 **c.** 10 **d.** 36

PRACTICE EXERCISES

See Extra Practice, page 461.

Solve and check.

• **1.** $2x + 1 = 7$ 3 • **2.** $3x + 7 = 19$ 4 **3.** $5x + 4 = 34$ 6 **4.** $7x + 2 = 16$ 2

5. $8x + 3 = 11$ 1 **6.** $6x + 2 = 38$ 6 **7.** $7x + 1 = 15$ 2 **8.** $9x + 6 = 15$ 1

9. $4x - 8 = 12$ 5 **10.** $9x - 5 = 13$ 2 **11.** $6x - 3 = 15$ 3 **12.** $3x - 7 = 14$ 7

13. $5x - 13 = 12$ 5 **14.** $8x - 4 = 4$ 1 **15.** $10x - 80 = 100$ 18 **16.** $9x - 8 = 19$ 3

17. $-3x + 2 = 11$ −3 **18.** $-5x - 8 = 2$ −2 **19.** $-4x + 10 = 50$ −10 **20.** $-2x - 15 = 3$ −9

21. $-8x - 6 = 2$ −1 **22.** $-9x + 3 = 3$ 0 **23.** $-12x - 9 = 27$ −3 **24.** $-6x + 14 = 26$ −2

• **25.** $\frac{x}{3} - 4 = 8$ 36 • **26.** $\frac{x}{7} + 5 = 12$ 49 **27.** $\frac{x}{11} - 2 = 8$ 110 **28.** $\frac{x}{13} + 10 = 11$ 13

29. $\frac{x}{9} + 3 = -5$ −72 **30.** $\frac{x}{4} - 7 = -8$ −4 **31.** $\frac{x}{6} + 9 = -1$ −60 **32.** $\frac{x}{5} - 14 = 6$ 100

33. $\frac{x}{6} - 5 = 1$ 36 **34.** $\frac{x}{8} + 7 = 19$ 96 **35.** $\frac{x}{3} - 19 = 1$ 60 **36.** $\frac{x}{5} + 7 = 16$ 45

MIXED REVIEW

Solve.

• **37.** 20% of what number is 17? 85

• **38.** 18 is what percent of 72? 25%

39. What is $37\frac{1}{2}$% of 56? 21

40. 45 is 50% of what number? 90

41. What percent of 36 is 12? $33\frac{1}{3}$%

42. $33\frac{1}{3}$% of 129 is what number? 43

43. 42 is what percent of 210? 20%

44. 2% of what number is 27? 1,350

45. 1,000% of 26.2 is what number? 262

46. 14% of what number is 28? 200

Solve. Write an equation as needed.

• **47.** Jim makes sodas and sundaes at the Frozen Fun Yogurt Stand. Last week the number of sodas he made was 4 fewer than 5 times the number of sundaes. If he made 96 sodas, how many sundaes did he make? 20 sundaes

48. Annette ordered 263 sugar cones for the Frozen Fun Yogurt Stand. The number of sugar cones was 8 more than 3 times the number of waffle cones. How many waffle cones were ordered? 85 waffle cones

49. There are 8 flavors of yogurt at the Frozen Fun Yogurt Stand. If you order a triple scoop cone with no two scoops alike, how many combinations of flavors can you choose from? 56 combinations

50. The Frozen Fun Yogurt Stand had a special last Thursday. For each double scoop cone you bought, you got a single scoop cone free. Raoul served 33 scoops of yogurt on the special. How many free scoops did he serve? 11 free scoops

CHECKPOINT The incorrect choices include common errors students make.

COMMON ERROR

Some students will forget to multiply or divide every term in the equation. For example, in item 1 of Checkpoint they would choose *c* as a result of the following computation.

$$\frac{x}{4} - 7 = 1$$
$$x - 7 = 4$$
$$x = 11$$

Encourage students to cover the variable, and then add or subtract to eliminate the other term.

ASSIGNMENTS

Level 1	Odd 1–27, 37–48
Level 2	Even 2–36, 37–49
Level 3	17–36, 43–50

FOLLOW-UP ACTIVITY

APPLICATION A chirping cricket can be used as a thermometer. Thirty-seven more than one-fourth the number of chirps in a minute is about equal to the air temperature in degrees Fahrenheit, $t = \frac{n}{4} + 37$.

Have students find the temperature for 20, 40, 60, 80, and 100 chirps per minute. (42°, 47°, 52°, 57°, 62°) Have students find a pattern to explain the relationship between an increase in the number of chirps and a rise in temperature. (For every 20 chirps, the temperature rises 5°.) Have them use the pattern to find the temperature at which crickets chirp 120, 140, and 200 times a minute. (67°, 72°, 87°)

SUPPLEMENTARY MATERIALS

TRP Practice, p. 168

TRP Reteaching, p. 105

TRP Transparency 84

OBJECTIVE

- Graph equations in two variables.

TEACHING THE LESSON

WARM-UP Write the following exercises on the chalkboard or an overhead transparency.

Evaluate for $x = -1$, $y = 3$, $z = -5$.

1. $x - y$ (-4)

2. $z - x$ (-4)

3. $x(y - z)$ (-8)

4. $x + y + z$ (-3)

5. $2y - x$ (7)

6. $-3x + z$ (-2)

INTRODUCTION Write: $x + 2 = 8$. Have students solve the equation. Ask if 6 is the only possible answer.

Write: $x + 2 = y$

Discuss how this equation differs from the one above. Ask if 6 is the only possible value for x. Guide students to realize that the value of x depends on the value of y. Therefore, the solution (x, y) must contain two values such as (1, 3).

INSTRUCTION Use Example 1 to reinforce the concept that the value of one variable determines the value of the second variable. Remind students that the solution is expressed as (x, y) and that both values must be used in the check.

The choice of values for x is arbitrary, but encourage students to use small negative integers, zero, and small positive integers to ease computation.

As students complete the (x, y) column in Example 3, ask if they detect a pattern (as x increases by 1, y increases by 2). Explain that the change from one y value to the next remains the same if the change from one x value to the next remains constant.

14.11 GRAPHING EQUATIONS IN TWO VARIABLES

An equation with two variables has an ordered pair of numbers as a solution.

EXAMPLE 1 Find the solution of $x - 3y = 7$ when $x = -2$.

1. Substitute -2 for x.

$$x - 3y = 7$$
$$-2 - 3y = 7$$

2. Solve.

$$-2 + 2 - 3y = 7 + 2$$
$$\frac{3y}{-3} = \frac{9}{-3}$$
$$y = -3$$

3. Check.

THINK: $x = -2, y = -3$

$$x - 3y = 7$$
$$-2 - 3(-3) = 7$$
$$-2 + 9 = 7$$
$$7 = 7$$

So, $(-2, -3)$ is a solution of $x - 3y = 7$.

An equation with two variables usually has many solutions. You can make a **table of values** to show the solutions.

EXAMPLE 2 Make a table of values to find some solutions for the equation $y = 5 - x$.

1. Assign x different values.
2. Find the solutions.
3. Write the ordered pairs.

x	$5 - x$	y	(x, y)
−2	5 − (−2)	7	(−2, 7)
−1	5 − (−1)	6	(−1, 6)
0	5 − (0)	5	(0, 5)
1	5 − (1)	4	(1, 4)
2	5 − (2)	3	(5, 3)

You can draw the graph of an equation with two variables by graphing the ordered pairs and connecting them with a line.

EXAMPLE 3 Graph the equation $y = 2x - 1$.

1. Make a table to find the ordered pairs.

x	$2x - 1$	y	(x, y)
-1	$2(-1) - 1$	-3	$(-1, -3)$
0	$2(0) - 1$	-1	$(0, -1)$
1	$2(1) - 1$	1	$(1, 1)$
2	$2(2) - 1$	3	$(2, 3)$
3	$2(3) - 1$	5	$(3, 5)$

2. Graph each of the ordered pairs. Connect them with a line.

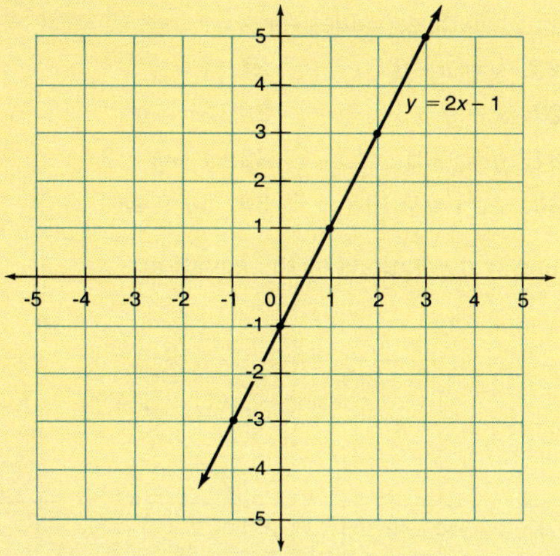

CHECKPOINT

Write the letter of the correct answer.

Find the value of the variable that is unknown.

1. $x + y = 8; x = 3$ c **a.** -3 **b.** 4 **c.** 5 **d.** -5

2. $7 - x = y; x = -2$ a **a.** 9 **b.** -9 **c.** 5 **d.** -5

3. $2x + y = 16; y = 10$ b **a.** -3 **b.** 3 **c.** 5 **d.** 13

4. $y - 3x = 12; y = 6$ d **a.** 6 **b.** -6 **c.** 2 **d.** -2

This can serve as a quick check on the y values and can also make it easy to write more sets of ordered pairs without substituting. Students can use -1, 0, and 1 for x to establish the pattern. If the pattern is correct, the points named by the ordered pairs will lie on a straight line. If one point is off the line, students should check that particular solution.

CHECKPOINT The incorrect choices include common errors students make.

TIME OUT Students should realize that since the sums of 62 and 176 are not great, neither age can be very great. Encourage students to start with the greatest square less than 62 (49) and work from there.

COMMON ERROR

Some students will still make computational errors with integers. For example, in item 4 of Checkpoint, they would choose *c*. Have these students review the rules for using signed numbers in the basic operations.

ASSIGNMENTS

Level 1	Odd 1–29
Level 2	Even 2–36, TO
Level 3	Odd 9–37, 38–40, TO

FOLLOW-UP ACTIVITY

APPLICATION Point out that graphs can be used to solve problems that involve two equations, each with two variables. Write the following problem on the chalkboard.

Two numbers add up to 7. The second number is 5 less than twice the first number. Find the numbers.

Instruct students to write two equations to find the missing numbers ($x + y = 7$ and $2x - 5 = y$). Using -1, 0, and 1 as values for x in both equations, graph each equation on the same set of axes. The solution is where both lines intersect. (4, 3)

SUPPLEMENTARY MATERIALS

TRP Practice, p. 169

TRP Reteaching, p. 106

TRP Lesson Aids, p. 12

TRP Transparency 85

PRACTICE EXERCISES

See Extra Practice, page 461.

Find the solution of the equation when $x = 0$, and then when $y = 0$.

• **1.** $x + y = 13$ 13, 13 **2.** $3 - x = y$ 3, 3 **3.** $5 + y = x$ -5, 5 **4.** $y - x = 9$ 9, -9

• **5.** $2x - 8 = y$ -8, 4 **6.** $5x + 3y = 15$ 5, 3 **7.** $3x - y = 12$ -12, 4 **8.** $-2y + 5x = 10$ -5, 2

9. $6 - 2x = y$ 6,3 **10.** $4 + 2x = y$ 4, -2 **11.** $y - 3x = 6$ 6, -2 **12.** $-4y + x = 8$ -2,8

Determine if the ordered pair is a solution of the equation. Write *yes* or *no*.

•**13.** $x - y = 11$; (2, 9) no **14.** $2 - x = y$; (0, 2) yes **15.** $x + y = 9$; (3, 6) yes

•**16.** $y = 2x - 4$; $(-2, -4)$ no **17.** $x - 3y = 6$; (6, 2) no **18.** $2x + 5y = 20$; (5, 2) yes

19. $x - \frac{3y}{5} = 6$; $(6, -10)$ no **20.** $\frac{3x}{4} + y = 12$; (16, 10) yes **21.** $\frac{2x}{3} - 2y = 6$; (9, 3) no

See page 479 for answers to Exercises 22–37.

Graph the equation.
Make a table of values that has three values for x.

•**22.** $y = x + 2$ •**23.** $y = x - 3$ **24.** $x = y$ **25.** $y = 3x + 1$

26. $x + y = 7$ **27.** $x - y = 5$ **28.** $3x = y$ **29.** $-2x = y$

30. $y = 2x - 1$ **31.** $y = -3x$ **32.** $x - y = 3$ **33.** $x = 2y$

34. $3x - y = 8$ **35.** $2x + y = 10$ **36.** $3y = 4x$ **37.** $x - y = -2$

Check students' answers. Sample answers below.

Use the graph to determine two solutions of the equation.

38.

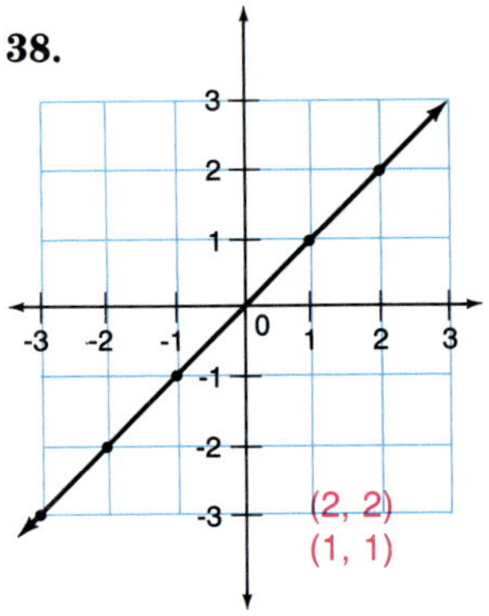

(2, 2)
(1, 1)

39.

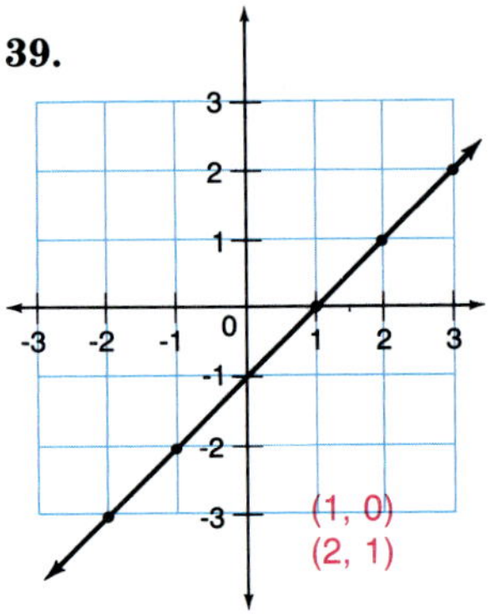

(1, 0)
(2, 1)

40.

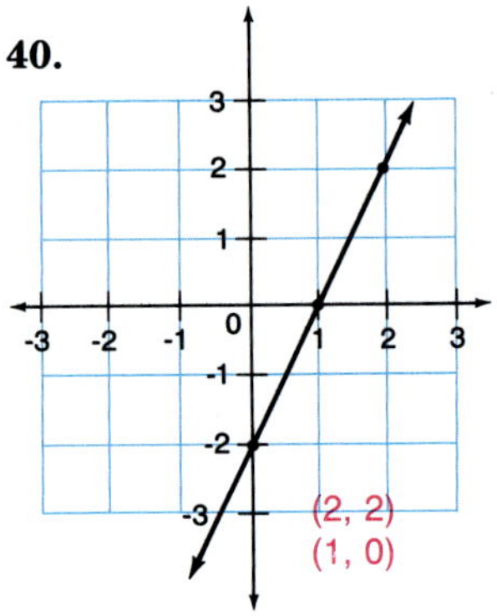

(2, 2)
(1, 0)

TIME OUT If you add the square of Tom's age to Mary's age, the sum is 62. If you add the square of Mary's age to Tom's age, the result is 176. What are their ages? Tom is 7, Mary is 13.

PROBLEM **Solving** APPLICATION

14.12 CONSUMER: ENERGY CONSERVATION

The Johnson family owns a 3-bedroom home that is heated by oil. Their furnace uses an average of 90 gal per month.
A consumer magazine provides a report about a new thermostat that can cut the use of home heating oil by 15%.
The Johnsons are interested in purchasing this new thermostat.

How many gallons of oil will be saved using this new thermostat for 4 months?

1. Write an equation. Use the formula $s = 0.15 \cdot 90 \cdot t$.
   Let s represent the amount of oil saved and t the amount of time.

   THINK: Amount of oil saved is 4 times 15% of 90 gal

   WRITE: $s = 4\ (0.15 \cdot 90)$

2. Solve. $s = 4\ (13.5)$
   $s = 54$

So, using the new thermostat for 4 months will save 54 gal of oil.

FOR DISCUSSION See TE side column.

The Johnson family uses an *average* of 90 gal of oil per month. What does this statement mean?

Solve. Use the formula $s = 0.15 \cdot 90 \cdot t$.

• **1.** How many gallons of oil would be saved in 36 months? 486 gal

2. About how many months would it take to save 400 gal of oil? 30 months

3. About how many months would it take to save 650 gal of oil? 48 months

4. How many gallons of oil would be saved in $4\frac{1}{2}$ years? 729 gal

Suppose oil costs $1.10 per gallon. Solve.

• **5.** How much less money would you pay for oil in 12 months with the new thermostat? $178.20

6. How much money would be saved in 3.5 years with the new thermostat? $623.70

OBJECTIVE

- Solve problems by writing and solving equations.

TEACHING THE LESSON

WARM-UP Write the following exercises on the chalkboard or an overhead transparency.

Solve.

1. $5a = 160$ **2.** $11.6 = 2d$

3. $32.2 = 4.6e$ **4.** $b = 0.25(6 \times 12)$

(**1.** 32 **2.** 5.8 **3.** 7 **4.** 18)

INTRODUCTION Ask students during which months a furnace is likely to use more oil, in which months is it likely to use less oil, and why.

INSTRUCTION Discuss how the problem is translated into an equation. Make sure students understand how each expression is formed. Remind students to apply the order of operations to find the answer.

Problems 1–6 use an equation for conserving oil similar to the completed example. Note that Problems 5 and 6 reflect an increase in the cost of oil.

FOR DISCUSSION Students should explain that an average of 90 gal per month means that the Johnsons use more than 90 gal in some months and less than 90 gal in others.

ASSIGNMENTS

Level 1	1–4
Level 2	1–6
Level 3	1–6

SUPPLEMENTARY MATERIALS

TRP Practice, p. 170

TRP Transparency 85

OBJECTIVES

- Review vocabulary.
- Practice key chapter concepts and skills.

USING THE REVIEW

The Chapter Review is designed to help students prepare for taking the Chapter Test. The first section focuses on vocabulary. It requires that students select words to complete statements. The second section presents practice exercises of key mathematical skills. Under each directive there is a sample exercise with the answer.

Each item of the review is referenced to the page on which the topic is taught in the Pupil's Edition. You may wish to have students refer to these pages to help review any concepts or skills they have not yet mastered.

It is suggested that students work in small-sized heterogeneous cooperative learning groups. Some cooperative learning methods that may be used are as follows:

1. After each student has independently completed the entire Chapter Review, a discussion should follow within each group about the solutions to the practice exercises.
2. The group can complete the entire Chapter Review by working together to discuss the sample exercises and then to answer the practice exercises.

End the lesson with an entire class discussion in which any questions brought up in group discussions are presented and answered.

CHAPTER REVIEW

Vocabulary Choose the letter of the word that completes the statement.

1. An algebraic expression uses a ■ for an unknown number. [400] c
   a. question mark **b.** blank **c.** variable **d.** dash

2. A mathematical sentence that uses $<$ to show the relationship between two expressions is an ■. [406] a
   a. inequality **b.** equation **c.** equality **d.** graph

3. To solve the equation $5x - 3 = 7$, you would ■ first. [416] a
   a. add **b.** subtract **c.** multiply **d.** divide

Skills Simplify. [398]

$(18 \div 2) - 4$ *5*

4. $(3 + 5)4$ 32

5. $(12 - 2)^2 \div 5$ 20

6. $48 - (7 - 1) + 2$ 44

7. $(15 + 2 \cdot 6) - 3^2$ 18

8. $\frac{12}{18 - 6}$ 1

Write an algebraic expression for the word phrase. [400]

16 fewer than a number *$n - 16$*

9. 3 more than twice a number $2n + 3$

10. 17 less than a number $n - 17$

11. A number increased by 5 $n + 5$

12. Sue's age 7 years ago $n - 7$

13. Twice a number, decreased by 83 $2n - 83$

Evaluate for $x = -2, y = -5$. [402]

$3x - y$ *-1*

14. $x - y$ 3

15. xy 10

16. $\frac{2x}{y - 1}$ $\frac{2}{3}$

17. $\frac{y + 5}{-8x}$ 0

Write a mathematical sentence. [406]

2 more than $\frac{1}{4}$ of a number is 8. *$\frac{1}{4}n + 2 = 8$*

18. 6 more than a number is 17. $n + 6 = 17$

19. 3 times a number decreased by 4 is 8. $3n - 4 = 8$

20. A number divided by 7 is less than 35. $\frac{n}{7} < 35$

Solve. [412, 414, 416]

$2x - 5 = 3$ *4*

21. $x - 13 = 25$ 38

22. $x + 19 = 21$ 2

23. $x - 5 = 93$ 98

24. $-12x = -24$ 2

25. $4x + 3 = 15$ 3

26. $-3x - 4 = 14$ -6

27. $\frac{x}{5} + 2 = -6$ -40

Chapter Test

Simplify.

1. $8 - 4 + 3$ 7
2. $12 \div 6 \cdot 2$ 4
3. $8 \cdot 3 \div 6 - 2^2$ 0
4. $9 - 5 \cdot 2 + 1$ 0
5. $(18 - 4)2 - 5$ 23
6. $\frac{6(20-4)}{3}$ 32

Write an algebraic expression.

7. The sum of a number and 17 $n + 17$
8. 3 more than 6 times a number $3 + 6n$

Evaluate for $x = -1, y = 2, z = -3$.

9. $x + y + z$ -2
10. $x - y - z$ 0
11. $\frac{x+z}{-y}$ 2
12. $\frac{2z - 3y}{4x}$ 3

Write a mathematical sentence.

13. 8 fewer than a number is 41. $n - 8 = 41$
14. 3 times a number is less than 34. $3n < 34$

Graph the solution. See page 479 for answers to Exercises 15–18.

15. $x < -2$
16. $x = 4$
17. $x \geq -1$
18. $x \leq 6$

Solve.

19. $x - 6 = 12$ 18
20. $x + 13 = -4$ -17
21. $9 + x = 8$ -1
22. $-7 + x = -3$ 4
23. $-7x = 35$ -5
24. $\frac{x}{-2} = 16$ -32
25. $9x = -81$ -9
26. $\frac{x}{-5} = -8$ 40
27. $7x + 4 = 32$ 4
28. $-2x - 9 = 11$ -10
29. $\frac{x}{4} - 13 = -15$ -8
30. $\frac{x}{-11} + 8 = 7$ 11

Find the solution of the equation when $x = -3$.

31. $x + y = 7$ 10
32. $y = -2x + 4$ 10
33. $3x - 4y = -9$ 0
34. $-4x + 3y = 27$ 5

See page 479 for answers to Exercises 35–36.

Graph each equation. Make a table that has three values for x.

35. $y = x - 5$
36. $y = 2x - 1$

Solve. Use the formula $s = 0.20 \cdot 100 \cdot t$.
Let s stand for the amount of oil saved and t for a period of time.

37. The Garcia family uses an average of 100 gal of fuel oil per month. A new thermostat will cut the amount of oil used by 20%. How much oil will be saved by using the new thermostat for 3 months? 60 gal
38. The Garcias hope that the new thermostat will allow them to save at least 200 gal of oil in a year. Is this possible? Use the formula above to find out. Yes. 240 gal could be saved in 1 year.

Objective

- Evaluate achievement of the chapter objectives.

Using the Test

The Chapter Test may be used as a posttest to evaluate the achievement of all students. However, you may wish to use the Chapter Posttest offered in the Teacher's Resource Package or to design your own chapter test. If this page is not used as a test, you may wish to assign it as additional review or practice.

The test items are correlated to the chapter objectives in the table below.

Chapter objectives	Test items
A. Simplify numerical expressions.	1–6
B. Evaluate algebraic expressions.	9–12
C. Write algebraic expressions and sentences.	7–8, 13–14
D. Graph the solution to an equation or inequality.	15–18
E. Solve equations.	19–34
F. Graph solutions to equations with two variables.	35–36
G. Apply computational skills to real-life situations.	37–38

Supplementary Materials

TRP Ch. 14 Posttest Form A, pp. 1–2

TRP Ch. 14 Posttest Form B, pp. 1–2

TRP Final Test, pp. 1–8

OBJECTIVES

- Simplify expressions by combining like terms.
- Solve equations by combining like terms.

USING THE PAGE

Point out that combining like terms can often simplify complicated expressions. Write the following expressions on the chalkboard:

- $5c + 4e - 9d + 6c$
  $(11c + 4e - 9d)$
- $14y - 7z - 2z - 6y$
  $(8y - 5z)$
- $7a^2 + 6a - 5a^2 + 2a$
  $(2a^2 + 8a)$

For each expression, call forward a volunteer to identify the like terms and then combine them to simplify the expression.

Before assigning the exercises as independent work, guide students in simplifying an expression that contains negative integers such as Exercise 15.

Emphasize that solving equations by combining like terms is a straightforward, two-step process. After students master the first step of combining like terms, they can apply their knowledge of solving equations in the second step.

After a little practice, many students enjoy combining like terms to simplify expressions and to solve equations. You may want to extend the exercise set by having students solve equations of the type:

- $2x + x + 2 = 11$ $(x = 3)$
- $6x - 4 - 4x = 12$ $(x = 8)$

ENRICHMENT COMBINING LIKE TERMS

In the expression $5x + 7a + 4x$, $5x$ and $4x$ are **like terms** because they have the same variable, x. Here are some other expressions with like terms.

$4a + 5b - 2a$	$8x^2 - 3x - 4x^2$	$6 + 3b - 2$	$7c + 5d^2 - 3c + 8d^2$
Like terms	**Like terms**	**Like terms**	**Like terms**

You can simplify some expressions by combining like terms.

Example Simplify: $-5b + 7a - 7b + 12a$

1. Identify the terms. $-5b + 7a - 7b + 12a$
2. Combine like terms. $-5b - 7b + 7a + 12a$

THINK: Group like terms together. $-12b + 19a$

You can solve some equations by combining like terms.

Example Solve: $7a - 4a = 24$. Check.

1. Combine like terms. $7a - 4a = 24$
   $3a = 24$
2. Solve for a. $\frac{3a}{3} = \frac{24}{3}$
   $a = 8$
3. Check. $7(8) - 4(8) \stackrel{?}{=} 24$
   $56 - 32 = 24$ It checks.

Simplify by combining like terms. 9. $9c + 6$ 10. $9d - 2$ 11. $10y - 4z$ 12. $5e + 4f$

1. $6x + 5x$ $11x$	**2.** $9a - 6a$ $3a$	**3.** $15c + 12c$ $27c$	**4.** $18d - 9d$ $9d$
5. $3a + a + 8a$ $12a$	**6.** $9r - 2r - 3r$ $4r$	**7.** $9n + 5n - 6n$ $8n$	**8.** $15h + 6h - 8h$ $13h$
9. $7c + 6 + 2c$	**10.** $8d - 2 + d$	**11.** $7y + 3y - 4z$	**12.** $12e - 7e + 4f$
13. $-y^2 + 7y^2 - 2y^2$ $4y^2$	**14.** $-6ab + 6ab$ 0	**15.** $17rs - (-6rs)$ $23rs$	**16.** $8x^2 - 6x^2 + 5x^2$ $7x^2$

Solve by combining like terms. 17. $x = 2$ 18. $b = 3$ 19. $k = -3$ 20. $m = 5$ 21. $d = 3$ 22. $a = -3$ 23. $c = 5$ 24. $y = 4$

17. $3x + 2x = 10$	**18.** $4b + 3b = 21$	**19.** $7k + 2k = -27$	**20.** $9m + m = 50$
21. $12d + 3d = 45$	**22.** $25a + 5a = -90$	**23.** $6c - 2c = 20$	**24.** $9y - 5y = 16$
25. $z - 8z = 28$ $z = -4$	**26.** $-11t + 2t = 54$ $t = -6$	**27.** $20r - 5r = 45$ $r = 3$	**28.** $30x - 5x = 100$ $x = 4$

CALCULATOR DIFFERENCE BETWEEN CALCULATORS

Do you have a scientific calculator or a nonscientific calculator? Enter the following example to find out.

Enter: [3] [+] [4] [×] [5] [=]

A scientific calculator uses the rules for the order of operations. It gives the correct answer of 23.
A nonscientific calculator performs the operations in the order in which they are entered. It gives an incorrect answer of 35.

Compute: $38 + 91 \div 13$

Scientific Calculator
Enter the numbers and the operation symbols in the order given.

Enter: [3] [8] [+] [9] [1] [÷] [1] [3] [=]

Display: [45.]

Nonscientific Calculator
Find the quotient $91 \div 13$ first. then add 38.

Enter: [9] [1] [÷] [1] [3] [=] [+] [3] [8] [=]

Display: [45.]

Evaluate the expression $7.8 + 1.5n$ when $n = 6$.

Scientific Calculator
Mentally substitute 6 for n. Enter the numbers and operation symbols in the order given.

Enter: [7] [.] [8] [+] [1] [.] [5] [×] [6] [=]

Display: [16.8]

Nonscientific Calculator
Mentally substitute 6 for n.
Multiply 1.5×6 first.
Then add 7.8.

Enter: [1] [.] [5] [×] [6] [=] [+] [7] [.] [8] [=]

Display: [16.8]

Compute. Use a calculator.

1. $29 + 6 \times 53$ 347 **2.** $37 + 54 \div 6$ 46 **3.** $360 + 19 \times 82$ 1,918
4. $76 + 851 \div 37$ 99 **5.** $4.7 + 3.6 \times 4.5$ 20.9 **6.** $8.4 + 31.9 \div 29$ 9.5
7. $0.28 + 0.32 \times 140$ 45.08 **8.** $0.4 + 58.5 \div 6.5$ 9.4 **9.** $6.29 + 1.9 \times 2.7$ 11.42

Evaluate each expression for the given value of n. Use a calculator.

10. $14 + 8n; n = 12$ 110 **11.** $49 + n \div 2; n = 68$ 83 **12.** $16 + 25n; n = 11$ 291
13. $17 + n \div 14; n = 98$ 24 **14.** $3.6 + 39n; n = 2.1$ 85.5 **15.** $0.6 + n \div 8; n = 5$ 1.225
16. $17 + 12 \div n; n = 2$ 23 **17.** $22 + \frac{n}{19}; n = 152$ 30 **18.** $7 + \frac{396}{n}; n = 36$ 18
19. $12.8 + 7n; n = 4.2$ 42.2 **20.** $3.6 + n \div 1.5; n = 6$ 7.6 **21.** $20 + \frac{196}{n}; n = 7$ 48

OBJECTIVE

- Identify the differences between scientific and nonscientific calculators.

USING THE PAGE

The examples and exercises in this lesson are all expressions of the form $a + b \times c$ or $a + b \div c$. Since addition is commutative, students who have nonscientific calculators simply perform the multiplication or division first, then add.

With nonscientific calculators, the memory feature can be used with expressions such as those shown in the lesson as well as with expressions involving subtraction, which is not commutative. The following example illustrates this.

Compute: $12 - 5 \times 2$

1. Use the **store** (**STO** or **M+**) key to store the product 5×2 in the memory.
2. To subtract this product from 12, use the **memory recall** (**MR** or **RCL**) key.

Enter: [5] [X] [2] [=] [STO] [1] [2] [-] [MR] [=]

Display: [2]

Many types of calculators are on the market, and the features vary. Encourage students to read the instructions that came with their instruments.

Scientific calculators also perform the four basic operations with negative numbers. To enter a negative number, you enter it without the sign. Then you press the **change of sign (+/−)** key. For example, to multiply $(-9) \times 4$, you enter [9] [+/−] [X] [4] [=]. The correct answer, −36, will be displayed.

To multiply $(-9) \times 4$ on a nonscientific calculator, you enter the numbers without the signs: [9] [X] [4] [=]. The number 36 will be displayed, but you must determine the correct sign.

EXTRA PRACTICE

CHAPTER 1, PAGES 4–5

1. hundreds; 200 **2.** ten millions; 30,000,000 **3.** thousands; 4,000 **4.** hundred billions; 500,000,000,000 **5.** ten thousands 40,000

Find the place and the value of the underlined digit.

1. 221 **2.** 935,456,820 **3.** 84,618 **4.** 508,912,631,008

5. 743,117 **6.** 6,284,122 millions; 6,000,000 **7.** 286,193 tens; 90 **8.** 592,816,134,726 ten billions; 90,000,000,000

Write the number.

9. 491 thousand, 224 491,224

10. 93 billion, 641 93,000,000,641

11. 6 million, 803 6,000,803

12. 825 million, 187 825,000,187

13. Three hundred five thousand, nine hundred forty-three 305,943

14. Twenty-three million, seventy-five thousand, eight hundred twenty 23,075,820

15. Sixty-two billion, two hundred four million, forty-six 62,204,000,046

16. One hundred sixteen billion, eighty-six million, thirty-nine 116,086,000,039

CHAPTER 1, PAGES 6–7

1. tenths; 0.6 **2.** ten-thousandths; 0.0002 **3.** ten-thousandths; 0.0004 **4.** hundredths; 0.01 **5.** hundred-thousandths; 0.00009 **6.** thousandths; 0.003 **7.** millionths; 0.000001 **8.** thousandths; 0.009

Find the place and the value of the underlined digit.

1. 62.6 **2.** 25.061284 **3.** 0.97845 **4.** 628.314

5. 4.387591 **6.** 78.20362 **7.** 29.0402613 **8.** 55.699

Write the number.

9. 8 and 42 hundredths 8.42

10. 2,624 ten thousandths 0.2624

11. 6 hundred and 13 thousandths 600.013

12. 766 millionths 0.000766

CHAPTER 1, PAGES 8–9

Compare. Write >, <, or =.

1. 18.0023 ● 18.023 < **2.** 8,392 ● 8,382 > **3.** 0.8003 ● 0.80030 = **4.** 2.8734 ● 2.87343 <

5. 650,412,731.662 ● 650,412,731.652 >

6. 7.02386 ● 7.02836 <

7. 42,861.73 ● 42,861.23 >

8. 738.002 ● 737.111 >

Arrange in order from greatest to least.

9. 104; 73.52; 24.863; 17.6 **10.** 54,183; 541.8; 60.18; 54.183 **11.** 409; 93.04; 43.09; 30.49 **12.** 0.8302; 0.382; 0.2803; 0.0382

9. 24.863; 73.52; 104; 17.6

10. 54.183; 541.8; 54,183; 60.18

11. 43.09; 409; 93.04; 30.49

12. 0.0382; 0.382; 0.8302; 0.2803

13. 92; 430; 4,329; 29,034; 409,320 409,320; 29,034; 4,329; 430; 92

14. 0.417; 0.41; 70.14; 0.4 70.14; 0.417; 0.41; 0.4

CHAPTER 1, PAGES 14–15

Solve, if possible. If not, identify the needed information.

1. In John Adams High School there are 2,334 students. 1,620 students come to school by bus. The rest walk or ride bikes. How many students walk to school?
Need to know how many ride bikes.

2. Cal's odometer reads 1,920.8 mi when he begins his road trip. After the first day, it reads 2,295.0 mi. After the second day, it reads 2,801.3 mi. How far did Cal drive the second day? 506.3 mi

3. According to a survey, 861 people entered the park on Saturday. 222 picnicked and 149 used the 19 tennis courts. The rest played baseball. How many played baseball? 490

4. The camp cook at Lake Wakega knows he can make 50 pancakes out of 1 lb of pancake mix. He has 5 lb of pancake mix. Can he feed 204 campers? Need to know how many pancakes each camper eats.

CHAPTER 1, PAGES 16–17

Round to the nearest hundred and to the nearest thousand.

1. 5,642
5,600; 6,000
2. 3,299
3,300; 3,000
3. 48,201
48,200; 48,000
4. 71,045
71,000; 71,000
5. 756,892
756,900; 757,000
6. 201,638
201,600; 202,000
7. 231,575
231,600; 232,000
8. 615,123
615,100; 615,000

Round to the nearest tenth and to the nearest hundredth.

9. 18.103
18.1; 18.10
10. 10.305
10.3; 10.31
11. 33.4444
33.4; 33.44
12. 19.555
19.6; 19.56
13. 0.092
0.1; 0.09
14. 0.0167
0.0; 0.02
15. 1.567
1.6; 1.57
16. 7.00346
7.0; 7.00

CHAPTER 1, PAGES 18–19

Write in exponent form.

1. 8 squared 8^2
2. 9 cubed 9^3
3. 15 to the 4th power 15^4
4. 26 cubed 26^3
5. 100 squared 100^2
6. 10 to the 12th power 10^{12}
7. $7 \times 7 \times 7 \times 7 \times 7$ 7^5
8. $11 \times 11 \times 11$ 11^3
9. $2 \times 2 \times 2 \times 2 \times 2$ 2^5

Write as a product.

10. 3^4 81
11. 4^2 16
12. 25^0 1
13. 7^2 49
14. 2^3 8
15. 9^0 1

Write in exponent form.

16. 1,000 10^3
17. 100,000 10^5
18. 1,000,000 10^6
19. 10,000,000,000 10^{10}

Write as a product.

20. 1^7 1
21. 10^9 1,000,000,000
22. 1^{12} 1
23. 10^2 100
24. 10^6 1,000,000
25. 10^0 1

427

CHAPTER 2, PAGES 28–29

Estimate by rounding.

1. 482 + 27 530	**2.** 390 +226 600	**3.** 5,970 +4,001 10,000	**4.** 32,031 551 +56,264 90,600	**5.** 1,609 6,312 +2,049 10,000
6. 54 −27 20	**7.** 795 −362 400	**8.** 3,804 − 768 3,200	**9.** 4,809 −1,245 4,000	**10.** 26,274 −16,108 10,000
11. 6,483 + 612 6,600	**12.** 2,096 −1,245 1,000	**13.** 72,546 −18,609 50,000	**14.** 91,111 + 8,867 99,000	**15.** 76,452 2,367 +24,786 102,000

Estimate by using front-end estimation.

16. 342 +874 1,100	**17.** 2,590 28 +1,078 3,000	**18.** 3,619 +4,456 7,000	**19.** 343,332 141,608 + 9,009 400,000	**20.** 59,401 +34,822 80,000
21. 73,108 −18,945 60,000	**22.** 9,345 +5,777 14,000	**23.** 13,367 +91,862 100,000	**24.** 89,806 − 5,712 80,000	**25.** 319,450 74,433 +822,101 1,100,000
26. 567,981 −390,417 200,000	**27.** 39,560 −11,053 20,000	**28.** 3,815 − 762 3,000	**29.** 426 −193 300	**30.** 15,687 − 9,200 10,000

CHAPTER 2, PAGES 30–31

Add.

1. 166 +555 721	**2.** 892 +439 1,331	**3.** 458 +675 1,133	**4.** 3,689 + 796 4,485	**5.** 7,043 +8,907 15,950
6. 44,036 + 3,016 47,052	**7.** 9,933 +2,448 12,381	**8.** 534,856 +760,455 1,295,311	**9.** 745,912 +312,276 1,058,188	**10.** 949,624 + 62,083 1,011,707
11. 469 10,823 + 7,118 18,410	**12.** 16,127 528 123,045 + 8,003 147,703	**13.** 412,065 7,783 26,655 +607,027 1,053,530	**14.** 6,817 19,341 943 + 7,726 34,827	**15.** 23,747 178,019 6,913 + 666 209,345
16. 56,453 18,352 26,308 +31,325 132,438	**17.** 21,912 66,909 34,865 +37,809 161,495	**18.** 308,416 403,769 293,008 +185,444 1,190,637	**19.** 125,645 912,863 377,715 +714,543 2,130,766	**20.** 647,792 351,115 662,258 +523,087 2,184,252

21. 708 + 93 + 215 1,016 **22.** 87 + 1,257 + 38,409 39,753 **23.** 834,906 + 2,176 + 45,573 882,655

CHAPTER 2, PAGES 32–34

Subtract. Check your answer.

1. 460 − 183 = 277	**2.** 851 − 438 = 413	**3.** 12,804 − 8,691 = 4,113	**4.** 4,283 − 2,624 = 1,659	**5.** 2,820 − 543 = 2,227
6. 69,005 − 40,326 = 28,679	**7.** 36,112 − 18,813 = 17,299	**8.** 281,113 − 79,465 = 201,648	**9.** 110,311 − 106,413 = 3,898	**10.** 683,117 − 229,032 = 454,085
11. 45,082 − 8,762 = 36,320	**12.** 19,411 − 14,708 = 4,703	**13.** 612,734 − 589,037 = 23,697	**14.** 810,056 − 311,086 = 498,970	**15.** 300,000 − 134,082 = 165,918
16. 789,135 − 9,456 = 779,679	**17.** 70,003 − 874 = 69,129	**18.** 230,049 − 209,876 = 20,173	**19.** 5,123,452 − 358,118 = 4,765,334	**20.** 6,019,012 − 5,923,714 = 95,298

First estimate by rounding. Then find the exact answer.

21. 87 − 52 40; 35
22. 683 − 179 500; 504
23. 7,821 − 1,684 6,000; 6,137
24. 18,891 − 12,963 10,000; 5,928
25. 402,264 − 295,611 100,000; 106,653
26. 892,126 − 602,820 300,000; 289,306

CHAPTER 2, PAGES 36–37

Multiply.

1. 30 × 10 300
2. 4,000 × 80 320,000
3. 300 × 200 60,000
4. 268 × 1,000 268,000
5. 640 × 10 6,400
6. 3,260 × 1,000 3,260,000
7. 2,000 × 500 1,000,000
8. 6,000 × 8,000 48,000,000

Divide.

9. 90,000 ÷ 30 3,000
10. 35,000 ÷ 700 50
11. 8,000,000 ÷ 2,000 4,000
12. 30,000 ÷ 300 100
13. 429,000 ÷ 1,000 429
14. 1,612,000 ÷ 100 16,120

CHAPTER 2, PAGES 38–39

Estimate by rounding.

1. 72 × 8 = 560	**2.** 631 × 5 = 3,000	**3.** 382 × 9 = 3,600	**4.** 6,220 × 4 = 24,000	**5.** 1,892 × 7 = 14,000
6. 56 × 46 = 3,000	**7.** 29 × 73 = 2,100	**8.** 118 × 93 = 9,000	**9.** 745 × 81 = 56,000	**10.** 3,085 × 12 = 30,000
11. 4,780 × 57 = 300,000	**12.** 3,089 × 33 = 90,000	**13.** 22,135 × 68 = 1,400,000	**14.** 58,036 × 88 = 5,400,000	**15.** 73,586 × 52 = 3,500,000
16. 4)389 100	**17.** 7)664 100	**18.** 5)4,377 800	**19.** 3)8,672 3,000	**20.** 8)4,469 500
21. 9)86,800 10,000	**22.** 6)57,060 10,000	**23.** 49)964 20	**24.** 33)286 10	**25.** 58)5,710 100

Estimate by rounding. Which answer is reasonable?

1. 3 × 787 a **a.** 2,361 **b.** 231 **c.** 1,492

2. 24 × 96 c **a.** 225 **b.** 9,820 **c.** 2,304

3. 78 × 3,120 a **a.** 243,360 **b.** 120,000 **c.** 24,350

CHAPTER 2, PAGES 40–41

Multiply.

4. 132 × 5 = 660	**5.** 612 × 7 = 4,284	**6.** 2,391 × 6 = 14,346	**7.** 4,748 × 3 = 14,244	**8.** 1,955 × 9 = 17,595
9. 62 × 51 = 3,162	**10.** 392 × 12 = 4,704	**11.** 826 × 23 = 18,998	**12.** 5,431 × 89 = 483,359	**13.** 7,622 × 43 = 327,746
14. 9 × 18 162	**15.** 2 × 75 150	**16.** 7 × 63 441	**17.** 4 × 36 144	**18.** 8 × 781 6,248

First estimate using compatible numbers. Then find the exact answer.

19. 8,920 × 3	**20.** 2,384 × 6	**21.** 27 × 61	**22.** 114 × 77	**23.** 439 × 56
27,000; 26,760	12,000; 14,304	1,800; 1,647	8,000; 8,778	24,000; 24,584
24. 42 × 43	**25.** 513 × 60	**26.** 285 × 52	**27.** 9,130 × 19	**28.** 6,120 × 103
1,600; 1,806	30,000; 30,780	15,000; 14,820	180,000; 173,470	600,000; 630,360

CHAPTER 2, PAGES 42–43

1. 295,800 **2.** 417,956 **3.** 2,832,720 **4.** 382,872 **5.** 1,549,360 **6.** 3,242,008 **7.** 873,441 **8.** 1,001,300 **9.** 7,610,853 **10.** 127,530 **11.** 3,823,728 **12.** 4,260,480 **13.** 113,364 **14.** 883,782 **15.** 185,472 **16.** 6,321,414

Multiply.

1. 408 × 725	**2.** 826 × 506	**3.** 6,105 × 464	**4.** 516 × 742
5. 3,620 × 428	**6.** 6,094 × 532	**7.** 907 × 963	**8.** 2,635 × 380
9. 8,003 × 951	**10.** 218 × 585	**11.** 4,306 × 888	**12.** 6,720 × 634
13. 268 × 423	**14.** 7,962 × 111	**15.** 207 × 896	**16.** 10,082 × 627

CHAPTER 2, PAGES 46–47

6. 101 R2 **7.** 1,355 R2 **8.** 154 R2 **9.** 571 R5 **10.** 185 R1 **11.** 2,462 **12.** 2,396 R3 **13.** 3,333 R5 **14.** 23,302 R7 **15.** 22,333 R8 **16.** 153 **17.** 243 **18.** 966 R2 **19.** 4,120 **20.** 8,797 R1 **21.** 14,200 R3 **22.** 7,448 **23.** 71,049 **24.** 171,958 R2 **25.** 71,573 R2

Divide.

1. 5)495 99	**2.** 8)247 30 R7	**3.** 7)840 120	**4.** 4)7,592 1,898	**5.** 2)6,441 3,220 R1
6. 6)608	**7.** 3)4,067	**8.** 8)1,234	**9.** 7)4,002	**10.** 6)1,111
11. 5)12,310	**12.** 9)21,567	**13.** 6)20,003	**14.** 8)186,423	**15.** 9)201,005
16. 612 ÷ 4	**17.** 1,701 ÷ 7	**18.** 8,696 ÷ 9	**19.** 20,600 ÷ 5	**20.** 52,783 ÷ 6
21. 71,003 ÷ 5	**22.** 22,344 ÷ 3	**23.** 142,098 ÷ 2	**24.** 687,834 ÷ 4	**25.** 501,013 ÷ 7

CHAPTER 2, PAGES 48–49 **16.** 20; 18 R29 **17.** 40; 41 R26 **18.** 30; 37 R3 **19.** 800; 772 R60

Divide.

1. 26)364 14	**2.** 13)793 61	**3.** 52)936 18	**4.** 81)891 11	**5.** 34)820 24 R4
6. 77)462 6	**7.** 48)727 15 R7	**8.** 22)6,527 296 R15	**9.** 63)2,961 47	**10.** 39)4,875 125
11. 57)7,296 128	**12.** 98)8,140 83 R6	**13.** 40)33,800 845	**14.** 91)59,332 652	**15.** 42)84,378 2,009

First estimate using compatible numbers. Then find the exact answer.

16. 821 ÷ 44	**17.** 3,429 ÷ 83	**18.** 928 ÷ 25	**19.** 47,924 ÷ 62
20. 367 ÷ 51 7; 7 R10	**21.** 6,272 ÷ 12 600; 522 R8	**22.** 2,098 ÷ 41 50; 51 R7	**23.** 78,429 ÷ 79 1,000; 992 R61

CHAPTER 2, PAGES 50–51

Divide.

1. 1,320 ÷ 258 5 R30 **2.** 7,500 ÷ 314 23 R278 **3.** 61,679 ÷ 593 104 R7 **4.** 4,561 ÷ 291 15 R196

5. 41,535 ÷ 585 71 **6.** 970,062 ÷ 642 1,511 **7.** 154,921 ÷ 701 221

8. 244,089 ÷ 312 782 R105 **9.** 608,145 ÷ 450 1,351 R195 **10.** 824,567 ÷ 809 1,019 R196

11. 378)759,636 2,009 R234	**12.** 175)67,200 384	**13.** 147)802,428 5,458 R102	**14.** 198)402 2 R6
15. 174)29,600 170 R20	**16.** 305)6,557 21 R152	**17.** 183)11,520 62 R174	**18.** 983)99,156 100 R856
19. 211)5,087 24 R23	**20.** 437)40,970 93 R329	**21.** 256)302,145 1,180 R65	**22.** 710)2,304,568 3,245 R618

CHAPTER 2, PAGES 52–53

Solve.

1. Jennifer has saved $73.00. On lay-away, she can buy a pair of roller skates for $7.00 a week for 12 weeks. Does she have enough money to buy the skates? If not, how much more does she need?
No. She needs $11 more.

2. It costs the Seabrook Inn $3.50, on the average, to serve a meal. How much will it cost to serve 75 people? If they charge an average of $6.25 a meal, how much profit will they make? $262.50 to serve 75 people. $206.25 profit

3. Derrick needs to buy 4 shock absorbers and 4 tires. The shock absorbers each cost $17.50. The tires each cost $79.00. How much money does Derrick need to buy the shock absorbers and the tires? $386.00

4. Marta wants summer clothes that cost $252.00. She earns $4.50 per hour baby-sitting. How many hours must she baby-sit to buy the clothes? If she baby-sits 7 hours per week, how many weeks must she work to earn enough money? 56 hours; 8 weeks

CHAPTER 3, PAGES 62–63

Add.

1. 18.302 + 1.9 20.202	**2.** 24.0005 + 66.1 90.1005	**3.** 5.418 + 2.2334 7.6514	**4.** 0.76 + 1.69 2.45	**5.** 543.00 + 46.12 589.12
6. 513.04 + 6.38 + 0.93 520.35	**7.** 2.6 + 119.73 + 88.1 210.43	**8.** 734.61 + 3.289 + 76.43 814.329	**9.** 17.561 + 6.092 + 0.836 24.489	**10.** $4.83 + 2.99 + 0.57 $8.39

First estimate by rounding, then find the exact answer.

11. 26.2 + 17.38 50; 43.58	**12.** 404.89 + 316.27 700; 721.16	**13.** 0.8318 + 0.3634 1.2; 1.1952	**14.** 0.071 + 0.7329 0.8; 0.8039	**15.** 1.007 + 5.8261 7; 6.8331

Solve for *n*.

16. $6.12 + 21.4 + 8 = n$ 35.52

17. $0.563 + 0.91 = n$ 1.473

18. $48.2 + 7 = n$ 55.2

CHAPTER 3, PAGES 64–65

Subtract.

1. 8.2 − 6.3 1.9	**2.** 24.7 − 18.3 6.4	**3.** 7.639 − 1.217 6.422	**4.** 12.008 − 6.51 5.498	**5.** 89 − 0.673 88.327
6. 281.063 − 41.026 240.037	**7.** 3 − 0.57 2.43	**8.** 7.005 − 1.217 5.788	**9.** 0.9382 − 0.41 0.5282	**10.** 68 − 33.064 34.936

First estimate by rounding, then find the exact answer.

11. $601.00 − 2.89 $597; $598.11

12. 24.024 − 11.82 10; 12.204

13. 0.38 − 0.19 0.2; 0.19

Solve for *n*.

14. $66.814 - 53.107 = n$ 13.707

15. $0.4028 - n = 0.3241$ 0.0787

16. $8.603 - n = 5.94$ 2.663

17. $0.923 - 0.058 = n$ 0.865

18. $421.9 - 2 = n$ 419.9

CHAPTER 3, PAGES 68–69

Multiply.

1. 8.4 × 10 84 **2.** 0.89 × 10 8.9 **3.** 62.7 × 100 6,270 **4.** 5.3 × 100 530

5. 73.8 × 1,000 73,800 **6.** 5.36 × 1,000 5,360 **7.** 0.22 × 1,000 220 **8.** 25.3 × 10,000 253,000

Divide.

9. 86.4 ÷ 10 8.64 **10.** 5.3 ÷ 10 0.53 **11.** 320 ÷ 100 3.2 **12.** 0.03 ÷ 100 0.0003

13. 90.3 ÷ 1,000 0.0903 **14.** 22 ÷ 1,000 0.022 **15.** 56.9 ÷ 10,000 0.00569 **16.** 3.42 ÷ 10,000 0.000342

CHAPTER 3, PAGES 70–71

Solve the problem by using the guess-and-check method.

1. Tom spent \$2.72 on 14 stamps. He bought 22¢ stamps and 4¢ stamps. How many of each kind of stamp did Tom buy? twelve 22¢ stamps, two 4¢ stamps

2. Mrs. Brown bought clothing for her children. She bought Sean 4 more items than Mike. She bought 18 items in all. How many things did she buy for Mike? 7 things for Mike

3. Marie bought milk and juice at the store. Her change was \$1.87 in quarters, dimes, and pennies. She received 10 coins. How many were dimes? 1

4. Sue and Jan ran a total of 21 km. Sue ran 5 more kilometers than Jan. How many kilometers did each girl run? Sue ran 13 km; Jan ran 8 km.

CHAPTER 3, PAGES 74–75

Multiply.

1. $\begin{array}{r} 38.5 \\ \times\ 16 \\ \hline 616 \end{array}$ **2.** $\begin{array}{r} 79.32 \\ \times\ 2.08 \\ \hline 164.9856 \end{array}$ **3.** $\begin{array}{r} 1.09 \\ \times 0.86 \\ \hline 0.9374 \end{array}$ **4.** $\begin{array}{r} 6.044 \\ \times 0.432 \\ \hline 2.611008 \end{array}$ **5.** $\begin{array}{r} 95.19 \\ \times 0.002 \\ \hline 0.19038 \end{array}$

6. $\begin{array}{r} 10.003 \\ \times\ 7.2 \\ \hline 72.0216 \end{array}$ **7.** $\begin{array}{r} 594 \\ \times\ 0.3 \\ \hline 178.2 \end{array}$ **8.** $\begin{array}{r} 0.471 \\ \times 0.002 \\ \hline 0.000942 \end{array}$ **9.** $\begin{array}{r} 17.4 \\ \times\ 8.8 \\ \hline 153.12 \end{array}$ **10.** $\begin{array}{r} 6.863 \\ \times 0.027 \\ \hline 0.185301 \end{array}$

First estimate by rounding. Then find the exact answer.

11. $\begin{array}{r} 898 \\ \times\ 0.5 \\ \hline \end{array}$ 450; 449 **12.** $\begin{array}{r} 4.03 \\ \times 0.92 \\ \hline \end{array}$ 3.6; 3.7076 **13.** $\begin{array}{r} 90.22 \\ \times\ 0.74 \\ \hline \end{array}$ 63; 66.7628 **14.** $\begin{array}{r} 19.8 \\ \times\ 5.2 \\ \hline \end{array}$ 100; 102.96 **15.** $\begin{array}{r} 0.9203 \\ \times\ 0.471 \\ \hline \end{array}$ 0.45; 0.4334613

16. 6.3×0.074 0.42; 0.4662

17. 0.31×0.006 0.0018; 0.00186

18. 2.9×0.48 1.5; 1.392

CHAPTER 3, PAGES 78–79

Divide.

1. $5\overline{)26.55}$ 5.31

2. $8\overline{)64.88}$ 8.11

3. $7\overline{)18.2}$ 2.6

4. $42\overline{)14.196}$ 0.338

5. $6\overline{)0.0192}$ 0.0032

6. $3\overline{)0.171}$ 0.057

7. $18\overline{)1.9836}$ 0.1102

8. $2\overline{)0.2014}$ 0.1007

First estimate using compatible numbers. Then find the exact quotient.

9. $154.32 \div 24$ 7; 643

10. $4.6718 \div 71$ 0.07; 0.0658

11. $280.854 \div 42$ 7; 6.687

12. $27.339 \div 30$ 0.9; 0.9113

13. $64.944 \div 82$ 0.8; 0.792

14. $78.032 \div 20$ 4; 3.9016

Solve for n.

15. $51.66 \div 3 = n$ 17.22

16. $3.114 \div 4 = n$ 0.7785

17. $0.0456 \div 8 = n$ 0.0057

CHAPTER 3, PAGES 80–82 **19.** 50; 55.25 **20.** 8; 7.92 **21.** 1; 1.23 **22.** 400; 405

Name the power of 10 needed to make the divisor a whole number.
Write 10; 100; 1,000; or 10,000.

1. 2.78)0.434 100 **2.** 2.5)625 10 **3.** 0.0008)640.64 10,000 **4.** 0.73)9.83 100

5. 0.035)130.2 1,000 **6.** 9.5)76.2 10 **7.** 0.549)64.1 1,000 **8.** 0.084)726 1,000

Divide.

9. 0.3)21 70 **10.** 0.6)1.8 3 **11.** 0.25)5.75 23 **12.** 0.015)18.6 1,240

13. 29.1)6,896.7 237 **14.** 20.3)103.124 5.08 **15.** 7.2)17.28 2.4

Divide. Round to the nearest hundredth.

16. 1.032)6.40 6.20 **17.** 0.87)8.64 9.93 **18.** 61.2)7.15 0.12

First estimate the quotient by using compatible numbers. Then find the exact answer.

19. 0.18)9.945 **20.** 3.7)29.304 **21.** 5.1)6.273 **22.** 0.05)20.25

23. 0.32)58.4 200; 182.5 **24.** 7.5)6.9525 0.9; 0.927 **25.** 0.24)2.088 10; 8.7 **26.** 8.9)81.88 9; 9.2

CHAPTER 4, PAGES 96–97

Use the bar graph below to answer Exercises 1–3.

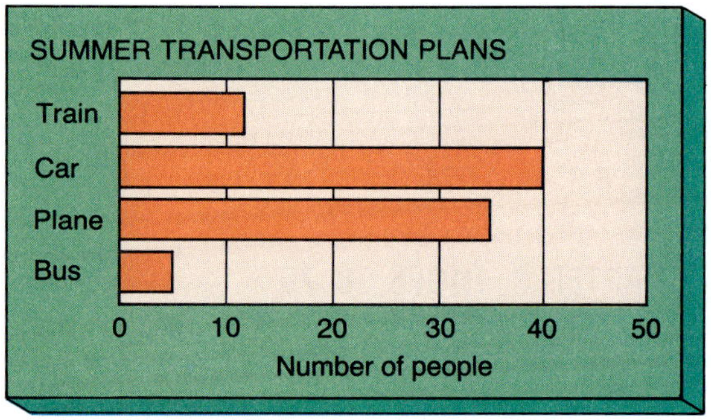

1. Which form of transportation are most people planning to use this summer? car

2. About how many more people are planning to travel by plane than by bus? about 30 more people

3. About how many people are planning to use ground transportation this summer? about 55 people

CHAPTER 4, PAGES 98–99

Find the mean. Round to the nearest tenth.

1. 2, 5, 10, 15 8

2. 28, 64, 37, 79, 88, 92 64.7

3. 208, 173, 180, 225, 158 188.8

4. 1,489; 2,045; 1,372; 3,004; 2,178 2,017.6

5. 3,567; 815; 2,003; 935; 267; 1,003 1,431.7

6. 4,004; 9,156; 10,487; 12,348 8,998.8

7. 24.6; 18.3; 7.9; 12.4; 9.04 14.4

8. 20.3, 19.5, 14, 26.2, 30.1 22.0

9. 326.8; 172.7; 419.5; 296.3; 509.7 345

10. 105.4, 267.3, 450.3, 336.4, 612, 708.3 413.3

CHAPTER 4, PAGES 100–101

Find the statistical measures for each set of data.
Round the answer to the nearest tenth.

Data	Range	Median	Mode	Mean
23; 19; 6; 19; 42; 3	**1.** 39	**2.** 19	**3.** 19	**4.** 18.7
7; 2.3; 9.2; 8.4; 9.2; 6.6; 9.2	**5.** 6.9	**6.** 8.4	**7.** 9.2	**8.** 7.4
64; 71.5; 58; 64; 62; 61; 70.8	**9.** 13.5	**10.** 64	**11.** 64	**12.** 64.5
600; 525; 743.6; 82.9; 310; 298.1	**13.** 660.7	**14.** 417.5	**15.** —	**16.** 426.6
730; 725.2; 730; 735; 734.6; 738	**17.** 12.8	**18.** 732.3	**19.** 730	**20.** 732.1
88; 123; 117.7; 121.5; 115; 102.3; 97	**21.** 35	**22.** 115	**23.** —	**24.** 109.2
315; 297.3; 304.5; 315; 299; 304.5	**25.** 17.7	**26.** 304.5	**27.** 315 and 304.5	**28.** 305.9

CHAPTER 5, PAGES 122–123

Write a fraction for the shaded part.

1.  3/4

2. 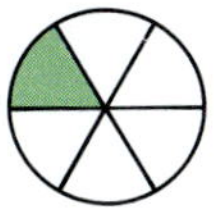 1/6

3. 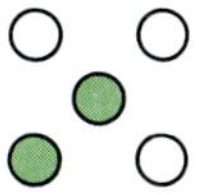 2/5

4.  7/16

5. 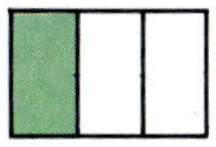 1/3

6.  3/7

7. 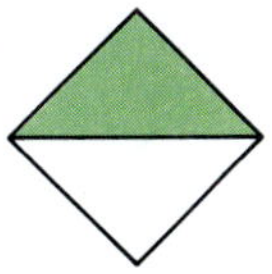 1/2

8. 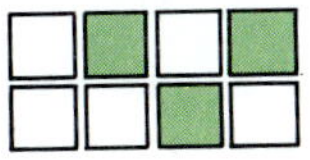 3/8

Write a fraction for the point.

9. 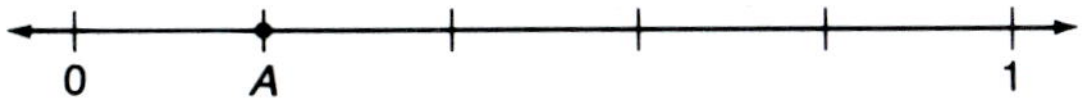

1/5

10. 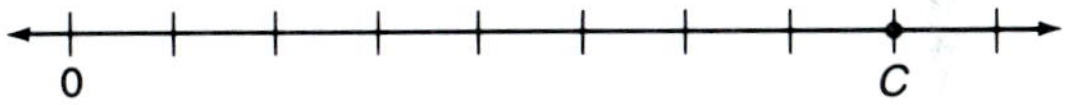

11/16

11. 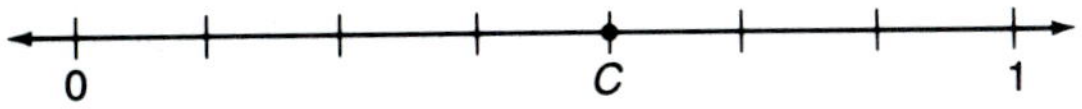

4/7

12. 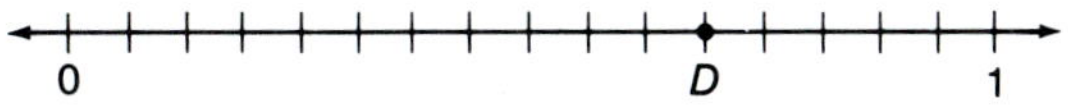

8/9

CHAPTER 5, PAGES 124–125

Find the greatest common factor (GCF).

1. 6, 25 1 **2.** 12, 20 4 **3.** 9, 36 9 **4.** 18, 27 9 **5.** 35, 49 7
6. 15, 40 5 **7.** 24, 81 3 **8.** 100, 95 5 **9.** 96, 64 32 **10.** 28, 12 4
11. 14, 21 7 **12.** 9, 33 3 **13.** 15, 22 1 **14.** 16, 22 2 **15.** 32, 48 16

Find the least common multiple (LCM).

16. 3, 8 24 **17.** 6, 10 30 **18.** 7, 9 63 **19.** 3, 12 12 **20.** 4, 5 20
21. 9, 21 63 **22.** 15, 4 60 **23.** 4, 16 16 **24.** 4, 10 20 **25.** 8, 20 40
26. 7, 8 56 **27.** 17, 4 68 **28.** 4, 12 12 **29.** 11, 12 132 **30.** 6, 14 42

CHAPTER 5, PAGES 126–127

Write the fraction in lowest terms.

1. $\frac{12}{20}$ $\frac{3}{5}$ **2.** $\frac{18}{24}$ $\frac{3}{4}$ **3.** $\frac{63}{81}$ $\frac{7}{9}$ **4.** $\frac{8}{28}$ $\frac{2}{7}$ **5.** $\frac{7}{21}$ $\frac{1}{3}$ **6.** $\frac{30}{54}$ $\frac{5}{9}$ **7.** $\frac{16}{66}$ $\frac{8}{33}$ **8.** $\frac{6}{9}$ $\frac{2}{3}$

9. $\frac{55}{95}$ $\frac{11}{19}$ **10.** $\frac{28}{38}$ $\frac{14}{19}$ **11.** $\frac{7}{49}$ $\frac{1}{7}$ **12.** $\frac{15}{90}$ $\frac{1}{6}$ **13.** $\frac{9}{21}$ $\frac{3}{7}$ **14.** $\frac{20}{90}$ $\frac{2}{9}$ **15.** $\frac{3}{27}$ $\frac{1}{9}$ **16.** $\frac{12}{42}$ $\frac{2}{7}$

Complete.

17. $\frac{1}{2} = \frac{■}{14}$ 7 **18.** $\frac{12}{27} = \frac{■}{9}$ 4 **19.** $\frac{2}{9} = \frac{■}{45}$ 10 **20.** $\frac{16}{24} = \frac{■}{3}$ 2 **21.** $\frac{5}{8} = \frac{25}{■}$ 40

22. $\frac{9}{60} = \frac{■}{20}$ 3 **23.** $\frac{20}{25} = \frac{4}{■}$ 5 **24.** $\frac{2}{7} = \frac{■}{35}$ 10 **25.** $\frac{3}{4} = \frac{■}{20}$ 15 **26.** $\frac{12}{48} = \frac{1}{■}$ 4

Using cross products, determine if the fractions are equivalent. Write $=$ or $\neq$.

27. $\frac{5}{6}$ ● $\frac{8}{9}$ $\neq$ **28.** $\frac{1}{7}$ ● $\frac{4}{28}$ $=$ **29.** $\frac{3}{9}$ ● $\frac{1}{3}$ $=$ **30.** $\frac{11}{16}$ ● $\frac{5}{8}$ $\neq$ **31.** $\frac{5}{12}$ ● $\frac{125}{300}$ $=$

CHAPTER 5, PAGES 128–130

Compare using equivalent fractions. Write $>$, $<$, or $=$.

1. $\frac{2}{5}$ ● $\frac{4}{5}$ $<$ **2.** $\frac{4}{7}$ ● $\frac{2}{7}$ $>$ **3.** $\frac{9}{14}$ ● $\frac{11}{14}$ $<$ **4.** $\frac{4}{13}$ ● $\frac{5}{26}$ $>$ **5.** $\frac{1}{2}$ ● $\frac{1}{3}$ $>$

6. $\frac{5}{8}$ ● $\frac{3}{4}$ $<$ **7.** $\frac{9}{10}$ ● $\frac{5}{6}$ $>$ **8.** $\frac{9}{27}$ ● $\frac{1}{3}$ $=$ **9.** $\frac{1}{8}$ ● $\frac{1}{7}$ $<$ **10.** $\frac{5}{10}$ ● $\frac{1}{2}$ $=$

11. $\frac{5}{7}$ ● $\frac{1}{2}$ $>$ **12.** $\frac{2}{30}$ ● $\frac{1}{10}$ $<$ **13.** $\frac{26}{50}$ ● $\frac{26}{100}$ $>$ **14.** $\frac{21}{40}$ ● $\frac{4}{20}$ $>$ **15.** $\frac{5}{25}$ ● $\frac{1}{5}$ $=$

Arrange in order from greatest to least.

16. $\frac{4}{7}, \frac{5}{7}, \frac{1}{7}, \frac{6}{7}, \frac{2}{7}$ $\frac{6}{7}, \frac{5}{7}, \frac{4}{7}, \frac{2}{7}, \frac{1}{7}$

17. $\frac{2}{8}, \frac{2}{4}, \frac{2}{10}, \frac{2}{6}, \frac{2}{12}$ $\frac{2}{4}, \frac{2}{6}, \frac{2}{8}, \frac{2}{10}, \frac{2}{12}$

18. $\frac{5}{12}, \frac{2}{7}, \frac{3}{8}, \frac{4}{5}, \frac{1}{10}$ $\frac{4}{5}, \frac{5}{12}, \frac{3}{8}, \frac{2}{7}, \frac{1}{10}$

19. $\frac{3}{4}, \frac{3}{9}, \frac{3}{8}, \frac{3}{6}, \frac{3}{10}$ $\frac{3}{4}, \frac{3}{6}, \frac{3}{8}, \frac{3}{9}, \frac{3}{10}$

20. $\frac{5}{8}, \frac{0}{8}, \frac{2}{8}, \frac{7}{8}, \frac{6}{8}$ $\frac{7}{8}, \frac{6}{8}, \frac{5}{8}, \frac{2}{8}, \frac{0}{8}$

21. $\frac{2}{5}, \frac{6}{7}, \frac{3}{4}, \frac{1}{3}, \frac{1}{6}$ $\frac{6}{7}, \frac{3}{4}, \frac{2}{5}, \frac{1}{3}, \frac{1}{6}$

CHAPTER 5, PAGES 132–133

Write as an improper fraction.

1. $4\frac{1}{9}$ $\frac{37}{9}$ **2.** $3\frac{1}{2}$ $\frac{7}{2}$ **3.** $7\frac{5}{6}$ $\frac{47}{6}$ **4.** $6\frac{2}{3}$ $\frac{20}{3}$ **5.** $3\frac{2}{9}$ $\frac{29}{9}$

6. $6\frac{5}{12}$ $\frac{77}{12}$ **7.** $2\frac{3}{4}$ $\frac{11}{4}$ **8.** $3\frac{7}{8}$ $\frac{31}{8}$ **9.** $1\frac{3}{4}$ $\frac{7}{4}$ **10.** $5\frac{2}{3}$ $\frac{17}{3}$

11. $3\frac{2}{3}$ $\frac{11}{3}$ **12.** $9\frac{1}{6}$ $\frac{55}{6}$ **13.** $5\frac{3}{16}$ $\frac{83}{16}$ **14.** $2\frac{1}{6}$ $\frac{13}{6}$ **15.** $3\frac{1}{12}$ $\frac{37}{12}$

16. $6\frac{3}{5}$ $\frac{33}{5}$ **17.** $9\frac{3}{7}$ $\frac{66}{7}$ **18.** $4\frac{5}{12}$ $\frac{53}{12}$ **19.** $10\frac{3}{10}$ $\frac{103}{10}$ **20.** $3\frac{9}{16}$ $\frac{57}{16}$

CHAPTER 5, PAGES 134–135

Solve.

1. Eric's lasagne recipe feeds 8 people and calls for 8 oz of cheese. He invited 22 friends for lasagne. How many pans of lasagne should he make? 3 pans

2. The Tygart City Police Academy accepts 75 recruits per class. 348 recruits are ready to begin training. How many must wait for a new class to form? 48 recruits

Frank's Auto Repair ordered 400 fan belts. The company sent 36 fan belts per crate, and the remainder were in a package.

3. How many crates of fan belts arrived? 11 crates

4. How many fan belts were in the package? 4 fan belts

CHAPTER 5, PAGES 136–137

Write as a fraction or mixed number in simplest form.

1. 4.2 $4\frac{1}{5}$	**2.** 0.05 $\frac{1}{20}$	**3.** 3.4 $3\frac{2}{5}$	**4.** 0.015 $\frac{3}{200}$	**5.** 1.54 $1\frac{27}{50}$	**6.** 6.125 $6\frac{1}{8}$
7. 0.3 $\frac{3}{10}$	**8.** 0.875 $\frac{7}{8}$	**9.** 2.44 $2\frac{11}{25}$	**10.** 5.25 $5\frac{1}{4}$	**11.** 0.065 $\frac{13}{200}$	**12.** 9.625 $9\frac{5}{8}$
13. 0.025 $\frac{1}{40}$	**14.** 10.8 $10\frac{4}{5}$	**15.** 30.78 $30\frac{39}{50}$	**16.** 7.03 $7\frac{3}{100}$	**17.** 0.55 $\frac{11}{20}$	**18.** 12.375 $12\frac{3}{8}$

Write the fraction or mixed number as a decimal.

19. $\frac{2}{5}$ 0.4	**20.** $\frac{3}{50}$ 0.06	**21.** $\frac{1}{10}$ 0.10	**22.** $\frac{5}{16}$ 0.3125	**23.** $\frac{2}{25}$ 0.08	**24.** $6\frac{7}{8}$ 6.875
25. $4\frac{1}{2}$ 4.5	**26.** $2\frac{1}{4}$ 2.25	**27.** $7\frac{9}{10}$ 7.9	**28.** $1\frac{3}{20}$ 1.15	**29.** $5\frac{5}{8}$ 5.625	**30.** $8\frac{3}{4}$ 8.75
31. $5\frac{3}{20}$ 5.15	**32.** $7\frac{1}{8}$ 7.125	**33.** $16\frac{1}{2}$ 16.5	**34.** $24\frac{5}{8}$ 24.625	**35.** $12\frac{7}{25}$ 12.28	**36.** $20\frac{17}{50}$ 20.34

Write the fraction or mixed number as a decimal. Round to the nearest hundredth.

37. $\frac{1}{14}$ 0.07	**38.** $\frac{4}{11}$ 0.36	**39.** $\frac{7}{9}$ 0.78	**40.** $\frac{6}{7}$ 0.86	**41.** $\frac{7}{12}$ 0.58	**42.** $\frac{1}{11}$ 0.09
43. $\frac{5}{13}$ 0.38	**44.** $\frac{7}{15}$ 0.47	**45.** $\frac{8}{11}$ 0.73	**46.** $\frac{5}{6}$ 0.83	**47.** $5\frac{4}{13}$ 5.31	**48.** $9\frac{9}{11}$ 9.82
49. $2\frac{1}{3}$ 2.33	**50.** $3\frac{4}{7}$ 3.57	**51.** $1\frac{1}{12}$ 1.08	**52.** $4\frac{2}{13}$ 4.15	**53.** $7\frac{2}{3}$ 7.67	**54.** $6\frac{3}{7}$ 6.43

Write the fraction or mixed number as a decimal. Use the bar notation.

55. $\frac{4}{9}$ $0.\overline{4}$	**56.** $\frac{7}{15}$ $0.4\overline{6}$	**57.** $\frac{49}{90}$ $0.5\overline{4}$	**58.** $\frac{2}{9}$ $0.\overline{2}$	**59.** $\frac{5}{6}$ $0.8\overline{3}$	**60.** $\frac{10}{11}$ $0.\overline{90}$
61. $2\frac{8}{9}$ $2.\overline{8}$	**62.** $1\frac{2}{3}$ $1.\overline{6}$	**63.** $1\frac{7}{11}$ $1.\overline{63}$	**64.** $3\frac{5}{9}$ $3.\overline{5}$	**65.** $40\frac{2}{3}$ $40.\overline{6}$	**66.** $19\frac{3}{7}$ $19.\overline{428571}$

CHAPTER 6, PAGES 146–147

Multiply. Write the product in simplest form.

1. $\frac{1}{4} \times \frac{3}{5}$ $\frac{3}{20}$ **2.** $\frac{2}{3} \times \frac{1}{6}$ $\frac{1}{9}$ **3.** $\frac{2}{3} \times \frac{3}{4}$ $\frac{1}{2}$ **4.** $\frac{12}{25} \times \frac{5}{6}$ $\frac{2}{5}$ **5.** $\frac{1}{16} \times \frac{1}{2}$ $\frac{1}{32}$ **6.** $\frac{1}{6} \times \frac{9}{10}$ $\frac{3}{20}$

7. $\frac{1}{3} \times \frac{5}{8}$ $\frac{5}{24}$ **8.** $\frac{1}{6} \times \frac{2}{3}$ $\frac{1}{9}$ **9.** $\frac{2}{4} \times \frac{2}{9}$ $\frac{1}{9}$ **10.** $\frac{1}{7} \times \frac{4}{5}$ $\frac{4}{35}$ **11.** $\frac{2}{5} \times \frac{1}{4}$ $\frac{1}{10}$ **12.** $\frac{2}{7} \times \frac{8}{9}$ $\frac{16}{63}$

Multiply. Use the shortcut when possible.

13. $\frac{7}{15} \times \frac{10}{21} \times \frac{1}{2}$ $\frac{1}{9}$ **14.** $\frac{2}{5} \times \frac{3}{4} \times \frac{5}{7}$ $\frac{3}{14}$ **15.** $\frac{6}{7} \times \frac{7}{9} \times \frac{1}{8}$ $\frac{1}{12}$ **16.** $\frac{1}{6} \times \frac{2}{3} \times \frac{9}{11}$ $\frac{1}{11}$

17. $\frac{5}{8} \times \frac{2}{3} \times \frac{2}{15}$ $\frac{1}{18}$ **18.** $\frac{6}{5} \times \frac{3}{18} \times \frac{1}{2}$ $\frac{1}{10}$ **19.** $\frac{6}{7} \times \frac{14}{24} \times \frac{1}{3}$ $\frac{1}{6}$ **20.** $\frac{3}{10} \times \frac{5}{8} \times \frac{4}{7}$ $\frac{3}{28}$

CHAPTER 6, PAGES 148–149

Multiply. Write the product in simplest form.

1. $7 \times \frac{3}{5}$ $4\frac{1}{5}$ **2.** $\frac{1}{8} \times 12$ $1\frac{1}{2}$ **3.** $10\frac{1}{3} \times \frac{5}{6}$ $8\frac{11}{18}$ **4.** $3\frac{1}{5} \times \frac{7}{8}$ $2\frac{4}{5}$ **5.** $6\frac{1}{3} \times 10\frac{1}{4}$ $64\frac{11}{12}$

6. $5\frac{5}{12} \times 3\frac{3}{4}$ $20\frac{5}{16}$ **7.** $4\frac{2}{3} \times 3\frac{1}{2}$ $16\frac{1}{3}$ **8.** $8\frac{2}{5} \times 6\frac{1}{3}$ $53\frac{1}{5}$ **9.** $3\frac{1}{3} \times 1\frac{1}{8}$ $3\frac{3}{4}$ **10.** $1\frac{4}{5} \times 5\frac{5}{8}$ $10\frac{1}{8}$

11. $3\frac{3}{4} \times \frac{1}{3} \times \frac{3}{5}$ $\frac{3}{4}$ **12.** $2\frac{1}{4} \times 1\frac{2}{3} \times 4\frac{4}{5}$ 18 **13.** $6\frac{1}{3} \times 2\frac{4}{10} \times 3\frac{3}{8}$ $51\frac{3}{10}$

14. $\frac{1}{2} \times 2 \times \frac{1}{3}$ $\frac{1}{3}$ **15.** $\frac{7}{8} \times 8\frac{1}{4} \times \frac{2}{7}$ $2\frac{1}{16}$ **16.** $3\frac{1}{7} \times 4\frac{2}{3} \times \frac{7}{8}$ $12\frac{5}{6}$

17. $3\frac{1}{7} \times 3\frac{1}{4} \times \frac{1}{3}$ $3\frac{17}{42}$ **18.** $\frac{5}{6} \times 2\frac{1}{6} \times \frac{3}{10}$ $\frac{13}{24}$ **19.** $\frac{4}{9} \times 12\frac{1}{4} \times 3\frac{2}{7}$ $17\frac{8}{9}$

20. $\left(3\frac{4}{5} + \frac{1}{10}\right) \times \frac{2}{3}$ $2\frac{3}{5}$ **21.** $\left(2\frac{3}{4} + 3\frac{1}{6}\right) \times \frac{4}{5}$ $4\frac{11}{15}$ **22.** $\left(1\frac{8}{10} + 2\frac{2}{5}\right) \times \frac{3}{7}$ $1\frac{4}{5}$

CHAPTER 6, PAGES 150–151

Find the reciprocal of the number.

1. 8 $\frac{1}{8}$ **2.** 14 $\frac{1}{14}$ **3.** 27 $\frac{1}{27}$ **4.** 100 $\frac{1}{100}$ **5.** $\frac{1}{9}$ 9 **6.** $\frac{5}{7}$ $\frac{7}{5}$

7. $\frac{8}{3}$ $\frac{3}{8}$ **8.** $\frac{10}{9}$ $\frac{9}{10}$ **9.** $4\frac{9}{10}$ $\frac{10}{49}$ **10.** $6\frac{7}{8}$ $\frac{8}{55}$ **11.** $2\frac{1}{5}$ $\frac{5}{11}$ **12.** $12\frac{2}{3}$ $\frac{3}{38}$

Divide. Write the quotient in simplest form.

13. $\frac{1}{2} \div \frac{1}{8}$ 4 **14.** $\frac{3}{4} \div \frac{3}{8}$ 2 **15.** $\frac{2}{3} \div \frac{4}{9}$ $1\frac{1}{2}$ **16.** $\frac{9}{10} \div \frac{5}{6}$ $1\frac{2}{25}$ **17.** $\frac{8}{17} \div 4$ $\frac{2}{17}$

18. $\frac{4}{5} \div \frac{8}{15}$ $1\frac{1}{2}$ **19.** $\frac{8}{3} \div \frac{4}{5}$ $3\frac{1}{3}$ **20.** $\frac{6}{7} \div 3$ $\frac{2}{7}$ **21.** $\frac{1}{3} \div \frac{1}{9}$ 3 **22.** $\frac{4}{9} \div \frac{2}{3}$ $\frac{2}{3}$

23. $7 \div \frac{1}{5}$ 35 **24.** $9 \div \frac{3}{4}$ 12 **25.** $\frac{1}{6} \div \frac{2}{3}$ $\frac{1}{4}$ **26.** $\frac{3}{5} \div \frac{1}{10}$ 6 **27.** $\frac{5}{9} \div \frac{1}{4}$ $2\frac{2}{9}$

CHAPTER 6, PAGES 152–153

Divide. Write the quotient in simplest form.

1. $4\frac{3}{4} \div 2$ $2\frac{3}{8}$ **2.** $8\frac{9}{10} \div 4$ $2\frac{9}{40}$ **3.** $3 \div 3\frac{1}{2}$ $\frac{6}{7}$ **4.** $1\frac{1}{3} \div 5$ $\frac{4}{15}$ **5.** $6\frac{3}{4} \div \frac{1}{8}$ 54

6. $\frac{1}{10} \div 9\frac{4}{5}$ $\frac{1}{98}$ **7.** $3\frac{7}{8} \div \frac{1}{32}$ 124 **8.** $2\frac{9}{25} \div \frac{1}{100}$ 236 **9.** $\frac{2}{3} \div 2\frac{2}{3}$ $\frac{1}{4}$ **10.** $2\frac{1}{5} \div \frac{3}{5}$ $3\frac{2}{3}$

11. $6 \div 2\frac{2}{3}$ $2\frac{1}{4}$ **12.** $12 \div 1\frac{3}{4}$ $6\frac{6}{7}$ **13.** $\frac{1}{9} \div 4\frac{1}{2}$ $\frac{2}{81}$ **14.** $6\frac{1}{5} \div \frac{1}{10}$ 62 **15.** $8\frac{1}{3} \div 5$ $1\frac{2}{3}$

CHAPTER 6, PAGES 154–155

Solve each problem by working backward.

1. On Tuesday Jan made one-half as many school safety posters as she made on Thursday. She made six on Tuesday. How many posters did she make on Thursday? 12 posters

2. Keith walked his dog four times as far on Saturday as he did on Sunday. He walked the dog 1.4 km on Sunday. How far did he walk on Saturday? 5.6 km

3. Bart had $25.00 on Friday. By Sunday, he had $9.05 left. He went to a movie on Saturday and then out to eat. His meal cost $7.75. How much money did Bart spend at the movies? $8.20

4. Ira worked $1\frac{1}{3}$ times as many hours last week than this week. This week he worked 39 hours. How many hours did Ira work last week? 52 hours

CHAPTER 6, PAGES 158–159

Add. Write the sum in simplest form.

1. $\frac{5}{8} + \frac{1}{8}$ $\frac{3}{4}$ **2.** $\frac{3}{8} + \frac{5}{32}$ $\frac{17}{32}$ **3.** $\frac{2}{5} + \frac{1}{10}$ $\frac{1}{2}$ **4.** $\frac{3}{4} + \frac{1}{6}$ $\frac{11}{12}$ **5.** $\frac{1}{3} + \frac{7}{12}$ $\frac{11}{12}$

6. $\frac{3}{5} + \frac{1}{3}$ $\frac{14}{15}$ **7.** $\frac{3}{5} + \frac{1}{2}$ $1\frac{1}{10}$ **8.** $\frac{1}{7} + \frac{8}{9}$ $1\frac{2}{63}$ **9.** $\frac{5}{8} + \frac{11}{12}$ $1\frac{13}{24}$ **10.** $\frac{2}{5} + \frac{5}{6}$ $1\frac{7}{30}$

CHAPTER 6, PAGES 160–161

Add. Write the sum in simplest form.

1. $2\frac{5}{6} + 1\frac{5}{8}$ $4\frac{11}{24}$ **2.** $9\frac{3}{5} + 3\frac{2}{20}$ $12\frac{7}{10}$ **3.** $10\frac{3}{10} + 8\frac{9}{10}$ $19\frac{1}{5}$ **4.** $12\frac{7}{8} + 11\frac{9}{16}$ $24\frac{7}{16}$

5. $7\frac{1}{4} + 8\frac{7}{16} + 6\frac{3}{16}$ $21\frac{7}{8}$

6. $7\frac{1}{4} + 8\frac{1}{3} + 5\frac{1}{6}$ $20\frac{3}{4}$

7. $3\frac{2}{3} + 6\frac{1}{2} + 8\frac{2}{5}$ $18\frac{17}{30}$

8. $4\frac{1}{4} + 2\frac{5}{6} + 10\frac{1}{2}$ $17\frac{7}{12}$

9. $6\frac{1}{2} + 4\frac{2}{5} + 9\frac{3}{10}$ $20\frac{1}{5}$

10. $3\frac{5}{8} + 6\frac{1}{4} + 2\frac{1}{5}$ $12\frac{3}{40}$

CHAPTER 6, PAGES 162–163

Subtract. Write the difference in simplest form.

1. $\frac{7}{8} - \frac{5}{8}$ $\frac{1}{4}$ **2.** $\frac{3}{4} - \frac{2}{5}$ $\frac{7}{20}$ **3.** $\frac{1}{4} - \frac{1}{9}$ $\frac{5}{36}$ **4.** $\frac{9}{10} - \frac{5}{6}$ $\frac{1}{15}$ **5.** $\frac{7}{10} - \frac{3}{8}$ $\frac{13}{40}$

6. $\frac{5}{6} - \frac{1}{6}$ $\frac{2}{3}$ **7.** $\frac{7}{9} - \frac{3}{5}$ $\frac{8}{45}$ **8.** $\frac{2}{3} - \frac{3}{10}$ $\frac{11}{30}$ **9.** $\frac{1}{2} - \frac{1}{4}$ $\frac{1}{4}$ **10.** $\frac{15}{36} - \frac{2}{9}$ $\frac{7}{36}$

CHAPTER 6, PAGES 164–166

Subtract. Write the difference in simplest form.

1. $3\frac{1}{2} - 1\frac{1}{4}$ $2\frac{1}{4}$ **2.** $6\frac{1}{8} - 4\frac{3}{4}$ $1\frac{3}{8}$ **3.** $23\frac{5}{8} - 12\frac{1}{2}$ $11\frac{1}{8}$ **4.** $19\frac{5}{16} - 5\frac{1}{4}$ $14\frac{1}{16}$

5. $9\frac{7}{12} - 6\frac{11}{12}$ $2\frac{2}{3}$ **6.** $14\frac{2}{5} - 8\frac{1}{10}$ $6\frac{3}{10}$ **7.** $8\frac{2}{3} - 5\frac{1}{2}$ $3\frac{1}{6}$ **8.** $11 - 9\frac{1}{4}$ $1\frac{3}{4}$ **9.** $13\frac{1}{3} - 10\frac{3}{4}$ $2\frac{7}{12}$ **10.** $5\frac{3}{10} - 4\frac{3}{4}$ $\frac{11}{20}$

11. $5\frac{3}{4} - \frac{5}{8}$ $5\frac{1}{8}$ **12.** $18\frac{5}{9} - 8\frac{5}{6}$ $9\frac{13}{18}$ **13.** $14 - 6\frac{1}{2}$ $7\frac{1}{2}$ **14.** $20 - 15\frac{3}{16}$ $4\frac{13}{16}$ **15.** $6\frac{1}{12} - \frac{2}{3}$ $5\frac{5}{12}$ **16.** $27\frac{3}{8} - 6\frac{2}{3}$ $20\frac{17}{24}$

CHAPTER 6, PAGES 168–169

Round to the nearest whole number.

1. $56\frac{4}{5}$ 57 **2.** $12\frac{1}{2}$ 13 **3.** $23\frac{4}{9}$ 23 **4.** $17\frac{1}{6}$ 17 **5.** $45\frac{5}{8}$ 46

6. $11\frac{1}{3}$ 11 **7.** $25\frac{7}{9}$ 26 **8.** $24\frac{9}{16}$ 25 **9.** $49\frac{11}{12}$ 50 **10.** $18\frac{3}{8}$ 18

Estimate.

11. $8\frac{3}{8} + 5\frac{1}{4}$ 13 **12.** $7\frac{3}{5} + 5\frac{1}{4}$ 13 **13.** $15\frac{3}{8} + 14\frac{4}{5}$ 30 **14.** $10\frac{1}{3} + 8\frac{3}{4}$ 19

15. $9\frac{1}{10} - 5\frac{3}{4}$ 3 **16.** $20\frac{1}{20} - 10\frac{1}{40}$ 10 **17.** $17\frac{39}{50} - 8\frac{3}{4}$ 9 **18.** $32\frac{1}{5} - 19\frac{2}{3}$ 12

19. $6\frac{1}{6} \times 4\frac{4}{5}$ 30 **20.** $2\frac{3}{4} \times 5\frac{2}{3}$ 18 **21.** $3\frac{3}{4} \times 1\frac{1}{8}$ 4 **22.** $7\frac{5}{6} \times 2\frac{1}{3}$ 16

23. $10\frac{3}{10} \div 5\frac{1}{5}$ 2 **24.** $4\frac{4}{9} \div 1\frac{2}{3}$ 2 **25.** $5\frac{3}{5} \div 2\frac{4}{15}$ 3 **26.** $2\frac{7}{8} \div \frac{15}{16}$ 3

27. $19\frac{3}{5} - 4\frac{1}{3}$ 16 **28.** $31\frac{2}{3} \div 7\frac{8}{9}$ 4 **29.** $4\frac{1}{5} \times 2\frac{1}{10}$ 8 **30.** $12\frac{3}{4} + 8\frac{7}{9}$ 22

CHAPTER 7, PAGES 180–181

Complete.

1. 9 in. = ■ yd $\frac{1}{4}$
2. 7,920 ft = ■ mi $1\frac{1}{2}$
3. $5\frac{1}{4}$ ft = ■ in. 63
4. 3 mi = ■ ft 15,840
5. 540 in. = ■ yd 15
6. 69 yd = ■ ft 207
7. 90 in. = ■ ft $7\frac{1}{2}$
8. $14\frac{1}{2}$ yd = ■ in. 522
9. 126 in. = ■ ft $10\frac{1}{2}$
10. 7,040 yd = ■ mi 4
11. 40 ft = ■ in. 480
12. 18 yd = ■ in. 648
13. 3 in. = ■ yd $\frac{1}{12}$
14. 2,640 ft = ■ mi $\frac{1}{2}$
15. 3 mi = ■ yd 5,280
16. 6 mi = ■ ft 31,680
17. 96 yd = ■ ft 288
18. 936 in. = ■ yd 26

CHAPTER 7, PAGES 182–183

Complete.

1. 7 pt = ■ c 14
2. 8,000 lb = ■ T 4
3. 6 gal = ■ qt 24
4. $\frac{1}{2}$ qt = ■ fl oz 16
5. $4\frac{3}{4}$ gal = ■ pt 38
6. $24\frac{1}{2}$ lb = ■ oz 392
7. 400 fl oz = ■ gal $3\frac{1}{8}$
8. 13 pt = ■ qt $6\frac{1}{2}$
9. 17 c = ■ pt $8\frac{1}{2}$
10. 10 c = ■ qt $2\frac{1}{2}$
11. 3 gal = ■ fl oz 384
12. 48 oz = ■ lb 3
13. 18 qt = ■ pt 36
14. 30 fl oz = ■ c $3\frac{3}{4}$
15. $2\frac{1}{4}$ T = ■ lb 4,500
16. 13 pt = ■ c 26
17. 1,000 lb = ■ T $\frac{1}{2}$
18. 10 gal = ■ qt 40
19. 3 qt = ■ fl oz 96
20. 7 gal = ■ pt 56
21. 18 lb = ■ oz 288

CHAPTER 7, PAGES 184–185

Add or subtract.

1. 2 pt 1 c +3 pt 1 c 6 pt	**2.** 6 yd 1 ft −2 yd 2 ft 3 yd 2 ft	**3.** 8 mi 1,680 ft +2 mi 5,000 ft 11 mi 1,400 ft	**4.** 12 lb 5 oz − 6 lb 9 oz 5 lb 12 oz
5. 9 ft 6 in. −1 ft 8 in. 7 ft 10 in.	**6.** 7 gal −3 gal 3 qt 3 gal 1 qt	**7.** 6 T 1,000 lb + 1,500 lb 7 T 500 lb	**8.** 3 c 7 fl oz +2 c 6 fl oz 6 c 5 fl oz
9. 3 mi 900 yd +6 mi 900 yd 10 mi 40 yd	**10.** 10 qt − 6 qt 1 pt 3 qt 1 pt	**11.** 17 lb 4 oz −10 lb 12 oz 6 lb 8 oz	**12.** 6 gal 2 qt +11 gal 3 qt 18 gal 1 qt
13. 6 pt 1 c +7 pt 1 c 14 pt	**14.** 9 yd 2 ft −7 yd 1 ft 2 yd 1 ft	**15.** 4 mi 340 ft +10 mi 920 ft 14 mi 1,260 ft	**16.** 18 lb 3 oz −16 lb 14 oz 1 lb 5 oz
17. 15 ft 1 in. − 12 ft 4 in. 2 ft 9 in.	**18.** 13 gal − 2 gal 1 qt 10 gal 3 qt	**19.** 5 T 500 lb + 1,700 lb 6 T 200 lb	**20.** 1 c 6 fl oz +4 c 2 fl oz 6 c

CHAPTER 7, PAGES 186–187

Solve.

Your plane leaves New York at 12:48 P.M. How long does it take you to reach Atlanta if you arrive at:

1. 2:04 P.M.? 1 hour 16 minutes
2. 1:42 P.M.? 54 minutes
3. 4:32 P.M.? 3 hours 44 minutes
4. 3:12 P.M.? 2 hours 24 minutes

The baseball game began at 3:30 P.M. How long did the game last if it ended at:

5. 5:54 P.M.? 2 hours 24 minutes
6. 6:49 P.M.? 3 hours 19 minutes
7. 6:09 P.M.? 2 hours 39 minutes
8. 7:50 P.M.? 4 hours 20 minutes

The Saturday double feature at the Plaza Theater begins at 1:50 P.M. How long will it last if it ends at:

9. 5:15 P.M.? 3 hours 25 minutes
10. 6:02 P.M.? 4 hours 12 minutes
11. 6:35 P.M.? 4 hours 45 minutes
12. 5:53 P.M.? 4 hours 3 minutes

CHAPTER 7, PAGES 192–193

Choose the most reasonable estimate.

1. The height of a coffee mug b **a.** 4 km **b.** 6 cm **c.** 6 mm
2. The length of a football field a **a.** 91 m **b.** 91 cm **c.** 91 mm
3. The height of a grandfather's clock a **a.** 3 m **b.** 3 cm **c.** 3 mm
4. The length of a dog's tail b **a.** 15 m **b.** 15 cm **c.** 15 mm
5. The width of a piano c **a.** 120 m **b.** 120 mm **c.** 120 cm
6. The thickness of a sandwich c **a.** 20 m **b.** 20 km **c.** 20 mm
7. The thickness of a pencil c **a.** 5 m **b.** 5 cm **c.** 5 mm

CHAPTER 7, PAGES 194–196

Choose the most reasonable estimate.

1. The mass of a tube of lipstick b **a.** 100 kg **b.** 100 g **c.** 100 mg
2. The mass of a table lamp a **a.** 7 kg **b.** 30 g **c.** 10 g
3. The capacity of a thermos bottle b **a.** 5 kL **b.** 1 L **c.** 6 mL
4. The capacity of an eyedropper c **a.** 1 kL **b.** 3 L **c.** 5 mL
5. The amount of water in a bathtub b **a.** 10 kL **b.** 100 L **c.** 1,000 mL
6. The mass of a bird feather c **a.** 10 kg **b.** 30 g **c.** 20 mg
7. The mass of an apple b **a.** 25 g **b.** 250 g **c.** 25 kg

CHAPTER 7, PAGES 198–199 12. 2,000,000

Complete.

1. 4 m = ■ mm 4,000	**2.** 2 cm = ■ mm 20	**3.** 800 cm = ■ m 8
4. 6 km = ■ m 6,000	**5.** 300 mm = ■ m 0.3	**6.** 70 m = ■ km 0.07
7. 3 kg = ■ g 3,000	**8.** 73,000 L = ■ kL 73	**9.** 5,000 mg = ■ g 5
10. 160 g = ■ kg 0.16	**11.** 7 L = ■ mL 7,000	**12.** 2 kL = ■ mL
13. 6 kg = ■ mg 6,000,000	**14.** 820 mL = ■ L 0.82	**15.** 82.3 mm = ■ cm 8.23
16. 60 mg = ■ g 0.06	**17.** 0.434 kg = ■ g 434	**18.** 6.888 km = ■ m 6,888

CHAPTER 7, PAGES 200–201

Choose the most reasonable temperature for each of the following.

1. A radiator	c	**a.** 32°F	**b.** 230°F	**c.** 55°C
2. Beef stew cooking	a	**a.** 350°F	**b.** 37°C	**c.** 60°C
3. A sick baby	c	**a.** 2°C	**b.** 40°F	**c.** 40°C
4. A day for sledding	b	**a.** 60°F	**b.** 28°F	**c.** 20°C
5. Steam from boiling water	b	**a.** 50°F	**b.** 95°C	**c.** 75°F
6. A freezer	b	**a.** −100°F	**b.** −5°F	**c.** 22°F

CHAPTER 7, PAGES 202–203

Solve each problem by drawing a diagram.

Jason's room measures 6 m × 5 m. Two windows take up a total of 2.5 m on the south wall which is 5 m in length. The door takes up 1.5 m on the north wall. His desk measures 2 m across, his bed is 2.2 m, and his dresser is 1.25 m.

1. Will all his furniture fit along the west wall? yes

2. How much space will be left along the wall with the door? 3.5 m

Lincoln Airport is built in the shape of a square. It measures 6 km on each side. A fuel truck enters the airport at the center of the north side and travels 2 km south. Then it travels 2.5 km west, then 0.5 km south.

3. How far is the truck from the western border of the airport? 0.5 km

4. How far is the truck from the southern border of the airport? 3.5 km

CHAPTER 8, PAGES 210–211

Complete to form an equivalent ratio.

1. $\frac{4}{5} = \frac{\blacksquare}{20}$ 16 **2.** $\frac{20}{150} = \frac{2}{\blacksquare}$ 15 **3.** $\frac{3}{13} = \frac{9}{\blacksquare}$ 39 **4.** $\frac{1}{6} = \frac{\blacksquare}{72}$ 12 **5.** $\frac{22}{14} = \frac{\blacksquare}{7}$ 11

Use these denominators to write a ratio equivalent to $\frac{2}{3}$.

6. 18 12 **7.** 36 24 **8.** 48 32 **9.** 96 64 **10.** 24 16 **11.** 108 72 **12.** 72 48 **13.** 54 36

Use these numerators to write a ratio equivalent to $\frac{40}{160}$.

14. 25 100 **15.** 16 64 **16.** 80 320 **17.** 7 28 **18.** 22 88 **19.** 15 60 **20.** 6 24 **21.** 32 128

Write the ratio in simplest form.

22. $\frac{3}{21}$ $\frac{1}{7}$ **23.** $\frac{36}{63}$ $\frac{4}{7}$ **24.** $\frac{6}{84}$ $\frac{1}{14}$ **25.** $\frac{50}{75}$ $\frac{2}{3}$ **26.** $\frac{25}{45}$ $\frac{5}{9}$ **27.** $\frac{6}{66}$ $\frac{1}{11}$ **28.** $\frac{42}{54}$ $\frac{7}{9}$ **29.** $\frac{48}{144}$ $\frac{1}{3}$

CHAPTER 8, PAGES 212–213

Write as a rate in simplest form.

1. 14 balls in 3 innings $\frac{14}{3}$

2. 95 words in 1 minute $\frac{95}{1}$

3. 2 mi in 6 minutes $\frac{1}{3}$

4. 16 apples for \$3.16 $\frac{4}{\$0.79}$

Compare the rates. Write >, <, or =.

5. $\frac{12\text{ pens}}{\$0.90} \bullet \frac{2\text{ pens}}{\$0.15}$ =

6. $\frac{8\text{ players}}{4\text{ games}} \bullet \frac{5\text{ players}}{7\text{ games}}$ >

7. $\frac{6\text{ eggs}}{\$0.59} \bullet \frac{12\text{ eggs}}{\$0.89}$ <

8. $\frac{21\text{ cats}}{14\text{ dogs}} \bullet \frac{32\text{ cats}}{23\text{ dogs}}$ >

9. $\frac{60\text{ mi}}{2\text{ hours}} \bullet \frac{30\text{ mi}}{1\text{ hour}}$ =

10. $\frac{5\text{ shows}}{2\text{ weeks}} \bullet \frac{14\text{ shows}}{20\text{ weeks}}$ >

CHAPTER 8, PAGES 216–217

Determine if the ratios form a proportion. Write *yes* or *no*.

1. $\frac{4}{2} \stackrel{?}{=} \frac{6}{3}$ yes **2.** $\frac{6.8}{3.4} \stackrel{?}{=} \frac{24.2}{8.6}$ no **3.** $\frac{12}{3} \stackrel{?}{=} \frac{116}{29}$ yes **4.** $\frac{67}{4} \stackrel{?}{=} \frac{33}{2}$ no

5. $\frac{5}{9} \stackrel{?}{=} \frac{7}{13}$ no **6.** $\frac{1.2}{2.5} \stackrel{?}{=} \frac{2.0}{4.5}$ no **7.** $\frac{25}{10} \stackrel{?}{=} \frac{100}{40}$ yes **8.** $\frac{16}{56} \stackrel{?}{=} \frac{30}{105}$ yes

Find n.

9. $\frac{n}{32} = \frac{5}{8}$ 20 **10.** $\frac{34}{20} = \frac{17}{n}$ 10 **11.** $\frac{19}{11} = \frac{n}{44}$ 76 **12.** $\frac{2}{3} = \frac{200}{n}$ 300

13. $\frac{6.3}{2} = \frac{n}{8.2}$ 25.83 **14.** $\frac{90}{1.5} = \frac{n}{6}$ 360 **15.** $\frac{0.5}{50} = \frac{4.5}{n}$ 450 **16.** $\frac{6}{n} = \frac{9}{30.6}$ 20.4

17. $\frac{16}{3.2} = \frac{n}{20}$ 100 **18.** $\frac{3.9}{n} = \frac{4.5}{7.5}$ 6.5 **19.** $\frac{n}{8.2} = \frac{20}{0.5}$ 328 **20.** $\frac{20}{35} = \frac{6.2}{n}$ 10.85

CHAPTER 8, PAGES 218–219

Which is the better estimate?

1. Suzanne wants to buy a skirt for \$24.99, shoes for \$42.84, socks for \$1.99, and a shirt for \$18.56. How much money should she have? b

a. \$24 + \$42 + \$1 + \$18 = \$85 **b.** \$25 + \$43 + \$2 + \$19 = \$89

Solve.

2. Jim wants to buy 4 avocados and 6 apples. Estimate the cost. \$6.00

3. Carol buys 12 cucumbers and 10 avocados. Estimate the cost. \$13.00

4. Wanda's mother sends her to Ray's for 5 avocados, 6 cucumbers, and 6 apples. Estimate the cost. \$8.50

RAY'S PRODUCE	
Apples	6/\$1.75
Cucumbers	12/\$2.99
Avocados	\$0.89

CHAPTER 8, PAGES 224–225

A model airplane was made to the scale of 1 in. : 3 ft. Find the actual measures.

Airplane	Length	Wingspan	Length of one wing	Height of tail fin
Actual measure of model	24 ft **1.** ■	18 ft **2.** ■	9 ft **3.** ■	3 ft **4.** ■
Scale measure of model	8 in.	6 in.	3 in.	1 in.

CHAPTER 9, PAGES 234–235

Write as a decimal.

1. 5% 0.05 **2.** 8% 0.08 **3.** 66% 0.66 **4.** 23% 0.23 **5.** 535% 5.35

6. 16% 0.16 **7.** 162% 1.62 **8.** 325% 3.25 **9.** 2.3% 0.023 **10.** 1.5% 0.015

11. $\frac{1}{4}$% 0.0025 **12.** $\frac{7}{8}$% 0.00875 **13.** $\frac{3}{5}$% 0.006 **14.** $37\frac{1}{2}$% 0.375 **15.** $40\frac{3}{4}$% 0.4075

Write as a percent.

16. 0.14 14% **17.** 0.61 61% **18.** 0.98 98% **19.** 0.8 80% **20.** 0.887 88.7%

21. 0.125 12.5% **22.** 0.333 33.3% **23.** 0.0725 7.25% **24.** 0.125 12.5% **25.** 0.00346 0.346%

26. 4.3 430% **27.** 3.9 390% **28.** 6.8 680% **29.** 7.25 725% **30.** 4.375 437.5%

CHAPTER 9, PAGES 236–237

Write as a fraction, whole number, or mixed number in simplest form.

1. 81% $\frac{81}{100}$ **2.** 3% $\frac{3}{100}$ **3.** 19% $\frac{19}{100}$ **4.** 63% $\frac{63}{100}$ **5.** 29% $\frac{29}{100}$

6. 15% $\frac{3}{20}$ **7.** 20% $\frac{1}{5}$ **8.** 90% $\frac{9}{10}$ **9.** 56% $\frac{14}{25}$ **10.** 78% $\frac{39}{50}$

11. 200% 2 **12.** 300% 3 **13.** 600% 6 **14.** 4,000% 40 **15.** 7,000% 70

16. 850% $8\frac{1}{2}$ **17.** 594% $5\frac{47}{50}$ **18.** 330% $3\frac{3}{10}$ **19.** 944% $9\frac{11}{25}$ **20.** 120% $1\frac{1}{5}$

21. 222% $2\frac{11}{50}$ **22.** 1,470% $14\frac{7}{10}$ **23.** 108% $1\frac{2}{25}$ **24.** 3,240% $32\frac{2}{5}$ **25.** 125% $1\frac{1}{4}$

26. 675% $6\frac{3}{4}$ **27.** 549% $5\frac{49}{100}$ **28.** 6,680% $66\frac{4}{5}$ **29.** 8,635% $86\frac{7}{20}$ **30.** 9,766% $97\frac{33}{50}$

CHAPTER 9, PAGES 238–239

Write as a percent.

1. $\frac{1}{10}$ 10% **2.** $\frac{4}{25}$ 16% **3.** $\frac{1}{2}$ 50% **4.** $\frac{3}{50}$ 6% **5.** $\frac{1}{4}$ 25% **6.** $\frac{7}{20}$ 35%

7. $\frac{3}{8}$ 37.5% **8.** $\frac{5}{12}$ $41\frac{2}{3}$% **9.** $\frac{4}{5}$ 80% **10.** $\frac{1}{3}$ $33\frac{1}{3}$% **11.** $\frac{2}{9}$ $22\frac{2}{9}$% **12.** $\frac{5}{6}$ $83\frac{1}{3}$%

13. 5 500% **14.** 2 200% **15.** 14 1,400% **16.** 30 3,000% **17.** 3 300% **18.** 17 1,700%

19. $1\frac{1}{5}$ 120% **20.** $2\frac{4}{5}$ 280% **21.** $4\frac{5}{6}$ $483\frac{1}{3}$% **22.** $9\frac{3}{8}$ 937.5% **23.** $3\frac{3}{5}$ 360% **24.** $1\frac{7}{8}$ 187.5%

25. $5\frac{7}{20}$ 535% **26.** $7\frac{3}{4}$ 775% **27.** $6\frac{2}{3}$ $666\frac{2}{3}$% **28.** $3\frac{14}{25}$ 356% **29.** $2\frac{13}{20}$ 265% **30.** $2\frac{1}{3}$ $233\frac{1}{3}$%

CHAPTER 9, PAGES 240–241

Find the percent of the number.

1. 3% of 90 2.7 **2.** 90% of 900 810 **3.** 100% of 50 50 **4.** 5.2% of 14 0.728

5. 4.4% of 600 26.4 **6.** $\frac{1}{4}$% of 36 0.09 **7.** $\frac{3}{8}$% of 48 0.18 **8.** $12\frac{1}{2}$% of 64 8

Compute.

9. What is 4% of 500? 20

10. What is 10% of 20? 2

11. What is $8\frac{1}{2}$% of 40? 3.4

12. What number is 0.17% of 92? 0.1564

13. What is 6% of 450? 27

14. 8.3% of 60 is what number? 4.98

15. What is 200% of 45? 90

16. 50.8% of 520 is what number? 264.16

17. What is $7\frac{1}{4}$% of 500? 36.25

18. What number is 130% of 800? 1,040

CHAPTER 9, PAGES 246–247

Find the missing number.

1. What percent of 20 is 12? 60%
2. What percent of 16 is 5.6? 35%
3. 8 is what percent of 10? 80%
4. 9 is what percent of 45? 20%
5. What percent of 400 is 3.2? 0.8%
6. What percent of 72 is 70.2? 97.5%
7. 75 is what percent of 300? 25%
8. 65 is what percent of 32.5? 200%

Estimate.

9. What percent of 10 is 5.4? 50%
10. What percent of 8 is 6.2? 75%
11. 19.4 is what percent of 59.8? $33\frac{1}{3}\%$
12. What percent of 20 is 1.1? 5%
13. 3.2 is what percent of 12.4? 25%
14. 3.9 is what percent of 5? 80%

CHAPTER 9, PAGES 248–249

Find the number. Change the percent to a decimal.

1. 7% of what number is 28? 400
2. 90 is 18% of what number? 500
3. 15% of what number is 30? 200
4. 3% of what number is 1.8? 60

Find the number. Change the percent to a fraction.

5. 12% of what number is 9? 75
6. 15 is 6% of what number? 250
7. 30 is $37\frac{1}{2}\%$ of what number? 80
8. 30 is $83\frac{1}{3}\%$ of what number? 36

Find the number.

9. 20% of what number is 20? 100
10. 160 is 400% of what number? 40
11. 125 is 125% of what number? 100
12. 2 is 0.4% of what number? 500
13. 2.5% of what number is 30? 1,200
14. 130 is 65% of what number? 200

CHAPTER 9, PAGES 250–251

Use a proportion to solve.

1. What number is 4% of 15? 0.6
2. 63% of 300 is what number? 189
3. 44 is what percent of 66? $66\frac{2}{3}\%$
4. 3 is 40% of what number? 7.5
5. 25 is what percent of 2.5? 1,000%
6. What number is 8.5% of 60? 5.1
7. 180 is what percent of 120? 150%
8. 45% of 280 is what number? 126

CHAPTER 9, PAGES 252–253

Solve by supplying the missing information.

1. Jenny needs 6 ft of material to make a skirt. How many yards of material should she buy? 2 yd

2. Paul uses 3 oz of yeast for every two loaves of bread he bakes. How many pounds of yeast does he need for 32 loaves of bread? 3 lb

3. It takes the Decorating Committee 45 minutes to decorate $\frac{1}{8}$ of the gym for the school dance. How many hours will it take to finish the job? 6 hours

4. Brian goes to visit his grandfather every summer. This year he leaves on June 28 and stays through August 31. How many days will he be at his grandfather's? 64 days

5. Dan is buying refreshments for the school dance. He expects 300 students to come to the dance and for each of them to drink three 8-oz cups of punch. How many quarts of punch does he buy? 225 qt

6. Leslie figures that she can earn $45 per week working part time in the bookstore. If she works for an entire year, how much money will Leslie earn? How much will she earn per month? $2,340/year; $195/month

CHAPTER 9, PAGES 258–259

Complete.

	Original amount	New amount	Percent change	
1.	$100	$150	■	50% increase
2.	8	5	■	$37\frac{1}{2}$% decrease
3.	15	20	■	$33\frac{1}{3}$% increase
4.	25	15	■	40% decrease
5.	6	24	■	300% increase
6.	5,600	4,200	■	25% decrease
7.	25	75	■	200% increase
8.	2,000	1,200	■	40% decrease
9.	420	■	10% increase	462
10.	300	■	5% decrease	285
11.	700	■	25% increase	875
12.	850	■	16% decrease	714

CHAPTER 10, PAGES 268–270

A clothes dryer contains 20 socks. Socks are picked at random. Refer to the chart at the right and find the probability.

PAIRS OF SOCKS

	Large	Small
Blue	3	2
White	2	3

1. P(blue) $\frac{1}{2}$

2. P(white) $\frac{1}{2}$

3. P(large blue) $\frac{3}{10}$

4. P(large white) $\frac{1}{5}$

5. P(small blue) $\frac{1}{5}$

6. P(small white) $\frac{3}{10}$

CHAPTER 10, PAGES 272–273

City Park Officials surveyed 500 park users on a Saturday in May. Here are the resuts. For the attendance figure, predict how many people will choose each activity listed at the right.

ACTIVITY	
Jogging	60
Baseball	210
Picnic	120
Other	110

Attendance	Jogging	Baseball	Picnic	Other	
1,000	**1.** ■	**2.** ■	**3.** ■	**4.** ■	120; 420; 240; 220
1,200	**5.** ■	**6.** ■	**7.** ■	**8.** ■	144; 504; 288; 264
1,500	**9.** ■	**10.** ■	**11.** ■	**12.** ■	180; 630; 360; 330

A spinner is divided into eight equal sections. Two sections are green, three are red, and three are blue. It is spun 240 times.

13. Predict how many times a red section will be spun. 90

14. Predict how many times a blue section or a green section will be spun. 150

CHAPTER 10, PAGES 276–277

Solve.

1. Jim has 3 shirts in his closet and 5 pairs of pants. He is going to choose one shirt and one pair of pants. How many different outfits can he choose? 15 outfits

2. Ava has 4 skirts and 6 blouses in her closet. She is going to choose one skirt and one blouse. How many different outfits can she choose? 24 outfits

3. There are 3 routes from Milltown to Sawville and 3 routes from Sawville to Elmont. How many routes are there from Milltown to Elmont? 9 routes

4. There are 4 routes from Ellis to Danton and 5 routes from Danton to Lincoln. How many routes are there from Ellis to Lincoln? 20 routes

CHAPTER 10, PAGES 278–279

Write *independent* or *dependent* to describe the series of events.

1. Two players' moves in a chess game dependent

2. Tossing a penny five times independent

3. Drawing numbers in a bingo game and replacing them independent

4. Dialing a telephone number correctly dependent

5. Tossing a basketball through a basket 10 times independent

Linda put 10 marbles in a can. There are 4 white ones, 3 blue ones, 2 red ones, and 1 yellow one. Without looking, she will pick marbles and not replace them. Find the probability.

6. *P*(yellow) $\frac{1}{10}$

7. *P*(white, then white) $\frac{6}{45}$

8. *P*(blue, then yellow, then red) $\frac{1}{120}$

9. *P*(red, then white or blue) $\frac{7}{45}$

10. *P*(white, then blue, then red, then yellow) $\frac{1}{210}$

CHAPTER 11, PAGES 292–293

Draw and label the figure. Check students' answers.

1. $\overleftrightarrow{DC}$ **2.** $\angle ABC$ **3.** $\overline{CD}$ **4.** $\overrightarrow{XY}$ **5.** $\angle MNO$ **6.** plane *XYZ*

Identify the geometric figure.

7. $\angle L$ angle *L*

8. $\overline{EF}$ line segment *EF*

9. $\angle CDE$ angle *CDE*

10. $\overleftrightarrow{ST}$ line *ST*

11. *Q* point *Q*

12. $\overrightarrow{MR}$ ray *MR*

Name the ray in three ways. Answers may vary.

13.

14.

Name the six line segments.

15.

$\overline{JK}, \overline{KL}, \overline{LM}, \overline{JL}, \overline{JM}, \overline{KM}$

Name the angle in four ways.

16.

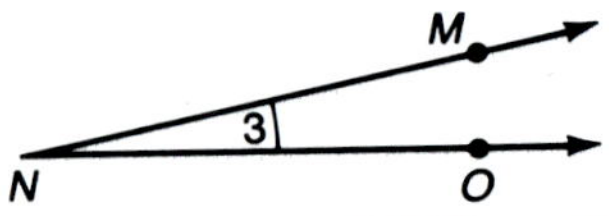

$\angle 3, \angle MNO, \angle ONM, \angle N$

Write *true* or *false*.

17. The length of a ray can be measured. False

18. A line segment can be named in at least two ways. True

CHAPTER 11, PAGES 296–297

Classify the angle as *acute, right, obtuse,* or *straight.*

1. 30° acute **2.** 43° acute **3.** 98° obtuse **4.** 90° right **5.** 9° acute **6.** 180° straight

7. 104° obtuse **8.** 25° acute **9.** 120° obtuse **10.** 145° obtuse **11.** 12° acute **12.** 16° acute

Identify the angle pair as *complementary, supplementary,* or *neither.*

13. 70°, 110° supp. **14.** 45°, 30° neither **15.** 60°, 30° comp. **16.** 15°, 165° supp.

17. 40°, 60° neither **18.** 65°, 25° comp. **19.** 72°, 18° comp. **20.** 100°, 75° neither

Find the complement of the angle.

21. 49° 41° **22.** 70° 20° **23.** 5° 85° **24.** 82° 8° **25.** 45° 45° **26.** 35° 55°

27. 10° 80° **28.** 33° 57° **29.** 28° 62° **30.** 75° 15° **31.** 56° 34° **32.** 15° 75°

Find the supplement of the angle.

33. 160° 20° **34.** 16° 164° **35.** 39° 141° **36.** 8° 172° **37.** 168° 12° **38.** 115° 65°

39. 125° 55° **40.** 6° 174° **41.** 27° 153° **42.** 140° 40° **43.** 89° 91° **44.** 175° 5°

CHAPTER 11, PAGES 298–300

Solve. Check students' constructions.

1. Use a ruler to draw a 6-cm line segment. Label it $\overline{CD}$. Construct $\overline{EF}$ so that $\overline{EF} \cong \overline{CD}$.

2. Use a protractor to draw a 30° angle. Label it $\angle SRO$. Construct $\angle TUV$ so that $m\angle TUV \cong m\angle SRO$.

CHAPTER 11, PAGES 301–303

Solve.

1. Draw a 6-in. line segment. Label it $\overline{RS}$. Use a compass and straightedge to bisect $\overline{RS}$ at point T. Bisect $\overline{RT}$ and $\overline{TS}$ at points U and V. The length of $\overline{RV}$ is what fractional part of the length of $\overline{RS}$? $\frac{3}{4}$

Trace the line segment and then construct the perpendicular bisector.
Check students' constructions.

2.
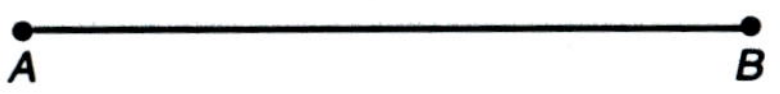

3.
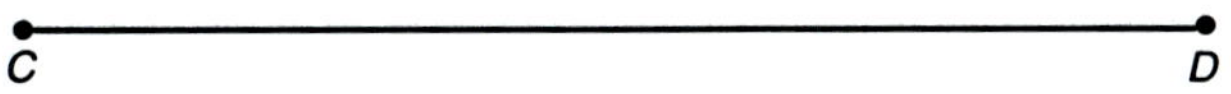

4.

5.

CHAPTER 11, PAGES 304–305

Identify and name the polygons.

1. 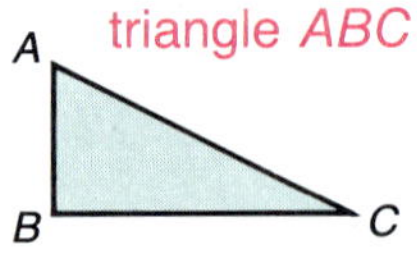

triangle *ABC*

2. 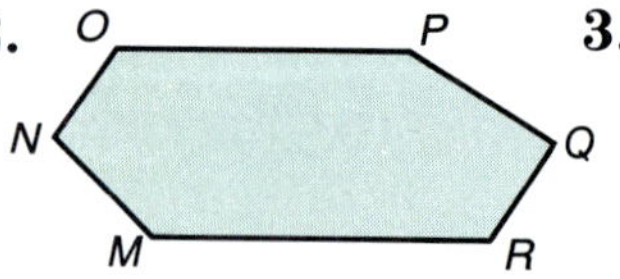

hexagon *MNOPQR*

3. 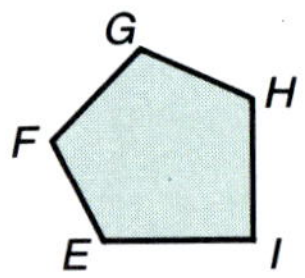

pentagon *EFGHI*

4. 

quadrilateral *STUV*

Find the missing length.

5. $r = 2.5$ cm, $d = $ ■ 5 cm

6. $r = 2\frac{3}{4}$ in., $d = $ ■ $5\frac{1}{2}$ in.

7. $r = 1\frac{1}{2}$ ft, $d = $ ■ 3 ft

8. $d = 9$ in., $r = $ ■ $4\frac{1}{2}$ in.

9. $d = 3.2$ m, $r = $ ■ 1.6 m

10. $d = 6\frac{1}{4}$ ft, $r = $ ■ $3\frac{1}{8}$ ft

11. $d = 6.3$ cm, $r = $ ■ 3.15 cm

12. $d = 7\frac{1}{2}$ in., $r = $ ■ $3\frac{3}{4}$ in.

13. $d = 10.4$ ft, $r = $ ■ 5.2 ft

CHAPTER 11, PAGES 312–313

Find each square root.

1. $\sqrt{16}$ 4 **2.** $\sqrt{25}$ 5 **3.** $\sqrt{36}$ 6 **4.** $\sqrt{49}$ 7 **5.** $\sqrt{121}$ 11 **6.** $\sqrt{144}$ 12

7. $\sqrt{17}$ 4.123 **8.** $\sqrt{20}$ 4.472 **9.** $\sqrt{42}$ 6.481 **10.** $\sqrt{53}$ 7.280 **11.** $\sqrt{74}$ 8.602 **12.** $\sqrt{96}$ 9.798

Find each square root. Round to the nearest tenth.

13. $\sqrt{14}$ 3.7 **14.** $\sqrt{19}$ 4.4 **15.** $\sqrt{24}$ 4.9 **16.** $\sqrt{31}$ 5.6 **17.** $\sqrt{46}$ 6.8 **18.** $\sqrt{52}$ 7.2

19. $\sqrt{58}$ 7.6 **20.** $\sqrt{67}$ 8.2 **21.** $\sqrt{75}$ 8.7 **22.** $\sqrt{88}$ 9.4 **23.** $\sqrt{92}$ 9.6 **24.** $\sqrt{98}$ 9.9

Find each square root. Use the table on page 313.

25. $\sqrt{324}$ 18 **26.** $\sqrt{729}$ 27 **27.** $\sqrt{1{,}156}$ 34 **28.** $\sqrt{1{,}764}$ 42 **29.** $\sqrt{2{,}209}$ 47

30. $\sqrt{3{,}136}$ 56 **31.** $\sqrt{4{,}356}$ 66 **32.** $\sqrt{5{,}329}$ 73 **33.** $\sqrt{7{,}569}$ 87 **34.** $\sqrt{8{,}281}$ 91

35. $\sqrt{2{,}601}$ 51 **36.** $\sqrt{5{,}041}$ 71 **37.** $\sqrt{8{,}649}$ 93 **38.** $\sqrt{4{,}624}$ 68 **39.** $\sqrt{5{,}776}$ 76

CHAPTER 11, PAGES 314–315

Determine whether each set of numbers can form a right triangle.

1. $a = 13$, $b = 12$, $c = 5$ yes

2. $a = 6$, $b = 7$, $c = 10$ no

3. $a = 9$, $b = 12$, $c = 15$ yes

4. $a = 30$, $b = 40$, $c = 50$ yes

5. $a = 5$, $b = 5$, $c = 10$ no

Find the missing length in the right triangle.

6. $a = 12$, $b = 16$, $c = $ ■ 20

7. $a = 10$, $b = 24$, $c = $ ■ 26

8. $a = 6$, $b = $ ■ 8, $c = 10$

9. $a = 3$, $b = 4$, $c = $ ■ 5

10. $a = $ ■ 20, $b = 21$, $c = 29$

CHAPTER 12, PAGES 330–331

Find the perimeter.

1. rectangle: $l = 4\frac{1}{2}$ ft, $w = 6\frac{3}{8}$ ft $21\frac{3}{4}$ ft
2. square: $s = 41\frac{7}{10}$ ft $166\frac{4}{5}$ ft
3. rectangle: $l = 10$ in., $w = 3\frac{1}{4}$ in. $26\frac{1}{2}$ in.
4. square: $s = 7.85$ m 31.4 m
5. regular pentagon: $s = 223$ yd 1,115 yd
6. regular octagon: $s = 14.2$ mm 113.6 mm
7. regular hexagon: $s = 5$ m 30 m
8. regular octagon: $s = 8.92$ mm 71.36 mm

CHAPTER 12, PAGES 332–333

Find the circumference of the circle. Use 3.14 for π with whole numbers and decimals. Use $\frac{22}{7}$ for π with fractions and mixed numbers.

1. $d = \frac{1}{4}$ yd $\frac{11}{14}$ yd
2. $d = \frac{1}{2}$ mi $1\frac{4}{7}$ mi
3. $d = 12\frac{3}{4}$ ft $40\frac{1}{14}$ ft
4. $d = 7$ km 21.98 km
5. $d = 11.2$ m 35.168 m
6. $d = 24.8$ cm 77.872 cm
7. $r = 6\frac{5}{8}$ yd $41\frac{9}{14}$ yd
8. $r = \frac{1}{6}$ mi $1\frac{1}{21}$ mi
9. $r = \frac{14}{15}$ ft $5\frac{13}{15}$ ft
10. $r = 8.5$ cm 53.38 cm
11. $r = 6$ m 37.68 m
12. $r = 0.2$ km 1.256 km

CHAPTER 12, PAGES 334–335

Find the area of the rectangle. **4.** 7.2 mm^2 **5.** 9.18 m^2

1. $l = 11$ ft, $w = 3$ ft 33 ft^2
2. $l = 3\frac{4}{5}$ ft, $w = 6$ ft $22\frac{4}{5}$ ft^2
3. $l = 1\frac{2}{3}$ yd, $w = \frac{1}{3}$ yd $\frac{5}{9}$ yd^2
4. $l = 3.6$ mm, $w = 2$ mm
5. $l = 5.1$ m, $w = 1.8$ m
6. $l = 0.9$ m, $w = 4$ m 3.6 m^2
7. $l = 4.25$ m, $w = 6.4$ m 27.2 m^2
8. $l = 0.3$ km, $w = 1.1$ km 0.033 km^2
9. $l = 17.6$ m, $w = 6.3$ m 110.88 m^2

Find the area of the square.

10. $s = 23$ in. 529 in.^{2}
11. $s = \frac{7}{8}$ in. $\frac{49}{64}$ in.^{2}
12. $s = \frac{2}{5}$ yd $\frac{4}{25}$ yd^2
13. $s = 14$ m 196 m^2
14. $s = 0.3$ cm 0.09 cm^2
15. $s = 1.1$ km 1.21 km^2
16. $s = 0.45$ km 0.2025 km^2
17. $s = 20.8$ m 432.64 m^2
18. $s = 3.12$ m 9.7344 m^2

Find the area of the parallelogram. **20.** $36\frac{1}{4}$ yd^2 **22.** 213.2 m^2 **23.** 1.5 cm^2

19. $b = 8$ ft, $h = 4$ ft 32 ft^2
20. $b = 7\frac{1}{4}$ yd, $h = 5$ yd
21. $b = 9\frac{1}{3}$ m, $h = 2$ m $18\frac{2}{3}$ m^2
22. $b = 52$ m, $h = 4.1$ m
23. $b = 0.5$ cm, $h = 3$ cm
24. $b = 6.4$ m, $h = 7$ m 44.8 m^2
25. $b = 7.25$ m, $h = 3.6$ m 26.1 m^2
26. $b = 5.83$ m, $h = 12.4$ m 72.292 m^2
27. $b = 41.2$ m, $h = 7.5$ m 309 m^2

CHAPTER 12, PAGES 336–337

Find the area of the triangle.

1. $b = 10$ in., $h = 7$ in. 35 in.^{2}
2. $b = 12.3$ cm, $h = 5$ cm 30.75 cm^2
3. $b = 1\frac{2}{7}$ m, $h = 4$ m $2\frac{4}{7}$ m^2
4. $b = 2\frac{1}{2}$ ft, $h = 4\frac{2}{3}$ ft $5\frac{5}{6}$ ft^2
5. $b = 8.7$ mm, $h = 4.5$ mm 19.575 mm^2
6. $b = \frac{11}{12}$ ft, $h = \frac{5}{6}$ ft $\frac{55}{144}$ ft^2

Find the area of the trapezoid. 7. 28 ft^2 8. 99 yd^2 9. 46.125 m^2 10. 19.36 m^2 11. 0.84 km^2 12. 4.2 m^2

7. $b_1 = 6$ ft, $b_2 = 8$ ft, $h = 4$ ft
8. $b_1 = 3$ yd, $b_2 = 15$ yd, $h = 11$ yd
9. $b_1 = 2.7$ m, $b_2 = 17.8$ m, $h = 4.5$ m
10. $b_1 = 2.6$ m, $b_2 = 9.5$ m, $h = 3.2$ m
11. $b_1 = 0.6$ km, $b_2 = 0.8$ km, $h = 1.2$ km
12. $b_1 = 2$ m, $b_2 = 0.4$ m, $h = 3.5$ m

CHAPTER 12, PAGES 338–339

Find the area of the circle. Use 3.14 for π with whole numbers and decimals. Use $\frac{22}{7}$ for π with fractions and mixed numbers.

1. $r = 16$ in. 803.84 in.^{2}
2. $r = 6$ mi 113.04 mi^2
3. $r = 12$ ft 452.16 ft^2
4. $r = 9$ yd 254.34 yd^2
5. $r = \frac{1}{5}$ ft $\frac{22}{175}$ ft^2
6. $r = 4\frac{1}{2}$ yd $63\frac{9}{14}$ yd^2
7. $r = 0.6$ in. 1.1304 in.^{2}
8. $r = 0.4$ ft 0.5024 ft^2

Find the area of the circle. Remember that the measure of the diameter is twice the measure of the radius.

9. $d = 13$ yd $132\frac{11}{14}$ yd^2
10. $d = 0.2$ m 0.0314 m^2
11. $d = 48$ in. 1,808.64 in.^{2}
12. $d = 32$ mm 803.84 mm^2
13. $d = 14$ m 153.86 m^2
14. $d = 4\frac{1}{2}$ in. $15\frac{51}{56}$ in.^{2}
15. $d = 6.8$ cm 36.2984 cm^2
16. $d = \frac{7}{12}$ ft $\frac{77}{288}$ ft^2

CHAPTER 12, PAGES 340–341

Solve by writing a simpler problem.

1. Barbra Hunter has 187 chickens. Each chicken eats $\frac{1}{4}$ lb of feed each day. Each pound costs $0.20. How much does Barbra spend on feed each day? $9.35
2. Chris Hunter has 6 corn fields. He uses 3,000 lb of fertilizer on each one. If fertilizer costs $59.10 per ton, what will his total cost be? $531.90
3. Christy raises evergreen trees. Each tree needs 6.4 m by 4.5 m to grow. She wants to plant 115 trees. How many square meters does she need? 3,312 m^2
4. Christy charges $29.95 for a Scotch pine tree, $19.95 for a Balsam tree, and $39.95 for a Douglas fir tree. She has an order for 24 balsams, 18 pines, and 48 firs. How much will she charge? $2,935.50

CHAPTER 12, PAGES 344–346

Identify the figure.

1.
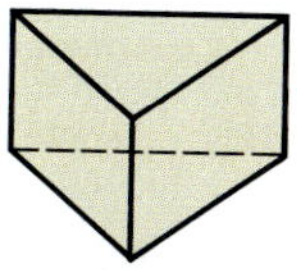
triangular prism

2.
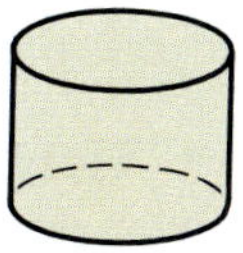
cylinder

3.
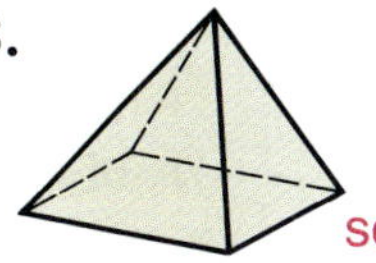
square pyramid

4.
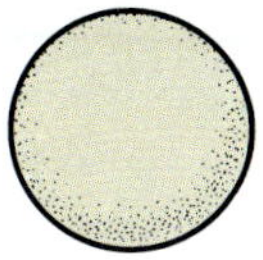
sphere

5.
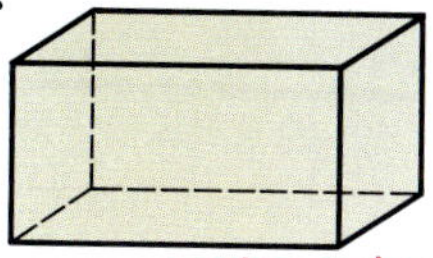
rectangular prism

6.
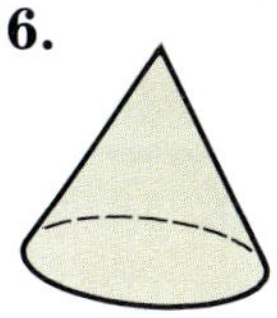
cone

CHAPTER 12, PAGES 347–348

Copy and complete the chart. The set of measurements describes a rectangular prism.

Length	Width	Height	Surface area	
1. 6 ft	8 ft	9 ft	■	348 ft^2
2. $2\frac{1}{2}$ yd	$6\frac{1}{2}$ yd	3 yd	■	$86\frac{1}{2}$ yd^2
3. 3.2 cm	5.8 cm	7 cm	■	163.12 cm^2
4. 0.9 m	0.9 m	0.9 m	■	4.86 m^2

CHAPTER 12, PAGES 350–351

Find the volume of the rectangular prism.

	Length	Width	Height		Length	Width	Height
1.	10 ft	4 ft	15 ft 600 ft^3	**2.**	8 ft	8 ft	8 ft 512 ft^3
3.	2.5 cm	7 cm	5.6 cm 98 cm^3	**4.**	11 m	3 m	6 m 198 m^3
5.	0.1 m	0.1 m	0.1 m $\frac{1}{1,000}$ m^3	**6.**	10 m	6.2 m	0.5 m 31 m^3
7.	$\frac{1}{2}$ ft	$3\frac{1}{2}$ ft	$\frac{3}{4}$ ft $1\frac{5}{16}$ ft^3	**8.**	$2\frac{1}{3}$ in.	$\frac{1}{2}$ in.	$4\frac{2}{3}$ in. $5\frac{4}{9}$ $in.^3$

Find the volume of the rectangular pyramid.

	Length	Width	Height		Length	Width	Height
9.	10 ft	4 ft	15 ft 200 ft^3	**10.**	13 yd	4 yd	8 yd $138\frac{2}{3}$ yd^3
11.	$3\frac{1}{2}$ in.	5 in.	4 in. $23\frac{1}{3}$ $in.^3$	**12.**	$\frac{1}{2}$ ft	$\frac{1}{2}$ ft	14 ft $1\frac{1}{6}$ ft^3
13.	9 m	2 m	5 m 30 m^3	**14.**	7 cm	7 cm	7 cm $114\frac{1}{3}$ cm^3
15.	6.2 cm	4 cm	5.1 cm 42.16 cm^3	**16.**	0.4 m	1.2 m	5 m 0.8 m^3

CHAPTER 12, PAGES 352–353

Find the volume of the cylinder. Use 3.14 for π with whole numbers and decimals. Use $\frac{22}{7}$ for π with fractions and mixed numbers.

	Radius	Height		Radius	Height		Radius	Height
1.	2 ft	8 ft 100.48 ft^3	**2.**	1 ft	4 ft 12.56 ft^3	**3.**	6 m	8 m 904.32 cm^3
4.	$\frac{1}{2}$ yd	6 yd $4\frac{5}{7}$ yd^3	**5.**	2 yd	$5\frac{1}{4}$ yd 66 yd^3	**6.**	0.3 m	4 m 1.1304 m^3
7.	3 m	9.1 m 257.166 m^3	**8.**	9 km	3.2 km 813.888 km^3	**9.**	8 cm	2.4 cm 482.304 cm^3

Find the volume of the cone. Use 3.14 for π for whole numbers and decimals. Use $\frac{22}{7}$ for π with fractions and mixed numbers.

	Radius	Height		Radius	Height		Radius	Height
10.	$4\frac{3}{4}$ ft	4 ft $94\frac{23}{42}$ ft^3	**11.**	$\frac{1}{4}$ yd	4 yd $\frac{11}{42}$ yd^3	**12.**	22 ft	$2\frac{1}{2}$ ft $1{,}267\frac{13}{21}$ ft
13.	6 m	9.2 m 346.656 m^3	**14.**	8 m	5.4 m 361.728 m^3	**15.**	0.5 cm	3 cm 0.785 cm^3

CHAPTER 13, PAGES 364–366

Write the opposite of the number.

1. +18 −18 **2.** −55 55 **3.** +37 −37 **4.** +6 −6 **5.** −21 21 **6.** +74 −74 **7.** −49 49

Compare the integers. Write > or <.

8. + 2 ● +6 < **9.** −88 ● +92 < **10.** −45 ● −76 > **11.** −1 ● +5 < **12.** −9 ● +3 <

13. −4 ● 0 < **14.** −7 ● −8 > **15.** +19 ● −22 > **16.** −36 ● +53 < **17.** −2 ● −3 >

Write the absolute value.

18. $|-2|$ 2 **19.** $|-8|$ 8 **20.** $|+75|$ 75 **21.** $|-15|$ 15 **22.** $|+134|$ 134

Use >, <, or = to make a true statement.

23. $|-4|$ ● $|+4|$ = **24.** $|+2|$ ● $|-6|$ < **25.** $|-7|$ ● $|0|$ > **26.** $|+10|$ ● $|+9|$ >

27. $|-18|$ ● $|+18|$ = **28.** $|-31|$ ● $|-31|$ = **29.** $|+14|$ ● $|-11|$ > **30.** $|0|$ ● $|1|$ <

CHAPTER 13, PAGES 367–369

Add.

1. 18 + 12 30 **2.** 43 + (−7) 36 **3.** −33 + (−35) −68 **4.** −52 + (−82) −134

5. −64 + 64 0 **6.** −19 + 9 −10 **7.** 22 + (−6) 16 **8.** 152 + (−50) 102

9. −70 + 44 −26 **10.** −14 + 25 11 **11.** −98 + (−30) −128 **12.** 86 + (−11) 75

CHAPTER 13, PAGES 370–371

Complete each statement.

1. $-5 - 3 = -5 + ■$ -3
2. $10 - 6 = 10 + ■$ -6
3. $-32 - (-47) = -32 + ■$ 47
4. $22 - (-73) = 22 + ■$ 73
5. $-64 - (-50) = -64 + ■$ 50
6. $-86 - (-21) = -86 + ■$ 21
7. $73 - (-59) = 73 + ■$ 59
8. $-116 - (-65) = -116 + ■$ 65
9. $-253 - 304 = -253 + ■$ -304
10. $-876 - 91 = -876 + ■$ -91

Subtract.

11. $6 - 3$ 3
12. $4 - 5$ -1
13. $1 - 8$ -7
14. $-7 - 5$ -12
15. $-2 - 2$ -4
16. $-8 - 3$ -11
17. $0 - (-12)$ 12
18. $1 - (-5)$ 6
19. $9 - (-7)$ 16
20. $-7 - (-2)$ -5
21. $-10 - (-3)$ -7
22. $-6 - (-4)$ -2
23. $17 - 30$ -13
24. $-21 - 7$ -28
25. $0 - 29$ -29
26. $-8 - (-31)$ 23
27. $-19 - (-45)$ 26
28. $-44 - 72$ -116
29. $-83 - (-98)$ 15
30. $4 - (-13)$ 17
31. $-23 - 37$ -60
32. $-56 - (-77)$ 21
33. $92 - (-87)$ 179
34. $-36 - 57$ -93

CHAPTER 13, PAGES 372–373

Multiply.

1. 3×9 27
2. -3×11 -33
3. $-8 \times (-5)$ 40
4. $-46 \times (-61)$ $2{,}806$
5. $4 \times (-14)$ -56
6. $-35 \times (-5)$ 175
7. 82×6 492
8. -27×9 -243
9. $15 \times (-100)$ $-1{,}500$
10. $55 \times (-14)$ -770
11. -176×23 $-4{,}048$
12. -16×52 -832
13. $2 \times (-3) \times 6$ -36
14. $-4 \times (-12) \times 1$ 48
15. $-15 \times (-3) \times (-2)$ -90
16. $6 \times 13 \times 5$ 390
17. $-9 \times (-3) \times 2$ 54
18. $10 \times (-1) \times (-7)$ 70
19. $-5 \times 3 \times (-4)$ 60
20. $12 \times 2 \times (-7)$ -168
21. $-8 \times 0 \times -17$ 0

CHAPTER 13, PAGES 374–375

Divide.

1. $7 \div 3$ $2\frac{1}{3}$
2. $8 \div 4$ 2
3. $16 \div 4$ 4
4. $20 \div 5$ 4
5. $-9 \div (-3)$ 3
6. $-64 \div (-16)$ 4
7. $-100 \div (-25)$ 4
8. $-46 \div (-2)$ 23
9. $60 \div (-3)$ -20
10. $36 \div (-6)$ -6
11. $63 \div (-7)$ -9
12. $51 \div (-17)$ -3
13. $-90 \div 5$ -18
14. $-44 \div 11$ -4
15. $-81 \div 9$ -9
16. $-75 \div 15$ -5
17. $84 \div (-12)$ -7
18. $-224 \div (-14)$ 16
19. $176 \div 22$ 8
20. $-420 \div (-28)$ 15
21. $-392 \div 7$ -56
22. $-168 \div (-21)$ 8
23. $255 \div (-3)$ -85
24. $-540 \div (-45)$ 12

CHAPTER 13, PAGES 378–379

Find the missing exponents.

1. $62 = 6.2 \times 10^{■}$ 1

2. $3{,}121 = 3.121 \times 10^{■}$ 3

3. $0.005 = 5 \times 10^{■}$ -3

4. $0.00085 = 8.5 \times 10^{■}$ -4

5. $7{,}000{,}000 = 7 \times 10^{■}$ 6

6. $76{,}400 = 7.64 \times 10^{■}$ 4

7. $0.000055 = 5.5 \times 10^{■}$ -5

8. $0.000000932 = 9.32 \times 10^{■}$ -7

Write in scientific notation.

9. 36 3.6×10^1

10. 230 2.3×10^2

11. 8,500 8.5×10^3

12. 0.5 5×10^{-1}

13. 0.62 6.2×10^{-1}

14. 0.019 1.9×10^{-2}

15. 1,500,000 1.5×10^6

16. 0.0000001 1×10^{-7}

Write as a standard numeral.

17. 4.1×10^3 4,100

18. 5.85×10^8 585,000,000

19. 3.71×10^{-7} 0.000000371

CHAPTER 13, PAGES 384–385

Compare. Write >, <, or =.

1. $\frac{3}{8}$ ● $\frac{5}{8}$ <

2. $-\frac{2}{7}$ ● $-\frac{4}{7}$ >

3. $-\frac{5}{20}$ ● $-\frac{1}{4}$ =

4. $-5\frac{4}{8}$ ● $-5\frac{1}{2}$ =

5. -0.8 ● -0.80 =

6. -0.27 ● 0.27 <

7. -83 ● -0.349 <

8. 6.28 ● -6.8 >

Write the absolute value.

9. $\left|\frac{4}{5}\right|$ $\frac{4}{5}$

10. $\left|\frac{1}{15}\right|$ $\frac{1}{15}$

11. $\left|-2\frac{7}{8}\right|$ $2\frac{7}{8}$

12. $|-0.6|$ 0.6

13. $|1.6|$ 1.6

CHAPTER 13, PAGES 386–387

Add or subtract.

1. $\frac{1}{6} + \left(-\frac{7}{12}\right)$ $-\frac{5}{12}$

2. $-2\frac{1}{3} + \frac{8}{9}$ $-1\frac{4}{9}$

3. $-0.80 + \frac{4}{5}$ 0

4. $-7\frac{1}{8} - \left(1\frac{1}{4}\right)$ $-8\frac{3}{8}$

5. $-48 + 23.14$ -24.86

6. $-83 - 13.07$ -96.07

7. $10.2 + (-0.08)$ 10.12

8. $5.89 + (-14.37)$ -8.48

9. $4 - \left(-1\frac{2}{3}\right)$ $5\frac{2}{3}$

10. $-17.04 + (-0.258)$ -17.298

11. $-60.05 - 23.47$ -83.52

12. $-2\frac{1}{8} + \frac{3}{4}$ $-1\frac{3}{8}$

Multiply or divide.

13. $\frac{2}{3} \times \frac{5}{6}$ $\frac{5}{9}$

14. $6\frac{1}{8} \times 3$ $18\frac{3}{8}$

15. $-1.2 \div \frac{3}{5}$ -2

16. $60 \times \left(-\frac{1}{4}\right)$ -15

17. $-70 \times (-35)$ 2,450

18. $24 \div (-8)$ -3

19. $1{,}000 \div 300$ 3.33

20. $-16 \times (-0.02)$ 0.32

21. $0.2 \times (-3.7)$ -0.74

22. $-2\frac{1}{5} \div \left(-\frac{1}{3}\right)$ $6\frac{3}{5}$

23. $1.96 \div (-0.04)$ -49

24. $-\frac{3}{8} \times \frac{2}{5}$ $-\frac{3}{20}$

CHAPTER 13, PAGES 388–389

Solve each problem by using one or more problem-solving strategies.

1. The Special Events Office advertises horseback rides through the park for 15 people at a time. They must have exactly 15 people in order to form a ride. If 72 people sign up, how many rides can they form? How many people will not ride? 4 rides; 12 people

2. The Parks Department rents out rowboats for $12 per hour. Ten boats are available between 10 A.M. and 5 P.M. Two people must ride in each boat. 140 people want to rent boats the first day. How much money does the Parks Department earn in boat rentals that day? $840

3. Fly-Hi Airlines sold 1,024 tickets for flights from Los Angeles to New York. Each plane holds 320 passengers. How many planes can Fly-Hi fill? How many passengers are on the unfilled flight? 3 planes; 64 passengers

4. Marion needs to make 2 sawhorses to hold up her table top. It takes 5 pieces of lumber each 95 cm long to make one sawhorse. How much lumber will be left over if she buys ten 1-m lengths? 10 pieces 5 cm each

5. Calvin works in a bottling plant. His boss has told him to increase his output to 1,500 bottles a day. Calvin bottles 525 bottles the first day, 600 bottles the second day, 675 bottles the third day, and so on. At this rate, how long will it take Calvin to reach his goal of 1,500 bottles per day? 14 days

6. An airport was built in the shape of a square. It measures 4 km on each side. A fuel truck entered the airport at the center of the north side and traveled 1.5 km south. It then traveled 1.5 km west, then 0.5 km south. At that point how far was the truck from the western border of the airport? 0.5 km

CHAPTER 14, PAGES 398–399

Write the operation and the number(s) you would compute first to simplify.

1. $2 \cdot 4 + 3$ $2 \cdot 4$
2. $5(7 + 8)$ $(7 + 8)$
3. $2 + 1 \cdot 4$ $1 \cdot 4$
4. $(9 + 6)7$ $(9 + 6)$
5. $11 - 2 - 8$ $11 - 2$
6. $(12 - 5) - 3$ $(12 - 5)$
7. $8 - 5 + 4$ $8 - 5$
8. $13 - 1 \cdot 7$ $1 \cdot 7$
9. $11 - 7 + 3$ $11 - 7$
10. $15 \div 3 + 6$ $15 \div 3$
11. $15 + 8^2$ 8^2
12. $(10 - 2)^2$ $10 - 2$

Simplify.

13. $(5 - 2) \times 4$ 12
14. $(7 - 3) - 1$ 3
15. $10 \div (9 - 4)$ 2
16. $8 - 2 + 6$ 12
17. $22 - 4^2 + 1$ 7
18. $40 + 9 - 3$ 46
19. $4^2 - (15 - 5)$ 6
20. $(2 + 6) \div 4$ 2
21. $\frac{10 - 3 + 5}{6}$ 2
22. $4 + (16 - 8)^2$ 68
23. $\frac{(2^3 + 4)7}{4}$ 21
24. $\frac{96 - (2 + 4)^2}{15}$ 4

Write parentheses to make each sentence true.

25. $(32 \div 8) - 5 - 4 = -5$
26. $(15 + 5) \div (4 + 1) = 4$
27. $(144 \div 12 - 3) \cdot 2 = 18$
28. $(12 + 96) \div (6 - 4) = 54$

CHAPTER 14, PAGES 400–401

Write an algebraic expression for the word phrase. Let n represent the number.

1. The quotient of a number divided by 70 $\frac{n}{70}$

2. 5 added to a number $n + 5$

3. 83 decreased by twice a number $83 - 2n$

4. 44 minus a number $44 - n$

5. The product of a number and 15 $n \times 15$

6. A number decreased by 20 $n - 20$

7. The ratio of a number to 34 $n \div 34$

8. A number divided by 18 $\frac{n}{18}$

9. The product of 35 and the difference of a number minus 12 $35(n - 12)$

10. The quotient of a number plus 8, divided by 21 $\frac{n + 8}{21}$

CHAPTER 14, PAGES 402–404

Evaluate for $a = -3$ and $b = 4$.

1. $b - a$ 7

2. $5 + a - b$ −2

3. $4b + a$ 13

4. $2(a - b)$ −14

5. $6a - b$ −22

6. $4 + (a - b)$ −3

7. $a - 2b$ −11

8. $3b \div a$ −4

Evaluate for $x = 2$, $y = -2$, and $z = 3$.

9. $(x - y)z$ 12

10. $\frac{8(z - y)}{4x}$ 5

11. $\frac{y + z}{x}$ $\frac{1}{2}$

12. $\frac{xz}{y}$ −3

13. $\frac{2(y - z)}{x}$

The formula for the area of a triangle is $A = \frac{1}{2}bh$, where b is the length of the base and h is the height. Find the area.

14. $b = 12$ m; $h = 2$ m $12\ m^2$

15. $b = 3.2$ ft; $h = 6.4$ ft $10.24\ ft^2$

16. $b = 5$ m; $h = 8.7$ m $21.75\ m^2$

17. $b = 4.5$ cm; $h = 2.3$ cm $5.175\ cm^2$

Some salespeople are paid a commission based on their total amount of sales. The formula for commission is $C = R \times A$, where C is the commission, R is the rate, and A is the amount of sales. Find the commission.

18. $R = 5\%$; $A = \$30,000$ \$1,500

19. $R = 7\%$; $A = \$50,000$ \$3,500

20. $R = 2\%$; $A = \$10,000$ \$200

CHAPTER 14, PAGES 408–409

Match the graph to the correct number sentence.

1. 

c

2. 

b

3. 

d

4. 

e

5. 

a

a. $x \leq -2$

b. $x = -1$

c. $x > 2$

d. $x < 0$

e. $x = 3$

CHAPTER 14, PAGES 412–413

Solve and check.

1. $x + 3 = 10$ 7 **2.** $x + 8 = 14$ 6 **3.** $x + 5 = 20$ 15 **4.** $x + 4 = 17$ 13

5. $x + 6 = 33$ 27 **6.** $9 + x = 18$ 9 **7.** $42 + x = 56$ 14 **8.** $x + 37 = 62$ 25

9. $x + 55 = 73$ 18 **10.** $x + 47 = 98$ 51 **11.** $x + 66 = 89$ 23 **12.** $54 + x = 75$ 21

13. $76 + x = 79$ 3 **14.** $32 + x = 69$ 37 **15.** $5 + x = 99$ 94 **16.** $9 + x = 123$ 114

17. $-4 + x = 8$ 12 **18.** $-7 + x = 9$ 16 **19.** $-5 + x = 13$ 18 **20.** $-6 + x = 14$ 20

21. $x - 3 = 16$ 19 **22.** $x - 5 = 18$ 23 **23.** $x - 33 = 13$ 46 **24.** $x - 74 = 38$ 112

25. $2 + x + 5 = 15$ 8 **26.** $8 + x + 9 = 22$ 5 **27.** $11 + x + 7 = 277$ 259

CHAPTER 14, PAGES 414–415

Solve.

1. $4x = 16$ 4 **2.** $3x = 30$ 10 **3.** $5x = 10$ 2 **4.** $8x = 32$ 4 **5.** $2x = 18$ 9

6. $-3x = 48$ -16 **7.** $9x = -27$ -3 **8.** $-7x = 56$ -8 **9.** $4x = 64$ 16 **10.** $-2x = 86$ -43

11. $-\frac{y}{2} = -22$ 44 **12.** $\frac{x}{6} = 16$ 96 **13.** $\frac{x}{2} = -38$ -76 **14.** $\frac{x}{5} = -13$ -65 **15.** $-\frac{x}{4} = 20$ -80

16. $0.01x = 2$ 200 **17.** $0.5x = 7$ 14 **18.** $0.02x = 16$ 800 **19.** $0.25x = 11$ 44 **20.** $0.9x = 81$ 90

CHAPTER 14, PAGES 416–417

Solve and check.

1. $3x + 2 = 11$ 3 **2.** $2x + 9 = 19$ 5 **3.** $4x + 3 = 27$ 6 **4.** $8x + 1 = 33$ 4

5. $2x - 4 = 14$ 9 **6.** $8x - 3 = 37$ 5 **7.** $5x - 6 = 29$ 7 **8.** $7x - 4 = 10$ 2

9. $2x - 4 = 196$ 100 **10.** $-3x - 13 = -64$ 17 **11.** $-6x - 12 = -102$ 15 **12.** $4x + 1 = 9$ 2

13. $\frac{x}{12} - 5 = 2$ 84 **14.** $\frac{x}{7} + 11 = 6$ -35 **15.** $\frac{x}{4} - 17 = 3$ 80 **16.** $\frac{x}{9} + 7 = 18$ 99

CHAPTER 14, PAGES 418–420

Find the solution to the equation when $x = 3$, then when $y = -2$.

1. $x + y = 14$ 11, 16 **2.** $x - y = 1$ 2, -1 **3.** $xy = 18$ 6, -9 **4.** $4x = y$ 12, $-\frac{1}{2}$

5. $6 - 2x = y$ 0, 4 **6.** $4x + 1 = y$ 13, $-\frac{3}{4}$ **7.** $2x - y = 4$ 2, 1 **8.** $-3y + x = 6$ -1, 0

Determine if the ordered pair is a solution of the equation. Write *yes* or *no*.

9. $x + y = 6\,(7, -1)$ yes **10.** $8 - x = y\,(6, 2)$ yes **11.** $\frac{3x}{y} = 10\,(10, 6)$ no

12. $4x - 2y = 0\,(2, 4)$ yes **13.** $\frac{x}{2y} = 6\,(-36, -3)$ yes **14.** $-3x + 2y = 7\,(-1, -2)$ no

GLOSSARY

absolute value The distance a number is from zero on the number line. [365]

acute angle An angle that measures more than 0° but less than 90°. [296]

acute triangle A triangle that has three acute angles. [306]

addends The numbers being added. In $3 + 6 = 9$, 3 and 6 are the addends. [30]

algebraic expression A mathematical phrase that uses symbols and variables to stand for one or more unknown numbers. [400]

angle Two rays with the same endpoint. [292]

angle bisector A ray that divides an angle into two parts. [299]

area The number of square units needed to cover a surface of a plane figure. [334]

bar graph A graph used to compare quantities by using horizontal or vertical bars. [106]

bisect To divide into two equal parts. [299]

broken-line graph A graph using points connected by line segments to show change over time. [108]

central angle An angle whose vertex is at the center of the circle. [304]

chord A line segment that has endpoints on the circle. [304]

circle The set of all points in a plane that are the same distance from the center. [304]

circle graph A graph using a divided circle to represent the parts of the whole. [317]

circumference The distance around a circle. [332]

complementary angles Two angles where the sum of their measures is 90°. [296]

compound interest Interest computed on the principal and on the interest previously earned. [261]

cone A three-dimensional figure with one circular base and a curved lateral surface that comes to a point. [345]

congruent Geometric figures that have the same shape and size are congruent. [298]

coordinates The ordered pair that corresponds to a point located on a grid. [380]

cube A rectangular prism with six congruent square faces. [344]

cylinder A three-dimensional figure with two congruent circular bases in parallel planes and a curved lateral surface. [345]

decimal A base 10 numeral that uses place value and contains a decimal point. [6]

degree A standard unit of measure for angles. [294]

denominator The bottom number of a fraction that identifies the total number of equal parts. In $\frac{1}{2}$, 2 is the denominator. [122]

dependent events When the outcome of the first event affects the outcome of the second event. [278]

diameter A chord that passes through the center of a circle. [304]

difference The result when subtracting numbers. In $5 - 3 = 2$, the difference is 2. [28]

digits The 10 symbols used to form numbers. 0, 1, 2, 3, 4, 5, 6, 7, 8, and 9 are digits. [4]

discount The amount that is subtracted from the list price of an item. [256]

dividend The number to be divided. In $40 \div 8 = 5$, the dividend is 40. [46]

divisible When one number divides another without a remainder. 20 is divisible by 4. [58]

divisor The number that is used to divide another number. In $20 \div 4 = 5$, the divisor is 4. [46]

edge A line segment in a polyhedron formed where two faces intersect. [344]

equation A statement of equality between two quantities using the "=" sign. [406]

equiangular triangle A triangle whose angles are of equal measure. [306]

equilateral triangle A triangle whose sides are of equal length. [306]

equivalent Having the same value. $2\frac{1}{2}$ is equivalent to $\frac{5}{2}$. [126]

estimate To approximate a value. [28]

exponent A number that tells how many times the base is used as a factor. [18]

face A side of a polyhedron. [344]

factors Numbers to be multiplied. In $4 \times 3 = 12$, 4 and 3 are factors of 12. [40]

formula An equation that expresses a relationship between two or more quantities involving more than one variable. [402]

fraction A part of a whole, or the expressed division of two whole numbers. [122]

greatest common factor (GCF) The greatest number which is a common factor of two or more numbers. [124]

hexagon A six-sided polygon. [304]

histogram A bar graph in which the bars represent the frequency of occurrence. The bars are placed next to each other to show changes in data in different intervals (periods of time). [92]

hypotenuse In the right triangle, the longest side which is also opposite the right angle. [314]

improper fraction A fraction with a numerator equal to or greater than the denominator. $\frac{22}{5}$ is an improper fraction. [132]

independent events When the outcome of the first event does not affect the outcome of a second event. [278]

inequality When two quantities do not have the same value. The symbols $>$, $<$, $\geq$, and $\leq$ designate this condition. [406]

integer A positive or negative whole number or zero. [364]

interest A charge for the use of money. Usually a percent of the amount borrowed, invested, or loaned. [260]

inverse operation Operations that "undo" each other. Addition and subtraction are inverse operations since $5 + 2 = 7$ and $7 - 2 = 5$. Similarly, multiplication and division are inverse operations since $4 \times 2 = 8$ and $8 \div 2 = 4$. [412, 414]

isosceles triangle A triangle with two sides of equal length. [306]

key On a map or graph shows the symbol, its meaning, and the amount it represents. [112]

least common denominator (LCD) The smallest number divisible by a set of denominators. The LCD of $\frac{1}{2}$, $\frac{1}{4}$, and $\frac{1}{5}$ is 20. [158]

least common multiple (LCM) The least number evenly divisible by two or more factors. [124]

legs In a right triangle, they are the two sides which are not the hypotenuse. [314]

line A set of points that is a straight path extending in opposite directions without end. [292]

line segment Part of a line, consisting of two endpoints and all the points in between. [292]

mean Statistical measure of the average of a set of data, found by dividing the sum of the items by the number of items. [98]

median Statistical measure of the middle value when a set of numbers is listed in order. [100]

mixed number A whole number and a fraction. [132]

mode Statistical measure of the value that occurs most often in a set of data. [100]

multiple The product of two factors. 6 is a multiple of 3 and 2. [124]

negative integers Numbers to the left of zero on the number line; the opposites of the positive integers. [364]

numerator The top part of the fraction that tells how many parts are being counted. In $\frac{3}{5}$, the numerator is 3. [122]

obtuse angle An angle that measures more than 90° but less than 180°. [296]

obtuse triangle A triangle that contains one obtuse angle. [306]

octagon An eight-sided polygon. [304]

odds A ratio describing the likelihood of an event; the ratio of favorable outcomes to unfavorable outcomes. [269]

opposites Any pair of integers whose sum is zero. [364]

463

order of operations To simplify an expression using the proper sequence of operations. First raise to a power, then multiply or divide from left to right. Finally add or subtract from left to right. [398]

ordered pair Two numbers that name a specific point in a coordinate plane. [380]

origin The point assigned to zero on the number line or the point where the horizontal and the vertical axes meet. [380]

outcome A result of an event. [268]

parallel lines Two lines in a plane that do not intersect. [301]

parallelogram A quadrilateral with opposite sides parallel. [318]

pentagon A five-sided polygon. [304]

percent Per hundred. The ratio of a given number to 100; "hundredths" or "out of 100." [234]

percent of change The amount of change divided by the original amount and renamed as a percent. [258]

perimeter The distance around a polygon. [330]

perpendicular bisector A line perpendicular to a line segment that divides the segment into two congruent parts. [302]

perpendicular lines Lines that intersect and form a right angle. [301]

pi (π) The number that is the ratio of the circumference of a circle to its diameter. It is approximately equal to $\frac{22}{7}$ or 3.14. [332]

pictograph A graph in which a symbol is used to represent a definite number. [112]

plane A set of points forming a flat surface that extends in all directions without end. [292]

point A location in space with no size. [292]

polygon A simple closed curve consisting entirely of line segments. [304]

polyhedron A three-dimensional figure with faces as polygonal regions, edges as line segments, and vertices as points. [344]

positive integers Numbers to the right of zero on the number line; the opposites of the negative integers. [364]

power of 10 The number 10 or any other number that can be formed by using 10 as a factor two or more times. [36]

prism A three-dimensional figure that has two parallel bases of identical size and shape and sides that are parallelograms. [344]

probability The chance or likelihood that an event will occur. It is expressed as a fraction between 0 and 1, including 0 and 1; the ratio of favorable outcomes to the total number of possible outcomes. [268]

product The result of multiplying two numbers. In $2 \times 3 = 6$, the product is 6. [38]

proportion Two equivalent ratios. [216]

pyramid A three-dimensional figure whose base is any polygon, and which has triangular lateral faces that meet at a common vertex. [344]

Pythagorean theorem The relationship among the legs of a right triangle where the square of the length of the hypotenuse is equal to the sum of the squares of the lengths of the legs ($a^2 + b^2 = c^2$). [314]

quadrilateral A four-sided polygon. [304]

quotient The result obtained by dividing one number by another. In $42 \div 7 = 6$, the quotient is 6. [38]

radius A line segment that goes from the center of the circle to a point on it. [304]

range A measure of spread or dispersion, found by taking the difference between the highest and lowest numbers in a given group of numbers. [100]

rate A ratio that compares different units. [212]

ratio A comparison of two like quantities by division. The ratio of a to b is written $\frac{a}{b}$ or $a{:}b$. [210]

rational number Any number that can be expressed as the ratio of an integer to a nonzero integer. 9, 0, $3\frac{1}{4}$, $-4.\overline{7}$ are all rational numbers. [384]

ray A part of a line with one endpoint that extends indefinitely in one direction. [292]

reciprocals Two numbers with products of 1. $\frac{1}{2}$ and 2 are reciprocals since $\frac{1}{2} \times 2 = 1$. [150]

rectangle A parallelogram with all right angles. [318]

regular polygon A polygon with sides of equal length and with angles of equal measure. [304]

remainder The amount left over in a division problem. In $23 \div 5 = 4$ R3, the remainder is 3. [46]

repeating decimal A decimal that repeats a digit *or* a group of digits without end. $0.\overline{3}$ and $7.\overline{142857}$ are repeating decimals. [136]

rhombus A parallelogram with sides of equal length. [318]

right angle An angle which measures exactly 90°. [296]

right triangle A triangle with a right angle. [306]

sample space A listing of all the possible outcomes in a given situation. [276]

scale drawing A drawing of an object drawn in proportion to the actual object it represents. [224]

scalene triangle A triangle with no two sides of equal length. [306]

scientific notation A number between 1 and 10 and the second factor is a power of 10. For example, $50{,}000{,}000 = 5 \times 10^7$ and $0.0000045 = 4.5 \times 10^{-6}$. [378]

similar figures Figures that have the same shape but may not have the same size. Their corresponding angles are congruent, and the ratios of the lengths of the corresponding sides are equal. [320]

solid figure A three-dimensional figure having length, width, and height. [344]

sphere A three-dimensional figure with a curved surface of points that are the same distance from the center. [345]

square A rectangle with sides of equal length. [318]

square root A factor which multiplied by itself yields a product. In $8^2 = 64$, the square root of 64 is 8. [312]

straight angle An angle that measures exactly 180°. [296]

sum The result of adding 2 or more numbers. In $2 + 3 = 5$, the sum is 5. [28]

supplementary angles Two angles where the sum of their measures is 180°. [296]

surface area The total area of the surface of a solid figure. [347]

transversal In a plane, a line that intersects two or more lines at a different point. [326]

trapezoid A quadrilateral with only one pair of parallel sides. [318]

triangle A three-sided polygon. [304]

tree diagram A drawing that shows all possible outcomes using a "branch" to represent each choice or possible outcome. [276]

variable A letter or symbol used to represent a number. [400]

vertex of an angle The common endpoint of the two rays of an angle. [292]

vertices of a polygon The points formed by the intersection of the sides. [304]

vertices of a polyhedron The points where the edges intersect. [344]

volume The amount of space inside a three-dimensional figure, measured in cubic units. [350]

whole number A number such as 0, 1, 2, ... Each successive number is formed by adding 1 to the previous number. [4]

SELECTED ANSWERS

CHAPTER 1

Page 3 **1.** How much? **2.** measure **21.** code **22.** location

Page 5 **1.** tens; 40 **2.** hundreds; 600 **21.** one hundred seventy-five **22.** three hundred fifty-eight **37.** 336,154 **38.** 893,201

Page 7 **1.** tenths; 0.1 **2.** hundredths; 0.06 **21.** five-tenths **22.** one and eight-tenths **37.** 4.27 **38.** 13.63 **53.** two thousand one hundred five and seventy-five hundredths

Page 9 **1.** $<$ **9.** $<$ **29.** b **33.** 9,742; 7,492; 4,279; 2,972 **37.** 36,081; 36,180; 38,016; 306,810 **40.** Paris and Istanbul

Page 11 **1.** 5,927 **5.** 2,749 **9.** 23,416 **10.** 23,436

Page 12 **1.** \$0.75 **7.** \$4.10

Page 15 **1.** a **2.** b

Page 17 **1.** 2,800; 3,000 **2.** 4,000; 4,000 **17.** 0; 0.02 **18.** 0.1; 0.08 **33.** \$5.00 **34.** \$40.00 **41.** \$0.86 **42.** \$1.90 **49.** 4,000 **50.** 49.5 **57.** \$0.36

Page 19 **1.** 2^2 **2.** 7^3 **13.** 4 **14.** 1 **20.** 10^2 **24.** 1 **25.** 1,000

Page 21 **1.** \$11.50 **4.** 3 items **10.** \$11.00

CHAPTER 2

Page 29 **1.** 1,100 **21.** 200 **31.** 1,100 **51.** 200

Page 31 **1.** b **5.** 110 **6.** 51 **40.** 710,700; 744,670 **44.** R: 8,000 mi; FE: 5,000 mi

Page 33 **1.** d **5.** 50 **6.** 14 **30.** 200; 190 **31.** 5,000; 5,172

Page 34 **39.** 700,000 **40.** 8,000,000 **45.** $>$ **46.** $<$ **51.** 15,524 cars

Page 35 **1.** 2 dimes, 1 penny = 21¢ **2.** 3 quarters, 1 nickel, 3 pennies = 83¢

Page 37 **1.** 200 **2.** 4,200 **21.** 80 **22.** 20

Page 39 **1.** 350 **21.** 40 **41.** 40 **47.** 70 **53.** 8,000 cans

Page 41 **1.** b **5.** 78 **30.** 2,241 **45.** 285 **49.** 1,924 **57.** 10,000; 11,540 **59.** 2,400; 2,166

Page 43 **1.** 96,228 **21.** 347,985 **41.** 193,824 **65.** \$65,403

Page 45 **1.** 1,840 **3.** 108

Page 47 **1.** 137 R2 **11.** 105 R2 **56.** 62 **57.** 224

Page 49 **1.** 8 R12 **6.** 6 R55 **46.** 10; 9 **47.** 30; 31

Page 51 **1.** 38 **13.** 20 R8 **41.** 1,062 **42.** 1,112 **49.** 95 boxes

Page 53 **1.** 1,488 cal **2.** 248 cal

Page 55 **1.** interested **2.** somewhat interested **10.** Band **11.** Dance Club, Future Teachers' Club, Science Club

CHAPTER 3

Page 63 **1.** 76.8 **6.** 91.583 **27.** 880; 873.704 **28.** 810; 810.94 **37.** 107.45 **41.** Snow

Page 65 **1.** 2.3 **6.** 11.629 **24.** 4; 3.57 **25.** 0; 0.0241 **29.** 7.847 **30.** 1.751 **35.** 1.62 seconds

Page 67 **1.** $280.00 **5.** $144.69

Page 69 **1.** 504 **2.** 72 **21.** 0.55 **22.** 0.9 **41.** 600.2 **42.** 0.006002 **53.** 17,000 **59.** $23.00

Page 71 **1.** b **2.** c

Page 73 **1.** $425.85 **2.** $370.60

Page 75 **1.** 44.24 **2.** 33.28 **31.** 480; 475.2 **32.** 1.4; 1.251 **44.** 2.002 **47.** $302.95 billion

Page 76 **1.** $84.00 **5.** $124.80

Page 77 **1.** 1.192 **3.** 0.139 L

Page 79 **1.** 6.83 **5.** 1.6 **37.** 6; 6.2 **38.** 1; 0.91

Page 81 **1.** 10 **2.** 100 **9.** 2 **10.** 12.008

Page 82 **39.** 7,000; 7,420 **40.** 40; 40.95 **59.** 11,458 **60.** 15,022 **75.** 38.7 weeks

Page 83 **1.** 26 **2.** 17.5 **9.** 13.5 mpg

Page 85 **1. moderate** **2. expensive** **13.** Discount **14.** Exclusive

CHAPTER 4

Page 93 **1.** 22 **2.** 31

Page 94 **6.** 2–3 **13.** 36.4 **14.** 12.22

Page 95 **1.** $12,009

Page 97 **1.** eggs **2.** milk **9.** 30 **10.** 35

Page 99 **1.** 7 **2.** 26 **13.** 118.88 **14.** 22.376

Page 101 **1.** 9 **2.** 6 **3.** 6 **4.** 6 **29.** 36 mph

Page 103 **1.** a **2.** b

Page 105 **1.** 26–50 **2.** over 50

Page 107 **1.**

Page 109 **1.**

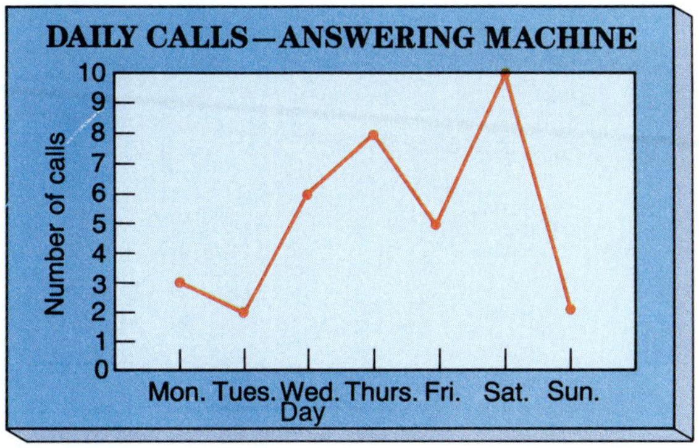

Page 111 **1.** $74.00 **2.** $41.00 **7.** $189.30 **9.** $100

Page 113 **1.** 10,000 **2.** green
7.

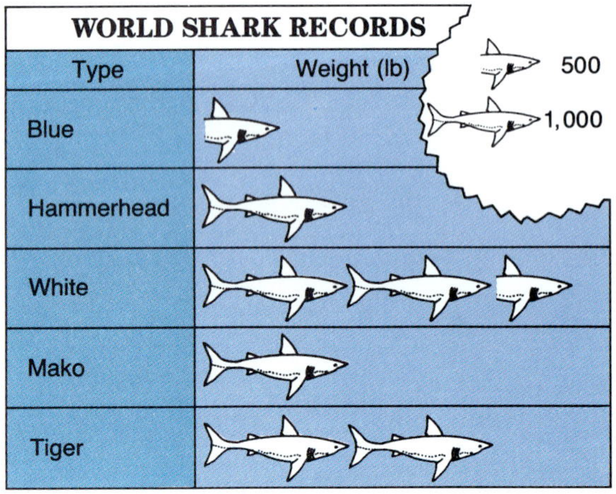

Page 115 **1.** \$355 **2.** \$404 **7.** \$4,108
11. \$3,232

CHAPTER 5

Page 123 **1.** $\frac{5}{6}$ **2.** $\frac{1}{4}$ **13.** $\frac{1}{4}$ **14.** $\frac{5}{12}$
19. $\frac{5}{6}$

Page 125 **1.** 2 **2.** 4 **11.** 2 **12.** 3
31. 6 **32.** 4 **41.** 12 **42.** 15

Page 127 **1.** $\frac{1}{3}$ **2.** $\frac{1}{2}$ **33.** 3 **34.** 6
53. = **54.** ≠ **63.** $\frac{2}{8}, \frac{1}{4}$

Page 129 **1.** < **2.** < **21.** < **22.** <

Page 130 **36.** $\frac{7}{8}, \frac{5}{8}, \frac{3}{8}$ **37.** $\frac{8}{10}, \frac{7}{10}, \frac{4}{10}$
44. $\frac{0}{6}, \frac{1}{6}, \frac{5}{6}$ **45.** $\frac{9}{24}, \frac{11}{24}, \frac{17}{24}$
52. 11,149 **53.** 15,039
68. blue paint **69.** No

Page 133 **1.** $\frac{7}{3}$ **2.** $\frac{5}{3}$ **33.** $1\frac{1}{2}$ **34.** $1\frac{1}{3}$
65. 14 **69.** 7

Page 134 **1.** b **2.** c

Page 137 **1.** $\frac{3}{10}$ **6.** $2\frac{7}{10}$ **25.** 0.5
33. 1.6 **49.** 0.33 **50.** 0.83
57. $0.\overline{3}$ **58.** $0.\overline{7}$

Page 138 **1.** $\frac{2}{5}$ **5.** transportation
10. housing

Page 139 **1.** 4 **2.** 5
7.

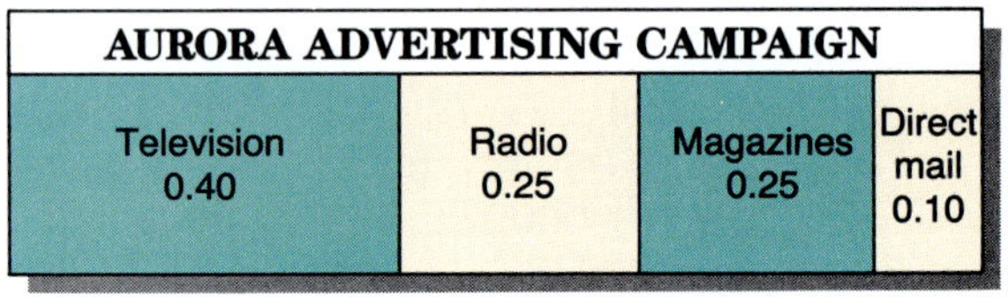

CHAPTER 6

Page 147 **1.** $\frac{1}{6}$ **13.** $\frac{3}{5}$ **37.** $\frac{1}{15}$ **38.** $\frac{1}{4}$
45. 415 **50.** 19

Page 149 **1.** 1 **11.** 3 **26.** 3
58. 27 mph

Page 151 **1.** $\frac{1}{3}$ **2.** $\frac{1}{7}$ **25.** $\frac{8}{9}$ **43.** $\frac{1}{4}$
55. 6 stakes

Page 153 **1.** 3 **6.** $\frac{1}{2}$ **11.** $\frac{4}{9}$ **21.** $\frac{3}{4}$

Page 155 **1.** a. 3; b. 27 **2.** a. \$55; b. \$85

Page 157 **1.** 12 **3.** $1\frac{1}{4}$ ft **7.** $\frac{3}{4}$ yd

Page 159 **1.** $\frac{3}{5}$ **11.** $\frac{9}{10}$ **16.** $1\frac{5}{12}$ **41.** $1\frac{1}{4}$
47. $\frac{11}{24}$

Page 161 **1.** $8\frac{1}{2}$ **6.** $7\frac{5}{8}$ **36.** $14\frac{3}{4}$
41. $5\frac{9}{10}$ mi

Page 163 **1.** $\frac{1}{3}$ **11.** $\frac{1}{4}$ **31.** $\frac{2}{3}$ **41.** 207
42. 7,480 **51.** 5 minutes

Page 165 **1.** $2\frac{1}{3}$ **6.** $3\frac{5}{8}$ **16.** $4\frac{11}{12}$ **26.** $3\frac{1}{4}$
40. 10.91 **41.** 48.374

Page 166 **49.** $2\frac{3}{8}$ lb **50.** $\frac{7}{8}$ lb

Page 167 **1.** \$3.15 **5.** \$2.25

Page 169 **1.** 5 **11.** 8 **16.** 1 **21.** 10 **26.** 3 **41.** 13 ft

Page 171 **1.** \$523.20; \$43.20 **9.** \$1,046.40; \$86.40

Page 173 **1.** about \$3,150 **2.** \$3,775 **9.** Ace; ourselves **10.** no

CHAPTER 7

Page 181 **1.** 24 **10.** 3 **28.** $\frac{3}{4}$ in. **29.** $2\frac{3}{16}$ in. **33.** Randolph

Page 183 **1.** 32 **10.** 5 **43.** $1\frac{1}{2}$ T

Page 185 **1.** 7 yd 2 ft **9.** 2 ft 1 in. **29.** 2 mi 1,110 yd

Page 187 **1.** 17 minutes **9.** 10:30 P.M. **15.** 3 hours 33 minutes **17.** 5:54 A.M. **21.** 12:43 P.M.

Page 188 **1.** 2 P.M.

Page 189 **7.** 8:30 A.M. **11.** no

Page 191 **1.** \$210.70 **2.** \$239.20 **9.** \$235.20

Page 193 **1.** b **2.** b **11.** 27 mm **12.** 66 mm **20.** centimeters

Page 195 **1.** b **7.** c

Page 196 **23.** $\frac{1}{6}$ **31.** $7\frac{7}{8}$ **39.** 45 kL is too much

Page 197 **1.** 9.5 cm **5.** 2 cm; 0.5 cm

Page 199 **1.** 2,000 **4.** 4 **37.** 32 **38.** 20 **39.** 17 **40.** 12 **53.** 74.4 kL

Page 201 **1.** 12°C **2.** 78°C **9.** c **10.** b **21.** no

Page 202 **1.** 67 cm **2.** 5

CHAPTER 8

Page 211 **1.** $\frac{6}{1}$ **2.** 6:1 **9.** 15 **10.** 16 **24.** 9 **25.** 18 **32.** 12 **33.** 4 **40.** $\frac{1}{3}$ **41.** $\frac{1}{6}$ **56.** $\frac{3}{1}$

Page 213 **1.** $\frac{\$20}{3}$ **2.** $\frac{7}{9}$ **19.** = **20.** > **28.** 8 pencils per pack **29.** 28 people per bus **38.** 3,280 **42.** 9.86 **46.** \$1.13

Page 215 **1.** \$0.11 per ounce **2.** \$0.065 per foot **7.** \$0.206 per ounce **8.** \$0.672 per can **12.** 3 lb for \$1.89 **13.** 6-oz jar for \$1.53

Page 217 **1.** no **2.** yes **9.** 15 **10.** 49 **25.** 0.25 **26.** 0.9 **31.** 14 cups

Page 219 **1.** b **2.** yes

Page 221 **1.** 290 mi **2.** 260 mi **7.** 5 **8.** 12.5 **13.** 378 mi

Page 223 **1.** 12.5 **7.** 9 in.

Page 225 **1.** 8 m **2.** 6 m **17.** 24 ft **18.** 18 ft **22.** 0.7 cm

Page 227 **1.** 5.0 cm **2.** 111 km **9.** 100 km **13.** 53 km **17.** 238 km

CHAPTER 9

Page 235 **1.** 0.28 **11.** 1.3 **16.** 0.0035 **26.** 0.0025 **31.** 0.035 **36.** 23% **46.** 370% **51.** $66\frac{2}{3}$% **56.** $8\frac{9}{11}$%

Page 237 **1.** $\frac{3}{100}$ **11.** $\frac{3}{25}$ **21.** 2 **31.** $1\frac{11}{100}$ **56.** 3.35 **57.** 2.16 **64.** 21 hours 48 minutes

Page 239 **1.** 50% **13.** $33\frac{1}{3}\%$ **25.** 300% **37.** 375% **55.** 2,776

Page 241 **1.** 1.2 **9.** 4 **17.** 4 **23.** 486 **31.** 20 **32.** 4 **39.** $\frac{5}{6}$

Page 243 **1.** $450 **2.** $1,050 **13.** $986.25; $130.25 **14.** $9,747.50; $3,897.50 **17.** $6,765.50

Page 245 **1.** $12.00; $36.00 **2.** $4.50; $19.50

Page 247 **1.** 40% **5.** 75% **9.** 200% **29.** 20% **33.** 50% **37.** $28\frac{4}{7}\%$

Page 249 **1.** 50 **2.** 20 **13.** 230 **14.** 200 **25.** 4.5 **26.** 12 **39.** 8,000 bits

Page 251 **1.** 7.2 **7.** 75% **13.** 40 **31.** $17,500

Page 253 **1.** b **3.** 108 ski poles

Page 255 **1.** $2,400 **7.** 3% **11.** $760

Page 257 **1.** $40 **2.** $9 **11.** $13.50; $76.50 **12.** $26.00; $104 **16.** $44

Page 259 **1.** 50% inc. **2.** $33\frac{1}{3}\%$ inc. **7.** 40% dec. **8.** 30% dec. **33.** 120 **34.** 28 **39.** 14%

Page 261 **1.** $60.00 **2.** $112.00 **21.** $522.84

CHAPTER 10

Page 269 **1.** $\frac{1}{4}$ **2.** $\frac{1}{12}$ **13.** 1:4 **14.** 3:2

Page 270 **25.** 18,000 **26.** 38 **37.** $\frac{1}{7}$

Page 271 **1.** 60.8 **5.** 76.6

Page 273 **1.** 25 **4.** 160 **5.** 60 **8.** 4,000 **9.** 3,000 **20.** 3,750 **21.** 8,750 **24.** $\frac{12}{25}$

Page 274 **1.** 250 **4.** 1,875 **7.** 0.25

Page 277 **1.** Aisle; Center; Window; Smoking, Center; Smoking, Window; Nonsmoking, Aisle; Nonsmoking, Center; Nonsmoking, Window **2.** 2 **3.** 3 **8.** 10,000 **9.** 625 **14.** 1,000 **15.** 17,576 **20.** $\frac{17}{20}$

Page 279 **1.** independent **4.** $\frac{1}{48}$ **5.** $\frac{1}{12}$ **13.** $\frac{2}{17}$ **14.** $\frac{1}{102}$ **23.** 27 **24.** 26.5

Page 281

1. Possible Combinations

1st digit	*2nd digit*
• 3	5
• 3	7
• 5	3
• 5	7
• 7	3
• 7	5

5. Tennis Matches

• Cathy	Ann
• Cathy	Wes
• Cathy	Glen
• Ann	Wes
• Ann	Glen
• Wes	Glen

Page 283 **1.** 60 **5.** 356 calculators **6.** 4,410 calculators

Page 285 **1.** $1,238 **4.** 0 **6.** 0 **9.** $1,238 **19.** car **20.** plane

CHAPTER 11

Page 293 **1.**

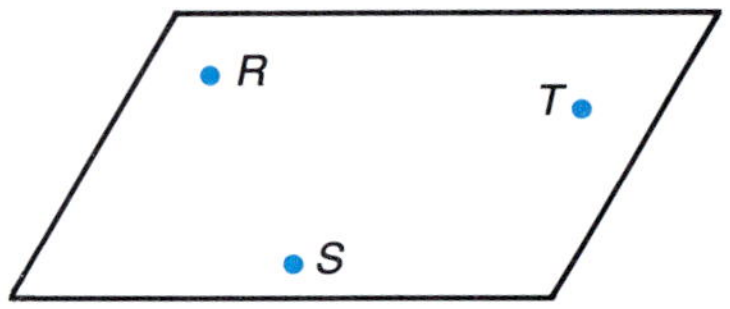

2.

7. angle T **8.** line segment BC **13.** line P **14.** angle 4 **21.** $\overrightarrow{AD}$, $\overrightarrow{AC}$, $\overrightarrow{AB}$ **25.** false

Page 295 **1.** 30° **2.** 45° **10.** 45°
11. 130°

18.

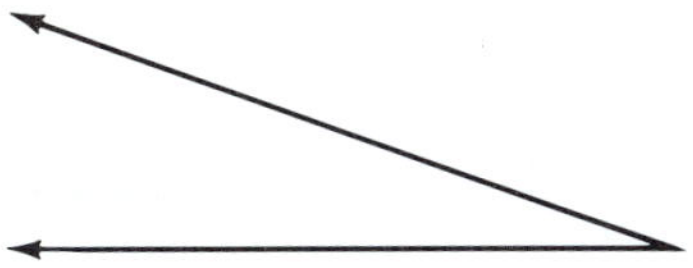

19.

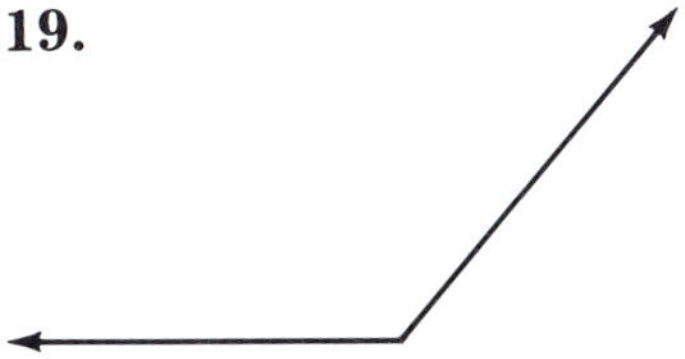

24. 110°

Page 297 **1.** obtuse **2.** acute **5.** acute
6. acute **11.** neither
12. complementary
16. complementary **20.** 70°
21. 35° **26.** 120° **27.** 40°

Page 299 **1.**

Page 300 **5.**

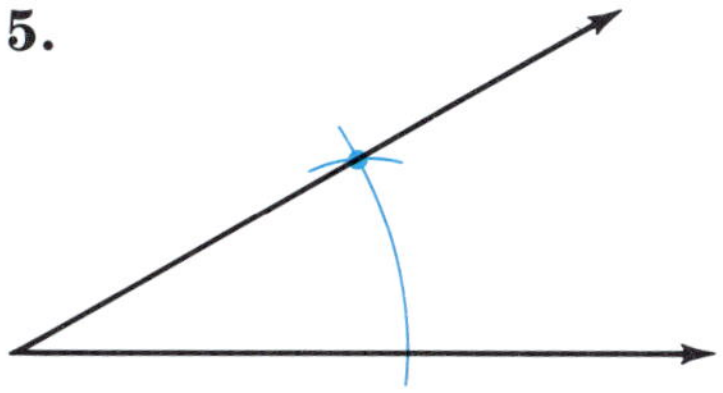

11.

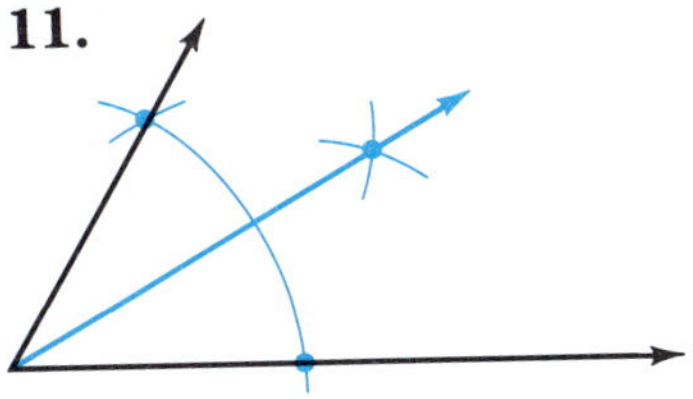

17. 36 **18.** 14

33.

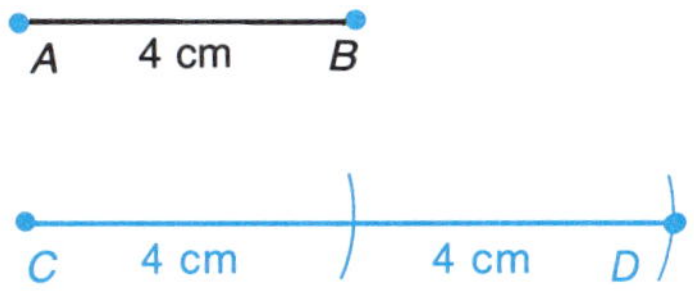

Page 303 **1.**

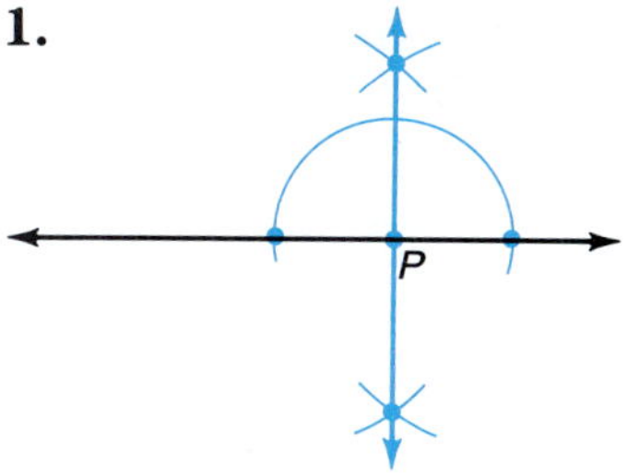

6.

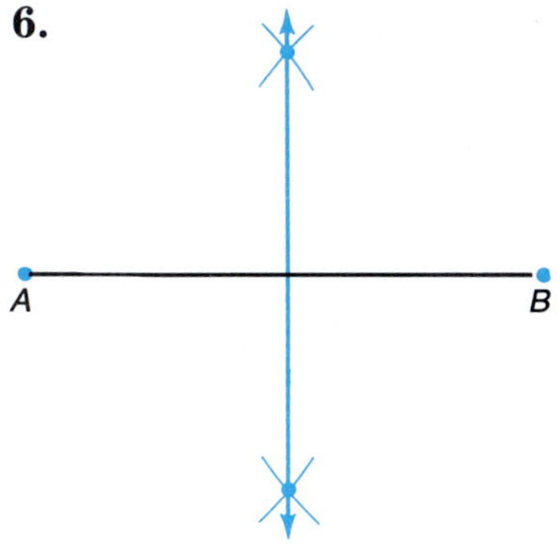

10.

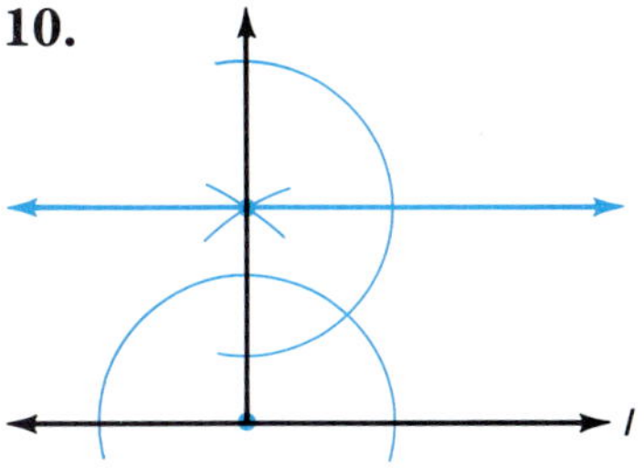

Page 305 **1.** triangle *ABC* **5.** diameter
9. 10 cm **15.** 4 ft **21.** 4

Page 307 **1.** isosceles **2.** equilateral
9. acute **10.** equiangular
17.

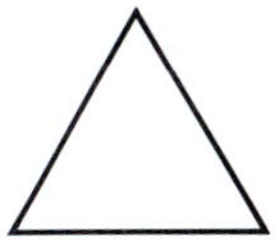

18.

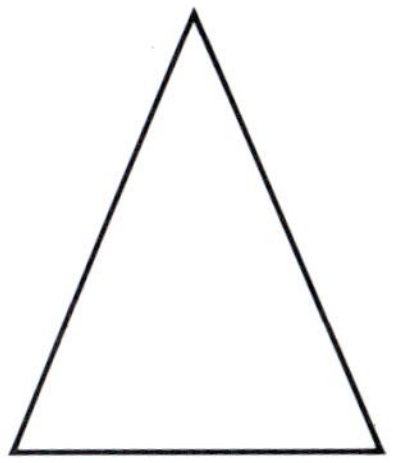

Page 308 **23.** 45° **27.** 65° **28.** 15° **35.** 4.2 **36.** 7.5 **43.** 35°; 110°

Page 311 **1.** multiply by 2 **2.** 128; 256

Page 313 **1.** 2 **2.** 3 **13.** 3.9 **14.** 4.2 **25.** 14 **26.** 24

Page 315 **1.** yes **2.** no **6.** 17 in. **10.** 25 **15.** 2 **16.** 4

Page 316 **1.** 27.5 ft **2.** 57 ft

Page 317 **1.** $86.25 **2.** $316.25

Page 319 **1.** parallelogram **2.** square **7.** all **8.** trapezoid

Page 321 **1.** $\angle X$ **2.** $\angle Y$ **7.** similar; congruent corresponding angles; ratio of sides equal **9.** 3 in.

Page 322 **1.** 9.1 m

Page 323 **5.** 12 m

CHAPTER 12

Page 331 **1.** 13.1 cm **4.** 56 mi **7.** 30 in. **10.** $11\frac{3}{4}$ ft **14.** 92 ft **18.** 9 m **20.** 104 m

Page 333 **1.** $1\frac{4}{7}$ ft **3.** $5\frac{1}{2}$ yd **9.** $2\frac{3}{4}$ mi **15.** $14\frac{2}{3}$ yd **21.** 11,100 yd

Page 335 **1.** 32 ft^2 **7.** 169 in.^{2} **13.** 12 ft^2 **19.** $10\frac{2}{3}$ yd^2

Page 337 **1.** 200 yd^2 **7.** 63 in.^{2} **14.** 20.4 **15.** 80 **22.** $24\frac{1}{2}$ ft^2

Page 339 **1.** $\frac{11}{14}$ yd^2 **2.** $1\frac{25}{63}$ mi^2 **7.** 3.14 in.^{2} **8.** 50.24 mi^2 **19.** 78.5 yd^2 **23.** 24 **24.** 3

Page 341 **1.** b **2.** a

Page 343 **1.** 20 yd^2 **5.** 26 yd^2

Page 345 **1.** rectangular prism **2.** sphere

Page 346 **7.** 8 **8.** 6 **9.** 12 **19.** false **20.** false

Page 348 **1.** 94 ft^2 **4.** 54 m^2 **10.** 18,400 ft^2 **11.** 4,600 cm^2

Page 349 **1.** 142 in.^{2} **4.** 1,728 in.^{2}

Page 351 **1.** 128 yd^3 **4.** 120 in.^{3} **5.** 729 ft^3 **12.** 3 m^3 **15.** 195 ft^3 **16.** 84 yd^3 **23.** $\frac{1}{4}$ gal

Page 353 **1.** 502.4 ft^3 **4.** 37.68 ft^3 **5.** $3\frac{15}{28}$ yd^3 **10.** 150.72 cm^3 **13.** 28.26 ft^3 **14.** $1\frac{43}{56}$ yd^3 **19.** 1.5 m

Page 355 **1.** 2.8 gal **5.** 1 gal **9.** 3 **10.** 8 **19.** $84.35

Page 357 **1.** $249 **2.** Limited **17.** Health Fair, Greenhouse, Trees **18.** Health Fair

CHAPTER 13

Page 365 **1.** -40 **2.** $-\$15$ **11.** -6 **12.** -9 **19.** -16 **20.** $+44$ **27.** $<$ **28.** $>$ **37.** 1 **38.** 2 **52.** $=$ **53.** $<$

Page 366 **60.** complementary **64.** 40° **72.** $+20,320$

Page 368 **1.** 3 **2.** -10 **13.** 28 **14.** 61 **29.** -2 **30.** 9 **39.** 37

Page 369 **43.** 57°F

Page 371 **1.** 7 **2.** -3 **17.** -2 **18.** -4 **23.** -2 **24.** -11 **39.** 20°

Page 373 **1.** 40 **2.** 88 **21.** 8 **22.** -18 **36.** -24 **37.** 1 **45.** 4 **46.** -32 **51.** $+210$

Page 375 **1.** 3 **5.** 2 **9.** -4 **57.** 100 **65.** 10^4

Page 376 **1.** 1°F **2.** -35°F **7.** -20°F

Page 379 **1.** 1 **2.** 2 **11.** 2.5×10^1 **12.** 3×10^2 **23.** 24 **27.** 0.31 **35.** 9.3×10^7

Page 381 **1.** $(-6, 6)$ **2.** $(1, 6)$ **9.** S **10.** O **25.** 0

Page 383 **1.** Pacific Ocean **2.** Indian Ocean **10.** 10°N, 130°W **14.** United States **15.** New Zealand

Page 385 **1.** $\frac{1}{2}$ **2.** -3 **7.** $<$ **15.** $=$ **23.** $=$ **35.** $\frac{3}{4}$ **36.** $\frac{2}{3}$

45. 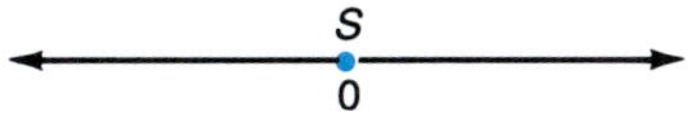

49. 116 **50.** 35

Page 387 **1.** -22.13 **2.** $-\frac{3}{8}$ **6.** -98.08 **11.** $-4\frac{1}{18}$ **21.** 1,794 **22.** 1,350 **30.** $-16\frac{2}{3}$ **31.** 6

Page 389 **1.** c **5.** 20 mL

Page 390 **1.** $8\frac{1}{4}$ **5.** increase

Page 391 **9.** Nostro

CHAPTER 14

Page 399 **1.** $3 \cdot 5$ **2.** $5 + 6$ **13.** 14 **14.** 19 **37.** $12 - (8 + 5) = -1$ **41.** 70°, acute

Page 401 **1.** $n + 6$ **2.** $n - 14$ **21.** 3 more than a number **22.** 9 minus a number **37.** $2n + 3$

Page 403 **1.** 11 **2.** 1 **17.** 3 **18.** -7 **33.** 4 **34.** 2 **53.** 40 m^3

Page 404 **57.** 21 m^2 **61.** 160 m^2

Page 405 **1.** 380 mi **2.** 338 mi **7.** 650 mph

Page 407 **1.** $18n = 36$ **2.** $n + 7 = 15$ **21.** $\frac{1}{6}$

Page 409 **1.** d **2.** e

6. 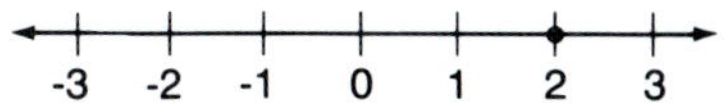

14. 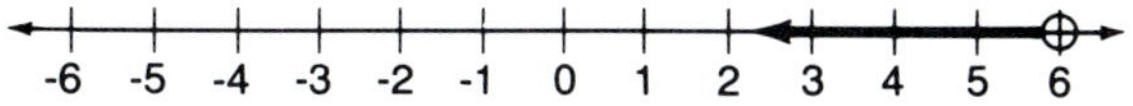

Page 410 **1.** 120 g **5.** 210 g

Page 413 **1.** 5 **2.** 7 **25.** 17 **26.** 11 **51.** $277.22

Page 415 **1.** 5 **2.** 7 **11.** 18 **12.** 32 **41.** 20 wpm

Page 417 **1.** 3 **2.** 4 **25.** 36 **26.** 49 **37.** 85 **38.** 25% **47.** 20 sundaes

Page 420 **1.** 13, 13 **5.** $-8, 4$ **13.** no **16.** no

22. 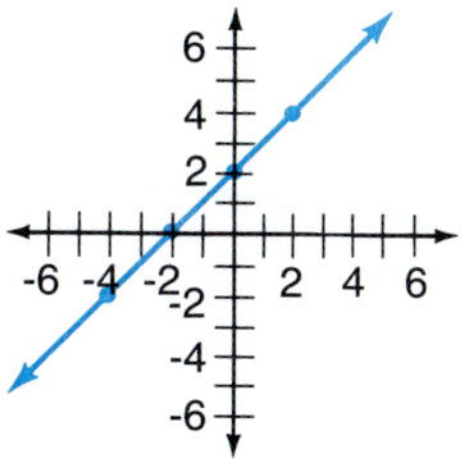

23. 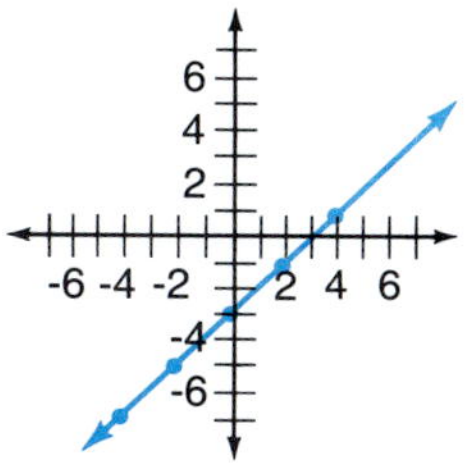

Page 421 **1.** 486 gal **5.** $178.20

TABLES

FORMULAS

Perimeter	Polygon	P = sum of the sides
	Rectangle	$P = 2l + 2w$
	Square	$P = 4s$
Circumference	Circle	$C = 2\pi r$ or $C = \pi d$
Area	Parallelogram	$A = bh$
	Rectangle	$A = lw$
	Square	$A = s^2$
	Triangle	$A = \frac{1}{2}bh$
	Trapezoid	$A = \frac{1}{2}(b_1 + b_2)h$
	Circle	$A = \pi r^2$
Volume	Rectangular prism	$V = lwh$
	Pyramid	$V = \frac{1}{3}Bh$
	Cube	$V = e^3$
	Cylinder	$V = Bh$, or $V = \pi r^2 h$
	Cone	$V = \frac{1}{3}Bh$, or $V = \frac{1}{3}\pi r^2 h$
Other	Diameter	$d = 2r$
	Pythagorean rule	$c^2 = a^2 + b^2$
Consumer	Distance	$d = rt$
	Interest (simple)	$i = prt$

METRIC SYSTEM OF MEASURES

Length
10 millimeters (mm) = 1 centimeter (cm)
10 centimeters = 1 decimeter (dm)
10 decimeters = 1 meter (m)
100 centimeters = 1 meter
1,000 millimeters = 1 meter
1,000 meters = 1 kilometer (km)

Capacity
1,000 milliliters (mL) = 1 liter (L)
1,000 liters = 1 kiloliter (kL)

Mass
1,000 milligrams (mg) = 1 gram (g)
1,000 grams = 1 kilogram (kg)
1,000 kilograms = 1 metric ton (t)

Temperature	
Freezing point of water:	0°C
Boiling point of water:	100°C
Normal body temperature:	37°C

U.S. CUSTOMARY SYSTEM OF MEASURES

Length
12 inches (in.) = 1 foot (ft)
3 feet = 1 yard (yd)
36 inches = 1 yard
1,760 yards = 1 mile (mi)
5,280 feet = 1 mile

Capacity
8 fluid ounces (fl oz) = 1 cup (c)
2 cups = 1 pint (pt)
2 pints = 1 quart (qt)
4 quarts = 1 gallon (gal)

Weight
16 ounces (oz) = 1 pound (lb)
2,000 pounds = 1 ton (T)

Temperature	
Freezing point of water:	32°F
Boiling point of water:	212°F
Normal body temperature:	98.6°F

SYMBOLS

$<$	is less than
$>$	is greater than
$\leq$	is less than or equal to
$\geq$	is greater than or equal to
$\neq$	is not equal to
$\approx$	is approximately equal to
$4 \cdot 8$	4 times 8
$12 \div 3$	12 divided by 3
4^5	4 to the fifth power
$0.\overline{16}$	0.16161616 . . .
$\sqrt{\ }$	square root
-8	negative 8
$\lvert -7 \rvert$	absolute value of negative 7
9^{-4}	9 to the negative fourth power
%	percent
7:2	the ratio of 7 to 2
$6/hour	the rate of $6 per hour
°	degree
A	point A
$\overleftrightarrow{AB}$	line AB
$\overrightarrow{AB}$	ray AB
$\overline{AB}$	line segment AB
$\angle ABC$	angle ABC
$m\angle D$	measure of $\angle D$
$\triangle XYZ$	triangle XYZ
$\perp$	is perpendicular to
$\parallel$	is parallel to
$\cong$	is congruent to
$\sim$	is similar to
π	pi (about 3.14)
(6, 4)	the ordered pair 6, 4
$P(2)$	the probability of the outcome 2
5!	5 factorial or $5 \times 4 \times 3 \times 2 \times 1$
${}_8P_5$	the number of permutations of 8 objects taken 5 at a time $8 \times 7 \times 6 \times 5 \times 4$

TABLE OF SQUARES AND SQUARE ROOTS

n	n^2	$\sqrt{n}$	n	n^2	$\sqrt{n}$	n	n^2	$\sqrt{n}$
1	1	1	35	1,225	5.916	68	4,624	8.246
2	4	1.414	36	1,296	6	69	4,761	8.307
3	9	1.732	37	1,369	6.083	70	4,900	8.367
4	16	2	38	1,444	6.164	71	5,041	8.426
5	25	2.236	39	1,521	6.245	72	5,184	8.485
6	36	2.449	40	1,600	6.325	73	5,329	8.544
7	49	2.646	41	1,681	6.403	74	5,476	8.602
8	64	2.828	42	1,764	6.481	75	5,625	8.660
9	81	3	43	1,849	6.557	76	5,776	8.718
10	100	3.162	44	1,936	6.633	77	5,929	8.775
11	121	3.317	45	2,025	6.708	78	6,084	8.832
12	144	3.464	46	2,116	6.782	79	6,241	8.888
13	169	3.606	47	2,209	6.856	80	6,400	8.944
14	196	3.742	48	2,304	6.928	81	6,561	9
15	225	3.873	49	2,401	7	82	6,724	9.055
16	256	4	50	2,500	7.071	83	6,889	9.110
17	289	4.123	51	2,601	7.141	84	7,056	9.165
18	324	4.243	52	2,704	7.211	85	7,225	9.220
19	361	4.359	53	2,809	7.280	86	7,396	9.274
20	400	4.472	54	2,916	7.348	87	7,569	9.327
21	441	4.583	55	3,025	7.416	88	7,744	9.381
22	484	4.690	56	3,136	7.483	89	7,921	9.434
23	529	4.796	57	3,249	7.550	90	8,100	9.487
24	576	4.899	58	3,364	7.616	91	8,281	9.539
25	625	5	59	3,481	7.681	92	8,464	9.592
26	676	5.099	60	3,600	7.746	93	8,649	9.644
27	729	5.196	61	3,721	7.810	94	8,836	9.695
28	784	5.292	62	3,844	7.874	95	9,025	9.747
29	841	5.385	63	3,969	7.937	96	9,216	9.798
30	900	5.477	64	4,096	8	97	9,409	9.849
31	961	5.568	65	4,225	8.062	98	9,604	9.899
32	1,024	5.657	66	4,356	8.124	99	9,801	9.950
33	1,089	5.745	67	4,489	8.185	100	10,000	10
34	1,156	5.831						

FRACTIONS AND PERCENTS

$\frac{1}{2} = 50\%$					
$\frac{1}{3} = 33\frac{1}{3}\%$	$\frac{2}{3} = 66\frac{2}{3}\%$				
$\frac{1}{4} = 25\%$	$\frac{3}{4} = 75\%$				
$\frac{1}{5} = 20\%$	$\frac{2}{5} = 40\%$	$\frac{3}{5} = 60\%$	$\frac{4}{5} = 80\%$		
$\frac{1}{6} = 16\frac{2}{3}\%$	$\frac{5}{6} = 83\frac{1}{3}\%$				
$\frac{1}{7} = 14\frac{2}{7}\%$	$\frac{2}{7} = 28\frac{4}{7}\%$	$\frac{3}{7} = 42\frac{6}{7}\%$	$\frac{4}{7} = 57\frac{1}{7}\%$	$\frac{5}{7} = 71\frac{3}{7}\%$	$\frac{6}{7} = 85\frac{5}{7}\%$
$\frac{1}{8} = 12\frac{1}{2}\%$	$\frac{3}{8} = 37\frac{1}{2}\%$	$\frac{5}{8} = 62\frac{1}{2}\%$	$\frac{7}{8} = 87\frac{1}{2}\%$		
$\frac{1}{9} = 11\frac{1}{9}\%$	$\frac{2}{9} = 22\frac{2}{9}\%$	$\frac{4}{9} = 44\frac{4}{9}\%$	$\frac{5}{9} = 55\frac{5}{9}\%$	$\frac{7}{9} = 77\frac{7}{9}\%$	$\frac{8}{9} = 88\frac{8}{9}\%$
$\frac{1}{10} = 10\%$	$\frac{3}{10} = 30\%$	$\frac{7}{10} = 70\%$	$\frac{9}{10} = 90\%$		

ADDITIONAL ANSWERS

CHAPTER 1

Page 3

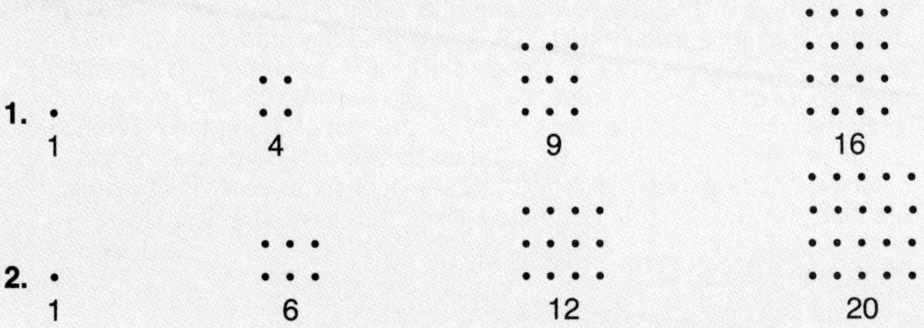

Page 5

1. tens; 40 **2.** hundreds; 600 **3.** hundreds; 800 **4.** thousands; 9,000 **5.** ten thousands; 30,000 **6.** thousands; 8,000 **7.** hundreds, 400 **8.** ten thousands; 90,000 **9.** ten thousands; 20,000 **10.** hundreds; 300 **11.** hundred thousands; 800,000 **12.** thousands; 6,000 **13.** ones; 8 **14.** hundred thousands; 300,000 **15.** ten millions; 70,000,000 **16.** millions; 3,000,000 **17.** hundred millions; 500,000,000 **18.** billions; 9,000,000,000 **19.** ten billions; 30,000,000,000 **20.** hundred billions; 900,000,000,000 **21.** one hundred seventy-five **22.** three hundred fifty-eight **23.** four thousand, five hundred sixty-eight **24.** eight thousand, six hundred four **25.** twenty-three thousand, nine hundred sixty-one **26.** forty-eight thousand, five hundred forty-three **27.** sixty-seven thousand, eight hundred one **28.** seventy-three thousand, four hundred ninety-six **29.** three hundred twenty-five thousand, four hundred six **30.** four hundred sixty-two thousand, one **31.** five hundred eighty thousand, one hundred two **32.** seven hundred six thousand, five **33.** one million, eight hundred ninety thousand **34.** two million, three hundred forty thousand, ten **35.** three million, four hundred thousand, nine **36.** eight million, sixty thousand, one hundred twenty-five

Page 7

1. tenths; 0.1 **2.** hundredths; 0.06 **3.** tenths; 0.4 **4.** hundredths; 0.05 **5.** hundredths; 0.04 **6.** thousandths; 0.003 **7.** tenths; 0.9 **8.** thousandths; 0.005 **9.** thousandths; 0.003 **10.** tenths; 0.8 **11.** ten thousandths; 0.0002 **12.** thousandths; 0.008 **13.** hundred thousandths; 0.00002 **14.** tenths; 0.1 **15.** thousandths; 0.003 **16.** hundredths; 0.04 **17.** ten thousandths; 0.0007 **18.** ten thousandths; 0.0003 **19.** hundred thousandths; 0.00002 **20.** millionths; 0.000006 **21.** five tenths **22.** one and eight tenths **23.** sixteen and nine tenths **24.** fifty-nine and six tenths **25.** four hundredths **26.** twenty-nine hundredths **27.** one and fifty-one hundredths **28.** two and eighty-nine hundredths **29.** seven thousandths **30.** thirty-nine thousandths **31.** two hundred forty-one thousandths **32.** one and three hundred seventy-two thousandths **33.** two ten-thousandths **34.** seventeen ten-thousandths **35.** three hundred fifty-six ten-thousandths **36.** one thousand two hundred forty-five ten-thousandths

Page 13

11. hundreds; 900 **12.** thousands; 7,000 **13.** hundred thousands; 300,000 **14.** millions; 7,000,000 **15.** tenths; 0.9 **16.** hundredths; 0.05 **17.** hundredths; 0.05 **18.** hundred thousandths; 0.00006 **19.** four hundred twenty-six **20.** two hundred seventy-nine thousand, three hundred **21.** ten and five hundred sixty-two thousandths **22.** one hundred twenty-nine ten-thousandths

Page 17

1. 2,800; 3,000 **2.** 4,000; 4,000 **3.** 5,300; 5,000 **4.** 6,100; 6,000 **5.** 8,600; 9,000 **6.** 17,900; 18,000 **7.** 63,800; 64,000 **8.** 74,200; 74,000 **9.** 90,000; 90,000 **10.** 93,200; 93,000 **11.** 354,200; 354,000 **12.** 489,600; 490,000 **13.** 666,100; 666,000 **14.** 899,500; 900,000 **15.** 942,200; 942,000 **16.** 1,000,000; 1,000,000 **17.** 0; 0.02 **18.** 0.1; 0.08 **19.** 1.5; 1.55 **20.** 4.1; 4.05 **21.** 17; 16.95 **22.** 39.9; 39.95 **23.** 85.6; 85.56 **24.** 0; 0.05 **25.** 0.5; 0.48 **26.** 2.2; 2.17 **27.** 11.2; 11.16 **28.** 55.1; 55.12 **29.** 0.4; 0.35 **30.** 5.1; 5.12 **31.** 23.7; 23.67 **32.** 49; 48.98

Page 23

7. twelve thousand-four hundred eighty-nine **8.** two hundred forty thousand, seven **9.** five and eight hundred forty-one thousandths **10.** four hundred eighty-seven ten-thousandths

Page 24

CHAPTER 2

Page 35

1. 2 dimes, 1 penny = 21¢ **2.** 3 quarters, 1 nickel, 3 pennies = 83¢
3. 2 quarters, 2 dimes, 2 pennies = 72¢
4. 1 quarter, 1 nickel, 4 pennies = 34¢ **5.** 3 quarters, 1 penny = 76¢
6. 1 quarter, 2 dimes, 2 pennies = 47¢
7. \$1 bill, 1 nickel, 4 pennies = \$1.09 **8.** \$10 bill, 2 \$1 bills, 1 quarter, 2 dimes, 2 pennies = \$12.47
9. 6 dimes, 8 pennies = 68¢
10. 1 half-dollar, 1 dime, 1 nickel, 3 pennies = 68¢

Page 39

21. 47 **22.** 107 **23.** 51 **24.** 99 **25.** 43 **26.** 1,975 **27.** 1,901 **28.** 439 **29.** 520 **30.** 1,055 **31.** 4,738 **32.** 10,554 **33.** 9,781 R2 **34.** 9,555 R1 **35.** 9,109 **36.** 43 R17 **37.** 20 R26 **38.** 21 R25 **39.** 498 R4 **40.** 2,591 R18

Page 43

1. 96,228 **2.** 137,935 **3.** 200,736 **4.** 224,895 **5.** 66,971 **6.** 357,576 **7.** 358,224 **8.** 452,120 **9.** 240,783 **10.** 280,966 **11.** 1,760,200 **12.** 805,968 **13.** 876,828 **14.** 1,281,638 **15.** 2,783,340 **16.** 459,486 **17.** 1,083,004 **18.** 2,057,376 **19.** 2,819,178 **20.** 3,720,332 **21.** 347,985 **22.** 208,800 **23.** 422,400 **24.** 106,981 **25.** 361,350 **26.** 327,800 **27.** 1,498,680 **28.** 2,288,415 **29.** 3,278,374 **30.** 2,419,956 **31.** 147,586 **32.** 73,023 **33.** 69,138 **34.** 387,032 **35.** 516,450 **36.** 640,491 **37.** 3,609,088 **38.** 4,165,700 **39.** 2,563,022 **40.** 3,537,534 **41.** 193,824 **42.** 169,608 **43.** 2,415,000 **44.** 3,266,340 **45.** 86,961 **46.** 86,064 **47.** 84,436 **48.** 635,508 **49.** 3,298,472 **50.** 226,074 **51.** 3,914,352 **52.** 5,675,028 **53.** 277,665 **54.** 131,493 **55.** 407,121 **56.** 768,456 **57.** 635,762 **58.** 135,642 **59.** 2,070,782 **60.** 2,151,478 **61.** 3,690,644 **62.** 640,256 **63.** 3,185,936 **64.** 1,321,320

ADDITIONAL ANSWERS

CHAPTER 7

Page 185

1. 7 yd 2 ft **2.** 11 lb 8 oz **3.** 2 ft 11 in. **4.** 24 gal 3 qt **5.** 4 yd 1 ft **6.** 14 gal 1 qt **7.** 3 pt **8.** 10 mi 70 ft **9.** 2 ft 1 in. **10.** 1 yd 1 ft **11.** 2c 4 fl oz **12.** 1 mi 880 yd **13.** 1 qt 1 pt **14.** 1 lb 15 oz **15.** 3 qt **16.** 1 T 1,670 lb **17.** 2 T 290 lb **18.** 1 pt 12 fl oz **19.** 3 pt **20.** 15 mi 623 yd **21.** 4 lb 13 oz **22.** 16 lb 12 oz **23.** 9 T 1,875 lb **24.** 2 mi 4,630 ft **25.** 2c 1 oz **26.** 6 T 1,600 lb **27.** 3 yd 2 ft **28.** 3 ft 6 in.

Page 196

Buckets

8L	5 L	2 L
8	0	0
3	5	0
3	3	2
5	3	0
5	1	2
7	1	0

CHAPTER 9

Page 261

1. \$60 **2.** \$112 **3.** \$54 **4.** \$250 **5.** \$562.50 **6.** \$349.50 **7.** \$371.25 **8.** \$5.00 **9.** \$2,000 **10.** \$96.25 **11.** \$6,000 **12.** \$4,851 **13.** \$56.25 **14.** \$562.50 **15.** \$5,625 **16.** \$56,250 **17.** \$3,744 **18.** \$10.36 **19.** \$1,771.81 **20.** \$217.81

CHAPTER 10

Page 281

1. 35 53 73
37 57 75

2.

Left Field	Right Field	Center Field
Joe	Al	Sam
Joe	Sam	Al
Al	Sam	Joe
Al	Joe	Sam
Sam	Al	Joe
Sam	Joe	Al

3. 246, 248, 264, 268, 284, 286
426, 428, 462, 468, 482, 486
624, 628, 642, 648, 684, 682
824, 826, 842, 846, 862, 864

4.

B, A, J, S	A, J, S, B	J, S, B, A	S, B, A, J
B, A, S, J	A, J, B, S	J, S, A, B	S, B, J, A
B, S, A, J	A, S, B, J	J, B, A, S	S, A, J, B
B, S, J, A	A, S, J, B	J, B, S, A	S, A, B, J
B, J, A, S	A, B, J, S	J, A, B, S	S, J, B, A
B, J, S, A	A, B, S, J	J, A, S, B	S, J, A, B

5. Cathy—Ann Ann—Wes Wes—Glen
Cathy—Wes Ann—Glen
Cathy—Glen

6. 1, 2 1, 3 1, 4 1, 5
2, 3 2, 4 2, 5
3, 4 3, 5
4, 5

CHAPTER 14

Page 401

21. 3 more than a number **22.** 9 decreased by a number **23.** 64 times a number **24.** a number minus 14 **25.** a number added to 17 **26.** 6 less than a number **27.** 5 times the sum of 3 and a number **28.** 4 times a number, minus 6, divided by 2 **29.** 2 less than the product of 11 and a number **30.** 13 times the difference of 9 minus a number **31.** 28 minus the product of 7 and a number **32.** a number minus 15 **33.** the quotient of a number minus 7, divided by 5 **34.** the product of 3 and the difference of a number minus 2, divided by 8 **35.** the quotient of 60 divided by the difference of 12 minus a number **36.** the quotient of 35 divided by the product of 2 and the sum of a number and 4

Page 409

6.
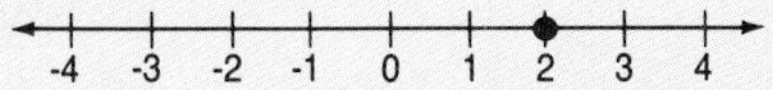

7.
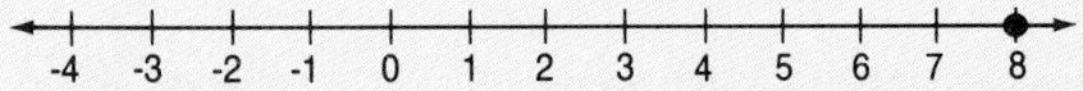

8.
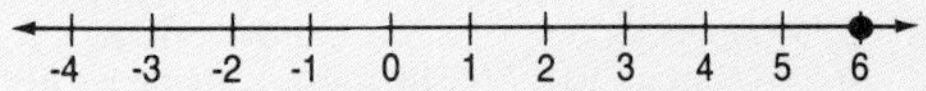

9.
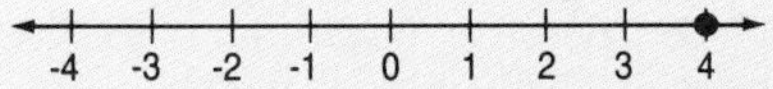

10.
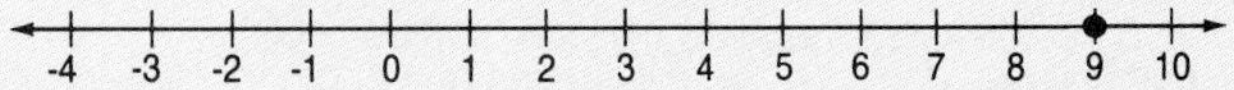

11.
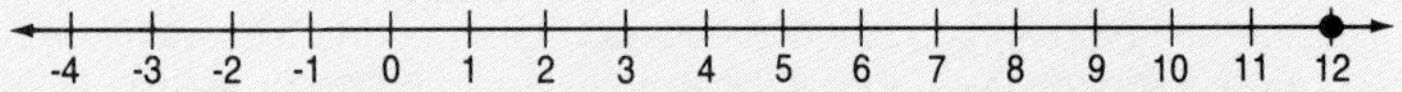

12.
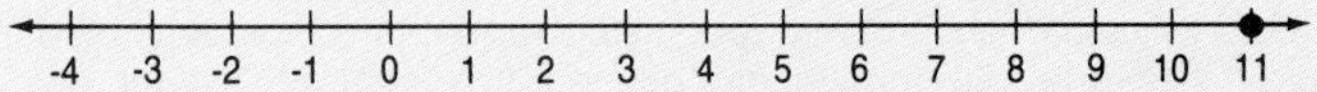

13.
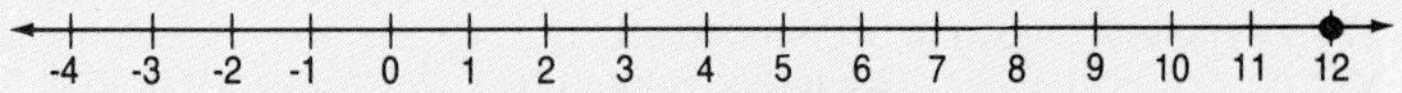

14.
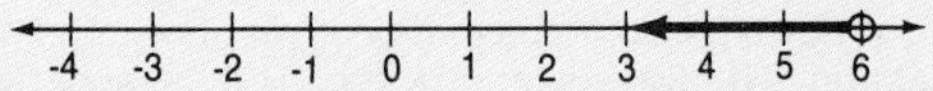

15.
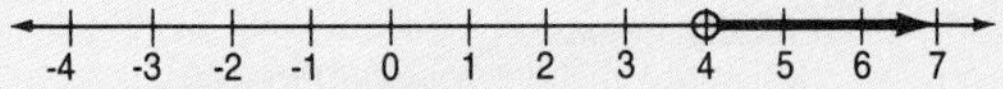

16.
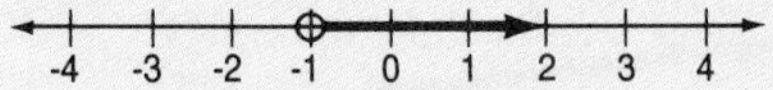

17.
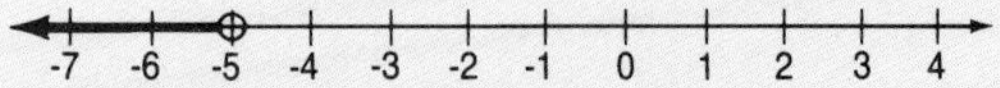

18.
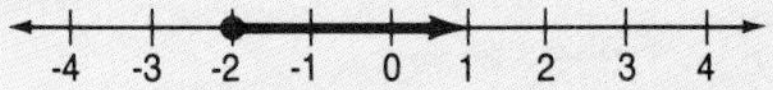

19.

20.
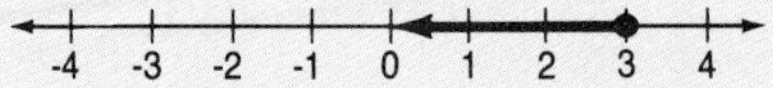

21.
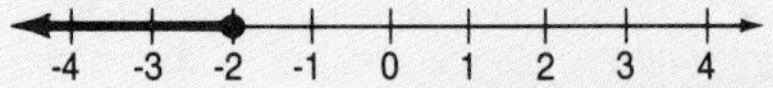

22.
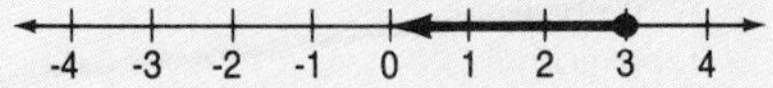

23.
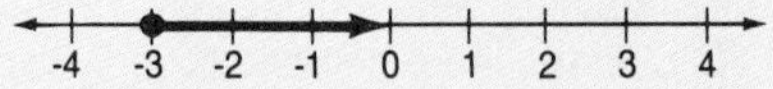

24.
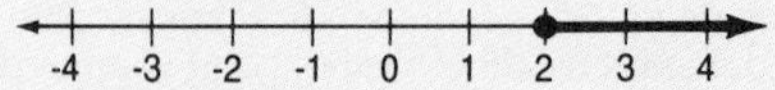

25.
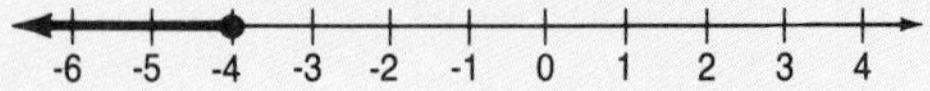

Page 411

19. 3 added to a number **20.** 12 less than a number **21.** 2 times a number minus 4 **22.** 73 minus a number **23.** the quotient of a number minus 7, divided by 5 **24.** the sum of the cube of a number plus 18 **25.** the sum of 16 times a number plus 9 **26.** the product of 12 times the difference of a number minus 5

ADDITIONAL ANSWERS

Page 411

39.
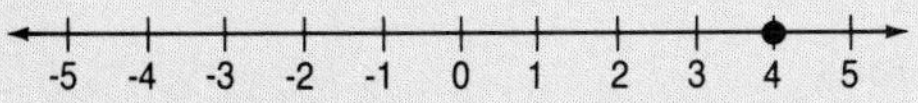

40.
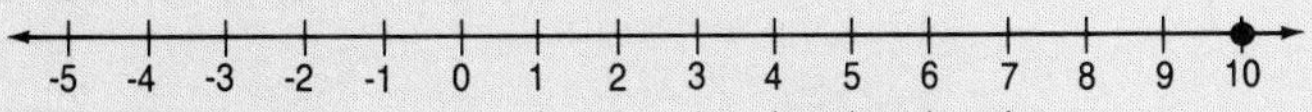

41.
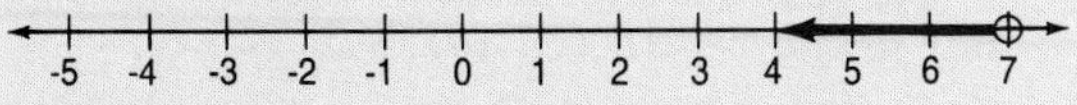

42.
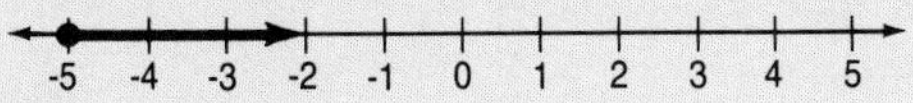

Page 420

22.
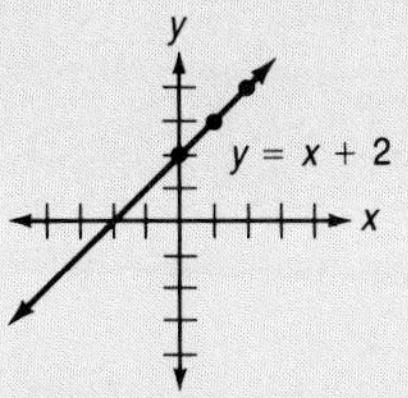

23.
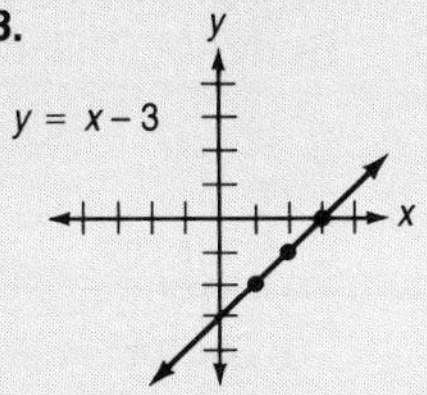

24.
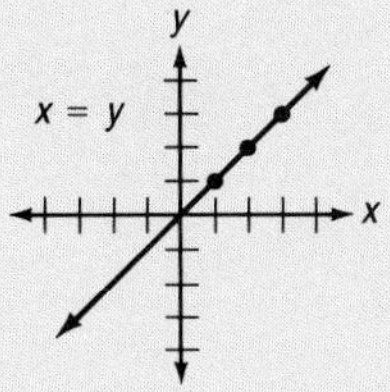

25.
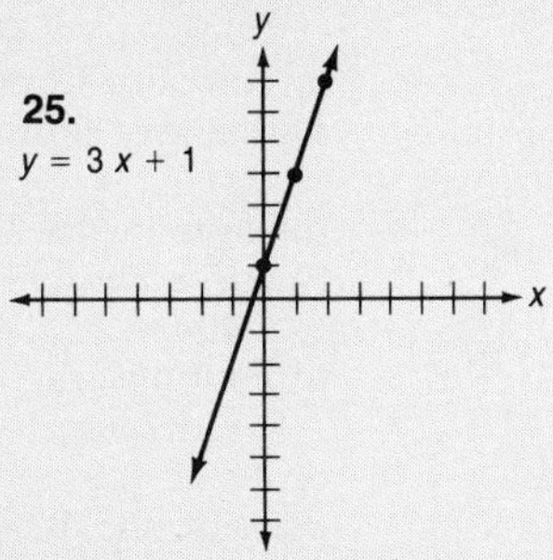

26.
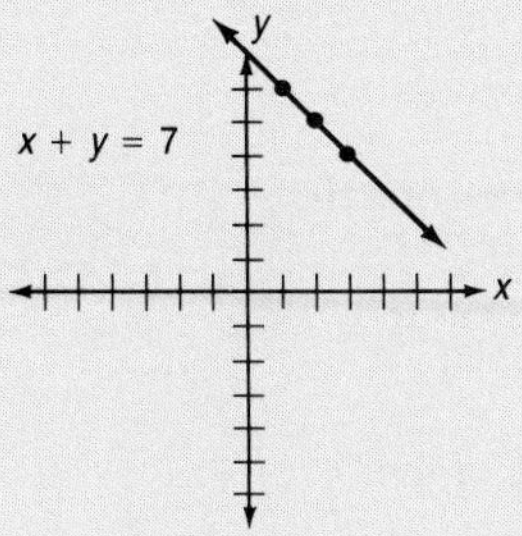

27.
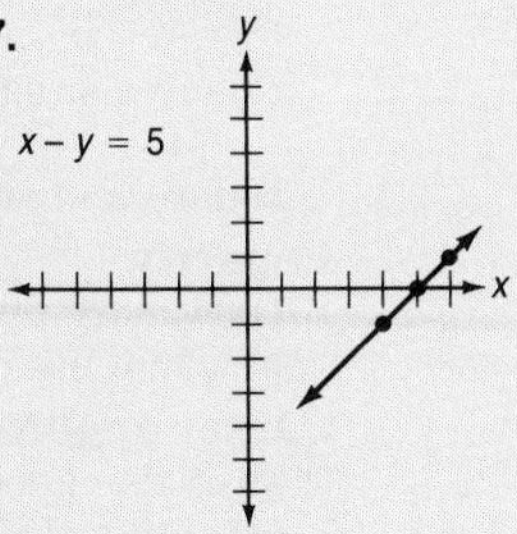

28.
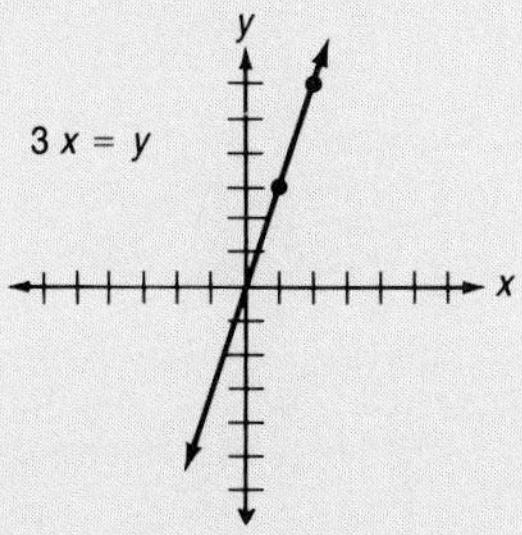

29.
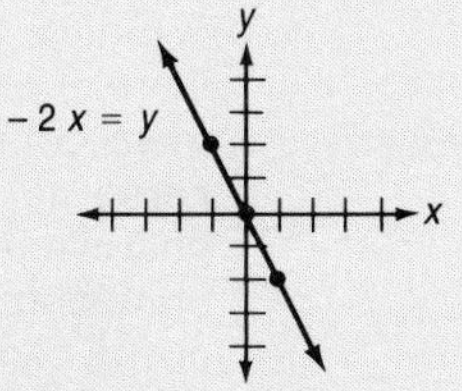

30.
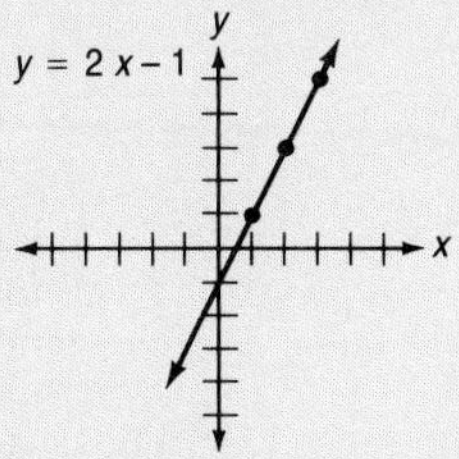

31.
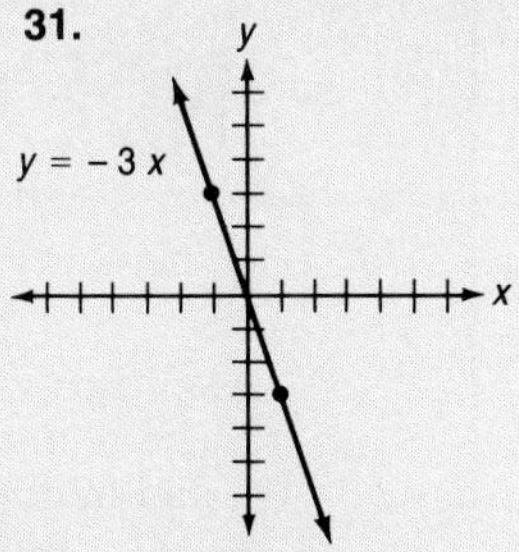

32.
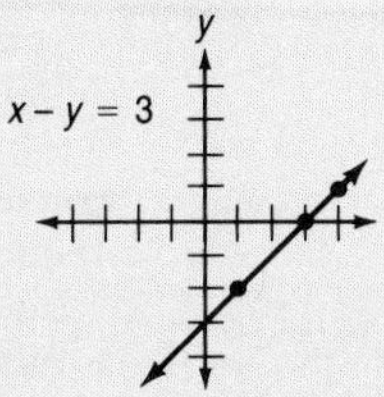

33.
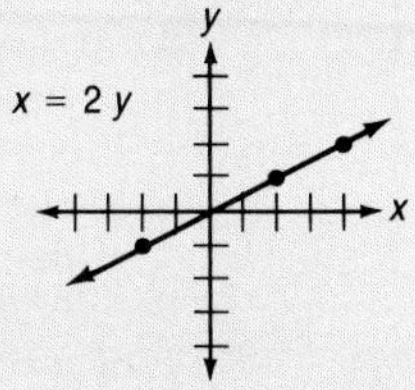

34.
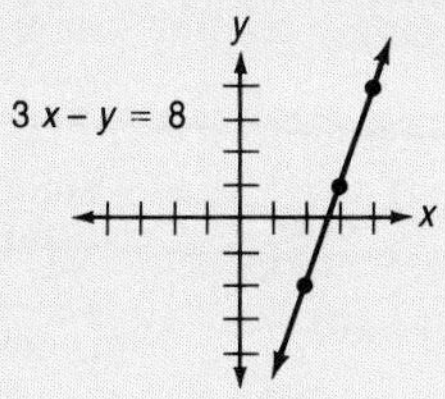

35.
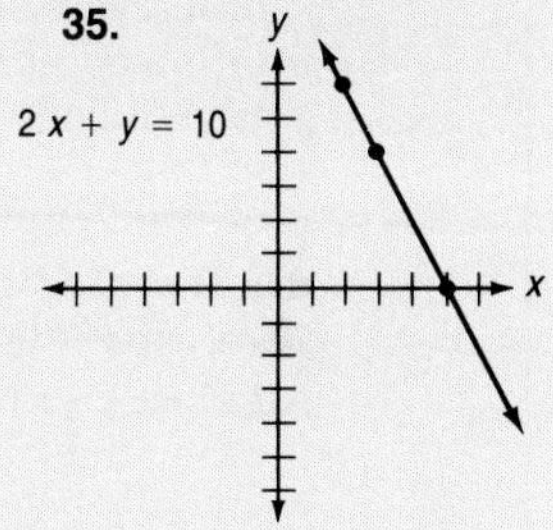

36.
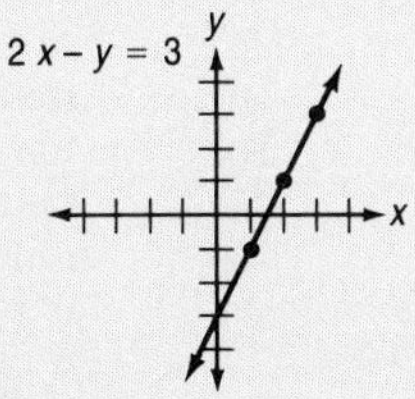

37.
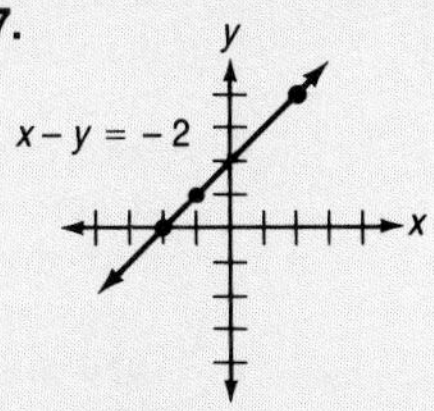

Page 423

15.
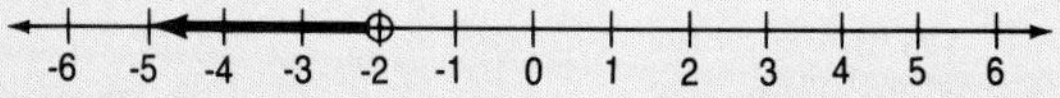

16.
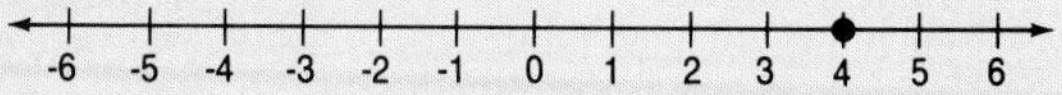

17.
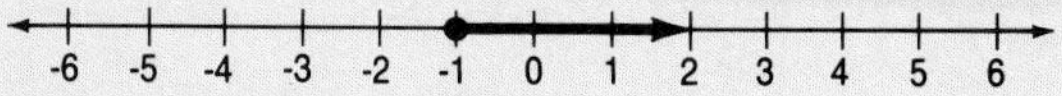

18.
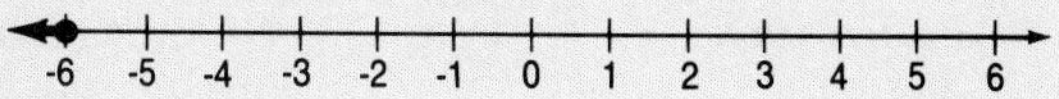

35.
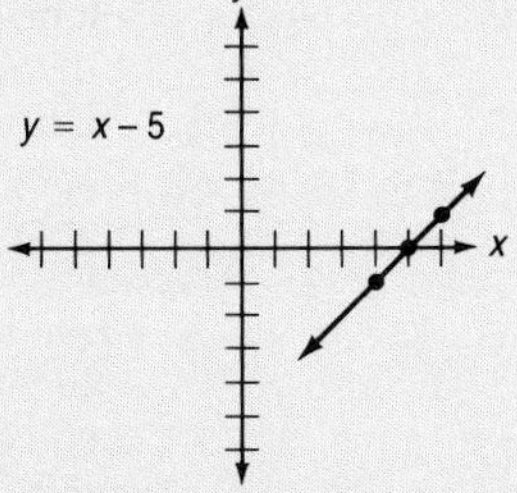

36.
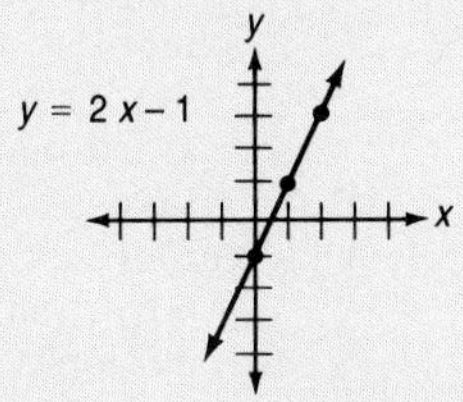

INDEX

D

E

Q

R

S

T

Credits

Book Design and Production: Textart Inc.
Cover Design: Textart Inc.
Cover Photo: Robert Fishman
Technical Art: Network Graphics

Illustrations: *Mark Giglio* 98, 110, 230, 234; *Steve Marchesi* 78, 127, 196, 202, 212; *Samantha Smith* 115, 180, 192, 194, 241, 273, 375

Picture Credits:
Contents Pages: III, IV–V, VI–VII, VIII–IX HRW Photo Ken Karp

Chapter 1 Opener: Page 1 HRW Photo Richard Haynes; **10–11** HRW Photo Ken Karp; **12** HRW Photo Richard Haynes; **15** James Blank/Archive Pictures Inc.; **17** Robert McElroy/Woodfin Camp & Assoc.; **21** Robert Ferrick/Odyssey Prod.
Chapter 2 Opener: Page 27 HRW Photo Richard Haynes; **35** HRW Photo Ken Karp; **45** Richard Hirneisi/Stock Shop; **52–53** HRW Photo Ken Karp; **55** HRW Photo Ken Lax
Chapter 3 Opener: Page 61 HRW Photo Michal Heron; **70–71** HRW Photo Ken Karp; **77** Sepp Seitz/Rainbow; **83** Jose Fernandez/Woodfin Camp & Assoc.
Chapter 4 Opener: Page 91 Thomas Sobolik; **92** Guido Alberto/Image Bank; **95** Richard Vadnai/Stock Market; **101** Philip Jon Bailey/The Picture Cube; **103** HRW Photo Ken Karp; **111** Bill Eppridge/DOT; **114** HRW Photo Ken Karp
Chapter 5 Opener: Page 121 Focus on Sports; **135** John W. Banagan/Image Bank; **138** HRW Photo Ken Lax
Chapter 6 Opener: Page 145 Sepp Seitz; **148** Bob Gelberg/Image Bank; **154–155** HRW Photo Ken Karp; **157** Dan McCoy/Rainbow; **167** Dan McCoy/Rainbow; **170–171** HRW Photo Ken Karp
Chapter 7 Opener: Page 179 HRW Photo Richard Haynes; **189** Aram Gesar/Image Bank; **191** Audrey Gottlieb/Monkmeyer Press; **200L** Bob Daemmrich/Stock Boston; **200R** Dede Hatch/The Picture Cube; **203** David Moore/Black Star
Chapter 8 Opener: Page 209 HRW Photo Michal Heron; **214–215** HRW Photo Ken Karp; **218–219** HRW Photo Ken Karp; **221** Steve Proehl/Image Bank; **223** Bob Woodward/The Stock Market; **227** HRW Photo Ken Lax
Chapter 9 Opener: Page 233 Dave Schaefer; **235** Robert J. Capece/Monkmeyer Press; **236** HRW Photo Ken Lax; **242–243** HRW Photo Ken Karp; **245** Alvis Uptis/Image Bank; **252–253** HRW Photo Ken Karp; **254–255** HRW Photo Ken Karp; **256** HRW Photo Ken Karp; **258** HRW Photo Ken Lax; **260** HRW Photo Ken Lax
Chapter 10 Opener: Page 267 Ed Kashi; **270** HRW Photo Ken Karp; **271** HRW Photo Ken Karp; **274** Abigail Heyman; **276** HRW Photo Ken Lax; **280–281** Dan Helms/Duomo; **282–283** HRW Photo Ken Karp
Chapter 11 Opener: Page 291 James Blank; **295** HRW Photo Ken Karp; **310–311** HRW Photo Ken Karp; **312** Richard Steedman; **316** HRW Photo Richard Haynes; **323** HRW Photo Ken Karp
Chapter 12 Opener: Page 329 Michal Heron; **333** Dave Bloch; **340–341** HRW Photo Ken Karp; **349** HRW Photo Ken Lax; **354** HRW Photo Ken Karp; **357** HRW Photo Ken Lax
Chapter 13 Opener: Page 363 Frank Fournier; **366** Johnny Johnson/DRK Photo; **369** Focus West; **376** Ken Karp; **379** NASA; **382** Armando Jenik/Image Bank; **389** Charles Gupton/The Stock Market; **391** Ken Karp
Chapter 14 Opener: Page 397 HRW Photo Michal Heron; **402** Paul Chauncey; **405** Hank de Lespinasse/Image Bank; **406** Ed Kashi; **410** Gabe Palmer/The Stock Market; **415** HRW Photo Ken Lax; **416** HRW Photo Richard Haynes; **421** HRW Photo Richard Haynes